Applied Geophysics for Karst and Sinkhole Investigation

The Dead Sea and other regions

Online at: https://doi.org/10.1088/978-0-7503-3635-2

Applied Geophysics for Karst and Sinkhole Investigation

The Dead Sea and other regions

Michael Ezersky

Geotec Engineering and Environmental Geophysics Ltd (Israel), Rishon Lezion, Israel

Lev V Eppelbaum

Department of Geophysics, Faculty of Exact Sciences, Tel Aviv University, Tel Aviv, Israel
and
Azerbaijan State Oil and Industry University, Baku, Azerbaijan

Anatoly Legchenko

Research Director in the Institute de Recherche pour le Développement (IRD) in Marceil (France), actually attached to the Université Grenoble Alpes (IGE laboratory in Grenoble), Grenoble, France

IOP Publishing, Bristol, UK

ISBN 978-0-7503-3635-2 (ebook)
ISBN 978-0-7503-3633-8 (print)
ISBN 978-0-7503-3636-9 (myPrint)
ISBN 978-0-7503-3634-5 (mobi)

DOI 10.1088/978-0-7503-3635-2

Version: 20230501

IOP ebooks

British Library Cataloguing-in-Publication Data: A catalogue record for this book is available from the British Library.

Published by IOP Publishing, wholly owned by The Institute of Physics, London

IOP Publishing, No.2 The Distillery, Glassfields, Avon Street, Bristol, BS2 0GR, UK

US Office: IOP Publishing, Inc., 190 North Independence Mall West, Suite 601, Philadelphia, PA 19106, USA

Contents

Preface

Historically, geophysicists have often been focused on the exploration aspects. This primarily comprises either oil or minerals. It is recognized that in both cases the use of geophysics allows a major return on investment.

In engineering and environmental geophysics (EEG), quantifying the cost benefits of using geophysical methods is much harder. Furthermore, geophysical information is difficult to convey in a format that is understandable by the civil engineer, engineer-geologist, hydrogeologist, or layperson (i.e. lawyer), who must use the information but is not well versed with the idiosyncrasies of geophysical methodology, data presentation. Interpretation of geophysical results is non-unique and requires specific knowledge and practical experience.

EEG specialists must have engineering thinking and be trained in engineering problems. Despite that, engineering and environmental geophysics is, often, the only remedy that can solve the desired problem quickly and reliably. However, the efficiency of using geophysics may be reduced if selecting methods are not suitable for the investigated problem. Every geophysical method is designed for a solution to specific problems under predefined conditions. Planning geophysical investigations for engineering purposes and their subsequent interpretation requires corresponding geological models. Every failure raises doubts about the capabilities of one or the other geophysical methods. In this book, we aim to show that the correct application of geophysical methods can be a powerful tool for solving complex physical-geological problems.

The idea of this book was formed in 2018 while we were finalizing the results of two international projects financed by the NATO Science for Peace program (SfP 981128) and MERC USAID program (M27-050) devoted to the development of the geophysical methods for resolving the Dead Sea (DS) sinkhole problem.

These projects were carried out in the western and eastern DS shores in the sinkhole development sites. Several teams from Institutions and Universities from Israel, Jordan, and France have taken part in field measurements and interpretation. Numerous geophysical methods were applied, tested, and developed aiming to understand the origin of the DS sinkholes and to suggest effective measures for the resolving this problem.

The DS karst has two specific features. Firstly, it is an evaporite nature caused by the buried salt layers dissolving by the very saline water (brine). The second one is the drop in the water level that takes place at a very high rate (~1 m/yr). Both these produce conditions at which processes of karst formation flow in real-time allowing observation and investigation of those during a time of human life. That is why the DS is considered a natural laboratory that despite its small size attracts great interest among scientists from different branches.

We try to avoid the complicated mathematical equations and theoretical calculations reducing them to the necessity for selecting the most suitable methods for the solution of karst-related problems in the world, and in the DS area in particularity.

We have drawn attention to new methods, especially those that proved efficient in the DS area, (characterized by the high groundwater salinity), such as transient electromagnetics (TEM), the surface nuclear magnetic resonance sounding method (SNMR) also known as magnetic resonance sounding (MRS), electrical resistivity tomography (ERT), and ground penetrating radar (GPR). We consider also seismic methods modified for sinkhole problem resolving, such as seismic refraction in its generalized reciprocal method technique, seismic reflection and diffraction methods in their multipass summation techniques, and multichannel analysis of surface waves (MASW). Microgravity and micromagnetic methods based on the high accuracy measurements, removing regional trends, advanced quantitative analysis, and 3D modeling also contributed to the detection of subsurface cavities - future sinkholes. Finally, we overview other methods such as borehole logging, remote sensing, and others applied by different researchers to investigate the Dead Sea sinkhole problem.

In our book, we collected a rich experience in the geophysical investigations of karst phenomena. It shows approaches to resolving different applied problems in the field of karst investigation and instruments for its resolution. Finally, geophysics is the part of geology that in turn is the key to a better understanding of the site model. If the geology of the area under study is properly understood, then all subsequent work can be done with more success.

We hope also that this book will assist the readers who wish to continue and elaborate on some of our ideas.

This book is divided into six chapters:

- Chapter 1 provides a short introduction to karst and its distribution throughout the world. It provides a brief overview of karst definitions and aspects.
- Chapter 2 is a brief overview of geophysical methods and examples of the application of geophysics worldwide for the karst problems solution.
- Chapter 3 describes the Dead Sea geology, hydrogeology, and other DS environment conditions.
- Chapter 4 is the core of the book. It overviews an application of different geophysical methods for the DS sinkhole problem studies.
- Chapter 5 considers a combination of different geophysical methods (based on both practice and information, probabilistic, and wavelet approaches) for resolving different aspects of karst problems.
- Chapter 6 contains some conclusions and recommendations.

Numerous bibliographical references provide readers with an overview related to the book's content. Numerous references to any books, papers, articles, equipment, software, etc do not constitute advertising or endorsement of any product, and no preference is implied, but reflect only the experience of authors

Michael Ezersky,
Geotec Ltd,
Ruzanski 18b, Rishon Lezion,
Israel

Lev Eppelbaum,
Tel Aviv University,
Department of Geophysics,
Faculty of Exact Sciences
Ramat Aviv, 6997801, Tel Aviv, Israel

Anatoly Legchenko,
University Grenoble Alpes,
Institute of Research for Development,
IGE, Grenoble, France, UGA – IGE,
CS 40700, 38058 Grenoble Cedex 9, France

Acknowledgments

We dedicate this book to our families and thank them for their immeasurable patience and support. We are grateful to Alla Ezersky (wife of Michael), Elina Eppelbaum (wife of Lev) and Tatiana Legchenko (wife of Anatoly). Our warm thanks go to Mr Amit Ronen, General Manager of the Geotec Company. We thank our publishers, Ms Emily Tapp and editor assistants for their patience, invaluable support, and constant attention. A special thank you is extended to our reviewers and contributors.

The authors are grateful to NATO SfP and USAID MERC programs for support of geophysical researches to resolve the DS sinkhole problem. We are grateful also to the Ministry of Energy and Water Resources of Israel and the Tamar Regional Council. Thanks go to the Geophysical Institute of Israel and Department of Geophysics, Tel Aviv University. We thank especially our colleagues for participation in the field studies and writing of research articles: A Al-Zoubi, A Abueladas, E Akkawi, C Camerlynck, L Bodet, S Keydar, M Rybakov, V Goldshmidt, and V Zheludev. We thank the scientists for assisting us in the geological interpretation of results, advising of geophysical data presentation of the material, and the content of this book. It first relates to P Milanovic, A Frumkin, and D Closson. We thank A Abelson, Y Yechieli, M Parise, J De Waele, and F Gutiérrez—well-known specialists in the field of evaporite karst. Also, thank you to U Basson.

You can communicate with us by e-mail:

Michael Ezersky: michael.ezersky10@gmail.com
Lev Eppelbaum: levap@tauex.tau.ac.il
Anatoly Legchenko: anatoli.legtchenko@ird.fr

Author biographies

Professor Michael G Ezersky

Michael G Ezersky received a professional engineering degree (MSc) in rock physics from Moscow Mining Institute (Russia) in 1972. In 1985 he received a PhD diploma in Physics and Mathematics from the Institute of Physics of the Earth named after O Yu Shmidt of the Russian Academy of Sciences (Moscow). He began his career with Designing, Surveying, and Research Institute Hydroproject named after S Y Zhuk (Moscow). His duties during his 22-year tenure included directing seismic-acoustic R&D projects and surveys at the construction of Hydro Power Stations in Russia and abroad. His geophysical researches were associated with investigations of the High Dam foundations, Large Underground Halls, and Tunnels. After immigrating to Israel in 1995 he accepted the position of researcher-geophysicist and from 2005 of Senior Researcher in Geophysical Institute of Israel. From 2016 Michael Ezersky has been Senior Resercher in Geotec Engineering and Environmental company Ltd. His professional interests are in engineering and environmental geophysics, and archaeology. During the ten last years, he led research projects connected with the development of geophysical methods and tools for investigation and mitigation of the Dead Sea sinkhole problem. In 2015 M Ezersky was awarded the title of professor (Grade A). M Ezersky is a reviewer in many scientific journals. He has written numerous papers dealing with underground mechanics, archaeology, environmental geophysics and hydrogeology. He is an active member of EAGE, IRMA, IGS.

Professor Lev V Eppelbaum

Lev V Eppelbaum received an MSc from the Azerbaijan State Oil and Industry University (ASOIU) in 1982 and a PhD from the Inst. of Geophysics (Azerbaijan), the Geophysical Inst. of Georgia, and the Geological Prospecting University (Moscow) in 1989. From 1982–90, he worked as a geophysicist, Researcher, and Senior Researcher at the Southern Branch of the All-Union Institute of Geophysics in Baku (Azerbaijan). From 1991–93 Eppelbaum completed postdoctoral studies in the Department of Geophysics and Planetary Sciences at Tel Aviv University; currently, he is a Research Professor at the Department of Geophysics. He is the author of about 420 publications, including ten books and more than 175 articles. His scientific interests cover potential (magnetic, gravity, thermal, and self-potential) and quasi-potential geophysical field analysis in complex geological–geophysical environments, integrated geophysical and geological data interpretation, tectonic, geodynamic, and paleomagnetic reconstructions. Eppelbaum's research interests also include analysis of nonlinear geophysical processes, archaeological and environmental investigations, and searching economic minerals. He was awarded 'The Knight of the Science and Arts', and V. Vernadsky medal of the Russian Academy of Natural Sciences in 2017 and 2018, respectively. In 2019, Lev Eppelbaum was awarded the Christian Huygens medal of the European Geosciences Union, and in 2020 he was awarded the Honorary Professor of the ASOIU. An International Scientific Base 'ResearchGate' shows the total number of his downloaded (read) publications as close to 210 000. In May 2022, he was elected a Foreign Member of the Georgian National Academy of Sciences.

Professor Anatoly Legchenko

Anatoly Legchenko is geophysicists working in the development of the Surface Nuclear Magnetic Resonance method. He graduated in 1979 as an engineer in computer sciences and electronics at the Electrotechnical Institute of Novosibirsk (NETI), Novosibirsk, Russia. In 1992, he received a PhD diploma in physics and mathematics of the Computer Centre of the Siberian Branch of Russian Academy of Sciences. In 2001, Anatoly was granted by a diploma of Habilitation à diriger les recherches (sciences de la terre) (HDR) of the University Paris XI, France.

Between 1981 and 1995, he was working in Novosibirsk (Russia), in the Institute of Chemical Kinetics and Combustion of the Siberian Branch of Russian Academy of Sciences. He started as an engineer in electronics, and then he focused his activities on the applied mathematics and geophysics being employed as a research scientist. In 1995, Anatoly joined the Bureau de Recherches Géologiques et Minières (BRGM) in Orléans, (France) and in 2003, he moved to the Institute de Recherche pour le Développement (IRD) in Marceil (France). He is actually attached to the Université Grenoble Alpes (IGE laboratory in Grenoble).

His professional interests are focused on the development of methods for non-invasive NMR water prospecting (conception, hardware and software) and a methodology of its practical application. He carried out numerous field studies of aquifers in different geological settings and at different scales, coupling geophysics and groundwater modeling. Groundwater-related natural hazards is one of the branches of his scientific activity.

Anatoly is one of animators of the international network for development and popularization of SNMR method (MRS workshops). He is an Associated Editor of European journal *Near Surface Geophysics* and reviewer in many scientific journals. Teaching of geophysics is one of his professional activities. Anatoly is an author and co-author of more than 100 scientific publications, including patents, research papers and books. In December 2021, his h-index (Web of Sciences) was 28.

Chapter 1

General introduction to the karst problem

This chapter is intended to introduce the readers to the worldwide karst problem. Because this book is devoted mainly to karstic geophysics, and especially to geophysics in the Dead Sea (DS) sinkholes we have to accent those problems that can be geophysically resolved. This Chapter is briefly composed based on glossaries and fundamental works of the well-known karst investigators (e.g., Monroe 1970, UNESCO 1972, U.S. Environmental Protection Agency 2000, Walthman 2003, Milanović 2005, Milanović *et al* 2019, Ford and Williams 2007, Gutiérrez *et al* 2014, De Waele *et al* 2017).

1.1 Definitions

Karst. Karst is the term used to describe a special style of landscape containing caves and extensive underground water systems that developed in especially soluble rocks such as *limestones, dolomites, gypsum, anhydrite halite, and their conglomerates* (Monroe 1970, Milanović 2005, Ford and Williams 2007). Karst is a result of very complicated processes that take place in soluble rocks in different climatic and hydrogeological conditions. Therefore, it is difficult to define the term 'karst' concisely. Nevertheless, by the term karst, we denote all the processes and phenomena connected with the *dissolution* of rocks by water. That takes place in areas with complex geological features and specific hydrogeological characteristics underlaid by soluble rocks. Karst areas are characterized by characteristic landforms that can be found both at the surface and below ground. Surface landforms commonly include enclosed depressions, sinkholes, sinking streams, caves, fluted rock outcrops, and large springs. Being a result of the solubility of the underlying bedrock, the karst areas, in general, are complicated by underground cavities and a high density of fractures. Note that rock solubility alone is not enough to form the karst phenomenon. There should be also developed secondary fracturing or porosity that together with solubility can induce karstic landforms. Also, the structure and lithology of rocks are important factors. For instance, in dense, massive, pure, and

doi:10.1088/978-0-7503-3635-2ch1

coarsely fractured rocks the best karst is developed. Thus, *karstic sinkholes require soluble rocks in the subsurface (limestone, dolomite, gypsum, salt, etc), whereas pseudokarst sinkholes require the existence of underground flow and physical erosion.*

1.2 Distribution of karst types around the world

1.2.1 Common background

The terrestrial Earth surface is covered by 64% sediments (a third of which, e.g. 20%–25%, are carbonates) (Hartmann and Moosdorf 2012). About 13% of the world's land surface is characterized by karst (Chen *et al* 2017). Figure 1.1 shows the distribution of the main karstic rocks. Their surface and near-surface outcrops occupy a total of 20% of the planet's dry ice-free land. Because of information availability, regional detail shown on these maps is of variable quality (Ford and Williams 2007). The mapping is also generalized and approximated; many very small outcrops are omitted, and possibly some large ones. The development of karst is a complex system driven by the dissolution of host rock and the subsequent removal of the dissolved matter by moving water. It is the process that, at various stages, initiates or triggers associated processes including erosion, collapse, and subsidence.

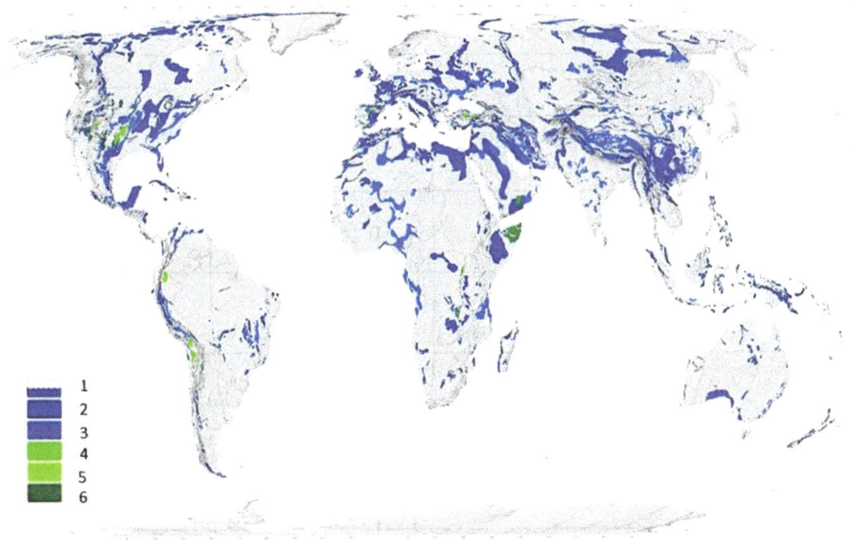

Figure 1.1. Map of the world distribution of the karstified rocks. Denoted: Karstifiable rocks/potential karst aquifers: (1) non-exposed/exposed karst aquifer; (2) continuous carbonate rocks; (3) discontinuous carbonate rocks; (4) continuous evaporites; (5) discontinuous evaporites; (6) mixed carbonate and evaporite rocks. (The map is modified by the Geological Survey of Ireland based on Chen *et al* (2017), with permission from Springer Nature). The map is presented at https://www.gsi.ie/en-ie/programmes-and-projects/groundwater-and-geo-thermal-unit/activities/understanding-irish-karst/Pages/What-is-karst.aspx.

Carbonates. Carbonate rocks are a class of sedimentary rocks composed primarily of minerals calcite or aragonite (limestone) and dolomite (dolomite rock named also dolostone. The formula of calcite is $CaCO_3$, whereas dolomite comprises also magnesium ($CaMg(CO_3)_2$). Calcite can be either dissolved by groundwater or precipitated, depending on several factors (the water temperature, pH, and dissolved ion concentrations). When conditions are right for precipitation, calcite forms mineral coatings that cement the existing rock grains together or it can fill fractures. Karst topography and cave development are usually in the form of carbonate rocks because of their solubility in dilute acidic groundwater. Cooling groundwater or mixing different groundwater flows will also create conditions suitable for cave formation. Marble is metamorphic carbonate rock. Rare igneous carbonate rocks exist as intrusive carbonatites and even rarer volcanic carbonate lava.

Not all carbonates display distinctive karst landforms and/or significant karst groundwater circulations because some are impure and their insoluble residues clog developing conduits. Taking into account an approximation of the karst distribution assessment, Ford and Williams (2007) estimate carbonate karst to occur over 10%–15% of the continental area.

Evaporites. Evaporite is the term for the water-soluble mineral sediments that result from the concentration and crystallization by evaporation from an aqueous solution (Jackson 1997). Two main types of evaporite deposits exist. First are marine ones, which can also be described as ocean deposits. The second group is non-marine evaporites, which are found in standing bodies of water such as lakes. Evaporites are formed by chemical sediments and consider sedimentary rocks. Because evaporites are formed by evaporation they should occur in a hot arid climate with small basins and a limited input of water. Evaporites occur in a hot arid climate with small basins and a limited input of water. When water becomes supersaturated by salts they precipitate. Evaporite karst is formed in evaporite rocks (gypsum, anhydrite, and salt). As follows from figure 1.2, evaporite karst has developed widely throughout the world. It can be found in all climatic zones Kozary *et al* (1968). At the same time, only seven karst areas in halite, two in anhydrite, and eleven in gypsum (figure 1.2(b)) have been studied for their cave minerals and speleothems, but most of these investigations are only preliminary ones (Forti *et al* 2017).

Anhydrite is an evaporite mineral that occurs in extensive layered deposits in sedimentary basins where large volumes of seawater have been evaporated. Anhydrite ($CaSO_4$), is also called *hydrous calcium sulfate*. Anhydrite consists of calcium, sulfur, and oxygen. The density of anhydrite varies between 2630 and 3000 kg m^{-3}, the solubility is 2.1 g L^{-1}, and the unconfined compressive strength ranges between 40 and 123 MPa. In comparison with gypsum, anhydrite is much denser and harder (on the Mohs scale, 3–3.5). It is typically interbedded with halite, gypsum, and limestone in accumulations that can be up to decades–hundreds of meters thick. Calcium sulfate (the compound that forms both gypsum and anhydrite) is the first to be deposited. Sodium chloride (halite compound) is next in the order of solubility. Sea-formed evaporates are generally thicker and are more widespread. Thus, (1) gypsum is hydrated while anhydrite does not contain water; (2) gypsum has a monoclinic crystal form, whereas anhydrite has an orthorhombic

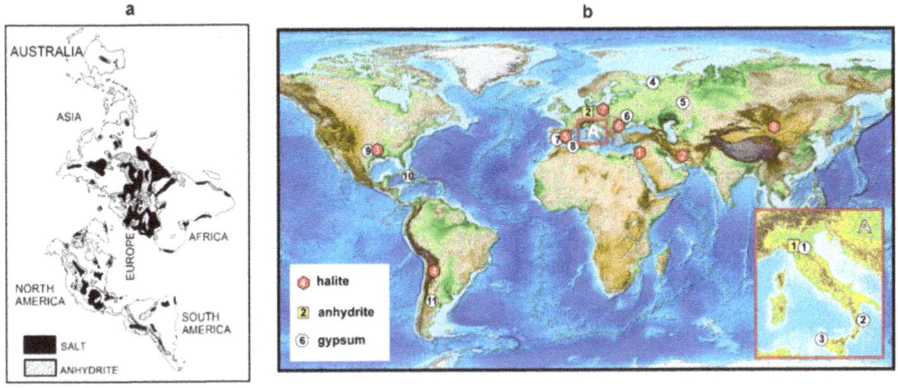

Figure 1.2. (a) Global distribution of the evaporite karst, (b) location of the evaporite karst areas where speleothems and cave minerals have been at least partially analyzed. **Halite**: (1) Mt. Sedom, (*also*, *DS buried karst*, M Ezersky) Israel; (2) Central Iran; (3) New Mexico, USA; (4) Atacama, Chile, (5) Cardona, Spain, (6) Mînzăleşti, Romania, (7) Wieliczka, Poland, and (8) Quinghai China. **Anhydrite**: (1) Upper Secchia Valley, Italy; (2) Central Germany. **Gypsum**: (1) Emilia Romagna, Italy; (2) Verzino, Calabria, Italy; (3) Santa Ninfa Sicily, Italy; (4) Pinega (Russia); (5) Kungur, Russia; (6) Podolia, Ukraine; (7) Almeria, Spain; (8) Darhedi, Algeria; (9) New Mexico, USA; (10) Punta Alegre, Cuba; and (11) Neuquén, Argentina. Panel (a)—reprinted from Kozary *et al* (1968), with permission of the Geological Society of America); panel (b)—reprinted with permission from Forti *et al* (2017), copyright *International Journal of Speleology*.

Figure 1.3. (a) Triassic anhydrite in the Upper Secchia Valley and (b) gypsum Hypogean karst form in Buco dei Buoi (Gessi Bolognesi) (Italy). Photos from Forti *et al* (2018), evaporite karst and caves of Emilia Romagna Region, with permission.

crystal structure; (3) whereas both minerals contain sulfur, anhydrite contains oxygen while gypsum does not contain it; (4) gypsum contains water while anhydrite does not, instead alters into gypsum when exposed to water (Frumkin and Forti 1997). Gypsum has a soft texture and can be easily scratched. Its crystals are very flexible and slim crystals can be slightly bent. Anhydrite is orthorhombic and does not react with hydrochloric acid. It is a rare mineral since many of its existing specimens have altered to become gypsum. Anhydrite usually occurs in arid places forming from the dehydration of gypsum. When exposed to water, anhydrite slowly turns into gypsum. An example of an anhydrite cave from the Emilia Romagna Region (Italy) is shown in figure 1.3(a).

Gypsum is defined as a soft sulfate mineral composed of calcium sulfate dihydrate, with the chemical formula $CaSO_4{\cdot}2H_2O$. Pure gypsum is white, but other substances found as impurities may give a wide range of colors to local deposits. Gypsum karst develops more rapidly than carbonate karst. Its distribution is limited neither by temperature (in the range from 0 °C to 40 °C) nor by the composition of natural waters. Gypsum karst is characterized by collapses, whose frequency can exceed two per km^2 per year, and a high density of sinkholes (up to 1000 per km^2) Gorbunova (1977).

It is widely mined and is used as a fertilizer and as the main constituent in many forms of plaster, blackboard/sidewalk chalk, and drywall. Surface outcrops of gypsiferous strata appear quite limited, evidently, because of the relatively low resistance of gypsum (a soft rock) to all types of denudation processes, rather than being a limited occurrence of sulfate rock deposits. De Waele *et al* (2017) report that gypsum has been distributed through all regions of Italy but comprises less than 1% of the country. These evaporite outcrops are very small and host-only mainly small caves. Therefore, in detail, the gypsum karst cavities were studied only in four regions: Piedmont, Emilia Romagna (Forti *et al* 2018, Caves of Emilia Romagna Region), Calabria, and Sicily. Gypsum belongs to the Permo-Triassic and Messinian ages, while anhydrite relates to Triassic sequences. An example of the cave in gypsum from the Emilia Romagna Region is shown in figure 1.3(b).

Recently, a massive quantity gypsum was discovered on the Dead Sea shores. This gypsum is connected with the rapid retreat of the Dead Sea during the past three decades. It is suggested that gypsum was formed by 'outsalting'—mixing of sulfate-bearing brines with variable salinities. The gypsum structures were formed during periods of regional aridity and low lake level (Weber *et al* 2021).

Sulfate rocks are widely extended at the shallow subsurface or the depth beneath them. Evaporites accumulated under carbonates or clastites are often removed by dissolution or are reduced by folding and thrusting. These rocks do not form outcrops (90% of anhydrite/gypsum and 99% of the salt).

The dissolution of sulfate rocks proceeds by different mechanisms and at different rates to those associated with the dissolution of carbonate rocks. The conversions of gypsum to anhydrite and back to gypsum are usual practices. As a rule, gypsum is converted to anhydrite and back. Gypsum forms at shallow depths, whereas at depths more than 450 m anhydrite is predominant. However, Klimchouk (1996) denotes numerous exceptions to this usual situation. The conversion of anhydrite to gypsum is a major significant process for karst development. It can be presumed that hydrogeological active karst within these evaporite rocks probably covers areas comparable to the active carbonate karsts. For example, Klimchouk (1996) mentions more than 35 countries and regions with recorded outcrops of gypsum.

Salt (halite), is the mineral (natural) form of sodium chloride (NaCl) (commonly known as rock salt). Halite typically occurs in arid climates as a result of ocean water evaporation. It is a sedimentary mineral. However, in some cases, halite is forming within inland lakes such as the Great Salt Lake of North America and the Dead Sea between Jordan and Israel (Abou Karaki *et al* 2016, Abelson *et al* 2017).

Belenitskaya (2018) describes several salt deposits that were formed when repeated episodes of seawater evaporation occurred in the restricted basins. Some of these deposits are thousands of meters thick. When these deposits are buried deeply, they can erupt to form a salt dome.

Many of the major sedimentary basins of the world contain salt (halite) strata. Saline deposits are distributed worldwide practically ubiquitously (figure 1.4).

Salt layers located at significant depths (>300 m) under impermeable and insoluble rocks are not subjected to dissolution by circulated groundwater. Such salt layers, because of their very low permeability, are often used for the disposal of radioactive waste (Rothfuchs *et al* 2004). However, there is no guarantee that even such layers will remain completely intact, especially at their peripheral parts. Milanović *et al* (2019) presume that the whole original ancient salt area will be more or less subjected to karstification.

Among all ordinary rocks, the evaporites are the most soluble. The karstification process is similar to that in other soluble rocks. The chemical dissolution added at a certain stage by the erosive acting of groundwater results in the formation of cavities and channels. With an increase in groundwater turbulence, the solute quantity is also increasing.

Evaporite karst rocks are often protected by less soluble or insoluble cover strata such as shales, carbonate, etc. Despite this protection, circulating waters can attack them and selectively remove them over large areas, even where they are buried as deeply as 300–1000 m. The phenomenon is termed interstratal karstification and

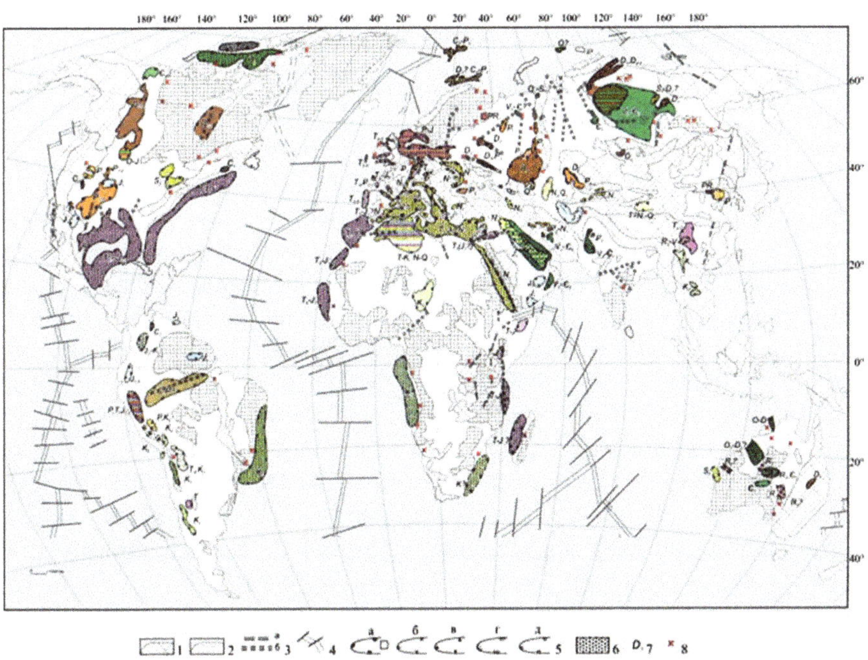

Figure 1.4. Map of salt bearing and salt dome basin of the world. After Belenitskaya (2018), copyright Scrivener Publishing LLC. With permission John Wiley & Sons.

may be manifested by collapse or subsidence of structures to the overlying rocks or at the Earth's surface (Ford and Williams 2007). An example of such karstification in Central Kansas has been described by Anderson *et al* (1998). The interstratal karstification occurs in carbonate rocks also but is of less significance. Intrastratal karstification refers to the preferential dissolution of a particular bed or another unit within a sequence of soluble rocks, e.g. a gypsum bed in a dolomite formation (Ford and Williams 2007).

Salt karst landscapes are limited to small patches in deserts. Interstratal solution of these rocks (with or without some surface expression) is widespread and very important (see an overview by Klimchouk 1996). The rocks are formed by homogeneous or heterogeneous precipitation in the marine, lake, or ponded waters that have been concentrated by partial evaporation, or as residues left by the complete evaporation. Seawater is quantitatively the most important source: restricted lagoons, plus sub-, inter-, and supratidal salt flats (known collectively by the Arabic term sabkha) are the main depositional environments (Ford and Williams 2007).

The karst formation in rock salt differently depends on the climatic factors—salt layers located above the local base level of erosion are strongly influenced by it, whereas layers below it are negligibly affected or even practically uninfluenced. That is why there are no salt outcrops in the humid climatic zones where they are quickly dissolved by the rains. On the other hand in an arid climate, we can see the presence of considerable surface salt (Korotkevich 1970).

1.2.2 Karst in the Mediterranean countries

This book is mainly devoted to consideration of the Dead Sea evaporite karst. Carbonate karst is widely described in the literature (Walthman 2003, Beck 2004, Milanović 2005, Ford and Williams 2007). We will consider here evaporite karst as an example of the Mediterranean region, where all types of karst are presented, e.g. gypsum and anhydrite karst in Italy. In Spain, most of the evaporitic formations are made of Ca-sulfates (gypsum/anhydrite) or Ca-sulfates and halite (Gutiérrez *et al* 2008a). Buried salt karst in the DS region causes the Dead Sea sinkholes. Evaporite deposits, such as gypsum, anhydrite, and rock salt, underlie about one third of the United States but are not necessarily exposed at the surface (Epstein 2005). At the same time, carbonate karst is also distributed in the Mediterranean region (Frumkin 2001, Bakalowicz 2015). Thus, our overview can be useful enough to understand the features of the karst in general.

Evaporite karst in Italy. Gypsum has been distributed in almost all Italian regions, which represents less than 1% of the country (De Waele *et al* 2017) As a rule, most evaporite outcrops are very small, outcropping rarely in Italy. Evaporite (gypsum and anhydrite) karst has been described in detail since the early 20th century. The gypsum cavities are mainly located along the northern border of the Apennine chain (Emilia Romagna and Marche), Calabria, and Sicily (figure 1.5).

More or less detailed studies on gypsum karst have been carried out in only four regions: Piedmont, Emilia Romagna (Klimchouk 1996), Calabria, and Sicily.

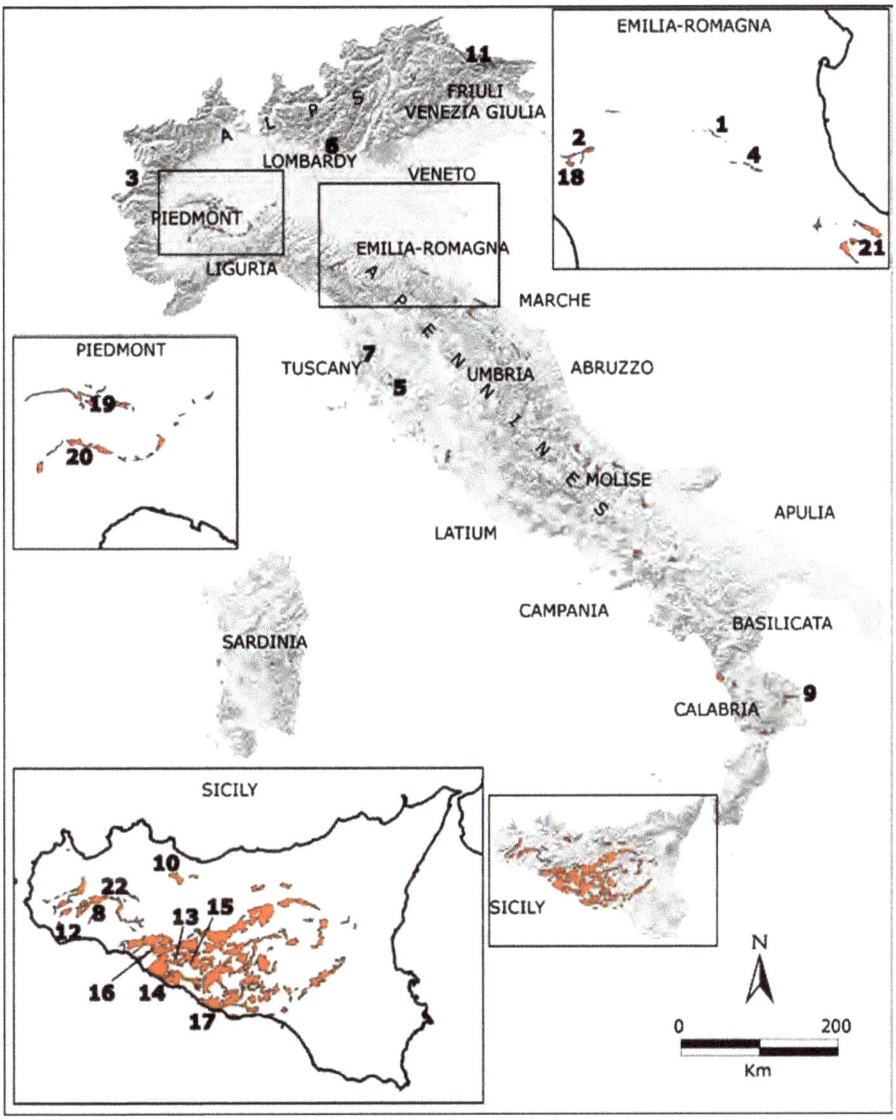

Figure 1.5. Map of distribution of the main gypsum outcrops in Italy. Digits denote main caves. Reprinted with permission from De Waele *et al* (2017), copyright *International Journal of Speleology*.

Gypsum is of Permo-Triassic and Messinian age, while anhydrite belongs to some Triassic sequences (Cucchi and Piano 2003). Recently, important caves have also been discovered in the underground gypsum mines in Piedmont.

Different Italian gypsum karst landscapes are shown in figure 1.6. They are different and complex and consider the various geological, environmental, and climatic conditions characterizing the country. From the morpho-structural

Figure 1.6. Large forms in Italian evaporite areas. (A) Truncated-conical solution drawdown doline in selenite gypsum (Monte Misciotto, NW Sicily). (B) Asymmetrical dolines with intermediate features between doline and blind valley (Ciminna Basin, NW Sicily). (C) Blind valley cut in clayey rocks ending at the foot of a selenite gypsum relief where the Grotta di Sant'Angelo Muxaro opens (Sant'Angelo Muxaro, central-western Sicily). (D) Gypsum polje developed in a faulted syncline depression (Siculiana, south-central Sicily). (E) The Vena del Gypsum ridge with its visible gypsum beds and the blind valley of Rio Stella (Romagna). (F) The Secchia River cutting the Triassic anhydrite (Reggio Emilia). (G) The Spipola Doline in the gypsum outcrops South of Bologna. (Photos A, C–D: Marco Vattano B, Antonino Lo Bello E, Piero Lucci F, Giovanni Bertolini G and De Waele J.) Reprinted with permission from De Waele *et al* (2017), copyright *International Journal of Speleology*.

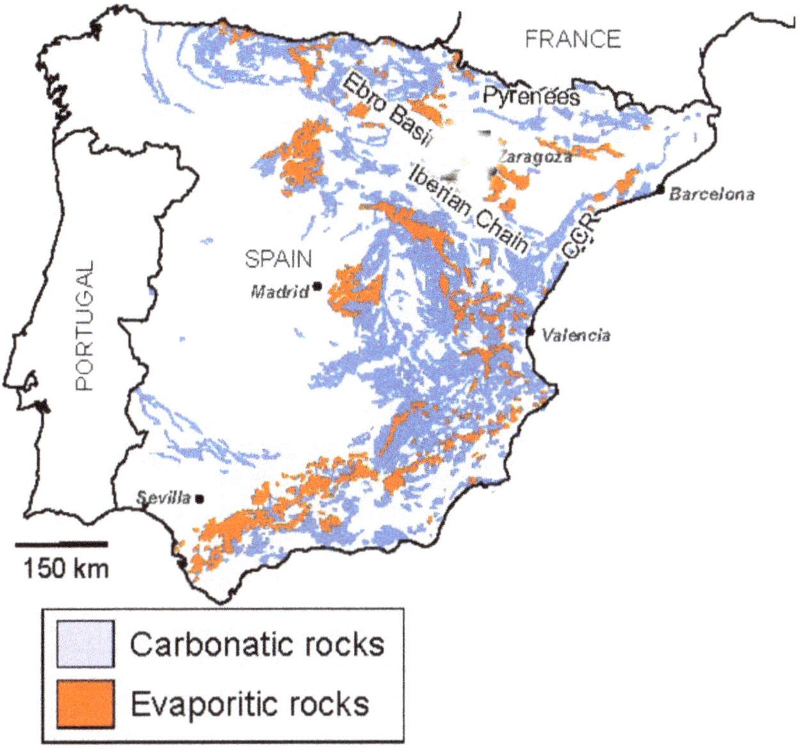

Figure 1.7. Distribution of the main evaporite outcrops in Spain and principal fluvial systems affected by the syn-sedimentary subsidence phenomena caused by evaporite dissolution. The small sketch shows the main geological domains in the Iberian Peninsula. Reprinted with permission from Pueyo Anchuela *et al* (2017), copyright *International Journal of Speleology*.

standpoint, the most common styles are tabular plateau, homoclinic ridges, fault scarps, folded relief, and isolated large gypsum blocks surrounded by clays (Forti and Sauro 1996).

Karst in Spain. In Spain (figure 1.7) evaporite outcrops cover more than 35 000 km^2, approximately 7% of the total country area (~500 000 km^2) (Macau and Riba 1962). Most of the evaporite structures consist of Ca-sulfates (gypsum/anhydrite) or Ca-sulfates and halite. Certain Paleogene marine evaporites also contain K–Mg-chlorides, and some tertiary continental formations bear substantial amounts of Na-sulfates in the subsurface (glauberite and thenardite). Evaporite dissolution leads to the hydrochemical degradation of surface waters.

Most of the gypsum outcrops are secondary gypsum produced by the hydration of anhydrite and, in the case of the continental evaporite bearing Na-sulfates, the incongruent dissolution of glauberite. Primary gypsum has been preserved only in some Neogene marine and terrestrial formations (Ortí 1989).

Karst in Israel (Frumkin 2001). The karst phenomena group are all a result of the subsurface water movement within carbonate formation. These phenomena are quite abundant in the central part of Israel, which is chiefly built in various types of carbonate rocks of the Late Cretaceous and early Cenozoic ages (Horowitz 1979). Karstic processes take place quite widely at present and most of the larger springs in Israel are fed by karstic aquifers.

In Israel, one can observe the variation of karst features: intensive karstification in Lebanon in the north and no karst in the Eilat region in the southern Negev desert. This is explained mainly by the corresponding climatological variations from the Alpine-Mediterranean climate in the Lebanon—Hermon Mts. in the north with precipitation more than 1000 mm yr^{-1} to the extremely arid southern Negev, with less than 50 mm yr^{-1} precipitation (Frumkin 2001). Evaporitic karst is characteristic of the Dead Sea area.

Mandel (1961) divided the active karstic processes into two main groups. The first comprises surface processes, which produce karst landscape and are presently active only in northern Israel. For instance, the first group is typical for Upper Galilee, where the annual rainfall exceeds 750 mm. These surface karstic landscapes were described in detail by Gerson (1974). The second group encompasses subterranean karst, which results in the formation and widening of caves and karstic aquifers. In Samaria, Judea, and Hebron hills, surface karst phenomena are presently not formed because of the insufficient rainfall, which is less than 750 mm yr^{-1}; this amount is noted by Mandel (1961) as the limit for the surface karst formation.

Carbonate rocks of Israel are predominantly of the Jurassic to Eocene age. The carbonate outcrops are seen in approximately 75% of the regions of Israel. Carbonates contain marls that hinder intensive karst development. Because of a lot of faults in the Hermon, Galilee, and Samaria regions the development of karst is also restrained. The faults constrain the continuous underground flow, limiting the development of large caves in these regions. It seems most limestone caves do not show any genetic relations to subaerial topography. Nowadays they are either dry or experience vadose drip water. They developed at moister conditions than those today. Frumkin (1994) suggests that these cavities can contain paleoclimatic records.

The massive underlying rocks are usually crossed by only a few bedding planes and fractures, in some places enlarged by dissolution (figure 1.8(a)). These are mostly presented by vertical shafts with rare horizontal sections (figure 1.8(b)).

Mentioned here the largest salt karst cave known in the world is the Malham cave 5685 m.

There are also some sea caves, formed by sea wave action with limited dissolution, attributed to chalk and the ‘kurkar’ sandstone.

1.2.2.1 Major karst regions

Major karst regions of Israel are shown in the map figure 1.9 (Frumkin 2001).

Hermon (1). There is Mt. Hermon at the north of Israel that in its higher part shows characteristic features of alpine karst (figure 1.10). Its elevation within Israel’s border is 2200 m. Several months per year the mountain is covered by snow.

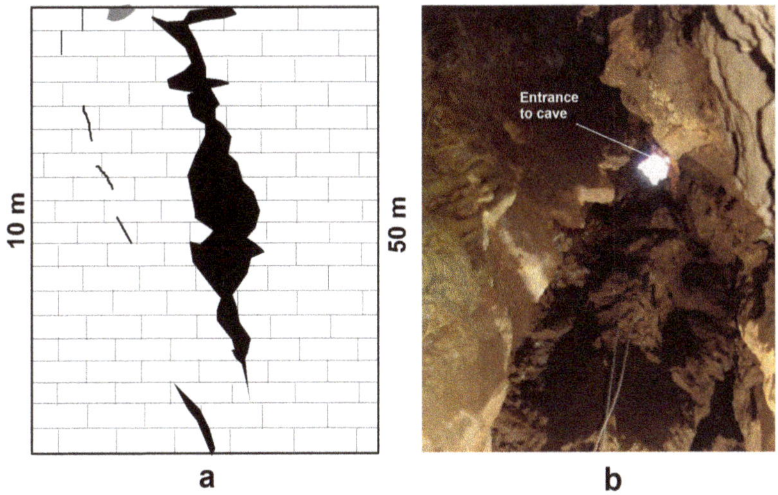

Figure 1.8. Karstification of carbonate in Israel. (a) A fracture enlarged by dissolution in limestone just under the epikarst. The height of the displayed area is 10 m Jerusalem, Israel. (b) A vertical active shaft developed along fractures is probably the most common vadose karst feature. The wall displays dissolution by films of descending water, condensation corrosion, and speleothems. David Cave, Ofra, Israel; the person is for scale. The photo is from A Frumkin, with permission and thanks.

Massive Jurassic limestone, several hundred meters thick, gives rise to development holokarst with the doline field and poljes. There are two known caves above 2000 m. One is Me'arat (e.g. Cave) Pitulim and the second one is Me'arat Mizpe Shelagim. Both are only 10 m long. On the Golan Heights (Ramat HaGolan) located south of Mt. Hermon composed of basalts can be found some pseudokarst features.

Galilee area (2). The Galilee Mts. are composed from Cretaceous to Eocene carbonates. This area is most karstic. The largest caves of the Galilee are Me'arat Sharakh and Me'arat Alma (shown in figure 1.11). The latter is a tectonic cave altered by the phreatic dissolution.

The small cave Me'arat Hakeshet is located in the western Galilee's Rosh HaNikra ridge, which is situated some 400 meters above sea level (figure 1.12). A geological rupture raised the ridge on its southern and south-eastern sides, creating steep slopes that look like a wall above the Nahal Betzet. However, on the northern side, the slope is gentler.

Carmel area (Mt. Carmel) (3). The Carmel ridge is composed mainly of Upper Cretaceous carbonates. The fluviokarst is commonly encountered, whereas the karstic closed depressions are rare.

An interesting example is the Misliya cave (figure 1.13), also known as Brotzen Cave, which is a collapsed cave at Mt. Carmel containing archaeological layers.

Large archaeological and extensive geophysical investigations were carried out here by Weinstein-Evron *et al* (2003). The performed excavations confirm the presence of relatively deep archeological deposits in the north-central portion of the

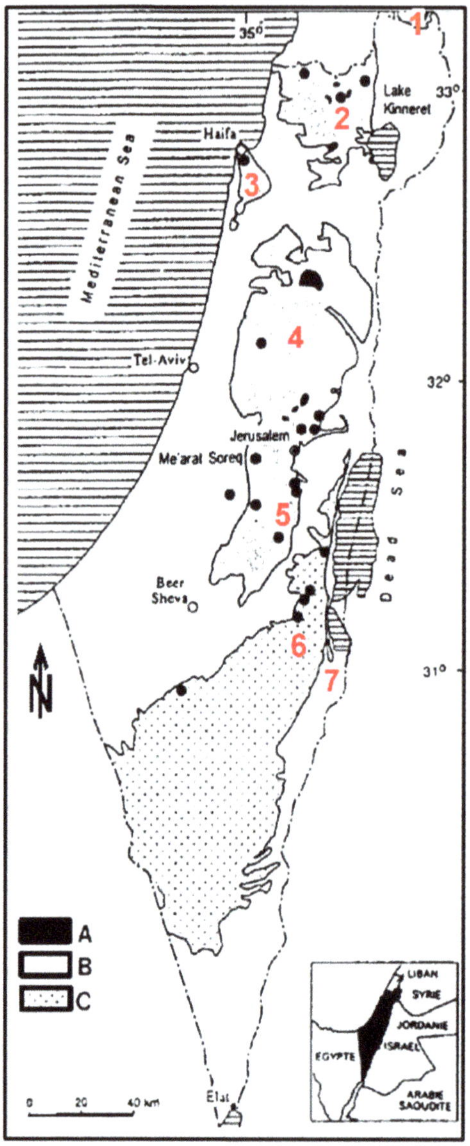

Figure 1.9. Karst regions of Israel. A—karst depressions, B—A Mediterranean type of karst, C—arid zone with carbonate rocks. Black circles—caves longer than 200 m, listed from north to south. (1) Mt. Hermon, (2) Galilee: caves Sharakh, Alma, Jermak, Bereniqi; (3) Karmel: Me'arat Ornit; (4) Samaria: caves Nahal Qana, Me'arat Na'ashneh, Yogev; (5) Judea: caves Makukh, Hameraglim; El-Gai, Atarot, Masaia, Pitria, Hariton, Tur Safa, Qneitra. To the west of Judea is Me'arat Niqbot Hamaim; (6) Judean Desert—Negev desert: caves Ha'igrot, Qann'im, Kidod, Zavoa, Ashalim; (7) Mt. Sedom. Reprinted from Frumkin (2001), copyright OCLc.

upper terrace of the site. Significantly, in places where these sediments were not covered with the talus, the high resistivity of the top layer may be related to the strongly lithified upper, occasionally exposed surface of the archeological sediments, themselves consisting of a rather compact breccia.

Figure 1.10. Mt. Hermon (north of Israel) (photo of Almog, with permission, CC0 Lisence, December 2005, https://commons.wikimedia.org/w/index.php?curid=2181987). This image has been obtained by the authors from the Wikimedia website, where it is stated to have been released into the public domain. It is included within this article on that basis.

Samaria area (Shomeron) (4). In Samaria, most karst landforms appear in Cretaceous limestone and dolomite. The Me'arat Nahal Qana is the largest cave. It is a tectonic maze with a phreatic chamber.

Judea area (Yehuda) (5). The karst features formed here in Cretaceous to Eocene carbonates. The absence of major faults has promoted the development of the largest and most abundant limestone caves in Israel.

There is the longest limestone cave Me'arat Hariton (Charitun) with 3450 m of mapped passages. Me'arat Soreq (named also Avshalom cave) is the geologically most studied cave (figure 1.14).

Me'arat Soreq is located in the western slopes of the Judean Hills, south of Nahal Soreq and approximately 2 km east of Bet Shemesh. The cave is 83 m long, 60 m wide, and 15 m high. The temperature and the humidity in the cave are constant year-round. The cave was discovered accidentally by workers blasting at the nearby quarry. This small cavern contains a wondrously varied world of stalactites and stalagmites. Some formations are at least 300 000 years old and can help scientists to track the climate changes in our region over millennia.

Figure 1.11. Alma Cave near Rehaniya (Upper Galilee, Israel): the way down to the inner entrance to the cave. This 'Alma Cave' image https://commons.wikimedia.org/wiki/File:Alma_Cave,_Upper_Galilee_-_02_-_descent_to_the_inner_entrance.jpg has been obtained by the authors from the Wikimedia website where it was made available by Dmitri Babin under a CC BY-SA 3.0 licence. It is included within this article on that basis. It is attributed to Dmitri Babin.

Figure 1.12. The Keshet Cave (a natural arch) located on a ridge near the Nahal Betzet. This 'Keshet Cave' image https://upload.wikimedia.org/wikipedia/commons/6/6f/PikiWiki_Israel_4518_Keshet_Cave.JPG (1600×1200) has been obtained by the authors from the Wikimedia website where it was made available by 'Ella Faust' under a CC BY 2.5 licence. It is included within this article on that basis. It is attributed to 'Ella Faust'.

Figure 1.13. Megadim Cliff, Mt. Carmel, in the center is the Misliya Cave. This image has been obtained by the authors from the Wikimedia website where it was made available by 'Hanay' under a CC BY 3.0 licence. It is included within this article on that basis. It is attributed to 'Hanay'.

Figure 1.14. Me'arat Soreq (named also Avshalom) stalactites and stalagmites. This image https://en.wikipedia.org/wiki/Avshalom_Cave#/media/File:Stalactite_Formation_in_Avshalom_Cave.jpg has been obtained by the authors from the Wikimedia website where it was made available by 'Sir_Joseph' under a CC BY-SA 3.0 licence. It is included within this article on that basis. It is attributed to 'Sir_Joseph'.

Judean Desert (Midbar Yehuda)—**Negev desert** (6). This region is arid, with precipitation lower than 200 mm yr^{-1}. Karstic caves are not known in the extremely arid southern Negev.

Mt. Sedom (7). Mt. Sedom is a salt diapir situated on the southwestern shore of the Dead Sea, which has been rising above the local base level throughout the Holocene. Karst's development within the salt body has kept pace with the rise

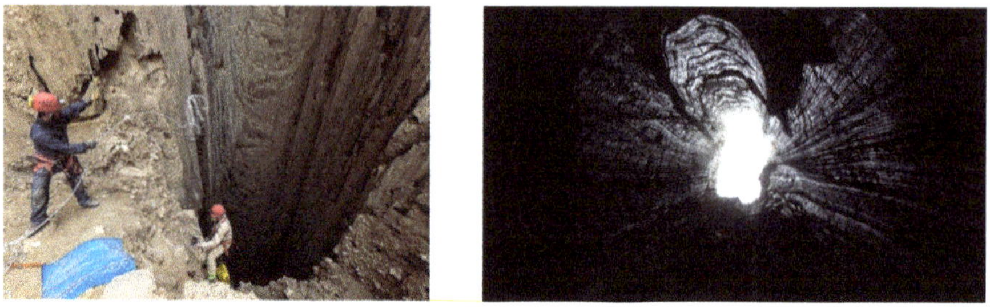

Figure 1.15. Me'arat Malcham in Mount Sedom has been discovered by the team of the Hebrew University of Jerusalem: (*Left*) entrance to cave; (*right*) View from within (Photo of Anton Chikishev/Hebrew University, with permission of NoCamel). salt-cave-israel-sedom-dead-sea.png (1822×1184) (greenprophet.com).

forming subhorizontal cave passages with vertical shafts. Salt diapir is extensively karstified despite its extremely arid climate due to the high solubility of rock salt. Some 100 caves have been studied, all of which were developed during the Holocene (Frumkin *et al* 1991).

The caves are branchwork vadose multilevel channels. The denudation and downcutting rate in the salt caves are some of the most rapid in karst (Frumkin and Ford 1995). The Me'arat Malham (figure 1.15) is the longest salt cave known in the world, with 5685 m of mapped passages. There are other caves longer than 200 m, Me'arat Sedom, for instance. Israeli researchers have just determined that a cave located in the Dead Sea southern point is the world's longest salt cave. Malham Cave, located on Mt. Sodom, has been discovered to span about 10 km, beating the earlier record agreed upon by scientists by almost half. The cave is composed of salt of unique composition like the Dead Sea salt. In most places in the world, the high solubility of salt caves has resulted in the existing primarily underground. They just dissolve by the rain and by the water. The unique conditions of Mt. Sodom in Israel are because the salt meets an arid climate with very low precipitation (50–100 mm yr^{-1}), then the salt can accumulate above the surface.

1.2.2.2 Dead Sea karst

The Dead Sea (DS) karst is a young phenomenon that occurred starting from the end of the 1980s to the present day that manifests itself as multiple sinkholes forming along the western and eastern shores (Taqieddin *et al* 2000, Yechieli *et al* 2006, Closson and Abou Karaki 2009, Abou Karaki *et al* 2016). The DS karst is caused by the dissolution of buried salt layers located at the depth of 20–50 m from the surface and the dropping of the DS level at the average rate of 1 m yr^{-1}. The DS sinkholes will be considered in continuation.

1.3 Some aspects of the karst problem

1.3.1 Karst classifications

In karst regions with complex geological structure, it is necessary to identify karst features that subdivide the regional karst settings into the types and subtypes of karst with some common characteristics. Several parameters can be used to make this division feasible (Milanović 2000).

Cvijić (1926) divided karst into three types basing on the karst topography that has been found in many areas of the world: holokarst, merokarst, and transitional type. It was the first karst classification.

Holokarst (complete karst). A karst area underlying thick carbonate rocks that is characterized by well-developed karst surface topography from the karren to poljes, extensive subsurface karst features like caves, caverns, galleries, chimneys, etc making feasible further development and creation of new karst phenomena (UNESCO 1972). In such areas we can see bare surfaces on thick deposits of limestone, well-developed karren, dolines, uvalas, poljes, and deep ponors, and extensive cave systems; they have small or no surface drainage. These areas give very specific appearances in the holokarst regions.

Merokarst (incomplete karst). Defined by Cvijić (1926) to indicate imperfect karst topography as found on thin, impure, or chalky limestone where surface drainage and dry valleys are presented in addition to some karstic features (Monroe 1970). On carbonate rocks (bituminous, marvelous, dolomitized), the karst phenomena are infrequent and the depth of karstification is limited.

The transitional type of the degree of karstification is between the holokarst and merokarst. The transitional type is usually formed in limestone isolated by impermeable and less soluble sediments (Milanović 2005).

Another classification includes two different types of karst: platform karst and geosyncline karst (Komatina 1973).

Platform karst is characterized by the horizontal or gently sloping strata and by platform relief. Carbonate rocks usually contain a higher percentage of marly material; therefore, their karstification is hindered.

Geosyncline karst develops in the distinctly folded and ruptured or faulted carbonate rocks. The synclinal fold regions with favorable climatic conditions are excellent environments for the maximal development of the karstification processes.

In this type of karst, its hydrogeological features are significantly more complex, and the karst processes are more active than these processes in latform karst. The solution activity of the carbonate rocks in the geosynclinal karst considerably exceeds the solution activity of the karst platforms (Komatina 1973).

By the period of formation, two types of karst can be separated: **paleokarst** (fossil karst) and **recent karst** (Milanović 2005).

Ford and Williams (1989) and Waltham and Fookes (2003) suggested the new 'full engineering description (classification) of carbonate karst'. These are as follows. (a) **Glaciokarst** that has extensive 'bare' rock surfaces with limestone pavements, rock scars, and deeply entrenched gorges. (b) **Fluviokarst** that has extensive dendritic systems of dry valleys, cut by rivers before they were captured by underground drainage into caves. (c) **Doline karst** that has a polygonal network of the interfluves separating closed depressions (dolines), each 100–1000 m across, it is a mature landscape, developed in temperate regions with the Mediterranean climates; e.g. the classical karst of Slovenia and the low-lying karst of Florida. (d) **Cone karst** (fencing karst) is dominated by repetitive conical or hemispherical limestone hills, 30–100 m high; it is a very mature landscape, largely restricted to inter-tropical regions. (e) **Tower karst** (fenglin karst) forms the most dramatic karst landscapes with isolated, steep-sided towers rising 50–100 m above-alleviated karst plains; it is the extreme karst type restricted to wet tropical regions with critical tectonic uplift histories.

There are recognizable subdivisions of these main karst types, and there are also additional landscape styles, e.g. formed in the arid regions where karst development is minimal. Construction practice is inevitably related to these geomorphological types, but an engineering classification of karst is more usefully based on the specific features that have the major influence on the ground conditions, namely caves, sinkholes, and rockhead morphology.

1.3.2 Karstification process

Karst is a synonym for carbonate rocks (limestones, dolomites), including all their conglomerates with carbonate matrix although the karstification process occurs also within the formations of gypsum and salt. **Karstification** is an epigenic (e.g. shallow subsurface) process of calcite dissolution in the course of infiltration of water including CO_2 from the atmosphere and soil (figure 1.16).

Karstification includes underground drainage systems with sinkholes and caves. The formation of caverns and channels is the direct result of a chemical dissolution, which at a certain stage of the process could be supported by the erosive action of water.

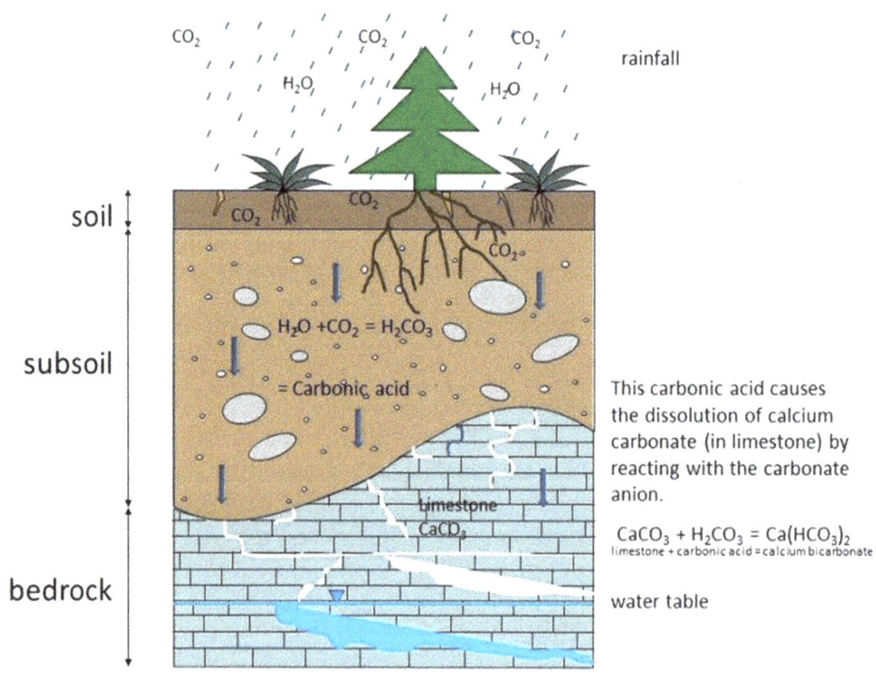

Figure 1.16. Karstification scheme in carbonates. Reprinted from the public site of Geological Survey of Ireland, with permission of Dr C Hickey.

1.3.3 Solubility of rocks

Soluble rocks dissolve the content of the formation in caves and cavities that may eventually collapse or act as holes into which the overlying materials funnel, causing subsidence at the surface. The collapse mechanism can be natural or man-made and its major trigger is infiltration of water (e.g. by heavy rainfall, burst pipes, or leaking drains). The rate of dissolution (limestone removal) depends on the chemical dissolution (or corrosion) and the mechanical erosion corrosion process. Both depend on several factors that make the exact rates difficult to determine. Benson and Yuhr (2016) denote that dissolution of carbonates is determined by many environmental factors. It is established that the natural dissolution of carbonate rock is highly variable, 1–3 cm/1000 years is an average range of values (Ford and Williams 2007). In general, the natural dissolution of limestone is a very slow geologic process. This process can be accelerated by man-made activity (leakage from the dam reservoirs, acids leaked from industrial facilities, etc).

The dissolution of gypsum is approximately two orders of magnitude greater than that of limestone. Under high groundwater gradients such as beneath dams, dissolution rates for gypsum may be as high as 10 mm yr^{-1} (Dreybrodt *et al* 2002).

Dissolution rates of gypsum and carbonates measured in Triest (Italy) have shown that average rates of dissolution of gypsum samples are roughly 30–70 times greater than the dissolution rates of carbonates, that after Klimchouk (1996) corresponds broadly with the theoretical expectations. Ford and Williams (1989) denote that there is relatively small information on the rates of rock salt dissolution. However, available field evidence shows that the development of solution features in rock salt is very rapid because of its great solubility. Frumkin and Ford (1995) measured the surface salt dissolution rate in Mt. Sedom (Israel) of 0.5–0.75 mm yr^{-1}. Salt (halite) solubility in water is 35% by weight at 25 °C and it increases at a higher temperature. Ezersky and Frumkin (2017) have shown that the DS salt is dissolved in the groundwater comprising in pores brine with 50–110 $g_{chloride}$ L^{-1} (or 22%–50% of that in saturated conditions, respectively). The high solubility of salt allows speleogenesis (i.e., cave development) to occur at a very short time scale, so salt caves are the fastest evolving karst caves (excluding ice cavities).

Generally, the relative solubility of limestone, gypsum, and rock salt under the same geologic conditions, is at the ratio of 1:100:10 000 (Reuter *et al* 1993). Thus, karst in the chloride rocks differs from the karst developed in the carbonate and sulfate rocks because of its much higher solubility. The rock salt is often located at a depth of 200–300 m and its dissolution rate is estimated as 0.3–0.5 mm yr^{-1} (Reuter *et al* 1993). Radiocarbon dates indicate that an entire cave or a single cave level may develop to accessible dimensions in a mere few hundred years (Frumkin and Ford 1995), compared to more than 10 000 years in limestone (Mylroie and Carew 1987). In accordance to Walkden (1974) karstification in carboniferous limestones, for instance, continued for 48 000 years and more. With that, the surface was lowered to 50 m. White (1988) evaluates the time of cave development in carbonates to 10 000–100 000 years.

Also, Milanović *et al* (2019) denote that many dams and reservoirs all over the world constructed in geological formations containing evaporites have encountered significant seepage or stability problems during their construction. All these problems are consequences of the high solubility of the common evaporite rocks, which are chiefly gypsum and salt.

1.3.4 Influence of tectonics in process karstification

Extensive joint systems within the very brittle limestones and dolomites are formed under the action of intensive tectonic processes. Because the karstification process is a result of the water penetrating permeable and soluble rock masses, fragmentation of the latter represents the most important factor of karstification, which operates both horizontally and vertically. That provides access to water that can migrate into deeper sections of the thick rock mass. As a consequence, the dissolution occurs within the rock since water moves within fractures. The result is the formation of a cave (dissolution cave). The surface collapses into the cave forming a sinkhole on the Earth's surface. Rock salt when deformed tectonically, behaves plastically resulting in no, or very few, joints. Therefore, groundwater can only contact the top of the rock salt. As a result, subsidence of the ground appears. If sinkholes exist, they are a function of the geomechanical properties of overlying rocks and processes after the actual dissolution (Reuter *et al* 1993, Milanović 2005).

1.3.5 Subsidence

Subsidence is the sudden sinking or gradual downward settling of the ground's surface with a small horizontal motion (sometimes it is absent). The terms *subsidence* and *collapse* usually are used as synonyms. The subsidence can be caused both by natural processes such as various karst phenomena, consolidation, normal faulting, etc, or human activities like subsurface mining or extraction of petroleum, natural gas, or groundwater (Neuendorf *et al* 2011).

Ground subsidence is of global concern to geologists, geotechnical engineers, surveyors, engineers, urban planners, landowners, and the public in general (National Research Council 1991). When the cave is formed within limestone and its roof is weakened it can collapse and the overlying rock and earth will fall into the cave, causing subsidence at the surface. This type of subsidence can cause the generation of sinkholes which can be many hundreds of meters deep. Subsidence is common when karstified rocks are covered with unconsolidated sediments.

In the salt, areas that occur under natural conditions the subsidence rate ranges from 1 mm yr^{-1} to much less than that. Subsidence can be greatly increased by the human cases the subsidence rate may greatly exceed 1 mm yr^{-1} up to a measured rate of 1 mm d^{-1} (Reuter *et al* 1993).

The role of water can be distinguished in four different categories:

- Underground water acting from below;
- Surface water acting as floodwater;
- Pore water within the alluvial deposits (overburden);
- Water acting indirectly, pressurizing the air in the aeration zone.

The genesis of subsidence is usually related to these categories of processes.

1.3.6 Depth of karstification, epikarst zone

Karstification depth is the depth to which the soluble rocks were exposed to the karstification process, which may vary widely. The transition zone represents the surface below which does not contain karstification. This is called the *base of valleys or (karstification)* (Monroe 1970).

Following a karst glossary (US EPA 2002), the upper part of the karstification zone is called *epikarst* (or epikarst zone) that is defined as a relatively thick (the thickness may vary significantly, but 15–30 m thick is a good generalization) portion of bedrock that extends from the base of the soil zone and is characterized by the extreme fracturing and enhanced solution. It is separated from the phreatic zone by an inactive, relatively waterless interval of bedrock that is locally breached by the vadose percolation. According to Milanović (1979), epikarst is the zone of greatest karstification, and at the same time the highest porosity in subsurface ranging from 10 to 20 m depth.

Based on the studying of karst porosity, Legrand and Striengfield (1975) (published in Milanović 1979) have determined that karstification decreases with depth. The 'progressive concentration of water in the upper part of the zone of saturation produces master conduits there, but causes no appreciable increase in flow and solution at greater depths' (figure 1.17). The solid graph in the figure (graph 1)

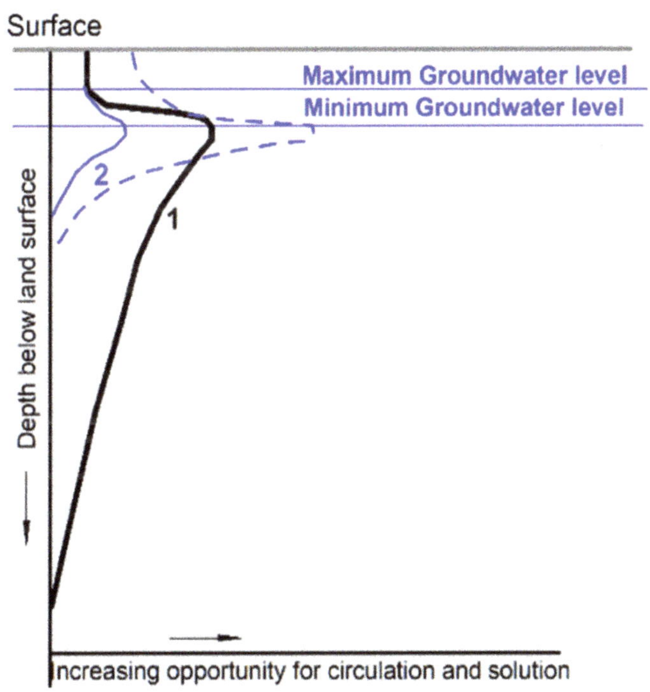

Figure 1.17. Graphs showing the degree variations of water circulation with depth below the land surface (Legrand and Striengfield 1975, copyright International Association of Hydrogeologists and UNESCO). (Was published in Milanović (1979) and modified by M Ezersky. With permission from CRC.)

represents the principal behavior, whereas the dashed and solid blue lines (graph 2) represent the less common but not unusual conditions.

Aquifers in carbonate rocks tend to have their permeability developed through the circulation of water and solution of the rock and to have an uneven distribution of permeability. The circulation of water and solution activity tends to be greatest in the upper part of the zone of saturation and tends to lessen with the depth increasing, i.e., karstification decreases with depth.

Ford and Williams (2007) suggested that the depth of active dissolution in the epigenic model is shallow. Water will accomplish 50%–80% of its solutional work within about 10 m from the surface.

The aforementioned authors have shown that generally karstification decreases with depth.

1.3.7 The porosity of karstified rocks

Porosity is one of the fundamental properties of karst rocks because it is closely connected with their hydraulic permeability (Schoen 2004). While the porosity of limestones varied generally from 1% to 36% permeability of such limestones varies from 10^{-6} to 10^{-2} (e.g., on 4 orders), respectively. Considering porosity and other physical properties of rocks we have to take into account that there is a scale effect in the rocks.

1.3.7.1 Carbonate rock porosity

Karst porosity consists of interconnected systems of fractures, karst channels, and caverns of various shapes, sizes, and fillings. According to its origin, karst porosity can be either primary or secondary.

Properties of carbonates can vary in the wide limits (Benson and Yuhr 2016) caves. Some types of porosity in rocks are shown in figure 1.18.

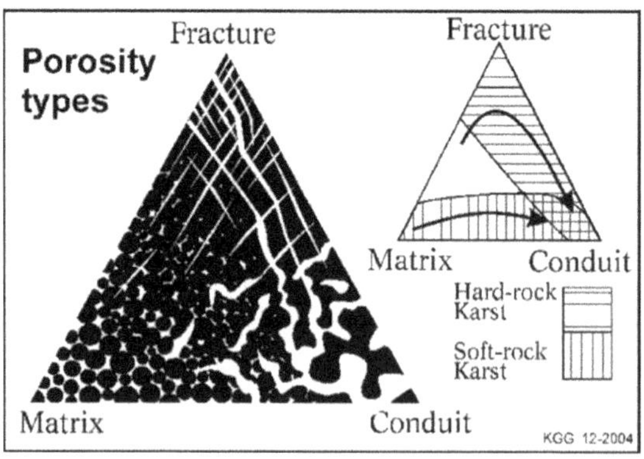

Figure 1.18. Limestone porosity types, (*left*) the three types of porosity and permeability which can each contribute to the overall karst porosity; and (*right*) evolution of the dominant permeability in hard rocks and soft-rock karsts. From Grimes (2006), with permission of Australian Speleological Federation Inc. and with gratitude to K Grimes and ASF.

Most caves have occurred in the hard rocks with low primary porosities. In such rocks, water flow is limited to fractures that then become enlarged by dissolution and erosion resulting in a cave system. The greatest karst development is generally seen in the massive thick-bedded limestone (Walthman *et al* 2005).

Primary porosity is of the syngenetic origin. It refers to the dissolutional development of karst features in newly-formed calcareous sediments, simultaneously with the cementation of the sediment into a rock. Syngenetic karst forms within a soft, porous, soluble sediment at the same time as it is being cemented into a rock (Jennings 1971). This type of karst porosity is not so distributed. It is found in the karst aquifers in Florida (Biscayne aquifer). It is also in the Miami region of Florida, at a depth of ~33 m (Stringfield 1966), and occurs on the Yucatan Peninsula in Mexico (Milanović 2005).

Vuggy porosity represents a specific type of karst porosity, consisting of small cavities lined with a crystalline, lenticular, or rounded shape. It varies in size from intercrystalline form to decimeter-size cavities. The vugs are partially connected, partly isolated, or disconnected. Larger vugs are often filled with clayey cave deposits. Sometimes they are completely plugged with calcite. This type of karstification often develops in single thick-bedded sequences. Karst aquifers with significant primary porosity are associated with the travertine masses.

Secondary porosity is that of some lithologic material that has developed after the rock was initially formed, such as faults and fractures, and may be capable of enlargement by the dissolution processes (Monroe 1970).

Secondary porosity consists of fractures, caverns, karst channels, and cave sediments (clay, sand, gravel, and terra-Rossa). This type of porosity is a more reliable characteristic of the karst terranes than the primary porosity. Different researchers calculated an average porosity ranged from 1.2% to 3.5% (Moor *et al* 1969, Torbarov 1976, Milanović 1979).

Effective porosity is the amount of interconnected pore space and fracture openings available for the transmission of fluids, expressed as the ratio of the volume of interconnected pores and openings to the volume of rock (Monroe 1970). Effective porosity varies in different rocks as 1%–2% (dolomites) and 1%–20% (sandy dolomite and dolomitic sandstone) in central Pennsylvania. At that, the average effective porosity of this sandy dolomite aquifer is estimated as 3%–5%. Bagrintseva (1977) reports the porosity of carbonate reservoirs of Central Asia in the range of 13%–23%. On this background porosity values for limestone (30%) and dolomite (26%) published by Morris and Johnson (1967) seems to be high enough.

There is also tertiary porosity, caused by solutional enlargement of the secondary porosity, and effective porosity that can be also the amount of interconnected pore space.

1.3.7.2 The salt porosity

Sedimentary salt rock is built of more than 95% halite and small amounts of concomitant minerals such as clay, calcite, etc. It has been well-studied because of its low permeability and suitability for the storage of radioactive waste (Baar 1977, Wittke and Schmitt 2000, Spiers *et al* 2005, WBI compilation 2006). The salt formations, as a rule, are formed at greater depth (300 m and more) at conditions of high confining pressures. That is why most of the salt studies were carried out at high pressure (Voronov and Grigor'ev 1976, Gevantman 1981, Popp and Kern 2000).

Sonnenfeld (1984) discussed the porosity of freshly precipitated evaporites, which he described as being initially very porous, loose, watery slush that gradually firms up, with a porosity of 10%–50%. The specific inclusion of halite in this group in the Dead Sea area is substantiated by the presently measured porosity values. Considerable diagenic changes are required to convert this material into a massive, compact deposit. After the burial and considerable compaction, the porosity of halite drops to less than 5%, and even to less than 0.5%. Hofrichter (1976) suggested that salt becomes tight (i.e., with effective zero porosity and permeability) under an overburden of about 100 m; Baar (1977) suggested that this salt occurs at depths exceeding 300 m.

Other authors have investigated a granular salt that was artificially compacted that is of importance for understanding porosity-dependent properties. Popp and Kern (1998) studied the dependence of ultrasonic velocities versus porosity and permeability in artificial dry salt specimens. Spangenberg *et al* (1998) have performed tests on wet granular salt samples to study interrelations between the porosity, permeability, and resistivity. In two latter studies, salt porosity was controlled and it could be considered as some reference. Several researchers have published articles devoted to the DS salt properties (Frydman *et al* 2008), dissolution and dynamics of permeability (Weisbrod *et al* 2012), and mechanical strength (Hatzor and Hayman 1997). All investigations were dealing with the laboratory testing of the salt samples. Ezersky and Goretsky (2014) have studied the porosity and permeability of the DS salt on the selected samples. The measured porosity varied from 3.2% to 21%. Properties of DS salt and its relationships with the elastic and electrical parameters will be considered in section 4.12.

1.3.8 Geomorphological features

There are numerous morphological forms associated specifically with karst regions. The most frequent are the gardens, dolines (sinkholes), dry valleys, shafts (potholes), caves, poljes, and karst plains (figure 1.19). The definitions of karst forms are given in Monroe (1970), Milanović (2005), and Ford and Williams (2007).

Enclosed (closed) depression. A general term for any enclosed topographic basin having no external drainage, regardless of origin or size. They are the most frequently occurring landforms of karst. They range in diameter from a few meters to tens of kilometers and in depth from a few centimeters to hundreds of meters. The smallest of these features are termed *dolines.*

Dolines (named commonly *sinks or sinkholes* in North America) are a basin- or funnel-shaped hollow in limestone. Some dolines are gentle grassy hollows; others are rocky cliff-bounded basins. A distinction may be made between those formed mainly by direct solution of the limestone surface zone, solution dolines, and those formed by collapse over a cave, collapse dolines, but it is generally not possible to establish the origin of individual examples. Dolines are the most common landforms

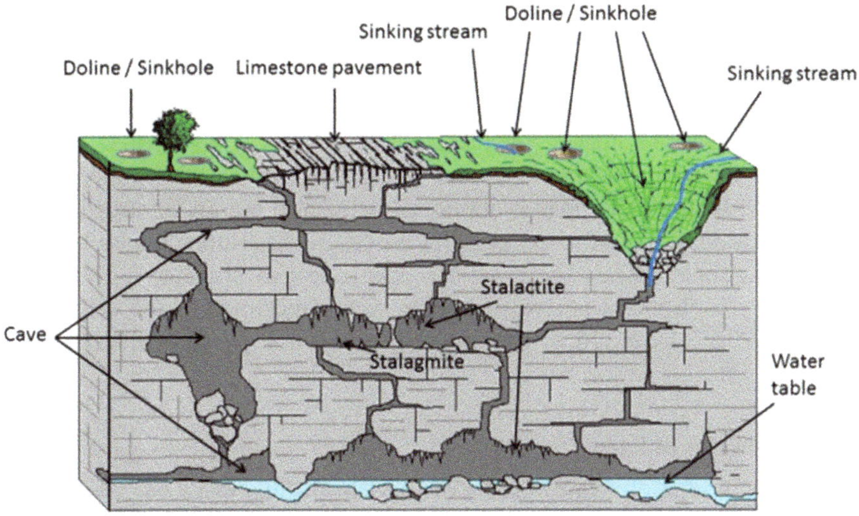

Figure 1.19. Karst landforms (Geological Survey of Ireland, Public domain. With permission Geological Survey of Ireland, Dr C Hickey).

in the karst areas. In most cases, they are results of the chemical action of water in soluble rocks. They are described as small-to-medium-sized closed depressions, ranging from meters to tens of meters in both diameter and depth. Sinkholes often start from the crossing of two and more faults. Another reason for sinkhole formation is the collapse of the cave roof and karst channel located at the shallow subsurface. The longer axis of the sinkhole is oriented along the longest fault, whereas a series of sinkholes denote the direction of flat or fault zone.

Karrens are channels or furrows by size from few millimeters to more than a meter, formed by solution on massive bare limestone surfaces; these channels are separated by ridges. Bögli (1980) describes the total complex of superficial solution forms found in the compact pure limestone. There are also known *cracked karrens* formed along cracks. They are rare in dolomite rocks and are quite small.

Dry valleys are valleys that lack surface water channels. Dry valleys are common in the carbonate rocks with expressed primary permeability and occur in other permeable rocks (such as sandstone) (Lowe and Waltham 1995). Lithological composition and tectonic activity have a considerable role in the formation of dry valleys. The active valleys can be transformed into the dry valleys where carbonate rocks are subject to tectonic uplift (Milanović 2000).

Caves and shafts are formed from solution channels 5–15 mm in diameter or width. This is an effective minimum opening for the turbulent flow (Ford and Williams 2007). *A cave* is a natural underground room or series of rooms and passages large enough to be entered by a human; generally formed by a solution of limestone.

Shafts are vertical or sub-vertical karst channels. They are located within zones of vertical fractures or systems that have been enlarged by karstification. Caves are less frequent karst forms than shafts. The shafts and caves with permanent and exclusive underground flows are considered to be the most important. They provide natural access to groundwater.

Polje (Slavic word means ‘field‘, whereas in English *polje* specifically refers to a karst plain or karst field). In areas of karst topography, a very large closed depression, in some places several kilometers long and wide, having a flat floor either of bare limestone or covered by alluvium, and surrounded by generally steep walls of limestone. The long axis of polje is developed parallel to the major structural trends

Superficial deposits tend to accumulate on the floor. Drainage may be either by surface watercourses (when the polje are open ones) or swallow holes (a ‘closed’ polje). Their development is encouraged by any impedance in the karst drainage (Sweeting 1977). It is the most prominent geomorphologic feature from a hydrogeological point of view.

1.3.9 Groundwater in karst

According to an often-cited estimation of Ford and Williams (2007), approximately 20%–25% of the global population depends largely or entirely on the groundwater obtained from karst. In some countries and regions such as Austria, the Dinaric region (Europe), and Southwest China, karst water contributes 50% or more to regional freshwater supplies (Klimchouk 1996, Wu *et al* 2009). There are large cities with more than a million inhabitants (San Antonio, Vienna, Rome, etc) that rely entirely or predominantly on the karst aquifers. However, about 25% of the world’s population lives in these regions and depends on the karst aquifers for its water supply. Horowitz (1979, p 161) reports that in the middle of the twentieth century karstic aquifers supplied more than 70% of water consumed in Israel. In some regions, such as the Dinaric region in Europe and Southwest China, karst water contributes to more than 50% of the water supply, and other areas, such as Damascus in Syria depend almost entirely on karst aquifers for water (Chen *et al* 2017).

1.4 Sinkholes, their classification, and nomenclature

Land subsidence can be defined as the sudden sinking or gradual downward settling of the Earth’s surface with small (or with absence of) horizontal motion, owing to the subsurface movement of earth materials (USGS 2013). Subsidence often occurs through the loss of subsurface support in karst terrain, which may result from several natural and human-caused occurrences. Karst is a distinctive topography in which the landscape is largely shaped by the dissolving action of water in carbonate bedrock (usually limestone, dolomite, or marble).

Sinkholes, a type of subsidence, are the natural and widely distributed geologic features in areas with underlying limestone, carbonate rock, salt beds, or other rocks that are soluble in water. Over periods measured in thousands of years, the carbonate bedrock can be dissolved through the acidic rainwater moving in fractures or cracks in the bedrock. This fact creates essential openings in the rock through which water and overlying soil materials will travel (figure 1.20).

Over time, the deposited soils compromise the strength of the bedrock, until it is unable to support the land surface above, and a collapse or sinkhole occurs.

Figure 1.20. A sinkhole opened in the Ben Shemen (Israel) near a farmer's house. The basement is composed of limestone (photo of M Ezersky).

Sinkholes or dolines are closed depressions with internal drainage that are characteristic features of karst landscapes. They display a wide range of morphologies (cylindrical, conical, bowl, or pan-shaped). Sinkholes vary in size from 1 to 1000 m in diameter and tens–hundreds of meters deep (Williams 2004).

The most frequently occurring and best-studied hazard in karst areas are dolines, which are often called sinkholes. Sinkholes very frequently occur in the evaporite terrains (Farrant and Cooper 2008, Gutiérrez *et al* 2008b), cause enormous economic losses in densely populated areas (e.g. the Ebro basin, Spain (Galve *et al* 2009)), and can also constitute significant hazards in limestone areas (e.g. Florida, USA, Brinkmann *et al* 2007).

Several researchers (Beck 2004, Williams 2004, Waltham *et al* 2005) have published similar classifications of sinkholes. These classifications consider two main categories of sinkholes: (1) those resulting from dissolutional lowering of the surface, and (2) those created by internal erosion and deformational processes caused by subsurface karstification (Gutiérrez *et al* 2008a). Correspondingly, the first group is represented by *dissolution sinkholes*. The second group includes four different types, depending on the type of material affected by downward gravitational movement (bedrock, caprock, and unconsolidated cover) and the subsidence mechanism (*collapse or suffosion*).

Walthman *et al* (2005) have compiled the classification available in the worldwide scientific literature (e.g., Sweeting 1972, Bögli 1980, Jennings 1985, Beck and Sinclair 1986, Culshaw and Walthman 1987, White 1988, Ford and Williams 1989, Lowe and Waltham 1995).

Analysis has shown that the most accepted classifications are solution sinkholes, collapse sinkholes, and suffosion ones (Ford and Williams 2007). Other classifications add some additional details associated with the type and properties of cover rocks. Bedrocks and caprocks can suffer brittle collapse, whereas cover deposits can be affected by both collapse and suffosion. This second group is the most important from an engineering perspective. In addition, the classifications include the term buried 'sinkhole' for those filled by sediments; compaction of the filling material may lead to the generation of compaction sinkholes (Waltham *et al* 2005).

United States Geological Survey suggests the following classification:

(a) Solution sinkholes (figure 1.21) are generated in the bare or uncovered karst by the differential dissolutional lowering of the ground where karst rocks are exposed at the surface. Dissolution of the limestone or dolomite is the most intensive where the water first contacts the rock surface and flow is focused in preexisting openings in the rock, such as along joints, fractures, and bedding planes. Dissolution also takes place in the zone of water-table fluctuation.

(b) *Cover-subsidence sinkholes (or suffosion sinkholes)* (figure 1.22) tend to develop gradually where the covering sediments are permeable and contain sand.

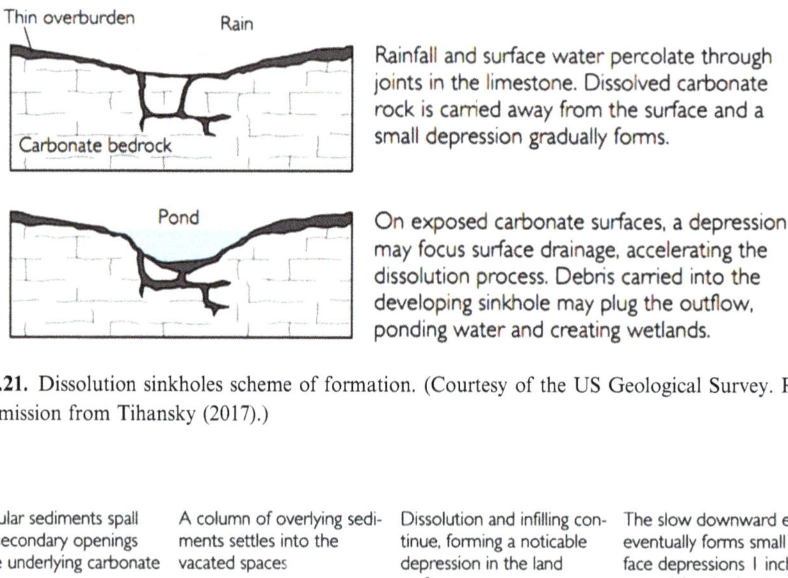

Figure 1.21. Dissolution sinkholes scheme of formation. (Courtesy of the US Geological Survey. Reprinted with permission from Tihansky (2017).)

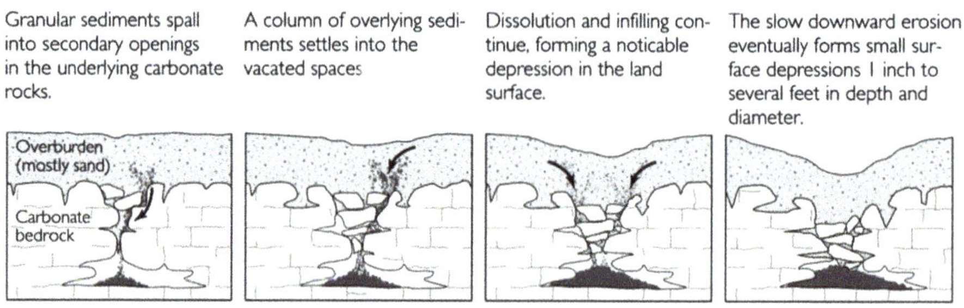

Figure 1.22. Cover-subsidence sinkholes. (Courtesy of the US Geological Survey. Reprinted with permission from Tihansky (2017).)

Figure 1.23. Cover collapse sinkholes. (Courtesy of the US Geological Survey. Reprinted with permission from Tihansky (2017).)

The process of cover-subsidence sinkholes is illustrated in figure 1.22. In areas where cover material is thicker, or sediments contain more clay, cover-subsidence sinkholes may go undetected for long periods.

If there are developed systems of fractures in carbonate rocks like figure 1.19, there can develop cover suffosion following the classification of Gutiérrez *et al* (2008a, 2008b) in figure 1.24. In this case, particles of sands can be washed out by descending water streams and remove through a system of fracture in carbonate.

(c) *Cover collapse sinkholes* (figure 1.23) may develop abruptly (over hours) and cause catastrophic damage. They occur where the covering sediments contain a significant amount of clay. Over time, the possibility of surface drainage, erosion, and deposition of sediment transform the steep-walled sinkhole into a shallower bowl-shaped depression. The stages of sinkhole formation are explained in figure 1.23.

1.4.1 Anthropogenic sinkholes

New sinkholes have been associated with land use, especially from groundwater pumping and construction (subway, for instance), etc. Sinkholes can also occur when natural water drainage is changed and new drainage systems are developed. Some sinkholes form when the land surface is changed, the result of creation of industrial and runoff-storage reservoirs. The water below ground is helping to keep the surface soil in place. Groundwater pumping for urban water supply can upset the natural balance to cause new sinkholes. The same can happen as a result of irrigation in sinkhole-prone areas. Also, pumping can result in a lowering of groundwater levels, which can trigger underground structural failure followed by sinkholes forming.

Walthman *et al* (2005) suggest classification explaining the main processes and mechanisms generating sinkholes in the carbonate karst terrains (dissolution, collapse, and suffosion).

However, we will use the classification of Gutiérrez *et al* (2008a) who suggested for evaporite karst areas that can involve a wider range of processes. In the proposed classification, types of sinkholes can be described using combinations of the different terms with a dominant material or process followed by a secondary one (e.g., cover and bedrock collapse sinkhole, cover sagging, and suffosion sinkhole).

Gutiérrez *et al* (2008a) suggests that every type of sinkhole formation is denoted by two terms: the first describes the material affected by downward gravitational movements (cover, bedrock, or caprock), while the second indicates the principal process involved (collapse, sagging or suffosion). This novel systematization follows the methodology of Beck's (2004) sinkhole classification.

An example of such classification is shown in figure 1.24 that presents a matrix where the horizontal line defines the process (collapse, sagging, and suffosion) and the vertical column defines the material (or layer) participating in sinkhole formation (cover, caprock, and bedrock).

Thus, the 'sagging bedrock' sinkhole denotes that the sinkhole formed because of the sagging of the bedrock. Figure 1.24 based on Gutiérrez *et al* (2008b) classification has been taken from Gutiérrez *et al* (2014).

Among different types of karst hazards (Gutiérrez *et al* 2014, Parise *et al* 2015), sinkholes are very typical for evaporite rocks, especially in consideration of the very rapid lithological dissolution/erosion (0.70 ± 0.17 mm yr^{-1}) (Cucchi *et al* 1998). Also, despite the natural processes, sinkholes may be originated, or indirectly induced, by human activities (anthropogenic sinkholes, Parise *et al* 2015).

Several tens of sinkholes have been identified at Marina di Lesina, even though the precise temporal reference is available for a dozen of the events only (Parise and

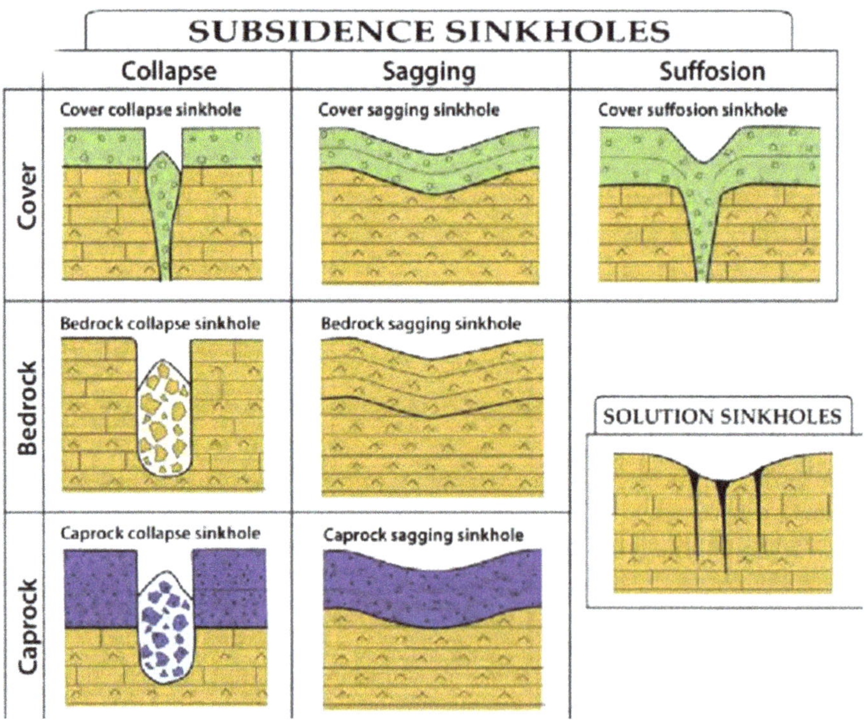

Figure 1.24. Modified from Gutiérrez *et al* (2008b) schema of sinkhole formation The main subsidence sinkhole types. Solution sinkholes generated by the corrosional lowering of the ground surface are not included (reprinted from Gutiérrez *et al* 2014. Copyright 2014, with permission from Elsevier).

Vennari 2013) (figures 1.25(A) and (B)). Another example of a sinkhole in the urban territory is shown in figure 1.25(C): sinkhole in the streets of Bologna, at the foot of Mt. Donato.

Figure 1.25 shows several sites composed of gypsum with sinkholes in the urban territories of the Emilia Romagna Region, in the Upper Secchia Valley (Italy). (A) Sinkhole in Triassic gypsum at Marina di Lesina, Apulia. (B) Road sign warning of the presence of sinkhole phenomena (Marina di Lesina). (C) Sinkhole in the streets of Bologna, at the foot of the Mt. Donato. (D) Sinkhole in the alluvial bed of Secchia River, Reggio Emilia; a big collapse of sinkholes formed in the Secchia riverbed (figure 1.25(D)), as documented by Colombetti and Fazzini (1976). After preliminary technical reports describing the phenomena, in recent years an escalation in the frequency of the events have been registered, with significant damage produced in buildings (Calligaris *et al* 2010, 2017, Zini *et al* 2015a, 2015b). Finally, figure 1.25(E) shows the leaning bell tower in the Quinis village. Its leaning is due to the presence of Triassic evaporites in the subsoil mantled mainly by alluvial deposits. The entire village is heavily affected by sinkholes (the three houses on the right side of the bell tower present stability problems and the church was demolished in the 1970s).

Figure 1.25. Hazards in the Italian gypsum areas (photos A and B: Jo De Waele C, Paolo Forti D, William Formella E, and Zini L). (Reprinted with permission from De Waele *et al* (2017).)

It can be added here sinkholes formed in the Zaragoza city area affecting the urban territory (figure 1.26).

Pueyo Anchuela *et al* (2017) presented images of surficial sinkholes formed within the central part of the Ebro Basin (NE Spain). These examples are diverse, including open collapses and subsidence zones (figure 1.27). In some cases, collapses relate to

Figure 1.26. Two sinkholes along with Highway #232 through Zaragoza City (Spain). (a) Collapse sinkhole generated as a result of the buried doline reactivation on the outskirts of Zaragoza (Gutiérrez *et a*l (2008a) with permission of Springer Copyright 2007, Springer-Verlag). (b) Oblique aerial view of sinkhole along the N-232 highway. Sinkhole S5 hosts a permanent lake 65 m long and 40 m wide and ~10 m depth (modified photos of Gutiérrez *et a*l (2011) with permission from Elsevier, copyright 2011 Elsevier, all rights reserved).

Figure 1.27. Sinkholes in the Zaragoza city. Field photographs of the surficial karstic evidence from the studied area. In these photographs, open collapses and concentric rings of cracks surrounding previous collapses can be identified. Note the shallow cavities due to the presence of a cemented gravel level. (Reprinted with permission from Pueyo Anchuela *et al* (2017), copyright *International Journal of Speleology*).

cavities located some centimeters from the surface. These cavities are stable at near-to-surface conditions due to the presence of cemented gravel deposits. In these cases there is no karst evidence of precluded collapses; in other cases, concentric cracks surrounding collapses can be mapped (note that it is significant for the geophysical mapping (Eppelbaum *et al* 2008)).

References

Abelson M, Yechieli Y, Baer G, Lapid G, Behat N, Calvo R and Rosensaft M 2017 Natural versus human control on subsurface salt dissolution and development of thousands of sinkholes along the Dead Sea coast *J. Geophys. Res.: Earth Surf.* **122** 1262–77

Abou Karaki N, Fiaschi S and Closson D 2016 Sustainable development and aAnthropogenic induced geomorphic hazards in subsiding areas *Earth Surf. Proc. Land.* **41** 2282–95

Anderson N L, Martinez A A, Hopkins J F and Carr T R 1998 Salt dissolution and surface subsidence in Central Kansas: a seismic investigation of the anthropogenic and natural origin models *Geophysics* **63** 366–78

Annan A P 1997 *Engineering and Environmental Geophysics: The Future* vol 12 (London: Geological Society, Engineering Geology Special Publications) pp 419–26

Baar C A 1977 *Applied Salt-Rock Mechanics 1: The In-situ Behavior of Salt Rocks* (Amsterdam: Elsevier) p 294

Bagrintseva K 1977 *Carbonate Oil and Gas Reservoir Rocks* (Moscow: Nedra) (in Russian)

Bakalowitcz M 2015 Karst and karst groundwater resources in the Mediterranean *Environ. Earth Sci.* **74** 5–14

Beck B F and Sinclair W C 1986 Sinkholes in Florida: an introduction *Florida Sinkhole Research Institute* Report 85-86-4 p 16

Beck B F 2003 *Sinkholes and the engineering and environmental impacts of Karst, Geotechnical Special Pub. No. 122* (Reston, VA: American Society of Civil Engineers) p 737

Beck B F 2004 Soil piping and sinkhole failures ed W B White *Encyclopedia of Caves* (New York: Elsevier) pp 523–8

Beck G (ed) 2005 Karst geology and hydrology *Proc. 8th Multidisciplinary Conf. on Sinkholes and the Engineering and Environmental Impacts of Karst* (Rotterdam: A.A. Balkema) p 437

Belenitskaya G A 2013 Tectonic aspects of the spatial and temporal distribution of saline basins of the world *Electronic Scientific Edition Almanac Space and Time 4* (in Russian)

Belenitskaya G A 2018 *Salt Systems of the Earth* (New York: Wiley)

Benson R C and Yuhr L B 2016 *Site Characterization in Karst and Pseudokarst Terranes. Practical Strategies and Technology for Practicing Engineers, Hydrologists and Geologists* (Berlin: Springer) p 421

Bögli A 1980 *Karst Hydrology and Physical Speleology* (Berlin: Springer) p 284

Bonacci O 1987 *Karst Hydrology with special references to the Dinaric Karst* (Berlin: Springer)

Brace W F 1981 The effect of size on mechanical properties of rock *Geophys. Res. Lett.* **8** 651–2

Brinkmann R, Wilson K, Elko N, Seale L, Florea L J and Vacher H L 2007 Sinkhole distribution based on pre-development mapping in urbanized Pinellas County, Florida, USA ed M Parise and J Gunn *Natural and Anthropogenic Hazards in Karst Areas: Recognition, Analysis, and Mitigation* vol 279 (London: Geological Society, Special Publication) pp 5–11

Calligaris C, Devoto S, Zini L and Cucchi F 2017 An integrated approach for investigations of ground-subsidence phenomena in the Ovaro village (NE Italy) ed P Renard and C Bertrand *Advances in the Hydrogeology of Karst and Carbonate Reservoirs* (Berlin: Springer) pp 71–7

Calligaris C, Zini L, Cucchi F and Stefanelli N 2010 Gypsum's role in the Friuli Venezia Giulia sinkholes. *2° I sinkholes: gli sprofondamenti catastrofici nell'ambiente naturale ed in quello antropizzato. ISPRA Roma 3–4 December 2009* pp 213–21

Chalikakis K, Plagnes V, Guevrin R, Valois R and Bosh F P 2011 Contribution of geophysical methods to karst-system exploration: an overview *Hydrol. J.* **19** 1169–80

Chen Z *et al* 2017 The world karst aquifer mapping project—concept, mapping procedure and map of Europe *Hydrogeol. J.* **25** 771–85

Closson D and Abou Karaki N 2009 Salt karst and tectonics: sinkholes development along tension cracks between parallel strike-slip faults, Dead Sea, Jordan *Earth Surf. Process. Landf.* **34** 1408–21

Colombetti A and Fazzini P 1976 L'alimentazione e l'origine della sorgente salso solfata di Poiano (Reggio Emilia): fenomeni di dissoluzione nella valle del fiume Secchia *Boll. Soc. Geol. Ital.* **95** 403–21

Cooper A H and Calow R C 1998 Avoiding gypsum geohazards: guidance for planning and construction. British geological survey *Technical Report WC/98/5 Overseas Geological Series*

Cucchi F and Piano C 2003 Inquadramento geografico e geologico dei gessi in Italia ed G Madonia and P Forti *Le aree carsiche gessose d'Italia*vol 2 (Memorie dell'Istituto Italiano di Speleologia) pp 17–26

Cucchi F, Forti P and Finnochiaro F 1998 Gipsum degradation in Italy with respect to climatic textural and erosional conditions *4th Int. Conf. on Geomorphology Suppl. Geogr. Fis. Dinam. Quat. III* **4** 41–9

Culshaw M G and Walthman A C 1987 Natural and artificial cavities as ground engineering hazards *Quar. J. Eng. Geol.* **20** 130–50

Cvijić J 1926 Geomorfologija (Morfologie Terrestre), Tome Secondo, Beograd, Yugoslavia.

De Waele J *et al* 2017 Evaporite karst in Italy: a review *Int. J. Speleol.* **46** 137–68

De Waele J, Guttierez F, Parise M and Plan L 2011 Geomorphology and natural hazards in karst areas: a review *Geomorphology* **134** 1–8

Dreybrodt W, Romanov D and Gabrosek F *et al* 2002 Karstification below dam sites: a model of increasing leakage from reservoirs *Environ. Geol.* **42** 518–24

Dursun A E 2022 Risk analysis of natural sinkholes hazards in Karapınar basin (Konya, Turkey) *Arab. J. Geosci.* **15** 279

Eppelbaum L V, Ezersky M G, Al-Zoubi A S, Goldshmidt V I and Legchenko A 2008 Study of the factors affecting the karst volume assessment in the Dead Sea sinkhole problem using microgravity field analysis and 3D modeling *Adv. GeoSci.* **19** 97–115

Epstein J B 2005 National evaporite karst—some western examples (https://pubs.usgs.gov/sir/2005/5160/PDF/Part2_4B.pdf)

Ezersky M and Frumkin A 2017 Evaluation and mapping of Dead Sea coastal aquifers salinity using Transient Electromagnetic (TEM) resistivity measurements *C. R. Geosci.* **349** 1–11

Ezersky M G and Goretsky I 2014 Velocity-resistivity versus porosity-permeability inter-relations in Dead Sea salt samples *Eng. Geol.* **183** 96–115

Ezersky M and Goretsky I 2015 The scale effect of Dead Sea salt velocities based on seismic-acoustic measurements *The Mechanical Behavior of Salt VIII* ed L Roberts, K Mellegard and F Hansen Rapid City SD, 26–28 May 2015 (London: Taylor and Francis Group) pp 137–44

Ezersky M G and Frumkin A 2020 Identification of sinkhole origin using surface geophysical methods, Dead Sea, Israel *Geomorphology* **364** 107225

Ezersky M G and Legchenko A 2014 Quantitative assessment of in-situ salt karstification using shear wave velocity, Dead Sea *Geomorphology* **221** 150–63

Ezersky M G and Frumkin A 2021 Subaerial morphology affected by groundwater aggressiveness: sinkhole susceptibility above karstified salt *Dead Sea, Geomorphology* **375** 107525

Farrant A R and Cooper A H 2008 Karst geohazards in the UK: the use of digital data for hazard management *Quart. J. Eng. Geol. Hydrogeol.* **41** 339

Ford D and Williams P 2007 *Karst Hydrogeology and Geomorphology* (New York: Wiley)

Ford D C and Williams P F 1989 *Karst Geomorphology and Hydrology* (London: Unwin Hyman)

Forti P and Sauro U 1996 The gypsum karst of Italy *Int. J. Speleol.* **25** 239–50

Forti P 2017 Chemical deposits in evaporite caves: an overview *Int. J. Speleol.* **46** 109–35

Forti P *et al* 2018 Evaporite karst and caves of Emilia Romagna Region *Tentative List—Supplementary Description* (https://whc.unesco.org/en/tentativelists/6317/)

Frumkin A and Forti P 1997 Liquid crystal cave, Israel ed C A Hill and P Forti *Cave Minerals of the World* (Huntsville, AL: National Speleological Society) pp 319–22

Frumkin A 1994 Morphology and development of salt caves *NSS Bull.* **56** 82–95

Frumkin A 2001 Karst and caves of Israel ed C Juberthie and V Decu *Encyclopaedia Biospeleologica: Moulis* vol 3 (Béziers: Société de Biospéologie) pp 1840–2 OCLC Leadership www.oclc.org ©2021 OCLC, 10 May 2021, (https://www.oclc.org/en/about/leadership.html)

Frumkin A 2013 Salt Karst *Treatise on Geomorphology vol 6 Karst Geomorphology* ed A Frumkin (San Diego, CA: Academic) pp 407–24

Frumkin A and Ford D C 1995 Rapid entrenchment of stream profiles in the salt caves of Mount Sedom, Israel *Earth Surf. Process. Landf.* **20** 139–52

Frumkin A, Magaritz M, Carmi I and Zak I 1991 The Holocene climatic record of the salt caves of Mount Sedom Israel *Holocene* **1** 191–200

Frumkin A, Zaidner Y, Na'aman I, Tsatskin A, Porat N and Valfson L 2014 Sagging and collapse sinkholes over hypogenic hydrothermal karst in a carbonate terrain *Geomorphology* **229** 45–57

Frydman S, Charrash J and Goretsky I 2008 Geotechnical properties of evaporate soils on the Dead Sea area *Eng. Geol* **101** 236–44

Galloway D, Jones D R and Ingebritsen S E 2013 *Land Subsidence in the United States* (Reston, VA: USGS) (https://pubs.usgs.gov/circ/circ1182/)

Galve J P, Gutiérrez F, Remondo J, Bonachea J, Lucha P and Cendrero A 2009 Evaluating and comparing methods of sinkhole susceptibility mapping in the Ebro Valley evaporite karst (NE Spain) *Geomorphology* **111** 160–72

Gerson R 1974 Karst processes of the eastern Upper Galilee, northern Israel *J. Hydrol.* **21** 131–52

Gerson R and Inbar M 1974 The field study program of the Jerusalem–Eilat symposium, 1974. Reviews and summaries of Israeli research projects *Z. Geomorphol. N.F. Supplement Band* **20** 1–40

Gevantman L H (ed) 1981 Physical properties data for rock salt National Bureau of Standards *Monograph 167* (Washington, DC: U.S. Government Printing Office) p 20234

Gil-Márquez J M, Juan A B, Mudarra M, Andreo B, Prieto-Mera J, Sánchez D, David Rizo-Decelis L, Argamasilla M, Nieto J M and de la Torre B 2017 Karst development of an evaporitic system and its hydrogeological implications inferred from GIS-based analysis and tracing techniques *Int. J. Speleol.* **46** 219–35

Glazek J 1989 Tectonic conditions for karst origin and preservation ed P Bosak, D C Ford, J Glazek and I Horacek *Paleokarst, A Systematic and Regional Review* (Amsterdam: Elsevier Academia) pp 569–74
Gorbunova K A 1977 *Karst in Gypsum of the USSR, 83* (Perm: Perm University) (in Russian)
Grimes K G 2006 Syngenetic karst in Australia: review *Helictite* **39** 27–38
Grimes K G 1975 Pseudokarst: definition and types ed A W Graham *Proc. 10th Biennial Conf.* (Brisbane: Australian Speleological Federation)
Gutiérrez F, Gurrero J and Lucha P 2008c Quantitative sinkhole hazard assessment. A case study from the Ebro Valley evaporite alluvial karst (NE Spain) *Nat. Hazards* **45** 211–33
Gutiérrez F, Calaforra J and Luch P 2008b A genetic classification of sinkholes illustrated from evaporite paleokarst exposures in Spain *Environ Geol* **53** 993–1006
Gutiérrez F, Parise M, De Waele J and Jourde H 2014 A review on natural and human-induced geohazards and impacts in karst *Earth-Sci. Rev.* **138** 61–88
Gutiérrez F, Benito-Calvo A, Carbonel D, Desir G, Sevil S and Guerrero J I 2019 Review on sinkhole monitoring and performance assessment of the performance of remediation measures by high-precision leveling and terrestrial laser scanner in the salt karst of the Ebro Valley Spain *Eng. Geol.* **248** 283–308
Gutiérrez F, Galve J P, Lucha P, Castañeda C, Bonachea J and Guerrero J 2011 Integrating geomorphological mapping, trenching, InSAR, and GPR for the identification and characterization of sinkholes in the mantled evaporite karst of the Ebro Valley (NE Spain) *Geomorphology* **134** 144–56
Gutiérrez F, Calaforra J M, Cardona J M, Orti F, Duran F and Garay, P. J J 2008a Geological and environmental implications of the evaporite karst in Spain *Environ. Geol* **53** 951–65 .
Hartmann J and Moosdorf N 2012 The new global lithological map database GLiM: a representation of rock properties at the Earth surface *Geochem. Geophys. Geosyst.* **13** 1–37
Hatzor Y H, Wainshtein I and Mazor D B 2010 Stability of shallow karstic caverns in blocky rock masses *Int. J. Rock Mech. Min. Sci.* **47** 1289–303
Hatzor Y H and Heyman E P 1997 Dilation of anisotropic rock salt: evidence from Mount Sedom diapir *J. Geophys. Res* **102** 14853–68
Herak M 1976 Geological approach to classification of terranes for consolidation treatments *Proc., I JUSIK 76* (Zagreb: Yugoslavia)
Higgins C 1984 Piping and sapping ed R LaFleur *Groundwater as a Geomorphic Agent* (Winchester: Allen & Unwin) pp 19–58
Hiltunen D R and Cramer B J 2008 Application of seismic refraction tomography in karst terrane *J. Geotech. Environ. Eng.* **134** 938–48
Hofrichter E 1976 Zur frage der porositat und permeabilitat von saltzgesteiner *Erdol Erdgas* **93** 77–80
Hollingthworth E, Brahanna V, Inlander E and Slay M E 2008 Karst regions of the world (KROW): global karst datasets and maps to advance the protection of karst species and habitats worldwide *(U.S. Geological Survey Scientific Investigations Report 2008-5023) U.S. Geological Survey Karst Interest Group Proc.* pp 17–24
Horowitz A 1979 *The Quaternary of Israel* 1st edn (New York: Academic)
Jackson J A 1997 *Glossary of Geology* 4th edn (Alexandria, VA: American Geologic Institute)
Jennings J N 1968 Syngenetic karst in Australia *Contributions to the Study of Karst*, ed P W Williams and J N Jennings, Research School of Pacific Studies, Australian National University Department of Geography Publication G/5: 41

Jennings J N 1971 *Karst* (Canberra: Australian National University Press)

Jennings J N 1985 *Karst Geomorphology* (Oxford: Blackwell) p 293

Klimchouk A 2004 Towards defining, delimiting, and classifying epikarst: its origin, processes, and variants of geomorphic evolution *Speleogene. Evol. Karst Aquifers* **2** 1–13

Klimchouk A B 1996 The dissolution and conversion of gypsum and anhydrite *Int. J. Speleol.* **25** 21–36

Knapp R W, Steeples D W, Miller R D and McEllwee C D 1989 Seismic-reflection surveys at sinkholes in central Kansas *Kansas Geophys. Surv. Bull.* **226** 95–116 https://www.kgs.ku.edu/Publications/Bulletins/226/Knapp/

Komatina M 1973 Conditions of Karst Development and Zoning of Karst Terranes, Vesnik Geozavoda X/XI, Beograd, Yugoslavia

Korotkevich G V 1970 *The Salt Karst* (Leningrad: Nedra) p 256 (in Russian)

Kozary M T, Dunlap J C and Humphrey W E 1968 Incidence of saline deposits in global time *Geol. Soc. Am. Spec. Pap.* **88** 43–57

Legrand H E and Striengfield V T 1975 Hydrogeology and hydrology of karst *Hydrogeology of Karst Terrains* (Paris: International Association of Hydrogeologists with the assistance of UNESCO) (Republished in Milanović, 1979)

Livne E 2012 Evaluation of the liquefaction potential along the Dead-Sea western shores—regional screening and geotechnical analysis of two boreholes *MSc Thesis* (Ben Gurion University of the Negev) p 118

Lowe D and Waltham T 1995 *A Dictionary of Karst and Caves: A Brief Guide to the Terminology and Concepts of Cave and Karst Science. Cave Studies Series Number 6* (London: British Cave Research Association) p 41

Lowe D and Waltham T 2002 *A Dictionary of Karst and Caves* vol 10 (London: British Cave Research Association Cave Studies) p 40

Macau F and Riba O 1962 Situación, características y extensión de los terrenos yesíferos en España *I Coll. Int. sobre las Obras Públicas en los terrenos yesíferos. Madrid* **5** 157–84

Mandel S 1961 Properties and genesis of the Turonian—Cenomanian aquifer in western Israel as example of a large limestone aquifer *Unpublished DSc Thesis* Technion, Haifa (in Hebrew, English summary)

Mendonca A F, Pires A C B and Barros J G C 1993 Pseudo-sinkholes in lateritic terrains, Brasilia, Brazil *Applied Karst Geology* ed Beck (Rotterdam: Balkema) pp 43–9

Mihevć A, Prelovshek M and Hajna N Z 2010 Introduction to Dinaric karst *2010 Karst Underground Protection Project* (Postojna) (https://izrk.zrc-sazu.si/sites/default/files/Dinarski%20kras_normalna%20locljivost.pdf)

Milanović P T 1981 *Karst Hydrogeology* (Littleton, CO: Water Resource Publications)

Milanović P T 2000 *Geological Engineering in Karst: Dams, Reservoirs, Grouting, Groundwater Protection, Water Tapping, Tunneling* (Belgrade: Zebra Publishing Ltd) p 347

Milanović P 1979 *Karst Hydrogeology and Methods of Investigation* (Trebinje: Institute for Karst Water Utilisation) (in Serbo-Croatian)

Milanović P 1990 Influence of construction on hydrogeological and environmental conditions in the karst region, eastern Herzegovina, Yugoslavia *Environ. Geol. Water Sci.* **15** 5–15

Milanović P 2002 The environmental impacts of human activities and engineering constructions in karst regions *Episodes* **25** 13–21

Milanović P 2005 *Water Resources Engineering in Karst* (Boca Raton, FL: CRC Press) p 328

Milanović P 2021 *Karst istoćne Hercegovine I Dubrovaćkog priobalja (Karst of Eastern Hercegovina and Dubrovnic Littoral)* (NET, Banja Luka: Bina)
Milanović P, Maksimovich N and Meshcheriakova O 2019 *Distribution of Evaporite Karst* (Berlin: Springer)
Mokhtari L G, Velayati S and Dadashzadeh Z 2012 Investigation of the relationship between tectonics and karstification *Int. Conf. on Environmental, Biomedical and Biotechnology IPCBEE* vol 41 (Singapore: IACSIT Press)
Monroe W H 1970 *A Glossary of Karst Terminology. Geological Survey Water-Supply Paper 1899-K. U.S. Geological Survey* (Washington, DC: U.S. Government Printing Office) p 26
Moore G K, Bruchette C R and Bingham R H 1969 *Limestone hydrology in upper stones River Basin, Central Tennessee, Tennessee Department of Conservation* (Washington, DC: Division of Water Resources and USGS)
Moore H and Beck B 2006 Karst terrane and transportation issues *Encyclopedia of Sustainability Science and Technology* (Cham: Springer) pp 5645–71
Morris D A and Johnson A I 1967 Summary of hydrologic and physical properties of rock and soil materials, as analyzed by the hydrologic laboratory of the U.S. Geological Survey, 1948–1960. USGS Water Supply Paper: 1839-D.
Mylroie J E and Carew J L 1987 Field evidence of the minimum time for speleogenesis *Natl. Speleol. Soc. Bull.* **49** 67–72
National Research Council 1991 *Mitigating Losses from Land Subsidence in the United States* (Washington, DC: National Academies Press) p 58
Neuendorf K K E, Mehl J P and Jackson J A 2011 *Glossary of Geology* 5th edn (Berlin: Springer)
Nof R N, Abelson M, Raz E, Magen Y, Atzori S, Salvi S and Baer G 2019 SAR interferometry for sinkhole early warning and susceptibility assessment along the Dead Sea, Israel *Remote Sens.* **11** 89
Ortí F 1989 Evaporitas marinas *Sedimentología* ed A Arche (Madrid: Consejo Superior de Investigaciones Cientificas) pp 89–177
Parise M 2019 Sinkholes *Encyclopedia of Caves* 3rd edn (New York: Academic) ch 110
Parise M and Vennari C 2013 A chronological catalogue of sinkholes in italy: the first step toward a real evaluation of the sinkhole hazard *Conf. Paper 13th Multidisciplinary Conf. on Sinkholes and the Engineering and Environmental Impacts of Karst* 383–92
Parise M, Parise M, Closson D, Gutiérrez F and Stevanović Z 2015 Anticipating and managing engineering problems in the complex karst environment *Environ. Earth Sci.* **74** 7823–35
Popp T and Kern H 1998 Ultrasonic wave velocities, gas permeability and porosity in natural and granular rock salt *Phys. Chem. Earth* **23** 373–8
Popp T and Kern H 2000 Monitoring the state of microfracturing in rock salt during deformation combined measurements of permeability and P- and S-wave velocities *Phys. Chem. Earth* A **25** 149–54
Pueyo Anchuela Ó, Pocoví Juan A, Casas Sainz A M, Gracia Abadías J and Liesa Carrera H D 2017 Integrated approach for sinkhole evaluation and evolution prediction in the Central Ebro Basin (NE Spain) *Int. J. Speleol.* **46** 237–49
Quinlan J F 1972 Karst-related mineral deposits and possible criteria for the recognition of paleokarst: a review of preservable characteristics of Holocene and older karst terranes *24th IGC, Section 6, Montreal, Canada*
Quinlan J F, Smith R A and Johnson K S 1986 Gypsum karst and salt karst of the United States of America *Atti Symp. Int. sul carsismo nelle evaporiti* vol 4 Le Grote d-Italia pp 73–92

Reuter F, Dietrich S and Stoyan D 1993 Sinkholes in carbonate, sulphate, and chloride karst regions: principles and problems of engineering geological investigations and predictions, with comments for the construction and mining industries ed Beck *Applied Karst Geology* (Rotterdam: Balkema) pp 3–25

Rodrigues V, Gutiérrez F, Green A G, Carbonel D, Horstmeyer H and Schmelzbach C 2014 Characterizing sagging and collapse sinkholes in a mantled karst by means of ground penetrating radar (GPR) *Environ. Eng. Geosci.* **20** 109–32

Rothfuchs T, Bollingerfehr W and Bechthold W 2004 Lessons learned from large-scale experiments at the Asse mine Germany *RADWASTE'04 6th European Conf. on Management and Disposal of Radioactive Waste (Luxemburg, 29 March–1 April 2004)* https://slideplayer.com/slide/741833/

Sanchez J A, Perez A, Coloma P and Martinez-Gil J 1998 Combined effects of saline depressions of Europe: north-east Spain *Hydrol. Process.* **12** 813–20

Savich A I 1984 Generalized relations between static and dynamic indices of rock deformability *Hydrotechn. Constr.* **18** 394–400

Savich A I and Kujundjić B D (ed) 1990 *Complex Engineering-Geophysical Explorations at the Hydrotechnical Construction* (Moscow: Nedra) (in Russian)

Schoen J H 2004 Physical properties of rocks ed K Helbig and S Treitel *Seismic Exploration* (Amsterdam: Elsevier) p 18

Sevil J, Gutiérrez F, Carnizer C, Carbonel D, Desir G, Garcia-Arnay and Guerro J 2020 Characterizing and monitoring a high-risk sinkhole in an urban area underlain by salt through non-invasive methods: detailed mapping, high-precision leveling and GPR *Eng. Geol.* **272** 105641

Sharma P V 1997 *Environmental and Engineering Geophysics* (Cambridge: Cambridge University Press)

Shorer Y 1992 *Israel Excursions* (North Hollywood: Keter Publishing House)

Sonnenfeld P 1984 *Brines and Evaporites* (Orlando, FL: Academic) p 613

Sowers G F 1996 *Building on Sinkholes, Design and Construction of Foundations in Karst Terrain* (Reston, VA: ASCE Press)

Spangenberg E, Spangenberg U and Heindorf C 1998 An experimental study of transport properties of porous rock salt *Phys. Chem. Earth* **23** 367–71

Spiers C J, Grupa J, Salzer K, Popp T and Stuhrenberg D 2005 Buffer/backfill in salt (long-term compaction, additives, precompacted bricks) *NFpro* pp 1–33

Stagg K G 1968 In situ tests on the rock mass ed K G Stagg and O C Zienkiewicz *Rock Mechanics in Engineering Practice* (New York: Wiley) pp 125–56

Steeples D W, Knapp R W and McElwee C D 1986 Seismic reflection investigations of sinkholes beneath Interstate Highway 70 in Kansas *Geophysics* **51** 295–301

Stringfield V T 1966 Artesian water in Tertiary limestone in the southern states *Professional Paper 517* (Washington, DC: USGS Press) p 65

Sweeting M M 1972 *Karst Landforms. Selected Glossary. Compiled by K Addison* (New York: Columbia University Press) p 362

Taqieddin S A, Abderahman N S and Atallah M 2000 Sinkhole hazard along the eastern Dead Sea shoreline area, Jordan: a geological and geotechnical consideration *Environ. Geol.* **39** 1237–53

Theron A, Engelbrecht J, Kemp J, Kleynhans W and Turnbul T 2017 Detection of sinkhole precursors through SAR interferometry: radar and geological consideration *IEEE Geosci. Remote Sens. Lett.* **14** 871–5

Tihansky A B 2017 Water Science School, Dissolution sinkholes scheme (USGS) https://www.usgs.gov/special-topics/water-science-school/science/sinkholes

Torbarov K 1976 Estimation of permeability and effective porosity in karst on the basis of recession curve analysis *Hidrologija i vodno bogatstvo krša* (Sarajevo: Zavod za hidrotehniku) p 97

US EPA 2002 A lexicon of cave and karst terminology with special to environmental karst hydrogeology *EPA/600/R-02/003* U.S. Environmental Protection Agency, Office of Research and Development, National Center for Environmental Assessment, Washington Office, Washington, DC https://cfpub.epa.gov/ncea/risk/era/recordisplay.cfm?deid=54964

U.S. Environmental Protection Agency 2000 *National Center for Environmental Assessment, A Lexicon of Cave and Karst Terminology with Special Reference to Environmental Karst Hydrogeology* (Washington DC: US EPA)

UNESCO 1972 *Glossary and Multilingual Equivalents of Karst Terms* (Paris: United Nations Educational. Scientific and Cultural Organization) p 72

Veress M 2016 *Covered Karsts* (Berlin: Springer) p 536

Vincent C, Thibert E, Gagliardini O, Legchenko A, Gilbert A, Garmbois S, Condom T, Baltassat J M and Girard J F 2015 Mechanisms of subglacial cavity filling in Glacier de Tête Rousse, French Alps *J. Glaciol.* **61** 1–15

Vlahović V 1975 Karst Nikšićkog Polja i njegova hidrogeologija (Hydrogeology of Karst Region of the Nikšićko Polje) *Društvo za nauku i umjetnost Crne Gore* (Serbo-Croatian: Podgorica, Yugoslavia)

Voronov F F and Grigor'ev S B 1976 Influence of pressure up to 100 kbar on elastic properties of silver, sodium, and cesium chlorides *Sov. Phys. Solid State* **18** 325–8

Walkden G M 1974 Paleokarstic surfaces in upper Visean (Carboniferous) limestones of the Derbyshear Block, England *J. Sedimen. Petrol.* **44** 1236–48 Menasha

Walthman A C and Fookes P G 2003 Engineering classification of karst ground conditions *Quart. J. Eng. Geol. Hydrogeol.* **36** 101–18

Walthman T 2003 Karst terrains ed P G Fooks, M Lee and G Milligan *Geomorphology for Engineers* (Caithness: Whittles Press)

Walthman T, Bell F and Culshaw M 2005 Sinkholes and subsidence . *Karst and Cavernous Rocks in Engineering and Construction* (Chichester: Springer, Praxis Publishing Ltd) p 382

Warren J K 2006 *Evaporites: Sediments, Resources, and Hydrocarbons* (Berlin: Springer)

WBI Compilation 2006 *Salt Permeability versus Porosity Palette* ed W Wittke (http://www.wbionline.de/).

Weber N, Antler G, Lazar B, Stein M, Yechieli Y and Gavrieli I 2021 Hydrological and thermodynamic controls on late Holocene gypsum formation by mixing saline groundwater and Dead Sea brine *Geochim. Cosmochim. Acta* **316** 363–83

Weinstein-Evron M, Beck A and Ezersky M 2003 Geophysical investigations in the service of Mount Carmel (Israel) prehistoric research *J. Archaeol. Sci.* **30** 1331–41

Weisbrod N, Alon-Mordish C and Yechieli Y 2012 Dynamic dissolution of halite rock during flow of diluted saline solutions *Geophys. Res. Lett.* **39** L09404

White W B 1988 *Geomorphology and Hydrogeology of Karst Terrains* (New York: Oxford University Press) p 464

White W B and Culver D C (ed) 2012 *Encyclopedia of Caves* 2nd edn (Amsterdam: Elsevier)

Williams P 2004 Dolines ed J Gunn *Encyclopedia of Caves and Karst Science* (New York: Taylor and Francis Group) pp 304–10

Wittke W and Schmitt B 2000 Permeability of salt rock—theory and experiment *EUROCK 2000, Tagungsband des 14. Nationalen Symp. fur Felsmechanik und Tunnelbau* pp 361–7

Wu P, Tang C, Zhu L, Liu C, Cha X and Tao X 2009 Hydrogeochemical characteristics of surface water and groundwater in the karst basin, Southwest China *Hydrol. Process* **23** 2012–22

Yang R Y, Jin S, Zhang T, Luo Y, Meng Z and Deng L 2017 Seismic response characteristics and identification of different types of karst collapse in Tahe oil field, Tarim Basin, NW China *Int. Geophysical Conf. (Qingdao, China, 17–20 April 2017)* The Society of Exploration Geophysicists and the Chinese Geophysical Society 2017 CGS/SEG

Yechieli Y, Abelson M, Bain A, Shtivelman V, Crouvi O, Wacks D, Bar G, Kalbo R and Lyachovski V 2004 Sinkhole phenomena along the Dead Sea shoreline *GSI Report No. GSI/21/04* (in Hebrew)

Yechieli Y, Abelson M, Bein A and Crouvi O 2006 Sinkhole 'swarms' along the Dead Sea coast: reflection of disturbance of lake and adjacent groundwater systems *GSA Bull.* **118** 1075–87

Yechieli Y, Ronen D and Berkowitz B 1995b Are sedimentary salt layers always impermeable *Geophys. Res. Lett.* **22** 2761–4

Yechieli Y 1993 The effects of water level changes in closed lakes (Dead Sea) on the surrounding groundwater and country rocks *PhD Thesis* (Israel: Weizmann Inst. Sci., Rehovot)

Zini L, Calligaris C, Devoto S, Zavagno E, Forte E, Petronio L, Boccali C and Cucchi F 2015a Fenomeni di sprofondamento nella piana di Enemonzo (UD) *Memorie Descrittive Carta Geologica d'Italia* **119** 101–10

Zini L, Calligaris C, Forte E, Petronio L, Zavagno E, Boccali C and Cucchi F 2015b A multidisciplinary approach in sinkhole analysis: the Quinis village case study (NE-Italy *Eng. Geol.* **197** 132–44

IOP Publishing

Applied Geophysics for Karst and Sinkhole Investigation
The Dead Sea and other regions
Michael Ezersky, Lev V Eppelbaum and Anatoly Legchenko

Chapter 2

An overview of geophysical methods suitable for karst problem resolving

Generally, sinkholes are surface (named also, subaerial) manifestations of karst or pseudokarst. The term karst includes a wide range of surface and subsurface landforms that develop by the dissolution of soluble bedrocks and the development of subsurface drainage (e.g., De Waele *et al* 2011, Ford and Williams 2007, Frumkin 2013, Benson and Yuhr 2016). On the other hand, some sinkholes were not formed by the natural dissolution of bedrock despite their resembling or acting like karst. These sinkholes are termed 'pseudokarst' (e.g., Grimes 1975, Eberhard and Sharples 2013, Benson and Yuhr 2016). During the pseudokarst processes, the material is physically removed in the solid-state (like suffosion), together with effects resulting from changes of phase and removal of liquid material. They are termed 'false-karst' or, sometimes 'analogous karst' (McGee 2012). Thus, for the formation of karstic sinkholes soluble rocks are required in the subsurface (limestone, dolomite, gypsum, salt, etc), whereas pseudokarst requires the existence of underground flows and physical erosion. To secure agriculture, construction, tourism, etc, site characterization is carried out.

Site characterization is the process of studying the geomorphologic, geologic, hydrologic, and engineering properties at the site, including the soil, rock, groundwater. It includes also human-modified conditions in the subsurface (e.g., utilities, structures, mines, and tunnels) that can impact site conditions (Benson and Yuhr 2016, Ezersky and Frumkin 2021).

Geophysical methods are used worldwide for a decision of karst problems. Taking into account the presence of excellent handbooks and monographs of leading worldwide geophysicists (Telford *et al* 1990, Khesin *et al* 1996, Sharma 1997, Reynolds 2011, Kirsch 2006, etc) and new books by different geophysical disciplines (Zhdanov and Keller 1994, Daniels 2007, Legchenko 2013, Dal Moro 2015, etc) we did not aim to repeat the known geophysical knowledge. We aim to overview the principles and new knowledge that appeared in the last decades.

doi:10.1088/978-0-7503-3635-2ch2

The geophysical methods can be used during two stages of the characterization of site conditions. At the start of the surveys, when the goal is to detect site anomalous conditions, the density of sampling increases. At this stage, the use of geophysics is often qualitative. Later, during the subsequent stages of investigations (designing and construction), geophysical data may be used more quantitatively to determine the specific engineering or hydrologic parameters. For instance, P-wave velocity is used for rippability (Sharma 1997). P-wave and S-wave velocities are used for calculating elastic constants (Reynolds 2011). TEM method is used for locating the fresh–seawater interface (Kafri *et al* 1997, ISRM 2004, Greenhouse *et al* 1995).

Two important principles of the karst investigations using geophysical methods deserve to be noted.

1. The geophysical scale of measurements should correspond to the scale of considered objects. It means that the method scale and target scale must coincide (or, in other words, the void must be of the same scale as the method). It should be taken into account in the planning of the geophysical surveys (Benson and Yuhr 1993).
2. Any geophysical measurements must be in close corresponding with geological data. Such measurements should help in visualizing the hidden geological peculiarities. Geophysicists have to use any geological data known from outcrops, boreholes, or any other features such as 'benchmarks'. Geophysical measurements can be very effective if one is to use interpolation between such sources of geological information or extrapolation away from them. In every case, the knowledge derived from geophysical measurements must be eventually reconfirmed against the known geological condition (McNeil 1980b).

2.1 Aspects of sinkhole hazards that can be resolved by geophysical methods

Surface geophysics is widely used in the karst problem in most of the carbonate karst regions of the USA: GPR and microgravity, geoelectric methods, seismic tomography, etc. A comprehensive review of the application of the geophysical method in the karst problem is given in Benson and Yuhr (2016).

2.1.1 Detectability of voids by geophysical methods

It is believed that the subsurface cavity can be detected using surface geophysical methods. The theoretical modeling of the cavities by a round hole within a medium states that cavities can be detected at a depth no more than 2–3 times their 'effective diameter' (Militzer *et al* 1979). If to take into account that a cavity is a geological object affected by different processes, the depth of object detection can be increased by up to 10 times (figure 2.1) (Benson and Yuhr 1993). For instance, rock strain, secondary permeability, and/or chemical alteration cause the 'halo' effect around the cavity effectively making it appear geophysically larger than it is (dashed circle in figure 2.1(b)) and allow it to be detected at greater depths, possibly up to 1.5–2 times the theoretically calculated depth. Similarly, the associated network of dissolution

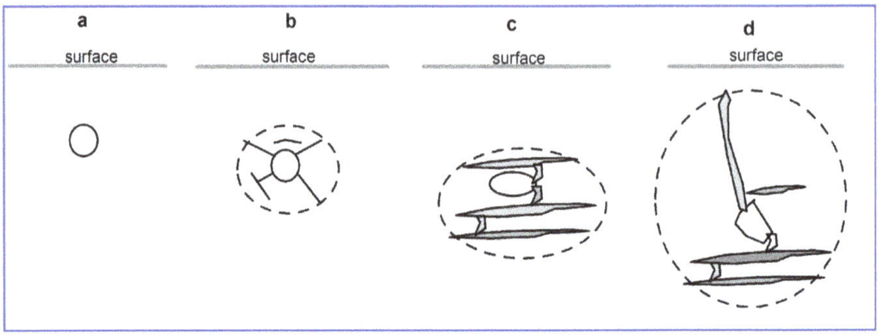

Figure 2.1. Explanations of the investigation depth estimation (a) primary spherical cavity. (b) Halo effect allows most cavities to be detected deeper. (c) Most cavities are part of a complex cave system and can be detected at even greater depths. (d) Near-surface indicators often allow cavity system to be detected beyond the range and resolution of most geophysical methods. Dashed contours in figures are an effective square of the detected object (suggested by Benson and Yuhr 1993).

enlarged fractures, cavities, and bedding planes involved with a natural cave system (dashed contours in figure 2.1(c)) will allow the cavity to be detected deeper than just the spherical approximation by itself. Finally, near-surface indicators often enable detecting the cavity system beyond the range and resolution of most geophysical methods (dashed circle in figure 2.1(d)). Because of these geological factors and improved techniques, the limits of detection have been significantly revised. 'Spherical cavities' can now be detected to depths of up to 6–8 times exceeding their diameter and linear cavities—up to 10 times exceeding their diameter by the microgravity method (for instance, Butler *et al* 1997).

Moreover, analyses carried out in several studies show that several indications can help mitigate this effect. The study of Yang and Drumm (2002), Maimon *et al* (2005) predict the presence of some mechanical indications at the surface when a large cavity has been opened at depth. The signs can be fractures or heightened porosity.

2.1.2 How a subsurface cavity can manifest itself on the surface

The planning and interpretation of results of geophysical investigations in the engineering–geologic zone require corresponding engineering–geologic models. Following Ogilvy *et al* (1980), *a priori* geologic–geophysical information on the main features of engineering–geologic and hydrogeological conditions in the region under investigation is used as input in the first stage. Based on this information an initial geophysical model of the subsurface is generated. The size, shape, and location depth are also important. Therefore, models of the sinkhole formation should be considered.

Yang and Drumm (2002) have suggested a simplified method for estimation of the soil stability concerning two potential modes of failure within the residual soil. The stability of the first failure mode depends on the development of arching in the residual soil, while the second failure mode is caused by the yielding and plastic flow of the soil into the soil dome and/or rock void. The study is based on the finite element method allowing the parametric studies of both stability modes for a typical residual clay soil over a range of cavity diameters. Stability Mode-1 is a potentially tensile failure. Shear stresses (and tensile cracks) develop between the cavity and surface. Mode-II stability: soil dome systems that are stable concerning Mode-I may still exhibit distress due to the excessive surface deformation

High shear stresses in the residual soil may lead to yielding or plastic flow. Authors suggest that sinkhole problems are more likely to be identified in the field of mode-I pattern, which is a shallow failure collapse mode. Stability Mode-II corresponds to dilatancy, e.g., increase of soil volume that, as was shown by Terzaghy *et al* (1996, pp 139, 147) takes place because of the accumulation of microcracks due to particle failure. It is a fact known from soil mechanics and explaining the difference of the actual friction angle from the theoretic angle of plastic flow. It means that from the geophysical point of view, the deformation of soil previous to collapse is accompanied by the variations of elastic velocity, resistivity, and other parameters characterizing soil properties.

Maimon *et al* (2005) have considered the stress distribution around underground cavities using 3D numerical models. The results obtained from this simulation show that tensile stresses formed above a cavity shaped like a slender ellipsoid disposed of an elastic medium (rock, salt) (figure 2.2). The most important thing here is that the tensile fractures formed above a cave can be a predictor of a future failure. Then the authors considered the stability and collapse mechanisms of cavities relating to sinkhole formation. The influence on the stress distribution and failure pattern

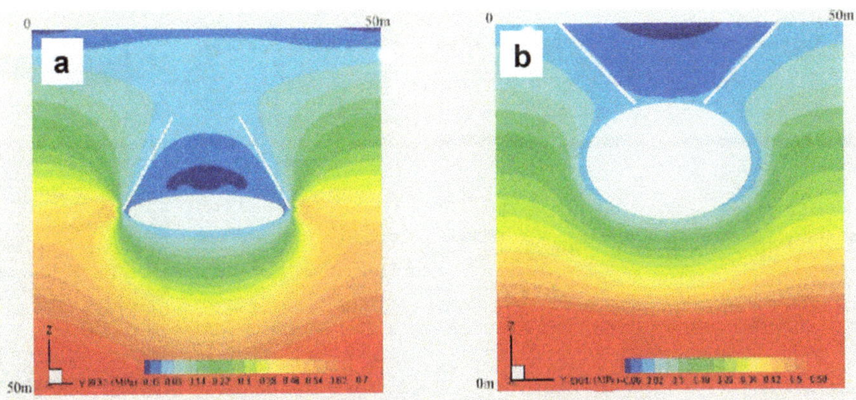

Figure 2.2. Stress distribution around the slender ellipsoid (cavity) at the size 7.5×7.5 m^2 within the elastic medium (salt, rock) calculated using 3D numerical modeling for tensile failure (Mode-I), the cavity is located at 25 m depth. Geometry of the cavity: (a) $7.5 \times 7.5 \times 1.5$ m^3; (b) $7.5 \times 7.5 \times 4$ m^3. *The Blue zone is tensile stresses.* White lines mark the predicted failure. The figure shows the last stage when tensile stress reached the surface. The cavity is unstable, but the failure will be along reverse (a) or normal (b) faults depending on cavity width and depth (reprinted with permission from Maimon *et al* 2005 copyright 2005 Geological Survey of Israel).

around a subsurface cavity is studied using numerical simulation of: (1) cavity depth-to-width ratio; (2) shape of cavity; (3) mechanical properties of the sediments; and (4) groundwater level lowering.

Maimon *et al* (2005), as in the study of Yang and Drumm (2002), consider two modes of failure. If tensile stress develops in an elastic medium (rock, salt) they likely result in surface cracks and sudden collapse (Mode I of the failure in figures 2.3(a) and (b)), whereas in clastic sediments—the shear zones are formed (Mode II of failure figure 2.3(c)).

Roche *et al* (2000) have carried out the two-dimensional elastic–plastic model for the study of fault and subsidence development in sand placed above the silicon-filled void. The obtained results have shown that the deformation of layers (subsidence) and development of fractures (faults) began at the bottom of the model and gradually grows up to the surface (figure 2.4).

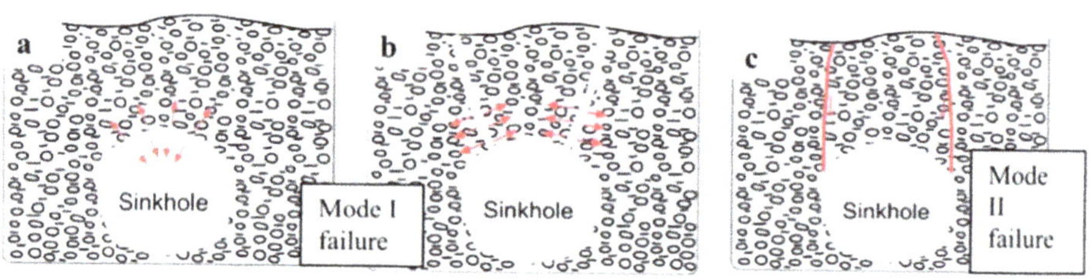

Figure 2.3. Failure modes by Maimon *et al* (2005) (reprinted with permission from Maimon *et al* 2005 with the permission from Geological Survey of Israel).

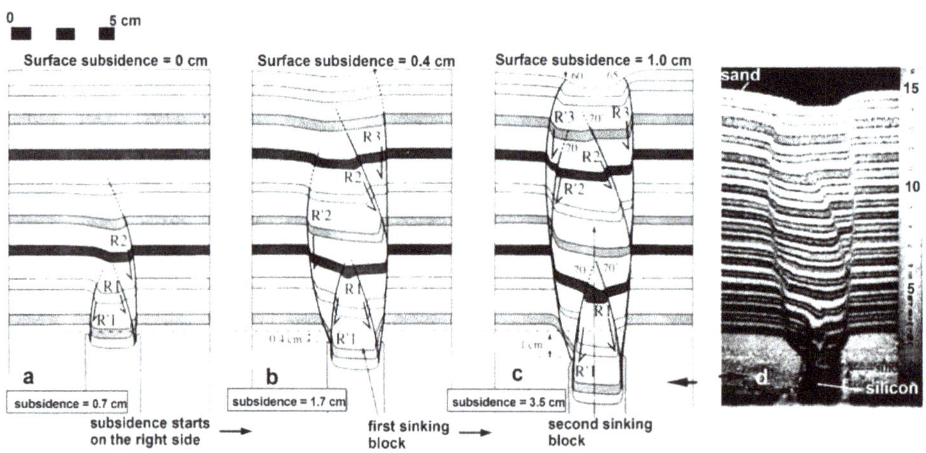

Figure 2.4. A two-dimensional elastic–plastic model for the study of fault and subsidence development in sand placed above silicon that fills the void. The experiments represent a temporal evolution of the system. The gradual growth of the subsidence from the bottom up to the surface. The fault formation stages are considered also. The study models the caldera collapse mechanism (reprinted with permission from Roche *et al* 2000, John Wiley & Sons, copyright 2000 by the American Geophysical Union).

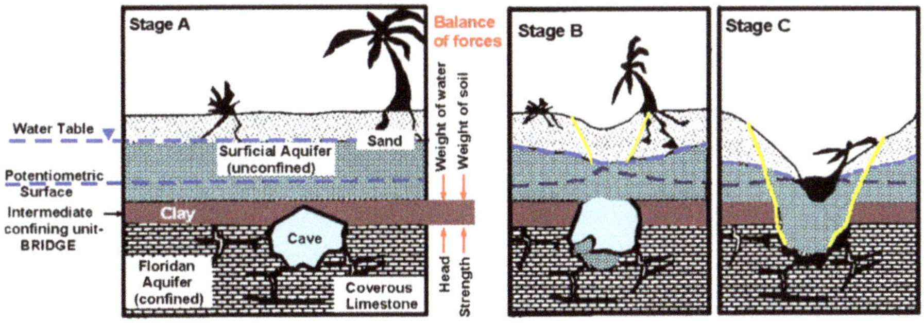

Figure 2.5. (a) Sinkhole formation hypothetic schema. (b) Three stages of the cover-collapse explaining the delay between the void formation within a carbonate rock and surface collapse. A similar process takes principally place in evaporates during a considerably shorter period (reprinted with permission from Lane 1986, copyright 1986 Florida Geological Survey).

The latter model shows that there can be a delay between the void formation and surface collapse (e.g., sinkhole formation) taking place either in subsidence mode or as collapse. Such a delay can be associated also with geological factors such as: (1) mechanical properties of the soil; (2) balance of the mechanical forces; and (3) geological structure of the subsurface. Such a case is described in the model of the 'cover-collapse sinkhole formation' modified by Lane (1986) for the carbonate karst (figure 2.5). The model explains the delay in surface collapse by the 'bridge effect' of the clay layer.

If aquiclude (clay layer, for instance) overlays the rocks, the balance will be composed of the clay layer strength, the pressure of the groundwater from below the clay layer (head of the confined aquifer), and the weight of the water (phreatic aquifer) from above the clay layer. The time factor serves in this case as a concurrent favorable factor permitting one to detect of the void before the surface will collapse.

Resumé. From the above analysis, two conclusions follow: (1) void formation at depth causes vertical and horizontal fractures and faults at the shallow subsurface; (2) between the void formation at depth and surface the collapse processes are developing over a long time. Both these conclusions allow being optimistic with respect to the possibility of detecting void and subsidence of the shallow subsurface before the collapse processes will happen.

2.1.3 Aspects of sinkhole hazards that can be resolved by geophysical methods

The use of geophysical methods in the karst problem is connected with some characteristic features of the subsurface caused by karstification. Currently, the use of geophysics in resolving karst problems is a usual practice for engineers and researchers dealing with the problem. However, many professionals are not familiarized with the numerous geophysical methods, measured parameters, and problems that can be resolved by those means. Geophysical methods, their advantages, and limitations, problems, common acquisition parameters, means of geophysical information presentation as well as their potential for successful resolving of karst problems are discussed in the normative documents (ASTM 6429 (2006), ASTM 5777 (2006)).

Understanding of both karst problems and the possibility of geophysical methods will lead to the more successful cooperation of geophysicists, geologists, and specialists in physical–geological modeling dealing with karst development (Hoover 2003).

Numerous publications have been devoted to this problem (e.g., Green 1974, Benson and Yuhr 1993, Hoover 2003, Chalikakis *et al* 2011). The American Society of Testing Materials (ASTM) suggests the selection of surface geophysical methods for resolving the different problems from unconsolidated soil thickness and elastic properties to studying archaeological sites. An example of the application of geophysics for cave detection is a single problem in the karst investigations (ASTM 6429 (2006), table 2.1). The Standard suggests a selection of geophysical methods based on their suitability ranking: either (A) or (B) depends on the possibility of geophysics to resolve the problem. Ranking of (A) or (B) means a primary (A) or secondary (B) selection of the method for a particular problem. Table 2.1 modified from Hoover (2003) has suggested only five methods with either an A or B suitability ranking for cave investigation. These recommendations conforming to ASTM are shown in table 2.1.

Only five methods are suggested for sinkhole and voids investigation. However, new methods are developed and old ones have been upgraded. Both of the aforementioned methods are used for resolving different aspects of the problem. For instance, the microgravity technique is considerably more sensitive in comparison to the gravity method (Debeglia *et al* 2006).

Electrical resistivity tomography (ERT) is a multi-channel analog of the vertical electric sounding (VES) method that employs up to 160 electrodes simultaneously and allows seeing results in the field during data acquisition (Elfakih *et al* 2020). Apart from this, ERT enables us to perform surveys in the three-dimensional (3D) mode. The magnetic resonance sounding (MRS also named surface nuclear magnetic resonance—SNMR) method is now relatively frequently applied to water-filled caves (Baucher *et al* 2006). Seismic reflection method using S-wave

Table 2.1. ASTM Consensus Standard Methods for assessing sinkholes and voids (reprinted with permission from Hoover 2003 copyright American Society of Civil Engineers).

Method	Consensus standard
Seismic refraction	B
Electrical (DS)	B (A)
Frequency domain electromagnetic (FDEM)	A
Ground-penetrating radar (GPR)	A
Gravity	A

reflection allowed increasing resolution in the uppermost subsurface (Krawzczyk *et al* 2012), whereas P-wave seismic reflection shows very impressive results in the deep parts of the subsurface (Anderson *et al* 1998). Seismic refraction tomography enables us to overcome the disadvantage of the conventional seismic refraction method, e.g., 'velocity inversion' (Sheenan *et al* 2005b). Multi-channel analysis of surface waves (MASW) is used for detecting subsurface cavities and stiffness of karstified bedrock (Park *et al* 1999a). Apart from this, other geophysical methods found application in the karst investigation. Moreover, new geophysics remote sensing methods have shown promising results in the sinkhole susceptibility assessment (Gutiérrez *et al* 2011, Oliver-Cabrera *et al* 2020). The goal of this book is to familiarize the reader with geophysical and other modern methods applied in the karst investigation. We do not intend to give a detailed description of these methods but present only a brief description and examples from the karst investigation worldwide and more detailed results of geophysical studies in the Dead Sea sinkhole problem.

The main methods used to study the subsurface geological, geomorphology, and hydrogeology features of the sinkhole sites are the following ones. The seismic refraction method (SRFR), based on compressional-wave velocity V_p, was applied to map hard layers (refractors) (Redpath 1973, Palmer 1986).

MASW was used for the shear wave velocity V_s determination, characterizes the rock rigidity (Park *et al* 1999b, Dal Moro 2015). The seismic reflection (SRFL) method based on V_p, was applied to detect the vertical and horizontal structure of deep subsurface (Yilmaz 2001, Anderson *et al* 1998); note also the recently developed SFRL method based on V_s for detection of a subsurface structure at shallow depth (Krawzczyk *et al* 2012, Polom *et al* 2013). The transient electromagnetic (TEM) method (termed also the time-domain electromagnetic (TDEM)) is used to measure the resistivity of aquifers. The method is applied either in sounding or mapping mode (Fitterman and Stewart 1986, Barsukov *et al* 2015).

To apply the geophysical method we have to understand what challenges exist in the problem and what can be resolved by geophysical means. Chalikakis *et al* (2011) select the following challenges in karst systems that can be resolved by geophysical methods: (1) the limits and extent of karst systems; (2) structural discontinuities; (3) preferential pathways and/or concentrated infiltration; (4) cavities: either empty (air-filled) cavities (4a) or water-filled cavities (4b).

This list can be extended: detection of soluble rocks, detection of structure and shape of the reflector, connected with sagging, salt dissolution, revealing of karstification zones within bedrock, caused by dissolution, detection of cavities in the shallow subsurface, connected with stoping, evaluation of water salinity and aggressiveness in evaporite karst (Ezersky and Frumkin 2021).

Chalikakis *et al* (2011) carried out an overview of the most effective geophysical researches published over the last 20 years in the scientific literature. Moreover, they evaluated the suitability and ratings of most surface geophysical methods used for karst-system exploration (table 2.2). Authors of this book have added to this table new methods developed in the last decades.

Table 2.2. Adequacy of ground-based geophysics for karst system exploration (reprinted, modified, and added from Chalikakis *et al* (2011), by the permission of Springer (GmbH) Verlag, copyright 2011).

		Suitability and rating of the method													
		With sedimentary covering					Without sedimentary covering					sedimentary covering			
Method	Measurement technique	Boundaries	Fractured zones	Preferential pathways	air-filled cavities	water-filled cavities	Boundaries	Fractured zones	Preferential pathways	air-filled cavities	water-filled cavities	Type of covering	Thickness of covering	Structural problems: Faults, subsidence	Soluble layers***
Electrical	VES	+++	++	+	0	0	+++	+	+	0	0	++	+++		
	ERT	++	++	++	++	++	+++	++	++	+++	++	+++	++		
	Mise-a-la-masse	0	0	++	0	++	0	+	++	0	++	0	0		
	SP	0	++	+++	0	++	0	++	+++		++	0	0		
EM	FDEM|Slingram	++	++	++	+	+	++	++	++	++	++	+++	++		
	TDEM (TEM)	+++	+	0	0	0	+++	+	0	0	0	+++	+++		
	GPR	0	+	+	+	+	+	+++	++	+++	++	+	0		
	CSAMT	+++	++	++	+	+	+++	+++	++	+	+	+++	+++		
	VLFres-RMT	+++	++	++	+	+	+++	+++	+++	+	+	+++	+++		
	VLF-EM	+++	+++	+++	+	+	++	+++	+++	+	+	++	+		
Seismic	Tomography	+++	++	++	++	++	+++	++	++	++	++	++	++		
	MASW	++	++	++	++	++	++	+++	++	++	++	++	++		
	Reflection P	++	+	0	0	0	+	+						+++	
	Reflection S													+++	
	Refraction				++			++		++			+++		+++
	Diffraction				+++					+++				+++	
Microgravity	Profiling or mapping	0	++	+	+++	++	0	+	+	+++	++	0	0		
Magnetometry	Profiling or mapping	0	0	0	0	0	0	0	0	0	0	++	0		
SNMR (MRS)	Sounding	0	0	0	0	+++	0	0	0	0	+++	+	0		

Suitability: +++ recommended; ++ appropriate but incomplete; + appropriate but limited; 0 not recommended.

CSAMT controlled source audio magnetotelluric; EM electromagnetic; ERT electrical resistivity tomography; ES electrical sounding; GPR ground-penetrating radar; MASW multi-channel analysis of surface waves; MRS magnetic resonance sounding; RF-EMF radiofrequency electromagnetic field; RMT radio magnetotellurics (enhanced VLF-R); SASW spectral analysis of surface; wave; SP self-potential; TDEM time-domain electromagnetic; VLF very low frequency; VLF-R very low frequency-resistivity, ***at favorable conditions.

2.2 An overview of geophysical methods used for karst problems worldwide

2.2.1 Surface geophysical methods

2.2.1.1 Seismic refraction method

2.2.1.1.1 General consideration

Introduction. The seismic refraction method is based on the measurement of elastic velocities and their attenuation. Seismic methods study the subsurface by generating seismic waves and observing their propagation through the Earth section. Various methods of field acquisition and data processing are used, mainly to produce cross-sections through the subsurface that can be interpreted in geologically meaningful ways. The methods are particularly widely employed in oil and gas industries (Bacon 2005). The velocity of wave propagation and amplitude (signal strength) of seismic waves depend on the dynamic elastic constants of the rock and their density. There are two groups of elastic waves: body waves (that penetrate the bulk of medium) and interface waves. The body waves are capable of traversing and 'probing' all depth levels of the subsurface. The second one is existing only near the boundary of layers such as the Earth's surface and is named 'surface waves'. Seismic surveying is mainly based on acquiring and analyzing signals generated by body waves and recorded at the Earth's surface. Two types of body waves: compressional and shear waves (named P- and S-waves, respectively) exhibit movement of particles parallel and orthogonal to the direction of wave propagation, respectively (figures 2.6(a) and (b)) (Reynolds 2011).

Reynolds (2011) selects two main seismic methods—*refraction* and *reflection*.

However, during the last two decades, these methods were added by MASW, which is a seismic exploration method for the determination of near-surface shear wave velocity profiles by analyzing Rayleigh and Love waves propagation over a wide range of wavelengths (Park *et al* 1999a, Dal Moro 2015). Also, the seismic reflection method based on S-wave application demonstrates high resolution at the shallow subsurface (Inazaki *et al* 2005, Krawczyk *et al* 2012). Apart from this, a method *of diffraction imaging* is developed aimed to detect underground heterogeneities (Landa *et al* 1987, Klem-Musatov 1994, Keydar and Shtivelman 2005).

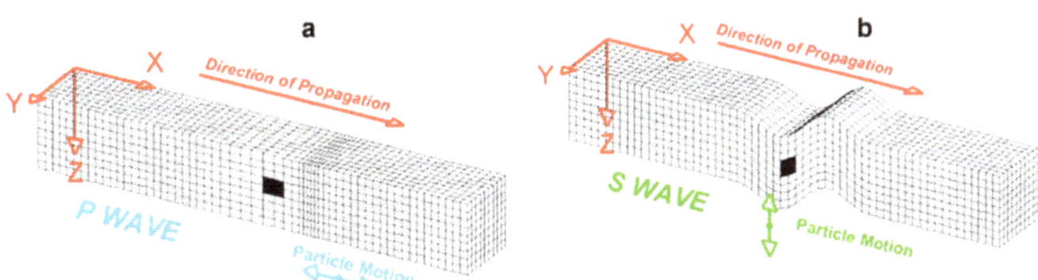

Figure 2.6. Graphical presentation of P- and S-waves explaining the relationship between the direction of propagation and displacement of medium particles for P- and S-body waves. The 3D grid created by L Braile allows the animation of the volume deformations (reprinted from Braile (2004), copyright 2004–10 L Braile).

Seismic refraction method. Seismic refraction surveys are carried out by measuring the travel times of refracted seismic waves as they propagate from the surface via subsurface layers and come back to the surface where their arrival times are recorded by a seismograph. The travel times depend on the layers' seismic velocity and the geometry of subsurface soil and rock layers.

The seismic refraction surveys are regulated by the ASTM D 5777-00 (2006).

The seismic refraction method is used for the calculation of depth and thickness of geological layers, for instance, depth to bedrock and water table, and to delineate geologic structure (Redpath 1973, Green 1974, Mooney 1984a). Velocity values of the longitudinal (compressional) P-waves (V_P) and the shear S-waves (V_S) are a measure of the layer properties and can be used for evaluating rock quality and rippability of rock (Sharma 1997, Reynolds 2011). If both compressional P- and shear S-wave velocities are measured, dynamic Poisson's ratio and elastic moduli of soil and rock can be determined *in situ*, when the rock density was measured.

Here i_c is the angle at which critical refraction occurs, and V_2 is greater than V_1. V_1 and V_2 are the seismic velocities of the upper and lower layers, respectively; i and r are the angles of incidence and refraction, respectively.

When a seismic wave propagates through layers with different elastic properties the direction of wave travel is changed according to Snell's law that states that the magnitude change of direction is governed by the contrast in the seismic velocity of layers:

$\sin i / \sin r = V1/V2$ for general refraction,

$\sin i_c = V_1/V_2$ for critical refraction.

Data acquisition of seismic refraction surveys. A typical seismic survey is performed using 12–24 geophones placed along a line at the surface (figure 2.7). The geophone separation, as a rule, is 1.5–6 m. The distance between geophones determines the

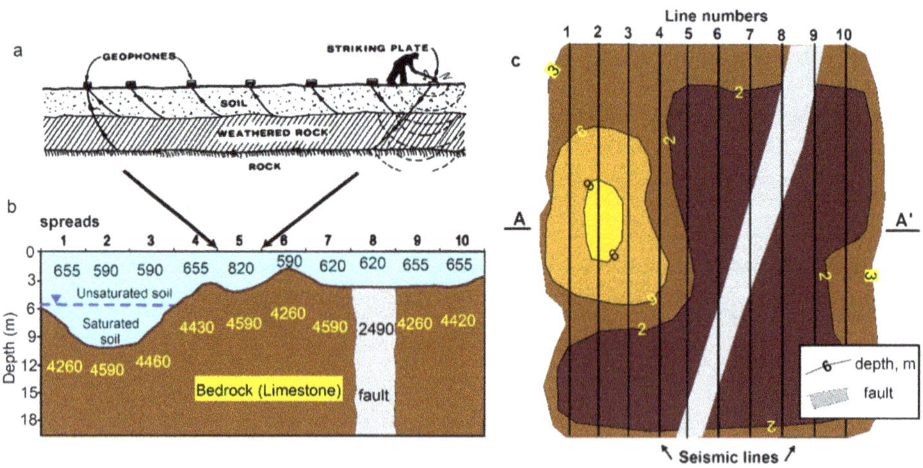

Figure 2.7. (a) Seismic refraction spread; (b) resulting cross-sections along line A–A′ developed from ten spreads; (c) map-to-bedrock (modified and added from Hoover 2003 by author).

resolution of the seismic survey. The less the separation, the less the penetration depth, and the less is the heterogeneity that can be detected (fractures, voids, layers, etc) and vice versa. Seismic energy sources, are commonly sledgehammer, dropping weight, hydraulic source placed on a vehicle (Digipulse, for instance), shot-gun and must excite longitudinal (P) waves or shear (S) waves. In the first case, a vertical stroke is carried out on a metallic plate, in the second one a horizontal stroke is performed on a horizontal beam. The place of stroke is named shot. The number of shots along a line is determined by the seismic processing technique selected for the survey. Usually, the number of shots is from 3 to 5 in the karst survey. More shots are applied in the case of seismic tomography. The number of shots determines the time and cost of a survey because both data determine acquisition and processing time (Hoover 2003)

The penetration depth of the seismic refraction survey is usually 1/3 length of the geophone spread (as a rule of thumb). The use of the far offsets (located out of the spread) increases the penetration depth of seismic waves. *The distance of the far offsets from the first or last geophones is approximately half of the seismic line length.*

The results of the seismic survey are presented as a velocity-depth section. Figure 2.7(b) presents a typical two-layered section with the denoted seismic velocity. In the studies of karst, velocity mapping or map-to-bedrock is carried out (figure 2.7(c)). In figure 2.7(a) a schema of the refraction along one line is shown. In this example, ten parallel spreads were shot. The section presented in figure 2.7(b) crosses these spreads along A–A′ line. Note, that the presented section has detected a fault zone characterized by low-velocity. The map in figure 2.7(c) shows isolines of depth to bedrock. The left part of the figure shows the zone of depressions in the bedrock, which may indicate the presence of a cavity.

A difference in refraction techniques. It is known, that there are different seismic refraction (SRFR) techniques based on the refraction of elastic waves shooting on a surface (Redpath 1973, Mooney 1984a, Palmer 1986, Reynolds 2011). The conventional reciprocal method (CRM) is probably the most used and one of the most convenient methods for processing and interpreting, known in Europe as ‘the plus–minus’ method (Hagedoorn 1959), in Russia the correlation method of refracted waves (CMRW) (Gamburtsev *et al* 1952) and in Japan as Hagiwara’s method (Kagawa *et al* 2010).

The advantage of using CRM is that it can conveniently accommodate and resolve simple departures from plane interfaces and homogeneous velocities. Besides, most of the computations are relatively simple arithmetic operations, which can be carried out to an extent with spreadsheets. Seismic refraction measurements are applicable for a wide diversity of problems. In the field of engineering geology, the method is applied to the mapping of geologic conditions including depth to bedrock, water table, stratigraphy, lithology, structure, and fractures. The characterization of the geological material (type of rock, degree of weathering, and rippability) is based on the seismic velocity, which depends on elastic and some other material properties (Mooney 1984a, Sharma 1997, Reynolds 2011).

The generalized reciprocal method (GRM) (Palmer 1986) is determined as the inversion technique for interpreting in-line seismic refraction data using travel times coming from both forward and reverse shots. The GRM allows resolving the geometry of subsurface refractors. Using the refraction migration one can calculate the detailed structure of a refractor and information about any localized lateral velocity variations within it. The distance of offset (e.g., migration) is the horizontal separation between a point on the refractor where a ray is critically refracted, and that at the surface where the ray emerges (Reynolds 2011). The field geometry includes 24 geophones, at least, (usually 48 ones) put along a straight line (spread) with equal distances (separations) that can be 1, 2.5, 5, 10, 20 m in depending on the required depth of investigation. Five shots, at least, along the seismic line should be carried out.

The time distance graph is the main output result of the seismic refraction measurement (figure 2.8).

Two shots performed near the first and last geophones (at −1 and 236 m distance) are named 'zero offsets'. The third shot carried out at the center of the line (at 117.5 m distance) is named 'split'. Finally, two shots 4 and 5 carried out ~100–120 m from every end of the spread mentioned above are 'far offsets' or simply 'offsets'. The distance between the zero offsets and the far offsets should satisfy the above-mentioned condition (marked by Italic). It depends on the refracting properties of the subsurface.

The reader can learn the GRM data processing in Palmer (1986) or Reynolds (2011). In our book, we will consider an application of the GRM method in sections 2.2.1.1.1 and 4.2.

The GRM method was intended for heterogeneous refractors (with varying lateral velocities) and inclined layers with dip <20°. Palmer's (1986) method provides accuracy in the refractor velocity determination of less than 5%. However, there can be concerns that should be considered using modeling.

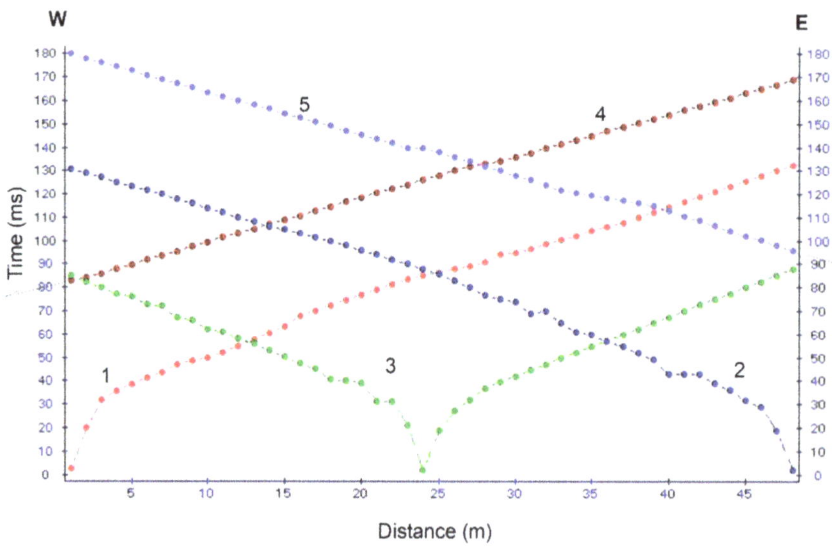

Figure 2.8. Time-distance graphs for 48-channel seismic refraction survey. The separation between geophones is of 5 m. 1, 2—zero-offsets; 3—split, 4, 5—far offsets.

Limitations of the method. The method is designed for a stratigraphic medium. It is sensitive to the velocity structure of the medium, e.g., the seismic velocity of layers must increase with the depth. The method cannot detect thin geological layers. The distance between seismic source and geophone should exceed three–five times the investigated depth.

Seismic refraction tomography (SRT) is a modification of the seismic refraction method, which uses a gridded, inversion technique to determine the velocity of individual 2D blocks (pixels) within a profile, as against the velocity of the layer in a layered model of the medium. SRT is an alternative to conventional seismic refraction analysis methods.

In a traditional delay time or reciprocal approach to seismic refraction, over-simplifying assumptions constrain the use of seismic refraction to layered 'cake geology' settings. In contrast, iterative modeling of numerous densely spaced raypaths, used in seismic refraction tomography, allows for the construction of seismic velocity sections accounting for velocity gradient, heterogeneity of short distance wave velocity, horizontal discontinuity, and even the presence of local velocity inversion—conditions frequent in the near-surface settings (Ostrowski et al 2016). An example of refraction tomography is shown in figure 2.9.

Software. The data processing of the seismic refraction method is carried out using commercially available software containing as a rule one–two methods. A thorough review of currently used programs is considered by Reynolds (2011).

Additionally, The Interpex software *IXRefraX* is intended for 2D interpretation using the GRM method (Interpex Ltd 2019). GRM has parameters that allow the method to be applied in more simplified ways. The forward and inverse modeling do give an interpretation with the constant horizontal velocities, but this is used primarily to remove the drudgery of assigning arrivals to layers as this can be done automatically by the software and edited by the user if desired (Stoyer 2011).

Other software, *WinSism*17 (W-Geosoft 2017), allows 2D data interpretation using four refraction techniques such as the intercept time method (SIM), the critical distance method, ABC/delay method, and the GRM method.

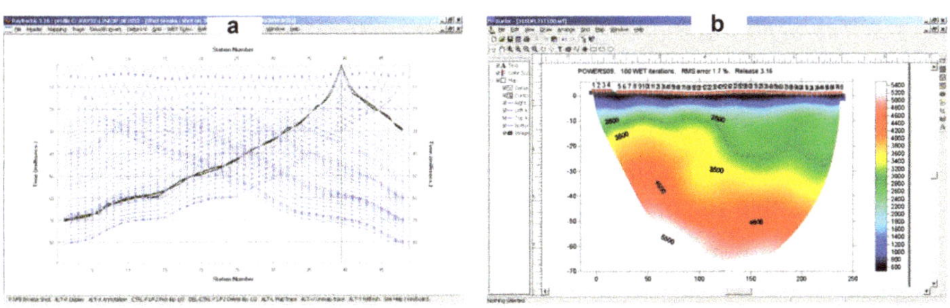

Figure 2.9. (a) Travel time curves and (b) smooth inversion with v.3.06, 100 WET (Wavepath Eikonal Traveltime) iterations, RMS error 1.7%. See the explanation in the text. (The following material from Rohdewald *et al* (2010), is published with the permission of the Environmental and Engineering Geophysical Society. All copyright privileges remain with EEGS. This material cannot be copied or used without the express written permission of EEGS.)

Recent software deserves to be mentioned that uses SRFR tomography.

There are now several different packages available on the market such as REFLEXW (Sandmeier 2020); SeisImagerTM (Geometrics and OYO 2003); ∗RAYFRACT TM (Rohdewald 2006); ∗GeoCT-II (GeoTomo 2007). Tomo has also produced a 3D tomography package (TOMO3D).

ReflexW v.9.5 (Sandmeier 2020) covers the complete range of wave data (seismic, GPR, ultrasound) and different geometry assembling (surface reflection and refraction, borehole, cross-hole, and tomography, and combination of borehole and surface measurements. There is a module for forward and inverse modeling. The software is working in the Windows environment. The software is organized in modular characters. The seismic refraction module allows 2D data interpretation, wavefront inversion, tomographic inversion using wavepath travel-time calculation, and forward raytracing.

The Rayfract® seismic refraction tomography software allows reliable imaging of subsurface velocity structure. The tomographic data interpretation is based on physically meaningful modeling of wave propagation with the wave paths instead of conventional seismic rays. This increases the numerical robustness of the inversion, especially in low coverage situations. An initial 1D gradient model is determined automatically directly from the travel-time data, with the Delta-t-V method (improved Herglotz–Wiechert method). This method delivers reliable models even in the case of velocity inversions. The initial model is then refined with the true 2D wavepath eikonal travel-time (WET) tomography processing. Also, the software allows the interpretation of travel times with the conventional plus–minus and wavefront seismic refraction methods (Rohdewald 2006).

TOMO3D is a 3D seismic refraction tomography software package designed for geophysical engineering applications (based on the Windows operational system). It includes travel-time picking, 3D travel-time QC and utilities, 3D velocity model viewer and utilities, 3D raytracer, 3D tomography, and tomography QC modules. TOMO3D allows users to perform the first-arrival travel-time tomography for three different survey geometries: refraction, cross-hole, and vertical seismic profiling (VSP). It has no restriction for any source/receiver layout.

Other software can be found on the corresponding Internet sites.

2.2.1.1.2 The application of the seismic refraction method for the karst problem

Following Hoover (2003), many seismic methods have been applied to the karst problem, but few have been successful. As a rule, a geologist does not know the position of hidden cavities when they are not written in the cave database list or if a map of cavities is not available. In these cases, geophysical methods can be an effective alternative for cavity detection. It at first relates particularly to ground-penetrating radar (GPR). It can be useful also for seismic methods, for which pertinent results have been recently obtained. Despite there being essential contrast in velocity and density between the cavity and surrounding rocks, the velocity contrast is diminished because of the diffusive character of P-wave interactions. Hickey *et al* (2009) conclude that the refraction seismic methods can be used with limited reliability. The inverted S-waves do not give enough resolution to detect cavity, but using backscatter it is possible to not only detect buried cavities but also to calculate the embedment depth using the diffraction hyperbolas in the common shot gathers (CSGs).

This phenomenon has been successfully used to detect the karst cavities (Breithaupt 2016, Xia *et al* 2007) and to discover tunnels (Sloan *et al* 2013).

Detection caves and voids (using raw data). Seismic refraction techniques can give us important information about the near-surface karst heterogeneities such as preferential pathways and near-surface cavities (Breithaupt 2016). The seismic refraction method is not able to delineate cavities when they are not included in horizontal layering (Sheenan *et al* 2005a). The inverse velocity limitation (an increase of velocity with a depth) leading to the fast layers hiding structures also limits the applicability of the method.

However, despite the disadvantages of seismic refraction, some phenomena can be applied to detect and approximate the geometry of subsurface heterogeneities.

Engelsfeld *et al* (2008) considered a model that suggests when a critically refracted up-going wavefield encounters a void, the latter will act as a boundary to wave propagation due to the large contrast in elastic parameters. The interaction between this boundary and wavefield (after Huygens principle) generates new sources, and a new wave field propagates from the void interface with a delay due to the added length of the wave path. This delay time depends on the depth and diameter of the cavity, as well as on the depth of the refractive interface.

The simple equations suggested by Engelsfeld *et al* (2008). to calculate the conduit embedment depth and diameter can be seen below in figure 2.10 and equations (1), (2), and (3).

Equation 1. The radius of the cavity: $2r = (x2 - x1)*\cos(\theta)$

Equation 2. Location of the cavity: $xp = (x1 + x3)/2 = (x2 + x4)/2$

Equation 3. Depth of the cavity: $z = (x2 - x3)/2\ \cos(\theta)$

where θ = critical angle = $(v1/v2)$.

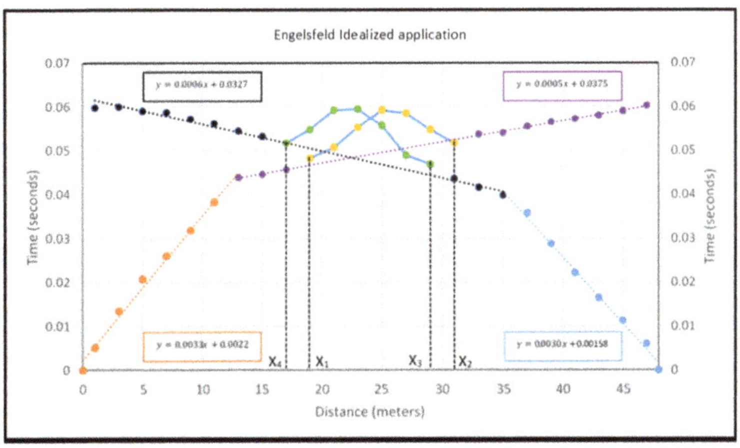

Figure 2.10. Application of Engelsfeld method from the time versus distance graph. The *x* location indicates the beginning and end of delays from forward and inverse shots. The velocity of each layer is taken from the inverse of the slope of each line but excludes the delay times caused by the void (reprinted from Breithaupt (2016), with permission).

And $v1$ and $v2$ are layers 1 and 2 velocities calculated from the *T–X* graph. The particular model has been designed for two layers where the conduit is in the first layer. Note also that there can occur a phenomenon known as 'wavefront healing' (Stein and Wysession 2003). In this case, a decrease in delay times can occur that is the result of the bending of waves around the void or low-velocity zone. The result can be the disappearance of some voids and conduits in the first arrivals depending on the size, depth, and properties of the materials contained in their boundaries. Breithaupt (2016) has tested the method at the Madison Blue Spring conduit (State Park, Madison county, Florida) where two caves are located at a depth range of 8.4–0.6 m. The study site is composed of karstified limestone located at an elevation of +22.2 m to −18.3. The water table is located at 15.5 m.

Seismic data were acquired over the conduit using a 24-channel seismograph Geode (Geometrics) with the 24 vertical 4.0 Herz geophones. All CSGs were collected with stacking (20 strokes) and were carried out using a 12-pound sledgehammer. A roll-along seismic data acquisition perpendicular to the conduit was used to define the lateral position and width of the conduit, and a varying azimuth survey to define the conduit's strike. The 46 m line was moved 12 m and the procedure was repeated until the line had a total length of 71 m with the first shot point at zero and the last at 71. An example of raw data is shown in figure 2.11. First arrivals with the picks are highlighted by red dots for both.

It is denoted that all delays are in the same location when the line is moved a magnitude of approximately 8 ms. The velocities are variable depending on the shot place. A depth to the bedrock (15–16 m) was calculated using the time intercept method. Engelsfeld's method of calculations show a depth mainly of 4–5 m. However, on some shots a depth of 16–39 m was calculated, which indicates that the method is only valid when perpendicular to the strike of the feature. The lateral position of the conduit was approximately right.

Thus, a survey has shown that seismic methods are effective for the detection of water-filled karst features even at depths of up to 25 m (Breithaupt 2016). Results testify that seismic methods can detect water-filled caves based on diffracted surface waves by viewing these data in the time and frequency domains. The combination of some seismic techniques (seismic refraction, reflection, and MASW) was ultimately

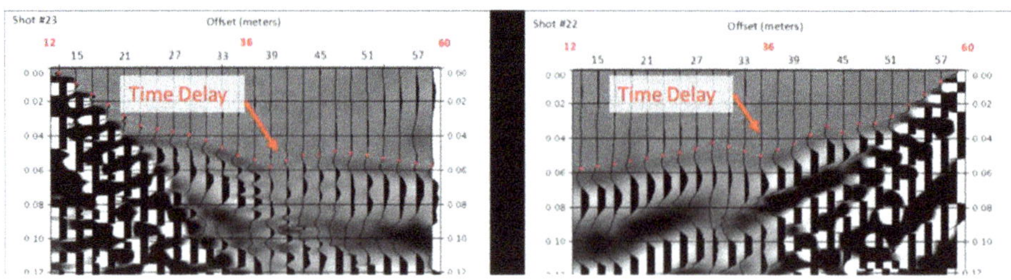

Figure 2.11. Example of roll-along line common shot gathers (Shots 23 and 22). First arrivals are indicated by the red dots. First break time delays are indicated with red arrows. Shot 23 displays maximum delay at an offset of 39 m and shot 22 at 35 m (reprinted from Breithaupt (2016), with permission).

successful in determining the location, depth, and approximate diameter of the hidden conduit.

Detection of the dissolved layer at shallow subsurface. Rucker *et al* (2005) using the seismic refraction method detected the gypsum layers. There is the layer of higher velocity ($V_P = 3300$ m s^{-1}) and S-wave velocity ($V_s = 1600$ m s^{-1}) for a thickness of about 4 m which was interpreted as dense gypsum. At a depth of 9–15 m, there is a softer layer with $V_s = 580$ m s^{-1}. Such a softer horizon is interpreted by the author as susceptible to significant dissolution consistent with the local topographic low at the project site. Below, at a depth of about 15 m, the S-wave velocity again increased to about 1600 m s^{-1} consistent with the massive gypsum/anhydrite (this layer was detected by the ReMe method). A similar problem of detection of the buried salt layers that was resolved using the seismic refraction method in the Dead Sea sinkhole area will be considered in continuation (e.g., Ezersky 2006). Note that the use of the seismic refraction method in the framework of the layered models is the most convenient and efficient. In this case, geological layers with different seismic velocities and their depths can be reliably detected.

Cave detection using seismic refraction tomography (SRT). Grandjean (2006) applied P-wave seismic tomography to detect air-filled cavities. The author has tested the method based on the tomography and Kirchhoff migration. The testing aimed to check the potential of the method in detecting and imaging buried heterogeneities in highly contrasted subsurface media.

The developed method includes processing first arrivals as well as later arrivals and uses them both in tomography and migration processes, sequentially. Tomography is used herein in interactive mode to reconstruct the field of background velocities from first arrival travel times. At first, the travel time inversion is carried out. Then, from the inverted velocity field, the seismic signals are migrated by a Kirchhoff method implemented in the space domain.

Figure 2.12 shows the different results obtained after the tomographic and imaging steps. The velocity field presumed constant equal to 200 m s^{-1}, is influenced by the local velocity anomalies ranging between 130 and 220 m s^{-1}. These anomalies can be interpreted as local compaction effects related to the building works (Grandjean and Leparoux 2004). Finally, the migrated section shows several seismic events, the strongest of which is related to the cylinder.

Another example. As a contradiction to refraction methods, in the SRT technique an initial model with the continuous velocity layers is not required. The SRT model consists of a multiplicity of small constant velocity grid cells or nodes. Then, inversion is performed automatically and involves raytracing through an initial model and comparing the modeled travel times to the field data, and adjusting the model grid-by-grid to match the modeled travel times to the field data. Then the procedure is iteratively repeated. The process is stopped when the number of given iterations is reached. Because there are no continuous velocity layers localized heterogeneity can be modeled (Sheenan *et al* 2005a). Authors have performed a refraction tomography line above a known mud-filled cavity centered on a well-designated GW-734. The cavity was encountered by the driller during the installation of well #GW-734 at a depth interval between 18 and 30 m. A

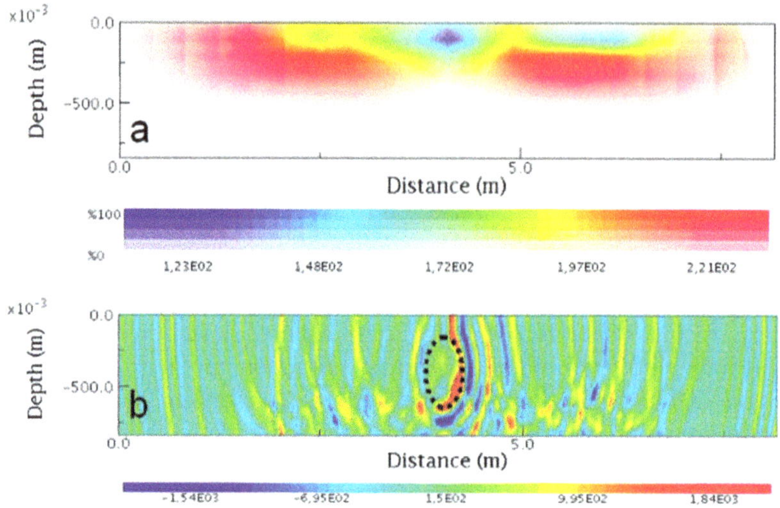

Figure 2.12. (a) The background velocity field is inverted from the first-arrival travel times (b) Anomaly reconstructed from the late arrivals using the velocity field shown in (a). The black dotted circle indicates the real position of the heterogeneity (reprinted from Grandjean (2006), with permission of EAGE, copyright 2006 European Association of Geoscientists and Engineers).

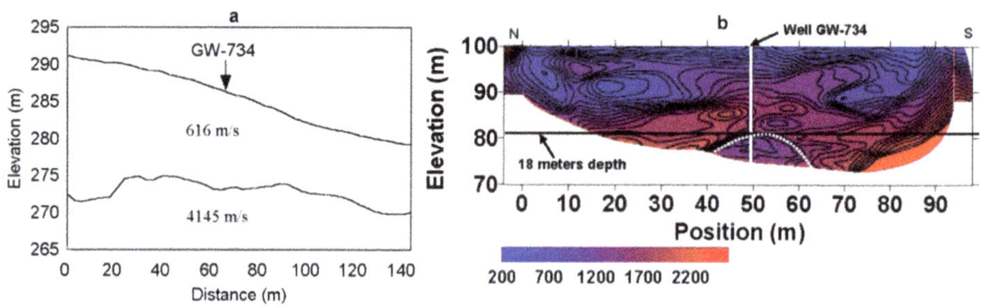

Figure 2.13. The known mud-filled cavity in Oak Ridge Reservation (Tennessee, USA) was discovered by the GW-734 well. (a) Conventional refraction analysis (arrow denotes well 734); (b) SRT result. The well indicated encountered weathered bedrock at 11 m, fresh bedrock at 13 m, and the cavity at 18 m (interpreted cavity shown by the dotted white line) (reprinted from Sheenan *et al* (2005a), with permission from the U.S. Geological Survey).

conventional refraction survey at this site did not detect the cavity (figure 2.13(a)). The SRT results for the line show a low-velocity feature with a location consistent with the drilling results (figure 2.13(b)). The velocity of the object is about 1000 m s^{-1}. The velocity of the surrounding area is about 2750 m s^{-1}, which coincide with the measured velocities for fractured and weathered carbonate at this locale.

Sheenan *et al* (2005a) conclude that the SRT method can image the buried cavities. Four seismic lines were shot in two different sites of the Oak Ridge Reservation mirror (Tennessee) known and possible (potential) cavities. One of the seismic lines was shot over a cavity revealed by drilling. The borehole discovered the top of the cavity at a depth of about 18 m and the bottom was at 30 m or deeper. The SRT has detected a cavity (as a low-velocity feature) at a depth consistent with the drilling. Thus, the SRT surveys can be effective tools to detect hidden cavities. However, sometimes there can be false detections (Sheenan *et al* 2005a, 2005b).

2.2.1.2 Multi-channel analysis of surface waves (MASW) method

2.2.1.2.1 Fundamentals of the MASW method

MASW method. During seismic surveys (seismic, reflection) when compressional-wave sources are used, tmost part (two-thirds) of the total seismic energy generated is converted into Rayleigh waves (Richart *et al* 1970). Therefore, in these methods surface waves are considered a hindrance that must be removed. Methods of surface waves developed in the 1980s–1990s such as the spectral analysis of surface waves (SASW) (Stokoe *et al* 1994) and MASW (Park *et al* 1999a) are based on the registration of the surface waves that are considered as useful signals.

The MASW method is a non-destructive seismic technique that uses surface waves propagating from the seismic source located on the Earth's surface to calculate shear wave velocities V_s of different polarization. The first type is vertically polarized waves, named Rayleigh waves propagating with velocity V_R. The second type is horizontally polarized Love waves propagating with velocity V_L (figure 2.14).

The Rayleigh wave travels in the direction of propagation inducing at that elliptical motion of particles whose amplitude exponentially decreases with the depth (figure 2.14(a)). Such elliptical motion is a result of a superposition of the vertical and horizontal components of oscillation. Love waves move only in a horizontal plane with a displacement of particles perpendicularly to the direction of wave propagation (figure 2.14(b)). Rayleigh and Love's waves are named surface waves because the amplitude of their oscillations is exponentially decreased with a depth. Therefore the depth of their penetration is limited to the wavelength λ.

The SASW method (Stokoe *et al* 1994) is based on using only two geophones with the consequent repetition of shots. The MASW method was suggested by Park *et al* (1999a) as the development of the SASW method to overcome its disadvantages: heightened noise and necessity to carry out numerous shots.

Active and passive MASW techniques. The method can be used in active and passive modes (Park 2006, Park *et al* 2007). In the active method, surface waves are generated by impulsive or vibrating seismic sources like falling weight, sledge-hammer, vibroseis. The geophones are placed along the line (figure 2.15(a)). In

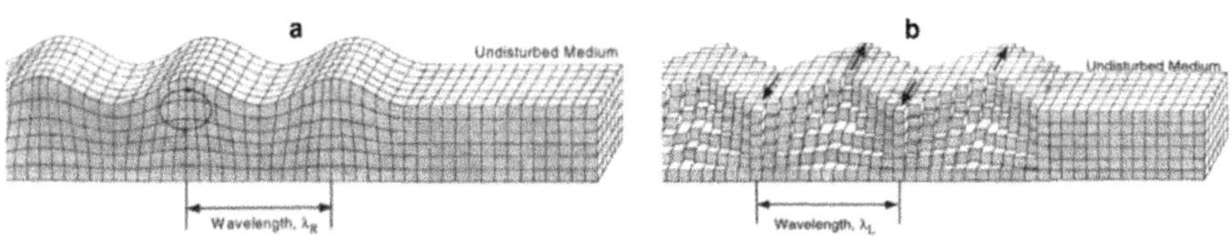

Figure 2.14. Scheme showing the particle oscillations near earth surface: (a) Rayleigh- and (b) Love-waves (reprinted with permission from Braile (2004)).

passive MASW the ambient noise is recorded using different arrays (linear, circle, orthogonal, etc) (figures 2.15(b)–(d)).

Dispersion of phase velocities. The surface waves have a dispersive nature, e.g., there is the dependence of the phase velocity from wavelength: the longer the wavelength, the higher is the phase velocity. Figure 2.16 explains this property of the surface waves.

When the Rayleigh wave propagates through the layered medium, phases with longer wavelength penetrate to deeper layers that, as a rule, are more compacted and have higher velocities. That is a reason for the dispersion. The propagation velocity of the Rayleigh wave depends on the shear-wave velocity. The Rayleigh-wave velocity propagating in a homogeneous half-space is slightly lower than ($0.87 < V_R < 0.96$), depending on the Poisson's ratio, and is not affected by dispersion (Richart *et al* 1970). In the vertically heterogeneous medium, i.e., a layered medium, V_R becomes frequency-dependent. Love velocity (V_L) is slightly higher than V_R.

The dispersion of the surface waves in a layered medium enables revealing the information regarding the properties of near-surface layers (Park *et al* 1999a). Being inverted, the dispersive Rayleigh wave phase velocity allows calculating the shear wave velocity profile of a multi-layered medium (Xia *et al* 1999). More about the surface waves reader can be found in Socco and Strobbia's (2004) tutorial.

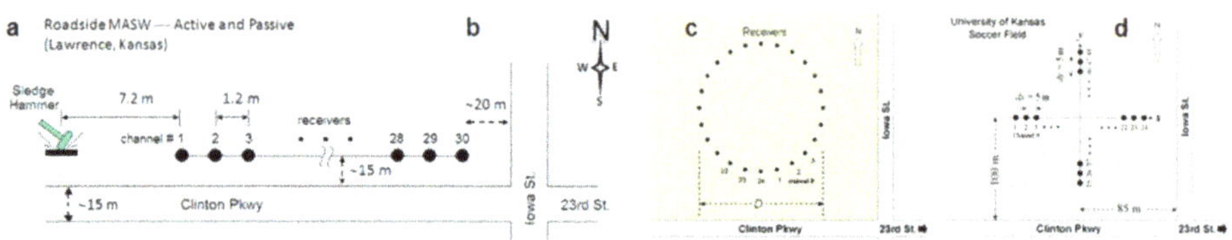

Figure 2.15. Different array modes are used in MASW surveys. (a) Linear active MASW using sledgehammer as a seismic source. (b) Linear passive array absorbing ambient noises. (c and d) Circular and cross layouts (respectively) arrays. Black points are geophones. (The source of this material is the Kansas Geological Survey website at http://www.kgs.ku.edu/. All Rights Reserved. Reprinted from Park (2006), with permission of KGS).

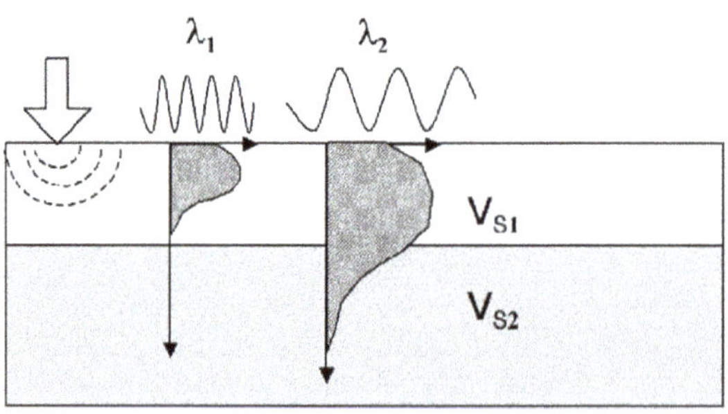

Figure 2.16. Schematic representation of geometric dispersion of Rayleigh waves: the vertical displacements associated with a short and a long wavelength (reprinted from Socco and Strobbia (2004), with permission of EAGE).

Multi-modes. In a layered medium having a finite number of layers and finite frequency range there exist a finite number of higher modes. Figure 2.17 shows multi-modal curves (Socco and Strobbia 2004).

Each mode, apart from the first one, exists only above a cut-off frequency, at which the modal phase velocity is the maximum shear-wave velocity of the system: higher modes have higher velocities and there can be common roots and apparent intersections of modes. The first one named fundamental mode is characterized by the lowest frequency.

Park *et al* (1999b) denote that the energy of higher modes depends on the source location: as the distance between the measurement line and source increases, the energy of the higher modes also increases.

MASW method includes three main parts: (1) data acquisition; (2) generation of the dispersion curves; (3) inversion of the dispersion curves.

1. Stage. Field data acquisition.

Active MASW is carried out on a horizontal or slightly tilted profile of 12–48 geophones with the eigenfrequency of 2.5–4.5 Hz linked to a multi-channel seismograph. Distance between the geophones is 1.5–10 m depending on the goals of the survey (penetration depth, resolution, etc). The MASW data acquisition of R- and SH-waves using dynamic DIGIPULSE source is presented in figure 2.18.

Passive MASW acquisition is based on the recording of ambient noise using different geophone arrays with the subsequent processing of the recorded data. Note that there are two different passive methods. The first method is named based on the array type (passive remote MASW, passive roadside MASW, etc) (Park *et al* 2007, Morton *et al* 2021). The second one (ReMi) is based on a linear array with increase electrode spacing (Louie 2001, Dal Moro 2015). It is a known fact that the passive

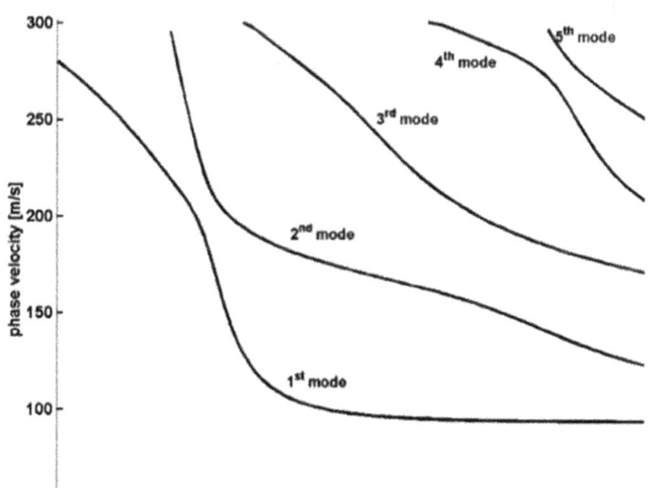

Figure 2.17. Modal curves for a three-layer model (3 m at 100 m s^{-1}, 3 m at 200 m s^{-1}, half-space at 300 m s^{-1}) (reprinted from Socco and Strobbia (2004), with permission of EAGE).

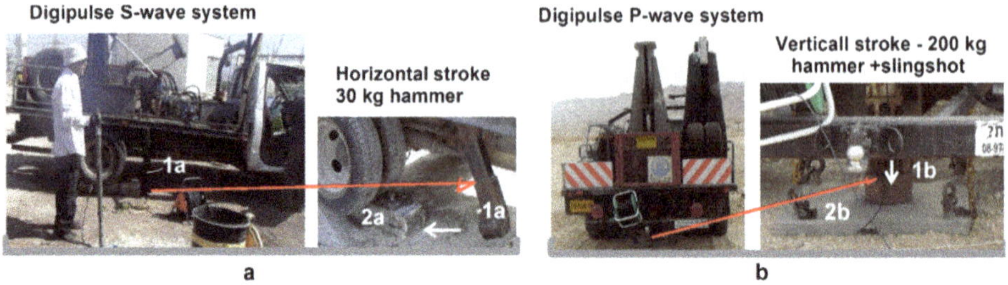

Figure 2.18. MASW data acquisition for the Rayleigh and Love waves. (a) Excitation of SH-waves using the horizontal stroke by a 30 kg hammer (1a) on the metallic beam (2a); (b) excitation of R-wave type using vertical stroke by a 200 kg hammer forced by slingshot (2a) on a metallic plate (2b).

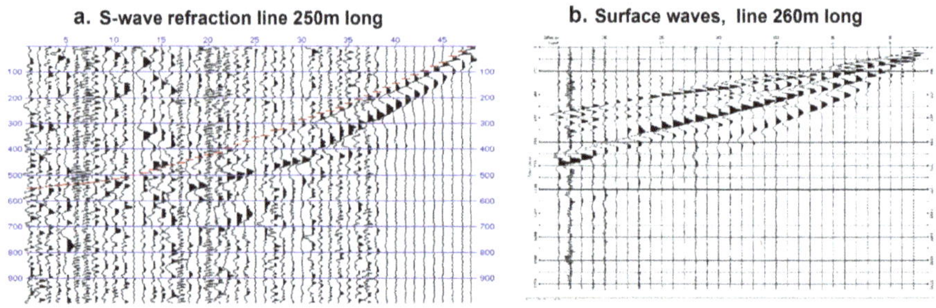

Figure 2.19. Comparison of field records for S-wave acquisition. (a) Raw data obtained during SH-refraction survey (records are complicated by strong noise); (b) field records of surface SV-waves.

method allows getting more penetration depth at a lower frequency than an active one. That is why using two acquisition modes together allows widening the frequency and penetration depth range (Park *et al* 2007).

Field records. A comparison of MASW and SH-wave refraction data is shown in figure 2.19. The record lengths of field data are 1000–4000 msec. Surveys were carried out along the same line of the 250 m length by different hammers. A comparison of surface waves (MASW) and SH-wave refraction data is shown in figure 2.19(a). The data acquisition of SH-waves using the refraction method was carried out using the horizontal stroke of a 30 kg sledgehammer on a horizontal metallic beam (figure 2.18(a)) allowing the excitation of clearest SH-waves at the background of other waves. Such a source allows the excitation of clearest SH waves at the background of other waves.

The records are shown in figure 2.19(a). Because of a weak source, SH-wave records are complicated by ambient noise. Such records allow penetration of SH-wave, as a rule, up to 25–30 m. On the other hand, the surface waves, because of strong vertical stroke by the 200 kg hammer (in our case, forced by slingshot) (figure 2.19(b)), are excited by the strongest energy, and field records are free of the noise. The propagation velocities of surface waves are the velocities of SV waves.

The quality of records and strong stroke allow penetration of surface waves down to the depths determined by wavelength (that is usually some tens to one hundred meters).

Note that records presented in figure 2.19(a) characterize SH-waves, whereas these in figure 2.19(b) relate to SV-waves. Velocities of SH and SV waves are seldom the same (Crice 2002). The latter denotes that horizontal anisotropy of 5%–15% is typical. Jackson and McCann (2015) conclude that 10%–20% anisotropy does not provide any significant changes to the tomography results. From our experience, the velocity of SH waves, as a rule, is somewhat higher (0%–10%) than SV-wave velocities. That should be taken to account.

Stage 2. Generation of dispersion curves.

The generation of a dispersion curve is one of the most critical steps for eventually generating an accurate shear-wave velocity profile. Recorded data are subjected to the generation of dispersion curves. For that, the wavefield-transformation' method converts the multi-channel record directly into an image, where a specific dispersion pattern is recognized in the transformed energy distribution (Park *et al* 1999a, 2007). As a rule, dispersion curves are displayed as the phase velocity versus frequency (figure 2.20). A dispersion curve (comprised of frequency–velocity points) summarizes the surface wave's dispersive properties (Dal Moro 2019).

Stage 3. Inversion of dispersion curves.

To calculate the V_S vertical profiles inversion should be carried out. There are two approaches to carrying out the dispersion. One is the least-squares approach which allows the automation of the process (Xia *et al* 1999). The calculation is carried out using the iterative inversion process. The dispersion curve, start model, V_S, density, and Poisson's ratio are requested to carry out an inversion. During the inversion, only V_S is updated whereas, Poisson's ratio, density, and model thickness remain unchanged through the inversion (Park *et al* 1999a). Xia *et al* (1999) have shown that among these four parameters used in the inversion, V_S has the most significant effect on the reliable convergence of the algorithm. This inversion was implemented in the

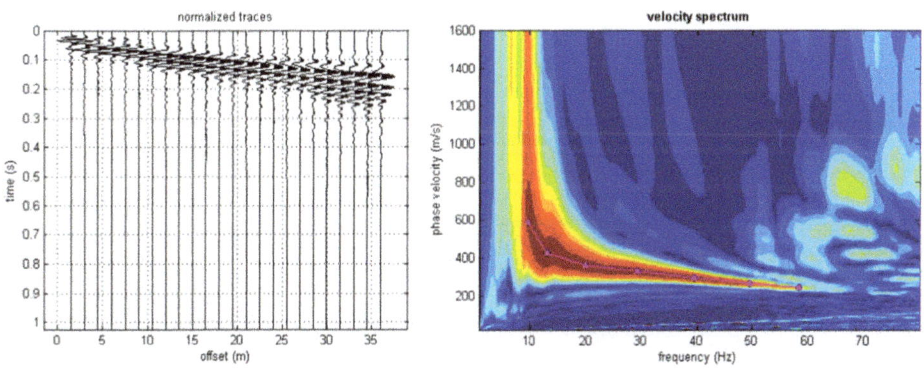

Figure 2.20. Surface wave records (left) and dispersion image with the picked curve (right). One can see only the fundamental mode. The spectrum type was picked up to about 6 Hz. At higher frequencies, a trend reveals intervention of the first higher mode (reprinted from Dal Moro (2019), the winMASW manual with permission (courtesy of www.winMASW.com)).

SurfSeis software (see in continuation). Dal Moro (2015) uses an automatic inversion of the dispersion data that is carried out via Genetic algorithms or forward modeling. Generic algorithms represent an optimization procedure belonging to the classification of the heuristic algorithms (or also global-search methods of soft computing). To carry out an inversion number of layers, limits of layer thickness and V_s values should be inserted into the initial model. These parameters are constraints of the calculation. Another way is the selection of the model using forward modeling. Then, this selected model is inserted into the inversion procedure. This approach has been implemented in the WinMASW software. An example of inversion using WinMasw software is shown in figure 2.21.

The upper left window in figure 2.21 shows picked dispersion curves of the fundamental (lowest frequency range), first, and second modes (all picked curves are denoted by circles). Above these curves, two inverted curves are shown: best fit model (solid, blue) and mean model (solid, green). The mean model is the average of different models considered during inversion, the best fit model is closest to the dispersion curve. The lower left window shows the development of the misfit evolution during the inversion procedure. One can see that 65 models were considered during the iteration process. The best fit curve (solid blue graph) after 20 iterations reached a 4% misfit and then was not changed. The average value of the mean model (dashed red graph) was varied in limits of 25%–35% during all iteration processes. Finally, the right window submits V_s vertical five-layered profile, as follows: the blue graph is the fittest model, the red dashed line is the average of considered mean models with a horizontal bar showing standard deflection of average from different considered models, the green area shows limits of calculated models inserted by the interpreter, the grey area shows the considered models during iterations.

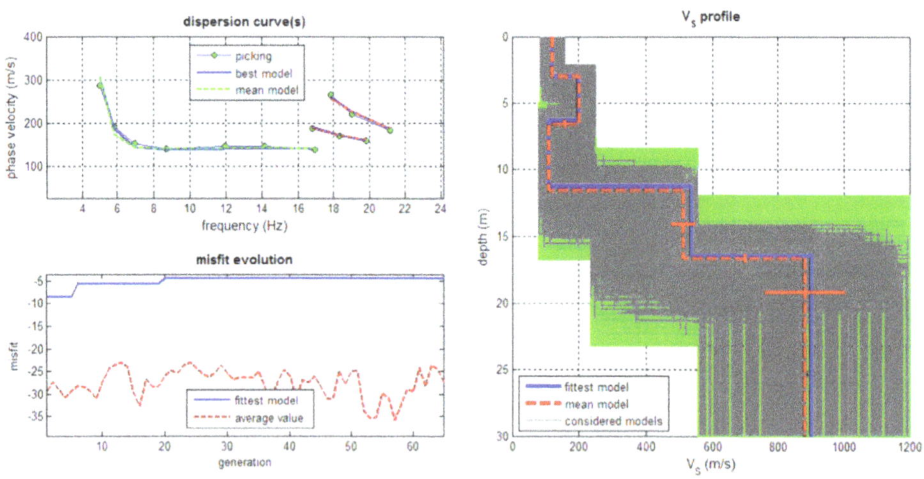

Figure 2.21. Results of inversion with WinMASW software (reprinted from Dal Moro (2015), the winMASW manual with permission (courtesy of www.winMASW.com)).

As a resumé, the blue best fit graph (right window) is the searched V_s model that fitted the inverted curve to the dispersion one with a misfit error of 4%.

Software. There are two commercially available software packages: SurfSeis and WinMASW.

The *SurfSeis* has been developed by the Kansas Geological Survey (Park *et al* 2008). The surveys are carried out using active, passive remote, and passive roadside MASW (figure 2.15(a)). Linearized inversion with a gradient-based iterative method is applied to dispersion data to reveal surface wave model (Miller *et al* 1999, Park *et al* 1999a, Xia *et al* 1999, 2003). The root mean square error (RMSE) between the theoretical dispersion curve and the measured one is usually used as an indicator of the closeness between the measured and iteratively calculated dispersion curves. The latest versions of the software (SurfSeis v.6.7) allow inversion of both Rayleigh and Love waves.

The *WinMASW* has been developed by Eliosoft geophysical software (Dal Moro 2015). The surveys are carried out using active methods (the Rayleigh waves, Love waves, and joint analysis of both) and a passive method named Refraction Microtremor (ReMi) (Louie 2001). An automatic inversion of the dispersion data is carried out via genetic algorithms or forward modeling (Dal Moro 2015). The software also allows one to carry out the joint processing (inversion) of shear wave velocity and HVSR (horizontal-to-vertical spectral ratio) (SESAME user guide 2006).

2.2.1.2.2 Application of MASW to karst investigation

The MASW method is a relatively new technique that is denoted as a method for cave detection neither in the ASTM 6429 nor in publications of researchers dealing with the problem (Benson and Yuhr 1993, Butler 1984, Hoover 2003). However, some investigators of the MASW method try to look for the way to use surface waves in the karst problem (Xia *et al* 2001, 2007, Park *et al* 1999a, 2009). Finally, Chalikakis *et al* (2011) include the MASW method in the list of surface geophysical techniques suitable for such karst-system exploration as detection of fractured zones, a search of karst boundaries, detection of preferential pathways, and cave detection. It is based on the sensitivity of the MASW method to the soil stiffness expressed in the shear wave velocity and shear modulus.

Debeglia *et al* (2006) have carried out a study of microgravity and a high-resolution MASW survey in the Orleans vicinity composed of siliceous and argillaceous alluvia, 5–10 m thick overlie the calcareous Beauce formation of Aquitanian age (Miocene) (chalky limestone), in which significant karst dissolution features occur.

The profile is shown in figure 2.22 as follows: (a) surface wave records and their dispersion images in the karst and out of the karst zones; (b) microgravity (Bouguer and residual anomaly profile); (c) MASW section and (d) surface wave image stack.

The MASW line revealed two perturbed zones coinciding with the negative gravity anomalies in figure 2.22(b). Features of MASW records in the northern zone are shown in figure 2.22(a). Left and right figures in (a) characterize the non-karstic area, whereas the central picture is characteristic of the karstified zone. There are several dispersion modes in this zone (figure 2.22(a), middle picture), and velocity

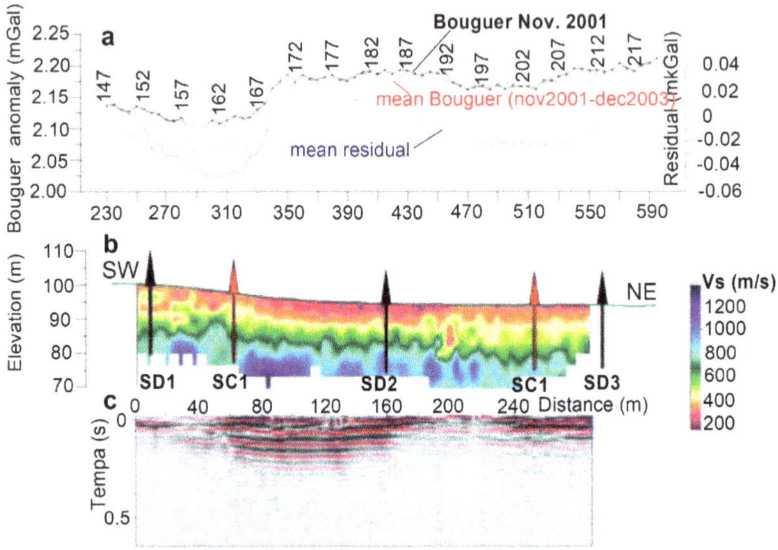

Figure 2.22. High-resolution MASW and microgravity reconnaissance along the central part of Profile 3: (a) MASW shots and the dispersion diagram recorded at the Parc Floral site in the southern (left), the central (middle), and the northern (right) parts of the high-resolution profile. (b–d) High-resolution MASW and microgravity reconnaissance along the central part of Profile 3: (b) gravity anomalies; (c) V_s velocity section, put attention on the color scale: blue zones are high velocity; (d) surface-wave image stack (reprinted from Debeglia *et al* (2006), with permission from EAGE publisher).

inversion in figure 2.22(c). The surface wave imagery section in these zones shows attenuation of the continuity of seismic markers (figure 2.22(d)). Such data characterizes fracturing and drainage pathways in the karstic zone.

One can see in figure 2.22(c) that the shear-wave velocity V_s varies in the range of 200–1000 m s^{-1}. The low-velocity values are consistent with the anthropogenic fill, whereas higher V_s corresponds to the massive limestone. The difference in MASW data agrees with the negative gravity anomalies.

One can see in figure 2.22(c) that the shear-wave velocity V_s varies in the range of 200–1000 m s^{-1}. The low-velocity values are consistent with the anthropogenic fill, whereas higher V_s corresponds to the massive limestone. The difference in MASW data agrees with the negative gravity anomalies.

This study has shown that the combination of microgravity and MASW measurements is an excellent approach to detect and characterize karst structure in the urban environment. Both methods are complementary in the scale and depth of investigation. Note that microgravity detects the density anomalies directly related to the cavities (in this case down to 20 m). However, its weakness is the ambiguity of its interpretation. The MASW, sensitive to the soil stiffness allows reducing ambiguity connected with the microgravity data interpretation. In this case, MASW has shown the weakness of zones where hidden cavities were detected.

Nasseri-Moghaddam *et al* (2005) carried out computer modeling of Rayleigh waves propagating through a medium with a buried cavity. They considered wavefront and its interaction with the cavity wall. A non-filled cavity is considerably less stiff than the surrounding medium. Therefore, the R-wave cannot propagate via the cavity.

Then, the cavity will affect phase velocities of these frequencies that are comparable with the size and embedment of the cavity (Park *et al* 1999a). figure 2.23(a) shows that the group velocity in the homogeneous medium (the S-model) is constant at all distances from the source. In the anomalous medium (V-model) the group velocity has a minimum, right after the void, and increases at farther distances from it. It is seen in figures 2.23(b)–(e), that both horizontal (x) and vertical (y) displacements sharply increase before cavity wall and attenuate after cavity passing. It is not shown in the figure that as the embedment depth of the void increases the difference in the group velocity decreases, and for embedment depths larger than 0.1λmax it is negligible.

The modeling has shown that before the cavity arises, significant ripples are caused by the reflection of the incident waves from the cavity wall. For the deep voids, the amplitude of reflections is so small that they cannot be seen in the response. Right after the cavity, a high attenuation of passed waves is observed. Thus, the cavity acts as a barrier that attenuates the incident front.

Based on the results of modeling, the authors suggest a quantitative method to locate a cavity.

(1) The underground cavity interacts with the incident Rayleigh waves. Part of the incident energy is reflected, diffracted, and transmitted further as Rayleigh waves. Residual energy is converted to body waves. Thus, the cavity plays the role of a damper and phase shifter. In the presence of a cavity, the homogeneous medium becomes dispersive for the Rayleigh-wave propagation.

(2) The transmitted Rayleigh wavefront is attenuated as a result of reflection and diffraction on the cave walls. Interaction of reflected, diffracted and incident waves can take place constructively or destructively. Then fields with amplified or attenuated waves can be observed between the cavity and source.

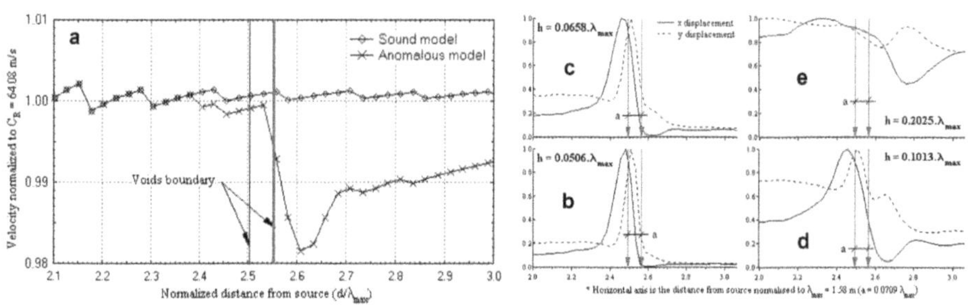

Figure 2.23. (a) Normalized group velocities measured at the surfaces in the homogeneous medium (the S-model) and in the medium with a cavity (anomalous V-model) (b)–(e) Variation energy–distance (ED) parameter with normalized distance from the source (h is the normalized embedment depth; h increases from (b) to (e)). Vertical lines are the limits of the cavity. See text for explanations. (Reprinted from Nasseri-Moghaddam *et al* 2005, is published with the permission of EEGS. All copyright privileges remain with EEGS. This material cannot be copied or used without the express written permission of EEGS.)

(3) The acquired data can be processed in both the time and frequency domains. The latter is more sensitive to the presence of an anomaly. The power spectral density and logarithmic decrement values are the most sensitive parameters.
(4) Finally, the ED parameter and amplified logarithmic decrement (ALD) values are computed easily from the Fourier spectrums. These parameters after normalization provide a quantitative method for cavity detection and evaluation of the embedment depth. This method has been named attenuation analysis of surface waves (AASW) (Nasseri-Moghaddam *et al* 2005).

One more approach has been presented by Orfanos *et al* (2020) who have suggested the methodology of constructing high-resolution frequency-dependent group and phase Rayleigh velocity maps in a 3D mode based on the surface wave tomography.

The implementation of the latter is based on the automated surface-wave phase-velocity measuring system (ASWMS) (Jin and Gaherty 2015). Originally, it is a *MATLAB*®-based package intended for analyzing and measuring the phase and amplitude of surface waves derived by regional and teleseismic earthquakes, to generate surface-wave tomography maps.

The acquisition layouts are shown in figure 2.24(a). The measurement system consists of a rectangular dense grid of 60 vertical 4.5 Hz geophones with an inter-station distance of 3 m, crossing as perpendicular as possible the tunnel. Shots at 34 positions were carried out with a weight drop, around and inside the rectangular grid, to have an adequate azimuthal coverage of the sources concerning the various geophone pairs. During data acquisition, the number of shots is carried out (figure 2.24(a)) and phase and group time-delays between all the available stations are calculated for each shot.

The automatic selection of the signal of interest in the time domain is carried out, and the cross-correlation function between the windowed signals of the analyzed stations is calculated. This automatic procedure is repeated for each shot, to estimate

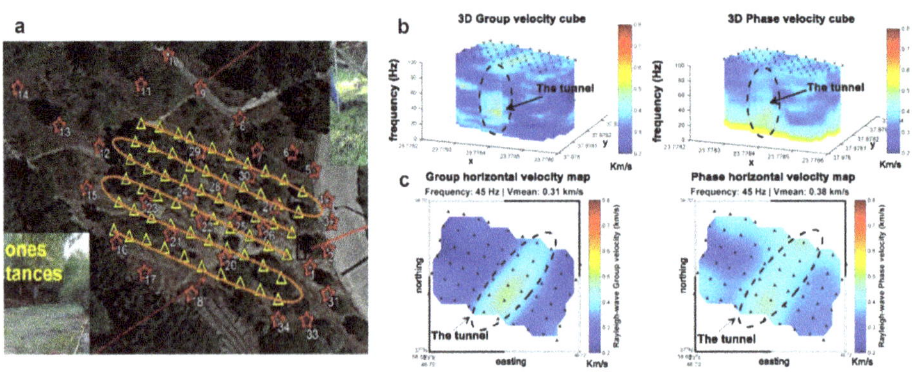

Figure 2.24. (a) Surface seismic wave tomography acquisition layouts at the controlled test site; (b) 3D frequency-dependent group and phase velocity images (c) group and phase velocity horizontal maps for the 45 Hz frequency (reprinted from Orfanos *et al* 2020, with permission from EAGE).

the phase and group delays that will be used for tomographic analysis. Then, the frequency-dependent phase and group velocities are inverted, to calculate spatial variations in apparent velocity using the Eikonal equation. The topographic maps of phase and group velocities are drawn with approximately 1×1 m grid size for each period.

The 3D frequency-dependent group and phase velocity images are shown in figure 2.24(b) in the frequency domain. It was possible to construct 3D velocity images with adequate density coverage in a frequency range of approximately from 10 to 100 Hz. The 3D maps present the spatial distribution of phase and group velocity for different frequencies. Taking into account that penetration depth of surface waves is defined by the wavelength (frequency) one can see that each frequency relates to different depths and characterizes different depth sections of the subsurface. On the other hand, suggesting that the depth of investigation is equal to one-third of the measured wavelength, maps of group and phase velocity at the 45 Hz corresponds to an approximately pseudo depth of 3 m. These maps shown in figure 2.24(c) define the limits of the tunnel with great accuracy.

This methodology, as well as presenting a method of detection of the tunnel, can also be used for the detection of cavities using surface waves.

Finally, we have to denote that although the MASW method can measure reliable shear wave velocity V_s in different geological conditions it cannot detect caverns directly looking for V_s anomalies because of limitations on the resolution of S-wave velocity profiles inverted from surface wave phase velocity (Bitri *et al* 2016). However, Park *et al* (2009) suggest that it is possible to image a near-surface anomaly through the simplest form of data acquisition using a swept-source, and through the simplest form of data processing without intensive analysis. Caves can be detected employing the MASW method using some characteristic features of propagation and interaction with cavities such as diffractions (Xia *et al* 2007, Breithaupt 2016). Interaction of surface wave with the cavity wall demonstrates some features such as an increase of the signal amplitude before the cave and its sharp attenuation after cavity passing. Based on these features, Nasseri-Moghaddam *et al* (2005) suggested the AASW method. Orfanos *et al* (2020) use ASWMS based on the calculation of the cross-correlation function between the windowed signals of the analyzed stations. The authors use 3D data acquisition. Finally, the MASW method can be also used in combination with other methods, for instance, microgravity (Debeglia *et al* 2006), with the electric resistivity tomography (ERT) method. In the latter case, the additional use of MASW and ERT results allows revealing multiple correlated V_s and resistivity anomalies on the segments, which may correspond to different stages of sinkhole propagation from the deep karst network to the surface through the alluvial cover.

2.2.1.3 Seismic reflection (SRFL) method

2.2.1.3.1 General consideration

Since the end of the 20th century, high-resolution seismic reflection surveying began to be applied in shallow investigations (i.e., to depths less than 200 m and especially less than 50 m).

Seismic reflection (SRFL) method. The purpose of the SRFL method is to determine the spatial configuration of underground geological units. The method is used to obtain important details about the geometry of the geological structures and also about the physical properties of the subsurface. In the hydrocarbon exploration, SRFL aims to find oil, or rather structural deformations like faults, folds that could form features capable of trapping migrating petroleum (Selley 2003). In engineering applications, the target is essentially less in size, and typical depths of penetration are of the order of the first decades of meters.

The SRFL technique is intended for measuring the time that is needed for a seismic wave to travel from a source (at a known location at/or near the surface) down into the ground where it is reflected from the interface between layers and then is recorded by a receiver at or near the surface at a known position. The time of a seismic wave traveling is named *two-way travel-time* (TWT).

Principles and practice of reflection surveying are thoroughly described in several textbooks and monographs (Waters 1978, Dobrin 1976, Telford *et al* 1990, Sharma 1997, Yilmaz 2001, Reynolds 2011) and therefore they will be only briefly discussed to provide the background for application to shallow-depth targets.

The first important (even, fundamental) problem in SRFL surveying is the conversion of TWTs (in the time domain) to depths (in the space domain). Here, the most important parameter that most affects the conversion to depth is seismic velocity. Following Yilmaz (2001) there is a slight relationship between the seismic data and geological peculiarities. That weak parameter is a seismic velocity. As a result, the depth section is never entirely reliable. Another fundamental problem is differences in the data processing performed by different interpreters using various means for data processing. These variances often follow from differences in the choice of parameters and the implementation of the detailed aspects of processing algorithms. That has happened even when starting with the same raw data. That is because each contractor, for instance, to estimate residual statics, most likely selected another correlation window, the traces used for the cross-correlation with the plot trace, and statistically treating correlation peaks. One more aspect of seismic data processing is the generation of artifacts while trying to improve the obtained signal. An optimal seismic program apart from its main destination also generates minimum numerical artifacts. Thus, data processing is a very subjective one. Some experts even consider it an art (Yilmaz 2001). These features of the SRFL method should be taken into account.

Seismic reflection principles. The SRFL method is governed by Huygens' principle (Fresnel 1900). With such knowledge, it is easier to understand how reflection, refraction, and diffraction occur. It is not necessary to consider the wavefront of every wave, but easier to consider a line normal to the wavefront along which the energy is propagated. Therefore, seismic wave propagation is considered in terms of rays and raypasses. Every time a wave falls upon an interface dividing two layers with the contrast in elastic properties, some of the energy is reflected off the interface, and the remainder passes through the boundary and is refracted upon entering the second medium. The reflection and transmission wave are considered in terms of the ratio of acoustic impedance of two layers (see Telford *et al* 1990, p 155, for a detailed discussion). The acoustic impedance (Z) is the product of the density (ρ) and the seismic velocity (V) for each respective layer. When a wave with energy falls normally to the interface between layers (1 and 2) with the energy of A_0 and those of the reflected and transmitted energy are, respectively, A_R and A_T (figure 2.25(a)), then (in the presumption of no energy loss) the sum of reflected and transmitted waves must be equal to the energy of the incident wave (i.e., $A_0 = A_R + A_T$).

On the other hand, for a seismic wave to be reflected from the interface between layers 1 and 2 to the surface there must be a contrast in acoustic impedances $Z = V\rho$, where V is a velocity of longitudinal waves and ρ is the density (i.e., $Z_i = V_i\rho_i$ for the ith layer). Then, the amplitude of the reflected wave from the second layer is described by the *reflection coefficient* (Telford *et al* 1990, p 156):

$$R = (Z_2 - Z_1)/(Z_2 + Z_1) \tag{2.1}$$

In other words, when we have more compacted rock, its acoustic impedance will be higher. To propagate seismic energy most efficiently across a boundary, the acoustic impedance contrast should be small (in the limit, no interface exists).

Figure 2.25(b) shows resulting reflection ray paths reflected from a horizontal interface across which there is a contrast in acoustic impedance Z (Z_2>Z_1). The shot point is located in the center of the seismic line and geophones are laid out in a line on each side of the shot. It is presumed that reflection takes place from the point. It takes place from some area of reflector defined by the first Fresnel zone. Second and subsequent Fresnel zones can be ignored. The first Fresnel zone at just one location is indicated in figure 2.25(b).

Thus, the incident waves are reflected at each point of incidence on a subsurface interface, above an area corresponding to the first Fresnel zone. The following geometry is visible from figure 2.25(b): (a) the point of reflection is located halfway between the source and the detecting geophone; (b) the spacing between reflection

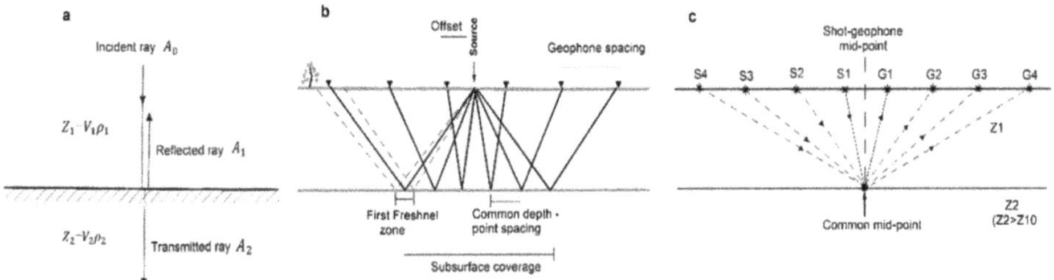

Figure 2.25. (a) Partitioning of energy from a normally incident ray with amplitude A_0 into reflected and transmitted rays with respective amplitudes A_1 and A_2. Z_1 and Z_2 are the acoustic impedances of the two layers. (b) Schematic view of reflection raypaths in single layer case for a six-channel seismograph, the first Fresnel zone at just one location is indicated, (c) Concept of Common mid-point, Si—shot, Gi—geophone. A horizontal interface divides layers between which there is a contrast in acoustic impedance Z ($Z_2 > Z_1$) (reprinted from Steeples and Miller (1990), with permission from SEG).

points on the interface is always half that of the geophone spacing; finally (c) the total subsurface coverage of an interface is half of the total geophone spread length.

There is one more concept important for reflection data processing that is variously called a common-reflection point (CRP) (Mayne 1962), a common-depth point (CDP), or a common mid-point (CMP) depending upon the preference of the author (figure 2.25(c)). The power of the CDP method is in the multiplicity of data that come from a particular subsurface location. Passing via the same point (common mid-point) from different shots, the reflection signal is enhanced. However, the data must be corrected for differences in travel time for the reflected waves caused by the differences in source-to-geophone distance. The degree of a multiplicity of data from a particular location is known as the 'CDP fold.' A 12-channel seismograph, for example, is commonly used to gather 6-fold CDP data. From a theoretical standpoint, the signal-to-noise (S/N) ratio of reflections improves proportionally to the square root of the CDP fold.

Three-dimensional survey. 3D reflection surveys were previously used only in perspective areas from the point of view of hydrocarbon recovery. When the 3D data acquisition was improved, costs fell and the method became increasingly used for the estimation of new fields. 3D surveys are also used to evaluate trends in the quality of the reservoir. The 3D advantages in comparison with 2D surveys resulted in a sharp increase in 3D survey usage. Although the method is highly used in oil exploration, in engineering investigations, it can be used only where the high cost of the survey is justified by the needed safety of the designed structures, for instance, in designing nuclear power stations, sites of safe disposal of radioactive waste at great depth.

Resolution of P-seismic reflection method. Vertical resolution is a function of the frequency of seismic energy that is generated and propagates. At a frequency of 500 Hz resolution can be as good as 1 m. Saturated fine-grained soils can be the optimum conditions for shallow reflection surveys because of the good coupling of geophones with the ground. It enables the generation of the highest frequency energy. Lateral resolution is a function of geophone separation, which is commonly 0.3–3 m. Common practice is a 2D high-resolution reflection cross-section of soil or rock layers along a profile line. However, recently 3D reflection surveys have also been conducted (ASTM 6429 (2006)).

S-wave reflection. The P-wave high-resolution seismic reflection technique is based on the excitement of a seismic disturbance at or near the ground surface and measuring the arrival times of compressional (P-waves) reflected from subsurface interfaces. To date, this method has developed into a powerful tool for subsurface characterization as part of groundwater or geotechnical studies (Yilmaz 2001). The main difficulty of the P-wave reflection method is the resolution at shallow depths. The use of shear (S) waves with low velocities enables increasing resolution in comparison with a conventional survey. In hard rocks, V_s/V_p the ratio is close to 0.5. For instance, V_s/V_p ratios for shale, sandstone, and limestone are 0.5, 0.62, and 0.56, respectively (Neidell 1985). The predominant frequency is also about half, and implying the S-wave will not increase resolution (Johnson and Clark 1992).

In unconsolidated saturated sediments, V_s/V_p ratio is close to 0.1 (e.g., Suyama *et al* 1986) and has a similar frequency content, in this case implying the S-wave can sufficiently increase resolution.

Pugin *et al* (2008) have shown different resolution and penetration depths for P-, SH- and SV waves in the case of water-saturated unconsolidated sediments with the P-wave velocities of 1500 m s^{-1} and the S-wave velocities of 150 m s^{-1}, if exploration is done by a vibratory source. In such a case, average structural resolutions are up to 1.67 m for P-waves, 0.63 m for SH-waves, and 0.27 m for SV-waves at the main frequencies of 220, 60, and 140 Hz, respectively.

The S-wave reflection method is actively developed by German researchers (Krawczyk *et al* 2012, Polom *et al* 2013, Wadas *et al* 2017). Wadas *et al* (2017) present a seismic reflection system working on the S-wave reflections. The seismic system, (figure 2.26), consists of a small vibrator source exciting transverse SH-waves (ELVIS, v. 7.0), and mobile land streamer of 120 m long, with mounted 120 one-component SH-geophones (Polom *et al* 2013). The system enables one to carry out the complete shear-wave seismic survey. When the system surveys on a sealed surface it allows reaching 10-times higher resolution than a compressional-wave survey. This takes place because of the absence of suppressed (Love) surface waves and interferences connected with them.

Figure 2.26. The photograph shows the seismic system that includes the micro-vibrator exciting transverse SH-waves (ELVIS, v. 7.0), the land streamer, and the recording system used for the SH-wave reflection-seismic surveys (reprinted from Wadas *et al* (2017), with permission from EGU, SS-BY 4.0 license).

The time-to-depth converted, uninterpreted reflection depth section as an example is shown in figure 2.27 (the whole depth of the section is 80 m).

2.2.1.3.2 The application of the seismic reflection method to the karst investigation
The seismic reflections can provide structural information on the subsurface. Indeed, reflections can reveal the vertical and sub-vertical boundaries such as faults and geologic contacts. Horizontal interfaces could be detected if, for example, there is an impedance contrast at the top or the bottom of the karst medium. Greenfield (1979) expressed doubts about how the SRFL methods will be used in karst areas for engineering applications. There are two reasons for this. First, the target of interest is quite shallow, being at most a decades of meters below the surface and often, only a few meters. The reflections from such shallow depths arrive with the large-amplitude surface wave (ground roll) arrivals, which make the reflections difficult to identify. For various reasons, it is difficult to sufficiently reduce the interfering energy. The second reason is that it is expensive to run and process these data.

The first difficulty is resolved today by SRFL based on S-wave use (Krawczyk *et al* 2012, Pugin *et al* 2019) or by high-resolution SRFL method using a short-spacing type land streamer (Inazaki, *et al* 2005, Inazaki 2006). The combining of methods can be used jointly with the high-resolution reflection survey (Busetti *et al* 2020).

The second is right, but we know sounding reflection surveys down to hundreds of meters (Steeples *et al* 1986, Anderson *et al* 1998).

High-resolution imaging of subrosion structures using S-wave seismic reflection.
To determine the key factors which have affected the development of a large sinkhole 26–30 m in diameter and 12–17 m depth, discovered in the urban area of Schmalkalden (Germany) and correspondingly the dissolution of subsurface structures, several shear-wave reflection-seismic profiles around the sinkhole were carried out (Wadas *et al* 2017). An example of such a profile S1 located north of the sinkhole is shown in figure 2.28(a).

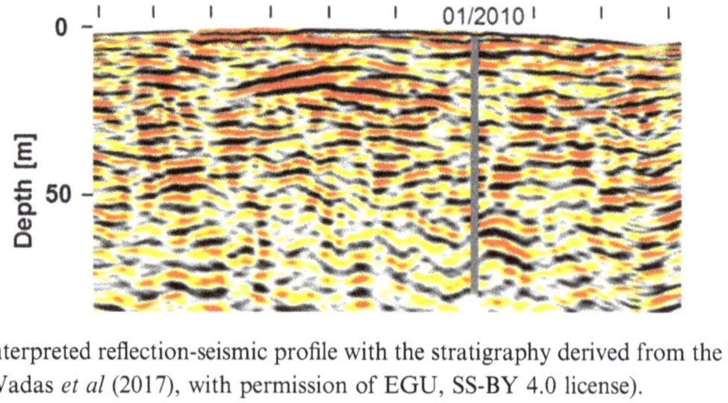

Figure 2.27. Uninterpreted reflection-seismic profile with the stratigraphy derived from the borehole #01/2010 (reprinted from Wadas *et al* (2017), with permission of EGU, SS-BY 4.0 license).

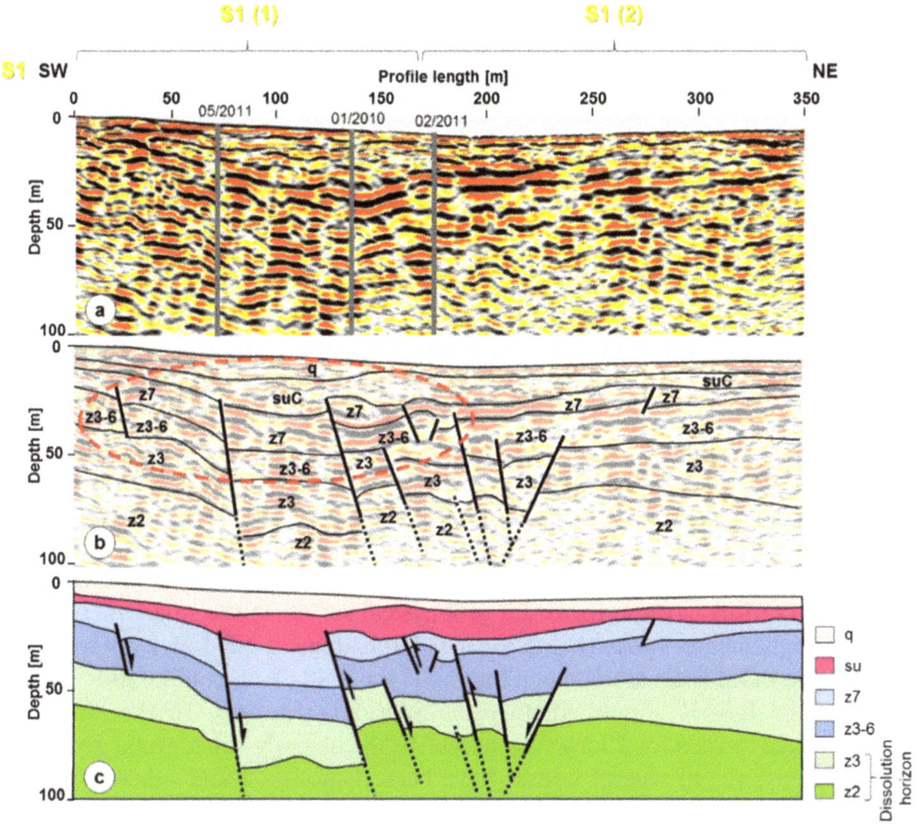

Figure 2.28. (a) Uninterpreted reflection-seismic profile S1; (b) with stratigraphy derived from boreholes ##05/2011, 01/2010, and 02/2011; and (c) with interpretation. Stratigraphy includes units Z3 and Z2 presented by gypsum and anhydrite that are subjected to dissolution. The profile was surveyed north of the sinkhole (reprinted from Wadas *et al* (2017), with permission of EGU, Creative Commons Attribution 4.0 license).

The profile 350 m length, was carried out north of the sinkhole. At 10–25 m depth, a continuous reflector with high amplitude can be traced throughout the entire profile (figure 2.28(a)). The strong impedance contrast represents the boundary between the Triassic sandstones (suC) and the Permian claystone (z7). This reflector, which can be found in all SH-wave reflection-seismic profiles, was used as a marker horizon.

In the southwest section, a bowl-shaped structure is visible (red dashed circle in figure 2.28(b)), which was interpreted as a dissolution-induced depression. A low-reflectivity zone (LRZ) can be observed at 70–100 m depth between 175 and 275 m profile length. This zone shows an almost transparent reflection pattern compared to the neighboring reflectors at the same depth. The LRZ correlates with the dissolution horizon within the z1–z3 unit composed of evaporitic rocks. The near-surface in all profiles is represented in detail, with a resolution of fewer than 1 m at depths down to 15 m and a resolution of 1–3 m at 50 m depth.

The results show that areas affected by phases of tectonic deformation are prone to enhanced dissolution. Around and in between the faults the deformations of the fault blocks are formed that lead to the generation of a damage zone with a dense fracture network, which enables the groundwater to flow through the subsurface and to leach soluble rocks. *The more complex the fault geometry and the more interaction between faults, the more fractures that are generated; e.g., dissolution in a strike slip-fault zone with steep normal and reverse faults will be enhanced more than in an area with just a simple normal or reverse fault.*

The faults do not only serve as fluid pathways through which groundwater can migrate upward; they also influence the entire groundwater flow. Wadas *et al* (2017) suggest that in a sinkhole area with a broad zone of parallel fault branches, steep dipping faults can act as barriers for the horizontal groundwater flow, but also serve as conduits for water to flow parallel to the fault strike, thus causing dissolution.

In Schmalkalden, the dissolution horizon in which the cavities are formed is located between the Permian Leine and Werra formations at 70–120 m depth. The long-lasting dissolution processes with rates that are probably slow, cause wide-spread dissolution. The processes are probably still ongoing.

High-resolution seismic reflection method in combination with InSAR and GPR

Busetti *et al* (2020) have performed a geophysical survey in the small village in the Alta Val Tagliamento Valley (Friuli Venezia Giulia region, NE Italy). An integrated non-invasive multi-scale study including the P-wave SRFL in combination with 3D GPR and PS-InSAR should identify the subsiding features. In the study area sinkholes were formed for a long time as well as the hazards linked to their presence.

During past years, several houses have been demolished and others have been damaged. The area is composed of an intensively karstified evaporite bedrock (gypsum) mantled by quaternary deposits. Also, the presence of regional thrust, the high amount of annual rainfall (1600–2000 mm y^{-1}), and the large fluctuations of the water table (often greater than 10 m) seem to be responsible for the above-mentioned phenomena. Cover-collapse and cover suffosion sinkholes are discovered in the area.

Ultra-high-resolution seismic reflection was conducted along two lines PS1 and PS2 through the village. Data acquisition was carried out using a Geode seismograph (Geometrics) connected to up to 48 channels. The seismic source was a 5 kg sledgehammer. The separation between geophones was 2 m, shot interval was 2 m, minimum and maximum offsets were 1 and 94 m, respectively, staking was 4, record length 1 sec and sampling interval was 0.25 ms. The whole data processing and analysis were performed by using the Seispace ProMAX suite (Landmark, Halliburton).

We show here results obtained in the line PS2 (figure 2.29).

The good quality of the field data allowed the drawing of the reflection section down at least to 60 m with a high resolution of some meters (figure 2.29(a)). The seismic interpretation was validated by data of some boreholes drilled in the past in an area. The top of evaporitic rocks (line TB in figure 2.29(c)) is imaged along with the profile. It dips towards S with an irregular trend, locally showing abrupt changes in depth, such as, for instance, in correspondence with the white arrow in

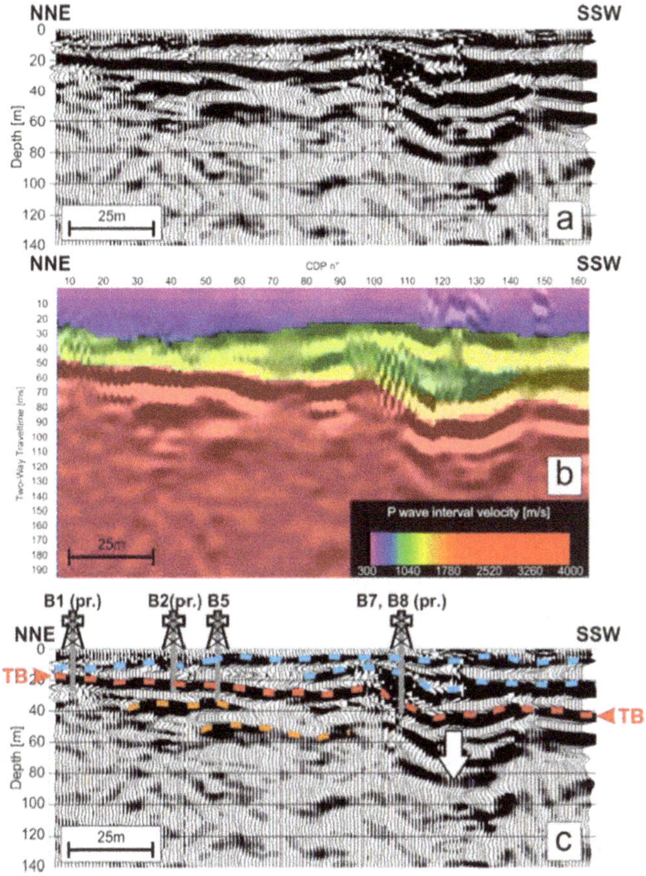

Figure 2.29. Seismic reflection sections along with profile PS2. (a) Processed depth converted section; (b) TWT (two-way travel-time) section with the interval velocity superimposed; (c) interpreted depth converted section. The red dashed line marks the top of the bedrock (TB), while light blue and orange lines highlight seismic horizons within the sediments and within the evaporites, respectively. The white arrow is located where the deformation reaches the maximum. The location and depth of boreholes B1, B2, B5, B7, and B8 have been used for validation (reprinted with permission from Busetti *et al* 2020, published by *Remote Sensing*, MDPI).

figure 2.29(c). The profile shows that the sinking is accommodated by the surface sedimentary layers, which are dipping less as they are closer to the topographic surface. The presence of the evaporites is clear due to their high velocity in comparison with the overlying sediments (figure 2.29(b)). That follows from the P-wave interval velocities analysis revealed from the integrated CMP and common-reflection surface (CRS) calculations Even more interesting is a local velocity inversion along with the N–S profile (figure 2.29(b)), located where the sinking is more evident. This enables us to conclude that there are lenses and intercalations of materials with poor geotechnical properties, and thus further increasing the instability of the buildings and the overall risk of the area.

Near-surface cavity detection by high-resolution seismic reflection method using short-spacing type land streamer
The experience of high-resolution SRFL is discussed in continuation.
Abandoned lignite (coal) mines at shallow depths often cause a collapse of the surface, which should be avoided in the area of renewal construction. To prevent collapse the exact positioning of abandoned mine galleries needs to be known. The microgravity and resistivity methods could not delineate subsurface cavities due to their poor spatial resolution. In this case, Inazaki *et al* (2005, 2006) suggested a high-resolution seismic reflection survey using the short-spacing Land Streamers combined with the GPR and surface wave surveys. The study site is located in the Nakashidami district, in the northeastern part of Nagoya (Japan). The area is bounded by incised valleys at two sides and a scarp at the northern side and was a lignite mining field. The lignite layer at the survey site is very thin (about 60 cm, 5–20 m deep) overlain by terrace deposits.

The Land Streamer tool (figure 2.30) consists of a 48 horizontal geophone array at 20–30 cm spacing, and an accelerometer array at 20 cm spacing. The sensors are mounted on a woven belt. Such an array can provide high-resolution images up to 30 m in depth. Streamers were used to detect the cavities.

As a result, distinct diffraction anomalies (figure 2.31) were detected in the unmigrated stack sections at the points of cavities that were checked by sounding or drilling from the surface.

Where the roof had collapsed, cavities had grown upward anomalies occurring at depths shallower than the inferred horizons of shelter tunnels or lignite mines. It is denoted that there is a good correlation between the diffraction on GPR sections and reflection ones.

A model investigation in detecting air-filled cavities
Yang *et al* (2017) have considered SRFL images of karst collapse of different origins. When the karst caves collapse, a lot of fractures would appear and change the structure of reservoirs. As a result, the permeability of reservoirs will also be changed. So the karst collapse body can be a new target of exploration and development.

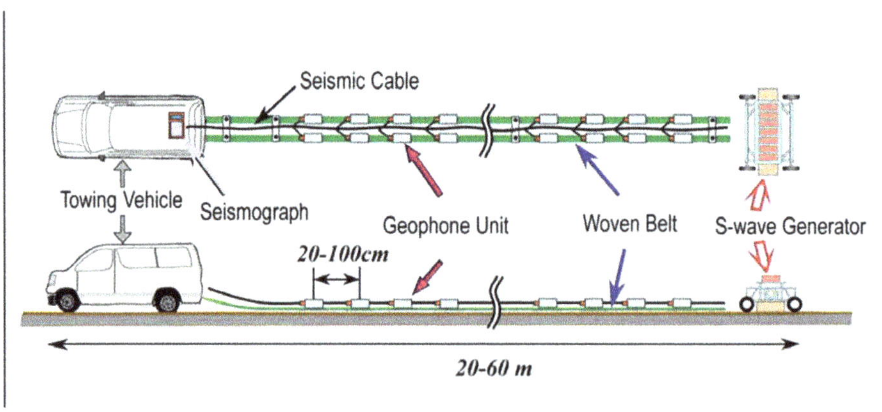

Figure 2.30. A woven belt type Land Streamer (reprinted from Inazaki 2006, with permission from EAGE).

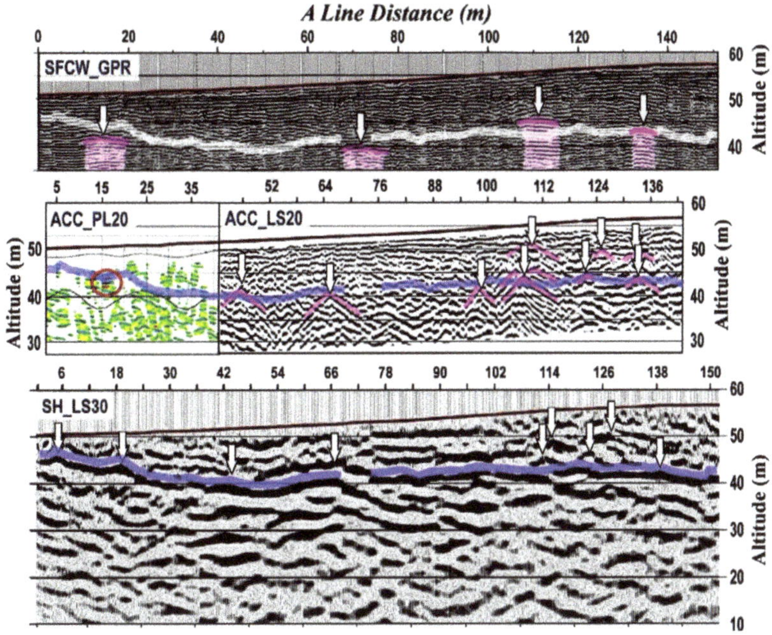

Figure 2.31. Comparison of seismic sections and GPR profile (top) along Line A. Clear diffractions are identified in the middle section (ACC_LS20). These diffractions are imaged roughly at the same points in the GPR profile and SH_LS30 section (bottom). The blue thick line shows the roof of the lignite layer. (Figure reprinted from Inazaki *et al* (2005), is published with the permission of EEGS. All copyright privileges remain with EEGS. This material cannot be copied or used without the express written permission of EEGS.)

As a rule, characteristics of karst collapse are more limited to the upper 'concave downward' of the seismic image. Dividing the karst collapse into different types and considering their SRFL characteristics using modeling methods that combine available geology with geophysics, it can be easier for us to identify the karst collapse on seismic sections and predict the reservoirs (Zheng *et al* 2011).

Into the basis of modeling, Yang *et al* (2017) put the theory of Loucks (1999), who suggested when karst caves survived: formation, transformation, and buried periods, and finally, become the karst collapse bodies. Moreover, most of the collapsed bodies are joint karst collapses which evolved based on several large single karst caves.

The main factors causing the collapse of karst caves are the gravity of the overlying ground, tectonic stress, dissolution of surface and groundwater, etc. Based on these factors we can divide the karst collapses into three types: load collapse, structural collapse, and sinkhole collapse (figure 2.32).

Based on field observations and the logging information, Yang *et al* (2017) designed the three different types of karst collapse seismic models, taking into account the characteristic features and shape of every type of collapse and seismic wave velocities in parts of models.

Seismic velocities in models are increased with a depth from 3500 m s^{-1} (uppermost layer) to 6000 m s^{-1} in bedrock (limestone). These values were put

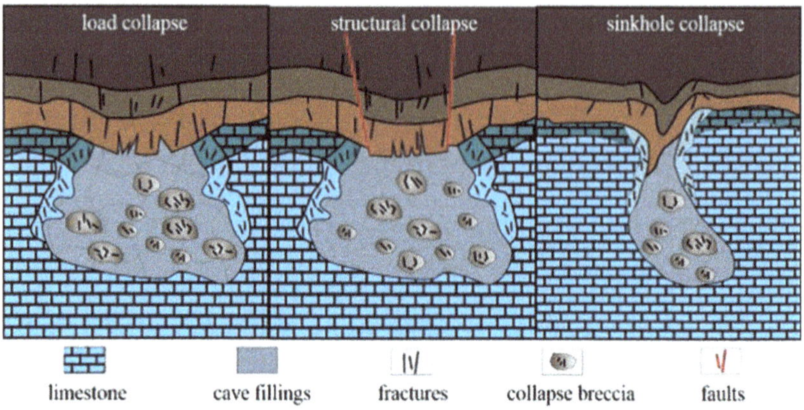

Figure 2.32. Schematic diagram showing three types of karst collapse: (a) load collapse; (b) structural collapse and (c) sinkhole collapse (reprinted from Yang *et al* 2017, with permission from SEG).

into the model calculation. The modeling technique Yang *et al* (2017) included 'the pre-stack depth migration' method (Ma and Li 2007). The results of the modeling are shown in figure 2.33. The combined seismic reflection sections are shown in figure 2.34.

Comparing real collapse structures with modeling results, one can distinguish different types of karst collapse in the seismic sections (figure 2.34). So, we can analyze the interior and boundary information on the structure of collapses and can distinguish between their different types, but not only their upper 'concave downward' shape.

On the seismic sections of different collapses, we can discern such features as strong 'beaded' or bright spot, strong reflection, and/or chaotic reflection characteristics at the internal of the karst collapse body.

We can see upper faults in the structural collapse. We are seeing the boundary of the karst collapse that is mostly 'beaded' reflection and faults. Finally, the sinkhole collapse indicates the characteristics of 'funnel-shaped' in the upper reflection events.

P-wave reflection for imaging of the buried depression

The problem of sinkhole development in the Kansas area was investigated by the P-wave seismic reflection for 25 years (Knapp *et al* 1989, figure 1). Sinkholes subsided gradually, although in some places catastrophic collapse took place. This section is a compilation of several works: (e.g., Knapp and Steeples 1986, Anderson *et al* 1998, Anderson *et al* 2017, Steeples *et al* 1986, Miller *et al* 1986). Several sinkholes have occurred progressively beneath Highway #70 (I-70) in Russel County, (Kansas) (figure 2.35).

The area of sinkhole development is underlain by the salt layers of the Permian formation of 20–90 m thick at the depth range varying from 150 to 500 m (figure 2.36). The sinking of the surface at a rate of 0.15–0.30 m per year continues since the late 1950s. The highway was twice repaired. In 1978 sinkhole (called Nielsen) has developed catastrophically in the saltwater disposal well.

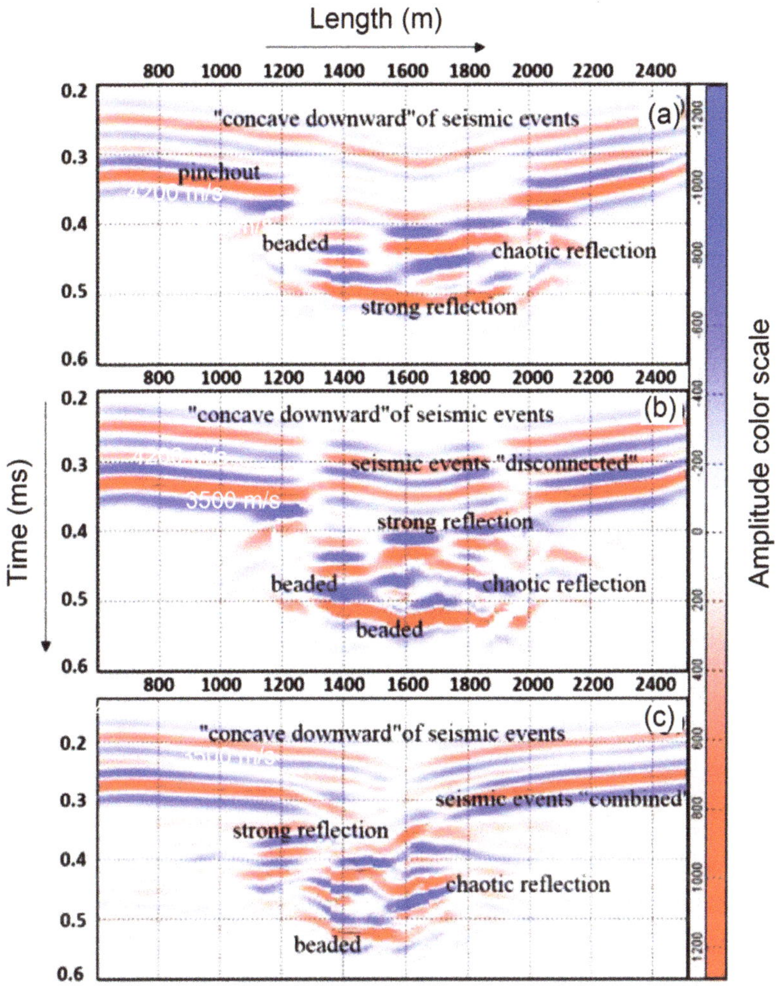

Figure 2.33. Seismic modeling results of three types of karst collapse corresponding to those in figure 2.32. Results are based on pre-stack depth migration (a) load collapse, (b) structural collapse, (c) sinkhole collapse (reprinted from Yang *et al* 2017, with permission from SEG).

Most of the studied sinkholes are associated with the Hutchinson salt layer dissolution by undersaturated brines or fresh-to-brackish water. Brines are introduced into leaky disposal wells, whereas brackish water leaks downward along the outside of the good casing (Knapp and Steeples 1986). Thus, sinkhole development is explained by either corrosion of the good casing or by not proper cement sealing to the salt beds.

After the formation of the Nielsen sinkhole, the Kansas Geological Survey (KGS) carried out SRFL studies along highway 70 (Steeples *et al* 1986). One of the aims was the safety of the highway in Russel County. About 4 km of six-fold CDP SRFL surveys were shot using the MiniSOSIE recording technique.

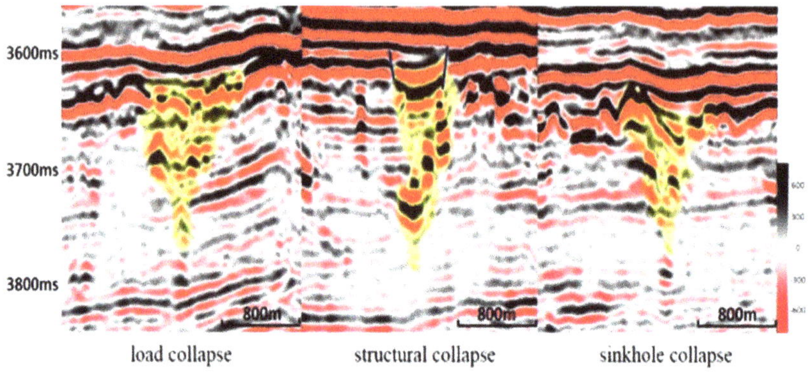

Figure 2.34. Seismic sections of three types of karst collapse (put attention to the vertical and horizontal scales of sections) (reprinted from Yang *et al* 2017, with permission from SEG).

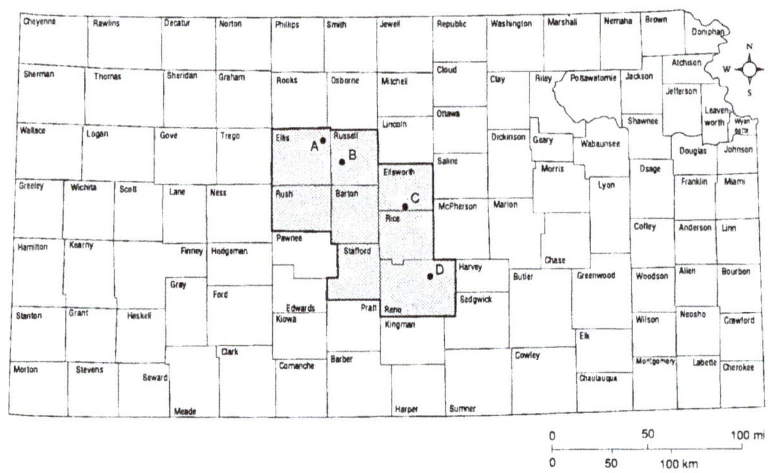

Figure 2.35. Map of Kansas showing locations of the (A) Nielsen sink (Ellis County), (B) I-70 sinks (Russell County), (C) Janssen sink (Ellsworth County), and (D) Punkin Center sink (Reno County). (The source of this material is the Kansas Geological Survey website at http://www.kgs.ku.edu/. All Rights Reserved. Reprinted from Knapp *et al* (1989) with permission of KGS.)

As an example, in figure 2.37 the paleosinkhole images centered at trace 1615 are shown. The clear sink is observed within the salt layer in reflection images at the time of 100–200 ms. The sink on the record corresponds to the sinkhole that occurred at the surface. The anomalous thinning of the residual rock salt is observed near the centers of the paleosinkhole, where the rock salt has been extensively (or totally) leached. Out of the center, the thickness of the rock salt increases rather abruptly by up to 40 m. Note that features characterizing the seismic reflection image are repeated through all sinkhole sites.

The interpreted record along the reflection line is presented in figure 2.38.

Recently, maximum surface subsidence has occurred between traces 1000 and 1160 and between traces 1320 and 1360. The western part of the reflection line is

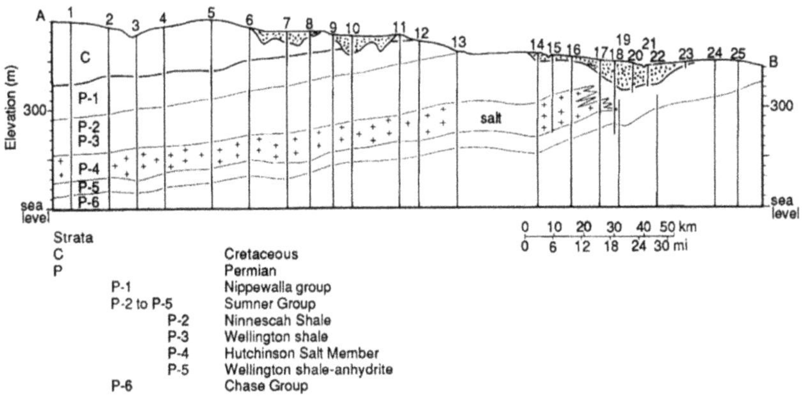

Figure 2.36. NW–SE geological section (AB) shows the salt layer in Kansas. Rock-unit names below the figure correspond to the location bets on the figure. The location of wells drilling for oil, gas, and water is depicted by cross-section at the top of the cross-section; modified from Waters (1978), where the well locations are listed and discussed. A—Nielsen sinkhole; B—I-70 sinkhole. (The source of this material is the Kansas Geological Survey website at http://www.kgs.ku.edu/. All Rights Reserved. Reprinted from Knapp *et al* (1989) with permission of KGS.)

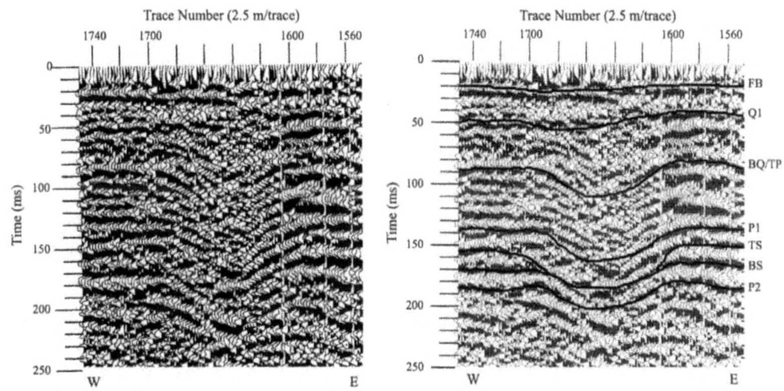

Figure 2.37. Typical seismic image, uninterpreted (left) and interpreted (right), of a secondary dissolution front (centered at trace 1615) (reprinted from Anderson *et al* (1998), with permission from SEG).

affected by subsidence. At these sites, the Hutchinson salt is up to 40 m thick (Anderson *et al* 1998).

Based on results and observations of SRFL surveys carried out in the sinkhole formation areas, the authors conclude that all documented dissolution/subsidence features (other than those related to mining) are situated on the active dissolutional salt margin, e.g., recent land subsidence in Kansas is mostly of natural origin.

It is important for the oil industry because previously such features were attributed to anthropogenic activity (usually related to the oil and gas exploration, Waters (1978)).

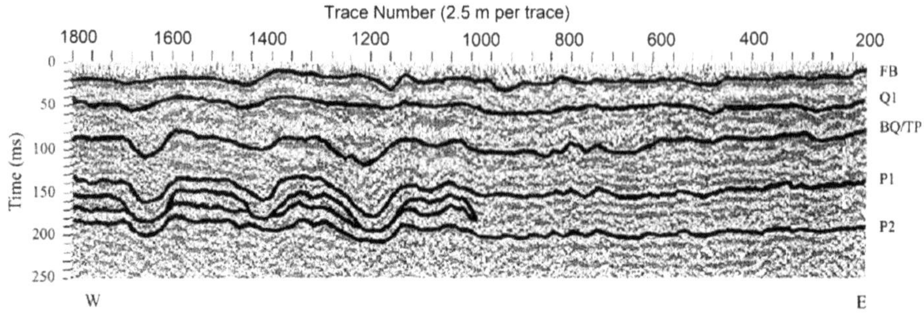

Figure 2.38. The interpreted seismic profile. The main dissolution front extends from the near-zero edge of continuous rock salt (trace 960) to the western edge of continuous dissolution (trace 1800). The secondary fronts (centered at traces 1200, 1415, and 1615) are located within the main front. These fronts developed along pre-existing fracture zones and are can be distinguished because of the anomalously high rate at which the associated salt dissolution occurs initially (reprinted from Anderson *et al* (1998), with permission from SEG).

2.2.1.4 Seismic diffraction method (SDFR)

One of the important problems in geophysical studies of the subsurface is the detection and localization of scattering objects such as karsts, mines, tunnels, and other natural and man-made local heterogeneities.

One of the methods that would be very effective for this purpose is active seismics, which is widely employed by the seismic exploration industry in its extensive and detailed studies of the subsurface. Adaptation of these methods to searching local underground heterogeneities is not a simple task. The seismic refraction and reflection methods are not directly applicable to a search for local underground heterogeneities of a small lateral extent (10–100 m).]

The energy reflected by such small objects would be very small, and it would be indistinguishable from other geological features of the subsurface. To be effective, the search for local heterogeneities has to use the features of a seismic wavefield, which is primarily associated with the presence of a local heterogeneity in the subsurface. Such features are known as scattered (diffracted) waves. In standard seismic prospecting, these waves are usually treated as noise. But for locating local objects these waves provide the most valuable source of information. Generally speaking, the wavefields arising from scattering objects are characterized by the presence of scattered or diffracted energy. *Diffracted wave* means edge diffraction, *the scattered wave* means a scattering from objects or velocity perturbations of about the finite size of the Fresnel radius (Landa and Keydar 1998).

Scattered (diffracted) waves are used in a developed approach called 'diffraction imaging' (Landa *et al* 1987, Kanasewich and Phadke 1988, Gelchinsky *et al* 1993, Klem-Musatov 1994, Landa and Keydar 1998, Belfer *et al* 1998, Keydar and Shtivelman 2005, Shtivelman and Keydar 2005). This approach has been originally proposed for the detection of geological fault location The object causing the diffraction acts as a secondary source of waves that spread out spherically from that point].

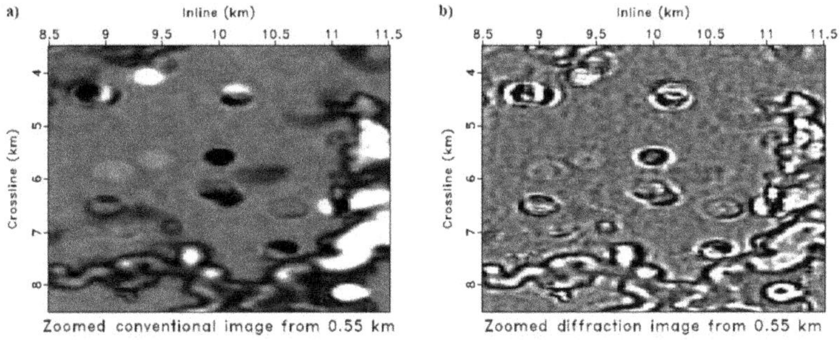

Figure 2.39. Zoomed Ordovician depth slices from 0.55 km: (a) conventional image and (b) seismic diffraction image (reprinted from Decker *et al* (2015), with permission from SEG).

The diffraction imaging is in development. There are studies where seismic diffraction was used to unmask the karstic features within the massive carbonates (Decker *et al* 2015, Barneff *et al* 2021). Figure 2.39(b) displays zoomed diffraction image within Ordovician carbonates in comparison with the conventional SRFL (Decker *et al* 2015).

In the framework of two international projects, Keydar *et al* (2010) have improved the existent diffraction-imaging technology for detecting underground heterogeneities and implemented it on the data that was acquired in the Dead Sea (see section 4.2.4).

2.2.1.5 Ground-penetrating radar (GPR)

2.2.1.5.1 Fundamentals of the GPR method

GPR method

Ground-penetrating radar (GPR) is a method using electromagnetic waves in the frequency range mainly 1–1000 MHz, to detect structures and features buried in the shallow subsurface (or in man-made structures) (Annan and Cosway 1992). The GPR is based on the transmission and receiving of electromagnetic waves at the Earth's surface (or boreholes) reflected from the objects or interfaces. At that, the elapsed time and amplitude of reflected waves are measured (Reynolds 2011).

The GPR is used for numerous applications, for instance, for the location of the shallow subsurface caves (located, as a rule, at depths of 10–20 m). This target is very important for the short-time prediction of surface collapse. That is proved by worldwide experience that GPR is the most suitable and successful method among all the high-resolution geophysical methods for the detection of karstic cavities and sinkholes, in a wide range of soil and rock (Benson and Yuhr 1993, Beres *et al* 2001, Casas *et al* 1996, Mellet 1995). The reader can find a thorough review of the theory and practice GPR method in Jol (2009).

GPR equipment principles. A radar system comprises a generator of electromagnetic pulses, transmission, and reception antennae. The system has an onboard or field computer that allows data processing both while acquiring data in the field and post-recording. As the antenna is moved on the Earth's surface, the receiver signals are displayed in the form of *radargram* as a function of their two-way travel time, i.e., the time taken from the instant of transmission to time of detection by the receiver. The radargram shows a 2D image along the line of radar motion. When numerous reflection profiles are collected, 3D images of buried materials in the subsurface can be constructed (McMechan *et al* 1997, Conyers 2013, p 166). The mutual arrangement of profiles can be different: parallel, parallel and perpendicular (grid), and arbitrary but located at the same surface ground (Sandmeier 2020). GPR's ability to produce 2- and 3D images of soils and sediments in shallow depths that are important for geology (10–15 m) and even within depths that are usually of importance for archaeology (a few centimeters to 3–4 m burial at most) means that complex images of geological deposits associated with archaeological features are possible (Conyers 2016).

The GPR main parameter determining the propagation and reflections of electromagnetic waves is the relative dielectric con*stant* ε_r, which varies from 1 for air to 81 for water (examples of the ε_r values, velocity, and electrical conductivity σ for different rocks are represented in table 2.3).

The degree of reflection from the interface between layers 1 and 2 is determined by reflection coefficient R, that is:

$$R = \frac{(V_1 - V_2)}{(V_1 + V_2)} \tag{2.2}$$

where V_1 and V_2 are the velocities of electromagnetic waves in 1 and 2 layers, respectively, and $V_1 > V_2$. The same equation can be written via ε_r:

$$R = \frac{\sqrt{\varepsilon_{r2}} - \sqrt{\varepsilon_{r1}}}{\sqrt{\varepsilon_{r2}} + \sqrt{\varepsilon_{r1}}} \tag{2.3}$$

In all cases, the magnitude of R lies in the range ± 1. The part of the energy transmitted through the interface is equal to $1 - R$.

In a low-loss medium the reflection depth of an electromagnetic wave (S) is:

$$S = \frac{t \times C}{2\sqrt{\varepsilon_r}} \tag{2.4}$$

where t is the two-way travel time, determined from the *radargram*, C is the speed of light in the free space (0.3 m ns^{-1}). The penetration depth versus TWT for different values is shown in figure 2.40.

Another important interrelation connects the attenuation (α) of the electromagnetic energy to the conductivity σ of the medium and dielectric constant ε_r

$$\alpha = \frac{1635 \times \sigma}{\sqrt{\varepsilon_r}}, \text{ [dB]} \tag{2.5}$$

Although there are not strict interrelations between the conductivity and dielectric constant, equation (2.5) indicates that such inverse correlation should exist.

Table 2.3. Approximate values of conductivity (σ), relative permittivity (ε_r), and velocity (V) for a variety of materials (after Ulriksen (1982), with permission).

Material	σ (mho m^{-1})	ε_r	V (m ns^{-1}) (*)
Air	0	1	~0.3
Pure water	10^{-4} to 3×10^{-4}	81	0.03
Seawater	4	81	0.01
Freshwater ice	10^{-3}	4	0.15
Granite (dry)	10^{-8}	5	0.134
Limestone (dry)	10^{-9}	7	0.113
Clay (saturated)	10^{-1} to 1	8–12	0.106–0.086
Snow (firn)	10^{-6} to 10^{-5}	1.4	0.253
Sand (dry)	10^{-7} to 10^{-3}	4–6	0.15–0.122
Sand (saturated)	10^{-4} to 10^{-2}	30	0.055
Silt (saturated)	10^{-3} to 10^{-2}	10	0.094
Seawater, ice	10^{-2} to 10^{-1}	4–8	0.15–0.106
Basalt (wet)	10^{-2}	8	0.106
Granit (wet)	10^{-3}	7	0.113
Shale (wet)	10^{-1}	7	0.113
Sandstone (wet)	4×10^{-2}	6	0.122
Limestone (wet)	2.5×10^{-2}	8	0.106
Copper	5.8×10^{7}	1	0.3
Iron	10^{6}	1	0.3
Soil			
Sandy (dry)	1.4×10^{-4}	2.6	0.186
Sandy (wet)	6.9×10^{-3}	25	0.06
Loamy (dry)	1.1×10^{-4}	2.5	0.189
Loamy (wet)	2.1×10^{-2}	19	0.069
Clayey dry	2.7×10^{-4}	2.4	0.193
Clayey wet	5.0×10^{-2}	15	0.077
Permafrost	10^{-5} to 10^{-2}	4–8	0.15–0.106

Data collected from Morey (1974), Keller and Frischknecht (1966), Von Hippel (1954). (*) Column is calculated from equation (2.4).

The velocity of electromagnetic waves within media V is expressed by equation (2.6):

$$V = \frac{C}{\sqrt{\varepsilon_r}}. \tag{2.6}$$

The selection of an antenna is the most important step to resolve the survey aim.

$$\lambda = \frac{1000 \times C}{f \times \sqrt{\varepsilon_r}}, \tag{2.7}$$

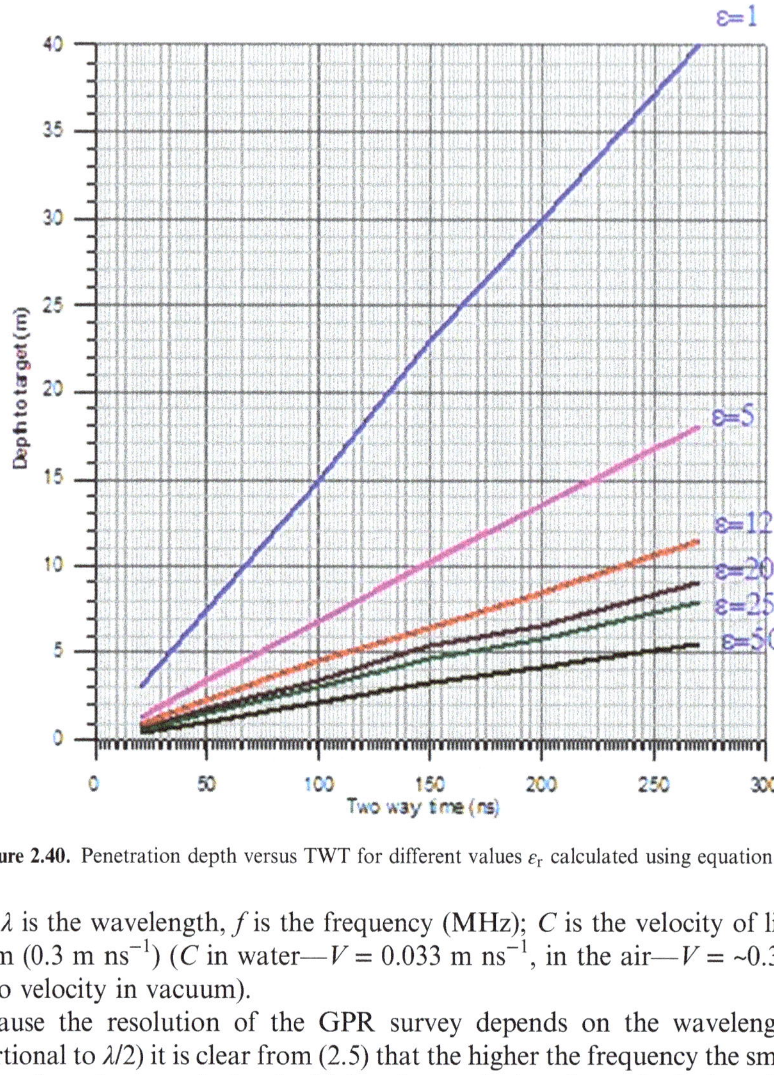

Figure 2.40. Penetration depth versus TWT for different values ε_r calculated using equation (2.4).

where λ is the wavelength, f is the frequency (MHz); C is the velocity of light in a vacuum (0.3 m ns^{-1}) (C in water—V = 0.033 m ns^{-1}, in the air—V = ~0.3 m ns^{-1} close to velocity in vacuum).

Because the resolution of the GPR survey depends on the wavelength (it is proportional to $\lambda/2$) it is clear from (2.5) that the higher the frequency the smaller the wavelength and, respectively, the higher the resolution. In geological applications, the depth penetration is often more important than the high resolution, antennas for frequencies of not more than 500 MHz are used.

The interpretation of the GPR study results is carried out using the radar section named radargram. The horizontal axis on the radargram is the distance along with the GPR profile. The vertical axis is usually TWT from the surface to the target and back. The time can be converted to the depth if the dielectric constant is known or was determined. The velocity of electromagnetic waves in the subsurface depends on the physical and chemical properties of the material. As the propagating radar waves pass through various materials in the ground, their velocity will also change (Conyers 2013, p 107). If the constituent differences at interfaces of the materials occur abruptly along the boundaries between very different materials in the ground, the radar waves' propagating velocity will also change when they pass across these contacts.

Depth of GPR measurements. The typical depth of radar penetration in rocks is 3–15 m, much less in clays, conductive soil, rock, or pore fluid, especially saline. For example, the penetration depth of electromagnetic waves is limited by high conductivity values of the soil, rock, and pore fluids. Conversely, if the surface sediments are highly resistive, such as quartz sands or dry clay-free limestone, GPR depths can be much greater (Benson and Yuhr 2016).

GPR systems. At present, there are a lot of GPR systems for surface and borehole measurements, principles of which are described in numerous monographs (e.g., Jol 2009, Reynolds 2011, Persico 2014, Utsi 2017). The experience of authors in geological and archaeological applications is associated mainly with the SIR-10 and SIR-20 systems (Production GSSI, USA) with antennae 100 MHz, 200 (or 250), 300 MHz, 400 (or 500) MHz, 900 MHz (Frumkin *et al* 2009, 2011). Another GPR system used in geological applications was the CU-II RAMAC GPR system (Production MALA, Sweden), which includes the CU-II control unit, a modern and upgraded unit known for its high versatility. The system allows us to perform ground-penetrating radar surveys, as well as surveys in boreholes and cross boreholes (tomography and reflection) (Ronen *et al* 2019). The system has several electronic control units, suitable for use with a variety of antennas—50 MHz, 100 MHz, 250 MHz, and 500 MHz antennas, both shielded and unshielded (Mala 2010).

Data processing. Comprises pre-processing (the suppression of corrupted and redundant traces); dewow and time-zero correction (removing very-low-frequency components); amplitude scaling based on moving average amplitude envelopes; deconvolution; noise removal using different kinds of filtering and enhancement of data quality; migration and time-to-depth conversion (Annan 2003).

GPR software. The main signal processing function is performed using the REFLEXW software package (produced by the K.J. Sandmeier Company, Germany) (Daniels 2007, Booth *et al* 2010, Sandmeier 2020). The software is employed for processing and interpretation of reflection and transmission data and enables 2D and 3D interpretation of the GPR data, as well as a 2D simulation of the electromagnetic wave propagation based on a finite difference approximation. Another software package, GPR-SLICE version 7.0 (2012), is used for 3D data processing and creating images of GPR reflection data. GPR-SLICE integrates radargram profiles taken over a survey grid to produce horizontal time slice maps of radar anomalies in the ground (Conyers 2013).

Interpretation of complicated GPR images

Typical GPR anomalies. We have considered models of sinkhole formation (section 2.1.). On one hand, it was shown that there should be observed some signs of the deep karst in the uppermost subsurface. They are heightened porosity, shear fractures, and deformations of the layers (subsidence) predicted by the theoretical models (Maimon *et al* 2005), and confirmed by geophysical mapping. Some models predict subsurface water conductive channels (Arkin *et al* 2000). These channels are observed in shallow subsurface (Frumkin *et al* 2009). Two examples of radargrams obtained above subsurface cavities are shown in figure 2.41. One of them was

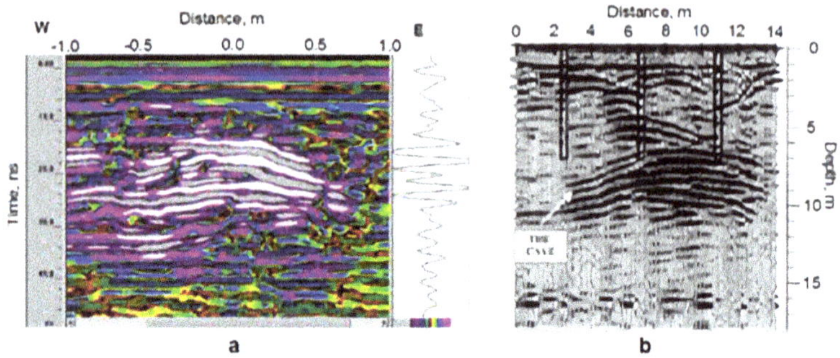

Figure 2.41. Typical images above the underground voids (a) The alluvial site at the south of the Dead Sea. Radargram has been obtained over a man-made void (horizontal tunnel). GPR signal is shown at right sides of radargram (Frumkin *et al* 2009). (b) Cave detected by the GPR in the Samaria Mountains of Israel in limestone bedrock (Ronen 2000, with permission from SPIE and A Ronen).

obtained in the DS shore (figure 2.41(a)), the second one belongs to A Ronen and was scanned in the Samaria Mountains of Israel in limestone bedrock (figure 2.41(b)). Note that both images are very similar despite different GPR systems, types of material, depth, sizes of cavities, etc. Voids and caverns of complicated forms that arise as a result of dissolution and floating up are also predictors of the sinkhole hazard important factors (Benson and Yuhr 1993). On the other hand, we have also analyzed the possibility of the GPR technique to be applied to the surveying of subsurface from point of view of soil conditions and signal penetration (figure 2.41).

The synthetic radar records over air-filled cylindrical voids (figure 2.42) confirm the hyperbolic shape of the anomaly in the subsurface (Munk and Sheets 1997).

Pre-evidences of karst: elements of sinkholes in the subsurface (review of literature)

We have considered models of sinkhole formation in the DS coastal area (section 2.2).

On one hand, it was shown that there should be observed some signs of the deep karst in the uppermost subsurface. They are heightened porosity, shear fractures, and deformations of the layers (subsidence) predicted by the theoretical models (Maimon *et al* 2005) and confirmed by geophysical mapping (Ezersky 2008). Some models predict subsurface water conductive channels (Arkin *et al* 2000). These channels are observed in shallow subsurface (Frumkin *et al* 2009). Voids and caverns of complicated forms arising as a result of dissolution and floating up (stoping) are also predictors of the sinkhole hazard important factors (Benson and Yuhr 1993).

On the other hand, we have also analyzed the possibility of the GPR technique to be applied to the surveying of subsurface from the point of view of soil conditions and signal penetration. In continuation, we have to consider is what GPR should distinguish in the subsurface from the point of view of sinkhole hazard. We will consider such elements of the subsurface disturbance as voids and caverns, subsurface subsidence, buried (refilled) sinkholes, vertical and horizontal faults, and fractures, etc.

Buried sinkholes. Sometimes buried sinkholes can be detected in the subsurface (Benson and Glaccum 1979, Batayneh *et al* 2002). The prominent radargram got from Benson and Glaccum (1979) is presented in figure 2.43. The figure shows an old

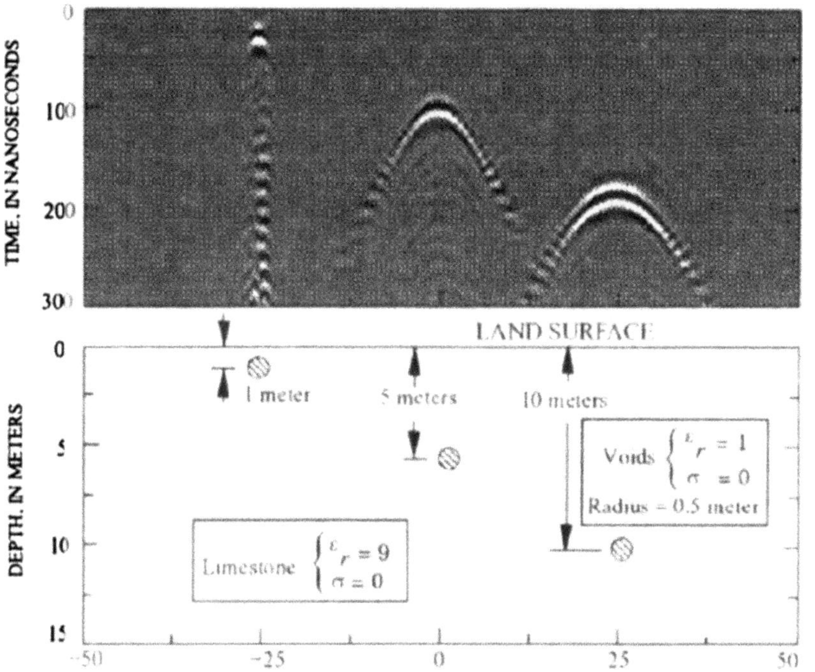

Figure 2.42. Synthetic radar records over air-filled cylindrical voids at different depths. ε_r and σ denote relative dielectric constant and electrical conductivity, respectively (reprinted with permission from Munk and Sheets (1997), courtesy of the US Geological Survey and Ohio Department of Transportation).

sinkhole that is filled with clean quartz sand with only a subtle depression at the surface. The cone-shaped profile indicates the extent of sinkhole activity and a zone of the potential hazard. The penetration depth, in this case, is 12–15 m (this example has been obtained from Lane 1986, figure 57, p 91).

Batayneh *et al* (2002) revealed refilled sinkholes at the Ghor Al-Haditha area in Jordan.

The survey revealed a newly formed sinkhole 18 feet (~6 m) wide. This sinkhole was associated with a 70-feet (~20 m) wide buried sinkhole that had a 27-feet (~9 m) wide throat extending through the subsurface clay to the limestone bedrock (SIR-2000 system, 500 MHz antennae, 102 nsec range).

Different elements of subsurface deformations can be detected by GPR. Benson (1995) reports an example of the apparent subsidence of subsurface layers detected by GPR.

The radargram indicates also two respective hyperbolic anomalies manifested on the void existence at the lower part of a section.

Vertical fractures in the upper part of the Earth's crust are the results of either local or regional tectonic stresses. Yalciner (2013) has used GPR to examine the subsurface geometry of the fissure-ridge travertine and the central fissure. GPR profiles were combined across the ridge to produce a 3D subsurface image of the travertine mass. The result has shown that the GPR profile can illustrate a vertical fault.

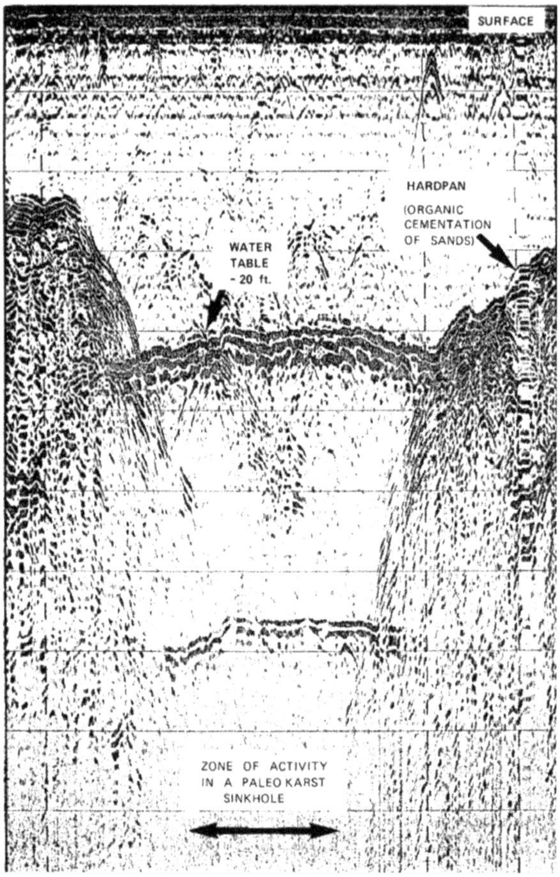

Figure 2.43. Ground-penetrating radar traverse across a paleokarst site (reprinted with permission from Benson and Glaccum (1979) referred to in Lane (1986)).

The example presented in figure 2.44 was obtained in Israel in the mountain region (Kofman and Herr 2001, Frydman and Kofman 2002). Two vertical faults can be seen in the figure. The right structure is seen at the surface, whereas the left one is covered by clay (white area).

Resonant oscillations GPR signal above the air-filled cavity. The results of the laboratory and field tests demonstrated that registration of the prominent resonant oscillations and/or the amplification of high-amplitude reflections with its horizontal extension during the towing antenna in the 'side' mode (e.g., GPR axis has been turned 90° to the direction of movement) may serve as reliable indicators of the existence of subsurface cavities (Kofman *et al* 2006, Frumkin *et al* 2009).

The radargrams obtained by towing 300 MHz antennae over air-filled cylinders of 1.5, 1.0, and 0.6 m diameter buried within sand are shown in figure 2.45. In this case we can observe the strongly expressed resonance oscillations, containing a series of high amplitude reflections with lowered frequency, when the 300 MHz antennae is passing over the center of the 1.5 m diameter cylinder. Two rules of thumb based on

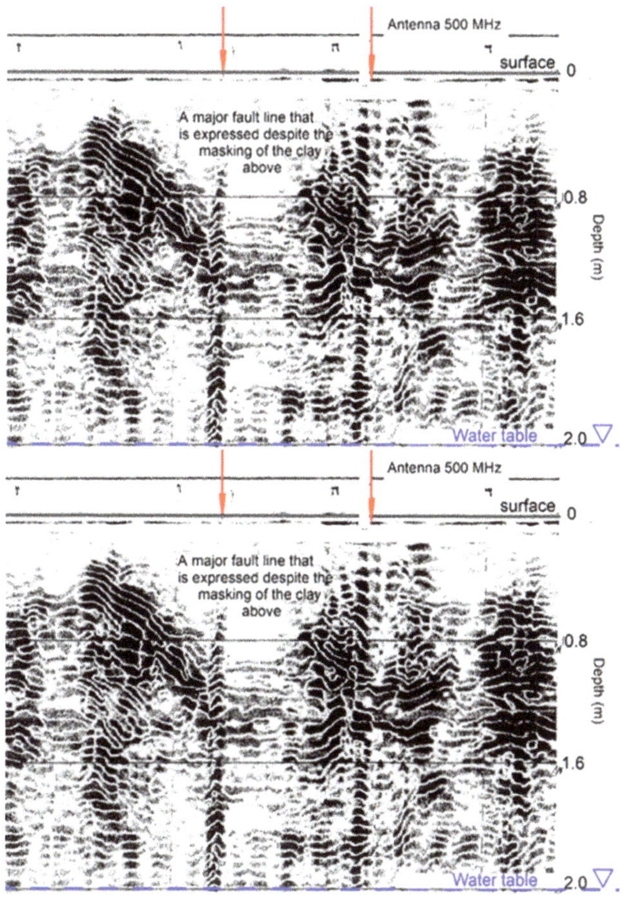

Figure 2.44. Two faults (denoted by red arrows) were detected by GPR in northern Israel (Alon Ha-Galil). The left fault is not seen at the surface, whereas the right one is visible here (reprinted with permission from Kofman).

practical GPR experience of void detection may be formulated for the prediction of the appearance of resonant radar pictures:

1. The diameter of the cavity should be more than the wavelength in the air of the antenna used.
2. The vertical size of the empty cavity should be not significantly smaller than its horizontal dimension.

It was found that strong reverberations are generated by the inner surface of the cavity. They approximate standing waves generated in cylindrical waveguides and waveguide resonators.

Resumé. Analysis of the worldwide literature has shown that there are some characteristic forms (anomalies) of the subsurface structure, which can relate to karst formation in the subsurface. As predicted by the above-considered model's deformations of the subsurface, the layers can accept characteristic forms

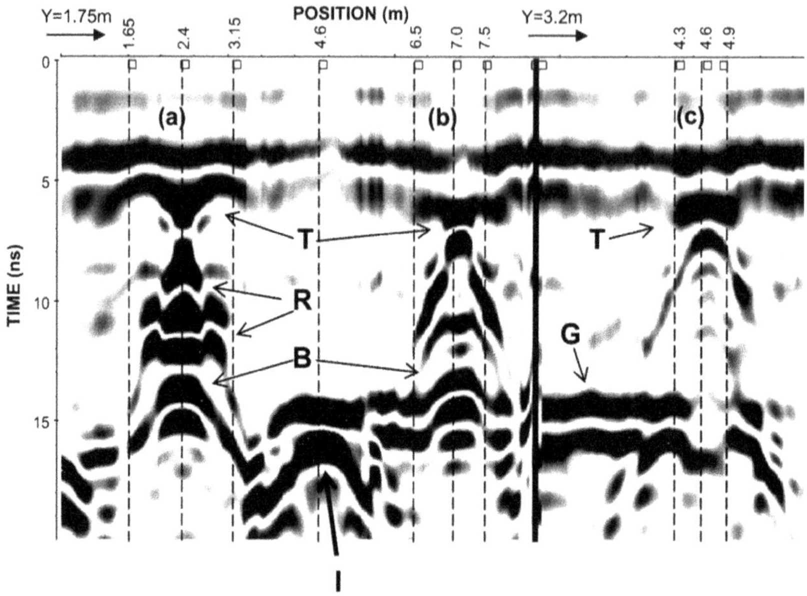

Figure 2.45. Data acquired with 300 MHz bistatic antennae towing over the centers above 1.5 m (a), 1.0 m (b) and 0.6 m (c) diameter cylinders, 20 ns range (reprinted from Kofman *et al* (2006), copyright 2006 with permission from Elsevier).

(voids, subsidence, fractures, etc), which can be detected by the GPR technique. Another question is: do these anomalies ever testify to the sinkhole hazard? This issue we will discuss in continuation.

Following the ASTM 6429-99 (2006), GPR provides the highest lateral and vertical resolution of any surface geophysical method. Antennas of different frequencies (10–3000 Hz) can be selected to optimize data to the project's needs. It should be taken to account that the lower the frequency of electromagnetic (EM) waves, the greater is the depth penetration. However, in this case, the resolution of the GPR data is less. Conversely, the higher frequencies provide less penetration with a higher resolution. The vertical resolution of GPR depends on antennae frequency and is ranged from a few centimeters to a meter. The horizontal resolution depends on the distance between the station measurements, or the sample rate, or both, and the towing speed of the antenna. The quality of GPR data is determined by the quality of the ground surface, its evenness, absence of jumps and dips. The major limitation of radar is its site-specific performance. The depth of penetration is limited by the presence of mineralogic clays, saline soil, or high conductivity pore fluids (Daniels 1989, Sharma 1997).

2.2.1.5.2 Application of GPR in sinkhole and karst investigations

Karstic void detection. We now consider a cavity in fractured limestone (Modi'in, Israel) (Frumkin *et al* 2009). The example of the karstic void detection in the Modi'in region is shown in figure 2.46. The karst was suspected because the subsurface is composed of carbonates and this area is traditionally karstic.

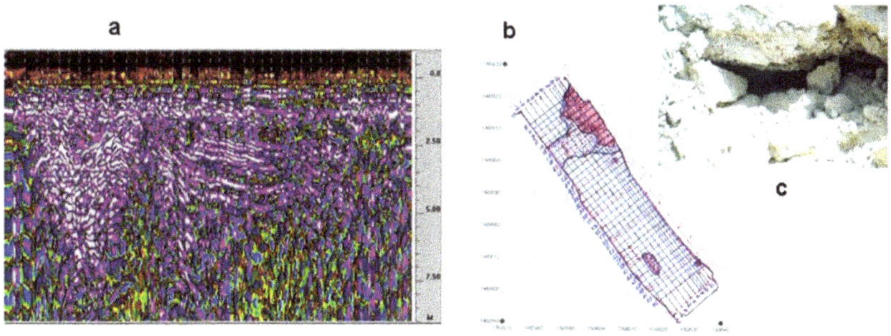

Figure 2.46. GPR section above the karst area (a), karst map (b), and (c) void opened during construction works (reprinted from Frumkin *et al* (2007) with permission from GII).

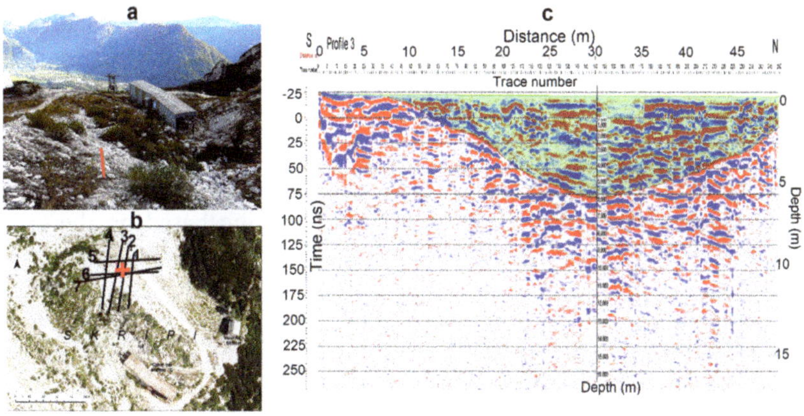

Figure 2.47. The cave S-19 in the Mt Kanin massif (NW Slovenia). (a) Place of study view. (b) Layouts of the GPR profiles (black solid lines). (c) Typical radargram along with profile 3. The red column (a) and red cross (b) are the locations of the detected entrance to the cave (b—Aerial images of Surveying and Mapping Authority of Slovenia) (reprinted with permission from Gossar and Čeru (2016)).

The study was performed with the SIR-20 system using a 200 MHz antenna, dielectric constant $\varepsilon_r = 6$, penetration depth is 9 m. The karst causes the dissipation of the electromagnetic waves on the small voids and fractures and it appeared at the radar section as strong reflections comprising a lot of hyperboles (white zone in figure 2.46(a)). The scanning of the building ground along the parallel and crossing lines permitted delineating the hazardous zone (red area in the map, figure 2.46(b)). Further construction works verified the karstic cave (figure 2.46(c)).

Karst detection in mountainous environments using a flexible antenna (Gossar and Čeru 2016). The known cave S-19 located in Mt Kanin massif (NW Slovenia) was lost to antennas in 1974 when a huge snow avalanche protection dyke was constructed across the cave entrance. To excavate the buried cave, the accurate location of the cave had to be determined first.

The GPR survey was carried out at the place of the suspected entrance to the S-19 cave (figure 2.47). The photo of the Scripi area (figure 2.47(a)) displays the study area Landschaft not suitable for the convenience of GPR antennae. Rough slopes and cobblestones impede smooth scanning of antennae. In these conditions, a Mala ProEx recording unit and 50 MHz special rough terrain antennas (RTAs) with a flexible snake-like design in the form of a long tube were used. The total length of a 50 MHz RTA is 9.25 m and the spacing between antennas is 4 m.

The study site is shown in figure 2.47(a). Gossar and Čeru (2016) hoped that it would be possible to find the cave entrance by locating the exact position of the local depression inside the rugged karstified terrain. The authors scanned seven GPR lines from 37 to 59 m in length following the scheme shown in figure 2.47(b). The typical radargram can be seen in figure 2.47(c). One can see in this radargram a visible difference in the character of reflections between the upper part of the section (marked by green) and its lower part. The green area in radargram is interpreted as an artificial debris infill. The lower part is interpreted as a bedrock. Reflections from the debris infill are almost absent or are very weak reflections. On the other hand, the reflections from the underlying bedrock are strong just below the boundary and are visible down to the maximum depth of around 13 m. In greater depths, the radargrams are characterized by an absence of any reflections due to the homogeneous limestone.

The cave entrance was visible on all radargrams as a topographic feature (a small depression) under debris. The exact location for digging based on the GPR data interpretation was determined (it is shown by the red column and red cross in figures 2.47(a) and (b), respectively). The thickness of the debris was evaluated at 6.5–7 m. Subsequent excavation resulted in confirmation of the GPR results: location of the cave entrance and its estimated depth. Thus, the GPR method was proven to be successful in providing a solution to this specific case in karst exploration and an important cave was saved.

Characterization of sagging and collapse sinkholes

Subsurface dissolution, internal erosion and deformations accompanying sinkhole development change the properties of subsurface sediments that can be detected by geophysical methods (Ezersky *et al* 2006). Rodriguez *et al* (2014) investigated the possibility of the GPR method to characterize sinkholes in covered karst terrain (figure 2.48). The aforementioned authors carried out a GPR study across two active buried sinkholes of different genetic types, e.g., collapse and sagging (Gutiérrez *et al* 2008b) in the mantle evaporite karst of Zaragoza city (NE Spain). In figure 2.48 two types of sinkholes are shown on the left side (cover collapse) and right side (cover sagging). In particular, a cover collapse sinkhole is characterized by the vertical faults (fractures) formed during the collapse, whereas a cover sagging sinkhole is of boiler-shape.

The employed method is based on the suggestion that the dissolution-induced subsidence structures commonly display geometrical features related either to gravitational deformation (e.g., fault, planes, shear zones, fissures, colluvial wedges, synforms) or depositions controlled by differential settlement (e.g., pinch outs) that may also be imaged by GPR method (see above).

Data acquisition was carried out using the Pulse EKKO system with the 100 MHz and 50 MHz unshielded antennas (Sensors and Software) and RIS system with

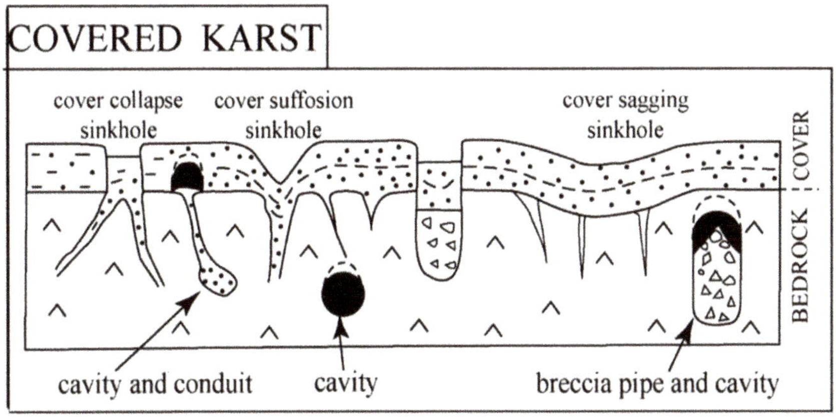

Figure 2.48. Diagram showing the main dissolution and subsidence features that may occur in covered karst settings (reprinted from Rodriguez *et al* (2014), with permission from The Geological Society of America).

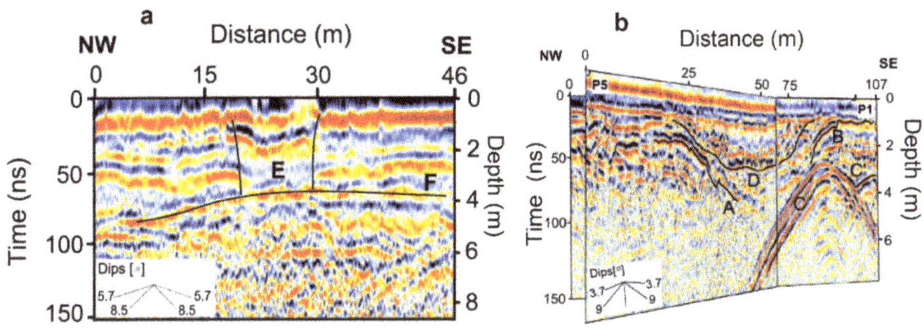

Figure 2.49. Two types of sinkholes GPR images: (a) collapse sinkhole in the Miraflores site; (b) pseudo-3D presentation of the sagging sinkhole in the Valdefierro site. in (a) E—an area of disturbed reflection patterns interpreted as the buried collapse sinkhole bounded by horizontal reflections; F—air reflection from the surface vegetation flanking the ditch 8 m to the southwest. in (b) A and B—lower boiler-shaped reflections of the sagging structure imaged in the GPR profiles; C—diffraction hyperbolas from above-surface objects; C9—air reflection from the terrace scarp; D—upper reflection indicating the general sinkhole internal geometry. Blues and reds define positive and negative pulses, respectively, and color intensity is a function of amplitude. Note vertical exaggeration 1:6 and the different horizontal scales (reprinted from Rodriguez *et al* (2014), with permission from The Geological Society of America).

the 180 MHz shielded antenna (IDS GeoRadar, Pisa, Italy). Data interpretation was carried out using the Landmark ProMAXH (Release 5000.0.3.3) seismic processing package.

After thorough data processing, the results obtained allowed reliably mapping the boundaries of the sinkholes, and to characterize their internal geometries (deformation style), infer the subsidence mechanisms, and estimate subsidence (figure 2.49).

The Miraflores is an active collapse sinkhole that was recognized in 1927. It has been presented as a scarp-edged depression. The depression is of sub-circular shape with a diameter of 12–14 m and 4 m deep (Gutiérrez *et al* 2008c). The sinkhole is located on an alluvial fan of the Ebro River. The geological section is composed of: alluvial fan deposits 1–2 m thick; approximately 28 m of terrace deposits comprising gravels with some sand and clayey silt beds, then (down to 30 m) a karstic residue consisting of grey and brown marls with scattered gypsum fragments; and finally, competent gypsum (at 38 m top). The sinkhole located at the uppermost layer is filled with clayey artificial deposits that reach at least 4 m in thickness (Gutiérrez *et al* 2008c). The groundwater table was measured at a depth of 15 m.

In figure 2.49(a), clear reflection disruptions (seen between 16 and 30 m) are related to a collapsed structure. The width of this zone is consistent with the 12–14 m-diameter depression in a topographic map. The disturbed zone (denoted as E) in the interpreted images can be traced to a depth of about 4 m, where noise begins to dominate the image. The F area in figure 2.49(a) is interpreted as a reflection coming from the surface vegetation flanking the ditch 8 m to the southwest.

Another case investigated by Rodriguez *et al* (2014) is the Valdeffierro sagging sinkhole (after Gutiérrez *et al* 2008b) located in the southwestern part of Zaragoza city. Its northeastern part is a semicircular collapse depression with a diameter of 36 m that has developed along the edge of an Ebro River terrace composed of gravels. This scarp-edged depression, resembling half-a-sinkhole, is partially filled by rubble and has a 3.2-m-deep nested collapse superimposed on the man-made fill. In the southwestern part of the structure, the only surface evidence of instability is a very gentle sag and open fissure 6 m long and 30–55 cm wide, the prolongation of which matches the scarp of the collapse

The pseudo-3D GPR combination of lines 1 and 5 that provide better insight into the geometry and dimensions of the sinkhole is presented (figure 2.49(b)). The fence diagrams enable us to trace and match reflections across different images that allow us to delineate the boundaries of the buried sinkhole in different directions to obtain a pseudo-3D visualization of the structure. Based on GPR information, it was concluded that the Valdefierro sinkhole is a basin structure with inward-facing dips typical of sagging sinkholes. The boiler-shaped reflection characterizes the subsidence. Different zones of the image are denoted by letters. Two main reflections A and B present the concealed sagging sinkhole in the study area. These reflections, probably coming from the same stratigraphic contact or unit on either side of the sinkhole, define a roughly symmetric form. Letters C and C′ denote diffraction hyperbolas. The first one (C) is caused by a power line that crosses the study area. The second one (C′) is a reflection from a terrace scarp that is parallel to the profiles. Finally, D in the GPR section shown in figure 2.49(b) is also a sagged stratigraphic marker similar to the above reflectors A and B. The reflectors A and B and can be traced across the entire structure, including the axial zone, where it is located at a depth of 2.25 m.

Like the Miraflores sinkhole, the GPR sections recorded at the Valdefierro site did not yield information on the maximum depth of deformation.

The GPR images at the Valdefierro sinkhole provide information on the subsidence structure in the sector of the valley floor where the acquisition of GPR data was feasible and the sinkhole barely shows any geomorphic expression.

GPR in the Tête Rousse glacier (Mont Blanc, France)

In July 1892 there occurred a rupture of a large water pocket at a height of 3150 m (a.s.l.) within the Tête Rousse glacier Mont Blanc area, French Alps). The sudden release of approximately 200 000 m^3 of water and ice caused deep erosion of soil and rocky materials and the formation of mudflow. Approximately, 20 min later a mudflow (~800 000 m^3 of a mixture of ice, water, and eroded rocks) reached the city of Saint-Gervais. It led to approximately 175 fatalities.

To prevent further outbursts of floods, two tunnels were drilled through the rock and ice, one having been maintained until the present without ever having drained any significant amount of water (Vincent *et al* 2010). With time the suspicion grew that the tunnel does not drain the groundwater. By the beginning of 21st century, the number of residents in the Sent-Gervaise had reached ~3000, and the problem of their safety was addressed to prevent floods similar to that of 1892. As a result, the decision was accepted by the authorities in charge of natural risks to carry out a complex of surveys to determine the status of the glacier safety. They included the following problems: (1) to study present conditions of the glacier; (2) to drain this subglacial water pocket, and (3) to monitor its evolution. To realize this plan, unique studies were carried out on the Tête Rousse glacier. A large number of GPR and SNMR measurements were carried out between 2007 and 2010 (Legchenko *et al* 2011, Vincent *et al* 2015, Garambois *et al* 2016).

These investigations aimed to check the potential existence of a subglacial water cavity in Glacier de Tête Rousse. Some results of the GPR survey through the glacier are presented here. The glacier photo is shown in figure 2.50(a) and GPR layouts are shown in figure 2.50(b).

The GPR study of 2014 was carried out using the IDS system with a 100 MHz shielded RTA. Lines P2, P3, P4, and P6 were acquired with the MALA system and unshielded RTA 100 MHz antennas. Seismic Unix software was used for data processing. Data acquired above the water-filled cavity indicates the 16.8 cm ns^{-1} velocity. Following equation (2.4) in this chapter dielectric constant $\varepsilon_r = 3.2$ is in correspondence with (Koh 1992). Each profile was first processed using a standard series of procedures (Garambois *et al* 2016). The filtering band was 50–300 MHz. The authors then applied static corrections to account for the topography and converted the two-way propagation times into depths. Figure 2.51 shows the longitudinal GPR profile P1 acquired along with the flow of the glacier in the W–E direction well illustrates the complexity of GPR data within the glacier and

Figure 2.50. (a) Photograph of the glacier taken at the east summit (credit: M Ezersky); (b) GPR line layouts. Pl a longitudinal GPR profile and P5 transverse one were recorded during the 2014 campaign. P2, P3, P4, and P6 transverse profiles were acquired during the 2010 campaign. The Universal Transverse Mercator coordinate system was used (zone 32) (reprinted from Garambois *et al* (2016), with permission from SEG).

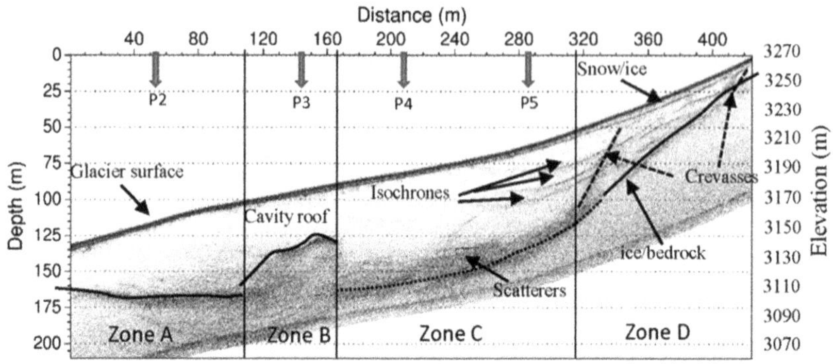

Figure 2.51. Processing and interpretation of the 100 MHz longitudinal profile P1 (W–E) after the time-to-depth conversion: data after migration in the t-k domain, and the associated interpretation. The arrows show crossing points with transversal profiles P2–P5 (reprinted from Garambois *et al* (2016), with permission from the Society of Exploration Geophysics).

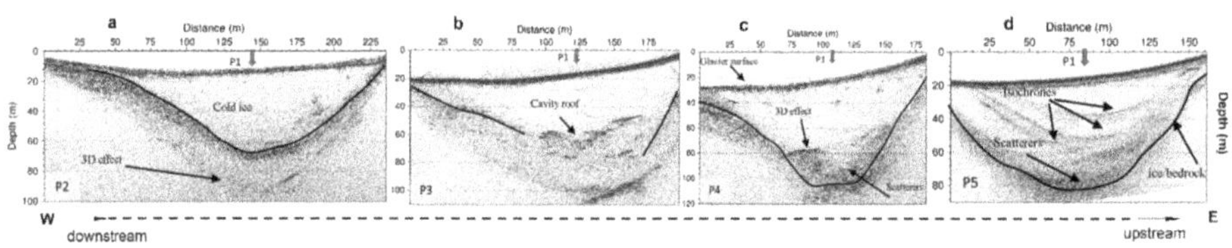

Figure 2.52. Processing and interpretation of the 100 MHz transversal profiles P2–P5 (E–W) (a) Zone A; (b) Zone B; (c) Zone C; (d) Zone D. Arrows show crossing points with longitudinal profile P1. For explanations see the text (reprinted from Garambois *et al* (2016), with permission from SEG).

the information they contain. The GPR section has indicated that the W–E glacier can be divided into four parts characterized by different reflectivity parameters. Transverse south to north profiles, respectively, acquired in these four parts from downstream (W) to upstream (E) (shown by arrows in figure 2.51) show the difference in structure. GPR radargrams along with these profiles are presented in figure 2.52.

The longitudinal GPR section of the glacier, presented in figure 2.51 shows characteristic boundaries of cavity roof, ice bedrock, etc. A large amount of scattered or reflected GPR energy detected in 2007 and of uncertain origin (i.e., rock blocks, water, and voids) was verified later, after SNMR that enabled evaluating the presence of water (Vincent *et al* 2015).

Zone A in the western part of the glacier (figures 2.51 and 2.52(a)), appears very homogeneous with slight internal scattering. The main deep-reflection event corresponds to the ice/bedrock interface, and scattering appears under this interface due to rock heterogeneities. This part of the glacier is interpreted by the authors as ice.

Zone B (figures 2.51 and 2.52(b)) is where a large reflection of the event appears, with the apex located at a depth of 40 m and in which the continuity of the ice/bedrock interface is lost.

Zone C in figures 2.51 and 2.52(c) (from 168 to approximately 320 m along longitudinal profile P1) is where the glacier appears relatively homogeneous in the first 50 m of depth followed by increased scattering in a 10–15 m thick layer near the glacier bed. In this part, it is still possible to trace the ice/bedrock interface but it is less clear than in the lower part.

Finally, Zone D (the uppermost part of the glacier) in figures 2.51 and 2.52(d) is characterized by a clear reflection at the snow/ice contact in the accumulation zone. There is a relatively clear interface between ice and bedrock, and a nearly total absence of basal scattering energy.

Thus, the migrated radar images, allow detecting the cavity roof. Depth zero corresponds to an altitude of 3189 m for images on the left and an altitude of 3192 m for those on the right. Then using all available longitudinal and transverse profiles the authors obtained the topography of the cave roof. Changes in this topography have shown that the thickness of the cavity roof strongly decreased after the first pumping (between May 2010 and May 2011), and amounted to >20 m at some locations during this period (Vincent *et al* 2015).

2.2.1.6 Transient electromagnetic method (TEM)

2.2.1.6.1 Fundamentals of the TEM method

TEM (also named the time domain electromagnetic method—TDEM) is sensitive specifically to bulk resistivity (conductivity) of the studied medium. It is right especially in the range of low resistivity (high conductivity) (Kaufman 1978). The TEM enables us to measure the electrical resistivity of the subsurface by inducing pulsed currents in the ground with a transmitter loop. The most frequent application of the method worldwide is the localization of the fresh-saline water interface in coastal areas and its stability with time (Fitterman and Stewart 1986, Hoekstra and Blohm 1990, Goldman *et al* 1991, Best and Todd 1996, Kafri *et al* 1997, Yechieli *et al* 2001, Siemon *et al* 2009, Kirkegaard *et al* 2011). Descloitres *et al* (2008) have used the TEM method for monitoring waste bio-degradation using resistivity changes with time in non-hazardous waste landfills. The TEM method can be used also for the detection and mapping of inorganic plumes, and seepage from brine pits.

Following modified Archie's law, the bulk electrical resistivity of the clay-free partially saturated soil with the ionic pore water conductivity depends on the porosity, pore volume occupied by the fluid, and the resistivity of the fluid filling the pore space (Archie 1942, Keller and Frischknecht 1966). In this case, the bulk resistivity is expressed as:

$$\rho_x = a\rho_w S^{-n}\phi^{-m} \tag{2.8}$$

where ρ_w is the resistivity of the solution filling the pores, ϕ is the porosity, $S = V_{el}/V_\Sigma$ is a fraction of the total pore volume filled with the same solution (referred to also as the *degree of soil pore filling*), V_{el} is the volume of the solution in the pores, V_Σ is the total pore volume per unit volume of soil, and n is an empirical parameter termed the *saturation exponent* (usually with a value of approximately 2). Parameters a and m appear to depend on the cross-section geometry of the pores along the flow path. The value of the parameter varies mostly within the range of 0.6–1.4. Parameter m varies within the range of 1.37–1.95. In the case of complete pore space saturation, parameter $S = 1$ and equation (2.8) becomes the conventional Archie's law:

$$\rho_x = a\rho_w\phi^{-m} \tag{2.9}$$

Equation (2.9) can be further simplified if we accept $a = 0.81$ and $m = 2$ for sandy DS aquifers (Kafri and Goldman 2005):

$$\rho_x/\rho_w = 0.81\varphi^{-2} = F, \tag{2.10}$$

where F is the *formation factor*, that must be constant for the given porosity. It should not depend on the resistivity of the water in the rocks.

It follows from equations (2.8) and (2.9) that there is direct dependence connecting bulk resistivity (ρ_x) measured by the TEM method with a water resistivity (ρ_w). The latter is the function of the water salinity—chloride concentration, TDS, etc (Keller and Frischknecht 1966).

Principles of the TEM method—general consideration

Electromagnetic measuring is carried out as follows: alternating current is emitting into the Earth via a square loop (insulated wire) located on the ground surface. This loop is termed the transmitter (Tx) (Fitterman 2005). The Tx emits pulses of the main frequency into the loop during turn-off and turn-on periods. That EM field occurs in the vicinity of the loop. Turning off the current within the loop is continued during some short time (~microseconds) that is termed 'the ramp time'. During ramp time, the magnetic field is time-dependent. The varying primary EM field induces a secondary EM field in the ground beneath the loop that immediately begins to decay generating additional eddy currents (figure 2.53). The latter propagate downward and outward from the loop into the subsurface like a series of smoke rings.

The secondary currents are received by an antenna linked to the receiver located in the center of the transmitter loop during the period of a current turn-off. The receiver senses eddy currents that are progressively removed away from the surface into the ground. Therefore, the depth of investigation depends on the time passed after the turn-off of the current. In turn, the intensity of the eddy currents at specific times and depths is determined by the properties of the subsurface such as the combined conductivity of the subsurface rock and fluid. The latter are a function of the soil and rock matrix, percentage of fluid saturation, and the conductivity of pore fluids (McNeill 1980c, Stewart and Gay 1986).

Main equations (McNeil 1994)

1. Apparent resistivity versus depth

We consider the case of the homogeneous earth in which the resistivity does not vary with either lateral distance or depth. The response of such earth (figure 2.54) can be divided into three segments: an early stage (where the response is roughly constant with time), an intermediate stage (response shape continually varying on a log–log plot), and a-late-stage (response is a straight line on a log–log plot). During the late

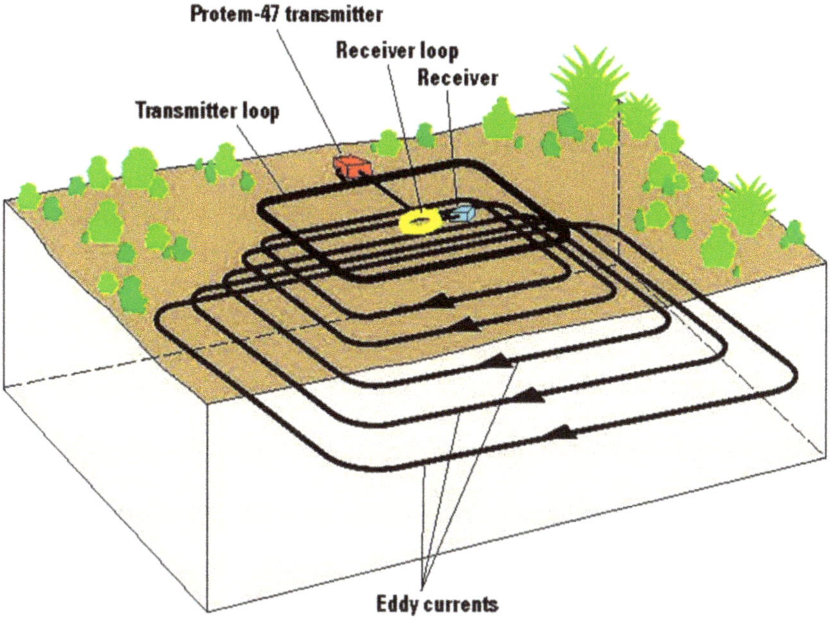

Figure 2.53. Diagram showing time-domain electromagnetic sounding layout and operation (reprinted from Fitterman (2005), modified from North Carolina Division of Water Resources (2004), courtesy of the U.S. Geological Survey).

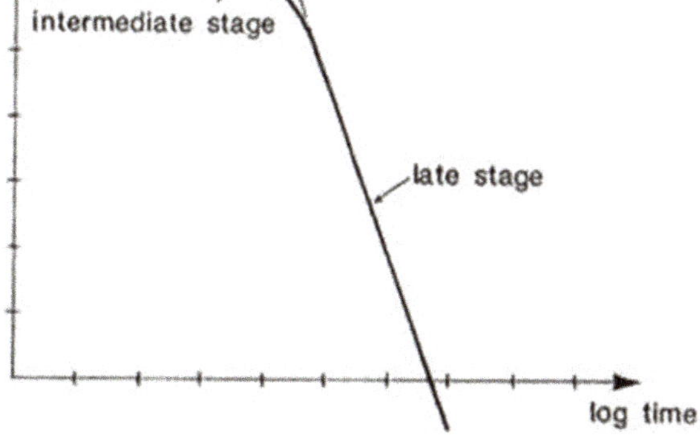

Figure 2.54. Log plot—receiver output voltage (*Y*) versus time (one transient) (*X*) showing stages of transient (reprinted from McNeill (1994), with permission from Geonics Ltd).

stage, the mathematics is considerably simplified (although generally it is a complex function of conductivity and time), and the response varies quite simply with time and conductivity as

$$e(t) = \frac{k_1 M \sigma^{3/2}}{t^{5/2}}, \tag{2.11}$$

where k_1 is the constant; M is the product of the Tx current (Amps), x is the area (m^2); σ is the terrain conductivity (S m^{-1}); t is the time (s) and $e(t)$ is the output voltage from a single-turn receiver coil from the area of one m^2.

If in the case for conventional resistivity measurement, when the relationship between the measured voltage and terrain resistivity is linear, in the case of TEM the measured voltage, $e(t)$ varies as $\sigma^{3/2}$, and it is more sensitive to small variations in the conductivity than conventional resistivity.

It also deserves to be mentioned that during the late stage, the measured voltage is decaying very quickly with a time at the rate $t^{5/2}$. The signal disappears into the system noise and further measurements are impossible. That is the maximum depth of exploration for our particular system.

We can now rewrite equation (2.11) (taking into account that ($\rho = 1/\sigma$) as)

$$\rho_a(t) = \frac{k_2 M^{3/2}}{e(t)^{2/3}\, t^{5/3}}, \tag{2.12}$$

where k_2 is the constant; M is the product of the Tx current (Amps), and x is the loop area (m^2).

Using this equation we can carry out sounding of resistivity with a depth (e.g., the depth of the measured resistivity increases with time).

To summarize, we see that, except for the early-time descending branch, and the intermediate-time anomalous region described above, the sounding behavior of TEM is analogous to that of conventional DC resistivity with the difference that in the TEM method depth of investigation is reached by an increase of the measurement time, while in DC method it is reached by increasing inter-electrode spacing.

2. Diffusion distance

Formation of the eddy current loop and its diffusion into the Earth we have considered above. McNeil (1980b) analyzed factors defining the penetration depth of the transient signal into the subsurface on an example of homogeneous half-space.

The surface current starts to diffuse into a homogeneous half-space near the transmitter, whereas at a great distance the current maintains the value dictated by the r^{-4} decrease.

The effect with time increase is that, due to the deficiency of current near the transmitter, the current seems to have to move out and down as a diffusing ring, as was described by Nabighian (1979). The diffusion distance is as

$$d = 2\pi \left(\frac{2t}{\mu_0 \sigma} \right)^{0.5} \tag{2.13}$$

where $\mu_0 = 4\pi \times 10^{-7}$, H m^{-1}, the magnetic permeability in a vacuum; $\sigma = 1/\rho$ is the conductivity of half-space. The significance of the diffusion distanced arises from the fact that it enables us to visualize the location of the maximum current density at any instant in time (McNeill 1980b).

As is seen from the above equation (2.13), in the ideal case the diffusion distance is a function of transient time (after transmitter turning-off) and conductivity of the subsurface. Determination of the diffusion distance is limited by ambient noise masking the useful signal, and by attenuation of the transient (McNeill 1994).

Data acquisition

There are two roughly TEM (TDEM) methodologies. The first is a conventional one intended for the shallow-to-deep investigation depth (McNeil 1994, Fitterman 2005). The second one is a fast methodology designed for non-deep relatively measurements with light equipment capable of carrying out preliminary data processing in the field (Barsukov *et al* 2015).

Conventional methodology (McNeill 1994, Fitterman 2005)

Usually, survey configuration consists of a square single-turn loop excited by the transmitter (Tx) (figure 2.55(a)). The length of the loop side is approximately equal to the desired depth of exploration, except that for depths less, than 40 m, the length can be as small as 5–10 m in the relatively resistive ground. The receiver coil (Rx) is a multi-turned (figure 2.55(b)), located at the center of the transmitter loop. It is connected to the receiver (figure 2.55(c)) through a cable of a short length. Such a transmitter–receiver configuration is termed as sounding with a central loop.

The method allows the depth measurements down to 6–1000 m or more. Shallow measurements can be made with a small transmitter loop of 1–10 m. Deeper measurements can require a large transmitter loop of 300–600 m. TEM methods can be used whenever there is a measurable resistivity contrast between the subsurface layers.

The equipment is autonomous, and heavy because of the 4–5 cable reels and an electric generator. The field team is 3–5 employees. During one day 4–6 stations are carried out.

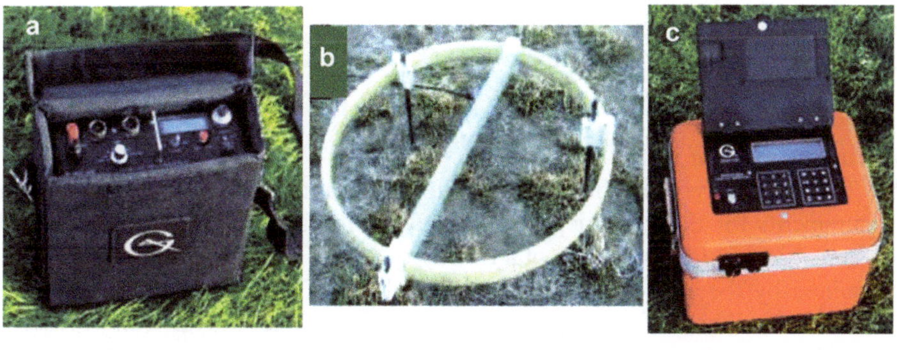

Figure 2.55. Time-domain electromagnetic sounding system: (a) Protem-47 transmitter, (b) transmitter loop, and (c) Protem receiver (reprinted from McNeill (1994) Geonics Limited, http://www.geonics.com/html/products.html).

TEM FAST methodology

The TEM FAST methodology is of high productivity. This methodology suggested by Barsukov *et al* (2015) is based on the TEM FAST equipment (AEMR, Netherlands) The equipment is portable and built into a light casing. It uses a single Rx/Tx loop. Such configuration Rx/Tx is termed a coincident, with the same loop acting both as a transmitter (Tx) and receiver (Rx) (Barsukov *et al* 2015). Using the coincident-loop configuration accelerates fieldworks. This configuration is much more sensitive to induced polarization (IP) and magnetic viscosity (the super-paramagnetic effect). The accumulator provides current 1–4 A. The TEM methodology can be used for vertical depth sounding, profiling, and mapping. A procedure commonly utilized for ground exploration using transient techniques is to lay a circle or square loop in the vicinity of the area to be examined.

The TEM FAST methodology was effectively employed in the Dead Sea Sinkhole problem and will be considered in chapter 4.

Data processing

Conventional TEM data are interpreted using the following software:

A computer program NTEMAVG v. 3.04, written by D V Fitterman (U.S. Geological Survey, written community 2006) for each sounding, averages the voltage data and evaluates these data statistically. Voltages that yielded a standard deviation greater than about 3% are not used (Fitterman 2005).

More updated GSPX7 software (Geonics Ltd): provides modeling and processing support for data collected with any PROTEM Time-Domain System Appropriate data acquisition and processing software. The program introduces minor changes that are provided without charge; new program versions that introduce major changes are subject to an upgrade charge.

Interpex Limited provides high-quality software (IX1D) for the processing, interpretation, and display of geophysical data. Programs are available to support the forward and inverse modeling of several geophysical techniques including both EM conductivity (frequency domain) and time-domain electromagnetics.

There are other software developers such as Geosoft Inc., HydroGeophysics Group, PetrosEiKon Inc.

TEM FAST methodology

Software for TEM FAST methodology IX1D, TEM Researcher, EM1dinv, 3D WIZARD will be considered in section 4.3.2.4.

TEM resolution and limitations

The TEM measurements have better lateral resolution than the DC resistivity measurements for the same depth of penetration because of the relatively smaller loop size compared to the length of a resistivity array. The TEM lateral resolution is determined by the efficient radius described above. The vertical resolution will be equal to the thickness of the ring tor. Typically, three to four layers can be resolved. The TEM measurements are subjected to interference from nearby metal pipes, cables, fences, vehicles, noise from power lines, and atmospheric storms.

2.2.1.6.2 Application of TEM in the karst problem

The application of the TEM method for void detection is related to the secondary choice or alternate method (ASTM 6429 (2006)). The application of the method is based on the detection of a low resistive zone (water-filled cavities) at the background of high resistivity zones (host rocks). It is clear that the issue of resolution arises, e.g., (1) what the difference in resistivity of the above-mentioned zones should be, and (2) what size of the heterogeneity to be detected should be. That is why there are not so many studies for the detection of karstic zones using the TEM method. However, researchers try to improve and use the TEM method to resolve some aspects of the problem. Such investigations are intensively developed in China and some other countries (Sun *et al* 2012, 2018, Zhu and Zhang 2008, Xue *et al* 2013). Another direction of investigation is the application of the airborne TEM measurements (Cox *et al* 2010, Reninger *et al* 2014).

Detection of water-filled cavities

Xue *et al* (2013) have carried out a feasibility study to check the potential of the transient electromagnetic (TEM) method in detecting and monitoring water-filled cavities. The results are analyzed using the 3D finite element method. The feasibility study has monitored subsurface conditions at the construction site of a power transfer station (PTS) location where many water-filled mine-out cavities exist. The construction site is located in the west of Shanxi province (China). From regional geological database and water analysis it is known that (1) voids in this area are water-filled, and (2) the water is more conductive than the surrounding host media. Moreover, before fieldwork, the numerical simulation was conducted to analyze the ability of TEM to detect these water-filled voids. Two stages of the research were carried out: (1) modeling, and (2) real field survey.

Stage 1: model study

According to the known properties of the survey area, models of coal and cavity have been designed. The principal nature model of subsidence in the mining area is shown in figure 2.56(a). The calculation model has been presented graphically in figure 2.56(b).

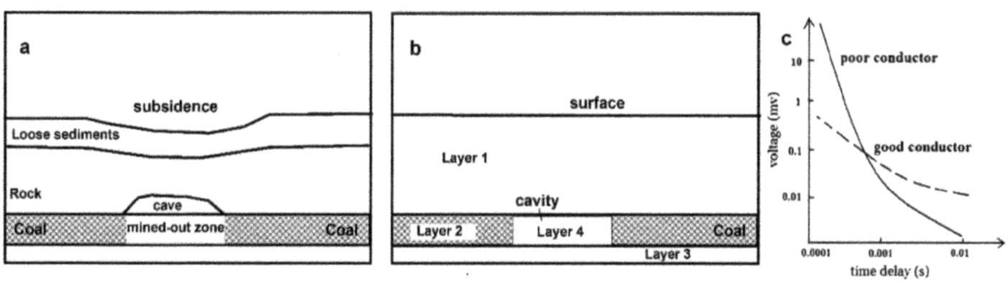

Figure 2.56. Model for calculation (a) a natural model of the mined area; (b) model for calculations; (c) expected response of TEM signal.

The TEM modeling has shown that the response expected at a coil receiver over a void filled with more conductive water (dashed line in figure 2.56(c)) has a slower decay than the response expected over the ground without a void (solid line).

Model parameters for the calculation were: layer 1 composed of limestone with a thickness of 240 m with resistivity $\rho_1 = 75\ \Omega m$; layer 2 composed of coal of 10 m thickness and resistivity ρ_2 of 400 Ωm, and layer 3 composed of rock characterized by resistivity $\rho_3 = 150\ \Omega m$. The water-filled cavity located within layer 2 is characterized by resistivity $\rho_4 = 15\ \Omega m$.

The model calculation has shown that the TEM is sensitive to the water-filled body in the case of resistivity value less than 40 Ωm when the apparent resistivity of surrounding rock is valued at about 75 Ωm.

Stage 2: field survey

The field survey was carried out in the Shanxi province located in the northeast of China. Resistivity of the rocks and cavities are as follows: air-filled cavity = >1000 Ωm, water-filled cavity = <100 Ωm, cracks <500 Ωm, coal 100–100 *n* Ωm, sandy clay <200 Ωm; sandstone = 100 *n*–1000 *n* Ωm; limestone = 100 *n*–1000 *n* Ωm. *n* = approximate times. The thickness of coal is 14 m maximum.

TEM measurements were carried out using a central loop arrangement.

Measurements were conducted around a section of the proposed location of PTS.

Sixteen lines were arranged around the PTS site with 312 survey points and a total section length of 2960 m (figure 2.57). The distance between the adjacent lines was 20 m, and the distance between the survey points is 10 m.

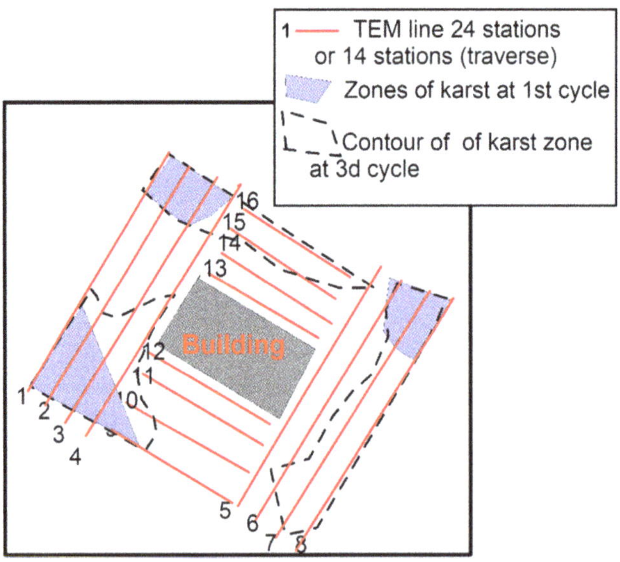

Figure 2.57. Layouts of TEM measurements around a power transfer station under construction. Blue zones are cavities Karstic zones detected in 2010 (1st cycle) and 2011 contours of these zones in 2011 (3rd cycle) are shown the area of the receiver coil is 2000 m^2, and the surveying grid of TEM is $20 \times 10\ m^2$.

Results

Results have shown that at first, the karstic zone at apparent resistivity pseudo-section appears as a low resistivity zone with the apparent resistivity less than 15 Ωm at a depth of 250–300 m (see lilac polygons in figure 2.57 for the location of these zones). The second one, the low resistivity zone, increased in size for more than 1 year (dashed lines in figure 2.57). The results were verified by the drilling. After the authors' opinion, the mined-out voids are enlarging with time.

A search of karstic water around the tunnel

Sun *et al* (2012) have developed a multi-component and multi-array detection method for tunnels based on the TEM technique for detecting karstic water around the tunnel. The method applies to inside-the-loop TEM surveys. The TEM measurement schema is shown in figure 2.58. In tunnels, a rectangular transmitter loop is used placed on the tunnel floor. Eddy currents occur in the geologic medium (because of Faraday's law) when a steady magnetic field is applied. If to decompose the arbitrary eddy current to vertical and horizontal components then in low resistivity geologic bodies with the regular shape placed just under the transmitter loop, one gets the maximum measured vertical EMF and the zero horizontal EMF in the central point (Xi *et al* 2010).

In reality, the situation is more complicated. We do not know where geological bodies are located: from the left side, right side, or just under the loop. The vertical component is significantly more accurate in the vertical direction, whereas the horizontal component is more accurate in the horizontal direction. Thus, using multi-component and multi-array measurements within the loop one can get more response data containing the distribution and shapes of the undesirable geological bodies.

Sun *et al* (2012) have tested the multi-component and multi-array system in a real tunnel complicated along its trace by karstic water. The test tunnel was located in a karst area in Chongqing, China. The host rocks along the tunnel are composed of limestones and their breaches characterized by high hydraulic conductivity. The field survey was carried out on a tunnel floor (figure 2.59). A 3×3 m^2 eighty turned coil was used as a transmitter. Within this loop, nine points were used for measurements, as shown in figure 2.59 (points (a) to (i)). Additionally, one line 10 m in length was put via the center of the loop along the tunnel axis. The transmitter frequency was 62.5 Hz; the current in the loop was 0.8 A; the receiver was a special high-frequency magnetic probe antenna with a resonant frequency of 250 kHz.

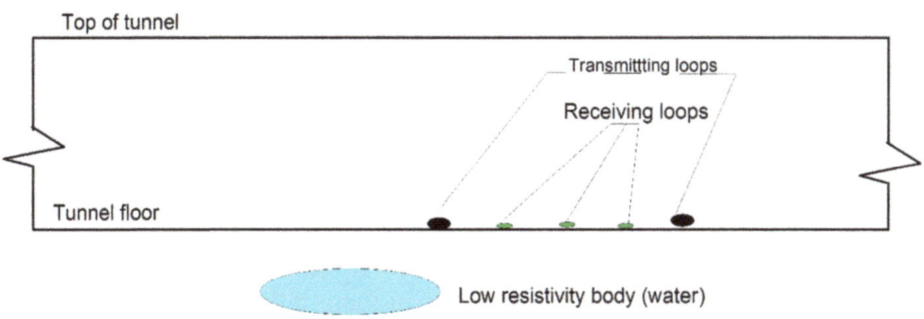

Figure 2.58. TEM measurement schema in the tunnel to detect low resistivity bodies under the tunnel floor.

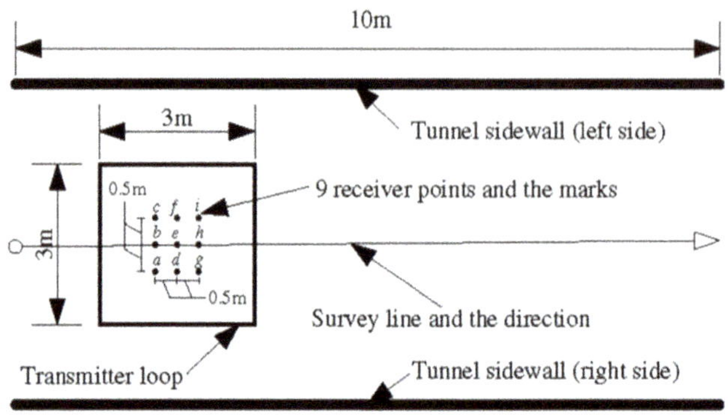

Figure 2.59. Top view of survey line and the transceiver arrangements (reprinted from Sun *et al* (2012), with permission from IOP Publishing).

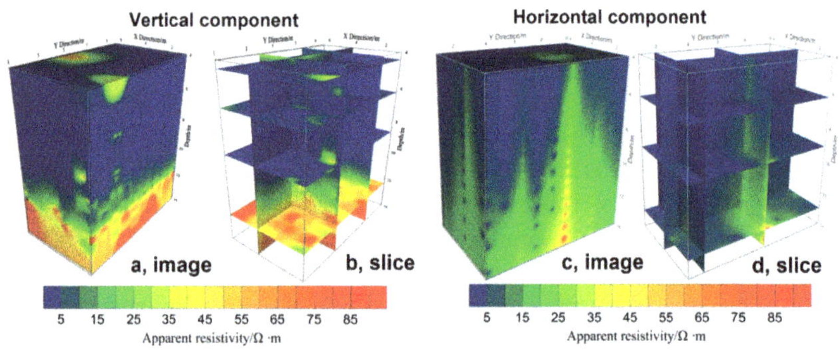

Figure 2.60. 3D interpolated apparent resistivity images of the vertical (a, b) and horizontal (c, d) components (a, c) the interpolated apparent resistivity images; (b, d) the slices of the interpolated apparent resistivity images. The depths of the 3D cubes are from 4 to 10 m (reprinted from Sun *et al* 2012, with permission from the IOP Publishing).

Data processing

Like conventional TEM data processing, the multi-channel profiles of the vertical and then horizontal components were drawn. Then, the apparent resistivities of all components were calculated. The moving conductive thin sheet theory is used for the time-depth conversion. In the beginning, maps of apparent resistivities were constructed for different receiver offsets ((a) to (i), see scheme in figure 2.59) for vertical and horizontal components. Finally, 3D apparent resistivity images were constructed for the vertical and horizontal components separately (figures 2.60(a), (b), (c), (d), respectively).

One can note the distribution difference of low resistivity anomalies in 3D apparent resistivity images and their slice images, respectively. The vertical component, for instance, shows lower resistivities at the bottom of the 3D cube than the horizontal component. Additionally, there are isolated resistivity anomalies.

To verify the TEM data, more than 10 drills were carried out in the test site. The drilling has shown that there are water inclusions, but they are not connected between themselves. There were found two water-saturated zones at 2 m and 7 m, respectively. The first one (2 m depth) is not so large and comprises mud. At a depth of 6 m, the water was emitted from the borehole under quite large pressure. The borehole has shown the thickness of this aquifer is rather large. The rocks are quite broken, weak, and hydraulically conductive (Sun *et al* 2012).

2.2.1.7 Frequency domain electromagnetic (FDEM)

2.2.1.7.1 General consideration

FDEM method. Frequency domain electromagnetic (FDEM) is termed also EM terrain conductivity measurement at low induction numbers (LINs) (McNeill 1980b). The method is used to measure the electrical conductivity of the subsurface using the magnitude and phase of the secondary field resulting from induced electromagnetic currents. The electrical conductivity of the earth material depends on the electrical properties of the soil and rock matrix, percentage of fluid saturation, and the conductivity of the pore fluids (McNeill 1980c). First systems for measurements of electromagnetic terrain conductivity (EM) were introduced in 1976. Nowadays, the EM method is being closely integrated with the global positioning systems (GPS). It allows determining the exact location of the measurement lines. The FDEM method intends resistivity (conductivity) surveys. The development of the EM31 and the EM34-3 (figure 2.61), allows mapping terrain conductivity at a rate of the operator's walk; furthermore, the reading of conductivity is averaged. It enables the investigator to provide high resolution in conductivity. Moreover, integrated with the GPS location of the measurements, variations, and the interpreted resistivity anomalies can be established with great accuracy.

Another type of EM exploration using two coplanar transmitters and receiver coils with horizontal orientation is named *Slingram methodology* (or HLEM—horizontal loop electromagnetic). With the Slingram methodology, one can provide three kinds of information simultaneously: (1) the apparent resistivity of the ground; (2) its apparent susceptibility; and (3) the characteristic signatures of the buried metallic objects. Both coils are constantly separated and operate at a fixed distance.

Figure 2.61. FDEM equipment (a) EM-31 system; (b) EM-34 system: two coplanar vertical loops (horizontal dipole), S-separation between transmitter–receiver loops that is 10, 20 or 40 m (McNeill (1980a), figures reprinted from the Geonics Limited Catalogue, with the permission from Geonics Limited).

Exploration depth can be varied by frequency or loop separation (Kirsch 2006). The devices used for the Slingram EM are EM31 and EM38 of the Geonics Limited (Technical Notes TN-31). There is described also a DualEM-421S multi-receiver EMI instrument operating at 9 kHz frequency (Dabas *et al* 2016). The system is designed for archaeological researches. The frequency usually has to be chosen in such a way that the skin depth does not limit the investigation determined primarily by coil spacing, and so that displacement currents are negligible. Therefore, a frequency lower than 100 kHz is required (Tabbagh 1986). Distance between the loops varies from 1 m up to 100 m.

Usually, the magnetic component of the superposition of the primary and secondary fields is measured. The measured field is split into in-phase and out-phase (= quadrature, 90° phase shift) components relatively the primary field. Both components are recorded. Software exists for modeling the forward response of EM fields over a known 3D subsurface, e.g., EMIGMA 9.5 (PetRos EiKon 2020). The application of electromagnetic techniques to the measurement of terrain resistivity, or more properly, conductivity is described by Keller and Frischknecht (1966), and McNeill (1980b), TN-31.

The principle of the FDEM method is presented in figure 2.62. A transmitter coil (Tx) excited by an alternating current of an audio frequency, is placed at the homogeneous earth. A receiver coil Rx is located a short distance S away. The alternating current in the transmitter coil generates the variable magnetic field that in turn induces very small currents in the Earth that generate a secondary magnetic field H. A secondary magnetic field H together with the primary field (HP) is felt by the receiver coil (Rx).

Secondary magnetic field depends on the ground conductivity (σ), the operating frequency (f), and intercoil separation (S). The secondary field is a simple function of these variables.

These constraints are incorporated in the design of the EM3I and EM34-3 whence the secondary magnetic field is shown to be:

$$\frac{H_s}{H_p} \cong \frac{i\omega\mu_0\sigma S^2}{4} \tag{2.14}$$

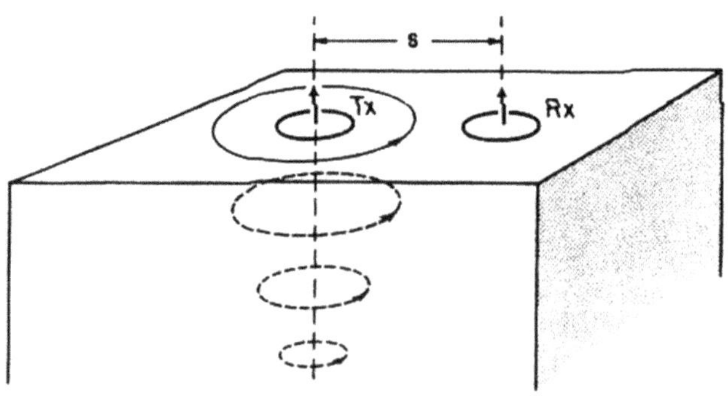

Figure 2.62. Induced current flow (homogeneous half-space) (reprinted from McNeill (1980a), with permission from Geonics Ltd).

Here, H_s is the secondary magnetic field at the receiver coil; H_p is the primary magnetic field at the receiver coil; $\omega = 2\pi f$; f is the frequency (Hz); μ_0 is the permeability of the free space; σ is the ground conductivity (mho m^{-1}); S is the intercoil spacing (m); $i = -1$.

From equation (2.14) is can be seen that the ratio $\frac{H_s}{H_p}$ is linearly proportional to conductivity σ and it is possible now to construct a direct reading terrain conductivity meter by measuring this ratio. It is possible because all parameters of the equation (1) are constants apart from the $\frac{H_s}{H_p}$ ratio and σ, e.g.

$$\sigma_a \cong \frac{4}{\omega\mu_0 S^2}\left(\frac{H_s}{H_p}\right) \tag{2.15}$$

The unit of conductivity is the mho (Siemen) or, more conventionally millimho (mmho).

The coils are normally carried with their planes either vertically (horizontal dipole) as shown in figure 2.61(b) or horizontally placed on the surface (vertical dipole). The same is shown in figures 2.63(a) and (c) schematically.

The corresponding graphs of the response versus relative depth of horizontal and vertical dipoles are shown in figures 2.63(b) and (d), respectively. Let us consider the vertical dipole (figure 2.63(c)) to understand how the system is working. There is selected a thin layer of the thickness dz located at depth of z, where $z = d/S$. It is possible to calculate the secondary magnetic field in the receiver coil arising from all of the current flow within this or any other horizontal thin layer. One can thus construct the function $\varnothing_V(z)$ shown in figure 2.63(d) which describes the relative contribution to the secondary magnetic field arising from a thin layer at any depth z. The material located at a depth $d = 0.4S$ (e.g., $z = 0.4$) contributes themaximum to the second magnetic field. The material located at a depth of $d = 1.5S$ still contributes significantly. At the same time, the ground at zero depth, i.e., the near-surface material, makes a very small contribution to the secondary magnetic field and therefore this coil configuration is insensitive to changes in the near-surface conductivity. For the horizontal dipole (figure 2.63(a)) the function $\varnothing_H(z)$ is significantly another (figure 2.63(b)). In this case, the maximum contribution to the secondary magnetic field is made at the zero-depth, e.g., by near-surface material. Then, the response falls off monotonically with a depth (McNeill 1980a).

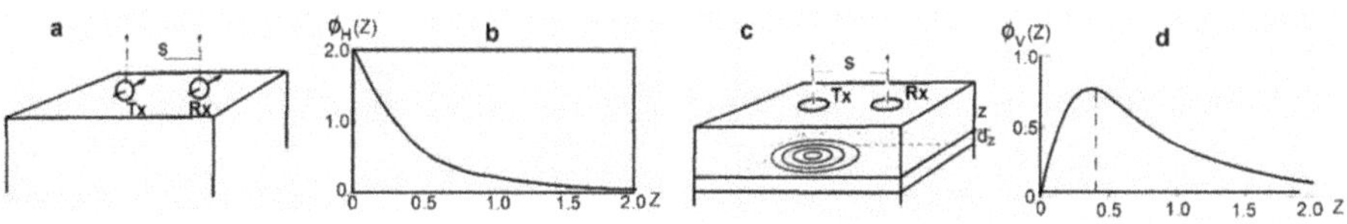

Figure 2.63. Schema (a, c) of the horizontal and vertical dipoles, respectively, and corresponding nomograms of response versus relative depth of horizontal (b) and vertical (d) dipoles (reprinted from McNeill (1980a), with permission from Geonics Ltd).

In figure 2.64(a) functions ∅ versus normalized depth (z) are compared for both coil configurations. This deserves to be emphasized in a different manner in which they respond to the material at different depths. The difference is important because either instrument can be rolled over so that the vertical dipole transmitter/receiver geometry becomes a horizontal dipole transmitter/receiver geometry and vice versa. This feature is useful in diagnosing and defining layered earth. The figure also shows that for regions greater than one intercoil spacing in depth the vertical transmitter/receiver dipole gives approximately twice the relative contribution of the horizontal transmitter/receiver dipole.

The functions shown in figure 2.64(a) are useful for describing the relative sensitivity of either of the two coil configurations to the material at various depths. However, a function derived from them is more useful for performing calculations. It is defined as the relative contribution to the secondary magnetic field or apparent conductivity from all material below a depth z and is given by

$$R_{\mathrm{VorH}}\,(z) = \int_{z}^{\infty} \varnothing_{\mathrm{VorH}}(z)dz \tag{2.16}$$

Equation (2.16) is termed the cumulative response and it is illustrated in figure 2.64(b) for vertical coplanar transmitter/receiver dipoles. It follows from the figure that, for example, that all material below a depth of two intercoils spacings (z = 2) yields a relative contribution of approximately 0.25 (i.e., 25%) to the secondary magnetic field at the receiver coil.

Exploration depth. The EM31 one-man instrument (figure 2.61(a)) has an intercoiled separation of 3.7 m that allows providing an exploration depth of about 6 m. The operating frequency is 9.8 kHz. When raising and lowering the instrument it is possible to achieve the identification of layering.

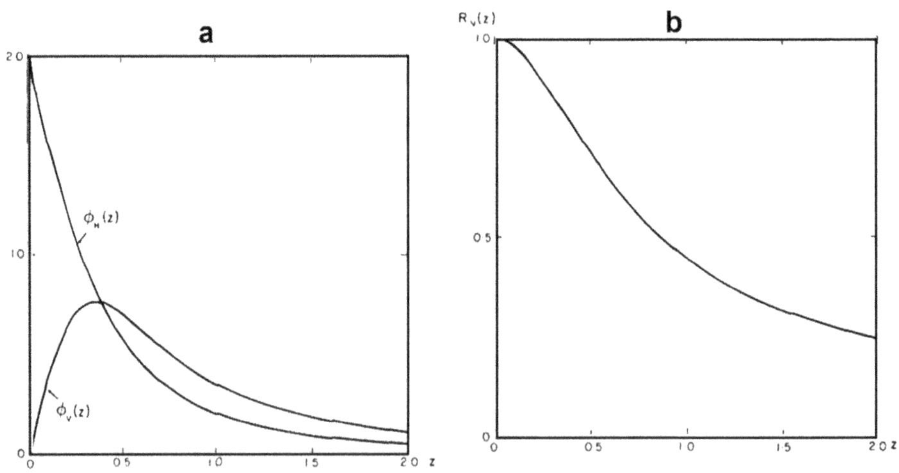

Figure 2.64. (a) Comparison of relative responses for the vertical and horizontal dipoles. (b) Cumulative response versus depth for vertical dipoles. $R_V(z)$ is the relative contribution to H_s from all materials below a (normalized) depth z (reprinted from McNeill (1980a), with permission from Geonics Ltd).

The EM-34 instrument (two-man) has two coils flexibly separated as 10, 20, 40 m. The intercoil separation is measured by the instrument meter and the receiver operator reads a meter to accurately set the coil to the correct spacing using cables of the corresponding length, which can be 10, 20, or 40 m. Conductivity is read by the transmitter operator from another meter. With the use of different separations and types of dipole, the different exploration depth is reached (table 2.4).

It is seen from table 2.4 that exploration depth is a function of the dipole type and separation between coils. However, investigation depth also depends on the site-specific conditions. For example, FDEM is limited by high conductivity values of the soil, rock, and pore fluids. The depth of resistivity measurements will be reduced when there is a highly resistive layer such as clean quartz sand, which would limit adequate current getting into the ground (Benson and Yuhr 2016).

Applications. Following the ASTM 6429-99 table 2.1, the FDEM method is that of primary choice for detection of the fractures and faults zones, landfill leachate, saltwater intrusions, soil salinity determination, drums, landfill, trench boundaries delineation, forensics, and archaeological features. In the detection of voids and sinkholes, following ASTM, FDEM is the method of secondary (B) choice or alternating method. However, Reynolds (2011, table 1.2) relates EM to that of primary (A) choice. Also, Hoover (2003, table 2.1) evaluates FDEM as the method of primary (A) choice for assessing sinkholes and voids.

FDEM measurements in the profiling mode are intended to detect and map the lateral changes in geologic and hydrogeologic conductivities. Sometimes, it is appropriate to carry out measurements with two coil orientations: coil axis vertical (vertical dipole) or coil axis horizontal (horizontal dipole). It can be performed in the EM-34 model. The vertical dipole provides a greater depth of measurement compared to the horizontal dipole with the same coil separation. Using three separations (10, 20, and 40 m) we can get different depths and some software allows the performing of inversion.

Data processing: FDEM relates to methods for which interpretation can be performed in the field (similarly to GPR, and DC). EM measurements are not intrusive, and are relatively easy to perform. The instruments can work both in manual and automatic mode. Field EM data can be plotted as a profile line or contoured. The data can sometimes be interpreted without any data processing.

Resolution. Frequency domain electromagnetics can provide excellent lateral resolution particularly when continuous measurements are used. Lateral resolution

Table 2.4. Exploration depth (McNeill (1980a), with permission of Geonic Ltd).

	Exploration depth (m)		
Intercoil spacing (m)	Horizontal dipoles	Vertical dipoles	Operating frequency (kHz)
10	7.5	15	6.4
20	15	30	1.6
40	30	60	0.4

is determined by the spacing between the transmitter and receiver coils and station spacing between measurements. Frequency domain EM methods have limited depth sounding capabilities. By using several intercoil spacings and both horizontal and vertical dipole configurations, interpretations of two or three contrasting layers can be performed.

Limitations of the method. FDEM measurements are sensitive to interference from the nearby metallic fences and pipes, vehicles, noise from power lines, and atmospheric distirbances. Apart from the EM distortions, the presence of utilities, fences, cars, or buildings limit the use of this method at developed properties and urban areas.

The effectiveness of the EM measurements decreases at very low conductivities and the measurements become nonlinear at high conductivity values.

The principal limitation of the method is the qualitative result, and depth estimates are normally limited to shallow or deep (Hoover 2003, ASTM 6429-99 2006).

Conclusion remark

We note and emphasise two factors that should be constantly in mind when performing FDEM surveys to map geological features.

(1) With the EM we map only the electrical conductivity. If the conductivity does not vary significantly with the geological environment, or if some artificial targets influence the conductivity, the survey results may be difficult to interpret.

(2) Measurement of terrain conductivity, must be connected with the available geological factors. It is always necessary to interpret geophysical data in conjunction with the known geology from outcrops, boreholes, or any other such 'benchmarks'. Geophysical measurements can be very effective by allowing interpolation between such sources, or careful extrapolation away from them. Finally, in every case, the knowledge derived from the geophysical measurements must be eventually reconfirmed against the known geological conditions (McNeill 1980b).

For further reading, we refer the reader to Keller and Frischknecht (1966), McNeill (1980b).

2.2.1.7.2 Application of FDEM to karst investigations

The application of the FDEM method can be effective in the study of such karst features as weathered depressions in the rock, open, water-filled, or sediment-filled sinkholes, and cavities, or larger cave systems. The ability to detect a given size cavity using FDEM decreases with increasing depth as for all surface geophysical methods (ASTM 6429-99 2006).

The results of an EM survey are typically presented in form of conductivity (resistivity map or profile), with a qualitative interpretation. Hoover (2003) suggests some examples of EM data interpretation using their calibration in the areas of known geological conditions. For instance, the interpretation can be calibrated to permit an estimate of depth if adequate information is available on the top-of-bedrock.

In continuation, we will consider some case histories of the FDEM application in soil and cavity detection.

Lange (1999) has mapped the karst structure and groundwater channels beneath sediments using three techniques: EM, gravity, and self-potential (SP). The EM method allowed the definition of the boundary between carbonate rock and alluvium.

The near-surface fracture density and extent of a karst system were mapped by Rey (2007) using the Slingram type EM method (EM-31 device) in southern France.

Jardani *et al* (2007) have reported on the detection of preferential infiltration pathways using the joint inversion of SP and Slingram EM source (EM-34 device) conductivity data.

EM conductivity mapping was combined with gravity, GPR, seismic refraction, and electrical sounding (ES) to identify sinkholes in the Oak Ridge reservations in Tennessee (USA) (Carpenter *et al* 1998).

The detection of cavities seems to be the most popular application for geophysical methods. The rock surrounding natural cavities is often disturbed, particularly in carbonate karst environments where a cave is formed by the chemical and physical interaction of groundwater with the rock. In such an environment, fractures and the dissolution of rock surrounding a karst cavity creates a larger bulk anomalous volume than the cave itself. The overview is given in Chalikakis *et al* (2011).

Soreq Cave area (Ezersky *et al* 2000).

The FDEM method was applied in the Soreq Cave area (Israel) with the main objective of testing the geophysical methods in a well-known cave and its vicinity.

The Soreq Cave is located at the western margins of the Judean Hills anticline within the Veradim Formation of the Cenomanian age. This formation is represented by a massive dolomite and limestone layer. The layers are covered by the Terra-Rossa soil 30–50 cm thick (figure 2.65(a)). The entrance to the cave is located at an elevation of +395 m AMSL. The maximum dimensions of the cave are 80 × 60 × 12 m and it is located at a depth of 8–35 m from the surface and deepens in a SE direction (figure 2.65(b)).

A total of 13 FDEM and eight CVES profiles were carried out in the area with dimensions of approximately 200×200 m both above the cave and outside it (figure 2.65(c)). The FDEM survey was carried out using a Geonics EM-34 XL instrument. All three transmitter/receiver (T/R) separations (10, 20, and 40 m) were used at six main profiles while only one separation, T/R = 20 m, was used at seven additional lines. The use of three separations facilitated the drawing of apparent resistivity pseudo-cross-sections along the appropriate lines while the use of separation T/R = 20 m on a relatively dense grid of points allowed plotting of the resistivity map of the whole test site at a depth of about 10–30 m (figure 2.65(d)). The measurement interval along the profiles was 5–10 m for separations 10 and 20 m and was as large as 20 m for the separation of 40 m. Since no quantitative interpretation of the FDEM measurements is possible, the results presented can be used mainly to locate resistivity anomalies in the lateral directions and for only a very rough estimation of their resistivities and depths.

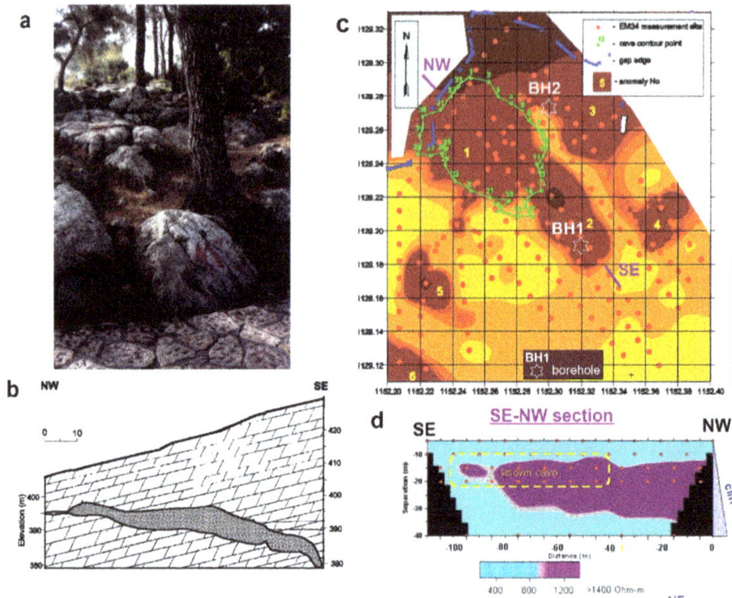

Figure 2.65. Soreq Cave area (a) the photo landform above the cavity; (b) NW–SE cavity section; (c) FDEM apparent resistivity map based on EM-34 profiling with separation 20 m between transmitter–receiver loops. High resistivity anomalies of more than 800 Ωm are located at the depth of the known Soreq cave and also out of that. Digits are the resistivity anomaly numbers; (d) FDEM resistivity pseudo-depth section along the SE-NW line. (The figure reprinted from Ezersky *et al* (2000) is published with the permission of the EAGE and EEGS. All copyright privileges remain with EAGE/EEGS. This material cannot be copied or used without the express written permission of EAGE/EEGS.)

FDEM results

A fairly strong resistivity anomaly (more than 800 Ωm as compared to approximately 200–400 Ωm of background resistivities) was detected above the known cave when measurements were carried out with 20 m offset (figure 2.65(c)).

No significant anomalies were detected with the 10 and 40 m offsets. Other strong anomalies were detected outside the cave.

FDEM method has shown a strong resistivity anomaly approximately at the cave depth (figure 2.65(c)). Another strong anomaly (2) was shown some 50 m to the southeast of the cave. This anomaly was interpreted as a possible unknown cave. Several smaller anomalies were detected at different locations (anomalies numbers 3–6). The possible origin of the two largest anomalies was investigated by follow-up drilling (Bar-Matthews *et al* 1999).

The borehole BH-1 drilled at the first anomaly (see figure 2.65(c) for location) penetrated the fractured dolomite from the surface to a depth of 9 m. From 9 m to 18 m was observed fractured dolomite, but fractures are filled with calcite and finally karstified fractured dolomite occurs from 18 m to a depth of 25 m. The cooling water has disappeared from a depth of 18 m. The borehole BH-2 was drilled at the high resistivity anomaly between the existing cave and the precipice. A geological cross-section here was presented as strongly fractured dolomite continues from the surface to a depth of 10–12 m, and strongly karstified dolomite occurs from 18 m to a depth of 28 m.

Thus, the investigations have shown that under the specific geohydrological conditions of the Soreq Cave site the EM methodology can detect caves as well as highly resistive dolomites with the calcite-filled fractures, but is unable to distinguish between them. Similar results were shown by the ERT method.

Electromagnetic prospecting for karst and fault systems

Vogelsang (1987) has reported an EM study of a massive Devonian limestone reef near the Lahn river in Germany. The EM profiling was carried out using the Slingram geometry, e.g., moving coplanar transmitter and receiver coils to create a vertically polarized inhomogeneous field of sine waves. In-phase and out-phase values of the frequencies 888 and 3555 Hz were used. The transmitters and receivers were usually placed 100 m apart. At first, modeling was carried out to estimate the EM response to low resistivity structures, such as cavities and faults. The electromagnetic response to such a low resistivity zone is shown in figure 2.66 (Ketola and Puranen 1967). If during the interpretation of field data, it was found that the geological strike direction and the direction of the survey line deviated more than 20′ from the perpendicular, a correction-as described by Ketola and Puranen (1967) was applied.

In modeling, mostly it was possible to determine the position of the fault plane and roughly its strike and dip, however, the thickness, or width of the fault, agreed with the model less frequently. The complete agreement between field and model curves was rarely reached.

The example of EM traverse through a massive Devonian limestone reef near the Lahn river in Germany is shown in figure 2.67 (Vogelsang 1987). The lower geological section has been generated based on a DC geoelectrical survey (VES) and borehole. There are two geological features in the section (a) karst pipe-filled clay and (b) fault with low resistivity clayey sediments in contact with limestone of high resistivity.

The upper section shows the EM profiles in-phase (A) and out-phase (B), both 3555 Hz. One can see that both plots denote geological features of the section by varying the resistivity. The karst pipe causes a local decrease of resistivity by 50%–60% (anomaly A) and contact between limestone and shale is detected by the step decrease of resistivity (anomaly B).

2.2.1.8 Very low frequency (VLF)

Investigation of the electromagnetic fields from distant VLF transmitters (mostly of military origin) is one of the most mobile and economical geophysical methods. At present, it finds frequent application in environmental studies, mining geophysics, and geophysical–geological mapping. VLF radiations are propagated in a spherical waveguide between the Earth and ionosphere where the energy of radiowaves is not attenuated even at great distances from the transmitter due to the channel effect. Because of the compact measuring equipment, the VLF method can be employed even in almost inaccessible regions (e.g., mountainous and foothill regions, deserts) (Eppelbaum 1991).

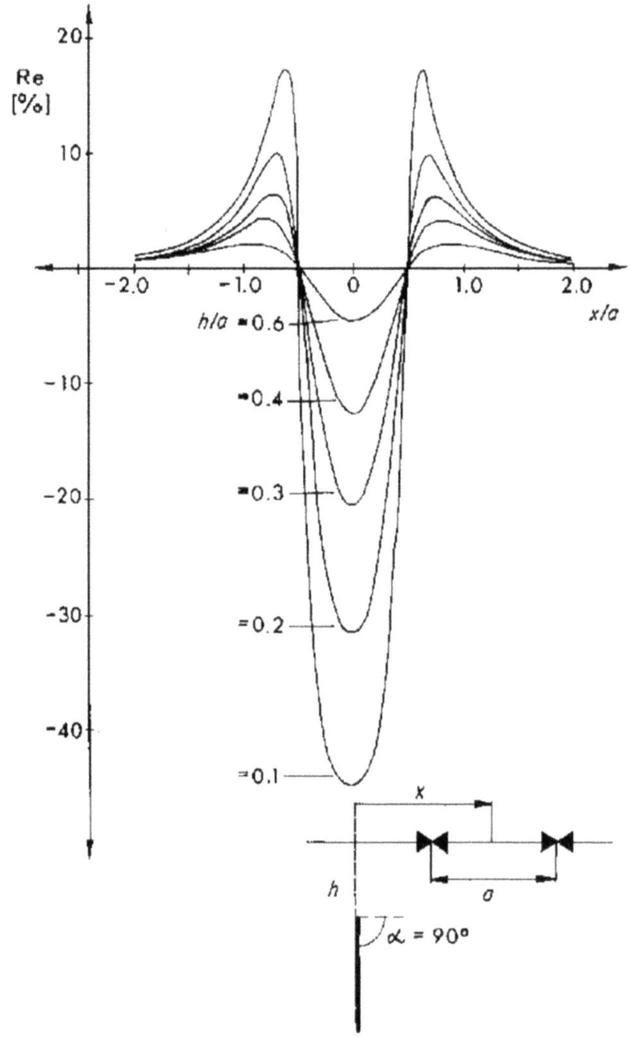

Figure 2.66. EM model curve of real component, measured across a vertical zinc sheet 1000×1000×0.82 mm^3. Half-plane material: zinc. a = 10 cm, f = 3600 cps, σ = 12.76 ×106 mho m^{-1}, W = 29.7 (reprinted from Ketola and Puranen (1967), with the permission of the Geological Survey of Finland).

The VLF equipment allows the measurement of the magnetic and electric components of the electromagnetic VLF field. However, the use of magnetic VLF components in many cases is a more efficient technology because of the comparatively low level of the various kinds of noise (Eppelbaum 1991). The VLF equipment developed in the former USSR, Russia, Czech Republic, and Poland (e.g., Gordeyev *et al* 1981, Eppelbaum and Khesin 1992) can be applied to obtain the values of magnetic components on coordinate axes: H_x, H_y, and H_z, as well as changes of radio-bearing Ψ to the employing VLF transmitter (as well as other parameters).

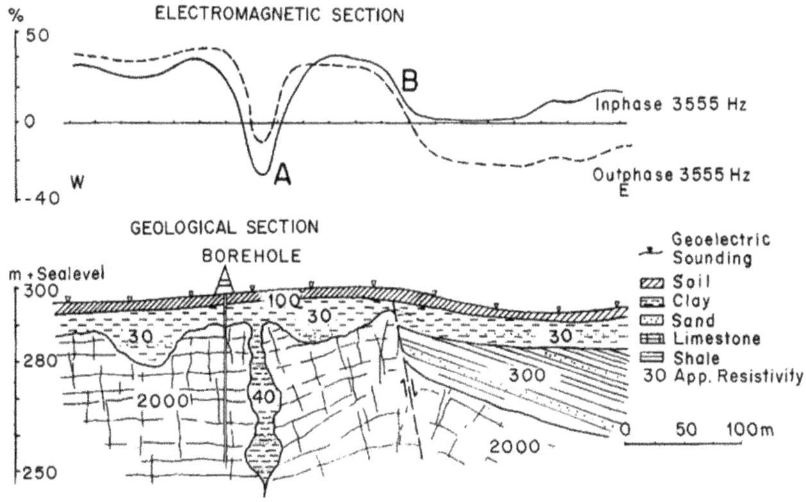

Figure 2.67. The geological cross-section of the Devonian limestone area near the River Lahn in Germany was obtained from d. c. soundings. A and B on the EM profile identify a karst pipe (anomaly A) and a fault with low resistivity clayey sediments in contact with limestone of high resistivity (anomaly B) (reprinted from Vogelsang (1987), with the permission from EAGE).

However, the most important factor is that this VLF equipment modification allows obtaining essential similarities with the magnetic field.

An important peculiarity of a VLF ROV survey at comparatively low altitudes (up to 100–150 m) is the slow attenuation of the VLF signal with increasing observation height over the Earth's surface (Eppelbaum and Mishne 2011).

Let us consider the developed advanced interpretation scheme for the VLF method in detail.

Elimination of temporal VLF variations

The problem of eliminating temporal variations in the VLF method is crucially important since the intensity of these variations is often compatible with the intensity of the averaged VLF signal during the time of the field survey. The following approaches have been suggested to overcome this problem.

In early VLF methodology, an approximate procedure was applied based on the field intensity measurements at a control point (CP) before and after fieldwork (Tarkhov 1961). The corresponding corrections were introduced by interpolation assuming that the field intensity change was linear during this period. However, the practical application of this method led to considerable errors.

A modification of this method is described in (Khmelevskoi and Bondarenko 1989). The time difference is that the period of supposedly linear variations is assumed to be 1 h. The VLF survey is rejected if the level of variation for 1 h exceeds 20%. This method, however, suffers from the following disadvantages. Experience in field investigations in various regions indicates that the VLF intensity (even over one hour) often cannot be approximated linearly with sufficient accuracy. In addition, the variation amplitudes during this period can exceed 20%. Besides this, the aforementioned methodology neglects the effect of various noise intensities on useful anomaly intensity. At the same time, it is obvious, that the VLF time variations affect the electromagnetic field passing from the air into the ground. The intensity of the secondary field tends to change (increase or decrease in the presence of anomalous targets of increased or decreased electric conductivity relative to the host media), and the temporal variations are non-uniform ones. When disregarded, this fact can distort the results of interpretation.

Some companies developed an automatic attachment to a VLF receiver, where temporal variations are eliminated by obtaining a synchronous ratio of the vertical component H_z to the total horizontal component H_φ of the VLF magnetic field. However, it was found experimentally that the H_z/H_φ ratio may vary even for observations at the same point. Therefore, quantitative interpretation of the H_z/H_φ ratio presents certain difficulties. Another approach consists of the selection of the frequency and time interval that is the most stable temporally for the area under study (e.g., Villee *et al* 1992). However, this approach would benefit from improvement.

The great distance from the utilized VLF transmitters (2000–10 000 km) accounts for the uniformity of the primary field (Eppelbaum and Khesin 1992) both at *CP* and in the investigated area. The VLF wavefront can be assumed to be a plane, whereas the primary field intensity is taken to be constant within the area of the detailed geophysical investigation (25–100 km^2). The following additive model of a geophysical field is applicable in practice for investigation by the VLF method (Eppelbaum and Mishne 2011):

$$F_j = \sum_j S_j\left[n_j(x), x\right] + \sum n_j(x), \tag{2.17}$$

where F_j is the field observed along with the profile, $\sum_j S_j\left[n_j(x), x\right]$ is the sum of effects from the anomaly-forming objects and geological inhomogeneities of the section (taking the dependence on field variation into account), $\sum n_j(x)$ represents the noise of temporal (time dependence is omitted in equation (2.17) for simplicity).

It is known that for the VLF field the conduction currents dominate the displacement currents (Gordeyev *et al* 1981, Khmelevskoi and Bondarenko 1989), which is the major condition for quasi-stationarity (Alpin *et al* 1985). The proportionality of the secondary electromagnetic fields (both magnetic and electric) to the primary fields has been confirmed in several publications (e.g., Dmitriyev *et al* 1977, Dmitriyev 1982). Hence given the physical effect of the sub-vertical VLF-field propagation in the ground (Gordeyev *et al* 1981, Olsson 1983). Taking also into account equation (2.17), the following simplified model of the VLF observations can be used in practice (Eppelbaum and Mishne 2011):

$$\left\{\begin{array}{l} H_0(t) = H(t) + B_0 H(t) \\ H_j(t) = H(t) + C_j(t) \end{array}\right\}, \tag{2.18}$$

where $H_0(t)$ is the VLF observations at the control point; $H_j(t)$ is the VLF observations on the profile, $H(t)$ is the primary VLF field intensity, B and C are the coefficients reflecting electromagnetic properties of the medium; indices 'j' and 'o' mark the observation point on the profile and CP, respectively.

We can consider that B_o = const, and this parameter, which is a basic value for the concrete area, can be assumed (for convenience) to be equal to zero. Solving equation (2.18) in the parametric form, we obtain the values H_{clear} cleared from the temporal variations for each profile point:

$$H_{\text{clear}} = \frac{H_j(t)}{H_o(t)} \overline{H}, \tag{2.19}$$

where $\overline{H}$ is the certain averaged value of the VLF field.

Thus, the proposed scheme of eliminating variations in the VLF method is based on a synchronous recording of the observations along with observation profile and temporal variations at CP with further specific processing of this data. Automatic recording of variations presents no difficulties and can be carried out with a discretization interval of 1–2 s. This method substantially eliminates observation distortions caused by the VLF temporal variations (both over a day and at different days in a survey) by reducing observation results to some common level (Eppelbaum and Finkelstein 1998).

Calculation of rugged relief influence

Various attempts have been made by various researchers to reduce the terrain relief effect. Tarkhov (1961) suggested an analytical approach. However, the analytical calculations proved to be too awkward even for the simple geological sections, above all, the obtained data were at variance with the experimental results. Sedelnikov (1983) analyzed in detail the theoretical problem of electric field distortions caused by a single mountain and a mountain range but no method was suggested for topographic effect elimination.

Karous (1979) studied topographic influences on the vertical component of the VLF magnetic field (H_z). It is clearly of interest, but no numerical results were reported. Eberle (1981) proposed simplified formulas to calculate relief corrections, however, their application is limited by several conditions, e.g., the disturbing object and the observation profile should be in the most favorable position for the primary VLF magnetic field excitation. This requirement is difficult to meet or seldom obtained.

The second approach is based on a physical (analog) modeling of VLF fields. Gordeyev *et al* (1981) developed models which describe quantitatively the distribution of the VLF field over different terrain relief forms. Baker and Myers (1980) basing on their model investigations suggested the VLF field 'reduction' method. This method has the same limitations as found in (Eberle 1981).

Gordeyev (1983) proposed a third approach. It is based on using a graticule (chart) method to reduce the topographic effect on H_φ, a component of VLF transmitters. It is difficult to apply at the strong rugged relief and can, in some cases, only impair the interpretation. This limitation is traceable to the 'build-up effect' of a rectangular graticule in a complicated topography, which in addition is often characterized by variable electric properties.

The effect of the Earth's surface topography on radio-wave propagation is a fairly important problem in radiophysics and has been investigated by many researchers. An overview of the literature indicates that this problem has yet to be adequately resolved. A numerical calculation of radio-wave scattering is a complex electrodynamic problem even for the simple case of diffraction at a homogeneous wedge with a finite conductivity (Zakharov and Pimenov 1982, Felsen and Markuvitz 1994). The difficulties increase progressively for statistically inhomogeneous (i.e., real) surfaces. Bass and Fuks (1978) emphasized that under natural conditions there is no need to find the detailed structure of a scattered field, and it is sufficient to employ some reflected-signal parameters averaged for some classes of surfaces and targets. In general, these researchers have concluded that wave scattering is determined largely by coordinate functions and represents a combination of wave diffraction on an arbitrary surface and an application of the theory of probability and mathematical statistics. An example of such a correlation is shown in figure 2.68 as a technique for reducing the terrain relief effect (northern Azerbaijan).

A review of the literature and personal results of VLF surveys (Eppelbaum 1991) leads to the following general rules:

(1) The plots of H_x (the x-axis is aligned with the observation profile) and the total horizontal magnetic field component H_φ substantially repeat the surface relief.
(2) The intensity of the recorded VLF magnetic field is directly proportional to the relative elevation of observation points. The topographic anomalies caused by local relief forms are similar to those from anomalous targets but have a smoother appearance.
(3) When the angle between the radio-wave arrival direction and that of a relief element is close to 0°, the relief effect is relatively large, and when this angle is close to 90°, it is relatively small, but in any case, the relief effect is always present.
(4) VLF field intensity is mainly dependent on the resistivity of the rocks composing the subsurface.

All the aforementioned items make it possible to employ the correlation method developed in magnetic prospecting (Khesin *et al* 1996) to the VLF method. A simple example of this approach application is shown in figure 2.69. Here is demonstrated using the correlation technique on a portion of the Katekh pyrite-polymetallic deposit in the southern slope of the Greater Caucasus. A frequency of 16.0 kHz was used in the investigation (Rugby VLF transmitted, Great Britain, was employed), and the profile azimuth was 60°. The ribbon-like band of the pyrite-polymetallic ore is located at a depth of about 60–80 m in the sandy argillaceous series of the Upper Aalenian. The average resistivity of the host medium is in the range of 700–900 Ωm; the resistivity of the massive ore body consists of a fraction of 1 Ωm. The calculated skin depth for the host medium is about 110 m. Since a small signal from a relatively deep-seated anomalous object is superimposed on a topographic anomaly, it was difficult to detect the anomaly from the hidden ore body.

Two correlations were made for the southwestern and the northeastern slopes $r_1 = 0.988$ and $r_2 = 0.85$, respectively. The approximation equation for the

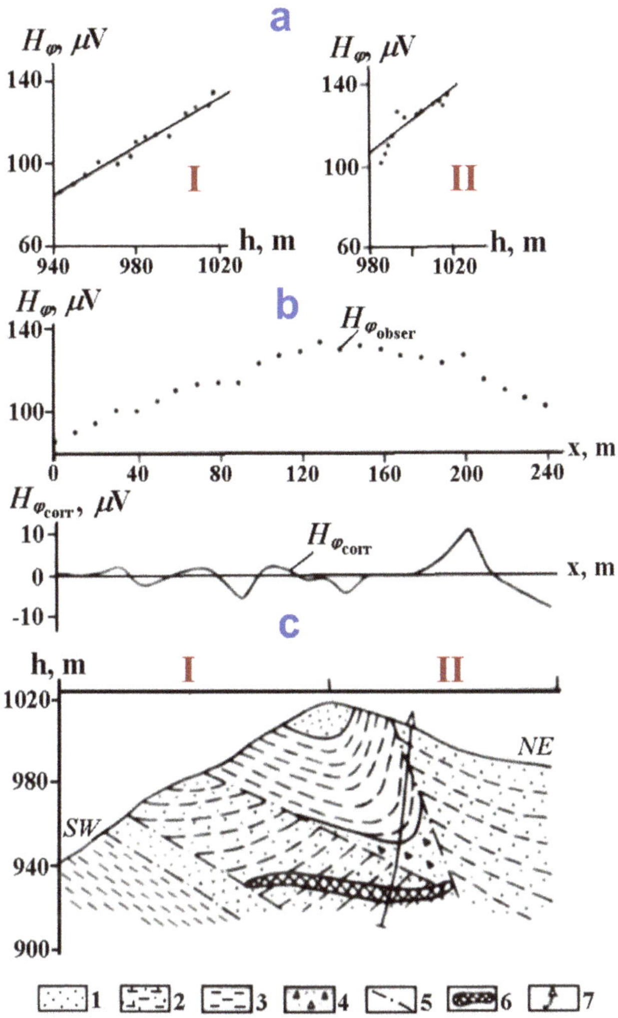

Figure 2.68. Application of the correlation technique for reducing the terrain relief effect on profile crossing the Katekh pyrite-polymetallic deposit (northern Azerbaijan). (a) Correlations, (b) graphs of the observed and corrected H_φ values, (c) geological section (1) fine and medium-grained sandstone, (2) alternating sandstone and clay shale strata, (3) clay shales, (4) brecciated zone, (5) fractures, (6) orebody, (7) prospecting borehole (reprinted from Eppelbaum (1991), copyright 1991, with permission from Elsevier).

southwestern slope is $H_{\varphi\ \text{appr}} = 84 + (0.6 \pm 0.9)h$, and for the northeastern slope—$H_{\varphi\ \text{appr}} = 105 + (0.79 \pm 0.49)h$. After removing the relief effect, a positive anomaly can be clearly seen on the northeastern slope in the corrected graph H_φ.

This anomaly may be due to the edge effect of a deep ore body, whereas the minimum on the southwestern slope may correspond to the opposite edge of the subhorizontal tabular ore body (Eppelbaum 1991).

VLF: a brief review of available methods of quantitative analysis

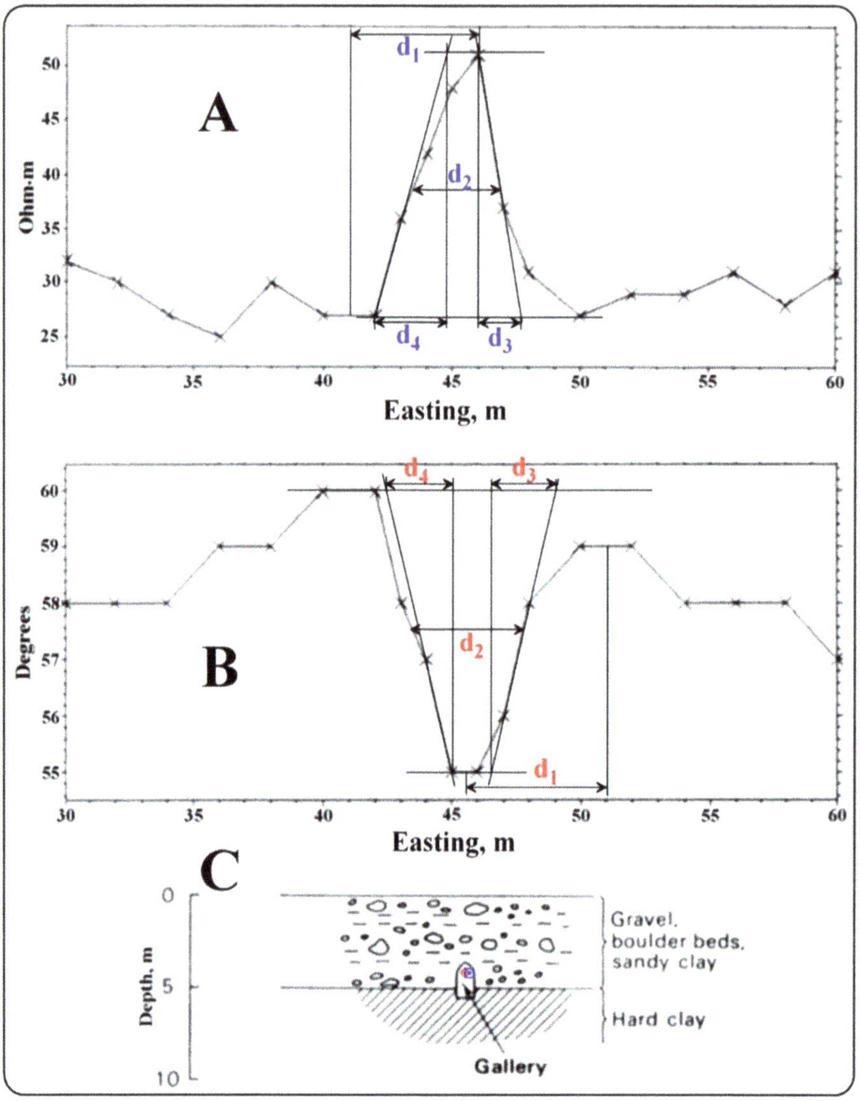

Figure 2.69. Quantitative interpretation of VLF-R apparent resistivity and phase profiles, NAA transmitter (Gutler ME, USA), 24 kHz (in the modern Reference Handbook the NAA transmitter is presented with the frequency of 25.3 kHz). A—apparent resistivity, B—phase anomaly, C—geological section. Observed curves and geological sections are from Ogilvy *et al* (1991), copyright 1991 with permission from Elsevier. Re-interpretation after Eppelbaum (2007). Symbols ▫ and ☒ designate the determined position of HCC center for the phase anomaly and apparent resistivity, respectively.

The performed overview indicates that there are practically no reliable techniques of quantitative interpretation in the VLF method.

The methods suggested by Tarkhov (1961) are very approximate. The procedure developed by Fraser (1969) is useful primarily only for revealing hidden anomalous objects but not for obtaining some quantitative parameters.

The techniques presented by Gordeyev and Sedelnikov (1974), based on calculating the VLF anomaly extrema ratio, require reliable knowledge of the zero line (implying knowledge of a normal field). The wrong choice of the zero line entails substantial errors in determining the anomalous object's quantitative parameters. Moreover, these techniques are not intended for interpretation in conditions of rugged terrain relief and oblique polarization.

Dmitriyev *et al* (1977) solved the direct problem for several electromagnetic methods (including the VLF method). These authors offer a numerical analysis of the EM anomalies from bed-like bodies as a function of their electric properties, dimensions, and position relating to the Earth's surface and the observation profile. However, computational difficulties restrict the range of models to simple ones.

Zhdanov and Keller (1994) noted that for the mathematical modeling of the electromagnetic field the most effective method is the finite differences. However, the practicality of its application strongly depends on the size of the design area and the desired accuracy of the computations.

Khesin *et al* (1996) pointed out that some methodologies developed for magnetotelluric sounding (Kaufman and Keller 1981, Zhdanov and Keller 1994) could be employed in the VLF method. Basokur and Candansayar (2003) suggested using the same approach (however, these authors concluded that this interpretation could only be qualitative).

Gordeyev and Sedelnikov (1974) and Gordeyev *et al* (1981) carried out the physical simulation of VLF curves from various models of a vertical inclined thin bed. The accuracy of the physical modeling and that of the numerical computations (for simple models) is at present approximately the same and is roughly 8%–10%.

Numerous other works, in many respects similar to those mentioned above, have dealt with anomaly interpretation in the VLF method. For instance, Karous and Hjelt (1979) and Olsson (1980) presented the calculation of anomalies from vertical and horizontal components of the VLF magnetic field using a model of an inclined thin bed with different dip angles. The depth of the upper edge of the bed varies as well. Olsson (1980), Tesmull and Crossley (1981), and Poddar (1982) computed components of the VLF fields for different types of geoelectric sections. Miecznik (1986) calculated the magnetic components of the VLF field caused by a conductive cylinder placed in a homogeneous medium (the cases of both **E**-polarization and **H**-polarization are discussed).

The characteristic VLF diagrams developed by Sinha (1998) can provide useful information about vertical and inclined conductive beds occurring in simple geological–geophysical conditions. The VLF first derivative method proposed by Djeddi *et al* (1998) is based on perspective (the authors suggest eliminating the topographic effect in the interpretation process).

Olsson (1983) solved the direct problem for an ideal conductive half-plane with different dip angles, when the half-plane is covered by variable-depth loose deposits. He proposed techniques for a simplified interpretation of VLF anomalies based on the employment of the extremum points in the anomaly plot. Pedersen and Oskooi (2004) provided synthetic models enabling a better understanding of the resolving power of the VLF data. The same authors propose applying a tensor variant of VLF measurements. Monteiro *et al* (2006) described a quantitative interpretation of VLF-EM data using 2D models (based on a 2D regularized modeling of VLF-EM data) that yields practical results about the average resistivity in the survey area. Kaikkonen and Sharma (2001) performed 2D modeling of noise-free and noisy media with single-body and complex geological models. The authors highlight the need to use reliable *a priori* geological–geophysical information before running the process.

VLF field: possible application of procedures developed for potential field analysis

The theory of alternating electromagnetic field used the following Maxwell's equations:

$$\mathrm{rot}\mathbf{H} = \mathbf{j} + \frac{\partial \mathbf{D}}{\partial t}, \tag{2.20}$$

$$\mathrm{rot}\mathbf{E} = \mathbf{j} + \frac{\partial \mathbf{D}}{\partial t}, \tag{2.21}$$

$$\mathrm{div}\mathbf{B} = 0, \tag{2.22}$$

$$\mathrm{div}\mathbf{D} = q. \tag{2.23}$$

Here **H** and **B** are the magnetic field vectors (**H** stands for intensity and **B** for induction); **E** and **D** are the electric field vectors (**E** stands for intensity and **D**—for induction); **j** is the conduction current density; q is the charge density.

Equation (2.20) can be written as (Alpin *et al* 1985):

$$\mathrm{rot}\ \mathbf{H} = \mathbf{C} = \mathbf{j}_{\mathrm{cond}} + \mathrm{j}_{\mathrm{disp}}, \tag{2.24}$$

where **C** is the total current density; $\mathbf{j}_{\mathrm{cond}}$ and $\mathbf{j}_{\mathrm{disp}}$ are the conduction and displacement current densities, respectively.

It has been found that at the frequencies of 10–30 kHz, $\mathbf{j}_{\mathrm{disp}} \approx 0$ (Dmitriyev 1982). Therefore,

$$\mathrm{rot}\ \mathbf{H} = \mathbf{j}_{\mathrm{cond}} \tag{2.25}$$

$$\mathbf{H} = \nu \mathrm{E}, \tag{2.26}$$

where ν is the electric conductivity.

An EM field can be considered quasi-stationary if it satisfies three conditions of quasi-stationarity. Alpin *et al* (1985) framed these conditions as applied to geophysics. These requirements are as follows:

(A) Slow temporal field variations;
(B) Closed currents;
(C) To avoid an appreciable delay in the magnetic field variations, the region including magnetic field generating currents and observation points must not be too large.

Let us consider these conditions in more detail.

(A) The study of temporal variations of the VLF fields has shown that sharp amplitude changes are not characteristic of these fields. Nevertheless, variations can be eliminated by the special procedure in fieldwork (see subsection 'Elimination of temporary VLF variations').

(B) This condition follows from equation (2.26), since div rot **H** = 0 and, therefore, div **j** = 0. Almost closed currents do not interrupt this condition.

(C) This condition can be expressed in the following way: the length of an electromagnetic wave in the ground should essentially exceed the length of the targets under study. This condition, as a rule, is fulfilled in the prospecting of mineral deposits, geological mapping, and object analysis in engineering and military geophysics.

The fundamental solution of the Laplace equation in the 2D case has the following form:

$$g_0(M, P) = \frac{1}{2\pi} \ln\left(\frac{1}{R_{MP}}\right), \tag{2.27}$$

where R_{MP} is the distance between points M and P (M is the observation point, and P is the point of the body).

The fundamental solution of the Helmholtz wave equation is Hankel function of the first kind with zero-order$H_0^{(1)}$:

$$g_0(M, P) = \frac{i}{4} H_0^{(1)}(kR_{MP}), \tag{2.28}$$

where k is the wavenumber, and i is the imaginary unit.

The function (2.28) is the analog of the function (2.27) for the Laplace equation.

Using the above correspondence, Hänl *et al* (1961) and Dmitriyev (1982) proved the relationship between Green's solutions of the Laplace and Helmholtz equations. This enabled them to conclude that the results obtained in the potential theory could be extended to the Helmholtz equation (EM field).

Dmitriyev *et al* (1977) indicated that the set of the singular points resulting from the VLF field anomalies from the anomalous hidden bed is analogous to the well-known behavior of singular points in the potential field anomalies. Zhdanov (1988) theoretically proved the application of a set of Cauchy-type integral analogs to the problems of electromagnetism. He proposed that the methods developed in the potential theory for analytical continuation, separation, and quantitative interpretation could be applied to the quasi-stationary electromagnetic anomalies. The quasi-stationarity of the electromagnetic field follows from condition (1).

An overview of publications (Gordeyev *et al* 1981, Poddar 1982, Sharma and Kaikkonen 1998, Beamish 2000, Oskooi and Pedersen 2006) indicates that a conductive thin bed (TB) is the most common model applied for the VLF technique. The analytical expressions of magnetic and VLF fields for this model (the 2D case) are presented in table 2.5.

Table 2.5. Comparison of analytical expressions for TB in magnetic prospecting and the VLF method.

Field	Analytical expression	
Magnetic	$Z_v = 2I2b\frac{z}{x^2 + z^2}$	$X_v = 2I2b\frac{x}{x^2 + z^2}$
VLF	$H_x = kH_0\frac{z}{x^2 + z^2}$	$H_z = kH_0\frac{x}{x^2 + z^2}$

In this table Z_v and X_v are the vertical and horizontal magnetic field components for vertical magnetization, respectively; H_x and H_z are the horizontal and vertical VLF magnetic field components, respectively; H_0 is the VLF primary field intensity; k is the coefficient reflecting geometry and conductivity of the anomalous TB.

Z_v is proportional to H_x, whereas X_v is proportional to H_z. Similar results were obtained for an HCC model (Eppelbaum and Khesin 1992).

Thus, the anomalies of magnetic fields of VLF transmitters can be interpreted by special methods developed in potential field theory (in particular, in magnetic prospecting, for conditions of rugged terrain relief, oblique magnetization (polarization), and an unknown level of the normal field). The **E**-polarization vector in the VLF field, in the first approximation, is the analog of the magnetization vector (Eppelbaum and Khesin 1992).

In this framework, the special techniques presented in (Eppelbaum *et al* 2000, 2001, Eppelbaum 2011a, 2019a, 2020, Eppelbaum and Mishne 2011) are of vital practical importance. In the VLF method can be applied the quantitative analysis methodologies developed in magnetic prospecting for models of inclined TB and horizontal circular cylinders, wherein the total horizontal component of the VLF magnetic field $H_\phi = \sqrt{H_x^2 + H_y^2}$ can be interpreted as the H_x component since the H_y component contribution is usually relatively small.

The essential distinctions reflecting the quasi-potential nature of the VLF EM field are as follows. In magnetic prospecting the induced magnetization vector of the anomalous target is approximately parallel to the geomagnetic field vector, irrespective of the bed dip direction, where the magnetic susceptibility does not exceed 0.1 *SI* unit. In EM prospecting by the VLF method under **E**-polarization for highly conductive objects, the equivalent polarization vector generating H_x and H_z anomalies approaches the body axis, with a slight deviation toward the vertical for gently sloping bodies (Gordeyev *et al* 1981, Poikonen and Suppala 1989). This makes it possible to employ the generalized angle θ derived from these interpretation methods and to estimate the angle of bed dip by the following empirical formulas (Eppelbaum and Khesin 1992):

for the H_x anomaly

$$\alpha = 3\theta + 90^\circ, \tag{2.29}$$

for the H_z anomaly

$$\alpha = 3\theta - 180^\circ, \tag{2.30}$$

where α is the anomalous bed inclination.

For the case of observations on a sloping relief, the angle α is calculated by the following way:

for the H_x anomaly

$$\alpha = 3(\theta - \omega_o) + 90^\circ \tag{2.31}$$

for the H_z anomaly

$$\alpha = 3(\theta - \omega_o) - 180^\circ, \tag{2.32}$$

(ω_o is derived in equation (2.26)).

Sometimes the position of the upper edge could be slightly shifted from the real upper edge downward along with the anomalous target. This is because the linear EM currents are concentrated in the upper portion of the conductive target. This portion may be situated below this target's upper edge.

Let us consider the following example of the VLF method application for hidden cave localization. A detailed VLF survey was carried out near Alcala de Henares, a small city 20 km east of Madrid, Spain (Ogilvy *et al* 1991). The main aim of this survey was to reveal air-filled underground galleries of relatively small sizes occurring in clastic sediments. One of the observed VLF profiles is shown in figure 2.70. The underground gallery was approximated by the HCC model inscribed to the upper part of this object. Results of quantitative examination of

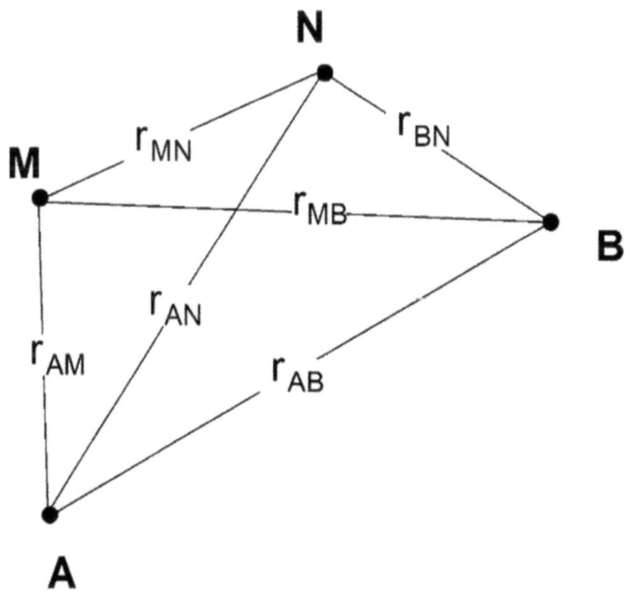

Figure 2.70. Four-point configuration for measurement of ρ_k. AB—current electrodes, MN—potential electrodes, r are distances between corresponding points.

the phase anomaly and apparent resistivity curves indicate that the obtained results have enough accuracy.

Development of the ROV technology and improvement of the VLF instrumentation make it possible to extensively monitor the underground karst phenomena by the use of ROV low-altitude surveys (Pedersen and Oskooi 2004, Pedersen *et al* 2009, Eppelbaum and Mishne 2011, Eppelbaum 2019c).

Application of different modifications of VLF surveys for underground cave localization can be found in Ogilvy *et al* (1991), Guérin and Benderitter (1995), Danişman, (2004), Bosch and Müller (2005), Eppelbaum (2007), Hamdan *et al* (2010), Sundararajan *et al* (2015), Singh *et al* (2020).

2.2.1.9 DC resistivity methods

2.2.1.9.1 Principles of the DC resistivity method—general consideration

Terms.

Apparent resistivity. In most of the resistivity methods, the normal electrical field is taken into account by the introduction of effective parameter ρ_k termed apparent specific resistivity. Let us presume that there are points A and B on the Earth's surface (figure 2.70). Grounding electrodes A and B are termed the current electrodes. By placing two electrodes M and N and measuring the difference of potentials between them, we can investigate the electrical field induced by the AB electrodes. These electrodes are termed potential ones.

Comment. In the scientific literature electrodes, A and B are denoted often as C1 and C2, respectively (e.g., current electrodes), whereas, potential electrodes M and N are denoted as P1 and P2, respectively (e.g., potential electrodes).

It can be shown that apparent specific resistivity

$$\rho_k = K\frac{\Delta U}{I}, \tag{2.33}$$

where $K=2\pi/\left(\frac{1}{r_{AM}} - \frac{1}{r_{BM}} - \frac{1}{r_{AN}} + \frac{1}{r_{BN}}\right)$ is the geometrical coefficient, ΔU is the difference of potentials between the electrodes MN, and I is the current introduced into the Earth via the electrodes AB (Yakubovski and Renard 1991).

Apparent resistivity depends on the location of the current and potential electrodes and the geoelectric section. If in the homogeneous half-space with a specific resistivity ρ_1 there is a high resistive layer (for instance, coal layer within limestone), then the distribution of current entering into the Earth via electrodes A and B will differ from that for a homogeneous medium. The current will be pushed up to the subsurface increasing the difference of potentials between potential electrodes and, as a result, ρ_k. Thus, $\rho_k > \rho_1$. If there is a conductive body in the geoelectrical section (for instance, ore deposit), then the current will be concentrated in the conductive body, and the latter will decrease in the vicinity of potential electrodes. It means, $\rho_k < \rho_1$. In the case of a homogeneous medium, $\rho_k = \rho_1$. The use of the method of resistivity for resolving geological problems is based on the dependence of the apparent resistivity and the character of the geoelectric section.

True resistivity

If electrical current propagates along the rod of length L and cross-sectional area A, then the resistance of this rod (R) is proportional to its length (L) and inversely proportional to cross-sectional area (A): This can be written as $R = \rho L/A$, where ρ is the true specific resistivity. On the other hand, resistance (R) from Ohm's law is given as

$$R = V/I, \tag{2.34}$$

where V and I are the potential difference across a resistor and the current passing through it, respectively. Then, true specific resistivity is defined by

$$\rho = (V/I)(A/L), \quad (\Omega\text{m}^2/\text{m}) \text{ or } \Omega\text{m} \tag{2.35}$$

This can be written alternatively as $\rho = \mathbf{E}/\mathbf{J}$ (Ωm), where **E** is the strength of the electric field (V m^{-1}) and **J** is the current density (A m^{-2}).

True resistivity is a property of the material and it does not depend on the location of electrodes concerning geological units.

The rocks and soils are characterized by the range of true resistivity dependently on water saturation, clay content, porosity, fracture density (rocks). Generally, soil resistivities range from 5 to 1000 Ωm (apart from the high resistive gravel). Bedrock resistivities range from 200 to 4000 Ωm (Ward 1990). Apparent resistivity depends on the location of electrodes concerning the geological units, the configuration of electrodes, conditions introducing the current into the Earth, and resistivity of the uppermost layer. That is why it can be said that the goal of geoelectrical prospecting is to define the distribution of true resistivity by measuring the apparent resistivity of the subsurface. Knowledge of the true resistivity can enable us to identify lithology of layers, weathering, water saturation, and detect cavities and signs of karstification.

2.2.1.9.2 Configuration of electrodes

To measure apparent resistivity, the different electrode installations are applied. These installations differ by the number of electrodes and their mutual arrangement. In dependence on a number of groundings, arrays can be two-electrode, three-electrode, four-electrode, etc.

The most useable among the different electrode installations are linear ones. In these installations, MN electrodes are placed usually in the middle third part of the AB segment, where the normal field is close to the homogeneous one.

In a special case of a straight line, the four-point installation is symmetrical. The four-point installation is called *Schlumberger* configuration (figure 2.71(a)). In this installation the potential electrodes, MN, are placed symmetrically to the center of AB such as $r_{\text{MN}} \leqslant r_{\text{AB}}$ (in what follows, we agree to denote these distances simply as MN and AB). Add here that the electrodes A and B are moved from MN as shown by arrows in figure 2.71(a). At that, the difference in potentials ΔU between MN decreases. When ΔU becomes comparable to the noise level separation, MN is increased and ΔU is increased too, allowing one to continue measurements. The increase of AB causes the penetration of the current to be deeper into the subsurface.

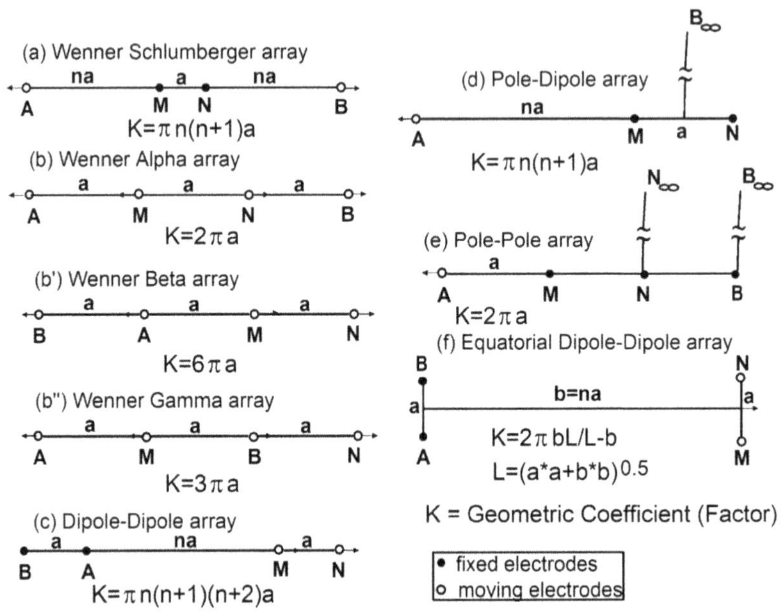

Figure 2.71. Different electrode configurations are most used for the DC resistivity method. Arrows show the direction of the electrode movement during measurement.

Thus, investigation depth is increased. The geometrical factor for the Schlumberger configuration is $K = \pi n(n + 1)a$, where n is (AM/MN = NB/MN).

A particular variant of the Schlumberger installation is the *Wenner* configuration (figure 2.71(b)) in which the distances between the electrodes are equal, e.g., AM = MN = NB = a. In this case, an increase in AB separation causes an increase in MN allowing ΔU constantly to increase and be at a measurable level. The geometrical factor for the Wenner configuration is $K = 2\pi a$.

Linear configuration *dipole–dipole* is the modification of the four-electrode installation. It is characterized by small separations of AB and MN electrodes that should be significantly less than the distance between their centers (figure 2.71(c)). The couple of electrodes MN is moving from AB. The geometrical factor $K = \pi n(n + 1)(n + 2)a$.

The three-point installation is obtained if one of the supply electrodes of the four-point installation is brought to infinity (that is, 10–15 AM distances). Such an installation is called a *pole–dipole* (figure 2.71(d)) or AMNB_∞. In this case, $K = \pi n(n + 1)a$. In this installation, the measurement point is AM/2.

Two-point installation termed *pole–pole* configuration or AM$N_\infty B_\infty$ (figure 2.71(e)) differs from three-electrode by two electrodes brought to infinity N and B. So, measuring ΔU between M and N, we, in reality, measure the potential of the electrode M. Geometrical factor $K = 2\pi a$.

Finally, there is *an equatorial dipole–dipole* configuration in which angle θ between dipoles AB, MN, and the axis of installation is equal to $\pi/2$ (figure 2.71(f)).

$$K = 2\pi bL/L - b, \text{ where } b = na,\ L = (a * a + b * b)^{0.5}$$

All electrode installations are characterized by the different sensitivity, resolutions, penetration depth, and used in dependence on the local conditions. Whitely (1973) has analyzed ~25 electrode configurations and characterized the advantages and disadvantages of each. Reynolds (2011) affirms that the choice of the array is often a function of the user's previous familiarity with the array type, availability of cabling and data acquisition software, and data processing and inversion software, as well as site-specific factors. Barker (1992) notes that the pole–dipole or dipole–dipole arrays which are normally employed are very sensitive to near-surface lateral changes in resistivity and the effects of these can be carried down into the back projection to cause erroneous results. Zhou *et al* (2002) compared the dipole–dipole, Schlumberger, and Wenner arrays using forward modeling. The authors clarified that all three configurations produced markedly different anomaly shapes for a conceptual model of the development of a cover-collapse sinkhole. In the real field, the measuring image from the dipole–dipole array was better than those from Wenner and Schlumberger for displaying the sinkhole collapse area. As denoted by the authors dipole–dipole array should be a more effective and less costly configuration.

See for further reading: Ward (1990), Kirsch (2006), Reynolds (2011).

2.2.1.9.3 The techniques of electrical prospecting

Resistivity measurements are carried out in two modes sounding and profiling. The goal of sounding is the determination of the depth and thickness of geological layers. The profiling is aimed to locate lateral changes in geologic conditions. At sounding measurements the spacing between the current electrodes AB is increased step by step and at that the depth of investigation (penetration of current into subsurface) increases.

Soundings generally apply to defining geologic layers where the geology is laterally homogeneous and layers are flat or gently dipping. At profiling (the profile measurements) electrode spacing is fixed. Profiling is used to locate and map areas of significant lateral variations in resistivity at a given depth (for example, clay areas).

Vertical electrical sounding (VES)

In its simplest concept, this method may be described as four electrodes laid in a line and inserted a few centimeters into the ground (figure 2.72(a)). An electric current is sent through one pair of electrodes (current or transmitting electrodes, A, and B); the potential difference established in the Earth by this current is measured using a second pair of electrodes (potential or measuring electrodes, M and N). While studying the variation of resistivity with depth, the spacing between the current and potential electrodes is gradually increased. The larger the spacing, the more pronounced the effect of material at depth. Usually, half-separation AB/2 increases in geometrical progression. It is collected equal ~ to 1.5; 2.2; 3; 4.5; 6; 8; 10; 15; 22; 30; 45; 60; 100; 150; 220; 300; 450; 600; 800; 1000 m. At AB/2 ranges 1.5–10; 15–100 and 150–1000 m, MN is collected 1, 10, and 100 m, respectively (Khmelevskoy 1980, Mooney 1984b).

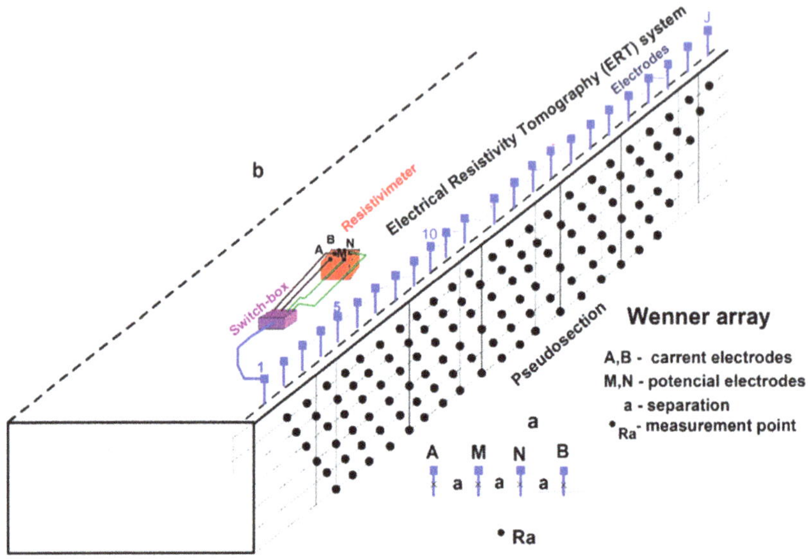

Figure 2.72. Scheme of geoelectrical data acquisition. (a) Vertical electrical sounding (VES) survey (Wenner electrode configuration). (b) Electrical resistivity tomography (ERT) (reprinted from Ezersky (2002) with permission of *Israel Journal of Earth Sciences*).

Since the ground is not generally homogeneous, equations (2.1) and (2.3) may be used to define the apparent resistivity of the Earth. The behavior of the apparent resistivity with separation AB roughly reflects the behavior of the true resistivity with a depth. To obtain a quantitative distribution of the resistivity with depth, the so-called inversion of the measured data is required.

VES data interpretation aims to determine the true resistivities and thicknesses of the successive strata below the different stations utilizing the measured field curves. The interpretation of VES curves was earlier usually made by curve-matching procedures in which albums of theoretical curves (Orellana and Mooney 1966) were used in conjunction with the auxiliary-point method of partial curve matching. This method requires a great deal of practice, and at times it can be frustrating to the inexperienced interpreter. Nowadays, there are a variety of PC-based programs, which allow computation for an arbitrary number of layers (IX1D by Interpex, IPI-1D by Moscow State University, 1Res, by Loke, etc). A comprehensive review of VES data processing software was performed in Badrinarayanan and Thiyagarajan (2018). Automatic inversion programs without any assumptions of the layering exist as well as programs that require the handling of the interpreter. There is a fundamental rule in data processing: to use all additional data available to generate an initial model with the known geological parameters (e.g., a standard thickness of stratigraphical units, depth to the groundwater table), available borehole data, or results of other geophysical measurements (Khmelevskoy and Shevnin 1988, Kirsch 2006).

Zohdy (1989) has proposed a fast iterative method for the automatic interpretation of Schlumberger-type VES curves which are obtained over the horizontally stratified, laterally homogeneous media. The initial model (the number of layers, their resistivity, and thickness) is automatically fixed as equal to the number of points on the digitized curve. Least-squares optimization is used in which an initial model is adjusted successively until the difference between the observed and model pseudo-sections is reduced to a minimum (Barker 1992).

The principle of the apparent resistivity curve interpretation is presented in figure 2.73.

Practically, the true resistivity-depth curve is unknown. We have only an 'observed' (measured) curve of apparent resistivity. The first stage is depth determination. The second stage is the resistivity arrangement.

Stage 1. Depth determination

To determine the true depth, it is necessary to shift AB/2 to the left to bring the assumed and unknown layering to the nearest 'in-phase' condition, and do the following iterative calculation:

(a) Let us replace the apparent resistivity ('observed') curve with an artificially layered ('digitized') curve, which assumes that the digitized electrode spacings are equal to the depths, at that the apparent resistivities are equal to the true resistivities at those depths (figure 2.73(a)).
(b) Compute a theoretical sounding curve by convolution for this multilayer model. The result is 'calculated' apparent resistivity curve (denoted by points in figure 2.73(b)).

Figures 2.73(c) and (d) show iterative steps of the fitting 'calculated' curve to the 'observed' one varying depth. When our goal is reached and the 'calculated' curve is matched 'in-phase' to the 'observed' one (figure 2.73(e)) it means that we have reached true depth.

The criterion of closeness between 'observed' and 'calculated' curves is the RMSE calculated by the equation:

$$\mathrm{RMS\%} = \sqrt{\frac{\sum_{J+1}^{N}\left(\frac{\bar{\rho}_{oj}-\bar{\rho}_{ej}}{\bar{\rho}_{oj}}\right)^2}{N}} \times 100, \tag{2.36}$$

where $\bar{\rho}_{0j}$ is the jth 'observed' apparent resistivity,

$\bar{\rho}_{ej}$ is the jth 'calculated' apparent resistivity, and

N is the number of digitized apparent resistivity points (with $j = 1$ to N).

Stage 2. Resistivity determination

At the next stage having 'calculated' the theoretical sounding curve using shifted depths and apparent resistivities, we must adjust the amplitudes of those resistivities to obtain a better fit between the 'observed' and 'calculated' sounding curves. This is done iteratively as follows. At each digitized electrode spacing on the 'observed' and 'calculated' curves, if the computed apparent resistivity at the jth spacing is less (or greater) than the corresponding 'observed' apparent resistivity, the corresponding true resistivity of the jth layer should be increased (or decreased) so that the 'calculated' apparent resistivity will rise (or fall) and approach the 'observed' resistivity (figure 2.73(d)). The amplitude of a layer resistivity is iteratively adjusted as:

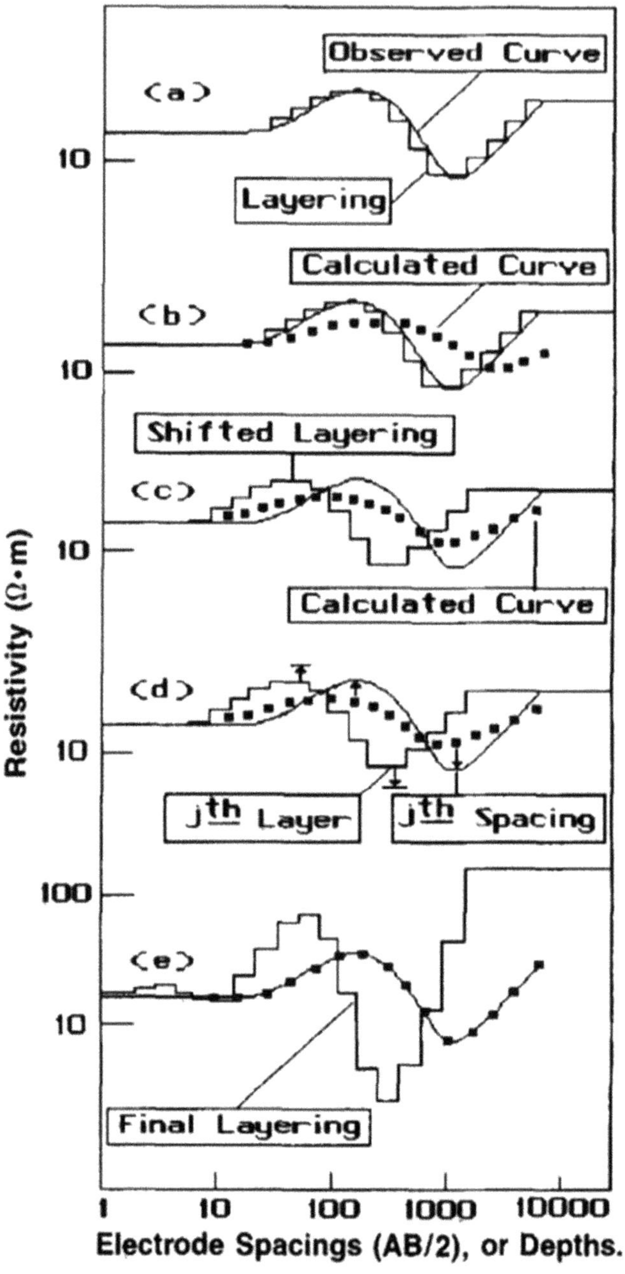

Figure 2.73. Graphs showing an iterative process of interpreting a sounding curve of unknown layering: (a) layering derived from a sounding curve, (b) calculated sounding curve, corresponding to (a), (c) optimally shifted layering and corresponding sounding curve for minimum RMS percent between the calculated and observed curve, (d) adjustment of each layer resistivity using logarithmic differences between calculated and observed sounding curves, and (e) final layering (the true resistivity digitized curve) after several iterations, with a calculated curve, matched observed curve (reprinted from Zohdy (1989), with the permission from SEG).

$$\rho_{i+1}(j) = \rho_i(j)\frac{\overline{\rho}_0(j)}{\overline{\rho}_{ci}(j)}, \tag{2.37}$$

where i is the number of iterations already carried out, j is the number of the layer (or electrode spacing), $\rho_i(j)$ is the resistivity of layer j after i iterations, $\overline{\rho}_0(j)$ is the 'observed' apparent resistivity at the jth spacing on the sounding curve, and $\overline{\rho}_{ci}(j)$ is the 'calculated' apparent resistivity at the jth spacing for the ith iteration.

The ratio of 'observed' to 'calculated' apparent resistivity is multiplied by the corresponding layer resistivity to adjust it to a higher or lower value. It is similar to the fitting of depths in iterative form. The iterative process is terminated when one of the following conditions is met: a prescribed minimum RMS percent (less than 2% for field data) is obtained; a slowness in further improvement in fit is detected (less than 5% reduction in successive RMS percent); a maximum number (30) of iterations is reached, or the RMS percent increases instead of decreases. The average number of iterations is about 10.

Electrical resistivity tomography (ERT)

In a heterogeneous medium, the electrical resistivity tomography (ERT) termed earlier continuous vertical electrical sounding (CVES) or 2D imaging method is used (Barker 1992). This method is also termed 'pseudo-section inversion'. Generally, ERT is based on the VES method. Several electrodes are laid out at similar intervals along the line (figure 2.72(b)), and the VES procedure is repeated successively on each pair of electrodes.

ERT data interpretation. The technique for transforming a pseudo-section from apparent resistivity to true resistivity (inversion) is similar to the above-mentioned Zohdy method, although a 2D pseudo-section modeling program must be used instead of the resistivity curve modeling algorithm. Firstly a finite difference network (Dey and Morrison 1979) is designed so that a single element or block of elements coincides with each point on the pseudo-section. The apparent resistivity of this point is used as the initial resistivity value of the associated element.

This initial model is then used to generate a theoretical pseudo-section data set.

The principle is based on a comparison of the observed and calculated pseudo-sections (figure 2.74).

Data processing and presentation comprised of three stages:

- The apparent resistivity pseudo-section is plotted (based on the field measurements) (figure 2.74(a));
- Apparent resistivity pseudo-section is calculated (figure 2.74(b));
- Inversion and plotting of the true resistivity sections using 2D smoothness-constrained inversion (figure 2.74(c)).

Finally, the apparent resistivity is calculated using the collected data.

Differences between the model and observed pseudo-section apparent resistivities are used to modify the true resistivity of each network element according to equation (2.37). The process is repeated until the observed data are modeled to within a predefined error limit or until the RMS difference reaches a minimum (see equation (2.37)) (Barker 1992).

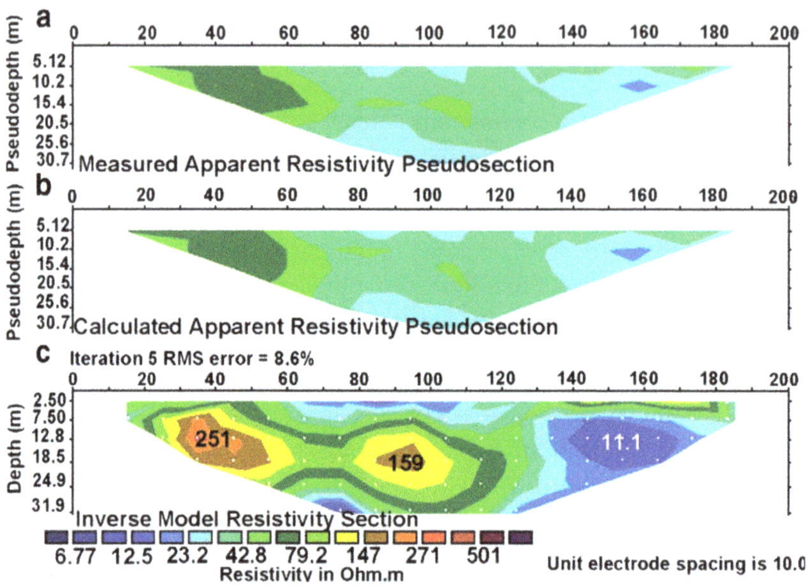

Figure 2.74. Principle of the ERT data inversion (a) measured apparent resistivity pseudo-section; (b) calculated apparent resistivity pseudo-section; (c) true resistivity inverse section.

One of the problems is how best to relate electrode spacing to depth to bring the computed apparent resistivities 'in-phase' with the measured data. For resistivity sounding, Zohdy (1989) resolved the problem by applying a range of shift factors and selecting that which gave the smallest discrepancy between the measured sounding curve and the initial model curve. Barker (1989) recognized that the shift applied by Zohdy was directly related to the depth of investigation and that for most resistivity distributions fast convergence could be obtained by applying a fixed shift factor equal to the median depth of investigation as defined by Edwards (1977) for the general case of the homogeneous geological section. For the Wenner array, this factor is 0.5 (i.e., layer depth = $0.5a$, where a is the spacing between adjacent electrodes). Although this value is correct only for the homogeneous section, the main effect of applying it to heterogeneous media is merely a reduction in the convergence rate.

The above-mentioned principles of the true resistivity section generation are released by the software Res2Dinv that divides the subsurface into several rectangular blocks, all at the start, with a measured apparent resistivity. The calculation is aimed to determine the resistivity of the rectangular blocks that will produce an apparent resistivity pseudo-section that agrees with the measured resistivity section. To reduce the difference between the calculated and measured apparent resistivity values the optimization method is applied that tries to 'adjust' the resistivity of the model blocks. A measure of this difference is given by the RMS error. Note that the lowest possible RMS is not always an absolute criterion. Sometimes, the model with the lowest RMS has shown unrealistic variations in the model resistivity values Also, the model with the lowest RMS error might not always be the 'best' model from a geological point of view. In general, the most prudent approach is to choose the model at the iteration after which RMS error does not change significantly (that takes place as a rule, between third and fifth iterations).

Inversion calculations are carried out using two new programs for ERT processing. RES2DINV Ver. 3.54 is a computer program for rapid 2D resistivity inversion using the least-squares method (Loke 2004). The program automatically determines a 2D resistivity model for the subsurface using data obtained from the electrical imaging surveys (Griffiths and Barker 1993). RES2DMOD is a 2D forward modeling program that calculates the apparent resistivity pseudo-section for a user-defined 2D subsurface model. This is intended for use as a utility program for the inversion program *Res2Dinv*. It is mainly intended to assist the user in selecting the appropriate array for different geological situations or surveys. The ERT data can be processed in 2D and 3D inversion and modeling mode using Geotomo Software.

Resistivity measurements are labor-intensive since the method requires ground contact. It takes time. To improve ground contact metal electrodes are derived into the ground and deployed connecting cables. Sometimes (in gravel, for instance) it requires adding water and/or bentonite into the electrode location to improve the contact with geological media. Lateral resolution depends on electrode spacing, as well as the spacing between station measurements. Processing of resistivity sounding data using computer software allows constructing the model with a number of layers, more than 3–4 usual for manual data processing. It can be said that the results of measurements are sensitive to interference from nearby metal pipes, cables, or fences, but less so than EM measurements.

The line length must extend at three to five times and sometimes more, the depth of interest, which results in long electrode spreads and cables. That is a disadvantage of the resistivity sounding in the laterally heterogeneous areas. Finding a space for long measurement lines can sometimes be a problem. Obtaining a good grounding can also be complicated in areas with high resistivity soils. In this case, the electrodes are placed into the bentonite.

2.2.1.9.4 An application of the DC resistivity method for karst detection

The application of the DC resistivity method (ERT) for water-filled cavity detection is related to the secondary choice or alternative method (ASTM 6429 2006) but can be used also as a method of the primary choice for air-filled cavities (Hoover 2003, Chalikakis *et al* 2011).

The air-filled cavity is characterized by an infinite resistivity and, therefore, it will increase the apparent bedrock resistivity. Contrarily, the mud-filled cavity has resistivity lower than host rocks. Therefore, it will decrease the apparent bedrock resistivity. That is why changes or anomalous apparent resistivity values are usually the features of interest.

The application of the method is based on the detection of a low resistive zone (water-filled cavities) at the background of high resistivity zones (host rocks) and contrarily. It is clear that the issue of resolution arises, e.g., (1) there should be a difference in resistivity of the above-mentioned zones, and (2) the size of the heterogeneity that must be detected should correspond to the scale of the method or more of that.

Werner (1984), notes despite theoretical considerations the standard Wenner array (collinear, semi-symmetric) cannot detect a cavity buried deeper than its radius (Militzer *et al* 1979). However, in practice, the weathered (destructed) rock around a cavity significantly enhances its detectability. The same reason is denoted by Benson and Yuhr (1993). Therefore, the potential of sinkhole hazards in many types of terrains can be delineated by the use of a combination of electrical resistivity methods. These determinations are usually less costly and more informative of such hazards than the conventional boring program. Variation of the methods exist and could be applied in various combinations to provide redundancy to improve the accuracy of the information derived from them (Dunscomb and Rehwoldt 1999). The 2D resistivity method is shown to have very good resolving abilities in karst terrain to image geological features such as pinnacled bedrock surfaces, 'over-hanging' rock ledges, fracture zones, and voids within the rock mass and the soil overburden. Performing the field survey with 28 electrodes can take approximately 2 h. The method allows data processing in the field and modeling can take 15–30 min. The method can be applied to detect fracture zones, the fault features, weathered rocks, voids, and pinnacles, identify large bedrock cavities below ground. Dunscomb and Rehwoldt (1999) apply a dipole–dipole array. All studies are carried out in limestone areas in the USA.

Roth *et al* (1999) carried out a case study to clarify the reliability of multi-electrode earth resistivity investigation for geotechnical applications in karst terrains.

A comparison between the boring and resistivity results has shown that resistivity study gives good information concerning the depth to the irregular surface of the bedrock.

However, the reliability of ERT is still under question with regard to locating and determining the size of possible voids.

A brief theoretical background of quantitative analysis

The electric field **E** is the electrostatic field (div **j** = 0, where **j** is the conductive current density). It is well known that the resistivity field is the potential one and satisfies the Laplace equation (Kaufman 1992):

$$\mathbf{E} = -\mathrm{grad}\Phi, \tag{2.38}$$

where Φ is the scalar electric potential.

There is an obvious similarity between the electric and magnetic fields (Eppelbaum 1999, 2007). Thus, it is possible to apply the advanced methodologies developed in magnetic prospecting (improved tangent, characteristic point, and areal methods) (e.g., Khesin *et al* 1996, Eppelbaum *et al* 2001, Eppelbaum 2011a, 2015a) to interpret the resistivity anomalies.

Examples of direct current anomaly interpretation

Figure 2.75 illustrates an application of these aforementioned methods for interpreting a theoretical anomaly over models of a sphere (a) and inclined bed (b). As can be seen from the figure, the application of interpretation methods developed in magnetic prospecting gave excellent results.

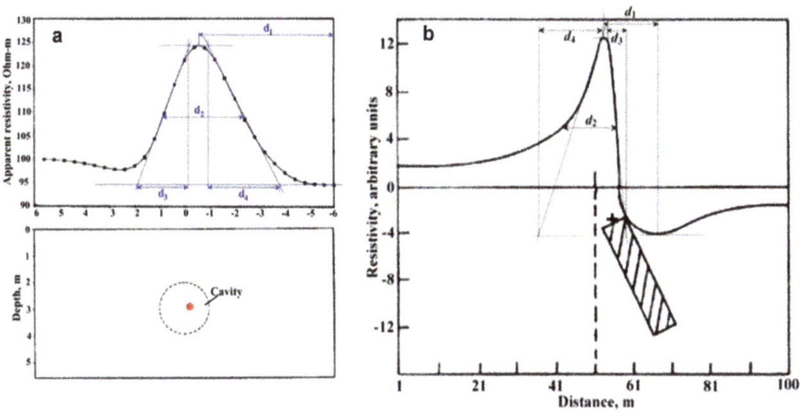

Figure 2.75. (a) Quantitative analysis of resistivity curve by the use of methods developed in magnetic prospecting (computed resistivity curve and geological model after Lowry and Shive (1990)). The small red circle shows the position of the center of the determined anomalous body, interpretation after Eppelbaum (2019a). (b) Interpretation of a theoretical resistivity anomaly over a model of an inclined thin bed (initial section after Sundararajan *et al* (2007)). The symbol '+' designates the position of the middle of the upper edge of the anomalous body. ((a) Reprinted from Lowry and Shive (1990), with permission from SEG. (b) Reprinted from Sundararajan *et al* (2007), with permission from EAGE.)

Figure 2.76 displays a real detailed physical–geological model of karst composed of the basis of drilling data, geological mapping, and different geophysical method interpretations. This karst domain is characterized by the following peculiarities: (1) at the contacts of limestone with the basic clays is recognized so-called carbonaceous-suffusion karst, (2) deepening of native rocks caused increasing of sand-alluvial deposits thickness, (3) unloading of subterranean waters is being under alluvial sediments. Results of quantitative analysis of resistivity anomaly (approximation model of HCC was applied) testify to a good agreement with the geological data (figure 2.77).

Application of the direct current surveys for detecting underground caves was demonstrated in Habberjam (1969), Parra and Owen (1986), Khmelevskoi and Bondarenko (1989), Lowry and Shive (1990), Hart and Rudman (1997), Eppelbaum (2007, 2019a, 2019c), Nuzzo *et al* (2007).

Analysis of electric resistivity tomography (ERT) images at different slices (levels) has shown that this methodology can be effectively applied to the ERT data interpretation.

We can note also that the Mise-à-la-masse method was successfully applied for mapping vadose zone karst at a site in eastern Pennsylvania, USA (Nyquist *et al* 2008).

Practical examples

VES Hydrosystem Trebyshniza (Yugoslavia) (Arandjelović 1969)

Mapping of Dinaric karst using VES curves

Generally, the VES method must be processed qualitatively and quantitatively. However, there is a great VES study based on qualitative processing only put into

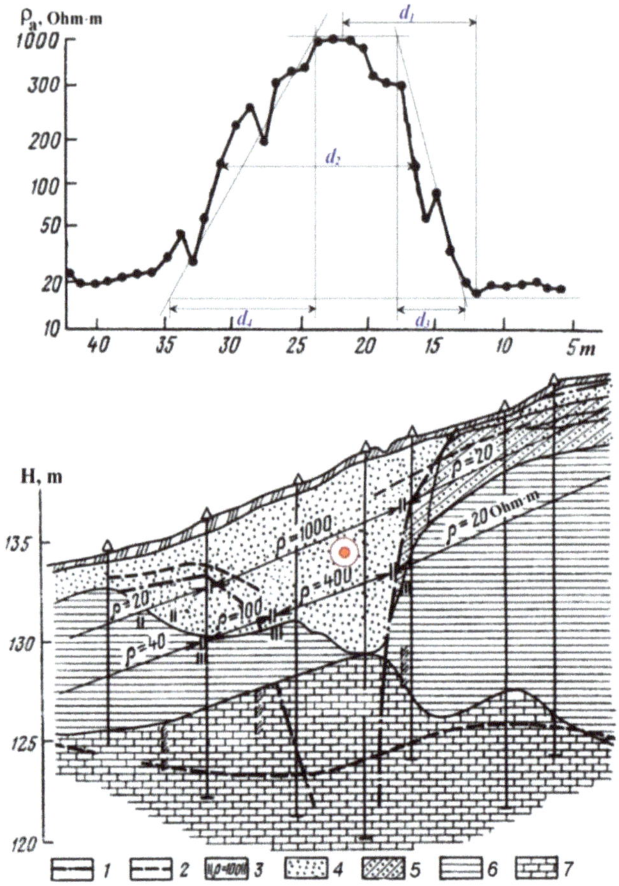

Figure 2.76. Quantitative interpretation of resistivity anomaly over the detail studied karst section (Russia). Curves of electrical profiling (above) and geological section (below) (1) faults, (2) geoelectrical boundaries obtained from the vertical electric sounding analysis, (3) vertical contacts of media with different electric characteristics, (4) alluvial sands, (5) sands, loams, and clay sands, (6) clays, (7) limestone. The ⊙ symbol marks the location of the center of a horizontal circular cylinder obtained from the analysis of the resistivity curve (observed curve and geological section after Khmelevskoi and Bondarenko (1989), with simplification; interpretation after Eppelbaum (2007)) (reprinted from Khmelevskoi and Bondarenko (1989)).

the design of the construction of the dam in the karstic region of eastern Herzegovina (Arandjelović 1976). The Dinaric Karst System is shared between northeast Italy, Slovenia, Croatia, Serbia, Bosnia and Herzegovina, Montenegro, Macedonia, and Albania. A comprehensive description of the Dinaric karst is presented in Mihevć *et al* (2010). The main watercourse in the region is the Trebišnjica River (figure 2.77), the largest sinking river in Europe (Milanović 1990, 2005). All the poljes in the catchment area are temporarily flooded, hydrologically separated, and closed karst entities. Enormous quantities of available water are not evenly distributed either in time nor in space.

The entire region was, from 1955 till the 2000s, a subject of detailed investigations. The Trebišnjica Hydrosystem (7 dams, 6 reservoirs, 5 tunnels 60 km long, 80 km of channels) is presented in figure 2.77. Large-scale investigations and sealing treatment have been successfully performed.

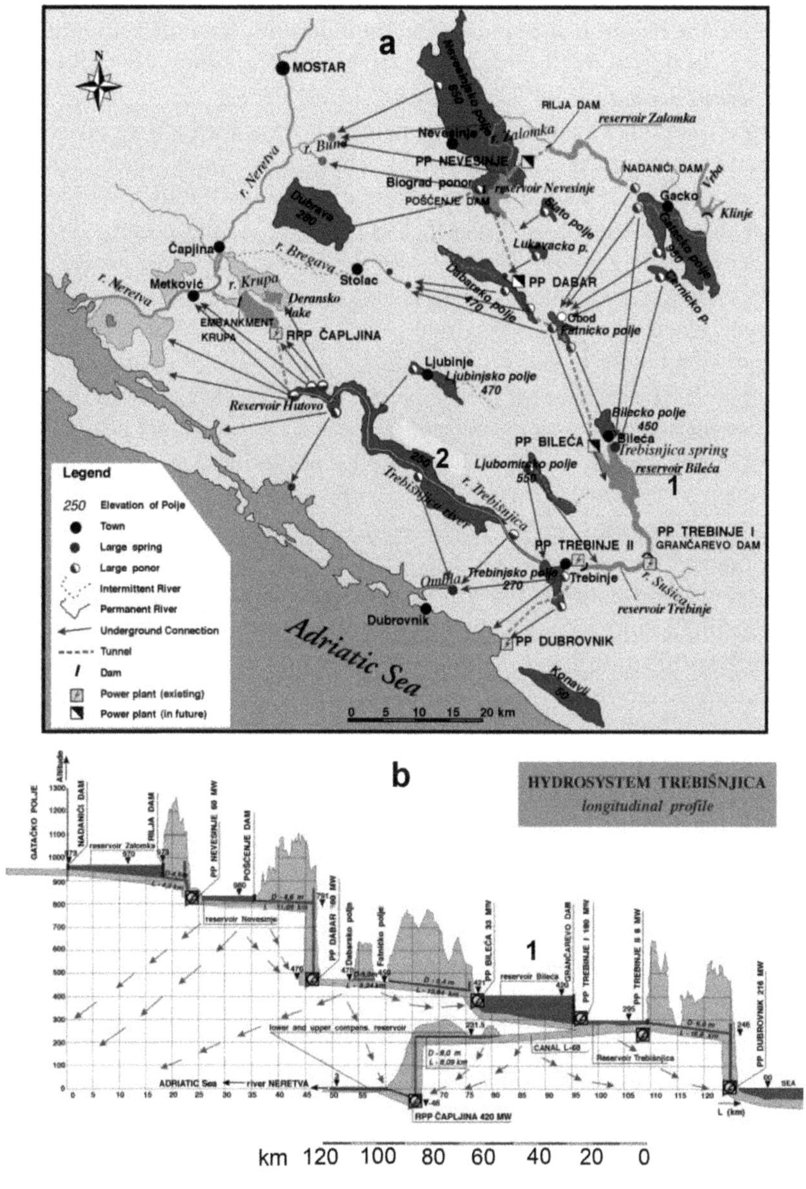

Figure 2.77. Hydro system Trebišnjica layout (a) plan view and (b) longitudinal profile. Arrows indicate the general direction of underground flows. 1—'Bileća' reservoir, 2—Popovo Polje (reprinted from Milanović (2002), with permission of *Episodes* and P Milanović).

The dam site is located in the canyon of River Trebišnjica. The concrete dam is 123 m high. The dam foundation is composed of fractured, karstified, and layered Lower Jurassic limestone. The entire reservoir is situated in karstified Mesozoic, limestones, and dolomites.

Geophysicists have made huge contributions to the study of karst and underground water karstic tunnels (Arandjelović 1976, Milanović 1979). A very effective VES method of electrical resistivity was applied to study the karstified layer and its thickness and extension in the space. The methodology included (a) calibration of the geological section at the first stage and (b) geoelectric surveys in the construction area in the second phase.

Geoelectric method. The geoelectrical method is described above. The method of electrical resistivity, first of all, limestone's, changes within a wide range (approximately from 100 to 1000 Ωm) and depends on many factors. This range was obtained in the karst region of Eastern Herzegovina, Dinaric karst (Arandjelović 1976).

VES and mapping surveys for the study area were carried out with a Schlumberger electrode array: maximal AB/2 = 550 m, MN = 50 m, and the measurement step l = 50 m. By these methods, for different reservoirs and tunnels of the Trebišnjica Hydrosystem, the total covered area consisted of more than 500 km^2. For the investigation of the right bank of 'Bileća' Reservoir, only a total of 160 km of cross-sections were mapped and 600 sounding points were used (Milanović 2002).

The problems that can be resolved with the resistivity method in karst areas include the determination of permeability of dam sites and reservoirs and delineation of underground watershed zones. They usually involve (Milanović 2000):

- Differentiation between the lithological members and their determination;
- Determination depth of the intensive karstified carbonate sediments, i.e., 'base karstification'; and
- Location of the zones of the tectonic and karst processes which may contain caves and conduits with the active underground karst flows.

Data processing (figure 2.78) is based on five-year measurements of 300 points in the same area. Note that the shape of curves of electrical sounding can differ at the same location at different times due to the different groundwater levels (Arandjelović 1976).

From diagrams in figure 2.78(a), one can see: four characteristic layers with different specific resistivities:

(a) the first layer (depth H_1) is characterized by electrical resistivity ρ_1. It is composed of intensively karstified and heavily disintegrated limestone with the limited overburden (predominantly Terra-Rossa);

(b) the second layer, of thickness H_2 and resistivity ρ_2 is composed of an intensively karstified limestone above the aquifer zone;

(c) the third layer, of thickness H_3 and electric resistivity of ρ_3, is located below the water table and is less karstified than the first and second media;

(d) finally, the fourth layer, with the thickness H_4 a resistivity of ρ_4, is characterized by the non-karstified and massive limestone.

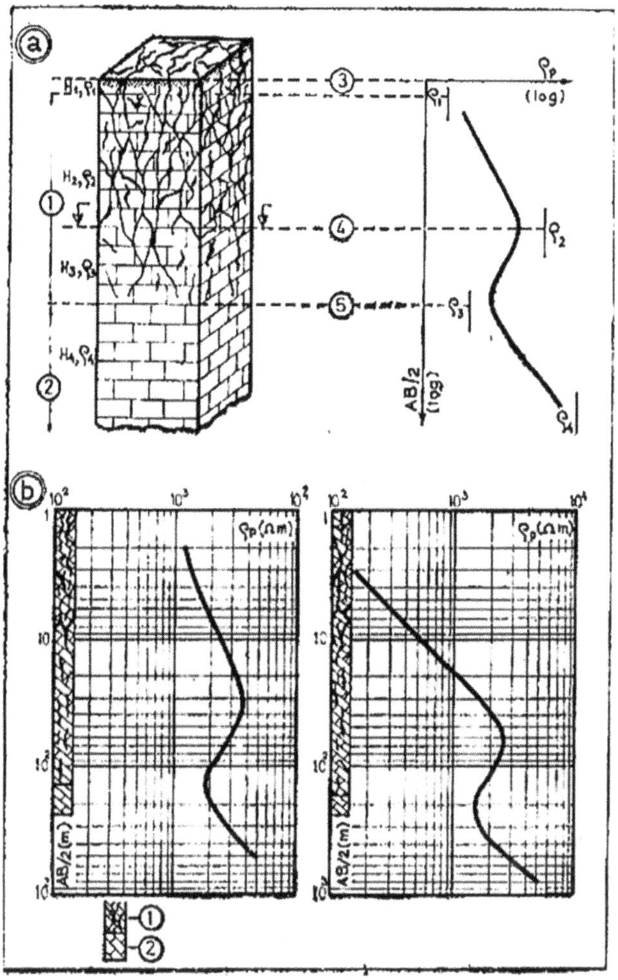

Figure 2.78. The geological column in karst with the schematic theoretical diagrams of electrical sounding (a) and diagram of electrical sounding near two boreholes in the karst of eastern Herzegovina, (b) examples of VES sounding curves, Axis *Y* is half of the electrode separation AB/2 that, in the first approach, is accepted as the depth of investigation. 1—karstified limestone, 2—nonkarstified limestone, 3—field surface, 4 6—water table, 5—base of karstification (figure from Arandjelović (1976), referred to in Milanović (1979) with gratitude).

It is evident from the general scheme (figure 2.78(a)) that electrical resistivity for karstified rocks above underground water table 4 in figure 2.78(a) is greater than the resistivity of karstified limestone saturated with water and smaller than or equal to the electrical resistivity of non-karstified limestones (Arandjelović 1976).

To estimate the complex weather tightness problem, hundreds of kilometers of VES were carried out in karstic areas. An example of such an investigation in the karst area of the Trebishnjica reservoir is shown in figure 2.79(a). There is seen a zone of relatively high electrical resistivity in graphs in figure 2.79(a) that was

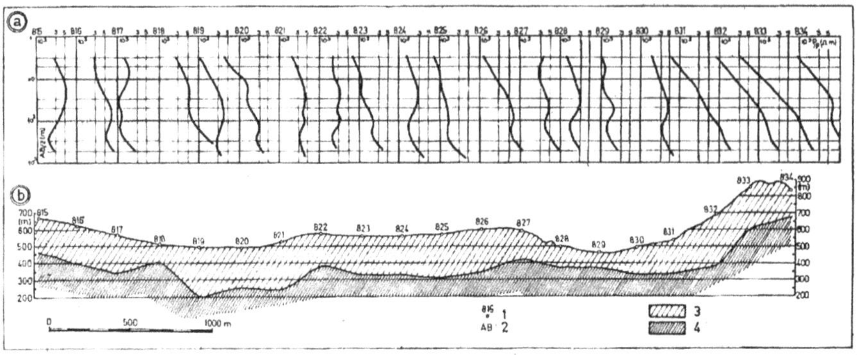

Figure 2.79. Results of electrical soundings with geological interpretation in the karst area of the Trebishnjica reservoir: (a) diagrams of electrical sounding scaled AB/2 along axis *Y*; (b) geological section along designed dam based on interpreting of soundings presented in figure 2.79(a), here axis *Y* is interpreted depth based on AB/2. 1—electrical sounding; 2—the separation between current electrodes; 3—heavily karstified limestone; 4 —non-karstified limestone (figure from Arandjelović (1976), referred to in Milanović (1979) with gratitude).

identified as a zone of non-karstified limestone and dolomite. This was also confirmed by drillings, permeability tests, and electrical logging. The geological cross-section along the dam (figure 2.79(b)) was obtained by interpreting results measurements that were used to prepare a map with a contour of the base of karstification (Arandjelović 1976).

Other special geophysical investigations concerning the predominant fracture direction were carried out using the circular electric sounding (figure 2.80). Such investigations were performed in the different points of the Dinarid karst to determine directions of the dominant flow direction coinciding with the minimum of electrical resistivity in the azimuthal diagram of fracturing.

The results of the geophysical studies were amazing. Before constructing the Hydrosystem the region was dying, suffering from a lack of water and electricity (figure 2.81 left). People, especially the youth, left the region in search of a better life. With the construction of a part of the Trebišnjica Hydrosystem, dramatic changes in the environment occurred (figure 2.81 right). In the excessively dry region, a large reservoir 'Bileća' (1 in figure 2.77) was formed and the permanent course of the Trebishnjica River was set at 65 km with a minimum discharge of 8 $m^3\ s^{-1}$. This arid region has been improved with more than 40 billion m^3 of the lake area. The natural intermittent flow of the Trebišnjica River has been changed to a permanent flow whose water surface area ranges from 12–19 million m^3, depending on discharge.

In Popovo Polje (2 in figure 2.77(a)), about 5000 hectares of soil has been protected from floods that, under natural conditions, make this polje almost useless for agriculture. The average duration of flooding (253 days per year) make it impossible to build any infrastructures in the polje. Inhabitants were forced to organize settlements in extremely unsuitable conditions on inaccessible steep rocky slopes above the flood level. Now, in one village alone, about 30 houses have been built at the same level as the polje. The old buildings on the rocky slope high above the flood level have been abandoned.

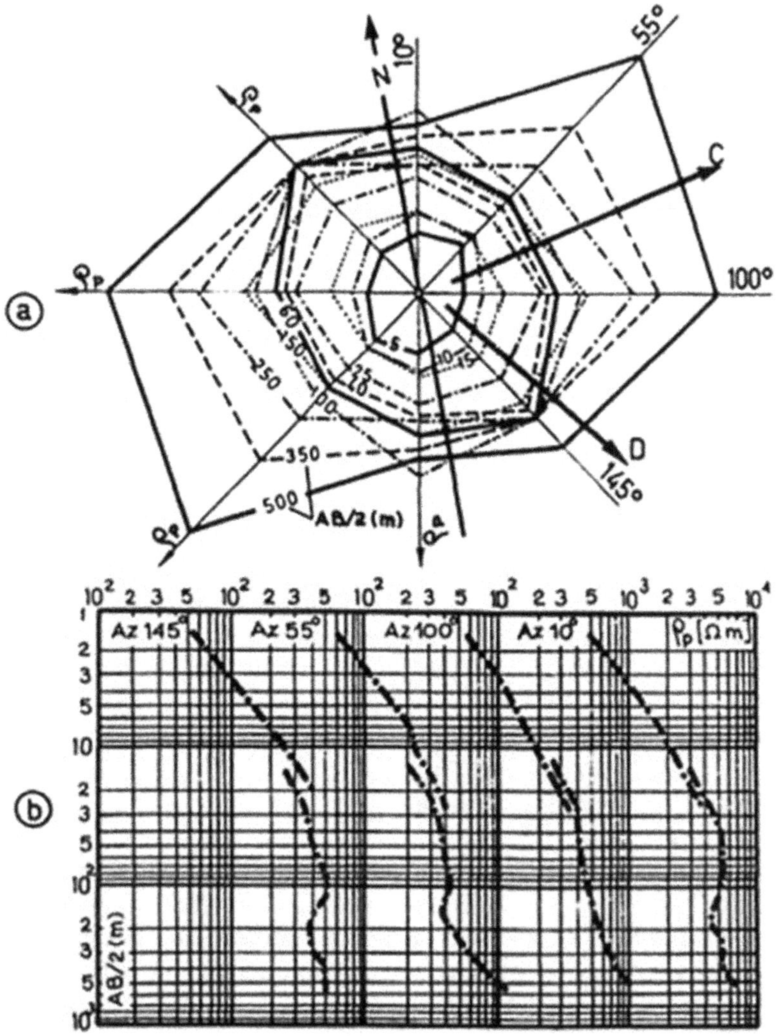

Figure 2.80. (a) Azimuthal curves of the circular electrical sounding show the predominant fracture direction (minimum resistivity) coinciding with the direction of maximum intensity flow. C and D—the directions of the maximum karstification of the Dinaric karst (b) VES curves are measured in the direction shown near every curve (Az 55°, for instance) (figure from Arandjelović (1976), referred to in Milanović (1979), with gratitude).

ERT Soreq Cave, Israel (Ezersky 2002)

Introduction. The large caves in Israel have been detecting still as a rule as the result of quarry explosion works. The geoelectric study was carried out at the Soreq cave area and its vicinity to split up into two different ones:

- To check if it is possible to detect the known cave located at the depth carrying out the surface measurements above it using the geoelectric methods.
- To locate the additional voids using the same methods under the known cave and in its vicinity.

Figure 2.81. (Left) is an old photo (1969) with Trebisnjica river bed in the dry period of the year. (Right) today Trebinje (Bosnia and Herzegovina): Trebišnjica River running through Trebinje City (both photos from Milanović (2021) with gratitude).

The investigation there was performed in two stages. The feasibility study was carried out in 1997 (Ezersky and Goldman 1998). The FDEM and ERT methods were applied in the Soreq Cave area with the main objective of testing the methods in a well-known cave and its vicinity (The FDEM survey is described in section 2.2.1.7). The second stage of the research includes:

- geological verification of the geophysical results including follow-up drilling.
- borehole electric and ultrasonic logging for a better understanding of the geological–geophysical model of the medium and the origin of false targets in cave locations using the ERT method.
- additional surface ERT measurements.

The Soreq Cave is one of a series of karstic caves situated within the steeply westward flank of the Judean Hills anticline and is located approximately 60 km inland and 400 m above sea level in the Veradim Formation of the Cenomanian age. This formation is represented by a massive dolomite and limestone layer. The cave is 80 m long, 60 m wide, and about 5.5 m in height with a total volume of about 25 000 m^3. Its location in the steeply dipping terrain results in a variation of its depth below the surface from less than 10 m at the western end to 35–40 m at the eastern end. A 30–50 cm thick soil cover above the dolomitic layer is relatively uniformly distributed and is composed of the Terra-Rossa formation. The climate in the cave area is typical of the Eastern Mediterranean semiarid conditions with about 70% of the rainfall occurring during the winter months (December–February), 10%–20% during the autumn (October–November), and 10%–20% during the spring (March–May). The annual average rainfall above the cave is 500 mm.

Survey. A total of 13 FDEM and 8 ERT profiles were carried out in the area with dimensions of approximately 200 × 200 m^2 both above the cave and outside it. The ERT profiles were measured using at the first stage by the standard set based Bison 2390 instrument (Canada) updated a specially designed 45-channel multi-electrode system with a handle switching unit for an ERT system and, at a second stage, an IRIS SYSKAL R2 Instrument (France) has been used for the data acquisition.

In this area, the pole–dipole electrode arrays were applied and the inter-electrode separation was 10 m. The ERT data were interpreted using commercially available 2D inversion software Res2Dinv v.3.45 (Loke and Barker 1996). The results of the inversions are presented in the form of 2D true resistivity cross-sections.

Two boreholes (bhl and bh2) both of 25–30 m depth were drilled by the Geological Survey of Israel for verification of high resistivity anomalies detected during a geoelectric survey of the first stage at the Soreq Cave area in 1998 (Bar-Matthews *et al* 1999).

Inverse resistivity cross-section along line 7 located above the well-known Soreq Cave is presented in figure 2.82.

The ERT method has shown a strong resistivity anomaly of 32 000 Ωm approximately above the cave. The lateral location of the known cave coincides roughly with the resistivity anomaly and the depth to the cave is accurately detected on its uplifted (NE) side.

The ERT method provides the possibility of mapping the karstic zones within the heterogeneous rock masses (Ezersky and Sobolevsky 2002). The resistivity map of the Soreq Cave area at a depth of 20 m is shown in figure 2.83. There is an evident interface between the karstified (brown zone) and non-karstified (green zone) dolomites.

It deserves mentioning the high background resistivity (1000–5000 Ωm) of massive limestones and dolomites of the Veradim Formation of the Cenomanian age. In all cases, in the ERT surveys above the known caves (at depths down to twice their characteristic size), high resistivity anomalies were detected. However, boreholes BH1 and BH2 (see figure 2.83 for location) have shown that a number of fractures are filled with the secondary calcite, and the refilled layers show very high resistivity.

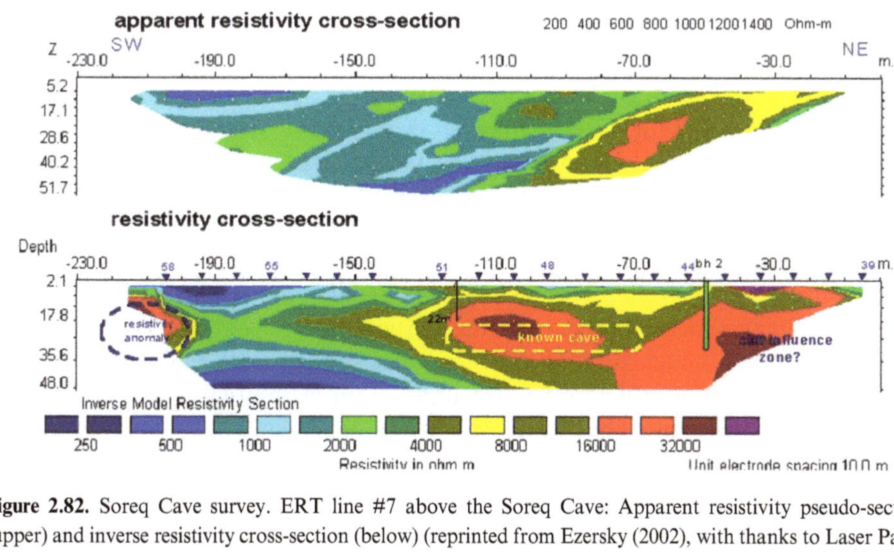

Figure 2.82. Soreq Cave survey. ERT line #7 above the Soreq Cave: Apparent resistivity pseudo-section (upper) and inverse resistivity cross-section (below) (reprinted from Ezersky (2002), with thanks to Laser Pages Publishing).

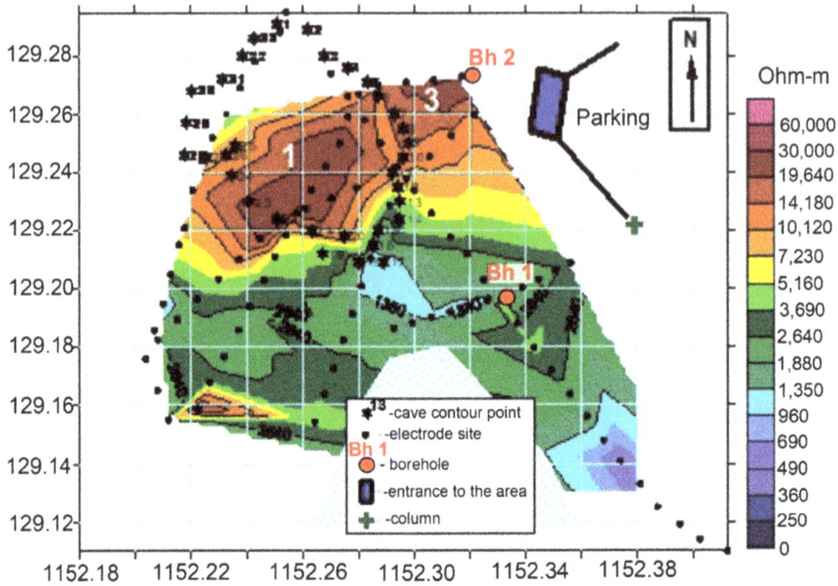

Figure 2.83. ERT resistivity map of the Soreq Cave area at a depth of 20 m (reprinted from Ezersky (2002), with thanks to Laser Pages Publishing).

Thus, the investigation showed that under specific geohydrological conditions (for instance, at the Soreq Cave site), the geoelectric methodology can detect caves as well as highly resistive rocks such as strongly karstified dolomites, dolomites with calcite-filled fractures, etc, but is unable to distinguish between them.

The equivalence problem then arises, which could be solved using the integrated technique (GPR, gravity survey, etc). A similar problem relates also to the FDEM method (see section 2.2.1.7)

Pseudo-3D ERT, Greene County Missouri (USA) (Kidanu *et al* 2020).

The ERT method is intended to generate 2D or 3D resistivity images depicting the distribution of the electrical resistivity values in the subsurface. It was shown by numerous researchers 2D ERT imaging is a suitable technique to map and characterize caverns and sinkholes in karst terrain (Barker 1992, Roth *et al* 1999, Zhou *et al* 2002, Carbonel *et al* 2014). However, in some situations, 2D ERT images are less accurate than desired because 2D ERT processing software cannot adequately address the lateral resistivity variations that occur outside of the vertical plane of the 2D ERT profile. True resistivity in the 3D mode which addresses lateral resistivity changes occurring outside the vertical plane of the 2D lines, can be acquired. 3D data acquisition involves the placement of current electrodes on the nodes of a rectangular grid and the measuring of all the mutual potentials between all electrodes. This approach is the most reliable, but it tends to be costly and time-consuming (Loke and Barker 1996, Pazzi *et al* 2018). An alternative to true 3D ERT imaging is the pseudo-3D imaging technique, which includes the acquisition of parallel and/or orthogonal 2D ERT profiles and the processing of the same as though they were true 3D data (Yi *et al* 2001, Chambers *et al* 2002).

Site and geology. The pseudo-3D ERT method was used to image the subsurface structure of an active sinkhole in Greene County Missouri.

The rock layers regionally dip gently towards the west with the minor faulting and folding. Bedrock in the study area is the Mississippian age limestone. This formation is characterized by layers of limestone interbedded with thin layers of chert and the presence of chert nodules within the limestone.

Data acquisition and processing. Conventional 2D electrical resistivity tomography (ERT) data were acquired along 16 parallel profiles. Pseudo-3D ERT images were generated using the same profiles spaced at 6.1 m across, which is four times the electrode spacing as recommended by Kidanu *et al* (2020). To generate the pseudo-3D ERT image, the 2D ERT data were combined into a 3D ERT data file and inverted using the RES3DINV program (Loke and Barker 1996).

The acquired ERT data were processed as both 2D data and pseudo-3D data taking into account the available core hole control, multi-channel analysis of surface waves (MASW) data, and field observations.

Results. The 2D and quasi 3D sections are shown in figure 2.84.

There are both the 2D ERT profile in figure 2.84(a) and the horizontal ERT depth slices, layer-11 to layer-16. The low resistivity anomalies in figure 2.84(b) representing the joint sets often tend to be wider and have a decrease of resistivity with depth (at least to a depth of 30 m). This typical phenomenon is most likely due to one or a combination of the following factors: (1) higher moisture concentrations at depth; (2) the widening of vertical seepage pathways through fractured rocks; and (3) more extensive solution-widening of fractures at the depth and increased concentrations of piped clay.

Conclusion. The interpreted increase in clay and moisture content with depth (at least to a depth of 30 m) along the linear joint sets implies increased joint width, increased moisture content, and/or increased clay content.

This conclusion is in agreement with the previous study (Kidanu *et al* 2016) which stated, that the subsurface structure of the sinkhole site is more reliably imaged on the pseudo-3D dataset than in the 2D dataset. The interpretation results of the pseudo-3D ERT indicated that the sinkhole developed at the intersection of three vertical solution-widened joint sets.

2.2.1.10 Induced polarization

2.2.1.10.1 Brief theoretical background

The induced polarization (IP) method is based on the study of secondary electric fields generated in the ground by electric currents. It is extensively applied due to its sensitivity mainly to electronic and ionic conductors (Komarov 1980).

Basing on Alpin *et al* (1985) and Smirnov (1975) and taking into account the work of Avdevich and Fokin (1978), IP field intensity can be written as follows (Khesin *et al* 1997):

$$\vec{E}_{IP} = \frac{1}{2\pi\varepsilon_o}\int_s \frac{2(\vec{P}, \vec{r})\vec{r} - Pr^2}{r^4} ds, \tag{2.39}$$

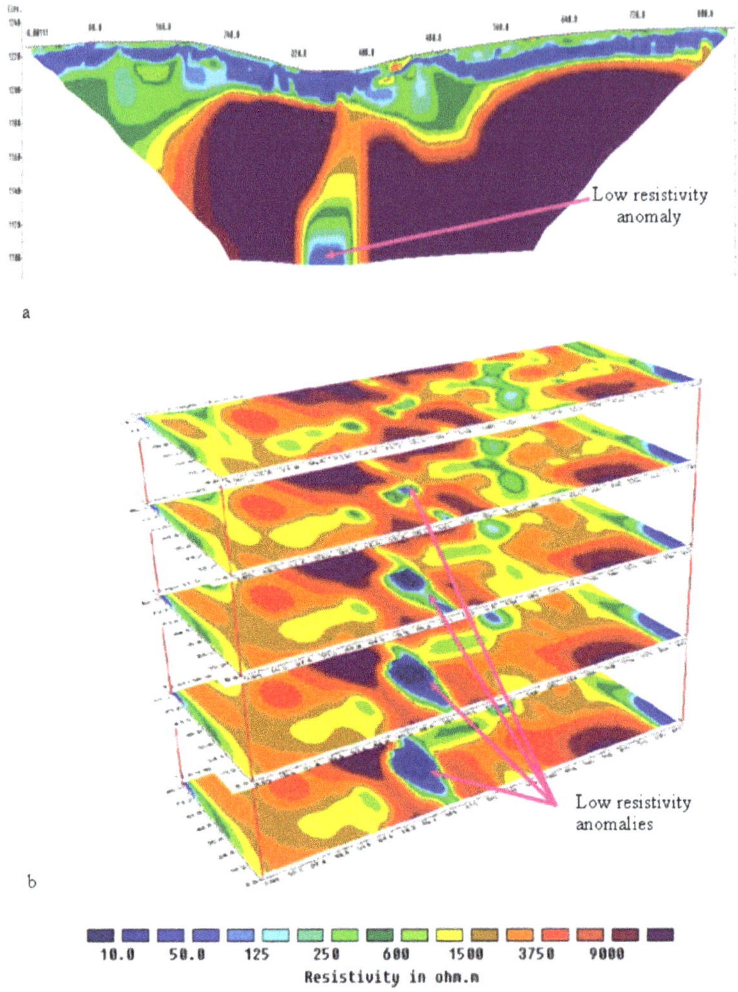

Figure 2.84. (a) 2D ERT profile. (b) Layers of the horizontal ERT depth slices showing the increase in width and the decrease in resistivity value with depth along a vertical low resistivity anomaly (reprinted from Kidanu *et al* (2020), with permission of S Kidanu and gratitude to AIMS geosciences).

where ε_0 is the dielectric constant of free-space; s is the cross-section area of the body; $\vec{P}$ is the polarization vector (dipole moment of a unit volume);

$r = \sqrt{(x_s - x)^2 + (z_s - z)^2}$ is the distance from the observation point M(x, z) to a certain point

$P(x_s, z_s)$, $\vec{r} = (x_s - x)\vec{i} + (z_s - z)\vec{k}$ is the radius-vector from the point M to point P; $\vec{i}$ and $\vec{k}$ are unit vectors directed along Ox and Oz axes, respectively.

The projections $\vec{P}_{IP}$ on the coordinate axes are:

$$P_{IP,\,x} = P_{IP}\cos\phi_{IP},\; P_{IP,\,z} = P_{IP}\sin\phi_{IP}, \tag{2.40}$$

where φ_{IP} is the angle between the vector $\vec{P}_{IP}$ and Ox-axis.

Then the components $\vec{E}_{IP}$ may be written as follows:

$$E_{IP,\,x} = -\frac{1}{2\pi\varepsilon_o}\int_s P_{IP}\frac{(z_d^2 - x_d^2)\cos\phi_{IP} - 2z_d x_d \sin\phi_{IP}}{r^4}ds, \tag{2.41}$$

$$E_{IP,\,z} = -\frac{1}{2\pi\varepsilon_o}\int_s P_{IP}\frac{(z_d^2 - x_d^2)\sin\phi_{IP} + 2z_d x_d \cos\phi_{IP}}{r^4}ds, \tag{2.42}$$

where $z_d = z_s - z$ and $x_d = x_s - x$.

The common expressions of the horizontal and vertical components of anomalous magnetic field take the form (Khesin *et al* 1996):

$$X_{an} = \frac{\mu_o}{4\pi}\int_s I\frac{-(z_d^2 - x_d^2)\cos\phi_m + 2z_d x_d \sin\phi_m}{r^4}ds, \tag{2.43}$$

$$Z_{an} = \frac{\mu_o}{4\pi}\int_s I\frac{(z_d^2 - x_d^2)\sin\phi_m + 2z_d x_d \cos\phi_m}{r^4}ds, \tag{2.44}$$

where μ_o is the absolute permeability of free-space, φ_m is the angle between the vector $\vec{I}$ and Ox-axis.

The expressions (2.41) and (2.43), (2.42) and (2.44), respectively, are proportional.

Taking into account that $E_{IP,z}$ is not measurable and X_{an} can be measured but rarely, it is necessary to find other relations. If we assume

$$\phi_{IP} = \phi_m + 90^o,\ P_{IP} = 4\pi k\varepsilon I, \tag{2.45}$$

(coefficient $k = 1\ \mathrm{V\ A^{-1}}$), then we obtain $E_{IP,}x = Z_{\mathrm{an}}$, since the right-hand sides of the expressions (2.42) and (2.44) (after some transformations) are the same. This fact has been noted earlier (Komarov 1980), but either for the vertical magnetization or the horizontal IP only.

To calculate the IP anomaly using equation (2.41), it is necessary to express the vector $\vec{P}_{IP}$ in the terms of the internal polarizing field and parameters of the host medium and the anomalous body. A well-known technique described in Komarov (1980) and Seigel (1959) has been used for this purpose. As a result (we omit here the detailed analysis of the problem), the theoretical analogy between the corresponding expressions for the main disturbing types of source bodies has been proved (Khesin *et al* 1997). The subsequent stage involves the application of advanced methods developed for quantitative interpretation of magnetic anomalies for analysis of IP data.

IP measurements were applied for searching and localization underground caves by the following investigators: Khesin *et al* (1997), Sheets (2002), Hughes and Karlson (2003), Al-Qufi *et al* (2012), Chen *et al* (2015), Moreno (2015), de Araújo *et al* (2019).

The basement for the application of interpretation methodology developed in magnetic prospecting for analysis of IP anomalies was given in Khesin *et al* (1997) and Eppelbaum and Khesin (2002).

2.2.1.10.2 Examples of IP data analysis

The following example (figure 2.85) demonstrates the quantitative analysis of IP anomaly observed over karst terrane composed of clay material (Yugoslavia). The application of the IP method is based here on the difference in polarizability of limestone (host medium) and clay (karst terrane) (figure 2.85). Obtained results illustrate enough accuracy of the performed analysis.

The next example (figure 2.86) presents the application of IP imaging for underground cave detection in the Dickson County Landfill, 20 km west of Nashville, Tennessee (USA) (after Hughes and Karlson 2003).

2.2.1.11 Self-potential (SP) method

2.2.1.11.1 Brief theoretical background

Self-potential (SP) method is one of the cheapest and technically simplest geophysical methods. Its widespread use limits the absence of a reliable interpretation scheme, first of all for complex physical–geological environments. The SP method is based on a study of natural electric sources (in some publications this method is named 'spontaneous polarization'). The term 'natural' denotes here that the SP field does not utilize any external artificial electric sources. The permanent electric fields arise as a result of the redox, filtration, and diffusion–adsorption processes in the upper part of the geological section. These fields are registered by the SP method, and the geophysical analysis of the targets generating these anomalies is the main goal of SP data investigation (Semenov 1980, Revil and Jardani 2013, Eppelbaum 2020).

The potential Π_i satisfies Laplace's equation everywhere outside the anomalous layer l (Zhdanov and Keller 1994).

$$E_{\mathrm{SP}} = -\mathrm{grad}\,\Pi. \tag{2.46}$$

The vector of the magnetic field **T** is determined as a gradient of magnetic potential U (Eppelbaum 2019a):

$$\mathbf{T} = -\mathrm{grad}\,U. \tag{2.47}$$

Table 2.6 presents the obvious similarity between the SP and magnetic fields for some interpreting models. Therefore, taking into account the theoretical similarity of

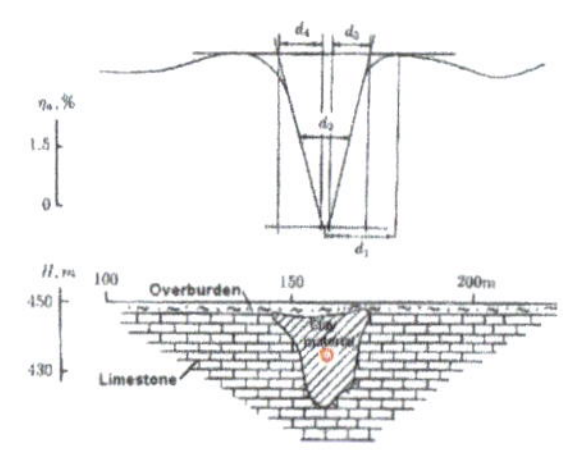

Figure 2.85. Quantitative interpretation of η anomaly over karst terrane composed of a clay material (initial data are taken from Šumi (1965)). The ⊙ symbol marks the location of the center of HCC obtained from interpretation of the anomaly curve (reprinted from Khesin *et al* (1997), with permission from EAGE).

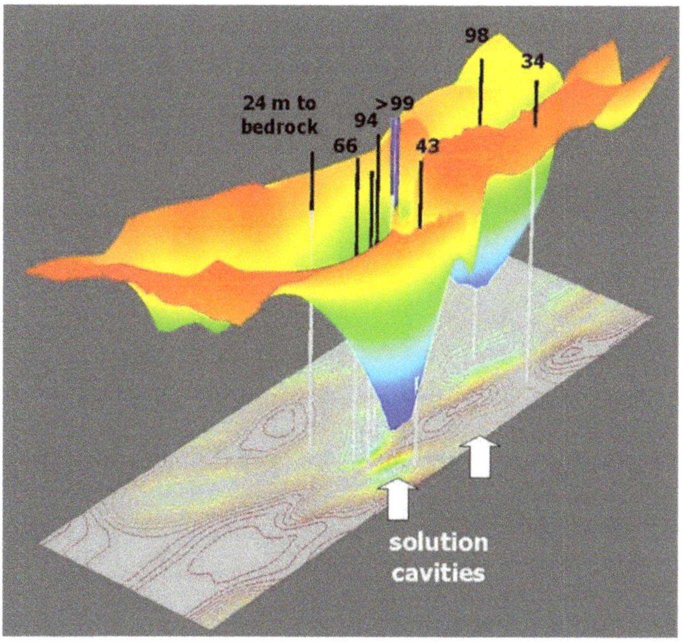

Figure 2.86. 3D representation of the IP data called by Hughes and Karlson (2003) a 'negative IP image' (Dickson County Landfill, Tennessee, USA). (Reprinted from Hughes and Karlson (2003), with permission of the Environmental and Engineering Geophysical Society. All copyright privileges remain with EEGS. This material cannot be copied or used without the express written permission of EEGS.)

the magnetic and SP fields (see equations (2.46) and (2.47) and corresponding analytical equations in table 2.6) (Eppelbaum and Khesin 2012), we can transfer the majority of interpretation methods (some uncertainty is for such methodologies as the upward and downward continuations) from the magnetic to SP field.

Here Z_v is the vertical magnetic field component at vertical magnetization; J is the magnetization; b is the horizontal semi-thickness of TB (thin-bed); m is the elementary magnetic mass; U_0 is the potential jump at the source body/host medium interface; r_0 is the polarized cylinder radius; R is the sphere radius; x is the current coordinate; z is the depth of the upper (DTB) edge (HCC or sphere center) occurrence.

2.2.1.11.2 Elimination of different kinds of noise

Different kinds of noise appearing in the SP measurements (electrode polarizability, calculation of the mean square error of SP observations, terrain relief effect, changing the water volume in the subsurface, uneven influence of solar radiation, and some others) and ways for their elimination are considered in detail in Eppelbaum (2019b, 2020).

2.2.1.11.3 Quantitative analysis of SP anomalies

Besides the geometric parameters of the anomalous target (corresponding similarity and methods of quantitative analysis are shown in detail in Eppelbaum (2019b,

Table 2.6. Comparison of analytical expressions for magnetic and self-potential fields.

Field	Analytical expression	
Magnetic	Thin bed $Z_v = 2J2b\frac{z}{x^2 + z^2}$	Point source (rod) $Z_v = \frac{mz}{(x^2 + z^2)^{3/2}}$
Self-potential	Horizontal circular cylinde $r\Delta U = 2\frac{\rho_1}{\rho_1 + \rho_2}U_0 r_0\frac{z}{x^2 + z^2}$	Sphere $\Delta U = \frac{2\rho_1}{2\rho_2 + \rho_1}U_0 R^2\frac{z}{(x^2 + z^2)^{3/2}}$

2020, 2021)), the self-potential moment can be also determined (Eppelbaum 2020). For the interpreting models of TB and horizontal circular cylinder (HCC), the self-potential moment can be calculated by the use of equations (2.48) and (2.49), respectively

$$M_{\Delta U} = \frac{1}{2}\Delta U_a h_0, \tag{2.48}$$

$$M_{\Delta U} = \frac{\Delta \mathrm{U_a h_c^2}}{(3\sqrt{3}/2)\cos(30^\circ - \theta/3)}, \tag{2.49}$$

where ΔU_a is the amplitude of SP anomaly (in mV), h_0 is the depth of the upper edge of TB (in meters), $h_c^2$is the squared depth to the center of the HCC (in m^2), and θ is some generalized angle (Eppelbaum 2020). The self-potential moment, by analogy with the magnetic field examination, can be employed for the arrangement of various SP anomalies.

SP measurements actively apply for the underground cave localization: (Corwin 1990, Wanfang *et al* 1998, Eppelbaum *et al* 2003, Eppelbaum 2007, 2020, Jardani *et al* 2007, Srigutomo *et al* 2010, Robert *et al* 2011, Chen *et al* 2018). Let us consider two typical examples: archaeological (1), and environmental (2) geophysics.

The prehistoric site of Sha'ar Hagolan is located south of the Sea of Galilee (Israel), on the bank of the Yarmuk River. SP measurements performed in the area of about 150 m^2 using grid 1 ×1 m enabled us to identify four underground caves associated with the ancient Neolithic village (figure 2.87). The anomalies 'C' and 'D' were selected for demonstration of the results of SP quantitative analysis (figure 2.88). Here interpretation models of the HCC were applied. For both targets, the depth of the center of HCC occurs at about 1.7 m. The interpretation results are in line with the available archaeological excavations in this area. The calculated value of electric polarization angle φ_p (see equation (2.48)) for anomaly 'C' is about 60°. Self-potential moment calculated for anomaly 'C' is $M_{\Delta \mathrm{U}} = \frac{6\ \mathrm{mV} \cdot 1.7\ \mathrm{m}^2}{2.6 \cdot \cos 20^\circ} = 4.18\ \mathrm{mV} \cdot \mathrm{m}^2$. Value of polarization angle φ_p for anomaly 'D' is about 65°, but given that parameters d_3 and d_4 are interchanged. Therefore, this angle is calculated from the other (opposite) side (Eppelbaum 2021) (figure 2.88). SP moment for the anomaly 'D' exceeds the previous

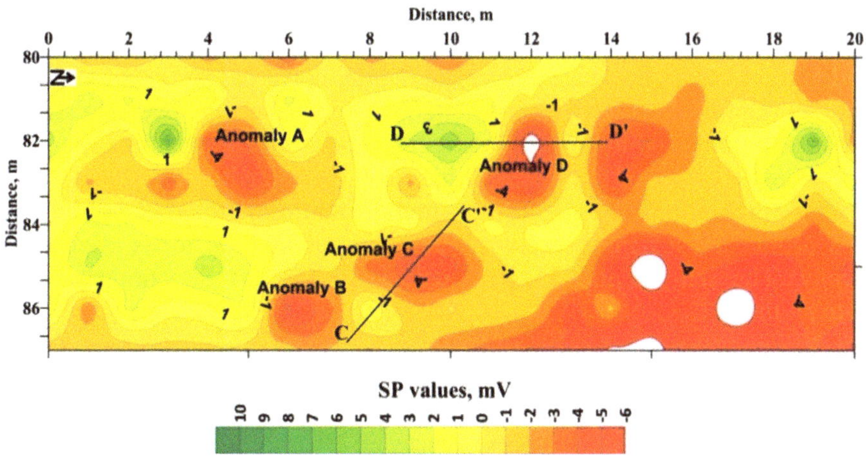

Figure 2.87. Self-potential map observed in the Sha'ar HaGolan site (northern Israel) and selected interpreting profiles (reprinted from Eppelbaum (2020), by permission from Springer-Verlag GmbH).

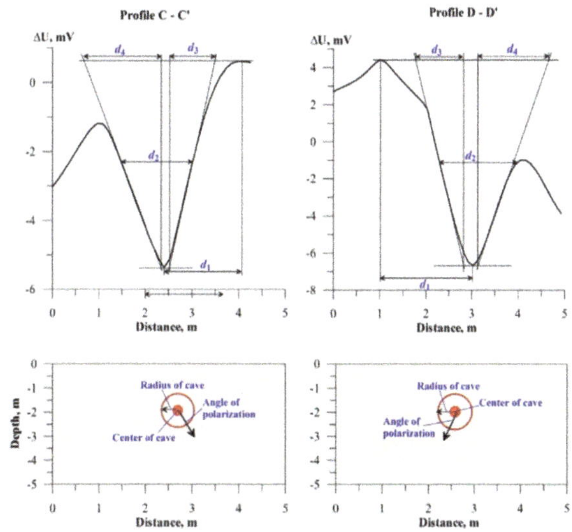

Figure 2.88. Quantitative analysis of anomalies 'C' and 'D' (see figure 2.88) in the site Sha'ar HaGolan (northern Israel) (reprinted from Eppelbaum (2020), by permission from Springer-Verlag GmbH).

calculated SP moment: $M_{\Delta U} = \dfrac{1.7\ \text{m}^2 \cdot 10.5\ \text{mV}}{2.6 \cdot \cos 22^{\circ}} = 7.41\ \text{mV} \cdot \text{m}^2$. Depth and size of the determined ancient targets agree with the available archaeological data (from the excavations carried out near this site). Interestingly, the previously performed ground penetration radar (GPR) survey (theoretically a much more exact geophysical method) gave ambiguous results due to the complexity of the medium.

Several impressive examples of SP application for detection of the underground cavities in southern Italy were displayed in Quarto and Schiavone (1996). Let us consider one of these field cases, where the buried karst cavity occurs in dolomitic limestones (figure 2.89). The cavity is horizontally extended and over it, a significant SP anomaly (up to 100 mV) was observed. Quantitative examination along profile A–B crossing a center of this anomaly has been performed (figure 2.90).

For interpretation models of TB (upper edge) and center of HCC were obtained depths of 6.0 and 9.5 m, respectively. The self-potential vector is oriented near-vertically. The self-potential moment calculated for this SP anomaly is about 3300 mV · m^2.

2.2.1.12 Surface nuclear magnetic resonance (SNMR) method

SNMR, also known as magnetic resonance sounding (MRS), was developed in the late 1970s (Semenov *et al* 1988) and since 1982 has been used for groundwater investigations. SNMR is a surface geophysical method recognized by the international community. The method was developed in the former USSR and now this development continues by geophysical teams in France, Germany, the USA, China, and Denmark.

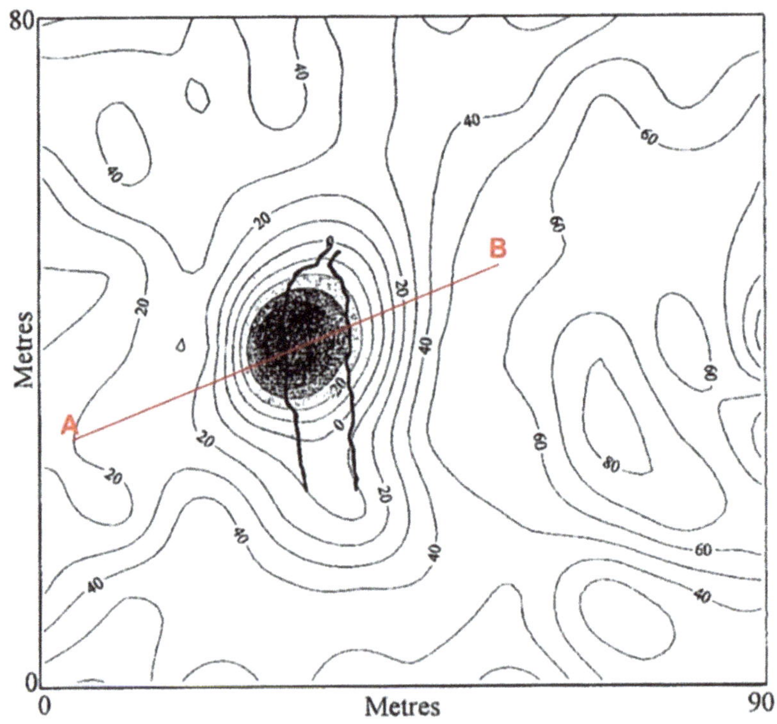

Figure 2.89. SP map over underground cave and position of interpreting profile A–B (SP map after Quarto and Schiavone (1996)) (reprinted from Eppelbaum (2021), with gratitude to MDPI).

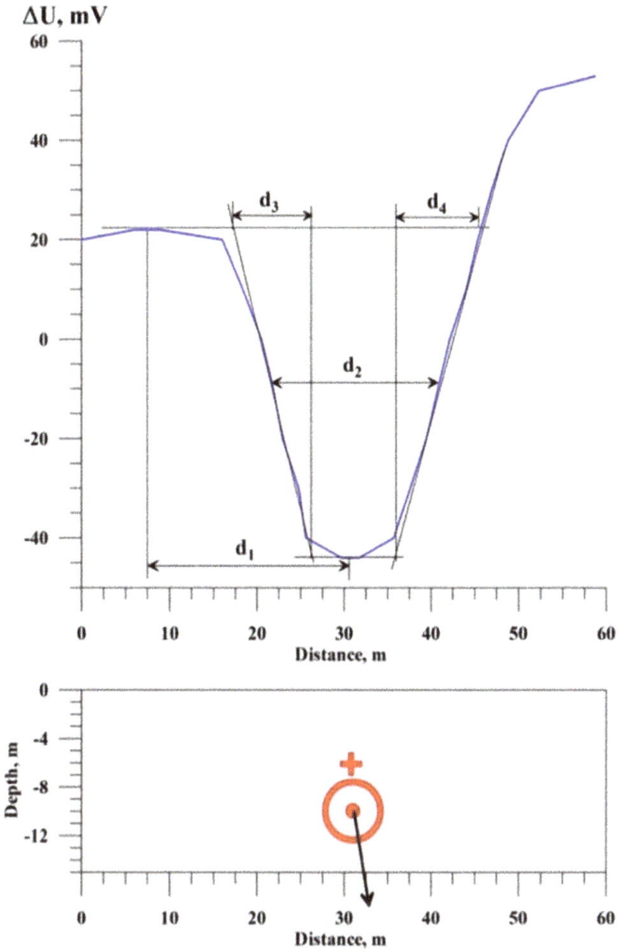

Figure 2.90. Quantitative analysis of SP anomaly along with profile A—B (location of the profile is shown in figure 2.17). Red cross designates the position of the middle of the thin-bed upper edge, and the circle—position of the center of the HCC (reprinted from Eppelbaum (2021), with gratitude to MDPI).

2.2.1.12.1 Background

SNMR method is based on the NMR phenomenon with a well-developed theory behind the technique. With NMR, we perform experiments on the nuclei of atoms. In some atoms, the nucleus possesses an overall spin. For example, hydrogen ^{1}H with an overall spin equal to 1/2 will have two orientations corresponding to two energy levels. When a static magnetic field is applied, the energy levels split. The initial populations of the energy levels are determined by thermodynamics and the lower energy level contains more nuclei than the higher one. Electromagnetic radiation allows one to excite these nuclei into the higher level with and observe exchanges between the energy levels (Farrar and Becker 1971).

For measuring, we irradiate samples by a pulse of EM energy, and when the pulse is cut off, and observe the EM field generated by the spin system returning to the previous state with the lower energy level. The frequency of radiation needed for exciting the nuclei is determined by the energy that separates the two levels (ΔE) defined by a static magnetic field. This frequency is termed the Larmor frequency. In a fixed static magnetic field, the Larmor is the signature of each type of nuclei possessing an overall spin. Figure 2.91 shows three principal steps necessary for measuring the NMR signal.

The spin system returns to equilibrium and generates an oscillating at the Larmor frequency magnetic field (the NMR signal). The spin system that we use with the SNMR method is protons of groundwater and without groundwater, the signal measured with SNMR is equal to zero. SNMR is a large-scale method investigating a volume ranging between $10 \times 10 \times 10$ m^3 and $150 \times 150 \times 150$ m^3 (Legchenko 2013) This volume is defined by the measuring setup. Figure 2.92 shows a typical

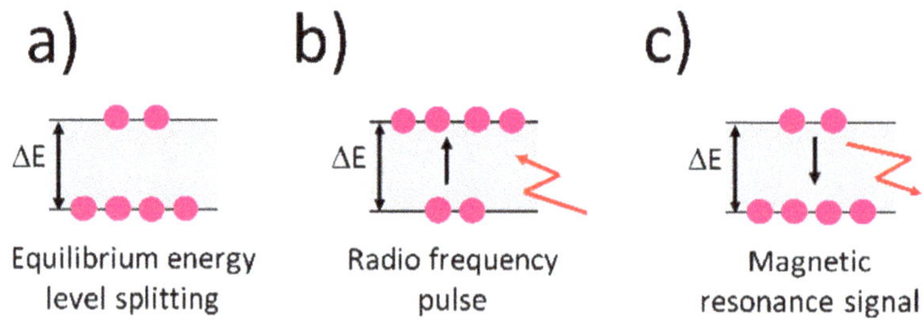

Figure 2.91. Schematic presentation of a pulsed NMR experiment: (a) in the equilibrium, the lower energy level is more populated; (b) during the pulse, the spin system absorbs electromagnetic energy and the higher level becomes more populated; (c) after the pulse cut-off, the spin system returns to equilibrium and releases absorbed EM energy.

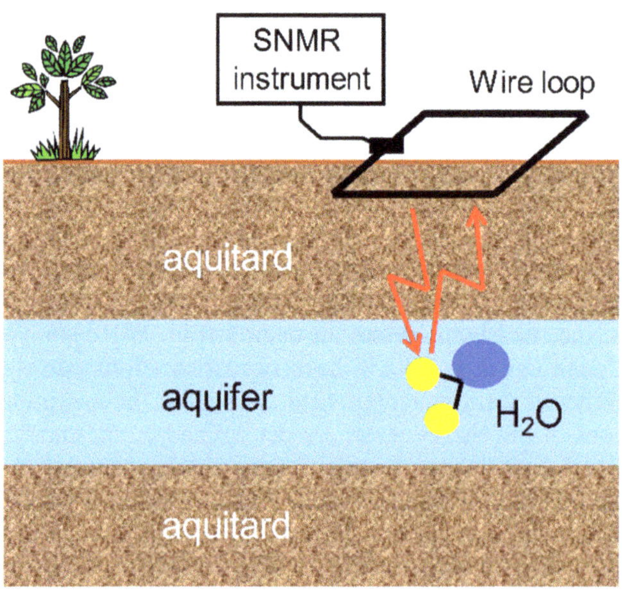

Figure 2.92. Typical SNMR field setup.

SNMR field setup. A large wire loop on the surface generates a pulse of the oscillating magnetic field. The same loop is used for measuring SNMR response from groundwater. The common loop size may vary between 50 × 50 m^2 and 100 × 100 m^2, which defines the area investigated with one SNMR station. Measuring the MRS signal can be also done using one or more additional receiving loops. The measuring procedure comprises the generation of one or more current pulses and measuring SNMR response after each pulse.

The simplest observation of the magnetic resonance phenomenon can be done using a one-pulse sequence, as shown in figure 2.93.

The measuring procedure starts by recording ambient electromagnetic noise. Then, a pulse of current oscillating at the Larmor frequency $f_0 = \omega_0/2\pi$ is generated in the transmitting loop

$$i(t) = I_0 \times \cos(\omega_0 t). \tag{2.48}$$

The pulse is characterized by the pulse moment

$$q = I_0 \times \tau_p, \tag{2.49}$$

where τ_p is the pulse duration. After the pulse is cut off, the instrument records an exponentially decaying free induction decay signal (FID1) oscillating at the Larmor frequency.

SNMR signals are characterized by the initial amplitude v_0 and the relaxation time constants T_1, T_2, T_2^* The amplitude v_0 depends on the amount of water below the loop and the aquifer geometry. The relaxation times allow estimating the saturated hydraulic conductivity of water-saturated formations.

The FID1 signal recorded following the first pulse and the instrumental delay Δt_{dead} allows measuring v_0 and T_2^*

$$v_1(t) = v_{01} \times \exp\left(-(t + \Delta t_{dead})/T_2^*\right), \tag{2.50}$$

where $v_1(t)$ is the envelope of oscillating at the Larmor frequency SNMR signal.

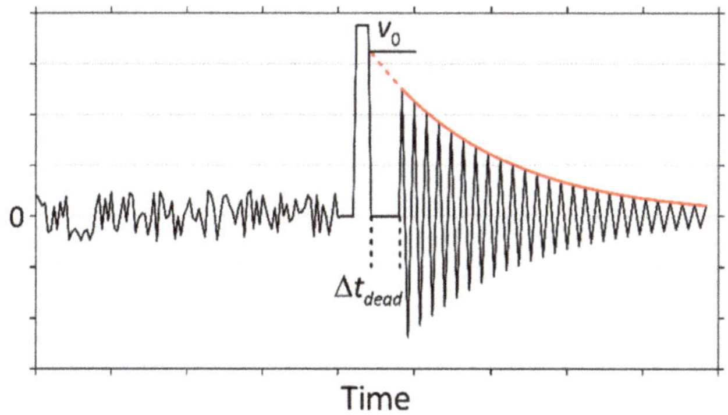

Figure 2.93. The one-pulse measuring sequence. Records before the pulse show the ambient EM noise and records after the pulse show the FID1 signal.

Two pulses separated by a variable delay τ_d allow measuring T_1. The free induction decay signal v_{02} is measured after the second pulse (FID2) and plotted versus τ_d

$$v_{02}(\tau_d) = \text{abs}\,(v_{01}) \times (1 - \exp(-\tau_d/T_1)). \tag{2.51}$$

Figure 2.94(a) shows two examples of the normalized FID1 signal decaying with different time constants T_2^* Figure 2.94(b) shows the normalized amplitude of the FID2 signal versus delay between the pulses. The black line corresponds to the signal with a short relaxation time T_1 and the red line to a long T_1.

For modeling the SNMR response using the Bloch model (Legchenko and Valla 2002), we assume a coincident transmitting/receiving loop and the FID1 signal. In the subsurface, the pulse of current in the loop creates an oscillating magnetic field $B_1(\mathbf{r})$. The nuclear magnetization of protons rotates at the flip angle

$$\theta_p = 0.5\,\gamma\, I_0^{-1} B_1\, q, \tag{2.52}$$

where γ is the gyromagnetic ratio of protons. One sounding comprises the SNMR signal with pulses of different pulse moments. The amplitude of the signal is given by the integral equation

$$v_0(q) = I_0^{-1} M_0 \omega_0 \int_V B_1 \sin(\theta_p) w(\mathbf{r}) dV, \tag{2.53}$$

where M_0 is the equilibrium nuclear magnetization of the protons, $0 \leqslant w(\mathbf{r}) \leqslant 1$ the water content, and $\mathbf{r} = r(x, y, z)$ the coordinate vector.

Equation 2.53 allows modeling the SNMR response considering 1D, 2D, and 3D geometry of the water-saturated formation. Figure 2.95 shows the amplitude versus pulse moment computed assuming a horizontally stratified subsurface and water-saturated layer ($w = 0.2$) at different depths (1D case). For modeling, we use a square loop of 80-m-side and the subsurface represented by a half-space of 100 Ωm (figures 2.95(a)) and 2 Ωm (figure 2.95(b)). These examples show a strong

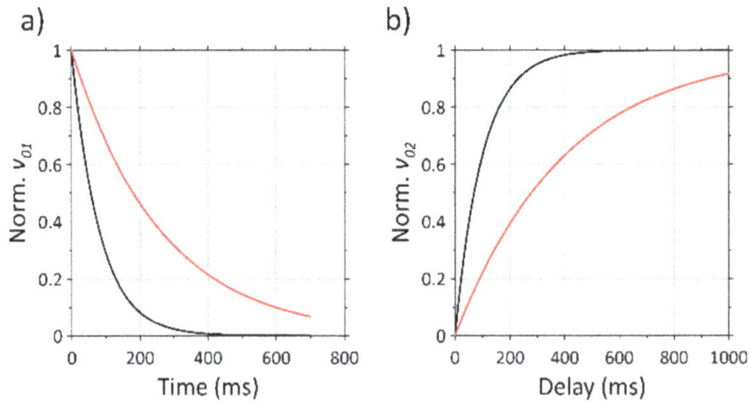

Figure 2.94. Examples of the SNMR signals: (a) FID1 signal with short (black) and long relaxation times T_2^*; (b) FID2 signal with short (black) and long relaxation times T_1.

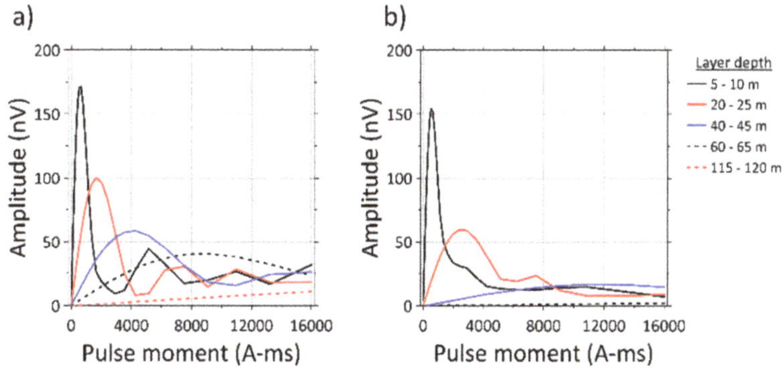

Figure 2.95. The amplitude of the SNMR signal was computed considering a 5 m-thick water-saturated layer (w = 0.2) at different depths. We assume an 80-m-side square loop and the half-space of: (a) 100 Ωm; and (b) 2 Ωm.

dependence of the SNMR signal of the layer depth and strong attenuation of the amplitude by electrically conductive subsurface.

The water content in the subsurface is the solution of equation (2.53). Usually, this linear integral equation is solved by approximating it by the matrix equation

$$\mathbf{Aw} = \mathbf{v}. \tag{2.54}$$

This SNMR inverse problem for water content is ill-posed and is often solved by the Tikhonov regularization method (Legchenko and Shushakov 1998). However, the use of the Tikhonov regularization is not exclusive and other methods have been reported in the literature (Guillen and Legchenko 2002a, 2002b, Mohnke and Yaramanci 2002, Grombacher *et al* 2017). Inversion can be carried out also in 2D (Girard *et al* 2007, Hertrich *et al* 2007) and 3D dimensions (Legchenko *et al* 2011, Chevalier *et al* 2014). The development of a time-lapse SNMR allows for more efficient use of SNMR data for hydrodynamic modeling (Legchenko *et al* 2020). Inversion for the relaxation time is also based on equation (2.53) (Legchenko and Valla 2002, Mohnke and Yaramanci 2005, Mueller-Petke and Yaramanci 2010, Müller-Petke *et al* 2013). Figure 2.96 summarizes an SNMR experiment.

Measurements of the FID1 signal provide the amplitude $v_{01}(q)$ and the relaxation time $T_2^*(q)$ versus pulse moment q. Measurements of the FID2 amplitude versus pulse $v_{02}(q, \tau_d)$ moment and the delay between the pulses τ_d allow measuring $T_1(q)$. Inversion of these data for the water content and for the relaxation time allows characterizing water-saturated formations

$$\begin{array}{c} v_{01}(q) \rightarrow w(\mathbf{r}),\ T_2^*(\mathbf{r}) \\ v_{02}(q, \tau_d) \rightarrow T_1(\mathbf{r}) \end{array}. \tag{2.55}$$

It is known that in water-saturated porous media T_1 and T_2^* can be used for estimating the mean grain size (Seevers 1966, Dunn *et al* 2002). The correlation between the aggregate class and the relaxation time T_2 was experimentally confirmed with SNMR (Schirov *et al* 1991). See table 2.7.

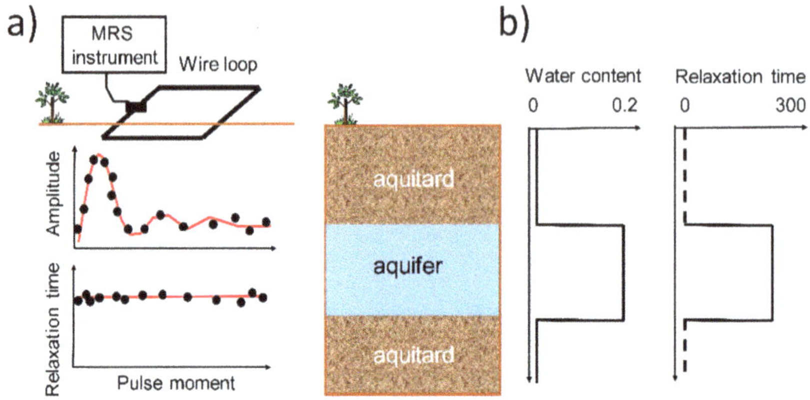

Figure 2.96. (a) Measured with SNMR data; (b) aquifer characterization after inversion (the dashed line shows that without water (w = 0) the relaxation time has no sense).

Table 2.7. Sediment grain size classification was defined by Wentworth (1922) and corresponding relaxation time T_2* reported by Schirov (Schirov *et al* 1991).

Wentworth size class	Grain size (μm)	Schirov aggregate class	T_2* (ms)
		Karst, bulk water	1000–2000
Very coarse sand	1000–2000	Gravel deposits	300–600
Coarse sand	500–1000	Coarse and gravelly sands	180–300
Medium sand	250–500	Medium sands	120–180
Fine sand	125–250	Fine sands	60–120
Very fine sand	63–125	Clayey sands, very fine sands	30–60
Coarse silt	31–63		
Medium silt	15.6–31		
Fine silt	7.8–15.6	Sandy clay	< 30
Very fine silt	3.9–7.8		
Clay	0.02–3.9		

The presented correlation shows that long relaxation times correspond to water in the geological material with large gran sizes and corresponding large pore spaces. Short relaxation times correspond to low permeable sediments with small gran sizes. Note that SNMR instruments allow reliable measurement of the SNMR signals with the relaxation time $T_2^* \geqslant 30$ ms and hence water in silt and clay cannot be detected with SNMR. A similar correlation between the relaxation time T_1 measured with the SNMR and the mean pore size was reported by Legchenko *et al* (2004).

In contrast to the water level measured in the borehole, SNMR results are not linked up to the hydraulic head and SNMR shows the location of a water-saturated formation.

Figure 2.97 shows expected SNMR results in typical geological settings.

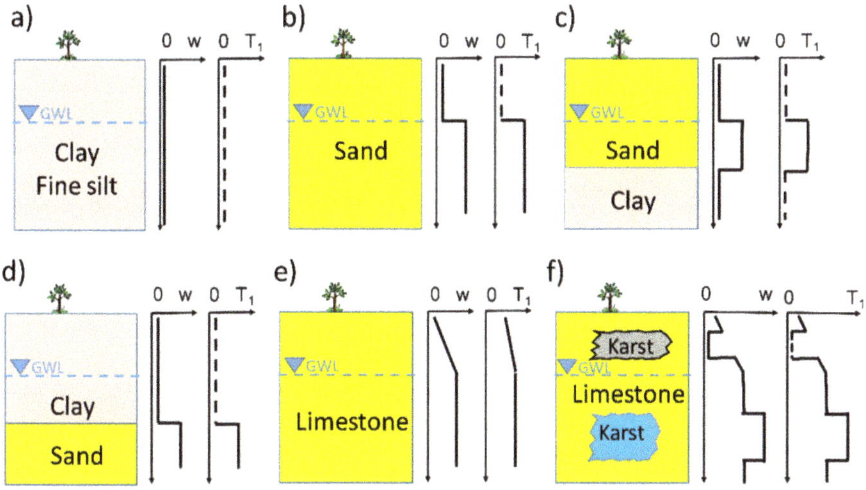

Figure 2.97. Expected SNMR results in typical geological setups. (a) In clay and fine silt SNMR does not show water. (b) In quartz sand water can be detected in the saturated part of the formation. (c) SNMR shows water only in sand formation. (d) In a confined aquifer SNMR does not show the water level, but the aquifer formation. (e) In non-magnetic material (chalk, limestone) SNMR shows water in both saturated and unsaturated zone. (f) Karst voids without water cause smaller SNMR water content and voids filled with water show increased values of both the water content and the relaxation time.

SNMR is a large-scale method and it can detect water in karst when the total volume occupied by karts below the measuring loop exceeds a few hundred cubic meters. The principal signature of water in karst caverns and channels is a long relaxation time (>400 ms). In unsaturated limestone, karst voids show no water. However, interpretation of SNMR measurements in the term of karst can be done only by taking into account the geological environment. The absence of water may be a clay deposit and not karst. A long relaxation time may point to a gravel formation and not to karst channels.

Karst investigation may be carried out using different approaches. Just one sounding is required if the goal is to confirm either the existence or the absence of karst below the loop. Bulk water in karst can be reliably identified by a long relaxation time. For example, this may be the case of a geotechnical study.

An approximate location of a karst zone can be found using measurements performed in a profile crossing the expected karst location (figure 2.98). Variations of the SNMR response along the profile can be interpreted as karst.

In two previous examples, interpretation of SNMR measurements does not require doing an inversion. Just analyzing raw signals can be sufficient. A more accurate study requires inversion. Inversion can be done considering a 1D, 2D, or 3D configuration. Measurements and inversion are two linked steps and the field setup must correspond to the inversion configuration. Figure 2.99(a) shows the 1D field setup. The measuring loops are separated over one length of the loop side. With a 1D measuring setup, 2D or 3D inversions cannot be carried out. A 2D (figure 2.99(b)) and a 3D (figure 2.99(c)) measuring setups assume that loops are

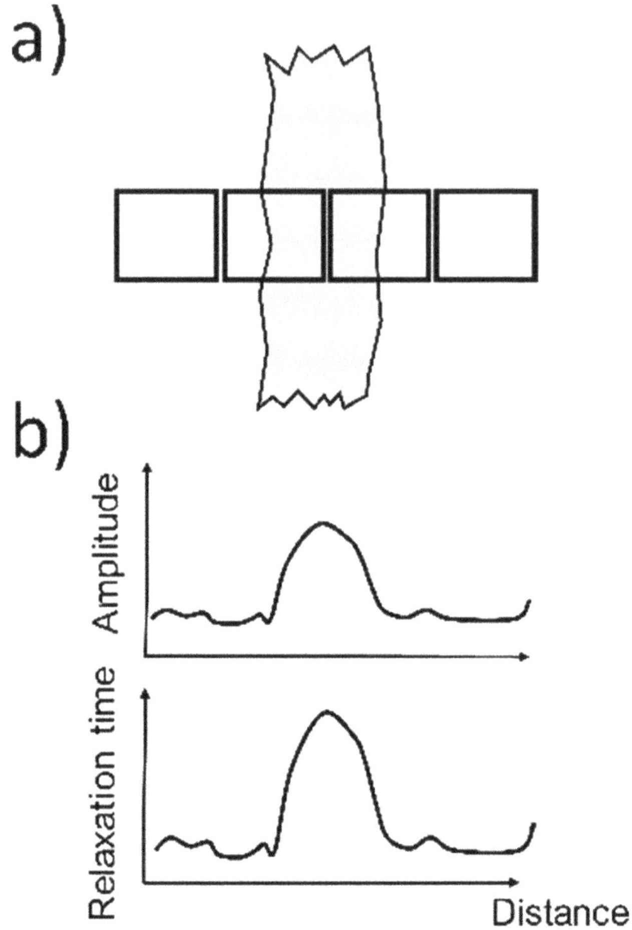

Figure 2.98. Example of profiling with SNMR. (a) A karst channel crossed by SNMR profile. (b) Expected SNMR response shows increased amplitude and relaxation time over the karst position.

set side by side or even overlapped. A 3D inversion cannot be done with 2D measurements. However, 3D measurements allow performing a 3D inversion or series of 1D and 2D inversions by selecting data corresponding to the inversion type.

Typical measuring loops used with the SNMR instrument are shown in figure 2.100. The area investigated with one loop is shown with dashed lines. Groundwater outside this area does not contribute to the measured SNMR signal. The square loop allows a larger depth of investigation but is sensitive to EM interference. In urban areas or close to electrical power lines a figure-of-eight loop should be considered.

SNMR equipment is commercially proposed by three companies. Vista Clara (USA) manufactures a large selection of magnetic resonance (NMR) geophysical tools for hydrogeological applications including SNMR equipment. IRIS Instruments (France) is the oldest manufacturer of SNMR instruments. Both

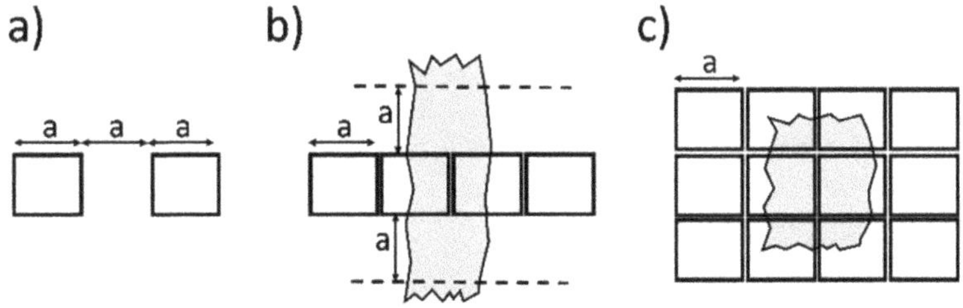

Figure 2.99. SNMR measuring setup around karst position: (a) 1D; (b) 2D; (c) 3D.

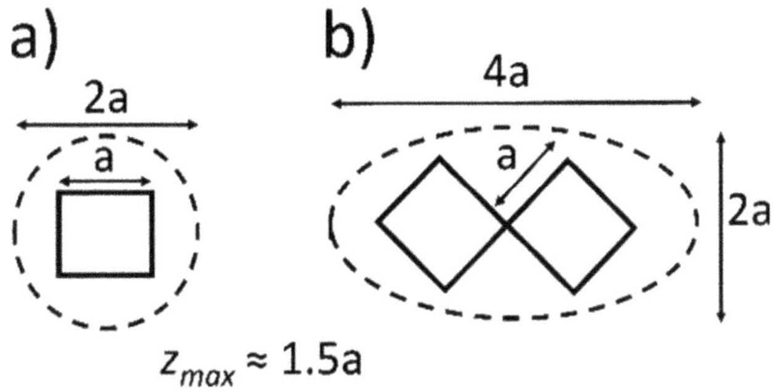

Figure 2.100. Typical SNMR loops: (a) square; (b) figure-of-eight. Dashed lines show an approximate area investigated with these loops.

companies propose different modifications of the instrument to sell or for rent. Vista Clara also offers flexible professional services. Radic-Research (Germany) manufactures a small-size SNMR instrument with the depth of investigation down to 10 m.

Processing and interpretation of SNMR data can be performed using software packages provided by the hardware manufacturers or using the MRSmatlab software (Müller-Petke *et al* 2016). MRSmatlab is a manufacturer-independent open platform that can be accessed by contacting the authors.

2.2.1.12.2 Example of karst investigation

Karst can develop in different geological environments as, for example, limestone, salt and ice, and karst structures are not always an easy target for investigation. It has been shown that the SNMR method is an efficient tool for investigating the water in karst caverns and channels (Vouillamoz *et al* 2005, Legchenko *et al* 2008a, 2014, Garambois *et al* 2016). Here, we discuss the principal steps to follow when investigating karst. We will use an example reported in the literature and reprocessing the field data using more recent software (Boucher *et al* 2006, Guérin *et al* 2009).

This SNMR survey has been carried out above a known karst system developed in the limestone geological environment in the Lot department of France (figure 2.101). The Poumeyssens karst shaft provides the possibility of visiting large interconnected conduits. One of them was selected for the methodological study. The water table is shallow, and the conduit is filled with water. For measuring, the $NUMIS^{PLUS}$ system

Figure 2.101. Location of the test site Poumeyssens.

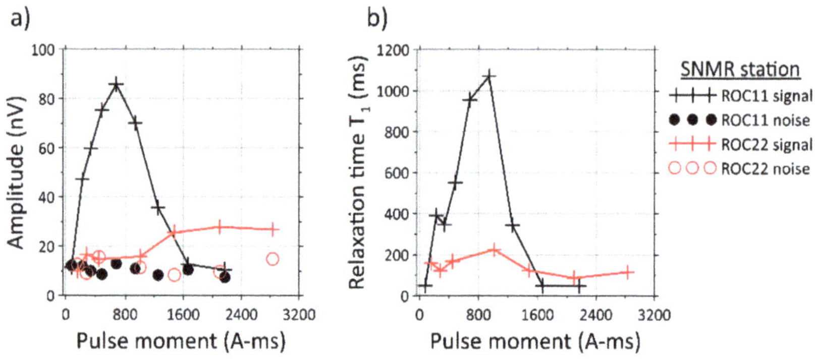

Figure 2.102. The amplitude and the relaxation time versus pulse moment were measured at two SNMR stations 50 m in between. (a) The amplitude (crosses) and the noise level (circles); (b) the relaxation time T_1.

of IRIS Instruments (France) with a 25 m-side square coincident loop (three turns) was used. The Larmor frequency was 1975 Hz. With this setup, water can be detected down to 40 m, and the resolution estimated with the singular value decomposition (SVD) method (Legchenko *et al* 2014, 2017) showed about 20 m.

At the beginning of the survey, the position of karst channels was known only approximately. For detecting the conduit, a few isolated SNMR soundings have been carried out. Figure 2.102 shows SNMR signals measured at two stations 50 m in between. In this geological environment, the increased amplitude and the relaxation longer than 400 ms observed at the station ROC11 undoubtedly point to a water-full karst formation.

The 1D inversion of this data set was not used because of the known 2D target. The second station (ROC22) showed no karst, and we used it for characterizing water in the limestone matrix (figure 2.103).

Using the results of these two stations, nine additional SNMR soundings for making a 2D profile were carried out. The orientation of this profile was expected perpendicular to the conduit. After the geophysical investigations were completed, the karst network was mapped by cave divers using a compass and a graduated Ariane's line. The inaccuracy of this method was estimated better than a few meters. The cross-section of the conduit was varying between 30 and 50 m^2. Figure 2.104

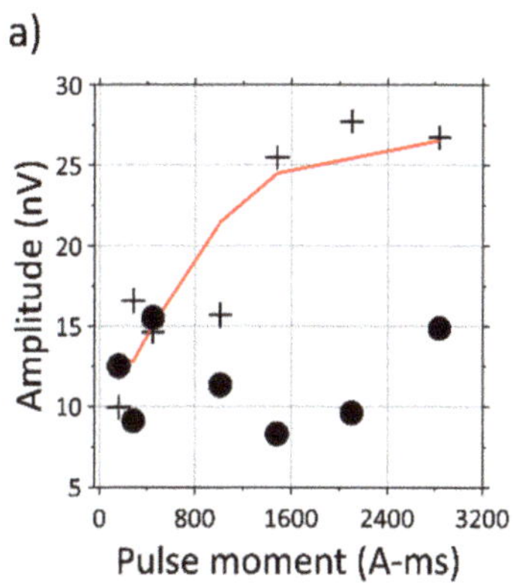

Figure 2.103. SNMR results at the station ROC22. (a) The measured amplitude (crosses) fitted by the inversion (red line) and the noise level (circles); (b) the water content versus depth.

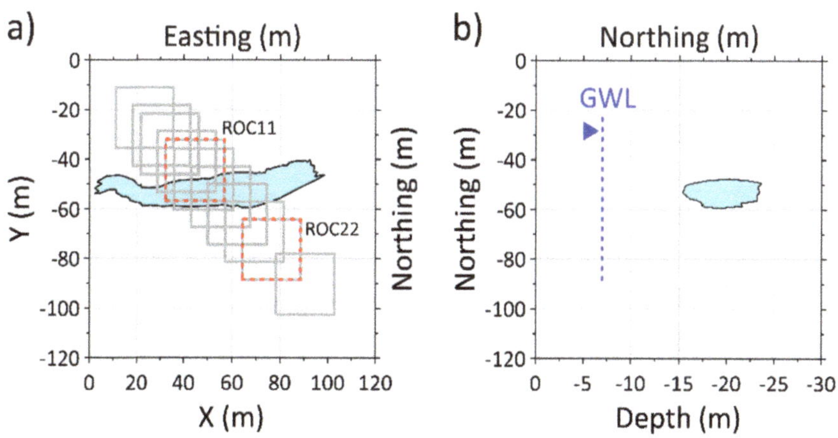

Figure 2.104. (a) Location map of the karst conduit and SNME loops; (b) the cross-section of the conduit. The *X* and *Y* coordinate axes correspond to the E–W and N–S directions. The red dashed lines show the positions of the stations ROC11 and ROC22.

shows the measuring loops and the conduit delineated by divers. One can see that the profile crosses the conduit with an angle of 45° relative to the perpendicular direction.

Red dashed lines in figure 2.105 show a band corresponding to large amplitudes and long relaxation times that may correspond to the conduit location. The resolution of karst localization using raw signals is limited and degraded by the non-perpendicular position of the profile direction and the conduit axis. However, the accuracy can be improved by using inversion results.

The data acquisition was carried keeping in mind a 2D inversion. However, the geometrical assumption for the 2D inversion was not respected, which may cause errors in the inversion results. On the other hand, the data set is not sufficient for accurate 3D inversion. We take into account variations of the sensibility within the SNMR loop (Girard *et al* 2007) and expect to get better results using the 3D inversion. Figure 2.106 shows discretization of the subsurface with homogeneous cells of 5 × 5 × 2 m^3.

3D inversion was done using the Tikhonov regularization method (Legchenko *et al* 2011). Measured amplitudes were fitted by the inversion fit with the RMSE corresponding to the mean noise level (figure 2.107).

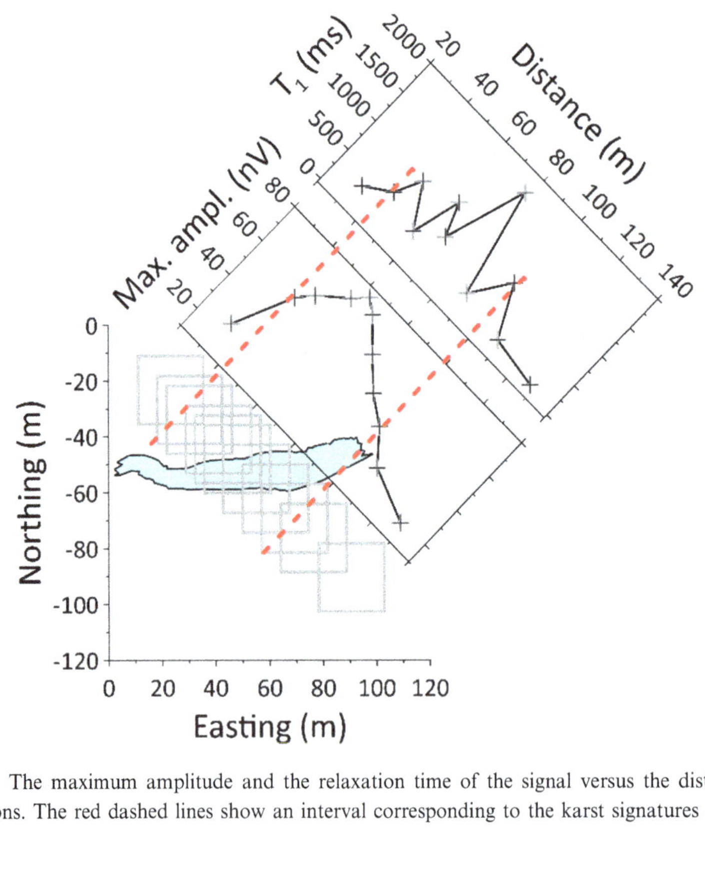

Figure 2.105. The maximum amplitude and the relaxation time of the signal versus the distance between SNMR stations. The red dashed lines show an interval corresponding to the karst signatures observed with SNMR.

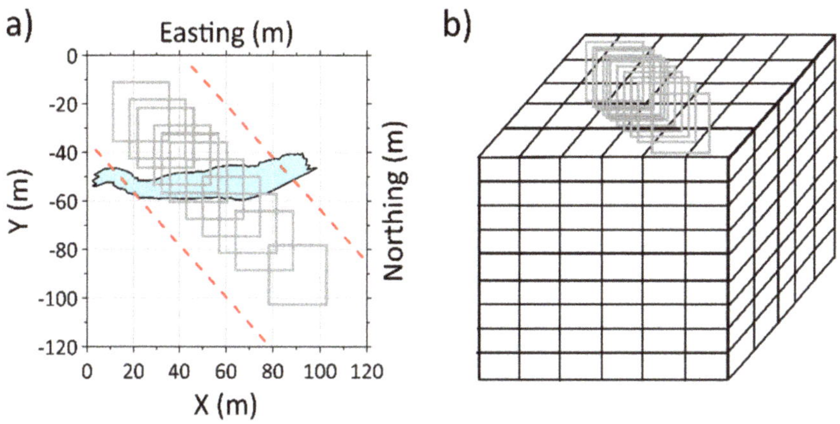

Figure 2.106. (a) Position of the SNMR profile relative to the karst conduit. The sensitivity zone is shown by the red dashed lines; (b) 3D discretization of the subsurface below SNMR field setup.

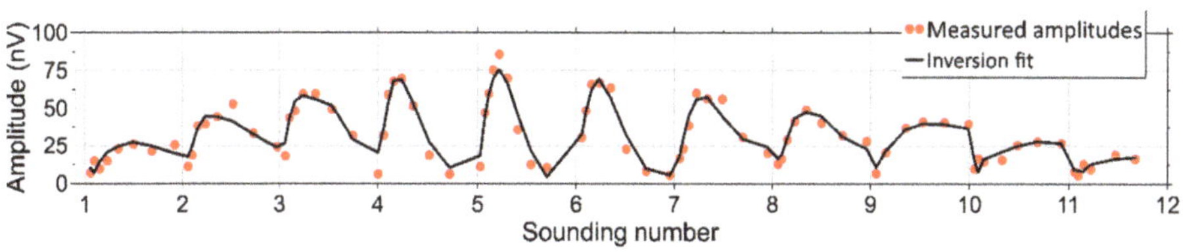

Figure 2.107. Measured amplitude of SNMR signal versus normalized pulse moment (red dots) and the inversion fit corresponding to the 3D inverse model (black line). The amplitudes corresponding to each sounding are plotted between the sounding numbers along with the profile (11 SNMR stations).

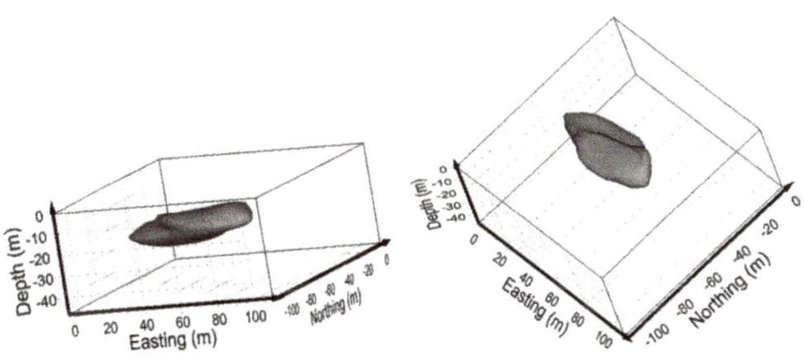

Figure 2.108. Position of the karst conduit determined by the 3D inversion.

Figure 2.108 shows the position of the karst conduit determined by the 3D inversion. The inverse model suggests a single conduit at a depth of 20 m, which is in good agreement with the ground truth.

Figures 2.109–2.112 show the detailed position of the conduit.

Results of the SNMR application presented in this example show that SNMR allowed locating the karst conduit filled with water. The limited resolution of 3D inversion causes an imprecise delineation of the conduit body and an

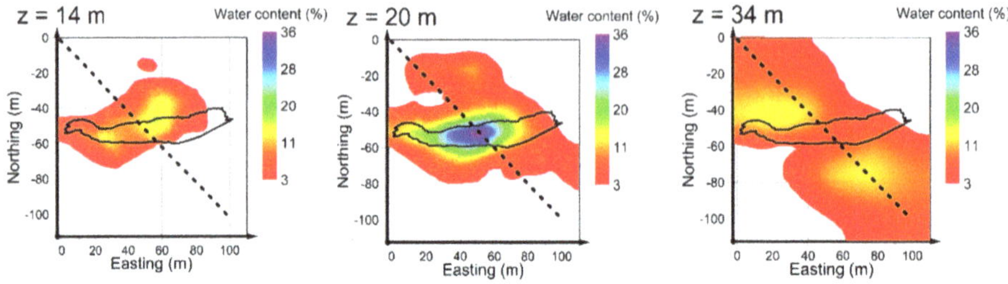

Figure 2.109. The water content in horizontal slices at different depths. The dashed line shows the profile axis and the black line shows the position of the conduit outlined by divers.

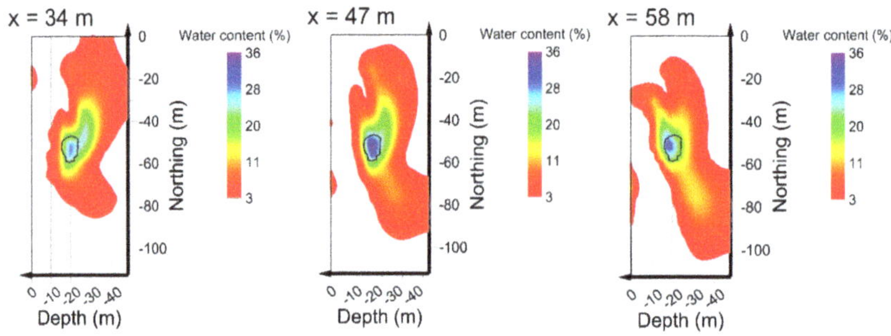

Figure 2.110. The vertical slices of the water content at different positions along the *X*-axis. The black line shows the true position of the conduit.

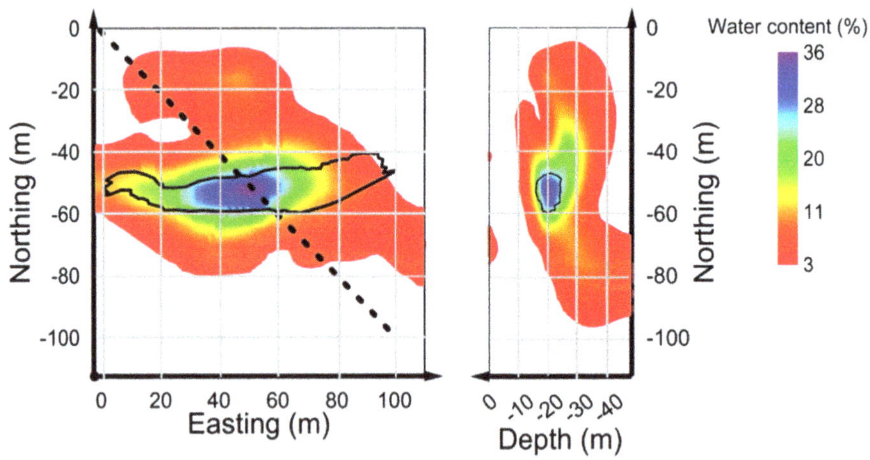

Figure 2.111. The water content image along the conduit.

underestimation of the water content (36% instead of 100%). Note that the resolution was additionally degraded by the field setup that provided insufficient coverage for the 3D inversion.

For this study, we did not use assumptions on the conduit position and geometry. Thus, the results represent a mathematical projection of field measurements onto

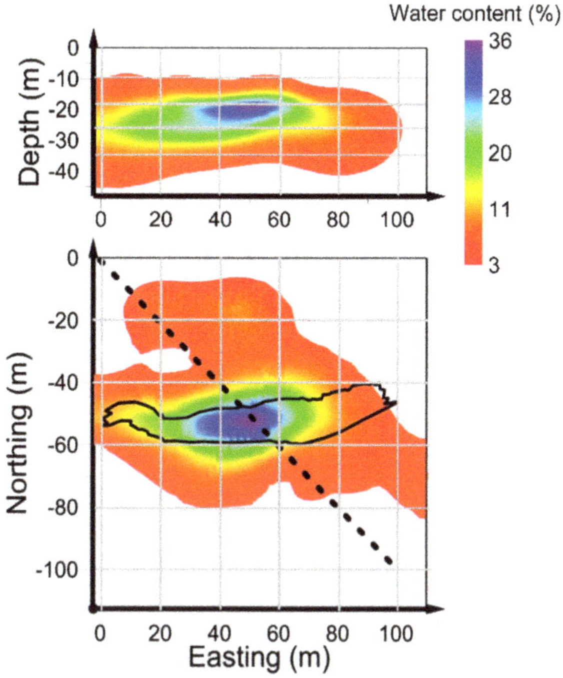

Figure 2.112. The water content image in the transverse to conduit direction.

solution space. SNMR alone cannot provide better knowledge about the karst, and various inversion procedures can only provide a variety of equivalent models. For selecting the best inverse model or improving the accuracy of the results, information about the conduit location derived from sources other than SNMR should be used. For that, additional geophysical or geological studies can be considered.

2.2.1.13 Microgravity

2.2.1.13.1 Brief theoretical background

The potential nature of a gravity field, which follows from the law of gravitation, is expressed by the following equations:

$$\text{rot}\,\mathbf{F} = 0 \tag{2.58}$$

and

$$\mathbf{F} = \text{grad}\,W, \tag{2.59}$$

where **F** is the gravity field intensity and W is the gravity potential.

The proportionality between the gravity field intensity and the density σ is determined by the expression:

$$\text{div}F = -4\pi G\sigma, \tag{2.60}$$

where G is the gravity constant.

As follows from equations (2.60) and (2.59), the gravity potential must satisfy Poisson's equation (Kaufman 1992):

$$\nabla^2 W = -4\pi G\sigma, \tag{2.61}$$

whereas points, where the mass is zero, ($\sigma = 0$)should satisfy Laplace's equation:

$$\Delta W = \frac{\partial^2 W}{\partial x^2} + \frac{\partial^2 W}{\partial y^2} + \frac{\partial^2 W}{\partial z^2}. \tag{2.62}$$

The potential character of these fields and proportionality of the main interpreting models allows application in the gravity method of advanced interpretation methodologies developed in magnetic prospecting (Eppelbaum 2007).

A key role in microgravity investigations is played by 3D gravity field modeling. Among several approaches of geological body approximation, most exact and optimal is the mathematical description of geological targets by a 3D horizontal polygonal prism (Khesin *et al* 1996, Eppelbaum and Khesin 2004).

An analytical expression for the first vertical derivative of gravity potential of (m − 1) angle horizontal prism has been obtained by integrating a common analytical expression:

$$W_{z'} = -\int_S \frac{z}{(R+y)R} dxdz \Big|_{y_1}^{y_2}, \tag{2.63}$$

where $R = \sqrt{x^2 + y^2 + z^2}$; S is the area of the normal section of the prism by the plane of XYZ.

$$W_{z'T} = \left\{ \begin{aligned} & -G\sigma \sum_{j=1}^{m-1} \left[V_j \sin\alpha_j \left(\text{In}\frac{R_{12j} + y_2}{R_{22j} + y_2} - \text{In}\frac{R_{11j} + y_1}{R_{21j} + y_1} \right) + \right. \\ & V_j \cos\alpha_j \left(\text{sgn}\left(y_2 V_j \right) \arccos\frac{V_j^2 R_{12j} + U_{1j}U_{2j}y_2^2}{r_{1j}r_{2j}\left(y_2^2 + V_j^2\right)} \right) - \\ & \left(\text{sgn}\left(y_1 V_j \right) \arccos\frac{V_j^2 R_{11j}R_{21j} + U_{1j}U_{2j}y_1^2}{r_{1j}r_{2j}\left(y_2^2 + V_j^2\right)} \right) + \\ & \left. \cos\alpha_j \left(y_2\,\text{In}\frac{R_{12j} + U_{1j}}{R_{22j} + U_{2j}} - y_1\,\text{In}\frac{R_{11j} + U_{1j}}{R_{21j} + U_{2j}} \right) \right] \end{aligned} \right\} \tag{2.64}$$

where G is the gravitational constant, σ is the density of the body, and α_j is the angle of the prism's side inclination.

Noise elimination and non-conventional calculation of terrain correction

Elimination of different kinds of noise appearing in microgravity investigation (and especially—corrections for the surrounding terrain relief) are considered in detail in Khesin *et al* (1996), Eppelbaum and Khesin (2004), Debeglia *et al* (2006),

Eppelbaum and Khesin (2012), Al-Zoubi *et al* (2013), Eppelbaum (2015a, 2017a, 2017b, 2019a), Pašteka *et al* (2017).

2.2.1.13.2 Examples of inverse problem solution and 3D modeling

A microgravity survey at the Medford Cave site (Florida, USA) was conducted to recognize the shallow subsurface air-filled cavities and tunnels (Butler 1984) (figure 2.113). For the Bouguer and terrain corrections calculation, the density of 1900 kg m^{-3} was employed. One of the profiles presented in Butler (1984) was quantitatively analyzed by the aforementioned methodologies (Eppelbaum 2007, 2011b). Figure 2.113 displays the results of interpretation of the most significant anomaly I (almost 70 microGals). Interpretation results have a good agreement with the geological data.

Planning of any microgravity investigations should be accompanied by the development of preliminary physical–geological models (***PAM***) (Eppelbaum *et al* 2000, Eppelbaum 2010) (results of 3D modeling by the use of the 3D software). The ***PAM*** presented in figure 2.114 reflects a real area with a buried cave located in the

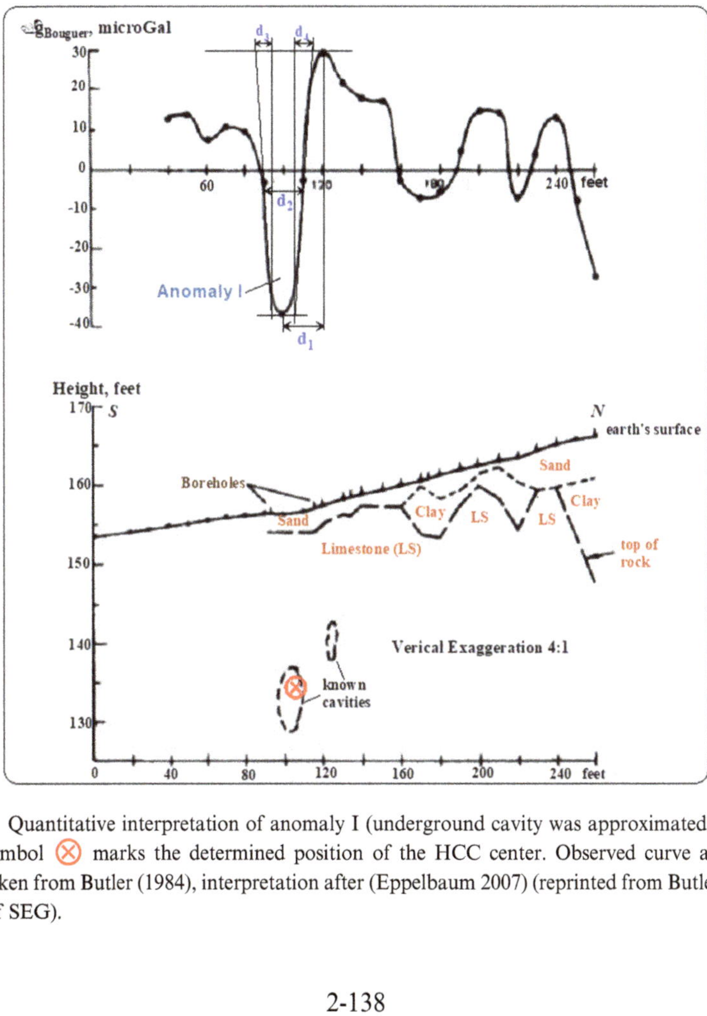

Figure 2.113. Quantitative interpretation of anomaly I (underground cavity was approximated by the model of HCC). Symbol ⊗ marks the determined position of the HCC center. Observed curve and geological section are taken from Butler (1984), interpretation after (Eppelbaum 2007) (reprinted from Butler (1984), with permission of SEG).

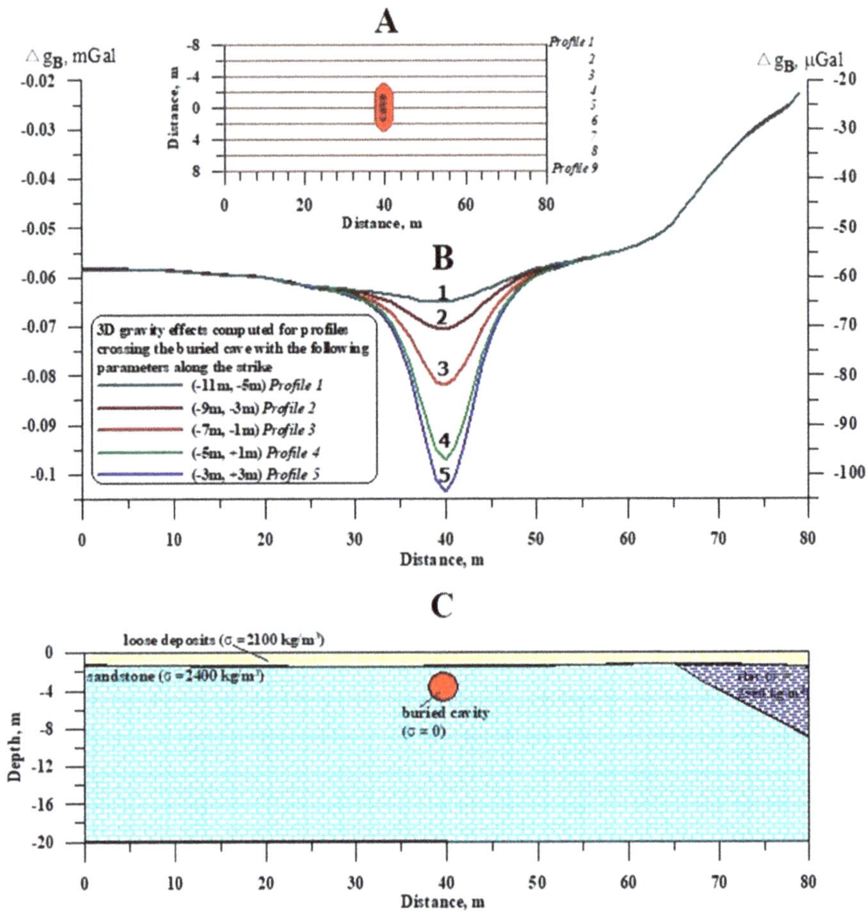

Figure 2.114. Physical–geological model of the buried cave and computed 3D gravity anomalies. (A) Location of the projected profiles and buried cave (view over). (B) Computed gravity effects along with the profiles 1–5. (C) Geological–archaeological sequence (reprinted from Eppelbaum (2011b), with modifications, with gratitude to Hindawi Publisher).

vicinity of Beit-Shemesh town (central Israel). The developed ***PAM*** was used for estimation of the expected gravity anomaly amplitudes, calculating the most optimal step of observations along proposed observation profiles and distances between profiles.

Even simple calculation of gravity field derivatives sometimes can also give important information (figure 2.115). This figure shows that 3D computing of the third derivative of gravity potential W_{xxz} (g_{xx}) makes it possible to reliably divide gravity effects from two closely located hidden caves and to obtain significant information about their geometric form (Eppelbaum 2011b).

Microgravity comparatively often is employed for the underground cave detection and localization (Bishop *et al* 1997, Beres *et al* 2001, Rybakov *et al* 2001, Di Filippo *et al* 2005, Styles *et al* 2006, Debeglia *et al* 2006, Eppelbaum *et al* 2008,

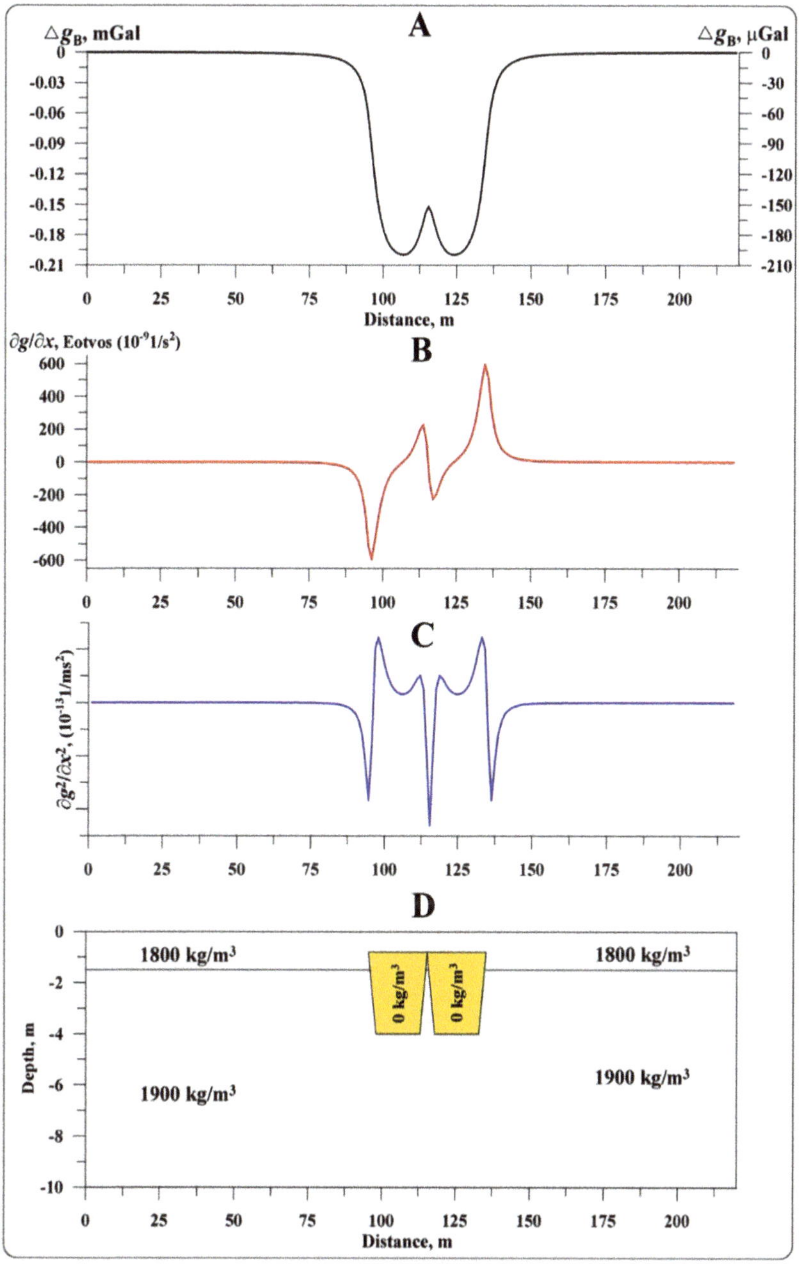

Figure 2.115. The computing of horizontal derivatives from models of two nearby occurring hidden caves. (A) Computed gravity curve. (B) Calculated first horizontal derivative of gravity field Δg_x. (C) Calculated second horizontal derivative of gravity field Δg_{xx}. (D) Physical–geological model (reprinted from Eppelbaum *et al* (2008), with permission from EGU).

Deroussi *et al* 2009, Leucci and De Giorgi 2010, Mochales *et al* 2008, Eppelbaum 2011b, 2017a, 2017b, Kaufman *et al* 2011, Al-Zoubi *et al* 2013, Pivetta and Braitenberg 2015, Howari *et al* 2016, Jacob *et al* 2020).

2.2.1.14 Magnetometry

2.2.1.14.1 Brief theoretical background

The Earth's magnetic field in the first approximation may be described by Maxwell's stationary equations:

$$\mathrm{div}\overrightarrow{B} = 0;\ \mathrm{rot}\overrightarrow{T} = 0, \tag{2.65}$$

where $\vec{B} = \vec{T} + 4\pi\vec{J}$ is the vector of magnetic induction, $\vec{T}$ is the vector of magnetic field strength, $\vec{J}$ is the vector of magnetization of the medium.

In the domain free from magnetic field sources, $\vec{J} = 0$ and vector of magnetic induction $\vec{B}$ is coincided with a vector $\vec{T}$ that allows using in the magnetic field strength a dimension of magnetic induction (Tesla). The unit used in the magnetic prospecting,

1 nanoTesla = 10^{-9} Tesla.

At the same time, the vector of the magnetic field strength is determined as a gradient of force function U named as magnetic potential:

$$\overrightarrow{T} = -\mathrm{grad}\,U. \tag{2.66}$$

It follows from (2.63) and (2.64) that by $\vec{J} = 0$ magnetic potential satisfies the Laplace equation

$$\mathrm{div}(\mathrm{grad}\,U) = \nabla^2 U = 0, \tag{2.67}$$

and by $\vec{J} \neq 0$satisfying the Poisson equation

$$\nabla^2 U = 4\pi \tag{2.68}$$

where ∇ is the Hamilton operator.

To study the Earth's magnetic field we usually use the rectangular Cartesian coordinate system: x-, y-, and z-coordinates (Ox, Oy, and Oz) are directed to the geographical north, to east, and to the Earth's center, respectively.

Projections of vector $\vec{T}$ to the coordinate axes (according to equation (2.64)):

$$X = -\partial U/\partial x;\ Y = -\partial U/\partial y;\ Z = -\partial U/\partial z. \tag{2.69}$$

Projection of vector $\vec{T}$ to the horizontal plane xOy is called a horizontal component H.

The main angle characteristics of the Earth's magnetic field are magnetic inclination I (angle between the horizontal component and vector $\vec{T}$) and magnetic declination D (angle between the axis Ox and horizontal component H).

The projections to coordinate axes and angle components have the following relationships:

$$\left.\begin{aligned} &X = H\cos D;\ H = T\cos I; \\ &Y = H\sin D;\ Z = T\sin I; \\ &T = \left|\vec{T}\right| = \sqrt{X^2 + Y^2 + Z^2}; \\ &H = \sqrt{X^2 + Y^2}. \end{aligned}\right\} \tag{2.70}$$

The total Earth's magnetic field $\overrightarrow{T_{\text{Earth}}}$ may be approximated by several components:

$$\overrightarrow{T_{\text{Earth}}} = \vec{T_1} + \vec{T_2} + \vec{T_3} + \vec{T_4} + \vec{T_5} + \vec{T_6} + \vec{T_7}, \tag{2.71}$$

where $\vec{T_1}$ is the field of uniformly magnetized Earth, $\vec{T_2}$ is the field caused by continental masses, $\vec{T_3}$ is the field from regional objects, $\vec{T_4}$ is the field from large-scale objects, $\vec{T_5}$ is the effect from local objects, $\vec{T_6}$ is the influence of small objects, and $\vec{T_7}$ is the effect of temporary magnetic variations due to various processes in the Earth's ionosphere and geodynamic processes inside the Earth.

The effect of temporal variations $\vec{T_7}$ (primary and sometimes—secondary components) is removed by conventional (Parasnis 1997) and non-conventional (Eppelbaum *et al* 2001) methods, respectively. Component $\vec{T_6}$ (occasionally component $\vec{T_5}$ should be also taken into account) is an object of our investigation.

2.2.1.14.2 Magnetic data processing and interpretation

Some conventional procedures of processing and interpretation are presented in popular textbooks—e.g., Telford *et al* (1990), Parasnis (1997), Kaufman *et al* (2008).

The developed advanced system of magnetic data analysis includes: (1) calculation of secondary effect of temporal magnetic variations (Eppelbaum *et al* 2001); (2) correlation method of elimination the rugged terrain relief influence (Khesin *et al* 1996); (3) calculation of the averaged magnetization of the upper part of geological section (Khesin *et al* 1996, Eppelbaum 2010); (4) application of information and wavelet methodologies for delineation of weak magnetic anomalies against the strong noise background (Eppelbaum *et al* 2011a, Eppelbaum and Khesin 2012, Eppelbaum 2014, 2017a); (5) advanced methods of magnetic anomaly quantitative interpretation under conditions of oblique magnetization, rugged terrain relief and unknown level of the normal magnetic field (Khesin *et al* 1996, Eppelbaum *et al* 2000, 2001, Eppelbaum and Mishne 2011, Eppelbaum 2015a, 2019a); (6) 3D magnetic field modeling by the use of GSFC program (Khesin *et al* 1996, Eppelbaum *et al* 2000, 2001); (7) development of final physical–geological model of the medium (Khesin *et al* 1996, Eppelbaum *et al* 2001, 2004).

The major principles of the quantitative interpretation are formalized for the conditions of vertical magnetization and do not work under conditions of oblique direct and oblique reverse magnetization (firstly in low and middle latitudes). Besides the influence of oblique magnetization, additional distortions of ΔT anomaly occur due to rigged (inclined) relief influence and superposition of magnetic anomalies of various orders.

In conditions of oblique magnetization, the 'reduction to pole'—calculation of pseudo-gravimetric anomalies procedure (e.g., Blakely 1995)—is often used. However, the procedure is suitable only when all interfering bodies in the studied area are magnetized parallel to the geomagnetic field (or to some known common magnetic field inclination angle) and simultaneously have sub-vertical dipping. Only in the case when the mentioned restrictions are observed can the magnetic fields be recalculated correctly; the obtained graphs would be symmetrical, and further interpretation using some conventional methods can be done. Similar approaches based on the specific transformations of the observed magnetic field (for instance, analytic signal (e.g., Roest *et al* 1992)), have the same limitations.

Our interpretation involves the application of methods developed especially for quantitative interpretation of magnetic anomalies in complex environments (oblique magnetization, rugged relief, and unknown level of the normal field) (Khesin *et al* 1996).

These methods employed improved modifications of the *characteristic point method*, the *tangent method*, and the *areal method*, utilizing the most commonly applied geometric models such as TB, HCC/sphere, thick bed, and horizontal plate (detailed explanation of these methods is given in Khesin *et al* (1996), Eppelbaum *et al* (2000, 2001), Eppelbaum (2011a, 2015a)). These four geometric models, with different modifications, may be used for approximation and corresponding quantitative interpretation of anomalies generated by various geological objects.

The following parameters are taken from the anomaly plot in the *characteristic point method*: d_1 = difference of semiamplitude point abscissae, d_2 = difference of extremum abscissae, d_5 = difference of inflection point abscissae.

In the *tangent method*, four tangents are employed: two horizontal lines concerning the anomaly extrema and two inclined lines passing through the points of the bend on the left- and right-hand branches of the anomaly plot. The following parameters are taken from the plot: d_3 = difference in abscissae of the points of intersection of an inclined tangent with horizontal tangents on one branch; d_4 = the same on the other branch (d_3 is selected from the plot branch with conjugated extremums), d_6 = interval between d_3 and d_4, d_8 = difference in abscissae between the point of intersection of a left inclined tangent with lower horizontal tangent and inflection point on the left branch (d_6 and d_8 are used for the model of thick bed only).

The *areal method* is based on the calculation of separate areas limited by the anomalous curve, horizontal line, and two vertical lines crossing some singular points at the anomalous curve (Eppelbaum and Mishne 2011).

When anomalies are observed on an inclined profile, the obtained parameters reflect not real, but a certain fictitious body. The transition from the fictitious body parameters to those of the real body can be realized using the following expressions (the subscript '*r*' stands for parameters of the real body):

$$\begin{Bmatrix} h_r = h + x_o \tan \omega_o \\ x_r = -h \tan \omega_o + x_o \end{Bmatrix}, \tag{2.72}$$

where h is the depth of body upper edge occurrence (or HCC center), x_o is the shifting of anomaly maximum from the projection of the center of the disturbing body to the Earth's surface (caused by oblique magnetization), and ω_o is the angle of the terrain relief inclination ($\omega_o > 0$ when the inclination is toward the positive direction of the x-axis).

It can be noted the following application of detailed magnetic investigations for underground cave localization (Rybakov *et al* 2005, Kis and Pustza 2006, Gibson *et al* 2004, Kim *et al* 2007, Mochales *et al* 2008, Saribudak 2010, Eppelbaum 2011a, 2019a, Balkaya *et al* 2012, Saad *et al* 2015).

The definite perspective of magnetic method evaluation is associated with increasing accuracy and frequency of measurements of magnetic survey and development of a new generation of remote operative vehicle (ROV) investigations (e.g., Eppelbaum and Mishne 2011, Eppelbaum 2016, 2019a).

2.2.1.14.3 Examples of inverse problem solution

An example of an inverse problem solution for the model of karst ($I = 0$) occurring in the magnetized medium ($I = 500$ mA m^{-1}) with oblique magnetization (45°) is presented in figure 2.116. A simple analysis of the magnetic curve testifies that a

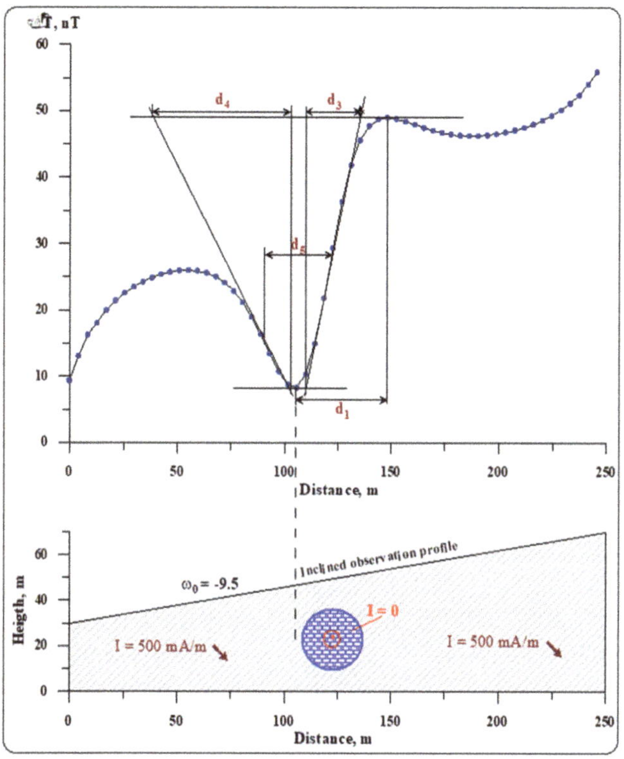

Figure 2.116. Quantitative analysis of model magnetic field generated by buried cave under complicated environments. Symbol ◉ designates the location of HCC center. (The following material from Eppelbaum (2007), is published with the permission of the Environmental and Engineering Geophysical Society. All copyright privileges remain with EEGS. This material cannot be copied or used without the express written permission of EEGS.)

borehole drilled at the minimal amplitude of magnetic field will not discover the hidden karst (due to the influence of oblique magnetization). Application of the developed methodology of inverse problem solution (Khesin *et al* 1996, Eppelbaum *et al* 2000, 2001, Eppelbaum 2015a) allows determining the position of HCC center with high accuracy.

The next example (figure 2.117) displays the application of ΔZ magnetic survey for detection of underground cavity occurring in strongly magnetized medium (basalts) in the Lava Beds National Monument (California, USA). The negative amplitude of the magnetic anomaly exceeds the value of 2000 nT. Results of the interpretation method application are displayed in figure 2.117.

The advanced methods of quantitative analysis developed in magnetic prospecting (Khesin *et al* 1996, Eppelbaum *et al* 2000, 2001, Eppelbaum 2015a) can be transferred to other potential and quasi-potential fields. For this aim similarity of fields and analytical expressions can be proved. The similarity of some analytical expressions of magnetic, gravity, self-potential, and temperature anomalies are shown in table 2.8.

Here Z_v is the vertical magnetic field component at vertical magnetization; J is the magnetization; b is the horizontal semi-thickness of TB; m is the magnetic mass; G is

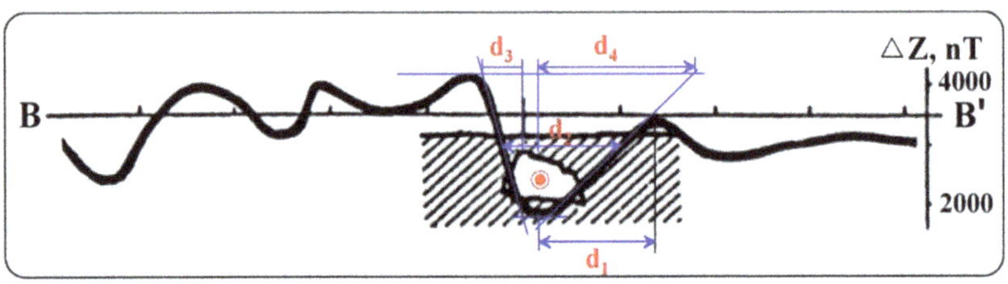

Figure 2.117. Interpretation of intensive negative magnetic anomaly along profile B–B′ observed over the buried void in basalt (Lava Beds National Monument, California). Symbol ⦿ marks the position of HCC center (owner of the magnetic curve and geological section is unknown) (after Eppelbaum (2007), with modifications). The following material from Eppelbaum (2007), is published with the permission of the Environmental and Engineering Geophysical Society. All copyright privileges remain with EEGS. This material cannot be copied or used without the express written permission of EEGS.

Table 2.8. Comparison of analytical expressions for natural potential fields.

Field	Analytical expression	
Magnetic	$\mathrm{TB}Z_v = 2J2b\dfrac{z}{x^2+z^2}$	Point source (rod)$Z_v = \dfrac{mz}{(x^2+z^2)^{3/2}}$
Gravity	$\mathrm{HCC}\Delta g = 2G\sigma\dfrac{z}{x^2+z^2}$	Sphere$\Delta g = GM\dfrac{z}{(x^2+z^2)^{3/2}}$
Self-potential	$\mathrm{HCC}\Delta U = 2\dfrac{\rho_1}{\rho_1+\rho_2}U_0 r_0\dfrac{z}{x^2+z^2}$	Sphere$\Delta U = \dfrac{2\rho_1}{2\rho_2+\rho_1}U_0 R^2\dfrac{z}{(x^2+z^2)^{3/2}}$
Temperature	—	Sphere$\Delta t = \dfrac{q}{\lambda_2}\dfrac{\mu-1}{\mu+2}\dfrac{R^3}{(x^2+z^2)^{3/2}}$

the gravity constant; σ is the density; M is the mass of the sphere; ρ_1 is the host medium resistivity; ρ_2 is the anomalous object (HCC or sphere) resistivity; U_0 is the potential jump at the source body/host medium interface; $r_0$is the polarized cylinder radius; R is the sphere radius; t is temperature; q is the heat flow density; $\lambda_2$is the thermal conductivity of the anomalous object; μ is the object-to-medium thermal conductivity ratio; x is the current coordinate; z is the depth of the upper DTB edge (HCC or sphere center) occurrence.

The proportionality of analytical expressions for different fields presented separately in columns 2 and 3 of this table is obvious.

A universality of correction methodology for geophysical observations observed at inclined (rugged) surface topography was presented in Alexeyev *et al* (1996).

2.2.1.15 Near-surface thermal investigations

2.2.1.15.1 Brief theoretical background

In solid bodies, heat flow according to Fourier's law (Tikhonov and Samarsky 1963):

$$\mathbf{q} = -\lambda \text{ grad } T = -\lambda \frac{\partial T}{\partial n}, \tag{2.73}$$

where T is the temperature, q is the heat flow density, λ is the thermal conductivity, and n is the outward normal to an isothermal surface.

If we assume that the heat transfer from the local sources is conductive and the temperature field is stationary, then the temperature distribution in a space without any sources satisfies the Laplace equation (Tikhonov and Samarsky 1963).

Taking into account the similarity of potential magnetic and thermal fields (equations (2.66) and (2.73)), and analytical expressions (table 2.8), we can apply for examination of temperature anomalies the interpretation methods (Eppelbaum *et al* 2014) developed in magnetic prospecting (Khesin *et al* 1996, Eppelbaum *et al* 2000, 2001). Eppelbaum *et al* (1996) proposed to use in thermal prospecting a new physical parameter—'thermal moment'—analogical to the 'magnetic moment' employed in magnetic prospecting. This parameter can be applied first of all for the temperature anomaly classification.

2.2.1.15.2 Elimination of various kinds of noise

Temperature measurements in the subsurface are usually carried out in subsurface (1.0–1.5 m depth) wells. This allows us to get rid of dependence on deep geological wells (usually, very expensive) and optimize the location of measurement points (Eppelbaum *et al* 2014). However, the near-surface measurements are accompanied by several kinds of noise (figure 2.118).

For the elimination of some noise components, the corresponding methodologies were developed (Eppelbaum 2009, Eppelbaum *et al* 2014). For instance, rugged relief influence can be reduced by application of the correlation method earlier developed in magnetic prospecting (Khesin *et al* 1996); for the elimination of the temporal temperature variations propagating from the Earth's surface to the

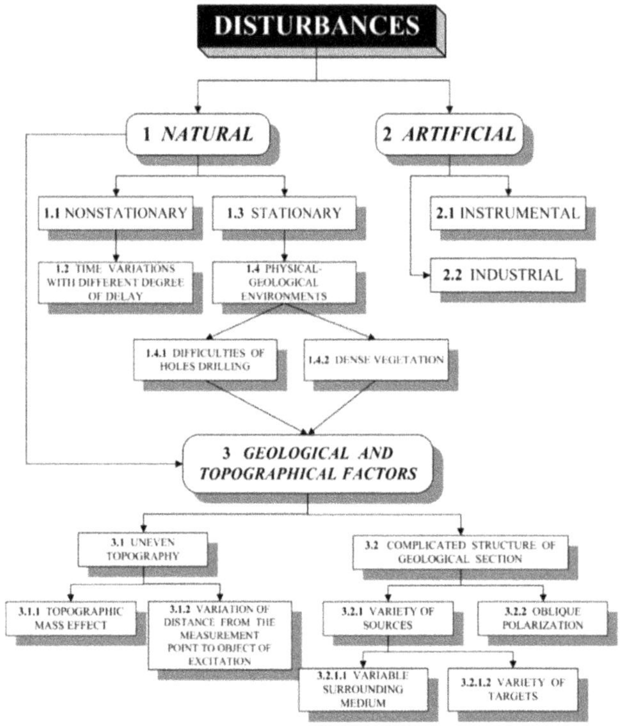

Figure 2.118. Near-surface temperature measurements: A general block-scheme of noise components (after Eppelbaum 2009, revised and supplemented) (reprinted from Eppelbaum 2009, with permission of Elsevier).

observation points, the special filtering procedure basing on the repeated measurements and some meteorological information was suggested (Eppelbaum 2009).

As in almost other geophysical methods, the thermal anomaly is a function of a ratio of the physical properties of the 'desired target/surrounding medium'. Values of the most typical thermal conductivities of rocks and environmental features are presented in table 2.9. As is seen from the above data, the existence itself of the temperature anomaly and its sign strongly depend on the type of host rocks (and in the case of sedimentary rocks, on their moisture saturation). However, it should be borne in mind that the presence of active faults can change the overall picture.

2.2.1.15.3 Thermal anomalies in the karst areas

Luetscher and Jeannin (2004) have analyzed temperature distribution in karst systems: the role of air and water fluxes. Karst systems display 'anomalous' temperature distributions which are frequently ignored or imprecisely understood. Bianchetti *et al* (1993) have carried out the temperature profile in the Simplon tunnel (Switzerland). Thermal anomalies have been recognized in carbonate environments in the range of 9.5 °C–35 °C. Hundreds of caves explored by speleologists to more than 500 m below the surface show temperatures close to those of the outside atmosphere (e.g., Choppy 1984, Badino 1992, Bitterli 1996, Audra 2001). As more

Table 2.9. Thermal conductivity of some rocks and environmental features (revised and supplemented after Eppelbaum (2009) and Eppelbaum *et al* (2014)).

Rock	Thermal conductivity ($W\ m^{-1} \cdot {}^{\circ}C$)	Environmental feature	Thermal conductivity ($W\ m^{-1} \cdot {}^{\circ}C$)
Dry soil	0.3	Air	0.26 (25 °C)
Sand	0.2–2.9 depending on water presence	Isolated cavity (sinkhole)	0.5
Gravel	0.3–0.5	Water	0.59 (25 °C)
Gypsum	0.5–1.5	Wood	0.09–0.14
Limestone	2.5–3.2	Show (fresh)	0.1–0.2
Clay	0.25–1.08	Snow (old)	0.5–1.8
Sandstone	2.4–2.5		
Dolomite	2.0–4.0		
Marls	1.5–1.54		
Shales	0.25–1.08		
Siltstone	1.55–1.60		
Quartz	3.0		
Rock salt	5.3–7.2		
Granite	2.2–3.5		
Granodiorite	2.5–3.0		
Basalt	1.5–2.2		
Gabbro	2.0–2.3		

than 10% of outcropping continental rocks are subject to intense karstic processes, interactions with human activities are frequent (Ford and Williams 2007). Air temperature at the surface of the karst massif is the result of a thermal balance between heat fluxes due to radiation, water precipitation, evapo-condensation processes, and thermal exchanges between the soil and the air.

Early attempts were made to predict cave temperature distribution theoretically (Cigna 1967, Andrieux 1969, Wigley and Brown 1971, Choppy 1984) but most authors were only concerned with the first hundred meters of karst systems, where cave temperature is still influenced by seasonal variations. Recent publications have tended to neglect the role of the geothermal heat flux in the energy balance of such systems (e.g., Badino 1995, Jeannin *et al* 1997, Lismonde 2002), however, the respective importance of heat fluxes transported by water and air is still a matter of debate.

Luetscher and Jeannin (2004) suggested a conceptual model of temperature distribution in natural karst environments that display thermal ‘anomalies’ as compared with other rocks.

In temperate regions, water circulation is usually high enough to ‘drawn-out completely the geothermal heat flux at the bottom of karst systems (phreatic zone). A theoretical approach based on temperature measurements carried out in deep caves and boreholes demonstrates, however, that air circulation can largely dominate water infiltration in the karst vadose zone, which can be as thick as 2000 m. Consequently, temperature gradients within this zone are similar to the lapse rate of humid air (~0.5 °C 100 m^{-1}). Yet, this value depends on the regional climatic context and might present some significant variations.

Another type of sinkhole is artificial ones caused by leakage from underground pipe mains in urban areas that may cause sudden ground subsidence or sinkholes. For a long time, researchers have been working on the hazard and risk assessment of sinkhole formation, especially natural sinkholes. However, much less work has been done on risk prediction and the mechanism of artificial sinkholes (Ali and Choi 2020).

Although many sinkhole detection methods have been developed, it is still difficult to predict sinkholes that occur in in depth areas. Also, conventional methods are inappropriate for scanning a large area because of their high cost. Unlike naturally occurring sinkholes, human-induced ones in urban areas are typically due to groundwater disturbances and leaks of water and sewage caused by large-scale construction.

Manohar *et al* (2000) describe the construction and properties of the thermal probe developed and calibrated. The probe is intended for the measurement of heat flux in laboratory samples or *in situ* conditions.

To study the distribution of thermal anomalies, thermal probes and thermal infrared measurements are used.

Frumkin *et al* (2015) used the thermal probe. A high-resolution infrared thermometer (model OS530HR: by OMEGA Engineering Inc.), operating in a wavelength range from 8 to 14 μm was used, to shed light on the hydrothermal plumes that dissolve the bedrock at present. Measurements were carried out in several sinkholes in the Nesher Quarry (Israel). Remote thermal infrared map showing variations of surface temperature (°C) relative to its mean value (17.4 °C) at the bottom of Nesher Quarry (27 October 2008). Red elliptical contours (center left) are 26 °C and 27 °C groundwater temperatures measured directly in boreholes at the central hydrothermal plume. A second plume is observed 100 m to SSE.

Gutiérrez *et al* (2008a, p 1011) write 'A complementary technique is the analysis of airborne and satellite multispectral and thermal images, which may be used to distinguish surface terrain patterns and to extract variations in moisture, vegetation, color, and temperature that may be related to sinkholes and subsidence areas'.

Calastrenc *et al* (2020) have successfully tested drone thermography for near-surface mapping of archaeological targets (including underground cavities) in areas with complex rugged relief. Implementation of integrated multi-source, multi-scale and multi-temporal approaches allowed the authors of this study to significantly increase the resolution of hidden object searches.

Benson and Yuhr (2016, p 140) denote that 'aerial thermography can detect variations in temperature as small as 0.05°, allows coverage over large areas, and can minimize laborious field efforts to identify these features on the ground'. Aerial thermography surveys are conducted to locate sources of springs discharging water on the bottom of lakes, creeks, and rivers. Davis (2007) reports that aerial thermography surveys provided a means of locating many unknown springs along the length of the river system in central Florida.

A thermal infrared device has been developed and tested by Bloomquist *et al* (2005).

Lee *et al* (2016) suggested early sinkhole detection using a drone combined with a thermal far-infrared (FIR) camera to detect potential sinkholes over a large area based on computer vision and pattern classification techniques and image processing.

2.2.1.15.4 Examples of thermal field analysis

Figure 2.119 shows the results of subsurface temperature measurements in southern Germany (Neckar valley near Kiebingen city) over deeply occurring karst terranes with comparatively hot water. The objective was to locate the small areas (spots) at which warm water rises from a depth of about 100–150 m. Kappelmeyer and Hänel (1974) suggested using the following expression to calculate the volume of heat flow:

$$Q = qF = \int_F \frac{\theta_m - \theta_b}{\lambda / a + z} \lambda df, \qquad (2.74)$$

where Q is the heat flow through anomalous area F, θ_m is the measured temperature at depth z inside the anomaly [°C], θ_b is the background temperature at depth z outside the anomaly [°C], z is the measured depth [m], λ is the thermal conductivity of the host rocks [W m^{-1}·°C], and α is the thermal diffusivity [10^{-7} m^2 s^{-1}].

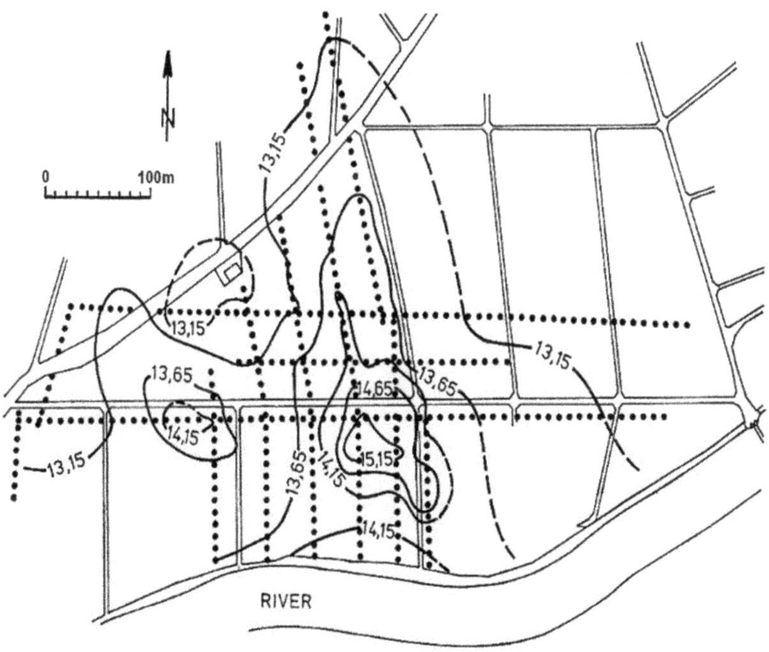

Figure 2.119. Temperature anomalies were observed at 1.5 m in depth in Kiebingen, Germany. These anomalies are caused by karst water emerging from a depth of 100–150 m in the groundwater at about 4–6 m in depth (reprinted from Kappelmeyer and Hänel (1974), with gratitude to Gebruder Borntrargen).

Applications of thermal methods for solving different geological–geophysical problems appearing in karst areas are presented in Kappelmeyer and Hänel (1974), Ravnik and Rajver (1998), Morat *et al* (1999), Stoessell *et al* (2002), Luetscher and Jeannin (2004), Luetscher *et al* (2008), Jalali *et al* (2009), Gui and Hu (2017), Liu *et al* (2017), Lei *et al* (2020), Zijártó *et al* (2021).

Methods for quantitative interpretation of temperature anomalies and their specific peculiarities were discussed in Khesin and Eppelbaum (1994), Eppelbaum (2010, 2019a), and Eppelbaum *et al* (2014).

A temperature anomaly with an amplitude of about 0.4 °C was registered over an underground cavity in Kracow's Planty Park, Poland (figure 2.120) (Moscicki 1987). This anomaly (an approximation model of the horizontal circular cylinder has been employed) has been interpreted using the above-mentioned methods (Khesin and Eppelbaum 1994). The center of the hidden cavity (approximated by the model of HCC) was calculated with satisfactory accuracy (figure 2.120).

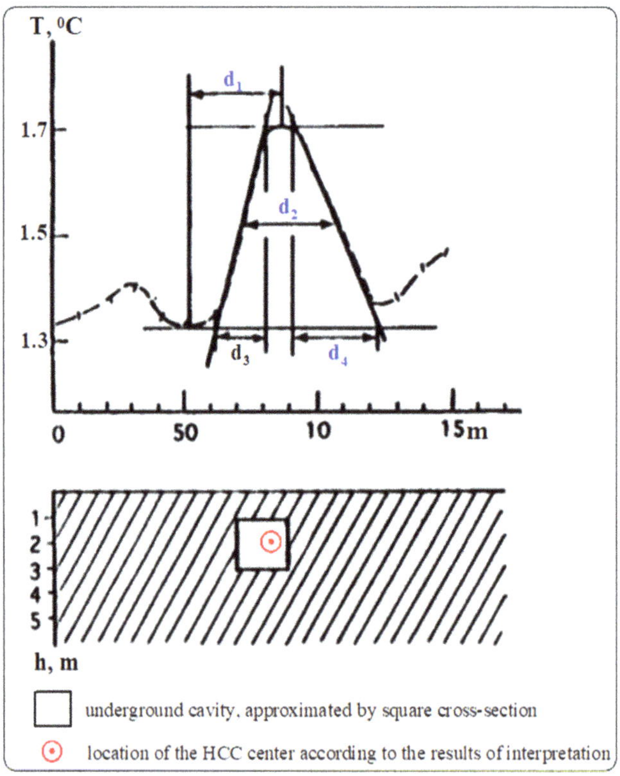

Figure 2.120. Quantitative interpretation of the temperature anomaly over an underground cavity (Kracow's Planty Park). The observed temperature and geological section are from Moscicki (1987). Interpretation after Khesin and Eppelbaum (1994) (reprinted from Khesin and Eppelbaum (1994), with permission from SEG).

2.2.1.16 Piezoelectric and seismo-electric investigations

2.2.1.16.1 Brief theoretical background

Systematic studies of piezoelectric and seismo-electrokinetic phenomena in the context of applied geophysics began in the former Soviet Union in the mid-1950s. These phenomena are manifested by electrical and electromagnetic processes occurring in rocks under the influence of elastic oscillations triggered by shots or mechanical impacts (hits). The following classification of piezoelectric and seismo-electrokinetic phenomena is suggested (Neishtadt *et al* 2006):

1. The shot-triggered phenomenon, which is observed in rocks in the vicinity of a shot or hit point. This effect is used in the piezoelectric method to mark the instant of elastic vibration generation.
2. The seismo-electric phenomenon *I*, manifested by the change of the electric current passing through rocks brought about by the passage of a seismic wave. This effect has not yet received a broad application in exploration geophysics.
3. High-frequency impulse electromagnetic radiation arises in polymetallic ore bodies under influence of elastic vibrations. The discovery of this phenomenon was utilized for the development of a radio-impulse method for searching and prospecting polymetallic deposits.

The electrokinetic phenomenon *E* is the most important one for buried karst studies. This phenomenon in rocks is manifested by the potential gradient due to the displacement of the liquid phase relative to the solid 'skeleton' of the rock under the influence of elastic waves. The similarity of this phenomenon to potential produced by water flow in rocks suggests a similar origin for these phenomena (Fridrichsberg 1995).

Ivanov (1940) first described the electrokinetic phenomenon *E* in sedimentary rocks. He offered a filtration model, based on the existence of a diffusion double layer at the solid/liquid phase boundary in sedimentary deposits. According to the filtration model, propagation of an elastic wave in the above media causes mutual displacement of the liquid phase (electrolyte) and solid phase, which upsets the equilibrium state of the diffusion double layer. This effect generates a so-called filtration potential in rocks. This is a flow potential, whose instantaneous value along a segment of an elastic wave can be calculated using the Helmholtz–Smoluchowski equation (Probstein 1994):

$$U = \frac{1}{4\pi}\int_{p_1}^{p_2} \frac{k\varepsilon\zeta}{\eta\sigma}dp, \tag{2.75}$$

where p_1 and p_2 are the instantaneous values of pressure at the studied site, respectively; k is the coefficient of dynamics of the elastic wave propagation; ε is the dielectric constant; ζ is the potential of the electric double layer; η is the viscosity of the solution; σ is the conductivity.

Frenkel (1944) offered a mathematical description of the electrokinetic phenomenon *E*, which was based on Ivanov's (1940) hypothesis regarding the electro-filtration nature of this phenomenon. Frenkel described compression and expansion of rock particles and pores, as well as the pore moisture under the influence of the elastic wave propagation. Moreover, he proposed an expression describing the propagation of the elastic waves in moisturized soil. For calculating electrokinetic potential, this formula uses the intensity of longitudinal electric field E for the stationary flow of liquid through the solid 'skeleton' pores:

$$E = i\frac{4\varepsilon\zeta\chi\omega^2}{\pi\sigma r^2\mu}f\rho_2\left(\frac{K_2\beta}{\rho_2\beta'\omega_0^2} - 1\right)u, \tag{2.76}$$

where μ is the coefficient of viscosity of the medium; i is the complex-valued electric field intensity; $\beta = 1/[f(1+\alpha)]$; α is the parameter that describes mechanical properties of the medium; $\beta' = 1+(\beta - 1)K_2/K_0$; K_0 is the coefficient of compressibility of the solid phase; $\alpha = K_1/f$ is the coefficient of permeability; r is the pore radius; f is the porosity; K_1 is the coefficient of permeability of the soil; ρ_2 is the true specific gravity of the pore moisture; K_2 is the coefficient of compressibility of the liquid phase; ω_0 is the propagation velocity of the longitudinal elastic wave; u is the displacement.

Equation (2.76) shows that the electric field intensity is proportional to the porosity of rock and is independent of the pore radius because the coefficient α is proportional to the square of this radius. With the assumed values of displacement, the electric field intensity is proportional to the square of the frequency of the elastic oscillations. In formula (2.76), it is assumed that the period required to produce the electrokinetic potential gradient is negligibly short compared to the oscillation period $2\pi/\omega$. This is why the value of E at any given moment practically coincides with the value that corresponds to the instantaneous meaning of the relative velocity (Neishtadt and Eppelbaum 2012).

2.2.1.16.2 Some peculiarities of practical application

Boulytchov (2000) observed the electrokinetic phenomenon E triggered by reflected seismic waves. The elastic waves were introduced into the ground by pneumatic hammer impacts (frequency of 1 kHz) and by a powerful piezo-ceramic vibrator. Stacking of the seismo-electric signals resulted in magnitudes of up to 200 μV. Measurements were conducted over sharp seismically reflecting boundaries, 0.1 m to 15 m deep. The observations enabled the determination of the overburden thickness in a 10–12 m topographic depression. The author maintains that the applied technique can be used for the delineation of caverns in the karsts (Boulytchov 1997, 2000). It would present an especially attractive choice in remote locations, due to the simplicity and portability of the required equipment.

Svetov and Gubatenko (1999) proposed a mathematical model describing the manifestation of the electrokinetic phenomenon E for low-frequency oscillations in porous and saturated rocks. The authors proposed an algorithm for calculating the electrokinetic component of the electromagnetic field: (1) determination of the displacement vector for the solid 'skeleton'; (2) solution of the fluid dynamic equations; (3) calculation of the electrokinetic current; (4) calculation of the magnetic field induced by the electrokinetic current.

Zhu *et al* (2000) reported that seismo-electric measurements could be an effective means of obtaining transport coefficients such as hydraulic permeability and other porous rock properties. This investigation experimentally confirms that the coupling between the seismic wave and the electromagnetic field in the kilohertz range is an electrokinetic one.

Peng *et al* (2017) have performed physical modeling which received the seismo-electric response from the two-cavity model. Comprehensive analysis of this model for different physical–geometrical conditions revealed some important peculiarities for this class of targets.

Eppelbaum (2017c) has proved that for quantitative piezoelectric/seismo-electric anomalies the advanced methods developed in magnetic prospecting for quantitative analysis of anomalies observed in complex physical–geological conditions (Khesin *et al* 1996, Eppelbaum *et al* 2001, Eppelbaum 2011a, 2015a) can be applied.

2.2.1.17 Other surface methods (HVSR, nanoseismicity)

2.2.1.17.1 Horizontal-to-vertical spectral ratio (HVSR) method

Ambient vibration recordings combined with the H/V spectral ratio technique have been proposed to help in characterizing local site effects (SESAME user guide 2006). Among the methods designed to assess the site effect, the method based on ambient noise recordings is a most practical and low-cost tool and has become more and more popular over the last decades (Kagami *et al* 1982, Yamanaka *et al* 1994). It offers a convenient technique, especially through urbanized areas (Bonnefoy-Claudet *et al* 2006). The method is intended to identify the main characteristics of site responses for soft deposits (i.e., resonance frequencies and amplification factor). In the last decades, the determination of H/V spectral ratios from ambient noise (HVSR) is supplemented with on-site geophysical, borehole, and geological information to derive the required models of the subsurface (Zaslavsky *et al* 2009, Gorstein and Ezersky 2015).

Nakamura (1989) hypothesized that site response could be estimated from the spectral ratio of the horizontal versus vertical components of noise observed at the same site (site of interest). Nakamura's method is based on the assumption that micro-tremors consist of body waves. Mucciarelli and Gallipoli (2004) claimed that the H/V spectrum of ambient noise is dominated by the upward propagation of SH waves through the layered media.

On the other hand, an explanation based on the opposite assumption that micro-tremors mainly consist of surface (Rayleigh) waves is also successful (e.g., Fäh *et al* 2001, Lachet and Bard 1994). Both models agree that the H/V spectra and the site response function for SH wave are the results of the velocity structure of the media, that both exhibit the same fundamental resonance frequencies with similar amplitudes at least when considering small motions. Malischewsky and Scherbaum (2004) demonstrated analytically the coincidence of H/V from the ellipticity of Rayleigh waves and the fundamental frequency for higher impedance contrasts (by a factor of 4).

It was demonstrated through many studies (Zaslavsky *et al* 2008, 2009) that, when noise measurements are made near boreholes and/or near refraction surveys, the fundamental frequency, and its corresponding H/V amplitude are practically the same as the fundamental frequency and its corresponding amplification level derived from the computed transfer function of SH-waves at low strains propagating through a relatively simple 1D model of the site, as known from geotechnical and geophysical surveying.

Computer code SHAKE (Schnabel *et al* 1972) is used to analytically evaluate the site response function. The specific parameters required for this analysis are:

- S-wave velocity, thickness, density, and damping of each layer in unconsolidated sediments, and
- S-wave and density of the hard rock (reflector).

Apart from the seismological problem the HVSR technique has become the primary tool of choice in many ambient noise related studies, such as studying the V_S structure of the subsurface (Stokoe *et al* 2006, Ezersky *et al* 2013, Lotti *et al* 2012, Kolesnikov and Fedin 2018)

HVSR data acquisition

The methodology of HVSR data acquisition and processing is presented in figure 2.122 (see Zaslavsky *et al* (2009) for details). Ambient noise measurements were made using portable instruments (Shapira and Avirav 1995).

Each seismograph station consists of three (one vertical and two horizontal) L4C velocity transducers (Mark Products) with a natural frequency of 1.0 Hz and a damping ratio of 65%–67% of critical (1 in figure 2.121(a)). The sample rate was 100 samples per second and the filter band-pass was between 0.2 Hz and 25 Hz. The recording system (figure 2.121(b)) consists of a multi-channel amplifier, a global positioning system (GPS) for timing, and a laptop computer with a 16-bit analog-to-digital conversion card to digitize and store the data (2 in figure 2.121(b)).

All the equipment: sensors, power supply, amplifiers, personal computer, and connectors were portable, allowing the performing of measurements in autonomous mode.

Nowadays HVSR equipment comprises the three-component geophone (6TD Guralp (USA), Tromino (Italy)) with a frequency range of 0.03–100 Hz.

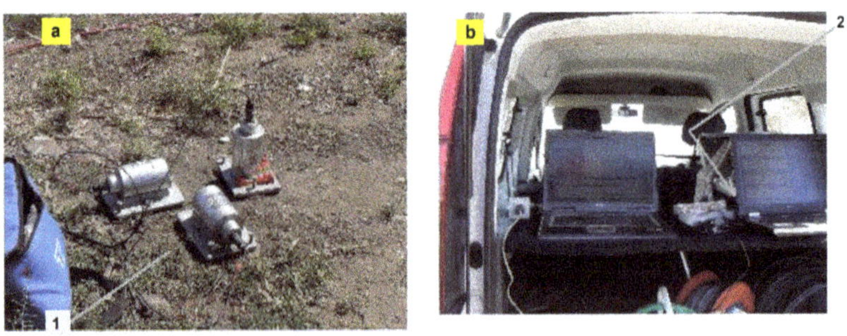

Figure 2.121. Photographs of noise measurements: (a) recording of seismic noise; (b) complect of the recording system. 1—horizontal sensor, 2—The analytical section (reprinted from Ezersky *et al* (2012), with gratitude to Ministry of Energy and water resources).

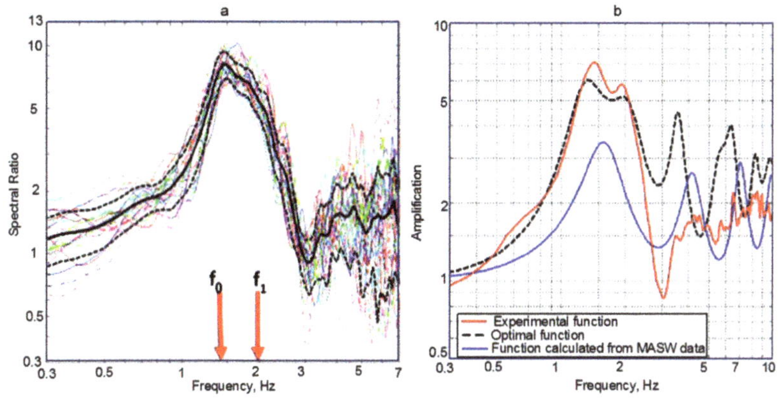

Figure 2.122. (a) Horizontal-to-vertical spectral ratios (individual, average, and standard deviation) were obtained in the Ramat Hakovesh site. The arrows show the fundamental f_0 and second resonance f_1 frequencies; (b) HVSR curve (red solid line) superimposed on analytical function calculated using MASW results only (blue line) and obtained by a combination of the HVSR and MASW methods (dashed line) (reprinted from Gorstein and Ezersky (2015), with gratitude to Camdemia).

HVSR data processing

To study the spectral characteristics of the ambient noise signals Fourier spectra and spectral ratios are computed. The record length (time window) used for spectral calculations is of at least 30 s. The selected time windows are Fourier transformed, using cosine-tapering (1 s at each end) before transformation and then smoothed with a triangular moving Hanning window. The H/V spectral ratios are obtained by dividing the individual spectrum of each of the horizontal components [$S_{NS}(f)$ and $S_{EW}(f)$] by the spectrum of the vertical component [$S_V(f)$]:

$$A_{NS}(f) = \frac{S_{NS}(f)}{S_V(f)};\ A_{EW}\ (f) = \frac{S_{EW}(f)}{S_V(f)}. \quad (2.77)$$

The average spectral ratio for each of two horizontal components is computed if the curves of average spectral ratios of the two components are similar to the average of the two horizontal-to-vertical ratios, defined as:

$$A(f) = \frac{1}{2n}\left[\sum_{i=1}^{n}\frac{S_{NS}(f)_i}{S_V(f)_i} + \sum_{i=1}^{n}\frac{S_{WE}(f)_i}{S_V(f)_i}\right]. \quad (2.78)$$

Routine analysis and data processing are performed using the software SEISPECT developed in the Seismology Division of GII (Perelman and Zaslavsky 2001).

SEISPECT is a MATLAB application for spectral analysis and processing of ground motion, including seismograms recorded by short-period and broadband seismic stations, as well as strong-motion accelerometers. The main modules of the program are visualizing and editing the input data; selecting time windows and computing fast Fourier transform and H/V spectral ratios; saving and displaying results.

Analytical response function computed using program SHAKE for a specified subsurface model should match the empirical response function. The program, based on the stochastic optimization algorithm (Storn and Price 1995), is applied to approximate the spectral ratio by theoretical function in the best way, considering the dominant frequency, its level, and the shape of the H/V curve. Within the chosen frequency interval $[w_1, w_2]$ we look for thickness (h_i) and S-velocity (v_i) that minimize the misfit function

$$F = \sum_{k=1}^{N} \left(f_{\text{SHAKE}}\,(\omega_k) - \left(\text{H/V}(\omega_k)\right)^2\right., \tag{2.79}$$

where ω_k are the points from the frequency interval $[w_1, w_2]$, $f_{\text{SHAKE}}(\omega_k)$ is the 1D theoretical transform function calculated by the SHAKE program, and H/V(ω) is the H/V spectral ratio.

Velocity and thickness are limited:

$$V_{1i} \leqslant v_i \leqslant V_{2i},\ i = \overline{1,\ M+\ 1} \text{ and } H_{1i} \leqslant h_i \leqslant H_{2i},\ \ i = \overline{1, M},$$

where M is the number of layers in the 1D model.

A field example of spectral ratios versus frequency from a test Israeli site Ramat Hakovesh is shown in figure 2.122(a). There are individual records (colored), standard deviation (black dashed lines), and average spectral ratio for this site (black solid line). The main frequencies are shown by arrows. Note the two site peak frequencies are inseparable. The analytical function (red line) calculated using a combination of the HVSR and MASW methods (black dashed line) is presented in figure 2.122(b). Separately, the transfer function calculated using only MASW results is shown by the blue line.

The analytical transfer function is fitted to the experimental spectral ratios by a selection of shear wave velocities (V_{S_i}) of layers and thickness h_i of these layers, where i is the number of the layer. The model of the subsurface corresponding to the fitted transfer function and experimental one is shown in table 2.10.

The prominent feature of the individual and average spectral ratios shown in figure 2.123(a) is two inseparable peaks at the frequencies of 1.4 and 1.9 Hz. Such distribution of peaks could be construed as follows: while the first resonance peak is related to the hard rock at depth, the second peak is caused by the intermediate hard layer directly overlying the deep reflector.

Table 2.10 contains two variants of the model calculations. The first is based on the MASW data only where four layers could be identified. The response function calculated using solely geophysical data and the broken dolomitic limestone at a depth of 73 m as a fundamental reflector is shown in figure 2.122(b) (blue solid line).

The second model is based on joint HVSR and MASW measurements in which the broken dolomite-limestone is an intermediate hard layer and overlays the fundamental reflector (dolomite) at a depth that should be estimated. The result of the optimization procedure is shown in figure 2.122(b). The optimal model providing the best fit between HVSR and the calculated response function is given in table 2.10 as well.

Table 2.10. Depth V_s profile derived from the MASW and optimal 1D subsurface model for determination of analytical site response function at the Ramat Hakovesh site. Reprinted with permission from Gorstein and Ezersky (2015).

	MASW		MASW+HVSR optimal model	
Lithology	Depth (m)	V_s (m s^{-1})	Depth (m)	V_s (m s^{-1})
Alluvium	0–5	205	0–5	170
Sand, loam, clay, calcareous sandstone	5–30	300	5–30	235
	30–50	500	30–55	450
	50–73	490	55–70	520
Broken dolomite, limestone	73–240	1480	70–260	1430
Dolomite	below 240 m			2400

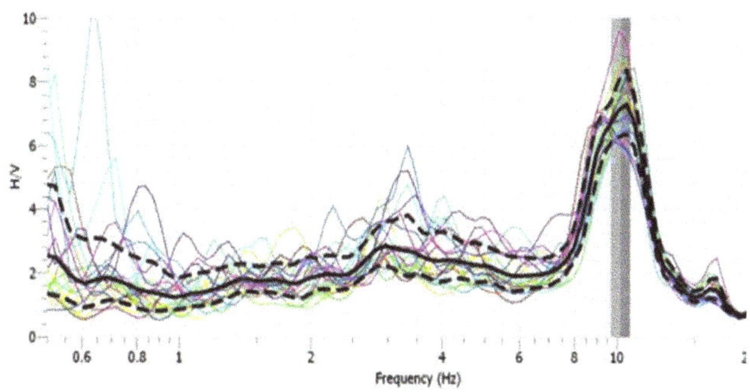

Figure 2.123. HVSR reveals the fundamental frequency of resonance *f*0, 10 Hz in this example (reprinted from Nehmé *et al* (2013)).

It should be pointed out that the H/V technique alone is not sufficient to characterize the complexity of site effects and in particular the absolute values of seismic amplification (SESAME user guide 2006). The method required the use of an additional method that can decrease the number of unknown parameters, such as the thickness of layers and their V_s values. Gorstein and Ezersky (2015) have shown that a successful addition to the HVSR is the MASW.

Application of H/V method in the karst problem

Pioneering work in this field was performed by Nunziata *et al* (1999). The authors investigated the changes of the local seismic response in natural cavities located in the historical center of Naples when affected by a seismic input such as the 1980 earthquake. They observed that the increments of the peak ground acceleration, as well as spectral amplification from numerical modeling, are more evident at the sites located away from the influence of the cavity itself concerning those obtained at sites placed over the grotto.

Sgarlato *et al* (2011) observed more complex effects coming from the ambient noise and earthquakes. The authors carried out some investigations in urban areas of Catania (Italy) in natural and artificial tunnels. The ambient noise and earthquake records were subjected to analysis of spectral ratio techniques, investigating also the possible directional effects. Performed measurements show that spectral peaks in the frequency range 3.0–7.0 Hz appear stable nearby the cavity, and show a tendency to be less pronounced at increasing distance from it. Moreover, the natural period of buildings erected nearby and over the considered cavity was evaluated to investigate the possible soil-structure interactions. The results so far obtained indicate that the height of the cavities plays an important role. Only cavities having a height greater than about 4 m show significant H/V amplification peaks. Moreover, spectral ratios deriving from the measurements performed inside the cavities show less pronounced spectral peaks than those obtained from measurements over and outside the cavities.

Approximately similar conclusions have been made by Lombardo and Rigano (2010).

Kolesnikov and Fedin (2018) investigated the possibilities of the use of microtremor records in the detection and delineation of near-surface underground cavities on the base of Barsukovskaya cave (Novosibirsk region, Russia). Moreover, the authors try to use parameters of noise to estimate the shape of the underground cavity horizontal projection. If the compressional-wave velocity in the enclosing rock is known, it is possible to evaluate the depth to the cavity top using the frequencies of the standing waves. The results of the field experiment confirmed the possibility of underground cavities detection using microtremor data.

Nehmé *et al* (2013) introduced the use of an HVSR technique to investigate the clay deposits and estimate their total volume in a Kanaan Cave, located in Metn District, Lebanon. The volume of unconsolidated clay sediments generally produced by the alteration of the calcareous rocks is one of the main factors determining cave development. The principle suggested by the authors is the determination of the frequency peak of the HVSR of ambient noise. The HVSR technique is applicable where a soft layer of material is located above a hard substratum. The ambient noise is recorded and H/V spectral ratio is calculated (figure 2.123). Then transfer function is matched to the spectral ratio record and the model of the medium ($V_{Si} = f(h_i)$) is selected as was described above. In this case, Nehmé *et al* (2013) use the Geopsy software to calculate the H/V spectra (SESAME user guide 2006). The thickness of the clay layer located above the cave bedrock is calculated.

Thus, the HVSR method allows deriving the clay volume, as well as its shape. This technique applied to the clay volume reveals some internal details, such as fallen blocks before the deposit accumulation, and helps one to understand the deposit evacuation dynamics. The study reveals new stages related to cave speleogenesis. The technique could be applied to constrain the clay volume and reconstruct the buried floor shape of the cave, underneath the clay deposit.

2.2.1.17.2 Microseismic activity

A microseismic (MS) activity named also nanoseismicity, acoustic emission (AE) is a worldwide known method used for the prediction of mining bursts (Hodson 1948, Vinogradov 1964), the problem of prediction of earthquakes (Scholtz 2002), the collapse of walls and arks in the underground tunnels and machine halls constructions (Hardy 1976, Ezersky *et al* 1993, Ezersky and Eppelbaum 2017, Philipp *et al* 2015), Monitoring of earth dam stability (Koerner, and Lord 1976) and slope stability (Smith *et al* 2017).

There are known attempts to use AE in the problem of karst (Hardy 1976, Hardy and Belesky 1986, 1987, Xingping and Meifeng 2004, Bakeev *et al* 2017).

Generally, the term AE denotes processes of radiation of elastic waves caused by inner sources located within the medium (Mazarchuk *et al* 2017). A necessary condition of AE occurrence is the locality and pulse character of the source. AE is determined as the process of radiation of mechanical waves by material caused by dynamic alteration of the inner structure of the medium material. The main source of AE in rocks is the formation of brittle fractures, their growing, interaction, sliding along plains of failure during which new segments of surface contact. In simple words, AE is associated with different modes of material destruction and radiation of elastic energy. Koerner *et al* (1981) have applied the term to processes taking place within soils and have used AE monitoring for the Earth dam stability control (Koerner and Lord 1976). This activity is seemingly well-founded, for AEs are indeed generated by deforming soil masses and technical feasibility is now firmly established.

AE activity in soils presents these findings based on fundamentals, small-scale laboratory tests, and large-scale laboratory tests. Furthermore, the technique has been applied to field situations in several cases. These include slope stability monitoring of dams and embankments, soil movements arising from horizontal and vertical deformations, seepage monitoring, and grout/hydrofracture monitoring. Specific case histories in each group are presented.

In sinkhole and karst development inner noise is induced by the inelastic failures within forming cavities (falling of stones, soil, conglomerates). The uncontrolled development of cavities near the Earth's surface, due to natural or human causes, pose a major engineering hazard. Regardless of the cause, these cavities can result in localized instabilities such as the collapse of highway or airport runway paving, foundation subsidence or collapse, and dam leakage.

Xingping and Meifeng (2004) have considered the characters of AE signals coming from rock damage in the subsiding area. These signals are synthetically analyzed and evaluated. These provide evidence to reversely deduce and accurately infer the position of rock fracture and damage in the subsiding area. Similar subsidence can be caused by karst processes as has been considered by Hardy and Belesky (1986).

This application is based on the principle suggested by Newton (1987) and shown in figures 2.124(a) and (b). The typical cavity in unconsolidated deposits such as residual clay is circular with the configuration of the top resembling a dome or an arch (figure 2.124(a)). The sides generally coincide with the occurrence of pinnacles or irregularities in the top of the bedrock. The wall is usually vertical as the opening grows toward the land surface. This configuration, however, can be modified by the configuration of the underlying openings in bedrock and by variations of the cohesion or competence of overlying beds.

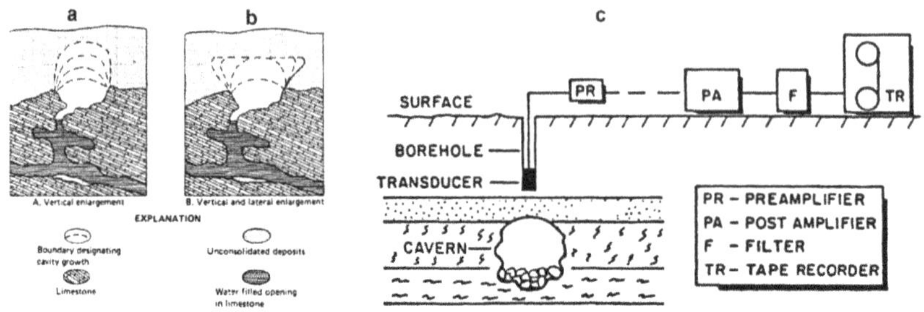

Figure 2.124. (a, b) Principle of sinkhole formation, and (c) measurement of AE/MS using a simple single-channel system for monitoring of an underground cavern system (a, b—Newton (1987) courtesy of the U.S. Geological Survey; c—Hardy and Belesky (1987), with permission of Japan Society of Applied Physics).

The configuration of a typical opening can change if its upward growth reaches the more competent bed. The roof flattens, the growth continues laterally, and walls taper toward the opening in bedrock (figure 2.124(b)).

The formation of the further sinkhole is accompanied by the falling of the stones and parts of the arch and walls. These fallings produce noise and pulses that can be recorded by the acoustic sensor. The principal scheme of acoustic noise monitoring is shown in figure 2.124(c) (Hardy and Belesky 1987). This system comprises a single acoustic sensor transferring acoustic pulse to an electrical signal, that then is amplified, filtered, and recorded by a tape recorder.

Hardy and Belesky (1987) investigate the feasibility of using AE and seismic tomography techniques to study the development of one type of near-surface cavity known as a 'sinkhole' as well as for monitoring such subsidence. Reference is made to the results of a recent sinkhole monitoring study undertaken at Capital City Airport, New Cumberland, Pennsylvania. These results are considered to be directly relevant to the problem of karst or mining-induced highway subsidence. The author's report summarizes the findings of the airport study and discusses the unique aspects of highway subsidence monitoring. A methodology to locate and monitor shallow cavities is contained within the report.

Systems developed in consequent years carried out multi-channel recording. The acoustic sensors were distributed throughout areas allowing localization of pulses (Abelson *et al* 2018). It is similar to the tracing of underground streams using the 'geo-bomb' method suggested by D Arangelovich (Milanović 2005, section 11.7)). The difference is the only source of underground pulses: in the first case that is falling of stones, in the second one—explosion of a bomb in the karstic stream.

Bakeev *et al* (2017) analyze the formation of a sinkhole and fracture pattern estimate depending on the rock properties. They consider the possibility of using geophysical methods for recording and analyzing AEs to monitor and predict the state of the medium. The problem of deformation of the sedimentary cover over the growing karst cavity is solved based on the elastoplastic model of the Drucker–Prager–Nikolaevsky and the equation of damage accumulation (Drucker, and Prager 1952). The specified kinetics of accumulation of damage allows authors to describe slow processes of degradation of the strength of the medium under stresses that are low for the development of inelastic deformations. The results are obtained for different values of the strength of karst rock. The authors show the influence of the kinetic parameters of damage accumulation on the shape of collapse depressions. It is also shown that AE is caused by a material fracture. One can follow different stages of karst development by looking at patterns of cells that fail at a given time. The observations have shown the relation between the intensity of material fracture and the intensity of seismic emission.

Joswig (2005) instead of the AE/MS term suggested another name 'nanoseismic monitoring' (NM). It deals with the other extreme: resolution at the limits of noise level, ambiguous phase readings, and just a few available stations. This method is based on noise occurring from the underground falling rocks in the shallow subsurface. Wust-Bloch and Joswig (2006) applied the NM method to predict sinkhole formation in the Nahal Hever north area in Israel. Recently Abelson *et al* (2018) tested the NM method to observe the process of sinkhole formation in the Mineral Beach site of the western DS shore. The nanoseismic tests we will consider in section 4.9.

2.2.2 Aerial and satellite methods

Aerial photography is a method intended for the overview of local and regional conditions through large areas. The method is cost-effective and it requires topographic and geologic maps for this type of information. An aerial photo is accepted as an additional powerful mapping tool, especially since it began to provide additional details on the regional and local levels. The aerial photographs are applied for many types of projects requiring spatial data from the same location at periodic intervals over a length of time (monitoring). The aerial photos are a valuable source of information about the studied area, but not a map. Gutiérrez *et al* (2019) denote that the most useful applications of imaging and remote sensing data are used in the karst projects to provide both spatial and temporal data for developing a karst inventory. Such application of aerial photos in the karst problem is caused by different manifestations of a sinkhole at the Earth's surface (subsurface) and by different sizes of karstified areas. The different manifestations include the surface sinkholes (bare karst), cavities mantled by unconsolidated deposits (covered or 'mantled karst'), or cavities overlain by non-karst rocks ('interstratal karst'). Ground deformations (associated with the sinkhole activity) typically affect spatially limited sub-circular areas a few meters or tens of meters in diameter. In addition, there is the dissolution of evaporites (sugging sinkholes) size of areas of a hundred meters or kilometers across. As a rule, karst induces displacements of the surface with dominated vertical downward strains. However, sugging can cause considerable horizontal movement.

Generally, the sinkhole-related subsidences are highly variable. The collapse sinkholes are frequently preceded by gradual ground subsidence (related to pre-failure plastic deformation) with subsequent sudden catastrophic displacement events. Baer *et al* (2018) have shown that in the Dead Sea conditions this preceding subsidence may begin as much as a few years or months before the occurrence of catastrophic collapse sinkholes, providing the possibility for the implementation of early warning systems.

Suffosion sinkholes forming usually in softcover deposits are relatively rare, do not show catastrophic behavior, and have quite a small size. Sagging subsidence is generally characterized by the progressive dissolution of highly soluble evaporitic rocks (for instance, halite) that may be located at a significant depth. In many cases, both sagging and collapse mechanisms operate simultaneously (Gutiérrez *et al* 2019). That is why the sinkhole susceptibility areas should be controlled by the aerial methodology, especially through the areas of intensive sinkhole development.

At present, many types of aerial photos or satellite imagery are available. They all have a lot of advantages and disadvantages (cost, coverage, scale, range of wavelengths, etc). However, it is a very important tool of karst area investigation (Benson and Yuhr 2016).

2.2.2.1 Aerial photos

There are qualitative and quantitative aerial photo applications. Black and white aerial photos have been the standard for analysis because of their availability and ease of qualitative and quantitative interpretation. Their continued acquisition over many decades provides a temporal record of the geologic, hydrologic, and cultural changes.

Qualitative interpretation of aerial photos allows estimation of the site access and features of the study area such as vegetation, lakes, rivers, roads, and trails. One can see also fences, areas of fill, landfills, structures, major pipelines, and transmission lines, along with both quarries and mines.

For karst, studies are very important to locate topographically (geographically) sinkholes including their size and spacing. By comparing older aerial photos (some go back to the 1930s) with more recent aerial ones one can determine sinkhole activity over long periods. Historic aerial photography can be used to identify old sinkholes where more recent land use has caused sinkholes to be completely covered, partially covered, or filled in.

From quantitative interpretation, one can obtain information as fracture pattern, geological mapping, hydrological conditions, drainage, and topography. This information is reached by looking at variations in location, size, shape, shadows, tone/color, texture, patterns, and other various characteristics (Clayton *et al* 1995). Aerials data are digitally acquired and analyzed using computer software.

Let us note the geological type of information that can be obtained from an analysis of aerial photos:

- Existing sinkholes are generally easy to identify on the aerial photos;
- Orientation of fractures and spacing along with the lineament analysis;
- The geology of a site can often be interpreted from landforms, drainage patterns, vegetation, and land use;
- Areas of subsurface instability (sinkholes and landslides) (Benson and Yuhr 2016).

The white aerial photos as a rule provide most of the visual information that is required for both the regional and site-specific scale site characterization. However, other methods can be used. There are other imaging and remote sensing measurements (Lillesand *et al* 2003, Paine and Kiser 2012).

Let us consider those imaging techniques that are most commonly used for karst investigations:

- Digital elevation models (DEMs) is a 3D computer graphics representation of elevation data to represent terrain, commonly of a planet (e.g., Earth), Moon, or asteroids (Hawker *et al* 2018). A DEM is a gridded digital representation of terrain, with each pixel value corresponding to a height above a datum.
- Thermal imagery is applied at the geologic mapping of vents, springs, seeps, cave openings, etc on land or water as long as a temperature contrast exists. It allows mapping variations in temperature with or without illumination and is usually flown at low altitudes (a few 100 m).
- Infrared (false color) film provides additional details of the vegetation and wet areas. It is used in forestry, agriculture, and vegetation studies (Clayton *et al* 1995).
- Color photos provide additional details of vegetation and wet areas despite they are limited by the Sun angle and cloud cover (Lillesand *et al* 2003).
- Synthetic aperture radar interferometry (InSAR) measures ground deformation data with centimeter precision. It can monitor ground movements down to 1 mm per year (Browitt *et al* 2007). Such measurements can be used night or day and penetrate the cloud cover and vegetation.
- LiDAR (light detection and ranging) is remote sensing (laser) technology that allows mapping topography through large areas using aircraft at a very high resolution and accuracy.

2.2.2.2 Remote sensing techniques (airborne/satellite measurements)

Generally, remote sensing is the process of detecting and monitoring the physical characteristics of an area by distant measuring reflected and emitted radiation, typically from satellites or aircraft. Imaging sensors collect remotely sensed data, which helps researchers to 'sense' things about the Earth (USCS). Some examples are:

- Electro-optic sensors on satellites and airplanes take images of large areas on the Earth's surface enabling to scan the landscape of the EM spectrum in the 'open windows';
- The Earth's surface enabling to scan the landscape of the EM spectrum in the 'open windows'
- Thermal sensors on satellites can be used to make images of temperature changes in the oceans;
- Tracking the growth of a city and changes in farmland or forests over several years or decades.

In application to this book, remote sensing can be used to map the sinkhole formations through great areas like Florida (Seale *et al* 2008), Minnesota (USA) (Wu *et al* 2016), Missouri (Kim *et al* 2019), salt karst of Ebro Valley (Spain) (Gutiérrez *et al* 2019), Konya Closed Basin in Turkey (Orhan *et al* 2020), Zhijin County, China (Chen *et al* 2018).

Remote sensing methods collect the data from the air or space using satellites, aircraft, or unmanned vehicles such as drones or balloons.

Recently geophysical methods, such as airborne TEM (Siemon *et al* 2009, Kirkeggard, *et al* 2011, Gondwe *et al* 2012), FDEM (Supper *et al* 2009), and GPR (Forte *et al* 2019) are also used as remote sensing methods.

Digital elevation models (DEMs) are gridded digital representations of terrain, with each pixel value corresponding to a height above a datum. We can distinguish between digital terrain models (ground without objects like houses and trees) and digital surface models. Since the pioneering work DEMs have grown to become an integral part of several scientific applications. DEMs can be created from the ground surveys, digitizing existing hardcopy topographic maps, or by remote sensing photogrammetry, radargrammetry, interferometry, LiDAR point (clouds processing techniques). DEMs are now predominantly created using remote sensing techniques with Smith (2002) observing the benefits that a large spatial area can be mapped by fewer people at a lower cost.

Photogrammetry is 'the art, science, and technology of obtaining reliable information about physical objects and the environment through processes of recording, measuring and interpreting photographic images and patterns of recorded radiant electromagnetic energy and other phenomena' (McGlone *et al* 2004). Photogrammetry is an analytical method primarily associated with making precise measurements of 3D objects and terrain features from 2D photographs. The method is based on the measuring of coordinates; the quantification of distances, heights, areas, and volumes. It prepares the topographic maps and generates digital elevation models and orthophotographs. There are two general types of photogrammetry aerial (with the camera in the air) and terrestrial (with the camera handheld or on a tripod). Terrestrial photogrammetry (termed also close-range) is dealing with object distances up to ca. 200 m (Wolf *et al* 2014).

Interferometric synthetic aperture radar (InSAR) is designed to measure the elevations of the Earth's surface from satellites. The principle of interferometry is based on the calculation of phase differences between two or more radar images. There can be two types of images difference. The first one is a difference of positions, e.g., two antennae onboard one vector acquire images simultaneously (single-pass interferometry such as SRTM mission). The second one is a difference of time, e.g., one antenna acquires images at two distinct times or multi-pass interferometry. Those can be compared after proper image coregistration. The resulting difference of phases is a new kind of image called *an interferogram* The latter is an interference pattern of fringes containing all the information on relative geometry. The resulting interferogram presents a contour map of the change in distance between the ground and the radar instrument The interferograms are sensitive to technical effects, such as relative variations in the radar's trajectory or variations in its frequency standard. Apart from this, interferogram records movements of the crust, perturbations in the atmosphere, dielectric modifications in the soil, and relief in the topography. All these phenomena affect an interferogram (Massonet and Feigle 1998).

The fact that the height of a point on the Earth's surface can be reconstructed from the phase difference between two signals arriving at two antennae. is the main idea of a space-borne radar interferometry technique. This is explained by fact that the phase difference is directly related to the difference in path lengths traversed by the signal between the point on the Earth's surface and the two antennae (figure 2.125).

InSAR is ideally suited for measurements of regional deformation and can be less expensive than obtaining from the sparse point measurements. The regional results can be used to optimize the location of point measurements or survey lines at a more local scale (Bawden *et al* 2003).

Many physical phenomena contribute to the phase measurement. They are phase variations within a pixel, contribution of the orbital trajectories, the contribution of the topography, contribution of displacements, atmospheric contribution, instrumental artifacts, etc (Massonet and Feigle 1998).

An example of the topography measured by InSAR on Mt Etna is shown in figure 2.126. This topographic contribution yields fringes which hug the topography

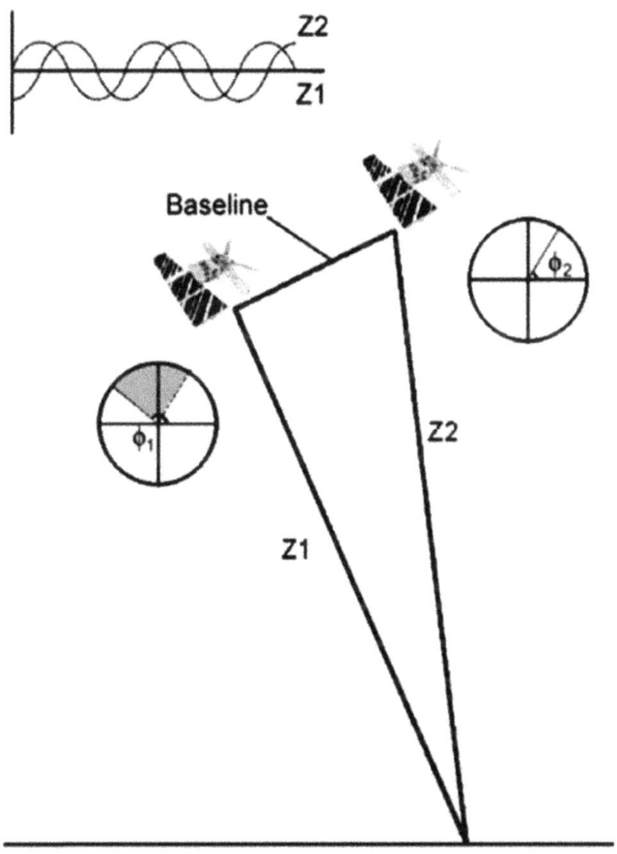

Figure 2.125. Global scheme of the interferometric acquisition. Z1 and Z2 are the distances between the antennae and a spot over the Earth surface (reprinted from Massonet and Feigle (1998), modified by Closson (2005), John Wiley & Sons, with permission of AGU).

Figure 2.126. Topographic fringes on Mt Etna. Here one fringe represents about 250 m of the topographic elevation. The shading represents the topographic relief from the DEM as if the illumination came from the west. The area is 40 × 40 km (reprinted from Massonet and Feigle (1998), John Wiley & Sons, with permission of AGU).

like contour lines ('topographic fringes'). Here, each fringe cycle in an interferogram corresponds to 250 m of the topographic elevation. To establish orders of magnitude, it is convenient to use the notion of altitude of ambiguity, or the shift in altitude needed to produce one topographic fringe.

As shown above, the InSAR techniques allow calculation of the topography of the surface (DSM). On the other hand, a variation of topography can provide information on the deformation of the surface. In this case, it will be differential synthetic aperture radar interferometry (DInSAR). The measurement of the Earth's deformations was applied in the Dead Sea area in association with the sinkhole problem to evaluate subsidence of DS shores in Israel and Jordan (Baer *et al* 2002, Abelson *et al* 2003, Closson *et al* 2003).

2.2.2.3 LiDAR

LiDAR (light detection and ranging) is a remote sensing technology. It is performed using the laser records of elevations at a rate of 2000–5000 pulses per second. The main advantage of LiDAR is that it maps the topography through large areas at very high accuracy by aircraft, helicopter, or drone (figure 2.127). Distance to a ground object depends on the time delay between the transmission and detection of a laser pulse. Lasers mounted on aircraft have a vertical precision of 15 cm (Stennett and Grusky 2008). The LiDAR measurements can be carried out in highly vegetated areas. In other cases, technology can be used to develop detailed bathymetric data in shallow water depths to 10 m (Finkl *et al* 2005).

LiDAR can take readings only through the openings or gaps in the tree cover but not through the canopy of the tree cover. Having repetition rates of 150 000 per second or more (around 10 shots m^{-2}), surface coverage can be achieved even in tree-covered areas (Stennett and Grusky 2008). It means LiDAR allows performing elevation measurements of the ground surface (DTM) through most tree covers and detailed surface topography can be reached. Concerning the karst problem, LiDAR can reveal the presence of sinkholes and depressions within the vegetated areas, which are not normally seen in aerial photography (Benson and Yuhr 2016). It is also an important application of the LiDAR method to monitor areas and map changes in topography over time. Such mapping can also be used to locate lineaments and possibly some cave entrances. One such example is the use of LiDAR revealing a fourfold increase in the number of sinkholes mapped for Winona County, Minnesota (Rahimi and Alexander 2013).

LiDAR-based topography maps and low sun angle shaded-relief images have the potential to highlight such features as subtle areas of subsidence as well as obvious sinkholes. As with all methods, there are inherent limitations.

Use of remote sensing techniques for karst investigation

Zhang *et al* (2019) developed a conceptual framework for sinkhole detection and mapping with airborne LiDAR. This concept lays the foundation for the future application of the airborne LiDAR for sinkhole detection and mapping. The proposed concept suggests using the airborne LiDAR data in the combination with auxiliary contexts to improve the accuracy of the morphological-based sinkhole detection methods and implement these by developing tools that can be used in the standard geographic information systems (GIS). Earlier, most of the research on LiDAR-based sinkhole detection postulated that morphological-based surface

Figure 2.127. (a) DJI Matrix 600 Pro drone, and (b) Yellowscan LiDaR Mapper II (Poirier *et al* (2020), with permission of ISTE OpenScience—Published by ISTE Ltd London, UK—openscience.fr) (reprinted from Poirier *et al* (2020), with gratitude to ISTE).

feature extraction methods can effectively detect sinkholes because of their geometric properties—sinkholes are oval-shaped concave depressions in the Earth's surface. The suggested conceptual framework allows for the development of the robust LiDAR-based sinkhole detection GIS toolset that provides an adequate degree of accuracy while maximizing the ability to assist inspectors with varying expertise.

Wu *et al* (2016) have suggested an automated approach for sinkhole extraction that may reduce inconsistencies and alleviate repeatability concerns associated with the visual interpretation methods. Also, the proposed method has contributed to improving the sinkhole inventory in several ways: (1) detection of non-inventoried sinkholes; (2) identification of previously inventoried sinkholes that have been filled; (3) delineation of sinkhole boundaries; and (4) characterization of sinkhole morphometric properties. The LiDAR data were acquired during 2008 in Fillmore County, which is an active karst area located in southeastern Minnesota that is part of the Upper Mississippi Valley Karst (USA). The study was based on the work of Wu *et al* (2015) who proposed the conceptual framework of the localized contour tree method for identifying surface depressions. The authors implemented the proposed conceptual framework and made the following improvements: (1) instead of using the whole DEM for the entire area to generate contours, they extracted a subset of the DEM representing depression areas, which can reduce the number of contours being generated and thus greatly reduce the computation time; (2) the computed morphometric properties of sinkholes provide critical information about sinkholes in addition to their locations; and (3) the improved algorithms have been implemented as an ArcGIS toolbox—Sinkhole Extraction Analyst. The automated procedure for the detection of sinkholes described in this study demonstrated its effectiveness and efficiency. In addition to detecting sinkhole locations, the localized contour tree method allowed for extracting sinkhole boundaries and quantifying sinkholes at different ranks across different spatial scales. Added to that, other machine learning-based methods such as decision tree and random forest could also be employed to facilitate sinkhole detection and susceptibility mapping, depending on the availability and quality of sinkhole training data.

Zhu *et al* (2014) (Kentucky cave) applied LiDAR in the Floyds Fork watershed to improve the sinkhole database for central Kentucky. The authors have developed a method of processing LiDAR data to identify sinkholes and successfully tested the method. The method consisted of four steps: creating a high-resolution digital elevation model (DEM) from LiDAR data, extracting surface depression features from the DEM, inspecting the depression features for probable sinkholes, and verifying the probable sinkholes in the field (figure 2.128).

A total of 1683 probable sinkholes were identified in the study area, compared to 383 previously mapped for the same area. Randomly selected probable 121 sinkholes were checked in the field and 106 of them were verified as karst sinkholes. This method increased the number of sinkholes by a factor of four with a success rate between 80% and 93% for the study area, demonstrating that the LiDAR sinkhole-mapping method is reliable and efficient. This method identified approximately 55% of the previously mapped sinkholes, and approximately 98% of the missed sinkholes appeared to be filled or covered for urban development and agriculture purposes.

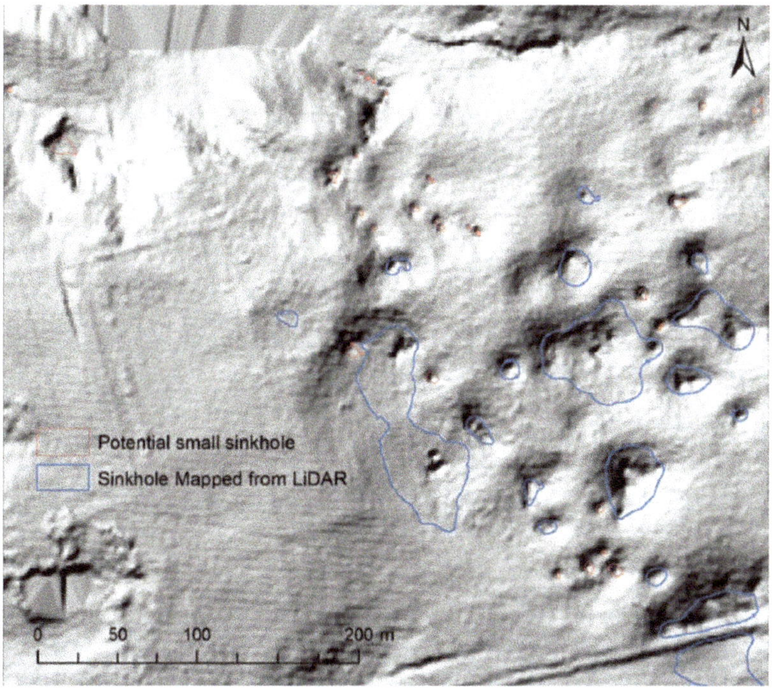

Figure 2.128. Shaded-relief map (5× vertically exaggerated) showing examples of potential small sinkholes not processed in this study (reprinted from Zhu *et al* (2014), with permission of the National Speleological Society (www.caves.org)).

Zini *et al* (2015) have applied multidisciplinary study in the Quinis, a small village sited in the Alta Val Tagliamento valley (Friuli Venezia Giulia The region, NE Italy). The region is characterized by the distribution of numerous sinkholes. It is a representative example of the complex regional geological situations typical of the northeastern Italian valleys, which at present is still not fully understood.

The data have been obtained from the oral witnesses, technical documents, field surveys, and LiDAR joined the Geodatabase. The sinkholes were classified following the scheme proposed by Gutiérrez *et al* (2008b, 2014), whereby sinkholes are described indicating the material affected by subsidence (cover, bedrock, and caprock) and the subsidence mechanism (collapse, suffosion, and sagging). The boreholes (12 of them have reached the bedrock of the gypsum unit), GPR, and seismic reflection surveys were conducted to define the bedrock morphology, characterize the mantling deposits, and identify the processes behind. What emerged from the study is a mandatory multidisciplinary approach to characterize the subsoil, because each technique is not able individually to take to a unique result. The data collected allowed the authors to draft a hydrogeological conceptual model of the Quinis village.

The lessons learned, albeit with some site-specific dependency, demonstrate the importance of broad-spectrum investigations, which are essential to understand the subsurface characteristics avoiding relevant socio-economic impact and supporting adequate future territorial planning.

Bernatek-Jakiel and Poesen (2018) suggested applying the LiDAR to detect piping areas with different vegetation types.

Busetti *et al* (2020) carried out a multidisciplinary study including InSAR in a small village in the Alta Val Tagliamento Valley (Friuli Venezia Giulia region, NE Italy In addition to conventional geophysical methods, the authors applied persistent scatterer interferometry (PSI) allowing construction of deformation time series maps including retrospective analyses. The bedrock in the study site is composed of gypsum. Sinkholes have been reported for a long time before. The result of the sinkhole presence was damaging and demolishing several houses. Busetti *et al* (2020) were able to identify and map the zones with higher vertical movements (multi-year interferometric data analysis) and to generate a summary map showing the highest vulnerable zone as a function of deformation thresholds. It was possible because of the data of interferometric analysis and peculiarities of the area.

DInSAR was applied by Galve *et al* (2015) to improve the sinkhole inventory in the Ebro Valley, NE Spain. There are three ways to reach this goal using DinSAR: (1) detection of non-inventoried sinkholes; (2) revision of sinkhole areas previously classified as inactive as active; and (3) refinement of underestimated sinkhole boundaries. The study has shown that the DInSAR results are suitable for analyzing active dissolution-induced subsidence. The application of DInSAR techniques may help in recognizing and better characterizing previously unknown karst subsidence problems and in preventing personal and property losses. However, the disadvantage of the method is that the available DInSAR maps combined overlook about 70% of the previously mapped active sinkholes mainly due to decorrelation.

Oliver-Cabrera *et al* (2020) have investigated small localized deformations using InSAR in Hernando Country in west-central Florida. The authors have applied the PSI technique. The obtained results enable us to reveal several areas of the localized subsidences at rates ranging from −3.7 to −4.9 mm yr^{-1}. The verification of subsiding locations confirmed the relationship of the subsidence with sinkhole presence.

Orhan *et al* (2020) investigated the middle of the Konya Closed Basin in Turkey, to evaluate sinkhole susceptibility using remote sensing and geographic information systems. A deformation map was produced using InSAR. Sentinel-1A SAR satellite images were employed to calculate the deformation of the study area using the small baseline subset (SBAS) algorithm. The Sentinel-1A satellite images, which are available for free and are open access, covered the entire study area. The complete surface deformation of the study area was obtained. The resulting map was divided into five classes with a natural break strategy. The deformation map showed up to 40 mm yr^{-1} subsidence in the line-of-sight (LOS) direction in some parts of the study area. The annual rate of deformation was considered the significant precursor deformation for information of the sinkholes expected to occur in the area. The produced map was in high agreement with the recent sinkholes in the district. The results of this study showed that the analytical hierarchy process can be a useful approach for susceptibility assessments when data on the main controlling factors have sufficient accuracy and spatial coverage

Remote measurements using geophysical methods. In the last decades, researchers have used conventional geophysical methods to investigate environmental problems by placing apparatus on helicopters.

Forte *et al* (2019) focus on helicopter-borne GPR surveys performed on rough topography typical of many mountain glaciers, which do not always exhibit a smooth surface and are often relatively small and fragmented. These surveys are indeed important because the mountain glaciers are so numerous in some areas that, at a regional scale, they can contain a significant fraction of ice volume, which must be taken into account to obtain meaningful global estimations. Such data are in turn necessary for reliable mass balance estimations, water equivalent assessments, and for both short and long-term forecasts. Garambois *et al* (2016) have shown the effectiveness of the GPR survey of an englacial water-filled cavity in the polythermal glacier Tête Rousse, French Alps. Another example is the survey of Gossar and Čeru (2016) in Mt Kanin massif (Slovenia and Italy) which discovered the entrance to the lost cavity using GPR with the special rough terrain antennas (RTA) and (Mala, Sweden). These examples testify that GPR is a really important method for mountain conditions. Moreover, shortly, due to global warming, several cirque glaciers and small ice caps will be divided into separated smaller fractions, while the topography will become more complex and rugged due to differential melting. Forte *et al* (2019) propose a few guidelines and optimization strategies to address several issues, including the choice of various data acquisition parameters, interpretation problems related to curvilinear or irregular flight paths, and trace positioning errors caused by the GPS malfunctioning or oscillating antennas. Such results have general validity and can be used for helicopter-borne survey planning, as well as for various data analysis and interpretation.

In the study of Supper *et al* (2009) submerged cave systems were mapped using airborne FDEM. There are two distinct flow domains in the karst aquifer. The first is limestone's fractures and the second one is conduits. To build a flow model of karst aquifers it is necessary to have detailed, spatially distributed information on the hydrological characteristics of the aquifer and flow paths. Such information can provide geophysical methods dealing with the distribution of electrical resistivities in the subsurface. Karstifiable rocks are a traditional focus of different geophysical research and partially electromagnetics (Nyquist *et al* 1999, Jardani *et al* 2007, Chalikakis *et al* 2011). Supper *et al* (2009) have investigated the potential of airborne EM mapping to provide such innovative information for improving groundwater modeling of karst aquifers. The airborne survey was carried out in the Sian Ka'an Biosphere Reserve located in Yucatán (Mexico), a coastal wetland of international importance. At first, the ground geoelectric and ground electromagnetic measurements were performed in March 2006 to determine the electrical properties of the study area subsurface environment. EM measurements using the Geonics EM-34 equipment were originally carried out along the highway Mex 307 between Chunyaxche and Tulum, crossing one branch of the cave system connecting Cenote Cristal and Escondido (Mayan Blue). The results (figure 2.129) show that there is a significant positive anomaly detected right above the cave.

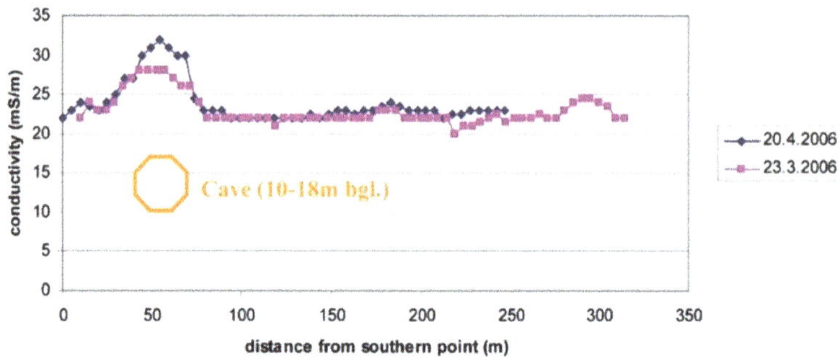

Figure 2.129. Results of EM-34 profiling (coil separation 20 m, frequency 1.6 kHz) over a known cave branch between cenote Cristal and cenote Escondido (apparent conductivities are given) (reprinted from Supper *et al* (2009), John Wiley & Sons, with permission of EAGE).

The cavity is located at 20 m depth and was surveyed using the EM-34 system with 20 m spacing between the antennas. One month later repeated measurements have produced almost the same results. The 3D forward modeling was carried out based on the ground EM results. That enables us to calculate the possible airborne electromagnetic response.

Then an airborne pilot survey was performed in 2007 to evaluate the applicability of the airborne electromagnetic methodology. This survey covers an area of 40 square kilometers above the well-mapped Ox Bel Ha cave system.

The Austrian airborne EM (AEM) system was used for the survey (Motschka 2001). The transmitting coils generate an EM alternating field with frequencies of 340 Hz, 3200 Hz, 7190 Hz, and 28 850 Hz.

A good representation of the cave system is shown in the pattern of the half-space resistivities map derived from the in-phase component of the 3200 Hz and 7190 Hz frequency (figure 2.130). The caves here can be seen as low resistivity anomalies.

The remarkable low resistivity areas along with the coastal line and lagoon can be caused also presumably by saline water intrusion into seashores or subsurface brackish water. This was verified by results from the geoelectric profile.

The results showed that the signature of the cave system could be detected. The study of Supper *et al* (2009) has testified to the ability of the airborne frequency domain EM measurements to localize and map karstic cave systems.

2.2.3 Borehole geophysics

2.2.3.1 Introduction

By measurements performed directly in the Earth, borehole methods can offer more enhanced results than surface methods due to closer distance to the desired objects. However, drilling boreholes is usually an expensive process: boreholes are not cheap, especially in the karst areas. Many tools require open holes (for instance, ultrasonic), while other tools can operate through the plastic casing (downhole, cross-hole). One instrument requires water in the borehole (suspended probe), another one can work only within the dry borehole (cross-hole). In addition, tomography measurements

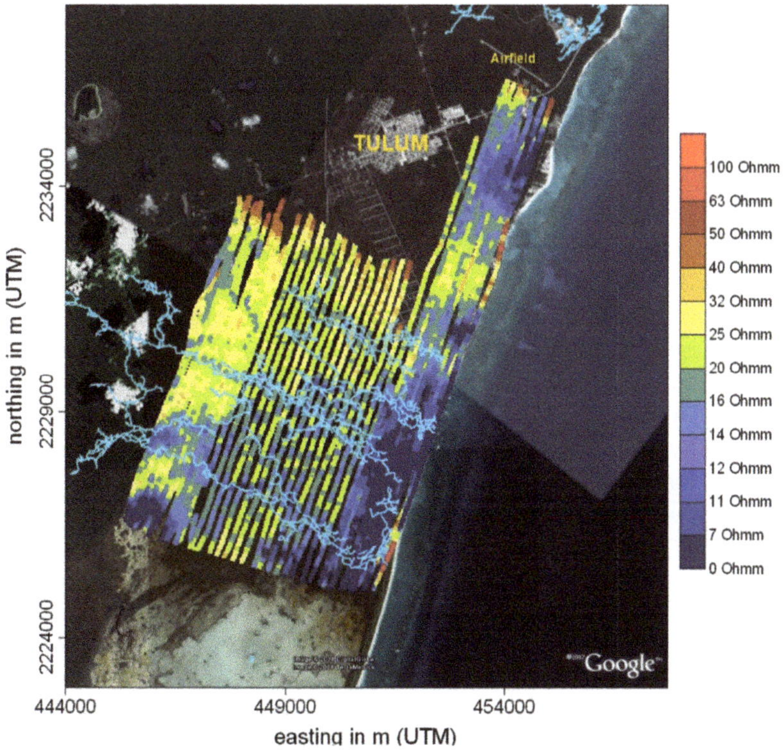

Figure 2.130. Resistivity map of the study area from the airborne EM survey. Results of homogeneous half-space inversion calculated from the in-phase component of 3200 Hz and 7190 Hz frequency and the bird altitude. Light blue lines show cave systems mapped by scuba divers (reprinted from Supper *et al* (2009), John Wiley & Sons, with permission of EAGE).

require several boreholes. Sensors located within the borehole are not sensitive to above-ground ambient noise, especially at large depths. There are strict standards established by the ASTM for performing tomography measurements and reporting. To identify high transmissivity zones it is necessary to use integrated approaches to borehole data (Brandon *et al* 2001).

There are conventional logging methods carried out within one borehole with the consequent use of different instruments (probes) measuring different parameters or of the borehole (itself) or characteristics of surrounding soil or rock (Serra 1984). Logs are measuring the electrical and magnetic properties of soil and rock (e.g., bulk electrical conductivity, resistivity, resistance, spontaneous potential (SP), magnetic susceptibility). Another log allows the determination of the flow characteristics. These relate to the fluid logs. Some logs measure the geometrical parameters of boreholes (diameter of boreholes or casing, inclination, and direction of boreholes in space). Another one produces visual/imaging logs that provide a photo of the borehole, casing, or rock surface Gamma radiation, relative moisture/fluid content in the unsaturated zone, and porosity in saturated soils can be characterized by nuclear kinds of logs, such as natural gamma, gamma–gamma, neutron, etc.

The radar, gravity, and full-waveform sonic logs are carried out in the open holes or PVC-cased boreholes. The goal of the sonic logs is *in situ* measurement of the P and S wave acoustic velocity in the frequency range of 3–15 kHz within a borehole. Note that there are ultrasonic logging measurements in the frequency range of 50–70 kHz and seismic (downhole or cross-hole, for instance) logs in the frequency range of 100–300 Hz. Two latter methods will be considered in the continuation. Because of the scale effect in the rocks, every one of these methods measures different values of wave velocities (Ezersky and Goretsky 2014; section 4.11.1 of this manuscript). We have to mention here that borehole radar has been used for a variety of applications including bedrock fractures, voids, and tracing tests. It will be considered in continuation). Density for large volumes of rock surrounding the borehole is carried out by the gravity measurements method. These data are not affected by borehole conditions (LaFehr 1983). Generally, borehole logging is used for the assessment of natural geological and hydrogeological conditions, measurements of the soil and rock properties within the distances of hundreds of meters, etc (Benson and Yuhr 2016). Paillet (2001) suggests several tools that may be applied to the characterization of heterogeneous karst aquifers: (1) Image and flow logs provide detailed information about the nature of hydraulically active zones intersected by boreholes; (2) geometric correlation of logs between the boreholes indicates possible connections between the zones in separate boreholes. (3) Water-quality and hydraulic-head data derived from logs can be used to identify subsurface flow paths; (4) cross-borehole flow experiments can be used to infer the properties of hydraulic connections among the subsurface conduits; (5) geophysical measurements can be made at local, intermediate, and large scales to infer the relation between scale and hydraulic conductivity.

Besides traditional geophysical logging in a single borehole, there is a range of other measurements that can be made between the ground surface and a borehole (downhole, uphole) or between boreholes (cross-hole). Although these measurements are carried out to significantly less depth (30–100 m) they significantly expand the quantity of material sampled from the borehole. These measurements would normally be made after site-specific geologic conditions had been defined so that the measurement locations can be selected to meet very specific objectives.

Add the method of borehole optical televiewer probe (BOTR) (depth to 30 m) allowing color pictures within boreholes. The camera can rotate 360° around the vertical axis and 220° around the horizontal axis. Such a study was carried out by Vadillo *et al* (2012) at an experimental karst site located near the Nerja cave (S. Spain), developed in Triassic dolomitic marbles. The images allow identification of major karstic elements, their depth, and approximate orientation.

The engineering boreholes are intended for site characterization. After drilling, the cased boreholes can be locked up for further study. When there is a chance of borehole collapse (soils, weathered rocks, salt, etc), the geophysical measurements may be carried out in the borehole just after drilling it. Often drilling equipment is located near the borehole up to completing the necessary measurements. Then the borehole is plugged.

In this section, we will consider main borehole investigations allowing measurements of longitudinal and shear wave velocities (V_P and V_S, respectively and resistivity of rocks surrounding the borehole, such as: (1) downhole measurements; (2) ultrasonic logging, and (3) resistivity logging by the method of apparent resistivity. Note: ultrasonic logging is carried out in the frequency range of 100–200 kHz, whereas downhole measurements are ranged from 80–300 Hz. There are also GPR measurements in the single borehole and between boreholes and similar ERT measurements.

2.2.3.2 Measurements in a single well (downhole, electric, and ultrasonic logging, GPR)

The downhole survey is aimed to determine both compressional and shear wave velocities of the subsurface. The measurement procedure follows ASTM D 7400-07 (2019). For elastic waves excitation, the Digipulse source or the handle sledge-hammer are used.

Data interpretation is carried out using one of the commercially available software packages (WinSism, ReflexW). The data acquisition parameters for downhole measurements (30–50 m) are as follows: record length: 250 ms for both P and S waves, sampling rate: 0.125 ms, Distance between geophone levels within the borehole during the downhole survey—1.0–2.0 m. The field records of the downhole and cross-hole surveys are shown in figure 2.131.

Note two features of the borehole survey. Figure 2.132(a) feature of S-wave that changes polarity when source polarity is reversed; and P-wave downhole measurements are accomplished by the tube wave that is a pressure pulse that propagates nearly unattenuated down (and up) the fluid column (figure 2.132(b)).

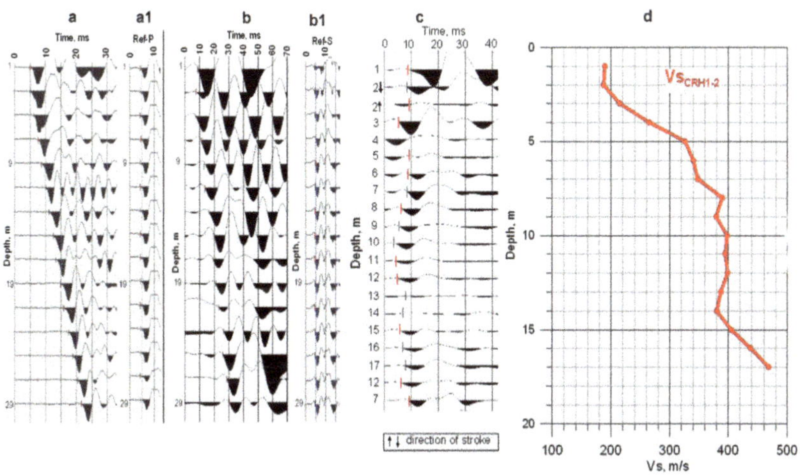

Figure 2.131. Examples of the field records of the P- and S-waves from the downhole surveys (a and b, respectively); (c) cross-hole surveys and (d) V_s-depth graph from records (c); a1 and b1 are control records of the trigger. In record (c) the second seismogram was received twice with stroke carried out in two directions, after completion of data acquisition repeatable shots were carried out at 12 m and 7 m to check the quality of records. Compare repeated strokes with those obtained at the same depth.

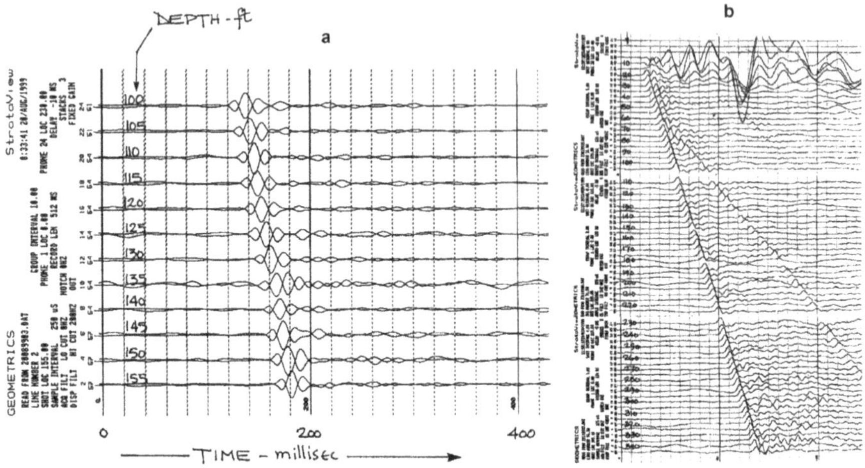

Figure 2.132. (a) Shear waves reverse polarity when the source polarity is reversed. (b) Composite downhole P wave record showing tube waves (reprinted from Crice (2011), with permission of SEG and D Crice).

The velocity of a tube wave is a function of the bulk modulus of the fluid, the elastic modulus of the casing, and the shear modulus of the surrounding material (Crice 2011). As a general rule, they are less of a problem for shallow surveys (<30 m). The tube waves are generally recognizable because their velocity is constant (see Crice (2011)) for details). S-wave changing polarity is used to identify S-wave records during field survey.

Ultrasonic testing (D2845-05) (Ezersky and Goretsky 2014, Ezersky and Eppelbaum 2017).

Method and equipment. In ultrasonic logging (USL) the elastic waves are spreading from a high frequency (10–100 kHz) transmitter to receivers along the borehole wall under investigation (Beck 1981). The 'dotty' (detailed) ultrasonic logging enables measuring elastic velocities within 0.1–0.2 m along the borehole (Savich *et al* 1969). Wave penetrates the rock approximately 0.1–0.2 m following the wavelength. The scheme of the dotted ultrasonic logging is presented in chapter 4.

Measurements by the method of apparent resistivity

The method is based on the measurement of the apparent resistivity of soils and rock around a borehole as a function of depth and their differentiation by specific electrical resistivity (Khmelevskoi 1979). A three-electrode potential sonde is used (see chapter 4).

*Measurements by the borehole-radar method (*Singha *et al* 2000).

The borehole radar method can be carried out either in a single borehole or from two boreholes. The borehole-radar reflection logging is similar to that of surface-radar profiling, except that antennae are connected and lowered as a single unit down the open or PVC-cased borehole (figure 2.133(a)).

The transmitted into the rock pulse propagates away from the borehole until it encountered material with different electromagnetic properties (fracture zone, change in rock type, cavity, etc). In this place, the radar pulse is reflected towards

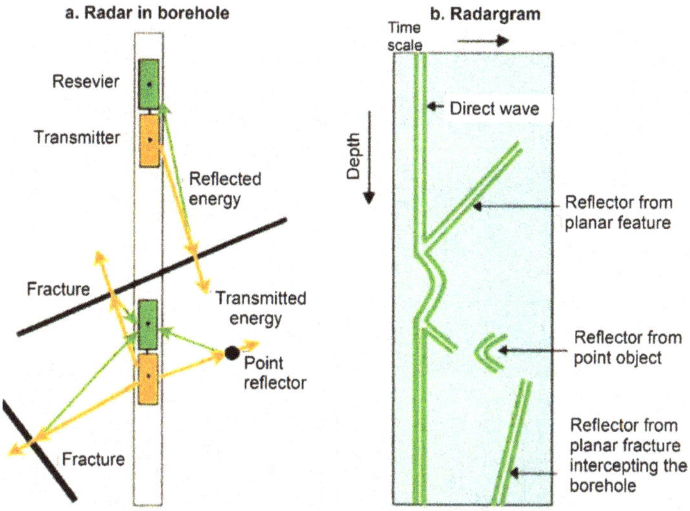

Figure 2.133. (a) The orientation of transmitter and receiver in a single borehole, and (b) resultant radar record (radargram) from a fracture and point reflector (reprinted from Singha *et al* (2000), courtesy of the U.S. Geological Survey).

the receiver. The radar reflection profile is recorded in radargram all characteristic features of the meeting in the pulse movement (figure 2.133(b)).

Borehole radar RAMAK (produced by MALA, Sweden) uses 100–250 MHz (completely digital antennas); designed for a maximum depth of borehole (optical fiber length): 1000 m, and an outer diameter of transmitter/receiver: 48 mm.

The radar range is dependent upon soil or rock type. In resistive rocks ranges of 10–40 m can be achieved. In conductive, clay-rich, or silty rock ranges can be limited to 5 m or less.

2.2.3.3 Measurements between a group of wells (tomography, cross-hole)

Cross-hole measurements are carried out following ASTM D4428/D4428M (2014).

Three boreholes (bh1, bh-2, and bh-3), about 3.0 m apart, center-to-center on the ground surface, are drilled in line (Azimuth 55°) in the central part of the site. ASTM D 4428 (2014) allows one to carry out cross-hole measurement in two boreholes: one is for source and the second one—for geophone. It is considered an optional method. In this case, the zero time determination is very important. The scheme of the cross-hole survey is shown in the diagram (figure 2.134). The boreholes are drilled using WIRE LINE technology guaranteeing borehole verticality no less than 2%. After drilling was completed, a PVC pipe (casing), 3 inches inside diameter, was inserted into the hole.

The space between the PVC pipe and the borehole wall is filled with a Portland concrete grout mixture. Verticality is therefore assumed in the borehole calculations.

A cross-hole test is carried out following the *preferred method* described in section 4.3.1 of ASTM D 4428. Note that D 4428 permits to use of two boreholes—one for source and the second for receiver geophone. In this case, the accuracy of distance

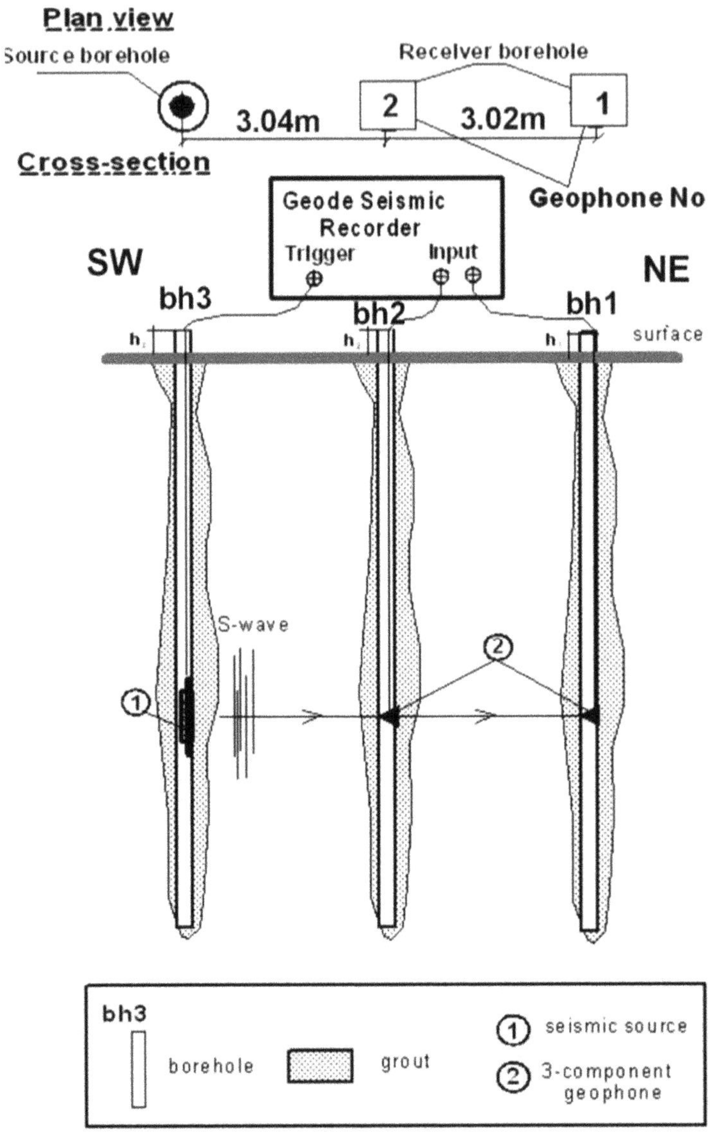

Figure 2.134. Scheme of the cross-hole measurements.

determining between points of measurement (source-geophone) becomes very important.

The source used to produce the transient pulse at one of the end holes (bh-3) is a special borehole source for SV-wave generation. This tool, illustrated in figure 2.134 (1), was designed, hand-made, and field-tested at the assembly point by the senior author of ASTM Standard D4428 and produced by Soils Engineering & Geophysics, Inc., USA. The source is a mechanical borehole tool consisting of a case (1), coupled to the pipe or borehole wall using an air-inflatable rubber inner tube (2). A single three-in-one, 0.5 inch diameter cable (3) houses controls for reversible spring-loaded hammer (4) activation, zero-time sensor (trigger) (5), and air for the clamping mechanism. The device is intended for use in dry-cased boreholes only.

Two BGH-3 borehole three-component geophones (v-l-t components) inserted into the other drill holes (figure 2.134 (2)) are used as receivers for the elastic waves. A special device is used to couple the geophone effectively to the borehole wall. The geophone is supplied with a compass permitting the orienting of its longitudinal component in a northerly direction so that the transversal component is oriented E–W while the vertical is parallel to the borehole axis. This permits accurate orientation of the strike for reliable selection of the P- and SH-waves. Cross-hole and downhole data are acquired using very high resolution.

The *cross-hole measurements* are carried out as follows: the cross-hole test is begun by placing the energy source in an end hole (bh-3) at a depth of 1.0 m from the surface and coupled to the wall. Two receivers (or one, for two borehole measurements) are placed at the same elevation in bh-2 (geophone No. 1) and bh-1 (geophone No. 2) boreholes, clamped and oriented (figure 2.134). The recording equipment is checked. The energy source is activated and records from both receivers are displayed on the recorder screen. The source activation (1–4 stacks) is carried out until a high-quality signal is accumulated. The source and receivers are lowered simultaneously and fixed at the next depth (1 m lower). Measurements are performed again until the maximum borehole depth has been reached. At some source and receiver positions, a reverse stroke is performed for the S-wave identification. During withdrawal of the energy source and receivers from the boreholes, the test is repeated at 5.0 m intervals until the ground surface is reached. Such measurement is carried out at depths of 5.0 and 1.0 m.

*Cross-hole measurements by the borehole-radar method (*Singha *et al* 2000).

In a cross-hole measurement, the transmitter and receiver are placed in different boreholes. Moving one antenna stepwise, one carries out the measurements and takes measurements at every stop until the bottom of the borehole is reached. The other antenna is then moved one step further, and the first antenna is moved up the hole, taking measurements again at every stop.

2.2.3.4 A brief overview of borehole geophysics in karst problem

As mentioned in sections 2.2.1.7 and 2.2.1.9, high resistive anomalies were detected by FDEM and ERT surface measurements above the existing Soreq Cave (figures 2.65 and 2.83, respectively) (Ezersky 2002). To verify these anomalies two boreholes (BH1 and BH2) by the 25–30 m depth everyone were drilled by the Geological Survey of Israel (see figure 2.83 for location) (Bar-Matthews *et al* 1999). Borehole BH-2 was drilled in the anomaly location and borehole BH-1 some 100 m south to BH-2. The ultrasonic loggings (USL) were conducted in both boreholes. Results of USL in borehole BH-1 are shown in figure 2.135.

Results obtained in both boreholes BH1 and BH 2 (see figure 2.83 for location) have shown that a number of fractures are filled with the secondary calcite, and the refilled layers show very high resistivity. Thus, the investigation has shown that

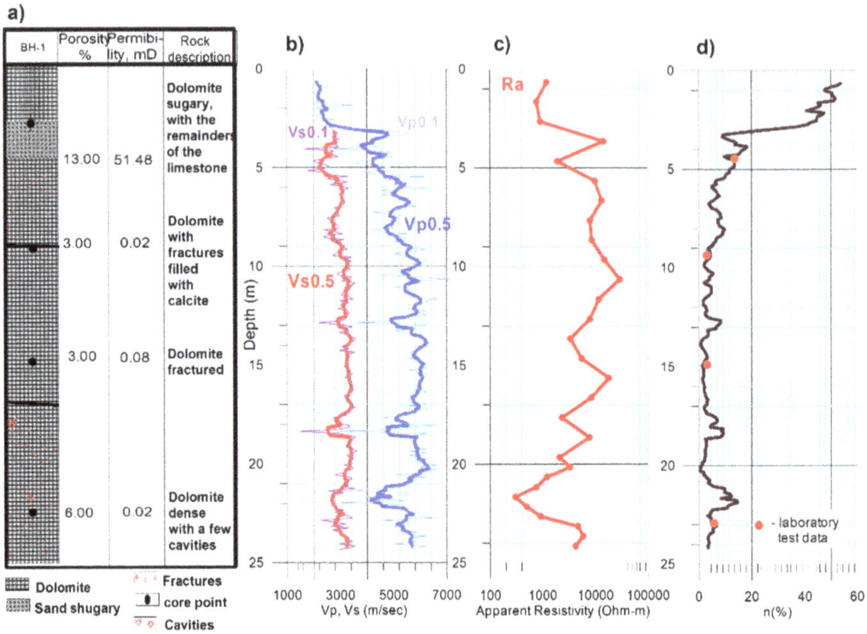

Figure 2.135. Soreq cave area, Israel. Results of geotechnical and geophysical investigations in borehole BH-1 (a) geological description and geotechnical parameters of rock samples (Bar-Mathews *et al* 1999); (b) ultrasonic logs (graphs V_p and V_s versus depth); (c) apparent resistivity logging graph; (d) porosity versus depth graph based on ultrasonic velocity V_p calculations (Dvorkin (2008), velocity in the rock matrix is accepted V_{pS} = 6390 m s^{-1}, and velocity in fluid filling pores V_{pf} = 1346 m s^{-1}). The red points designate the porosity values determined in the laboratory. (Reprinted from Ezersky *et al* (2000), with permission EAGE, the material is published with the permission of the Environmental and Engineering Geophysical Society. All copyright privileges remain with EEGS. This material cannot be copied or used without the express written permission of EEGS Permission.)

under specific geohydrological conditions (for instance, at the Soreq Cave site), the geoelectric methodology alone can detect caves as well as high resistive rocks such as strongly karstified dolomites, dolomites with calcite-filled fractures, etc, but is unable to distinguish between them. To resolve such an equivalence problem an integrated technique (GPR, gravity survey, etc) should be applied. A similar problem relates also to the FDEM method (see section 2.2.1.7).

Zhang *et al* (2019) have applied resistivity tomography (ERT) and cross-borehole seismic computed tomography (SCT) for the karst prospecting along the designed metro line between Hangzhou and Fuyang, China. The trace of the subway is composed of soluble rocks mainly of the Ordovician limestone that directly contact the Quaternary strata. These rocks are considered to cover the karst. The subsidence is affected by human influences and natural conditions. Therefore, there is a danger of the sudden collapse of the overlying strata into open karst. The stratigraphy of the site from shallow to deep parts consists of plain fill, various clays, clay-gravel layers, and the weathered limestone is located at 23 m (a borehole was drilled down to 30 m).

The developed karst water is hydraulically related to porous water. The weathered limestone is overlaid by soil layers of different thicknesses and poor mechanical properties. The weathered limestone is a typical structure of the karst covering that is prone to sudden and continuous collapse. The karst cavities are filled with a mix of clays and water. After analysis of possible methods of karst investigations along the trace of the future metro line, the existing boreholes were selected to be used for geophysical field application for identification of karst development conditions.

A total of 250 boreholes 20 m apart and 40–50 m depth for the karst prospecting along the designed metro line were proposed. Deep inside each borehole, both geophones and electrodes were arranged for cross-borehole SCT and ERT.

The following parameters of rock and cavity have been introduced into the model:

Medium	V_P (m s^{-1})	Density (kg m^{-3})	Resistivity (Ωm)
Limestone	4000	2500	500
Clay and water-filled cavity	1500	2000	10

Zhang *et al* (2019) present seismic, electrical, and electromagnetic prospecting operating system (SEMOS) for the karst investigation. The seismic source of cross-borehole tomography is an electric spark and its multiply repeated and excited waveform is of high quality.

An example of tomography results between two boreholes XK-2YYZ-D75 and XK-2YYZ-Z78 are presented in figures 2.136(a) and (b), geological interpretation of

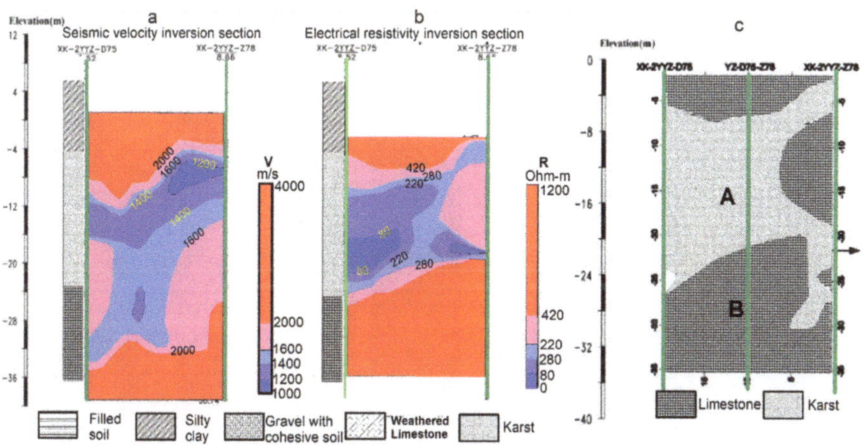

Figure 2.136. The cross-borehole prospecting results along the designed Hangzhou–Fuyang intercity railway. (a) Seismic velocity tomography section. (b) Electrical resistivity tomography section. (c) The section of combined interpretation of geophysical and geological results. (The material from Zhang *et al* (2019) is published with the permission of the Environmental and Engineering Geophysical Society. All copyright privileges remain with EEGS. This material cannot be copied or used without the express written permission of EEGS.)

tomography is shown in figure 2.136(c). The black–white section was modified to a color one by M Ezersky to better understand its velocity and resistivity structure.

The results obtained using the cross-borehole seismic velocities and resistivity profiles as well as results of drilling exploration were used for precise interpretation of the karst development.

Lane *et al* (1998). have used radar tomography (borehole-to-borehole) to monitor the effectiveness of a saline tracer test in sedimentary rocks at Belvidere, Tennessee (USA). Two hydraulically conductive zones were identified using borehole geophysical logs, flowmeter logs, and straddle-packer testing. After that, a saline tracer was injected into one of the permeable zones. The principle of this measurement is an increase in the radar signal attenuation when the saline tracer fills fractures. The difference in radar signal attenuation during the time was controlled. The conducted tomography has shown an increase in attenuation in the image plane after the salinetracer injection in one of the boreholes. This increase in attenuation has propagated between the boreholes for some period (figure 2.137).

Milanović (2005) refers to the borehole-radar investigation in the tunnel located behind the Ombla Spring (Croatia) where four boreholes were measured including a total of 663 m surveyed borehole length (Niva 1990). In the radargrams, the

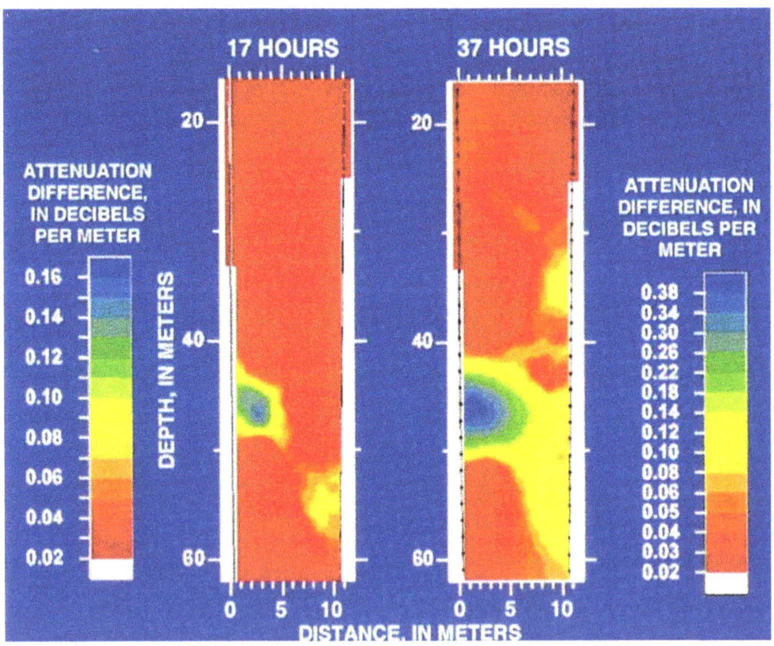

Figure 2.137. Borehole-to-borehole tomography was performed using sixty megahertz borehole radar. Two attenuation-difference tomograms were collected between two boreholes during 17 and 37 h after injection of tracer in one of the boreholes. (The material from Lane *et al* (1998), is published with the permission of the Environmental and Engineering Geophysical Society. All copyright privileges remain with EEGS. This material cannot be copied or used without the express written permission of EEGS.)

hyperbolic images are displayed testifying to the presence of cavities, voids, or water tunnels.

Deceuster *et al* (2006) using twenty-three boreholes beneath a four-floor public building have carried out 2D cross-borehole resistivity tomographies (ERT) where sinkholes were already collapsed These investigations were aimed to generate a model of the complex subsurface geometry to estimate and minimize the costs of remediation works. At the start, preliminary cross-borehole ERT field tests were performed to determine a suitable combination of arrays and to estimate the stability of these measurements on the site. The inversion was carried out using the Res2DInv® software. Weathered zones were identified as low resistivity zones ($\rho < 50\ \Omega$m) as against high resistivity solid limestone ($\rho > 200\ \Omega$m). That allowed building a 3D model of the ground conditions.

The 3D resistivity model built on ERT results has shown that:

(1) Depth-to-bedrock under building varies from 6 to more than 24 m;
(2) Inside the underbuilding area were selected four zones of different depth-to-bedrock ranges, and three vertically elongated weathered zones were also delineated.

Thus, the ERT study has allowed the designing of a remediation plan.

Subsequent monitoring of the remedial works displayed that tomographies gave a very accurate overview of the buried rockhead depths (with a precision of about 1 m).

Several works consider the disadvantages of the tomography methods.

Jackson and McCann (2015) consider the advantages and disadvantages of the application of seismic tomography in different geological conditions. A major problem can arise at the investigation stage when available natural cavities or mine shafts and adits are not always detected by the drilling program. The conventional approach requires a large array of closely spaced boreholes to provide an acceptable assessment of the site when this type of hazard is suspected. This approach was used in the location of mine workings at open-cast sites in Britain by the British Coal Opencast. The cross-hole seismic methods provide a viable approach to scanning the rock masses between the boreholes to locate these natural or artificial discontinuities.

The interpretation of cross-borehole seismic tomography surveys over three known geological structures has led to the identification of factors that can be important when interpreting tomograms, namely:

- the evenness of the angular ray coverage;
- velocity anisotropy;
- fast arrivals (e.g., heat waves) propagating 'outside'.

It is proposed that the cross-borehole tomographic surveys should not be taken immediately at their face value in the absence of geological control.

Deceuster *et al* (2006) also denote the disadvantage of the ERT tomography. As in any geophysical inversion, the non-uniqueness of the solution remains and could lead to erroneous reconstructions. Geophysicists prefer to interpret the models concerning the local geological context to exclude the improbable solutions. Moreover, when several panels are measured, a few control drillings inside the cross-borehole panels can give an insight into the confidence in the interpretations.

Authors of this book note that there is one important disadvantage of the borehole drilling in the karst areas associated with the risk of its collapse in the weathered karstified rocks. Our example from the Ramat Ha Kovesh (Ezersky 2007) has indicated that in such cases should be planned additional surface seismic measurements.

References

Abelson M, Baer, G, Shtivelman V, Wachs D, Raz E, Crouvi O, Kurzon I and Yechieli Y 2003 Collapse-sinkholes and radar interferometry reveal neotectonics concealed within the Dead Sea basin *Geophys. Res. Lett.* **30** 1545

Abelson M, Aksinenko T, Kurzon I, Pinsky V, Baer G, Nof R and Yechieli Y 2018 Nanoseismicity forecasts sinkhole collapse in the Dead Sea coast years in advance *Geology* **46** 83–6

Alexeyev V V, Khesin B E and Eppelbaum L V 1996 Geophysical fields observed at different heights: a common interpretation technique *Proc. of the Meeting of Soc. of Explor. Geophys. (Jakarta)* 104–8

Al-Qufi A, Al-Malabeh A and Al-Tarazi E 2012 Characterization of lava caves, using 2D induced polarization imaging, Umm Al Quttein area, NE Jordan *Proc. of the 15th Int. Symp. of Volcanospeleology (Amman, Jordan, March 15–22, 2012)* 71–83

Ali H and Choi J 2020 Risk prediction of sinkhole occurrence for different subsurface soil profiles due to leakage from underground sewer and water pipelines *Sustainability* **12** 310

Alpin L M, Dayev D S and Karinsky A D 1985 *Theory of Fields Employed in Geophysical Prospecting* (Moscow: Nedra) (in Russian)

Al-Zoubi A, Eppelbaum L, Abueladas A, Ezersky M and Akkawi E 2013 Removing regional trends in microgravity in complex environments: testing on 3D model and field investigations in the eastern Dead Sea coast (Jordan) *Int J Geophys.* **2013** 341797 (Special Issue)

Anderson N, Martinez A, Hopkins J F and Carr R T 1998 Case history. Salt dissolution and surface subsidence in central Kansas: a seismic investigation of the anthropogenic and natural origin models *Geophysics* **3** 366–78

Anderson N L, Martinez A A, Hopkins J F and Carr T R 2017 Salt dissolution and surface subsidence in Central Kansas: a seismic investigation of the anthropogenic and natural origin models *Geophysics* **63** 366–78

Andrieux C 1969 Contribution a' l'e'tude du climat des cavités naturelles des massifs karstiques *PhD Thesis* (Universite' de Bordeaux)

Annan A P 1999 *Practical Processing of GPR Data* (Mississauga, Canada: Sensors and Software Inc.) p 16

Annan A P 2003 *Ground Penetrating Radar Principles, Procedures, and Applications* (Mississauga, Canada: Sensors and Software Inc.)

Annan A P and Cosway S W 1992 Ground penetrating radar survey design *SAGEEP* 329–51

Arandjelović D 1969 A possible way of tracing groundwater flows in karst *Geophys. Prospect.* **XII** 404–21

Arandjelović D 1976 Geophysics in the karst. Institute for geological, geophysical, and geotechnical research *Geophys. Instit. Beograd* **17** 1976

Archie G E 1942 Electrical resistivity as an aid in determining some reservoir characteristics *Trans. Am. Inst. Mining Eng* **146** 54–62

Arkin Y, Gilat A and Kofman L 2000 Developing Dead Sea sinkholes revealed by ground-penetrating radar *GSI Current Res.* **12** 25–30

ASTM D 6429-99 2006 *Standard Guide for Selecting Surface Geophysical Methods* (West Conshohocken, PA: ASTM)

ASTM D 5777-700 2006 *Standard Guide for Using the Seismic Refraction Method for Subsurface Investigation* (West Conshohocken, PA: ASTM)

ASTM D 7400-07 2019 *Standard Test Methods for Downhole Seismic Testing* (West Conshohocken, PA: ASTM)

ASTM D 4428/D 4428M-14 2014 *Standard Test Method for Crosshole Seismic Testing* (West Conshohocken, PA: ASTM)

Audra P 2001 Valeur et répartition de la dissolution spécifique dans les karsts des montagnes Nakanaï *Nakanaï, 20 Ans d'Explorations* ed P Audra, P Coninck and J-P Sounier (Istres: Association Hémisphés Sud) 77–86

Avdevich M M and Fokin A F 1978 *Electric Modeling of Potential Geophysical Fields* (Leningrad: Nedra) (in Russian)

Bacon M 2005 *Encyclopedia of Geology* (Amsterdam: Elsevier) pp 151–8

Badino G 1992 Microclimi sotterranei *Grotte e Storie dell'asia Centrale: le Esplorazioni Geografiche del Progetto Samarcanda* ed T Bernabei and A De Vivo (Venice: Centro editorial Veneto) pp 151–3

Badino G 1995 *Fisica Del Clima Sotterraneo* (Bologna: Istituto Italiano di speleologia)

Badrinarayanan T S and Thiyagarajan G 2018 Software applications for interpreting geoelectric data in ground water exploration techniques *Sustainable Technologies for Intelligent Water and Environmental Management (STIWEM-2018) (Chidambaram)*

Baer G, Schattner U, Wachs D, Sandwell D, Wdowinski S and Frydman S 2002 The lowest place on earth is subsiding–an InSAR (interferometric synthetic aperture radar) perspective *Geol. Soc. Am. Bull.* **114** 12–23

Baer G, Magen Y, Nof R N, Raz E, Lyakhovsky V and Shalev E 2018 *InSAR Measurements and Viscoelastic Modeling of Sinkhole Precursory Subsidence: Implications for Sinkhole Formation, Early Warning, and Sediment Properties* (Washington, DC: American Geophysical Union)

Bakeev R A, Stefanov Y P, Duchkov A A and Miyasnikiov A V 2017 Modeling of karst deformation and analysis of acoustic emission during sinkhole formation *AIP Conf. Proc.* **1909** 020009

Baker H A and Myers S O 1980 A topographic correction for VLF-EM profiles based on model studies *Geoexploration* **18** 135–44

Balkaya Ç, Göktürkler G, Erhan Z and Ekinci Y L 2012 Exploration for a cave by magnetic and electrical resistivity surveys: Ayvacık Sinkhole example, Bozdağ, İzmir (western Turkey) *Geophysics* **77** B135–46

Barker R D 1989 Depth of investigation of collinear symmetrical four-electrode arrays *Geophysics* **54** 1031–7

Barker R 1992 A simple algorithm for electrical imaging of the subsurface *First Break* **10** 53–62

Bar-Matthews M, Ayalon A, Goldman M and Ezersky M 1999 The selection of boreholes for cave location at two sites above the Soreq Cave (in Hebrew)

Barneff W A, Hyslop C and Park J *et al* 2021 Viscoacoustic diffraction imaging to delineate karst features in a carbonate reservoir *First Break* **39** 39–43

Barsukov P, Fainberg E and Khabensky E 2015 Shallow investigations by TEM-FAST technique: methodology and examples *Electromagnetic Sounding of the Earth's Interior Methods in Geochemistry and Geophysics* ed V Spichak vol 40 (Amsterdam: Elsevier)

Basokur A T and Candansayar M E 2003 Enhancing VLF data for qualitative interpretation: an example of massive chalcopyrite exploration *Lead. Edge* **22** 568–70

Bass F G and Fuks I M 1978 *Wave Scattering from Statistically Rough Surfaces* (Oxford: Pergamon)

Batayneh A T, Abueladas A A and Moumani K A 2002 Use of ground-penetrating radar for assessment of potential sinkhole conditions: an example from Ghor al Haditha area, Jordan *Environ. Geol.* **41** 977–83

Baucher M, Girard J-F, Legchenko A, Baltassat J-M, Dorfliger N and Chalikakis K 2006 Using 2D inversion of magnetic resonance sounding to locate a water-filled karst conduit *J. Hydrol.* **330** 413–21

Bawden G W *et al* 2003 Measuring human-induced land subsidence from space. U.S. Geological Survey Fact Sheet 069-3

Beamish D 2000 Quantitative 2D VLF data interpretation *J. Appl. Geophys.* **45** 33–47

Beck A E 1981 *Physical Principles of Exploration Methods* (London: Macmillan)

Belfer I, Bruner I, Keydar S, Kravtsov A and Landa E 1998 Detection of shallow objects using refracted and diffracted seismic waves *J. Appl. Geophys.* **38** 155–68

Benson AK 1995 Application of ground penetration radar in assessing some geological hazards: examples of groundwater contamination, faults, cavities *J. Appl. Geophys.* **33** 177–93

Benson R C and Glaccum R A 1979 Radar surveys for geotechnical site assessment *Convention Exposition (Atlanta, October 23–25, 1979)* Preprint 37948

Benson R, Glaccum R A and Noel M R 1982 Geophysical techniques for sensing buried wastes and waste migration, Office of Research and Development, U.S. Environmental Protection Agency, Las Vegas, NV, published by National Water Well Association, Dublin, OH

Benson R C and Yuhr L 1993 Spatial sampling consideration and their applications to characterizing fractured rock and karst systems *Applied Karst Geology* ed B F Beck (Rotterdam: Balkema) pp 99–113

Benson R C and Yuhr L B 2016 Site characterization in karst and pseudokarst terranes *Practical Strategies and Technology for Practicing Engineers, Hydrologists, and Geologists* (Berlin: Springer) p 421

Beres M, Luetcher M and Paymond O 2001 Integration of penetrating radar and micro-gravimetric methods to map shallow caves *J. Appl. Geophys.* **46** 249–62

Bernatek-Jakiel A and Poesen J 2018 Subsurface erosion by soil piping: significance and research needs *Earth Sci. Rev.* **185** 1107–28

Best M E and Todd B J 1996 Electromagnetic soundings, pseudo-resistivity logs and implications for porosity and groundwater salinity *Proc., Symp. on the Application of Geophysics to Environmental and Engineering Problems (SAGEEP)* (Keystone, CO) 1061–74

Bianchetti G, Zuber F, Vuataz F-D and Rouiller J-D 1993 Hydrogeologische und geothermische Untersuchungen im Simplontunnel *Matériaux Pour la Géologie de la Suisse, Géotechnique* no 88 (Berne: Commission Géotechnique, ASSN) pp 1–75

Biot M A 1952 Propagation of elastic waves in a cylindrical bore containing a fluid *J. Appl. Phys.* **23** 997–1005

Bishop I, Styles P, Emsley S J and Ferguson N S 1997 The detection of cavities using the microgravity technique: case histories from mining and karstic environments *Geol. Soc. Lond. Eng. Geol. Spec. Publ.* **12** 153–66

Bitri A, Samyn K and Filippi C 2016 Can we really detect cavities using seismic surface waves? *22nd European Meeting of Environmental and Engineering Geophysics: Near Surface Geoscience 2016 (Barcelona, September 2016)* ffhal-01371628f

Bitterli T (ed) 1996 *Proyecto Cerro Rabon 1990–1994* (Oaxaca, Mexico: Speleo Projects, Basel)

Blakely R J 1995 *Potential Theory in Gravity and Magnetic Applications.* (Cambridge: Cambridge University Press)

Bloomquist D, Shrestha R L and Slatton C 2005 Early sinkhole detection and verification using airborne laser and infrared technologies *Technical Report* USABC-354-54 Department of Civil Coastal Engineering, University of Florida

Bonnefoy-Claudet S, Cotton F and Bard P Y 2006 The nature of the seismic noise wave field and its implication for site effects studies: a literature review *Earth Sci. Rev.* **79** 205–27

Bosch F P and Müller I 2005 Improved karst exploration by VLF-EM-gradient survey: comparison with other geophysical methods *Near Surf. Geophys.* **3** 299–310

Booth A D, Clark R and Murray T 2010 Semblance response to a ground penetrating radar wavelet and resulting errors in velocity analysis *Near Surf. Geophys.* **8** 235–46

Boucher M, Legchenko A, Girard J F, Baltassat J M, Dorflinguer N and Chalikakis K 2006 Using 2D inversion of MRS soundings to locate a water-filled karst conduit *J. Hydrol.* **330** 413–21

Boulytchov. A A 1997 Geophysically predicted and discovered large cave emptinesses in Siberia *Proc. l2th Int. Cong. of Speleology (Switzerland)* vol 5 89–92

Boulytchov A 2000 Seismic-electric effect method on guided and reflected waves *Phys. Chem. Earth* A **25** 333–6

Braile L 2004 Seismic Waves Demonstrations and Animations, website of Purdue University https://web.ics.purdue.edu/~braile/edumod/waves/WaveDemo.htm

Brandon W C, Behr R E, Blackey M E and Zay A 2001 Identification of discrete high transmissivity zones in sparsely fractured limestone; an integrated approach using fracture trace analysis and surface and borehole geophysical methods, Loring, Maine, in Abstracts with Programs *Geol. Soc. Am.* **33** 413

Breithaupt C 2016 Cave detection using seismic methods at Madison Blue Spring state park, County, Florida *MSc Thesis* (Michigan Technological University Madison)

Browitt C *et al* 2007 Terra not so firm *Geoscientist* **17** 18–22

Busetti A, Calligaris C, Forte E, Areggi G, Mocnik A and Zini L 2020 Non-invasive methodological approach to detect and characterize high-risk sinkholes in urban cover evaporite karst: integrated reflection seismics, PS-InSAR, Leveling, 3D-GPR, and Ancillary Data. A NE Italian Case Study *Remote Sens.* **12** 28

Butler K 1984 Microgravimetric and gravity gradient techniques for detection of subsurface cavities *Geophysics* **49** 1084–96

Butler K, Sjostrom K J and Llopis J L 1997 Selected short stories on novel applications of near-surface geophysics *Leading Edge* **16** 1593–600

Calastrenc C, Baleux F, Poirier N and Rendu C 2020 Thermographie aéroportée par drone. Nouvelle procédure pour la détection archéologique en haute montagne *ArcheoSci.* **1** 81–96

Carbonel D, Rodríguez V, Gutiérrez F, McCalpin J P, Linares R, Roqué C, Zarroca M, Guerrero J and Sasowsky I 2014 Evaluation of trenching, ground-penetrating radar (GPR), and electrical resistivity tomography (ERT) for sinkhole characterization *Earth Surf. Process. Landf.* **39** 214–27

Carpenter P, Doll W and Kaufmann R 1998 The geophysical character of buried sinkholes on the Oak Ridge reservation, Tennessee *J. Environ. Eng. Geophys.* **3** 133–45

Casas A, Lazaro R, Vilas M and Busquet E 1996 Detecting karstic cavities with ground penetrating radar at different geological environments in Spain *Proc. 6th Int. Conf. GPR'96, Sendai, Japan* 455–60

Chalikakis K, Plagnes V, Guerin R, Valoes R and Bosch F P 2011 Contribution of geophysical methods to karst-system exploration: an overview *Hydrol. J.* **19** 1169–80

Chambers J, Ogilvy R and Kuras O *et al* 2002 3D electrical imaging of known targets at a controlled environmental test site *Environ. Geol.* **41** 690–704

Chen Y, Gan F, Chen Y, Wu C, Ke P, Wei J and Lu C 2015 Types and genesis of induced polarization anomalies in the water-rich structures of the Taian karst area *Environ. Earth Sci.* **74** 1047–59

Chen Y, Qin X, Huang Q, Gan F, Han K, Zheng Z and Meng Y 2018 Anomalous spontaneous electrical potential characteristics of epikarst in the Longrui Depression, Southern Guangxi Province, China *Environ. Earth Sci.* **77** 1–9

Chevalier A, Legchenko A, Girard J-F and Descloitres M 2014 3D Monte Carlo inversion of magnetic resonance measurements *Geophys. J. Int.* **198** 216–28

Choppy J 1984 Températures de l'air *Phénomènes Karstiques, Série 1.1, Processus Climatiques, 3ème Partie* (Paris: Spéléo-Club de Paris) pp 1–73

Cigna A 1967 An analytical study of air circulations in caves *Int. J. Speleol.* **3** 41–54

Clayton C R I *et al* 1995 *Site Investigations, A Handbook for Engineers* 2nd edn (Cambridge: Blackwell Science)

Closson D, Karaky N A, Hassen H, Derauw D, Barbier C and Ozer A 2003 Space-borne radar interferometric mapping of precursory deformations of a dyke collapse, Dead Sea area, Jordan *Int. J. Remote Sens.* **24** 843–9

Closson D 2005 Cooccurrence between the Geo-Hazards induced by the Dead Sea level lowering and the geological setting *PhD Thesis* University of Liege

Conyers L B 2012 *Interpreting Ground-penetrating Radar for Archaeology* (Walnut Creek, CA: Left Coast Press)

Conyers L B 2013 *Ground-penetrating Radar for Archaeology* 3rd edn (Lantham, MD: Altamira Press, Rowman and Littlefield Publishers)

Conyers L B 2016 *Ground-penetrating Radar for Geoarchaeology.* 1st edn (New York: Wiley) p 160

Cook K L and Van Nostrand R G 1954 Interpretation of resistivity data over filled sinks *Geophys. Prospect.* **21** 716–23

Corwin R F 1990 The self-potential method for environmental and engineering application ed S H Ward *Geotechnical and Environmental Geophysics* vol 1 (Tulsa, OK: Society of Exploration Geophysicists) pp 127–45

Cox L H, Wilson G A and Zhdanov M S 2010 3D inversion of time-domain airborne electromagnetic data *20th IAGA WG 1.2 Workshop on Electromagnetic Induction in the Earth IAGA* Expanded Abstract

Crawford N C, Lewis M A, Winter S A and Webster J A 1997 Microgravity techniques for subsurface investigations of sinkhole collapses and detection of groundwater flow paths through karst aquifers *Hydrogeology and Engineering Geology of Sinkholes and Karst* ed B Beck, A Pettit and J G Herring (Rotterdam: Balkema) pp 203–18

Crice D 2002 Borehole shear-wave surveys for engineering site investigation *Geostaff.* 14 www.georadar.com/geostaff

Crice D 2011 Near-surface, downhole shear wave surveys: A primer *Leading Edge* **30** 113–240

Dabas M, Anest A, Thiesson J and Tabbagh A 2016 Slingram EMI devices for characterizing resistive features using apparent conductivity measurements: check of the dual EM-421S instrument and field tests *Archaeol. Prospect.* **23** 165–80

Dal Moro G 2015 *Surface Wave Analysis for Near-surface Applications.* (Amsterdam: Elsevier) p 244

Dal Moro G 2019 *Win MASW Manual* (Lorca: Eliosoft)

Daniels D J (ed) 2007 *Ground Penetrating Radar* 2nd edn (London: The Institution of Engineering and Technology)

Daniels J J 1989 fundamentals of ground penetrating radar *Proc., Symp. on the Application of Geophysics to Engineering and Environmental Problems, Society of Engineering and Mineral Exploration Geophysicists* 62–142

Danişman A 2004 Investigation of karstic cavities by using the VLF-EM method *J. Black Seal Mediterr. Environ.* **10** 113–22

Davis J 2007 Springs eternal *Prof. Surveyor Mag.* **27** 8–12

de Araújo O S, Hindi E C, Rigotti A and Rigoti F H 2019 Improved imaging of a karst aquifer using focused source electromagnetic and differentially normalized method: a qualitative analysis *Geophys. Prospect.* **67** 1965–76

Debeglia N, Bitri A and Thierry P 2006 Karst investigations using microgravity and MASW: application to Orléans, France *Near Surf. Geophys.* **4** 215–25

Deceuster J, Delgranche J and Kaufman O 2006 2D cross-borehole resistivity tomographies below foundations as a tool to design proper remedial actions in covered karst *J. Appl. Geophys.* **60** 68–86

Decker L, Janson X and Fomel S 2015 Carbonate reservoir characterization using seismic diffraction imaging *Interpretation* **3** SF21–30

De Waele J, Gutierrez F, Parise M and Plan L 2011 Geomorphology and natural hazards in karst areas: a review *Geomorphology* **134** 1–8

Deroussi S, Diament M, Feret J B, Nebut T and Staudacher T 2009 Localization of cavities in a thick lava flow by microgravimetry *J. Volcanol. Geotherm. Res.* **184** 193–8

Descloitres M, Laurent J P, Morra C, Clément R, Oxarango L and Gourc J P 2008 Monitoring resistivity in non-hazardous waste landfill using Time Domain Electromagnetism (Drôme, France) *Near Surface 2008 (Krakov, Poland, 15–17 September)* 5824

Dey A and Morrison H F 1979 Resistivity modeling for arbitrary shaped two-dimensional structures *Geophys. Prospect.* **27** 106–36

Di Filippo M, Santoro S and Toro B 2005 Microgravity survey of Roman Amphitheatre of Durres (Albania) *Trans. of 6th Archaeological Prospection (Rome)*

Djeddi M, Baker H A and Tabbagh A 1998 Interpretation of VLF-EM anomalies of 3D structures by using linear filtering techniques *Ann. Geofis.* **41** 151–63

Dmitriyev V I, Baryshnikova I A and Zakharov K V 1977 *Anomalous Electromagnetic Fields of Geological Bodies* (Leningrad: Nedra) (in Russian)

Dmitriyev V I (ed) 1982 Computing mathematics and computer methods in exploration geophysics *Geophysicist's Manual* (Moscow: Nedra) (in Russian)

Dobrin M B 1976 *Introduction to Geophysical Prospecting* (New York: McGraw Hill)

Drucker D C and Prager W 1952 *Quart. Appl. Math.* **10** 157–65

Dvorkin J P 2008 Yet another vs equation *Geophysics* **73** E35–9

Dunn K J, Bergman D J and Latorraca G A 2002 *Nuclear Magnetic Resonance Petrophysical and Logging Applications.* (London: Elsevier Science)

Dunscomb M B and Rehwoldt E 1999 Two-dimensional resistivity profiling; geophysical weapon of choice in karst terrain for engineering applications *Hydrogeology and Engineering Geology of Sinkholes and Karst* ed B F Beck, J Pettit and J G Herring (Rotterdam: Balkema) pp 219–24

Eberhard R and Sharples C 2013 Appropriate terminology for karst-like phenomena: the problem with 'pseudokarst' *Int. J. Speleol.* **42** 109–13

Eberle D 1981 A method of reducing terrain relief effects from VLF-EM data *Geoexploration* **19** 103–14

Edwards L S 1977 A modified pseudo section for resistivity and induced-polarization *Geophysics* **42** 1020–36

Elfakih T A, Bahi L, Akhssas A, Ouadif L and Benkmil R 2020 Electrical resistivity tomography contribution to the characterization of underground cavities in the region of Safi, Morocco *E3S Web Conf.* **150** 03023

EMIGMA 9.5 2020 https://petroseikon.com/

Engelsfeld T, Sumanovac F and Pavin N 2008 Investigation of underground cavities in a two-layer model using the refraction seismic method *Near Surf. Geophys.* 221–31

Eppelbaum L V 1991 Examples of terrain corrections in the VLF method in the Caucasian region, USSR *Geoexploration* **28** 67–75

Eppelbaum L V 1999 Quantitative interpretation of resistivity anomalies using advanced methods developed in magnetic prospecting *Trans. of the XXIV General Assembly of the Europ. Geoph. Soc. (Strasbourg)* vol 1 *166*

Eppelbaum L V 2007 Revealing of subterranean karst using modern analysis of potential and quasi-potential fields *Proc. 2007 SAGEEP Conf.* vol 20 *(Denver, USA)* 797–810

Eppelbaum L V 2009 Near-surface temperature survey: An independent tool for buried archaeological targets delineation *J. Cult. Herit.* **12** e93–103

Eppelbaum L V 2010 An advanced methodology for Remote Operation Vehicle magnetic survey to delineate buried targets *Trans. of the 20th General Meeting of the Intern. Mineralogical Association, CH30G: Archaeometry (general session): Composition, Technology, and Provenance of Archaeological Artifacts (Budapest, Hungary)* 103

Eppelbaum L V 2011a Study of magnetic anomalies over archaeological targets in urban conditions *Phys. Chem. Earth* **36** 1318–30

Eppelbaum L V 2011b Review of environmental and geological microgravity applications and feasibility of their implementation at archaeological sites in Israel *Int. J. Geophys.* **2011** 927080

Eppelbaum L V 2014 Geophysical observations at archaeological sites: Estimating informational content *Archaeol. Prospect.* **21** 25–38

Eppelbaum L V 2015a Quantitative interpretation of magnetic anomalies from bodies approximated by thick bed models in complex environments *Environ. Earth Sci.* **74** 5971–88

Eppelbaum L V 2015b High-precise gravity observations at archaeological sites: how we can improve the interpretation effectiveness and reliability? *Trans. of the 11th EUG Meet., Geophysical Research Abstracts 17 (Vienna, Austria)* 1–4 EGU2015-3012

Eppelbaum L V 2016 Remote operated vehicles geophysical surveys in air, land (underground) and submarine archaeology: General peculiarities of processing and interpretation *Trans. of the 12th EUG Meet., Geophysical Research Abstracts* vol 18 *(Vienna, Austria)* 1–7 EGU2016-10055

Eppelbaum L V 2017a Delineating buried karst terranes on the basis of modern wavelet theory and data mining *Proc. 2017 SAGEEP Conf. (Denver, CO)* vol 30 *343–6*

Eppelbaum L V 2017b From micro- to satellite gravity: understanding the Earth *Am. J. Geogr. Res. Rev.* **1** 1–34

Eppelbaum L V 2017c Quantitative examination of piezoelectric/seismoelectric anomalies from near-surface targets *Geosciences* **7** 1–12

Eppelbaum L V 2019a *Geophysical Potential Fields: Geological and Environmental Applications* (Amsterdam: Elsevier)

Eppelbaum L V 2019b Advanced system of self-potential field analysis in ore deposits of the South Caucasus *Proc. National Azerbaijan Academy of Sciences* **2** 21–35

Eppelbaum L V 2019c System of potential geophysical field analysis in complex environments and some results of its application (*Christiaan Huygens Medal Lecture*) *Trans. of the 14th EUG Meet., Geophysical Research Abstracts* vol 21 *(Vienna, Austria)* EGU2019-1775

Eppelbaum L V 2020 Quantitative analysis of self-potential anomalies in archaeological sites of Israel: an overview *Environ. Earth Sci.* **79** 1–15

Eppelbaum L V 2021 Review of processing and interpretation of self-potential anomalies: transfer of methodologies developed in magnetic prospecting *Geosciences* **11** 1–33

Eppelbaum L V, Ben-Avraham Z and Itkis S E 2003 Integrated geophysical investigations at the Halutza archaeological site *Proc. 64 EAGE Conf. (Florence, Italy)* vol P151 *1–4*

Eppelbaum L V and Finkelstein M I 1998 Radon emanation, magnetic and VLF temporary variations: removing components not associated with dynamic processes *Collection of Selected Papers of the XXVI General Assembly of the European Seismological Commission, Tel Aviv* 122–6

Eppelbaum L V and Khesin B E 1992 VLF-method: elimination of noises and quantitative interpretation. Collection of papers *Regional Symp. on Electromagnetic Compatibility '1992—From a Unified Region to a Unified World', section 'LF to ULF' Electromagnetics and the Earth, 5.2.1* (Tel Aviv: IEEE) 1–6

Eppelbaum L V and Khesin B E 2002 Some common aspects of magnetic, induced polarization and self-potential anomalies interpretation: implication for ore target localization *Collection of Selected Papers of the IV Intern. Symp. on Problems of Eastern Mediterranean Geology* 279–93

Eppelbaum L V and Khesin B E 2004 Advanced 3-D modelling of gravity field unmasks reserves of a pyrite-polymetallic deposit: A case study from the Greater Caucasus *First Break* **22** 53–6

Eppelbaum L V and Khesin B E 2012 *Geophysical Studies in the Caucasus* (Heidelberg: Springer)

Eppelbaum L V and Mishne A R 2011 Unmanned airborne magnetic and VLF investigations: effective geophysical methodology of the near future *Positioning* **2** 112–33

Eppelbaum L V, Modelevsky M M Jr and Pilchin A N 1996 Thermal investigation in petroleum geology: the experience of implication in the Dead Sea Rift zone, Israel *J. Petrol. Geol.* **19** 425–44

Eppelbaum L V, Itkis S E and Khesin B E 2000 Optimization of magnetic investigations in the archaeological sites in Israel *Special Issue of Prosperzioni Archeologiche 'Filtering, Modeling and Interpretation of Geophysical Fields at Archaeological Objects'* 65–92

Eppelbaum L, Ben-Avraham Z and Katz Y 2004 Integrated analysis of magnetic, paleomagnetic and K-Ar data in a tectonic complex region: an example from the Sea of Galilee *Geophys. Res. Lett.* **31** L19602

Eppelbaum L V, Khesin B E and Itkis S E 2001 Prompt magnetic investigations of archaeological remains in areas of infrastructure development: Israeli experience *Archaeol. Prospect.* **8** 163–85

Eppelbaum L, Ezersky M, Al-Zoubi A, Goldshmidt V and Legchenko A 2008 Study of the factors affecting the karst volume assessment in the Dead Sea sinkhole problem using microgravity field analysis and 3D modeling *Adv. Geosci.* **18** 1–2

Eppelbaum L V and Khesin B E 2012 *Geophysical Studies in the Caucasus* (Heidelberg: Springer) p 411

Eppelbaum L V, Kutasov I M and Pilchin A N 2014 *Applied Geothermics* (Heidelberg: Springer)

Erchul R A 1993 Effective geophysical methods for Virginia karst terrains *Applied Karst Geology* ed B F Beck (Rotterdam: Balkema) pp 131–4

Ezersky M G 2002 The CVES method as applied to geohydrological, environmental, and engineering surveys in Israel *Isr. J.Earth Sci.* **51** 29–34

Ezersky M G 2007 Seismic downhole survey of V_p and V_s velocities in the borehole 1A, Ramat Kovesh, Israel *GII report* 273/250/07 (in Hebrew)

Ezersky M 2008 Geoelectric structure of the Ein Gedi sinkhole occurrence site at the Dead Sea shore in Israel *J. Appl. Geophys.* **64** 56–69

Ezersky M G 2006 The geophysical properties of the Dead Sea salt applied to sinlhole problem *J. Appl. Geophys.* **58** 45–58

Ezersky M G, Bodet L, Akkawi E, Al-Zoubi A, Camerlynck C, Dhemaied A and Galibert P-Y 2013 Seismic surface-wave prospecting methods for sinkhole hazard assessment along the Dead Sea shoreline *J. Environ. Eng. Geophys.* **18** 233–53

Ezersky M G and Eppelbaum L 2017 Geophysical monitoring of underground constructions and its theoretical basis *Int. J. Georesour. Environ.* **3** 56–72

Ezersky M G and Frumkin A 2021 Subaerial morphology affected by groundwater aggressiveness: sinkhole susceptibility above karstified salt, Dead Sea *Geomorphology* **375** 107525

Ezersky M and Goldman M 1998 Detection of potentially dangerous subsurface karstic zones using electrical and electromagnetic methods in the Mishor Rotem Area *GII Report* 906/69/98 (in Hebrew)

Ezersky M G and Goretsky I 2014 Velocity–resistivity versus porosity–permeability interrelations in Dead Sea salt samples *Eng. Geol.* **183** 96–115

Ezersky M and Sobolevsky L 2002 Estimates of the rock mass quality using geophysical methods. Second stage *GII Report* 263/228/02

Ezersky M G, Rudyak M S and Zhdanov V V 1993 Forecast and monitoring of the stability of the tunnel Zhinvali Project on basis of geophysical data *Int. Symp. on Assessment and Prevention of Failure Phenomena in Rock Engineering (Istambul, Turkey, April 1993)* 485–92

Ezersky M, Bruner I, Keydar S, Trachtman P and Rybakov M 2006 Integrated study of the sinkhole development site on the western shores of the Dead Sea using geophysical methods *Near Surf. Geophys.* **4** 335–43

Ezersky M G, Goldman M, Bar-Matthews M and Ayalon A 2000 Geoelectric investigation of the Soreq Cave area in Israel *6th Annual EAGE-/EEGS Meeting (September 2000) (Bohum)* 4 cp-51-00126

Ezersky M, Gorstein M, Keydar S, Kalmanovich M and Perelman N 2012 Combining the H/V spectral ratio (HVSR) technique with a method using surface waves (MASW) to map the subsurface: implementation in the Dead Sea area *Unpublished GII Report* 025/759/13

Fäh D, Kind F and Giardini D 2001 A theoretical investigation of average H/V ratios *Geophys. J. Int.* **145** 535–49

Farrar E and Becker T 1971 Pulse and Fourier transform NMR *Introduction to Theory and Methods* (New York: Academic) p 115

Felsen L and Markuvitz N 1994 *Radiation and Scattering of Waves* (Englewood Cliffs, NY: Prentice-Hall)

Finkl C W *et al* 2005 Interpretation of seabed geomorphology based on spatial analysis of high-density airborne laser bathymetry (ALB) *J. Coast. Res.* **21** 501–14

Fitterman D V and Stewart M T 1986 Transient electromagnetic sounding for groundwater *Geophysics* **51** 955–1005

Fitterman D 2005 Time-domain electromagnetic soundings to characterize water quality within a freshwater/saline-water transition zone, Estancia Valley, New Mexico, July 2005—a reconnaissance study. In cooperation with the New Mexico Bureau of Geology and Mineral Resources, USGS

Ford D C and Williams P W 2007 *Karst Geomorphology and Hydrology.* (New York: Chapman and Hall) p 601

Forte E, Bondini M B, Bortolleto A, Dossi M and Colluchi R R 2019 Pros and cons in helicopter-borne GPR data acquisition on rugged mountainous areas: critical analysis and practical guidelines *Pure Appl. Geophys.* **175** 1103–22

Fraser D C 1969 Contouring of VLF-EM data *Geophysics* **34** 958–67

Frenkel Y I 1944 On the theory of seismic and seismoelectric phenomena in a moist soil *Izvestia Acad. Sci. USSR, Ser. Geogr. Geophys.* **8** 133–50 (in Russian, transl. to English)

Fresnel A 1900 Fresnel's prize memoir on the diffraction of light *The Wave Theory of Light: Memoirs* ed H Young and F H Crew (Woodstock, GA: American Book Co.) pp 81–144

Fridrichsberg D A 1995 *Course of Colloidal Chemistry* (St. Petersburg: Chemistry Publ.) (in Russian)

Frumkin A 2013 Salt karst *Karst Geomorphology: Treatise on Geomorphology* ed J F Shroder and A Frumkin vol 6 (San Diego: Academic) pp 407–24

Frumkin A, Kofman L and Ezersky M 2007 Improvement of the reliability of subsurface void detection, including sinkhole development, at the Dead Sea shore area using Ground Penetrating Radar (GPR) *GII Report* 283/31/07

Frumkin A, Kofman L and Ezersky M 2008 Improvement of the reliability of subsurface void detection, including sinkhole development, at the Dead Sea shore area by means of Ground Penetration Radar (GPR) *Intermediate Report* 283\311\07 Hebrew University—TECHNION Research and Development Foundation LTD—Geophysical Institute of Israel

Frumkin A, Kofman L and Ezersky M 2009 Improvement of the reliability of subsurface void detection, including sinkhole development, at the Dead Sea shore area by means of Ground Penetration Radar (GPR) *Report* MNI_ES-36-2008 Hebrew University—TECHNION Research and Development Foundation Ltd—Geophysical Institute of Israel

Frumkin A, Ezersky M, Al-Zoubi A, Akkawi E and Abueladas A-R 2011 The Dead Sea hazard: geophysical assessment of salt dissolution and collapse *Geomorphology* **134** 102–17

Frumkin A, Zaidner Y, Na'aman I, Tsatskin A, Porat N and Vulfson L 2015 Sagging and collapse sinkholes over hypogenic hydrothermal karst in a carbonate terrain *Geomorphology* **229** 45–57

Frydman S and Kofman L 2002 Improving the reliability of location and mapping of karst cavities with ground-penetrating radar *Final Report of the National Building Research Institute to the Ministry of Construction and Housing, Technion* (in Hebrew, English abstract)

Galve J P, Castaneda C and Gutierrez F 2015 Railway deformation detected by DInSAR over active sinkholes in the Ebro Valley evaporite karst, Spain *Nat. Hazards Earth Syst. Sci.* **15** 2439–48

Gamburtsev G A, Riznichenko Y V, Berzon L S, Yepinatyeva A M, Pasechnik J P, Kosminskaya J P and Karns E V 1952 Correlation method of refraction waves *Academic Science* (Moscow: USSR Press) (in Russian)

Garambois S, Legchenko A, Vincent C and Thibert A 2016 Ground-penetrating radar and surface nuclear magnetic resonance monitoring of an englacial water-filled cavity in the polythermal glacier of Tête Rousse *Geophysics* **81** WA131–46

Gelchinsky B, Keydar S and Belle H 1993 Studying salt dome structure by homeomorphic imaging methods *Seismo-Series* vol 59 (Norway: University of Bergen) p 87

Geomatrix 2010 (http://geomatrix.co.uk)

Geometrix software SeisImager/2D (https://geometrics.com/software/)

Geonics Limited Technical Note TN-31 1996 *Why Doesn't Geonics Limited Build a Multi-Frequency EM 31 or EM 38?* Mississauga, ON: Geonics Limited.

Gibson P J, Lyle P and George D M 2004 Application of resistivity and magnetometry geophysical techniques for near-surface investigations in karstic terranes in Ireland *J. Cave Karst Stud.* **66** 35–8

Girard J-F, Boucher M, Legchenko A and Baltassat J-M 2007 2D magnetic resonance tomography applied to karstic conduit imaging *J. Appl. Geophys.* **63** 103–16

GSPX7 software 2020 Geonics Limited (http://geonics.com/html/software.html)

GeoTomo 2007 *GeoCT-II* (Houston: GeoTomo) 2016

Goldman. M, Gilad D, Ronen A and Melloul A 1991 Mapping of seawater intrusion into the coastal aquifer of Israel by the time domain electromagnetic method *Geoexploration* **28** 153–74

Gondwe B R N, Ottowitz D, Supper R, Motschka K, Merediz-Alonso G and Bauer-Gottwein P 2012 Regional-scale airborne electromagnetic surveying of the Yucatan karst aquifer (Mexico): geological and hydrogeological interpretation *Hydrol. J.* **20** 1407–25

Gordeyev S G 1983 Terrain relief considerations in magnetic fields of remote transmitters *Proc. TSNIGRI* vol 179 56–60 (in Russian)

Gordeyev S G and Sedelnikov E S 1974 On the interpretation of the VLF method results on the basis of the conductive bed modeling *Proc. TSNIGRI* vol 116 88–99

Gordeyev S G, Sedelnikov E S and Tarkhov A G 1981 *Electric Prospecting by the Radio Comparing and Bearing Method (VLF)* (Moscow: Nedra) (in Russian)

Gorstein M and Ezersky M 2015 Combination of HVSR and MASW methods to obtain shear wave velocity model of subsurface in Israel *IJGE* **1** 20–41

Gossar A and Čeru T 2016 Search for an artificially buried karst cave entrance using ground-penetrating radar: a successful case of locating the S-19 Cave in the Mt. Kanin massif (NW Slovenia) *Int. J. Speleol.* **45** 135–47

Grandjean G 2006 Imaging sub-surface objects by seismic P-wave tomography: numerical and experimental validations *Near Surf. Geophys.* **4** 279–87

Granndjean G and Leparoux D 2004 The potential of seismic methods for detecting cavities and buried objects: experimentation at a test site *J. Appl, Geophys.* **56** 93–106

Green R 1974 The seismic refraction method-a review *Geoexploration* **12** 259–84

Greenfield R J 1979 Review of geophysical approaches to the detection of karst *Bull. Assoc. Eng. Geol.* **16** 393–408

Griffiths D H and Barker R D 1993 Two-dimensional resistivity imaging and modeling in areas of complex geology *J. Appl. Geophys.* **29** 21–6

Grombacher D, Fiandaca G, Behroozmand A A and Auken E 2017 Comparison of stabilizer functions for surface NMR inversions *Near Surf. Geophys.* **15** 533–44

Greenhouse J *et al* 1995 An introduction to application of surface geophysics in environmental investigations *A Short Course Presented at the Symp. on the Application of Geophysics to Environmental and Engineering Problems* (Environmental and Engineering Geophysical Society) *(Orlando, FL)*

Greenwood W 2015 *In situ* shear wave velocity measurements in rocks (https://geoengineer.org/education/web-class-projects/cee-544-soil-site-improve-winter-2015/assignments/in-situ-shear-wave-velocity-measurements-in-rocks)

Grimes K G 1975 Pseudokarst: definition and types ed A W Graham *Proc. 10th Biennial Conf.* (Brisbane: Australian Speleological Federation) pp 6–10

Guérin R and Benderitter Y 1995 Shallow karst exploration using MT-VLF and DC resistivity methods *Geophys. Prospect.* **43** 635–53

Guérin R, Baltassat J M, Boucher M, Chalikakis K, Galibert P Y, Girard J F, Plagnes V and Valois R 2009 Geophysical characterization of kart netwrks: application to the Ouysse system (Poumeyssen, France) *C. R. Geosci.* **341** 810–17

Gui H and Hu J 2017 A numerical simulation of impact of groundwater seepage on temperature distribution in karst collapse pillar *Arab. J. Geosci.* **10** 1–11

Guillen A and Legchenko A 2002a Application of linear programming techniques to the inversion of proton magnetic resonance measurements for water prospecting from the surface *J. Appl. Geophys.* **50** 149–62

Guillen A and Legchenko A 2002b Inversion of surface nuclear magnetic resonance data by an adapted Monte-Carlo method applied to water resource characterization *J. Appl. Geophys.* **50** 193–205

Gutiérrez F, Cooper H A and Johnson K S 2008a Identification, prediction, and mitigation of sinkhole hazards in evaporate karst areas *Environ. Geol.* **53** 1007–22

Gutiérrez F, Guerrero J and Lucha P 2008b A genetic classification of sinkholes illustrated from evaporite paleokarst exposures in Spain *Environ. Geol.* **53** 993–1006

Gutiérrez F, Guerrero J and Lucha P 2008c Quantitative sinkhole hazard assessment. A case study from the Ebro Valley evaporite alluvial karst (NE Spain) *Nat. Hazards* **45** 211–33

Gutiérrez F, Galve J P, Lucha P, Castañeda C, Bonachea J and Guerro J 2011 Integrating geomorphological mapping, trenching, InSAR, and GPR for the identification and characterization of sinkholes: a review and application in the mantled evaporite karst of the Ebro Valley (NE Spain) *Geomorphology* **134** 144–56

Gutiérrez F, Parise M, De Waele J and Jourde H 2014 A review on natural and human-induced geohazards and impacts in karst *Earth Sci. Rev.* **138** 61–88

Gutiérrez F, Benito-Calvo A, Carbonel D, Desir G, Sevil J, Guerrero J, Martinez-Fernandez A, Karamplaglidis T, Garcia-Arnay A and Fabregat I 2019 Review on sinkhole monitoring and performance assessment of the performance of remediation measures by high-precision leveling and terrestrial laser scanner in the salt karst of the Ebro Valley, Spain *Eng. Geol.* **248** 283–308

Habberjam G M 1969 The location of spherical cavities using a tripotential resistivity technique *Geophysics* **34** 780–4

Hagedoorn J G 1959 The plus–minus method of interpreting seismic refraction sections *Geophys. Prospect.* **7** 158–82

Hamdan H, Kritikakis G, Andronikidis N, Economou N, Manoutsoglou E and Vafidis A 2010 Integrated geophysical methods for imaging saline karst aquifers. A case study of Stylos, Chania, Greece *J. Balkan Geophys. Soc.* **13** 1–8

Hänl H, Maue A W and Westpfahl K 1961 *Theorie Der Beugung.* (Berlin: Springer)

Hardy H R 1976 Monitoring of stability of geologic structures using near-surface microseismic transducers *Proc. 1st Conf. of AE/MA* (Zurich: Trans Tech Publications)

Hardy H R and Belesky R M 1986 Potential of application of seismic and acoustic emission/ microseismic techniques to the monitoring of highway subsidence *Final Report* FHWA-PA-85-035+84-32, Contract # DOT-FA-01-84-C-0005 Federal Highway Administration (https:// trid.trb.org/view/269059)

Hardy H R and Belesky R M 1987 Use of acoustic emission and seismic tomography techniques to study sinkhole development *Jpn. J. Appl. Phys.* **26** 3–8

Hart D and Rudman A J 1997 Least-squares fit of an ellipse to anisotropic polar data: application to azimuthal resistivity surveys in karst regions *Comput. Geosci.* **23** 189–94

Hawker L, Bates P, Neal J and Rougier J 2018 Perspectives on Digital Elevation Model (DEM) simulation for flood modeling in the absence of a high-accuracy open access global DEM *Front. Earth Sci.* **6** 233

Hertrich M, Braun M, Günther T, Green A G and Yaramanci U 2007 Surface nuclear magnetic resonance tomography *IEEE Trans. Geosci. Remote Sens.* **45** 3752–9

Hickey C, Schmitt D, Sabatier J and Riddle G 2009 Seismic measurements for detecting underground high-contrast voids *Symp. on the Application of Geophysics to Engineering and Environmental Problems* 929–36

Hoekstra P and Blohm M 1990 Case histories of time domain electromagnetic sounding in environmental geophysics *Investigations in Geophysics No.5, Geotechnical and Environmental Geophysics* ed A Brown and S Ward vol II (Tulsa, OK: Society of Exploration Geophysicists) 1–14

Hodson E A 1948 The rockburst problem *Can. Min. Metall. Bull.* **440** 664

Hoover R A 2003 Geophysical choices for karst investigations *Ninth Multidisciplinary Conf. on Sinkholes and the Engineering and Environmental Impacts of Karst (Reston, VA)* B F Beck 529–38

Howari F M, Aldouri L and Sadiq A 2016 Gravity investigations of recent sinkholes and karst pits of Dahal Al-Hamam, State of Qatar *Environ. Earth Sci.* **75** 1–10

Hughes L J and Karlson N R 2003 Mapping structural pathways for DNAPL transport in Karst using induced polarization *Proc. SAGEEP 2003 Conf. (San-Antonio, TX)* 849–56

Inazaki T, Cawamura S, Tazawa O, Yamanaka Y and Kano N 2005 Near-surface cavity detection by high-resolution seismic reflection method using short-spacing type land streamer

18th EEGS Symp. on the Application of Geophysics to Engineering and Environmental Problems 959–70

Inazaki T 2006 High-resolution S-wave reflection survey in urban areas using a woven belt type land streamer *Near Surf. Geosci., Helsinki, Finland.* **A016** 4

Interpex Ltd 2019 IX1D v.3.6 Inversion Software (http://interpex.com/SoftwareIndex.htm.785)

Iris Instruments 2010 (http://iris-instruments.com)

ISRM 2004 Suggested methods for land geophysics in rock engineering *Final Report of ISRM Commission*

IPI-1D Program for 1D VES and IP (http://93.180.18.107/rec_lab3.htm)

Ivanov A G 1940 The electro seismic effect of the second kind *Izvestia Acad. Sci. USSR, Ser. Geograph. Geophys.* **5** 699–727 (in Russian, transl. to English)

Jackson P D and McCann D M 2015 Cross-hole seismic tomography for engineering site investigation *1997, Modern Geophysics and Engineering GeologySpecial Publication No. 12* ed D M McCann, M Eddleston, P J Fenning and G M Reeves (University of St Andrews: Geological Society Engineering Geology), 247–64 http://egsp.lyellcollection.org/

Jacob T, Pannet P, Beaubois F, Baltassat J M and Hannion Y 2020 Cavity detection using microgravity in a highly urbanized setting: a case study from Reims, France *J. Appl. Geophys.* **179** 104113

Jalali N, Saghafian B, Imanov N and Museyyibov M 2009 Recognition of shallow karst water resources and cave potentials using thermal infrared image and terrain characteristics in semi-arid regions of Iran *Int. J. Appl. Earth Obs. Geoinf.* **11** 439–47

Jardani A, Revil A, Santos F, Fauchard C and Dupont J P 2007 Detection of preferential infiltration pathways in sinkholes using joint inversion of self-potential and EM-34 conductivity data *Geophys. Prospect.* **55** 749–60

Jeannin P-Y, Liedl R and Sauter M 1997 Some concepts about heat transfer in karstic systems *Proc. 12th Int. Congress of Speleology* vol 1 *(Switzerland)* (Basel: Spéléo Projects), pp 195–8

Jin G and Gaherty J B 2015 Surface wave phase-velocity tomography based on multichannel cross-correlation *Geophys. J. Int.* **201** 1383–98

Johnson W B and Clark J C 1992 Improving subsurface resolution with the seismic reflection method. Geometrix.com

Jol H (ed) 2009 *Ground Penetrating Radar Theory and Applications* 1st edn (Amsterdam: Elsevier) p 544

Joswig M 2005 *Second Stuttgart Summer School on Nanoseismic Monitoring, Course tutorial and SparseNet Software* (http://geophys.unistuttgart.de/sss05.html).

Kafri U, Goldman M and Lang B 1997 Detection of subsurface brines freshwater bodies and the interface configuration in-between by the time domain electromagnetic method in the Dead Sea rift, Israel *Environ. Geol.* **31** 42–9

Kafri U and Goldman M 2005 The use of the time domain electromagnetic method to delineate saline groundwater in granular and carbonate aquifers and evaluate their porosity *J. Appl. Geophys.* **57** 167–78

Kagami H, Duke C M, Liang G C and Ohta Y 1982 Observation of 1- to 5-second microtremors and their application to earthquake engineering. Part II. Evaluation of site effect upon seismic wave amplification deep soil deposits *Bull. Seism. Soc. Am.* **72** 987–98

Kagawa T, Matsuoka T, Aizawa T and Hayashi T 2010 Analysis of refraction seismic survey using convolution product Butsari Tansa *Geophys. Explor.* **63** 311–20

Kaikkonen P and Sharma S P 2001 A comparison of performances of linearized and global nonlinear 2-D inversion of VLF and VLF-R electromagnetic data *Geophysics* **66** 462–75

Kanasewich E and Phadke S 1988 Imaging discontinuities on seismic sections *Geophysics* **53** 334–45

Kappelmeyer O and Hänel R 1974 Geothermics with special reference to application *Gebruder Borntrargen* (Berlin: Stutgart)

Karous M 1979 Effect of relief in EM-data methods with very distant source *Geoexploration* **17** 33–42

Karous M and Hjelt S E 1979 *Determination of Apparent Current Density from VLF Measurements.* (Oulu, Finland: Department of Geophysics, University of Oulu) pp 1–9

Kaufman A 1978 Resolving capabilities of the inductive methods of electroprospecting *Geophysics* **43** 1392–8

Kaufman A A 1992 *Geophysical Field Theory and Method, Volume 1: Gravitational, Electric, and Magnetic Fields* (San Diego, CA: Academic)

Kaufman A A, Hansen R O and Kleinberg R L K 2008 *Principles of the Magnetic Methods in Geophysics* (Amsterdam: Elsevier)

Kaufman A A and Keller G V 1981 *The Magnetotelluric Sounding Method* (Methods in Geochemistry and Geophysics vol 15) (Amsterdam: Elsevier)

Kaufman G, Romanov D and Nielbock R 2011 Cave detection using multiple geophysical methods: unicorn cave, Harz Mountains, Germany *Geophysics* **76** B71–7

Keller G V and Frischknecht F C 1966 *Electrical Methods in Geophysical Prospecting* (Oxford: Pergamon) p 519

Keydar S and Shtivelman V 2005 Imaging zero-offset section using multipath summation *First Break* **23** 21–4

Keydar S, Pelman D and Ezersky M 2010 Application of seismic diffraction imaging for detecting near-surface inhomogeneities in the Dead Sea area *J. Appl. Geophys.* **71** 47–52

Ketola M and Puranen M 1967 Type curves for interpretation of Slingram anomalies over tabular bodies *Report of Investigation No.1* Geological Survey of Finland

Khesin B E and Eppelbaum L V 1994 Near-surface thermal prospecting: Review of processing and interpretation *Geophysics* **59** 744–52

Khesin B E, Alexeyev V V and Eppelbaum L V 1996 Interpretation of geophysical fields in complicated environments *Ser.: Modern Approaches in Geophysics* (Boston: Kluwer Academic Publishers (Springer))

Khesin B E, Alexeyev V V and Eppelbaum L V 1997 Rapid methods for interpretation of induced polarization anomalies *J. Appl. Geophys.* **32** 117–33

Khmelevskoy V K 1980 Electrorazvedka/Spravochnik geophisika, Moscow (in Russian)

Khmelevskoy V K and Shevnin V A (ed) 1988 *Electrical Sounding of Geological Medium. Part 1.* (Moscow: Moscow University Press) p 176 (in-Russian)

Khmelevskoi V K and Bondarenko V M (ed) 1989 *Electric Prospecting* vol II (Moscow: Nedra) (in Russian)

Kidanu S T, Torgashov E V and Varnavina A V *et al* 2016 ERT-based investigation of a sinkhole in Greene County, Missouri *AIMS Geosci.* **2** 99–115

Kidanu S T, Varnavina A, Anderson N and Torgashov E 2020 Pseudo-3D electrical resistivity tomography imaging of subsurface structure of a sinkhole—a case study in Greene County, Missouri *AIMS Geosci.* **6** 54–70

Kirkegaard C, Sonnenbororg T O, Auken E and Jørgensen F 2011 Salinity distribution in heterogeneous coastal aquifers mapped by airborne electromagnetics *Vadose Zone J.* **10** 125–35

Kim J-H, Yi M-J, Hwang S-H, Song Y, Cho S-J and Synh J-H 2007 Integrated geophysical surveys for the safety evaluation of a ground subsidence zone in a small city *J. Geophys. Eng.* **4** 332–47

Kim Y J, Nam B H and Youn H 2019 Sinkhole detection and characterization using LiDAR-derived DEM with logistic regression *Remote Sens.* **11** 1592

Kirsch R (ed) 2006 Groundwater geophysics *A Tool for Hydrogeology* (Berlin: Springer)

Kis K I and Pustza S 2006 Application of magnetic field derivatives for locating Sarmatian graves *J. Appl. Geophys.* **60** 13–26

Klem-Musatov K 1994 *Theory of Seismic Diffractions* (Houston, TX: SEG)

Knapp R W and Steeples D W 1986 High-resolution common depth point seismic reflection profiling: field acquisition parameters design *Geophysics* **51** 283–94

Knapp R W, Steeples D W, Miller R D and McElwee C D 1989 Seismic reflection survey at sinkholes in central Kansas *Bulletin* **226** 95–116

Koerner R M and Lord A E 1976 Acoustic emission monitoring of earth dam stability *Water Power, Dam Constr.* **28** 45–9

Koerner R, McCabe W and Lord A 1981 Acoustic emission behavior and monitoring of soils *Acoustic Emissions in Geotechnical Engineering Practice* ed V Drnevich and R Gray (West Conshohocken, PA: ASTM International) pp 93–141

Kofman L, Ronen A and Frydman S 2006 Detection of model voids by identifying reverberation phenomena in GPR records *J. Appl. Geophys.* **59** 284–99

Kofman L and Herr N 2001 Detection and mapping of fault zones and karst caves by GPR and borehole radar system *IGS Annual Meeting*

Koh G 1992 The dielectric constant of ice at 26.5–40 GHz *J. Appl. Phys.* **71** 5119

Kolesnikov Y U and Fedin K V 2018 Detecting underground cavities using microtremor data: physical modeling and field experiment *Geophys. Prospect.* **66** 342–53

Komarov V A 1980 *Electric Prospecting by the Induced Polarization Method* 2nd edn (Leningrad: Nedra) revised and supplemented (in Russian)

Krawczyk C M, Polom U, Trans S and Dahm T 2012 Sinkholes in the city of Hamburg—new urban shear-wave reflection seismic system enables high-resolution imaging of subrosion structures *J. Appl. Geophys.* **78** 133–43

Lachet C and Bard P Y 1994 Numerical and theoretical investigations on the possibilities and limitations of Nakamura's technique *J. Phys. Earth* **42** 377–97

LaFehr T R 1983 Rock density from borehole gravity surveys *Geophysics* **48** 259–401

Landa E and Keydar S 1998 Seismic monitoring of diffraction images for detection of local heterogeneities *Geophysics* **63** 1093–100

Landa E, Shtivelman V and Gelchinsky B 1987 A method for detection of diffracted waves on common-offset sections *Geophys. Prospect.* **35** 359–73

Landau L D and Lifshitz E M 1960 *Electrodynamics of Continuous Media* (Oxford: Pergamon)

Lane E D (ed) 1986 *Karst in Florida. Special Publication No.29.* (Florida Geological Survey, Department of Natural Resources)

Lane J W , Jr, Joesten P K, Haeni F P, Vendi M and Yeskis D 1998 Use of borehole-radar methods to monitor the movement of a saline tracer in carbonate rock at Belvidere, Illinois

Proc. SAGEEP'98 (Wheat Ridge, CO, March 22–26, 1998) R S Bell, M H Powers and T Larson (Chicago, IL: Environmental and Engineering Geophysical Society) pp 323–32

Lange A L 1999 Geophysical studies at Kartchner Caverns State Park, Arizona *J. Cave Karst Stud.* **61** 68–72

Lee E J, Shin S Y, Ko B C and Chang C 2016 Early sinkhole detection using a drone-based thermal camera and image processing *Infrared Phys. Technol.* **78** 223–32

Legchenko A V and Shushakov O A 1998 Inversion of surface NMR data *Geophysics* **63** 75–81

Legchenko A and Valla P 2002 A review of the basic principles for proton magnetic resonance sounding measurements *J. Appl. Geophys.* **50** 3–19

Legchenko A, Baltassat J-M, Bobachev A, Martin C, Robin H and Vouillamoz J-M 2004 Magnetic resonance sounding applied to aquifer characterization *J. Ground Water* **42** 363–73

Legchenko A, Ezersky M, Camerlink C, Al-Zoubi A, Chalikakis K and Girard J-F 2008a Locating water-filled karst caverns and estimating their volume using magnetic resonance soundings *Geophysics* **73** 51–61

Legchenko A, Descloitres M, Vincent C, Guyard H, Garambois G, Chalikakis K and Ezersky M 2011 Three-dimensional magnetic resonance imaging for groundwater *New J. Phys.* **13** 025022

Legchenko A 2013 *Magnetic Resonance Imaging for Groundwater* (London: Wiley-ISTE) p 157

Legchenko A *et al* 2014 Monitoring water accumulation in a glacier using magnetic resonance imaging *Cryosphere* **8** 155–66

Legchenko A, Comte J-C, Ofterdinger U, Vouillamoz J-M, Lawson F M A and Walshd J 2017 Joint use of singular value decomposition and Monte-Carlo simulation for estimating uncertainty in surface NMR inversion *J. Appl. Geophys.* **144** 28–36

Legchenko A *et al* 2020 Time-lapse magnetic resonance sounding measurements for numerical modeling of water flow in variably saturated media *J. Appl. Geophys.* **175** 103984

Lei J, Pan B, Guo Y, Fan Y and Zhang L 2020 Depth prediction of karstic-fault reservoirs based on wellbore temperature measurements and numerical simulations *Geofluids* **2020** 1–11

Leucci G and De Giorgi L 2010 Microgravimetric and ground-penetrating radar geophysical methods to map the shallow karstic cavities network in a coastal area (Marina Di Capilungo, Lecce, Italy) *Explor. Geophys.* **41** 178–88

Lillesand T M *et al* 2003 *Remote Sensing and Image Interpretation* 5th edn (New York: Wiley)

Lismonde B 2002 Aérologie des systèmes karstiques *Climatologie du Monde Souterrain T2* (Isère: Comité D`partemental de Spéléologie) pp 1–362

Liu W, Zhou C, Liu Z, Yang C and Brancelj A 2017 The temperature variation in an epikarstic cave and its impact factors: a case from Velika Pasica Cave, Central Slovenia *Arab. J. Geosci.* **10** 1–9

Loke M H and Barker R D 1996 Rapid least-squares inversion of apparent resistivity pseudo-sections by a quasi-Newton method *Geophys. Prospect.* **44** 131–52

Loke M H 2004 Electrical imaging surveys for environmental and engineering studies, a practical guide to 2D and 3D surveys. Malaysia

Lombardo G and Rigano R 2010 Local seismic response evaluation in natural and artificial cavities *Proc. 3 Int. Symp. 'Karst Evolution in the South Mediterranean Area'* vol 14 *(Ragusa, Italy, 29–31 May 2009)* (Speleologia Iblea C.I.R.S.) 71–8

Lotti A, Saccorotti G, Fiaschi A, Matassoni L, Gigli G, Pazzi V and Casagli N 2012 Seismic monitoring of a rockslide: the torgiovannetto quarry (Central Apennines, Italy) ed G Lollino, A Manconi, F Guzzetti, M Culshaw, P Bobrowsky and F Luino *Engineering Geology for Society and Territory* vol 5 (Berlin: Springer) pp 537–40

Lowry T and Shive P N 1990 An evaluation of Bristow's method for the detection of subsurface cavities *Geophysics* **55** 514–20

Louie J N 2001 Faster, better: shear-wave velocity to 100 meters depth from refraction microtremor arrays *Bull. Seism. Soc. Am.* **91** 347–64

Loucks R G 1999 Paleocave carbonate reservoirs: origins, burial depth modifications, spatial complexity, and reservoir implications *AAPG Bull.* **83** 1795–834

Luetscher M and Jeannin P-Y 2004 Temperature distribution in karst systems: the role of air and water fluxes *Terra Nova* **16** 344–50

Luetscher M, Lismonde B and Jeannin P-Y 2008 Heat exchanges in the heterothermic zone of a karst system: Monlesi cave, Swiss Jura Mountains *J. Geophys. Res.* **113** 1–13

Ma S and Li Z 2007 Review of wave equation prestack depth migration methods *J. Prog. Explor. Geophys.* **30** 153–61

Maimon O, Lyakhovsky V, Agnon A and Abelson M 2005 Stability of cavities and formation of sinkholes along the Dead Sea coast *Geological Survey of Israel Report* GSI/19/05 65 (in Hebrew, Abstract in English)

Mala 2010 ProEx—professional explorer control unit *Operating Manual. Mala* 60

Malischewsky P G and Scherbaum P 2004 Love's formula and H/V-ratio ellipticity of Rayleigh waves *Wave Motion* **40** 57–67

Manohar K, Yarbrough D W and Booth J R 2000 Measurement of apparent thermal conductivity by the thermal probe method *J. Test. Eval., JTEVA* **28** 345–51

Massonet D and Feigle K 1998 Radar interferometry and its application to changes in the Earth's surface *Rev. Geophys.* **36** 441–500

Mayne W H 1962 Common reflection point horizontal data stacking trchnicues *Geophysics* **27** 927–38

Mazarchuk Z, Skalskyi V and Serhienko O 2017 Acoustic emission *Methodology and Applications* (Berlin: Springer)

McGee D K 2012 Hydrologic and microclimate characterizations of Thornton's Cave, West-Central Florida (USA) *Stud. UBB Geol.* **57** 53–61

McGlone J C, Mikhail E M, Bethel J S and Mullen R 2004 *Manual of Photogrammetry* 5th edn (Bethesda, MD: American Society for Photogrammetry and Remote Sensing)

McMechan G, Gaynor G and Szerbiak R 1997 Use of ground-penetrating radar for 3-D sedimentological characterization of clastic reservoir analogs *Geophysics* **62** 786–96

McNeill J D 1980a Electromagnetic terrain conductivity measurement at low induction numbers *Technical Note* TN-6 Mississauga, ON: Geonics Ltd 15

McNeill J D 1980b Applications of transient electromagnetic techniques *Technical Note* TN-7 Mississauga, ON: Geonics Ltd 17

McNeill J D 1980c Electrical conductivity of soils and rocks *Technical Note* TN-5 Mississauga, ON: Geonics Ltd 22

McNeill J D 1994 Principles and application of time domain electromagnetic method for resistivity sounding *Technical Note* TN-27 Mississauga, ON: Geonics Ltd

Mellet J S 1995 Ground-penetrating radar applications in engineering, environmental management, and geology *J. Appl. Geophys.* **33** 157–66

Miecznik S 1986 Ciala Cylindryezne w Polo Plaskey Fali Electromagnetycznei *Techn. Poszuk. Geol.* **25** 16–22

Mihevć A, Prelovshek M and Hajna N Z (ed) 2010 *Introduction to Dinaric karst* (Postojna: Karst Underground Protection Project) (https://izrk.zrc-sazu.si/sites/default/files/Dinarski%20kras_normalna%20locljivost.pdf)

Milanović P 1979 Hydrogeologija karsta I metode istrazivanja. Publisher 'HE Trebishnica' Institute za karstenje I zastitu voda na krsu, Trebinje (in Serbian)
Milanović P 1990 Influence of construction on hydrogeological and environmental conditions in the karst region, eastern Herzegovina, Yugoslavia *Environ. Geol. Water Sci.* **15** 5–11
Milanović P T 2000 *Geological Engineering in Karst: Dams, Reservoirs, Grouting, Groundwater Protection, Water Tapping, Tunneling* (Belgrade: Zebra Publishing Ltd) p 347
Milanović P T 2002 The environmental impacts of human activities and engineering construction in karst regions *Episodes* 25
Milanović P 2005 *Water Resources Engineering in Karst* (Boca Raton, FL: CRC Press)
Milanović P 2021 Karst istoćne Hercegovine I Dubrovaćkog priobalja. Karst of Eastern Hercegovina and Dubrovnic littoral. NET. Republike Srpske (in Serbian, extended abstract in English)
Milanović P, Maximovich N and Meshcheryakova O 2019 *Dams and Reservoirs in Evaporites.* (Berlin: Springer)
Militzer H, Rösler R and Lösch W 1979 Theoretical and experimental investigations for integration of geological and geophysical techniques **27** 640–52
Miller R D, Steeples D W and Veis T V 1986 Shallow seismic-reflection study of a salt dissolution subsidence features in Stafford County, Kansas *Kansas Geol. Survey Bull.* **237** 71–6
Miller R D, Xia J, Park C B and Ivanov J M 1999 Multichannel analysis of surface waves to map bedrock *Lead. Edge* **18** 1392–6
Mochales T, Casas A M, Pueyo E L, Pueyo O, Román M T, Pocoví A, Soriano M A and Ansón D 2008 Detection of underground cavities by combining gravity, magnetic and ground-penetrating radar surveys: a case study from the Zaragoza area, NE Spain *Environ. Geol.* **53** 1067–77
Mohnke O and Yaramanci U 2002 Smooth and block inversion of surface NMR amplitudes and decay times using simulated annealing *J. Appl. Geophys.* **50** 163–77
Mohnke O and Yaramanci U 2005 Forward modeling and inversion of MRS relaxation signals using multi-exponential decomposition *Near Surf. Geophys.* **3** 165–85
Monteiro S F A, Mateus A, Figueiras J and Concalves M A 2006 Mapping groundwater contamination around a landfill facility using the VLF-EM method—a case study *J. Appl. Geophys.* **60** 115–25
Morat P, Le Mouel J-L, Poirier J-P and Kosobokov V 1999 Heat and water transport by oscillatory convection in an underground cavity *C.R. Acad. Sci. Paris/Earth Planet. Sci.* **328** 1–8
Moreno F J M 2015 Detection and characterization of karstic caves: integration of geological and geophysical techniques *PhD Thesis* Universidad de Granada
Morton S L, Ivanov J, Peterie S L, Miller R D and Livers-Douglas A J 2021 Passive multichannel analysis of surface waves using 1D and 2D receiver arrays *Geophysics* **86** EN63–75
Mooney H 1984a Handbook of engineering geophysicsvol 1 *Seismics* (Minneapolis, MN: Bison Instrument Inc.)
Mooney H 1984b Handbook of engineering geophysicsvol 2 *Electrical Resistivity* (Minneapolis, MN: Bison Instrument Inc.)
Morey R M 1974 Continuous sub-surface profiling by impulse radar *Conf. on Subsurface Exploration for Underground Excavation and Heavy Construction* (Reston, VA: American Society of Civil Engineers) 213–32
Moscicki W J 1987 Temperature anomalies over underground cavities *Geophys. Prospect.* **35** 393–423

Motschka K 2001 Aerogeophysics in Austria *Bull. Geol. Surv. Japan* **52** 83–8

Mucciarelli M and Gallipoli M R 2004 The HVSR technique from microtremor to strong motion: empirical and statistical considerations *13th World Conf. of Earthquake Engineering (Vancouver)* 45

Mueller-Petke M and Yaramanci U 2010 QT inversion—comprehensive use of the complete surface NMR data set *Geophysics* **75** WA199–209

Müller-Petke M, Walbrecker J O and Knight R 2013 The inversion of surface-NMR T1 data for improved aquifer characterization *Geophysics* **78** 1–12

Müller-Petke M, Braun M, Hertrich M, Costabel S and Walbrecker J 2016 MRSmatlab—a software tool for processing, modeling, and inversion of magnetic resonance sounding data *Geophysics* **81** WB9–21

Munk J and Sheets R A 1997 Detection of underground voids in Ohio by use of geophysical methods *Report of U.S. Geological Survey 97-4221 in Cooperation with the Ohio Department of Transportation and U.S. Department of Transportation, Columbus, OH*

Nabighian M N 1979 Quasi-static transient response of a conducting permeable sphere in a dipolar field *Geophysics* **35** 303–9

Nakamura Y 1989 A method for dynamic characteristics estimations of subsurface using microtremors on the ground surface *Quart. Rep. RTRI Jpn.* **30** 25–33

Nasseri-Moghaddam A, Cascante G and Hutchinson J 2005 A new quantitative procedure to determine the location and embedment depth of a void using surface waves *J. Environ. Eng. Geophys.* **10** 51–64

Nehmé C *et al* 2013 The use of passive seismological imaging in speleogenetic studies: an example from Kanaan Cave, Lebanon *Int. J. Speleol.* **42** 97–108

Neidell N S 1985 Land applications of shear waves *Lead. Edge* **4** 1–96

Neishtadt N M and Eppelbaum L V 2012 Perspectives of application of piezoelectric and seismoelectric methods in applied geophysics *Russ. Geophys. J.* **51–52** 63–80 (in Russian)

Neishtadt N, Eppelbaum L and Levitski A 2006 Application of seismo-electric phenomena in exploration geophysics: review of Russian and Israeli experience *Geophysics* **71** B41–53

Newton J G 1987 Development of sinkholes resulting from man's activities in the eastern United States U.S. Geological Survey Circular 968

Niva B 1990 Results from Borehole Radar Tests in Ombla Tunnel, unpublished report, ABEM AB Borehole Geophysics, Mala, Sweden (Referred to in Milanović, 2005)

NRC 1997 *Satellite Gravity and the Geosphere* (Washington, DC: National Academy Press)

Nuzzo L, Leucci G and Negri S 2007 GPR, VES, and refraction seismic surveys in the karstic area 'Spedicaturo' near Nociglia (Lecce, Italy) *Near Surf. Geophys.* **5** 67–76

Nyquist J, Petruccione J and Roth M 1999 Characterization of shallow karst terrain using multifrequency electromagnetic induction: two examples from eastern Pennsylvania *Expanded Abstracts 69th SEG Meeting (Houston, TX)* 547–50

Nyquist J E, Mintz H E and Roth M J S 2008 Mise-à-la-masse and smoke tests for mapping vadose zone karst *Lead. Edge* **11** 1510–5

Nunziata C, Natale M and Panza G F 1999 Estimation of cavity effects on response spectra for the 1980 earthquake in the historical centre of Naples ed P Secoe *Earthquake Geotechnical Engineering.* (Rotterdam: Balkema) pp 9–14

Ogilvy A A, Bogoslovsky V A and Kuzmina E N 1980 Geoelectric models in engineering geophysics *Geophys. Prospect.* **28** 945–55

Ogilvy R D, Cuadra A, Jackson P D and Monte J L 1991 Detection of an air-filled drainage gallery by VLF resistivity method *Geoexploration* **28** 139–52

Oliver-Cabrera T, Wdovinsky S, Kruse S and Robinson T 2020 InSAR detection of localized subsidence induced by sinkhole activity in suburban West-Central Florida *Proc. IAHS* **382** 155–9

Olsson O 1980 VLF-anomalies from a perfectly conducting half-plane below an overburden *Geophys. Prospect.* **28** 415–34

Olsson O 1983 Computation of VLF response over half-plane and wedge models *Geophys. Prospect.* **31** 171–91

Orellana E and Mooney H M 1966 Master tables and curves for vertical electrical sounding over layered structures *Interciencia* Madrid 34

Orfanos C, Leontarakis K and Apostolopoulos G 2020 A surface wave tomography tool in geotechnical applications—a 3D Experiment at a controlled test site *Near Surface Geoscience Conf., and Exhibition 2020 (Belgrade, Serbia, 30 August–3 September 2020)*

Orhan O, Yakar M and Ekercin S 2020 An application on sinkhole susceptibility mapping by integrating remote sensing and geographic information systems *Arab. J. Geosci.* **13** 886

Oskooi B and Pedersen L B 2006 Resolution of airborne VLF data *J. Appl. Geophys.* **582** 158–75

Ostrowski S, Rybak-Ostrowska B and Lasoki M 2016 Fault zone imaging with seismic refraction tomography *Near Surf. Geosci.* Tu22 B06 4

Paillet F L 2001 Borehole geophysical applications in karst hydrogeology *U.S. Geological Survey, Karst Interest Group Proc. (St. Petersburg, FL, February 2001)* 116–23

Paine D P and Kiser J D 2012 *Aerial Photography and Image Interpretation* 3rd edn (New York: Wiley)

Palmer D 1986 Refraction seismics *The Lateral Resolution of Structure and Seismic Velocity* (London: Geophysical Press Ltd)

Parasnis D S 1997 *Principles of Applied Geophysics* 5th edn (London: Chapman and Hall) pp 104–76

Park C B, Miller R D and Xia J 1999a Multichannel analysis of surface waves (MASW) *Geophysics* **64** 800–8

Park C B, Miller R D, Xia J, Hunter J A and Harris J B 1999b Higher mode observation by the MASW methods *SEG Technical Program Expanded Abstracts* 524–7

Park C 2006 *SurfSeis 2. User's Manual v.2.0.* ed M Adkins-Heljeson (Lawrence, KS: Kansas Geological Survey)

Park C B, Miller R D, Xia J and Ivanov J 2007 Multichannel analysis of surface waves (MASW) —active and passive methods *Lead. Edge* **26** 60–4

Park C B, Ivanov J and Brohammer M 2008 *SurfSeis2, User's Manual v.2.0* ed M Adkins-Heljeson (Lawrence, KS: Kansas Geological Survey)

Park C B, Miller R D and Xia J 2009 Detection of near-surface voids using surface waves *22nd SAGEEP* (Houston, TX: Society of Exploration Geophysicists) 291–6

Parra J O and Owen T E 1986 Hole-to-hole resistivity signatures of cylindrical cavities: dipole–dipole electrode array *Geophysics* **51** 788–99

Pašteka R, Mikuška J and Meurers B 2017 *Understanding the Bouguer Anomaly* (Amsterdam: Elsevier)

Pedersen L B and Oskooi B 2004 Airborne VLF measurements and variations of ground conductivity *Surv. Geophys.* **25** 151–81

Pedersen L B, Persson L and Bastani M 2009 Airborne VLF measurement and mapping of ground conductivity in Sweden *J. Appl. Geophys.* **67** 250–8

Peng R, Di B, Wei J, Ding P, Zhao J, Pan X and Liu Z 2017 Experimental study of the seismoelectric interface response in wedge and cavity models *Geophys. J. Intern.* **210** 1703–20

Olsson O 1980 VLF-anomalies from a perfectly conducting half-plane below an overburden *Geophys. Prospect.* **28** 415–34

Olsson O 1983 Computation of VLF response over half-plane and wedge models *Geophys. Prospect.* **31** 171–91

Orellana E and Mooney H M 1966 Master tables and curves for vertical electrical sounding over layered structures *Interciencia* Madrid 34

Orfanos C, Leontarakis K and Apostolopoulos G 2020 A surface wave tomography tool in geotechnical applications—a 3D experiment at a controlled test site *Near Surface Geoscience Conf., and Exhibition (30 August–3 September 2020)* (Belgrade, Serbia)

Orhan O, Yakar M and Ekercin S 2020 An application on sinkhole susceptibility mapping by integrating remote sensing and geographic information systems *Arabian J. Geosci.* **13** 886

Oskooi B and Pedersen L B 2006 Resolution of airborne VLF data *J. Appl. Geophys.* **582** 158–75

Ostrowski S, Rybak-Ostrowska B and Lasoki M 2016 Fault zone imaging with seismic refraction tomography *Near Surf. Geosci.* Tu22 B06 4

Paillet F L 2001 Borehole geophysical applications in karst hydrogeology *Google Book of U.S. Geological Survey, Karst Interest Group Proc. (St. Petersburg, Florida, February 2001)* 116–23 https://books.google.com

Paine D P and Kiser J D 2012 *Aerial Photography and Image Interpretation* 3rd edn (New York: Wiley)

Palmer D 1986 Refraction seismics *The Lateral Resolution of Structure and Seismic Velocity.* (London: Geophysical Press Ltd)

Parasnis D S 1997 *Principles of Applied Geophysics.* 5th edn (London: Chapman and Hall) 104–76

Park C B, Miller R D and Xia J 1999a Multichannel analysis of surface waves (MASW) *Geophysics* **64** 800–8

Park C B, Miller R D, Xia J, Hunter J A and Harris J B 1999b Higher mode observation by the MASW method [Exp. Abs.] *Soc. Explor. Geophys.* 524–27

Park C 2006 *SurfSeis 2. User's Manual v.2.0.* ed M Adkins-Heljeson (Lawrence, KS: Kansas Geological Survey)

Park C B, Miller R D, Xia J and Ivanov J 2007 Multichannel analysis of surface waves (MASW) —active and passive methods *Lead. Edge* 60–4

Park C B, Ivanov J and Brohammer M 2008 *SurfSeis2, User's Manual v.2.0* ed M Adkins-Heljeson (Lawrence, KS: Kansas Geological Survey)

Park C B, Miller R D and Xia J 2009 Detection of near-surface voids using surface waves *22nd SAGEEP* (Houston, TX: Society of Exploration Geophysicists), 281–86

Parra J O and Owen T E 1986 Hole-to-hole resistivity signatures of cylindrical cavities: dipole–dipole electrode array *Geophysics* **51** 788–99

Pašteka R, Mikuška J and Meurers B 2017 *Understanding the Bouguer Anomaly.* (Amsterdam: Elsevier)

Pazzi V *et al* 2018 Assessing subsoil void hazards along a road system using H/V measurements, ERTs and IPTs to support local decision makers *Near Surf. Geophys.* **16** 282–97

Pedersen L B and Oskooi B 2004 Airborne VLF measurements and variations of ground conductivity *Surv. Geophys.* **252** 151–81

Pedersen L B, Persson L and Bastani M 2009 Airborne VLF measurement and mapping of ground conductivity in Sweden *J. Appl. Geophys.* **67** 250–8

Peng R, Di B, Wei J, Ding P, Zhao J, Pan X and Liu Z 2017 Experimental study of the seismoelectric interface response in wedge and cavity models *Geophys. J. Intern.* **210** 1703–20

Perelman N and Zaslavsky Y 2001 Analysis of seismic signals in frequency domain, (SEISPECT) *GII Report* 569/345/01. 2001

Persico R 2014 *Introduction to Ground Penetrating Radar: Inverse Scattering and Data Processing* (New York: Wiley-IEEE Press) p 392

Philipp J, Plankers K, Gartner G and Teihmann L 2015 On potential of in-situ acoustic emission (AE) technology for the monitoring of dynamic processes in salt mines *Proc. Conf. on Mechanical Behavior of Salt VIII* L Robert, K Mellegard and F Hansen (Boca Raton, FL/ Rotterdam: CRC Press/A.A. Balkema) pp 89–98

Pivetta T and Braitenberg C 2015 Laser-scan and gravity joint investigation for subsurface cavity exploration—the Grotta Gigante benchmark *Geophysics* **80** B83–94

Poddar M 1982 Very low-frequency electromagnetic response of a perfectly conducting half-plane in a layered half-space *Geophysics* **41** 1059–67

Poikonen A and Suppala I 1989 On Modeling airborne very low-frequency measurements *Geophysics* **54** 1596–606

Poirier N, Baleux F and Calastrenc C 2020 The mapping of forested archaeological sites using UAV LiDaR. Feedback from a south-west France ISTE OpenScience 1–24

Polom U, Bagge M, Wadas S, Winsemann J, Brandes C, Binot F and Krawczyk C M 2013 Surveying near-surface depocentres by means of shear wave seismics *First Break* **31** 67–79

Probstein R F 1994 *Physiochemical Hydrodynamics: An Introduction* 2nd edn (New York: Wiley)

Pugin A, Brever K, Cartwright T and Sargent S L 2019 Detection of tunnels and boulders using shallow SH-SH reflected seismic waves *Lead. Edge.* **38** 436–41

Pugin A, Pullan S E and Hunter J A 2008 SV-wave and P-wave high resolution seismic reflection using vertical impacting and vibrating sources *SAGEEP* **21** 585–94

Quarto R and Schiavone D 1996 Detection of cavities by the self-potential method *First Break* **14** 419–30

Rahimi M and Alexander E C Jr 2013 Locating sinkholes in LiDAR coverage of glacial-fluvial karst, Winona County, M.N; L Land, D H Doctor and B J Stephenson *Proc. 13th Multidisciplinary Conf. on Sinkholes and the Engineering and Environmental Impacts of Karst* (National Cave and Karst Research Institute, Symp.) **2** 469–80

Ravnik D and Rajver D 1998 The use of inverse geotherms for determining underground water flow at the Ombla karst spring near Dubrovnik, Croatia *J. Appl. Geophys.* **39** 177–90

Redpath B B 1973 Seismic refraction exploration for engineering site investigation *Technical Report* E-73-4 Explosive Excavation Research Laboratory Livermore, CA, Distributed by NTIS U.S. Department of Commerce

Reninger P A, Martelet G, Lasseur E, Beccaletto L, Deparis J, Perrin J and Chen Y 2014 Geological environment of karst within chalk using airborne time domain electromagnetic data cross-interpreted with boreholes *J. Appl. Geophys.* **106** 173–86

Reppert M, Vichabian Y, Sogade J and Morgan F D 1999 Cave detection using GPR *SAGEEP (March, 1999)* 287–92

Res1D Loke M N Geotomo Software, Malaisia

Res2Dinv v.32001.54 2004 Geotomo Software. geotomo@tm.net.my, (www.geoelectrical.com)

Res2Dmod V.3.01. Loke,M.H. Rapid 2D resistivity forward modeling using finite-difference and finite-element methods

Res3Dinvx64 v.3.14 2018 Geotomo Software (www.geoelectrical.com)

Res3Dmodx64. V3.04 2014 Geotomo Software (www.geoelectrical.com)

Revil A and Jardani A 2013 *The Self-Potential Method: Theory and Applications in Environmental Geosciences* (Cambridge: Cambridge University Press)

Reynolds J M 2011 2nd edn *An Introduction to Applied and Environmental Geophysics* (New York: Wiley)

Rey F 2007 Ressources en eau souterraine dans les chaînons béarnais (Pyrénées-Atlantiques, France) [Groundwater resources in the 'chaînons béarnais' (Western Pyrenees,. France) Geometry and hydrogeological functioning of four carbonated aquifers] *PhD Thesis* Université de Bordeaux I466

Robert T, Dassargues A, Brouyère S, Kaufmann O, Hallet V and Nguyen F 2011 Assessing the contribution of electrical resistivity tomography (ERT) and self-potential (SP) methods for a water well drilling program in fractured/karstified limestones *J. Appl. Geophys.* **75** 42–53

Roche O, Druitt H T and Merle O M 2000 Experimental study of caldera formation *J. Geophys. Res.* **105** 395–416

Roest W R, Verhoef J and Pilkington M 1992 Magnetic interpretation using the 3-D analytic signal *Geophysics* **57** 116–25

Rucker M L, Crum G, Meyers. R and Lommler J C 2005 Geophysical identification of evaporite dissolution structures beneath a highway alignment *Sinkholes Eng. Env. Impacts Karst* (Reston, VA: ASCE)

Richart F E, Hall J R and Woods R D 1970 *Vibrations of Soils and Foundations* (Englewood Cliffs, NY: Prentice-Hall)

Rodriguez V, Gutierrez F, Green A G, Carbonel D, Horstmeyer H and Schmelbach C 2014 Characterizing sagging and collapse sinkholes in a mantled karst by means of ground penetrating radar (GPR) *Environ. Eng. Geosci.* **XX** 109–32

Rohdewald S, Sheehan H J R and Burton B 2010 Processing of seismic refraction tomography data. Short course notes *Presented at SAGEEP, 2010. Symp. on the Application of Geophysics to Engineering and Environmental Problems Keystone (Colorado, April 10, 2010)*

Rohdewald S 2006 *RAYFRACTTM Manual* (Vancouver: Intelligent Resources Inc.)

Ronen A 2000 Cave detecting and mapping in construction site in Israel—a case study *Proc. of the 8th Int. Conf. GPR'2000 (Australia)* 812–5

Ronen A, Ezersky M, Beck A, Gatenio B and Simhiov R B 2019 Use of GPR method for prediction of sinkholes formation along the Dead Sea Shores, Israel *Geomorphology* **328** 28–43

Roth M J S, Mackey J R, Mackey C and Nyquist J E 1999 A case study of the reliability of multi-electrode earth resistivity testing for geotechnical investigations in karst terrains *Hydrogeology and Engineering Geology of Sinkholes and Karst* ed B F Beck, J Pettit and J G Herring (Rotterdam: Balkema) pp 247–52

Rybakov M, Goldschmidt V, Fleischer L and Rostein Y 2001 Cave detection and 4-D monitoring: a microgravity case history near the Dead Sea *Lead. Edge* **20** 896–900

Rybakov M, Rotstein Y, Shirman B and Al-Zoubi A 2005 Cave detection near the Dead Sea—a micromagnetic feasibility study *Lead. Edge* **24** 585–90

Saad R, Khan N, Saidin M, Shakhidan S and Karamah S S 2015 Magnetic reconnaissance at the Great Cave (Kashmir Smast), northern Pakistan *Electron. J. Geotechn. Eng.* **20** 11155–60

Sandmeier K J 2020 *REFLEXW Version 9.5* (Karlsruhe, Germany: Sandmeier Geo) https://sandmeier-geo.de/index.html

Savich A I, Koptev V I, Nikitin B N and Yaschenko Z G 1969 *Seismic-Acoustic Methods for Study of the Rock Masses.* ed A M Epinatieva (Moscow: Nedra) p 239 (In Russian)

Sedelnikov E S 1983 The effect of topography on the field of a distant source of alternating electromagnetic field *Izvestiya USSR Acad. of Sci., Ser: Phys. Solid Earth* 102–6

Seigel H O 1959 Mathematical formulation and type curves for induced polarization *Geophysics* **24** 547–65

Saribudak M 2010 Urban geophysics: a mapping of Mount Bonnell Fault and its karstic features in Austin, TX *Proceed. of the SAGEEP 2010 Conf. (Keystone, Colorado)* pp 482–97

Seale L D, Florea L J, Vacher H L and Brinkmann R 2008 Using ALSM to Map Sinkholes in the urbanized covered Karst of Pinellas County, Florida—1, methodological considerations *Environ. Geol.* **54** 995–1005

Schirov M, Legchenko A and Creer G 1991 A new direct non-invasive groundwater detection technology for Australia *Explor. Geophys.* **22** 333–8

Schnabel P B, Lysmer J and Seed H B 1972 SHAKE: a computer program for earthquake response analysis of horizontally layered sites *Report No. EERC72-12* (Berkeley: University of California)

Scholtz C H 2002 *The Mechanics of Earthquake and Faulting* 2nd edn (Cambridge: Cambridge University Press) p 471

Seevers D O 1966 A nuclear magnetic method for determining the permeability of sand-stones *Paper L. Annual Logging Symp. Trans.: Society of Professional Well Log Analysts*

Selley R C 2003 Petroleum geology *Encyclopedia of Physical Science and Technology* ed R Meyers 3rd edn (New York: Academic)

Semenov A S 1980 *Electric Prospecting by Self-Potential Method* 4th edn revised and supplemented (Leningrad: Nedra) (in Russian)

Semenov A, Burshtein A, Pusep A and Schirov M D 1988 A device for measurement of underground mineral parameters *USSR Patent* 1079063 (in Russian)

Serra O 1984 Fundamentals of well-log interpretation *Acquisition of Logging Data.* (Amsterdam: Elsevier) p 423

SESAME user guide 2006 Guidelines for the implementation of the H/V spectral ratio technique on ambient vibration measurements, processing and interpretation open file (http://sesame-sim.sourceforge.net/docs/SESAME/SESAME_Users_Guide.html)

Sgarlato G, Lombardo G and Rigano R 2011 Evaluation of seismic site response nearby underground cavities using earthquake and ambient noise recordings: a case study in Catania area, Italy *Eng. Geol.* **122** 281–91

Sharma V 1997 *Environmental and Engineering Geophysics* (London: Cambridge University Press)

Sharma S P and Kaikkonen P 1998 Two-dimensional non-linear inversion of VLF-R data using simulated annealing *Geophys. J. Int.* **133** 649–68

Sheenan J R, Doll W E, Watson D B and Mandell W A 2005a Application of seismic refraction tomography to Karst Cavities, In: US Geological Survey Karst Intersect group *Proc. Rapid City (South Dakota, 2005)* E L Kuniansky 29–38

Sheenan J R, Doll W E and Mandel W A 2005b An evaluation of methods and available software for seismic refraction tomography analysis *J. Environ. Eng. Geophys.* **10** 21–34

Sheets R A 2002 Use of electrical resistivity to detect underground mine voids in Ohio *U.S.G.S. Water-Resources Investigations Report* 02-4041 1–10

Shtivelman V and Keydar S 2005 Imaging subsurface inhomogeneities using multipath summation *First Break* **23** 39–42

Shapira A and Avirav V 1995 PS-SDA Operation Manual *Unpublished IPRG Report* No. Z1/567/79

Siemon B, Christiansen A V and Auken E 2009 A review of helicopter-borne electromagnetic methods for groundwater exploration *Near Surf. Geophys.* **7** 629–46

Singh A, Maurya S K and Sharma S P 2020 Forward modeling and inversion of very low frequency electromagnetic data over rugged topography using 2D triangular elements *Advances in Modeling and Interpretation in Near Surface Geophysics* ed A Biswas and S P Sharma (Cham: Springer) pp 97–120

Singha K, Kimball K and Lane J W Jr 2000 Borehole-radar methods: tools for the characterization of fractured rock. USGS Fact Sheet 054-0 May (https://water.usgs.gov/ogw/bgas/publications/FS-054-00/)

Sinha A K 1998 Recent developments in quantitative interpretation of VLF-EM data ed K K Roy *et al Deep Electromagnetic Exploration* (New Delhi: Narosa Publishing House) pp 599–606

Sloan S D, Peterie S L, Miller R D, Ivanov J, McKenna J R and Broadfoot S W 2013 *Tunnel detection using near-surface seismic methods* (Geophysical Society of Houston) pp 13–16

Smirnov A A 1975 *Introduction to the Theory of Electromagnetic Field* (Moscow: Nedra) (in Russian)

Smith A, Dixon N and Fowmes G J 2017 Early detection of first-time slope failures using acoustic emission measurements: large-scale physical modeling *Geotechnique* **67** 138–52

Smith L C 2002 Emerging application of interferometric synthetic aperture radar (InSAR) in geomorphology and hydrology *Ann. Assoc. Am. Geog.* **92** 385–98

Socco L V and Strobbia C 2004 Surface-wave method for near-surface characterization: a tutorial *Near Surf. Geophys.* **2** 165–85

Srigutomo W, Arkanuddin M R, Pratomo P M, Novana E C and Agustina R D 2010 Application of qualitative and quantitative analyses of self-potential anomaly in caves detection in Djuanda Forest Park, Bandung *Am. Inst. Phys. Proc.* 1325 164–7

Stein S and Wysession M 2003 An introduction to seismology *Earthquakes. and Earth Structure* (Oxford: Blackwell)

Steeples D, Knapp R and McElwee C 1986 Seismic reflection investigations of beneath Interstate Highway 70 in Kansas *Geophysics* **51** 295–301

Steeples D W and Miller R D 1990 Seismic reflection methods applied to engineering, environmental, and groundwater problems *Geotechnical and Environmental Geophysics, 1, Review and TutorialInvestigations in Geophysics no. 5* ed S Ward (Houston, TX: SEG) pp 1–30

Stennett T and Grusky S W 2008 Lidar: fact or fiction *Prof. Surv. Mag.* **28** 16–20

Stewart M and Gay M C 1986 Evaluation of transient electromagnetic soundings for deep detection of conductive fluids *Ground Water* **24** 351–6

Stilke G 1959 On elastic surface waves at a cylindrical hole in an infinite solid *Geophys. Prospect.* **7** 273–86

Stoessell R K, Coke J G and Easley D H 2002 Localized thermal anomalies in Haloclines of Coastal Yucatan Sinkholes *Ground Water* **40** 416–24

Stokoe K H, Wright G W, James A B and Jose M R 1994 Characterization of geotechnical sites by SASW method *Geophysical Characterization of Sites* ed R D Woods (Rotterdam: A A Balkema)

Stokoe I I K H, Kurtulus A and Park K 2006 Development of field methods to evaluate the nonlinear shear and compression moduli of soil *Earthquake Geotechnical Engineering*

Workshop, Canterbury 2006 (EGE 2006) (Christchurch, New Zealand, November 21–24, 2006)

Storn R and Price K 1995 Differential evolution: a simple and efficient adaptive scheme for global optimization over continuous spaces *Technical Report* TR-95-012 International Computer Science Institute, Berkeley

Stoyer C 2011 Interpreting Colin Zelt's 'Shootout' Seismic Refraction Data Set Using the Generalized Reciprocal Method

Styles P, Toon S, Thomas E and Skittrall M 2006 Microgravity as a tool for the detection, characterization and prediction of geohazard posed by abandoned mining cavities *First Break* **24** 51–60

Svetov B S and Gubatenko V P 1999 Electromagnetic field of the mechanical–electrical origin in the porous water-saturated mountainous rocks: statement of the problem *Izvestia Russ. Acad. Sci., Phys. Earth* **10** 67–73

Šumi F 1965 Prospecting for non-metallic minerals by induced polarization *Geophys. Prospect.* **13** 603–16

Sundararajan N A, Al-Gaami M A, Ramabrahmam G and Srinivas Y 2007 A real spectral analysis of the deformation of a homogeneous electric field over a thin bed—a Hartley transform approach *Geophys. Prospect.* **55** 901–10

Sundararajan N, Raef A and Al-Wardi M 2015 Integration of micro-gravity and Very Low Frequency (VLF-EM) electromagnetic (EM) data analyses to outline hazardous cavities and low rock-strength: coastal carbonates, Wadi Shab *Oman. Proc. SAGEEP 2015 Conf. (Austin (TX), USA)* 1–6

Sun H, Li X, Li S, Qi Z, Su M and Xue Y 2012 Multi-component and multi-array TEM detection in karst tunnels *J. Geophys. Eng.* 9 359–73

Sun H, Li K, Chen R, Li H, Ran Q, Liu G, Lin J, Li S and Su C X 2018 Experimental study on transient electromagnetic responses to shallow karst *Chin. J. Rock Mech Eng.* **37** (in Chinese, Abstract in English)

Supper R, Motschka K, Ahl A, Bauer-Gottwein P, Gondwe B, Alonso G M, Römer A, Ottowitz D and Kinzelbach W 2009 Spatial mapping of submerged cave systems by means of airborne electromagnetics: an emerging technology to support protection of endangered karst aquifers *Near Surf. Geophys.* **7** 613–27

SurfSeis v.6 2017 (www.KGS.ku.edu/software/surfseis)

Suyama K, Imai T, Ohtomo H, Ohta K and Takahashi T 1986 Delineation of structures in alluvium and diluvium using SH-wave reflection and VSP methods *54th Annual International Meeting, Society of Exploration Geophysicists* vol 1 *(Atlanta)*S H Danbom and S N Domenico 79–96

Tabbagh A 1986 Applications and advantages of the Slingram electromagnetic method for archaeological prospecting *Geophysics* **51** 576–84

Tarkhov A G 1961 *Principles of Geophysical Prospecting by Radio Comparing and Bearing Method (VLF)* (Leningrad: Gosgeoltekhizdat) (in Russian)

Technical Notes TN-31 Geonics Limited Application of 'dipole–dipole' electromagnetic system for geological depth sounding (http://geonics.com/pdfs/technicalnotes/tn3)

Telford W M, Geldart L P, Sheriff R E and Keys D A 1990 *Appllied Geophysics* 2nd edn (Cambridge: Cambridge University Press)

TEM FAST 48 (Manual) v.8a 2018 Applied Electromagnetic Research (AEMR) (http://aemr.net/tem_res.htm)

TEM-researcher v.8 2017 (http://aemr.net/tem_res.htm)

TEM-3D Wizard program (http://aemr.net/tem_res.htm)

Terzaghi K, Peck R B and Mesri G 1996 *Soil Mechanics in Engineering Practice* 3rd edn (New York: Wiley)

Tesmull F A and Crossley D I 1981 *Inversion of VLF-Data for Simple Lateral Inhomogeneities* (Ottawa: Geological Survey of Canada) pp 79–86

Tikhonov A N and Samarsky A A 1963 *Equations of Mathematical Physics* (Oxford: Pergamon)

Thomas B and Roth M J 1999 Evaluation of site characterization methods for sinkholes in Pennsylvania and New Jersey *Eng. Geol.* **52** 147–52

Ulriksen C P F 1982 Application of impulse radar to civil engineering *PhD Thesis* Lund University of Technology

USCS Remote sensing (https://usgs.gov/faqs/what-remote-sensing-and-what-it-used?qt-news_-science_products=3#qt-news_science_products)

Utsi E C 2017 *Ground Penetrating Radar: Theory and Practice* 1st edn (Oxford: Butterworth-Heinemann) p 224

Vadillo I, Benavente J, Neukum C, Grützner C, Carrasco F, Azzam R, Liñán C and Reicherter K 2012 Surface geophysics and borehole inspection as an aid to characterizing karst voids and vadose ventilation patterns (Nerja research site, S. Spain) *J. Appl. Geophys.* **82** 153–62

Villee M A, Chouteau M and Palacky G J 1992 Effect of temporal and spatial variations of the primary signal on VLF total-field surveys *Geophysics* **57** 97–105

Vincent C S, Garambois E, Thibert E, Lefèbvre E, Le Meur E and Six D 2010 Origin of the outburst flood from Tête Rousse glacier in 1892 (Mont-Blanc area, France) *J. Glaciol.* **56** 688–98

Vincent C, Thibert E, Gagliardini O, Legchenko A, Gilbert , Garambois S, Condom T, Baltassat J-M and Girard J-F 2015 Mechanisms of subglacial cavity filling in Tête Rousse glacier *J. Glaciol.* **61** 609–23

Vinogradov S D 1964 *Acoustical Observations of Rock Destroying Processes* (Moscow: Nauka)

Vogelsang D 1987 Examples of electromagnetic prospecting for karst and fault systems *Geophys. Prospect.* **35** 604–17

Von Hippel A R 1954 (1995) *Dielectrics and Waves* (New York: Wiley)

Vouillamoz J M, Descloitres M, Toe G and Legchenko A 2005 Characterization of crystalline basement aquifers with MRS: a case study in Burkina Faso *Near Surface Geophys.* **3** 205–13

Wadas S H, Tanner D V, Polom U and Krawczyk C M 2017 Structural analysis of S-wave seismics around an urban sinkhole: evidence of enhanced dissolution in a strike-slip fault zone *Nat. Hazards Earth Syst. Sci.* **17** 2335–50

Wanfang Z, Beck B F and Stephenson J B 1998 Investigation of groundwater flow in karst areas using component separation of natural potential measurements *Environ. Geol.* **37** 19–25

Ward S H 1990 Resistivity and induced polarization methods *Geotechnical and Environmental Geophysics* ed S H Ward vol I (Tulsa, OK: Society of Exploration Geophysicists) pp 147–90

Waters K H 1978 *Reflection Seismology* (New York: Wiley)

Wentworth C K 1922 A scale of grade and class terms for clastic sediments *J. Geol.* **30** 377–92

Werner E 1984 Review of electrical resistivity methods *Proc. 1st Multidisciplinary Conf. on Sinkholes (Orlando, Florida, 15–17 October)* 231–4

Wigley T and Brown M C 1971 Geophysical applications of heat and mass transfer in turbulent pipe flow *Boundary-Layer Meteor.* **1** 300–20

W-Geosoft 2017 (http://wgeosoft.ch/PDF/winsism_pub_us.html or http://wgeosoft.ch/menu_soft.html)

Wolf P R, Dewitt B A and Wilkinson B 2014 *Elements of Photogrammetry with Applications in GIS* 4th edn (New York: McGraw-Hill)

Wust-Bloch G H and Joswig M 2006 Pre-collapse identification of sinkholes in unconsolidated media at Dead Sea area by 'nanoseismic monitoring' (graphical jackknife location of weak sources by few, low-SNR records) *Geophys. J. Int.* **167** 1220–32

Wu Q, Deng C and Chen Z 2016 Automated delineation of karst sinkholes from LiDAR-derived digital elevation models *Geomorphology* **266** 1–10

Wu Q, Liu H, Wang S, Yu B, Beck R and Hinkel K 2015 A localized contour tree method for deriving geometric and topological properties of complex surface depressions based on high-resolution topographical data *Int. J. Geogr. Inf. Sci.* **29** 2041–60

Xi Z Z, Liu J, Long X and Hou H T 2010 Three-component measurement in transient electromagnetic method *J. Cent. South Univ (Sci. Technol.)* **41** 272–6 (in Chinese with English abstract)

Xia J, Miller R D and Park C B 1999 Estimation of near-surface shear-wave velocity by inversion of Rayleigh waves *Geophysics* **64** 691–700

Xia J, Miller R D and Park C B 2001 Feasibility of using the MASW method to define a sinkhole impact area at Calvert Cliffs Nuclear Power Plant, Maryland *Open-file Report* No. 2001-19 Lawrence, KS: Kansas Geological Survey

Xia J, Miller R D and Park C B 2003 Inversion of high-frequency surface waves with fundamental and higher modes *J. Appl. Phys.* **52** 45–57

Xia J, Nyquist J E, Xu Y, Roth M J S and Miller R D 2007 Feasibility of detecting near-surface features with Rayleigh-wave diffraction *J. Appl. Geophys.* **62** 244–53

Xingping L and Meifeng C 2004 *Analysis on Acoustic Emission Signal from Rock Damage in Subsiding Area Based on Non-Balanced Statistics* (https://en.cnki.com.cn/Article_en/CJFDTotal-YSLX200402018.htm)

Xue G Q, Cheng J L, Zhou W Y and Li H 2013 Detection and monitoring of water-filled voids using transient electromagnetic method *Env. Earth Sci.* **70** 2263–70

Yakubovski U B and Renard I B 1991 Electrical prospecting *Handbook for Universities* 3rd edn (Moscow: Nedra) (in Russian)

Yalciner C C 2013 Investigation of the subsurface geometry of fissure-ridge travertine with GPR, Pamukkale, western Turkey *J. Geophys. Eng.* **10** 035001

Yamanaka H, Takemura M, Ishida H and Niva M 1994 Characteristics of long-period microtremors and their applicability in exploration of deep sedimentary layers *Bull. Seism. Soc. Am.* **84** 1831–41

Yang M Z and Drumm E C 2002 Stability evaluation for the sitting of municipal landfills in karst *Eng. Geol.* **65** 185–95

Yang R, Jin S, Zhang T, Luo Y, Meng Z and Deng L 2017 Seismic response characteristics and identification of different types of karst collapse in Tahe oil field, Tarim Basin *Conf. paper (CGS/SEG Int. Geophysical Conf. Qingdao) (China, 17–20 April 2017)* 3

Yi M J, Kim J H and Song Y *et al* 2001 Three-dimensional imaging of subsurface structures using resistivity data *Geophys. Prospect.* **49** 483–97

Yechieli Y, Kafri U, Goldman M and Voss C I 2001 Factors controlling the configuration of the fresh-saline water interface in the Dead Sea coastal aquifers: synthesis of TDEM surveys and numerical groundwater modeling *Hydrol. J.* **9** 367–77

Yilmaz O 2001 *Seismic Data Processing* ed S M Doherty (Houston, TX: SEG)

Zakharov E V and Pimenov Y Y 1982 *Numerical Analysis of Radiowave Diffraction* (Moscow: Radio & Communication) (in Russian)

Zaslavsky Y *et al* 2008 Microzoning of site response parameters in the towns of Dimona and Bet Shean (Israel) *Boll. Geofis., Teor. Appl.* **49** 109–30

Zaslavsky Y, Shapira A, Ataev G, Gorstein M, Aksinenko T, Kalmanovich M, Perelman N and Hofstetter R 2009 Using ambient noise measurements in the process of assessing earthquake hazards in urban areas: examples from Israel ed T Miura and Y Ikeda *Earthquake Engineering: New Research* (Hauppauge, NY: Nova Publisher) pp 155–218

Zhang S, Bogus S, Baros S, Neville P and Dow R 2019 Karst sinkhole detecting and mapping using airborne LiDAR—a conceptual framework *2019 Tran-SET Annual Conf.* **271** 02005

Zhang J, Liu S, ChenZg Q, Wang B and Ren C 2019 Application of cross-borehole integrated geophysical methods for the detailed investigation of Karst in Urban Metro Construction *J. Environ. Eng. Geophys.* **24** 525

Zhdanov M S 1988 *Integral Transforms in Geophysics* (Berlin: Springer)

Zhdanov M S and Keller G V 1994 *The Geoelecrical Methods in Geophysical Exploration* (Amsterdam: Elsevier)

Zheng D, Zhang L and Shen F 2011 Characterization and modeling study of karst networks in the ordovician carbonate reservoirs, Tarim Basin *AAPG Annual Convention and Exhibition (Houston, TX, April 10–13, 2011)*

Zhou W, Beck B F and Stephenson J B 1999 Application of electrical resistivity tomography and natural-potential technology to delineate potential sinkhole collapse areas in a covered karst terrane ed B F Beck, J Pettit and J G Herring *Hydrogeology and Engineering Geology of Sinkholes and Karst* (Rotterdam: AA Balkema), pp187–93

Zhou W, Beck B F and Adams A L 2002 Effective electrode array in mapping karst hazards in electrical resistivity tomography *Environ. Geol.* **42** 922–8

Zhu Z, Haartsen M W and Toksöz M N 2000 Experimental studies of seismoelectric conversions in fluid-saturated porous media *J. Geophys. Res.* **105** 28055–64

Zhu J, Taylor T P, Currens J C and Crawford M M 2014 Improved karst sinkhole mapping in Kentucky using LiDAR techniques: a pilot study in Floyds Fork Watershed *J. Cave Karst Stud.* **76** 207–16

Zhu W D and Zhang Y G 2008 Study on the application of the transient electromagnetic method to the detection of.an underground cave *Build. Tech. Dev.* **11** 73–6

Zijártó M, Galsa A, Tóth Á and Mádl-Szönyi J 2021 Numerical analysis of the potential for mixed thermal convection in the Buda Thermal Karst, Hungary *J. Hydrol.: Reg. Stud.* **34** 1–20

Zini L, Calligaris C, Forte E, Petronio L, Zavagno E, Boccali C and Cucchi F 2015 A multidisciplinary approach in sinkhole analysis: the Quinis village case study (NE-Italy) *Eng. Geol.* **197** 132–44

Zohdy A A R 1989 A new method for the automatic interpretation of Schlumberger and Wenner sounding curves *Geophysics* **54** 245–53

Michael Ezersky, Lev V Eppelbaum and Anatoly Legchenko

Chapter 3

Dead Sea evaporite karst, its features, and methods of investigation

3.1 Introduction

More than six thousand sinkholes have been opened during the last 30 years along the Dead Sea (DS) coastal areas in Israel (DS sinkhole database 1998–2016, Abelson *et al* 2006, 2017, Yechieli *et al* 2016) and Jordan (Closson and Abou Karaki 2009, Abou Karaki *et al* 2016) (see figure 3.1).

Generally, sinkholes are a subaerial manifestation of karst or pseudokarst. The term karst is commonly used to describe a wide range of surface and subsurface landforms that develop by the dissolution of the soluble rock (Ford and Williams 2007, De Waele *et al* 2011, Frumkin 2013, Benson and Yuhr 2016). On the other hand, some sinkholes resemble or act like karst but were not formed by the natural dissolution of rock. Those are termed 'pseudokarst' (Grimes 1975, Eberhard and Sharples 2013, Benson and Yuhr 2016). Pseudokarst processes include a variety of mechanical agencies in which material is removed in solid-state, together with the effects resulting from the changes of phase and removal of liquid material. They are also known as 'false karst' or are sometimes referred to as 'analogous karst'. Thus, karstic sinkholes require soluble rocks in the subsurface (limestone, dolomite, gypsum, salt, etc), whereas pseudokarst sinkholes require the existence of underground flow and physical erosion.

To avoid problems in agriculture, industry, tourism, etc, site characterization is carried out. Site characterization is the process of successive understanding of the geomorphological, geological, hydrological, and engineering properties at the site including the soil, rock, groundwater, and in many cases, man-modified conditions in the subsurface (e.g., utilities, structures, mines, and tunnels) that can impact site conditions (Benson and Yuhr 2016). Many case histories have shown that most failures are caused by an improper understanding of site conditions that can impact the project. These failures could have been avoided by focusing attention on the

doi:10.1088/978-0-7503-3635-2ch3

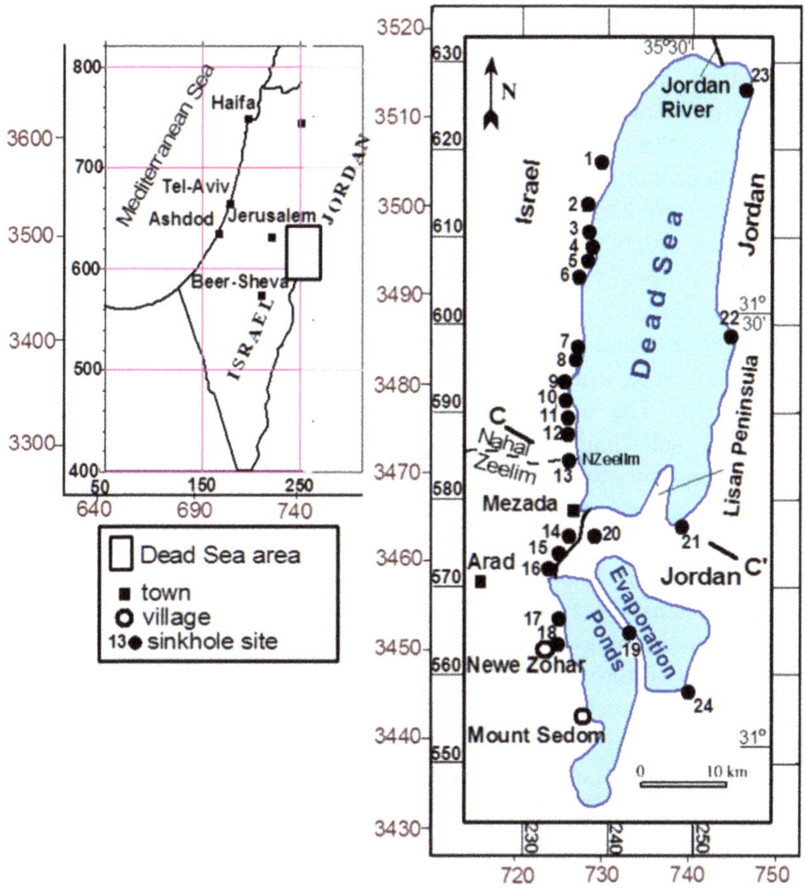

Figure 3.1. Sinkhole sites along Dead Seashore: 1—Zikim, 2—Avenat, 3—N.Qane, 4—Samar, 5—N. Deragot, 6—Mineral Beach, 7—Ein Gedi and Nahal Arugot, 8—Yesha, 9—Mazor, 10—Nahal Hever north, 11—Nahal Hever south, 12—Asa'el, 13—Nahal Zeelim, 14—Rahaf, 15—Lynch Straight, 16—Mor, 17—Ein Boqeq, 18—Newe Zohar, 19—Dam-2, 20—Lisan, 21-Ghor Al-Haditha, 22—Wadi Mujib, 23—Suweimeh, 24—unnamed; (20–24—after Closson and Abou Karaki (2009)). **CC′**—a line of the generalized geological section (the coordinates in the frame are new Israel Mercator grid in km, the outer coordinates are Universal Transverse Mercator grid, in km).

geomorphic conditions. Later these failures could be corrected based on drilling and geophysical investigations.

The problem of sinkhole formation is the main concern in the areas composed of limestones, gypsum, and other soluble rocks. It is associated with collapse or subsidence into subsurface cavities (e.g. Klimchouk 2005, De Waele *et al* 2009, Frumkin *et al* 2009a, Parise *et al* 2009, Gutiérrez *et al* 2014). In carbonate karst areas sinkhole clusters usually develop and modify during long geological periods. Different methods for the detection of sinkholes and other karst features in carbonate rocks (generally, limestone) were developed (Arandjelović 1969, Milanović 1981, 2000, Beck and Herring 2001, Beck 2003). However, these methodologies should be adapted, or new methods developed, for the unique environment of evaporite deposits along the Dead Sea coast in Jordan and Israel. It is also important for terrains composed of evaporites in Italy, Spain, and others. Two main factors have conditioned the difference of the karst areas in carbonate rocks and the sites composed of evaporates in the Dead Sea region: (1) subsurface dissolution of the salty rock is observed quite often in response to changes in environmental conditions; and (2) the presence of high salinity groundwater. The first factor is mainly related to sinkholes formed in evaporite areas, where dissolution rates are one to three orders of magnitude higher compared with limestone (Reuter and Stoyan 1993, Gutiérrez *et al* 2008). Salt is a very soluble material even in slightly saline (undersaturated) circulating groundwater. Therefore, the subsurface dissolution of salt by such groundwater and collapse of the overlying material (sinkhole formation) may also be relatively rapid, from days to some weeks (Frumkin 1995, Van Sambeek 1996, Stiller *et al* 2016). The second factor is that the Dead Sea groundwater contains essentially more salt than normal groundwater that is slightly slowing down the dissolution process (months, years) and allowing the observation of changes in physical properties of the subsurface on a human life scale.

Because of high solubility, salt outcrops are rare and produce the typical vadose karst morphologies. It is characterized by localized sinkholes, blind valleys, and underground streams (e.g. Frumkin 1994, Burthens *et al* 2009, De Waele *et al* 2009). Changes in geomorphology, including sinkhole development, can be extremely rapid, especially in areas of human activity altering the groundwater circulation (Lucha *et al* 2008). Researchers have suggested different approaches to the investigation and prediction of possible problems in karst areas. The sinkhole detection and study is a difficult task that should be resolved by using both surface and subsurface information (Gutiérrez *et al* 2008). The surface data may be obtained from remote sensing measurements, including high-resolution geodetic techniques (InSAR, photogrammetry, LIDAR, and DEMs), field surveys, building-damage maps, historical and recent topographical maps, evidence of people, exposed dissolution and subsidence features (paleokarst), etc. Subsurface data can be derived from geophysical surveys, boreholes, speleological exploration, and so on. Several researchers have suggested protective measures (Taqieddin *et al* 2000, Yechieli *et al* 2006, Gutiérrez *et al* 2014, Milanović *et al* 2019). Galve *et al* (2009a, 2009b) proposed probabilistic sinkhole modeling for hazard assessment and sinkhole susceptibility mapping for evaporite karst.

Following determination suggested by Gutiérrez *et al* (2014) 'Sinkholes are surface manifestations of subsurface dissolution and internal erosion and deformation, commonly hidden from direct observation and most subaerial geomorphological study methods'.

There is a strong contrast in the physical properties between different geological formations in the Dead Sea coastal area. It allows the use of surface geophysical methods (Frumkin *et al* 2011). Different geophysical methods have been applied to study the various aspects of sinkhole hazard evaluation (Eppelbaum *et al* 2008, Legchenko 2013, Ezersky *et al* 2017). To study the structure and properties of the shallow subsurface at the sinkhole development sites, authors have tested high-resolution seismic refraction, reflection, diffraction, electrical resistivity tomography (ERT), ground-penetrating radar (GPR), and microgravity methods. These methods have different resolutions and penetration depths. The methods are complementary to each other and reveal useful information about the subsurface. Each one investigates certain characteristics, but when combined they provide us with new information about the subsurface. The seismic refraction method is utilized to map salt layers and to detect dissolved zones. Surface nuclear magnetic resonance (SNMR) and microgravity methods allow detection of cavities (Rybakov 2005, Eppelbaum *et al* 2008, Legchenko *et al* 2008b). Transient electromagnetic method (TEM) is used for the determination of salt–fresh water interface. Moreover, the TEM method allows the determination of the DS water salinity and aggressiveness of salt and their mapping through the sinkhole-prone areas (Ezersky and Frumkin 2017, 2021). That is why geophysical methods can facilitate sinkhole susceptibility assessment in the DS coastal areas both in Israel and Jordan. Numerous geophysical methodologies have been developed and applied to explore the evaporite karst phenomenon. This chapter is aimed at summarizing the available experience and comparing predictions of sinkhole expansion through the DS coastal areas. We also tried to compare these predictions with the results of recent geological and geophysical observations. Such an analysis is important at present, for instance, for the construction of a channel between the Red and Dead seas (Beyth 2007) that is being decided.

The chapter will initiate discussing the presence of a salt layer both on the western (Israel) and eastern sides of the DS shores. Such analysis in the Ghor Al-Haditha area (eastern DS side) could be based on an integrated analysis of all available recently observed geophysical data.

3.2 Red Sea–Dead Sea channel

Nowadays, the southern part of the Dead Sea basin contains industrial areas both in Israel and Jordan. Industrial areas host the extractive industries which produce salts from the evaporation of the Dead Sea water. On May 9, 2005, Jordan, Israel, and the Palestinian Authority signed an agreement to go ahead with a feasibility study for the Two Seas Canal. The proposed canal would siphon seawater 230 m uphill from the Red Sea's Gulf of Aqaba in Jordan and then run down into the Dead Sea.

The goals of the Red Sea–Dead Sea conduit are as follows (Harza JRV Group, 1998):

- To desalinate water;
- To save the Dead Sea from environmental degradation;
- To generate hydroelectric power;
- To build a symbol of peace in the Middle East.

The project includes power plants, which will provide electricity for pumping the water from the Red Sea and water desalination plants. Desalination of ~2000 million cubic meters of seawater will allow producing up to 850 million cubic meters of fresh water per year with 20–300 mg L^{-1} TDS to the three countries (Beyth 2007).

Despite desalination being an important part of the project that will allow the production of fresh water for the arid region it will cause the water level in the Dead Sea to rise, creating serious concerns about the environmental impact of the channel (Angelakis *et al* 2021).

Sinkhole problems and other environmental changes may influence land-use practice, cause demographic changes and threaten the economic development of this region. These changes may also influence political stability and their impact must be investigated.

The construction of a conduit from the Gulf of Elat (Aqaba) to the Dead Sea (termed 'The Peace Conduit' or 'Red Sea–Dead Sea Canal'—RSDSC) should resolve several problems of the DS. However, such a complex unit to transfer some 2.0 billion m^3/year of water along an active segment of the Dead Sea Transform zone in immediate proximity to shallow and deeper groundwater resources should also be evaluated very cautiously (Shirav-Schwartz *et al* 2006). However, the potential impact on limnology, geochemistry, and biology of the DS is not well understood (Shirav-Schwartz *et al* 2006).

The Dead Sea occupies a morphotectonic valley that formed along the intra-land Dead Sea transform that extends some 1000 km from the Red Sea to the zone of plate convergence in southern Turkey. The Arava Valley (figure 3.2) is part of this

Figure 3.2. Location map of the Arava Valley (Landsat imagery). Reprinted from Shirav-Schwartz *et al* (2006), with permission Geological Survey of Israel.

system—an elongated depression of about 165 km, from the Dead Sea in the north to the Gulf of Elat (Aqaba) in the south (Amit *et al* 2002).

The area is covered by an extensive veneer of sediments that accumulated during several depositional cycles separated by periods of erosion between the Cambrian and the Late Eocene.

Five variants of the RSDSC were studied by researchers (The Harza JRV Group 1996) from the environmental, technical, and economic points of view, and the eastern alternatives passing along. The Jordanian side of the Arava Valley was selected as the most suitable. There will be sections along the RSDSC that are very close to the surface or are planned as open canals (figure 3.3), and these will be subjected to flash floods, both during construction and along with its continuous operation.

Due to numerous geo-environmental conditions: immediate proximity to shallow groundwater resources; flash-flood risks to open/near-surface sections of the conduit; a cluster of tectonic related features, such as seismicity, active faults, high amplification potential (site response); and recent surface displacements the trace of the proposed Red Sea–Dead Sea (RSDS) conduit along the Arava Valley creates

Figure 3.3. The eastern proposed alignment for the Peace Conduit. Marked in red are open canal/near-surface stretches. Reprinted from Shirav-Schwartz *et al* (2006), with permission from Geological Survey of Israel.

a variety of design problems. These factors will be thoroughly analyzed in the course of a substantial study (Beyth 2007).

In December 2013, the three parties signed an agreement for the implementation of a vastly scaled-back version of the RSDS Project. It is designed to accomplish two objectives: to provide new water to a critically water-scarce region; and to understand better, under scientific supervision, the consequences of mixing the Red Sea and the Dead Sea waters.

3.3 Geology and hydrogeology of the Dead Sea region: an overview

3.3.1 Geological background

The DS (more exactly—lake), with a water level (in 2017) of 431 m below mean sea level, is the deepest subaerial point on Earth, located in an extremely arid environment with annual precipitation of 50–100 mm. It is an example of a hypersaline lake (34.2% salinity in 2010), and it is also the deepest salt lake in the world (Goetz 1986, Ben-Avraham 2001). Usually, a hypersaline lake is a landlocked body water that contains significant concentrations of sodium chloride or other mineral salts. The DS water salinity surpasses the salinity of the ocean water by 10 times (3.5%; i.e. 35 g L^{-1}). As mentioned above, the DS lies within a transform that looks like an elongated depression, 150–165 km long and 15–17 km wide at its center, divided into two sub-basins (figure 3.4). At the deepest point in the northern sub-basin, the maximum water depth is ~300 m. The southern subaerial sub-basin is dry; it contains

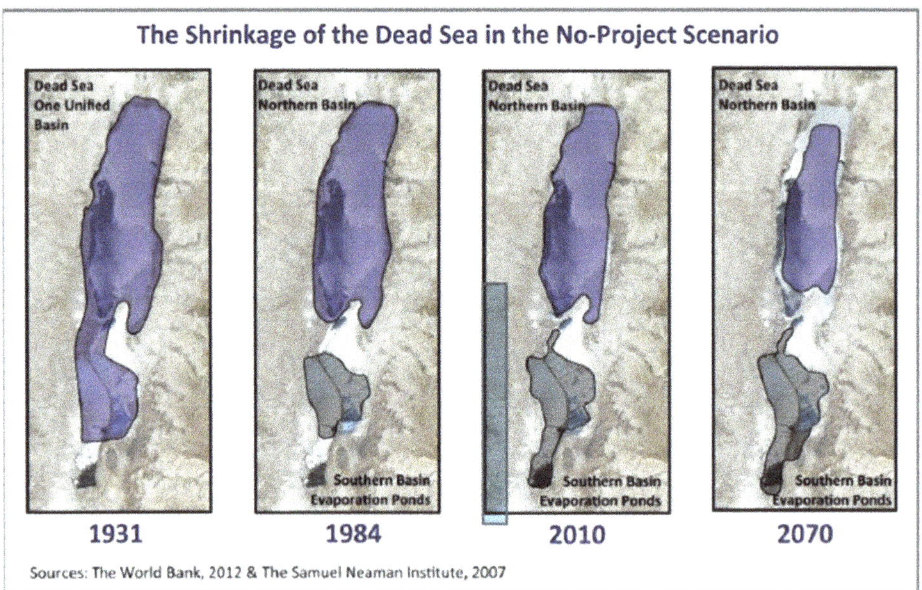

Figure 3.4. Shrinkage of the surface area of the Dead Sea as a result of the decline in its water level since 1931 through the 139 years, 1984 and 2010, and as projected for the year 2070 (reprinted from Willner *et al* (2013), with permission from Arava Institute for Environmental Studies, Kibbutz Ketura). (Permission of NASA Laboratory and EO SEOR.) https://earthobservatory.nasa.gov/images/77592/the-dead-sea.

commercial artificial evaporation ponds (EPs) both in Israel and Jordan and reaches a maximum elevation of about 405 m below the mean sea level (b.m.s.l.). Note that elevation of water in the southern basin rises annually by 20 cm.

During the Middle Miocene (18–14 Ma) the Dead Sea transform started separating Israel from the Arabian plate (Quennell 1958, Freund *et al* 1970, Garfunkel 1997). An elongated morphotectonic depression formed along the Dead Sea transform, between the Red Sea in the south and Lebanon in the north. As tectonic subsidence started to prevail over the filling of the basin, the depression began acting as an endorheic basin. The basin has been occupied by various lakes, the latest being the late Pleistocene Lake Lisan followed by the Holocene Dead Sea. The Lisan Formation contains alternations of detritic and aragonitic lamina, as well as conglomerates, clays, and evaporites (Begin *et al* 1974). Maximal Lake Lisan level was attained just before the last glacial maximum (Bartov *et al* 2002, Lisker *et al* 2009). Since then until the end of the Pleistocene, Lake Lisan receded abruptly, consequently depositing a salt layer around 11 000–10 000 years ago (Stein *et al* 2010). This latest Pleistocene salt layer ranges from a few meters to a few tens of meters in thickness, lying sub-horizontally tens of meters under the present surface.

Close to the Pleistocene–Holocene transition the basin was partly inundated by the Holocene lake, known as the Dead Sea. During most of the Holocene, the lake level was at ca. −400 m, with small oscillations (Frumkin 1997, Bookman Ken-Tor *et al* 2004). This was followed by a recent and still ongoing drop of ~40 m from the late 1980s to ca. −434 m below the mean sea level in 2019 due to anthropogenic use of water. Simultaneously, the lake shores suffer tremendous impact: the ground is collapsing and subsiding in hundreds of points along the lake, with people, roads, and the property being swallowed in the more catastrophic events. The collapse is believed to result mainly due to the dissolution of salt by aggressive groundwater, following the retreat of the Dead Sea level and the groundwater halocline. Geological evidence suggests that a previous major lake level fall occurred *naturally*. This may provide a new explanation for a curious historical–geological phrase in the book of Genesis (14:10), which suggested recording formation of the collapse sinkholes which occurred in response to the historic falling lake level, associated with the climatic desiccation.

The Holocene sediments of the Dead Sea basin include lacustrine and fluviatile facies (Migowski *et al* 2006): aragonite–detritic laminar alternations, clastics, and evaporites, mostly of low mechanical strength. The alterations of permeability in the layers divide the coastal aquifer into several sub-units. The coastal aquifer water is not fresh but is undersaturated and aggressive concerning salt. A halocline interface commonly underlies the coastal aquifer, with relatively fresh water overlying much saltier water (Frumkin and Raz 2001, Yechieli *et al* 2006). The Dead Sea level drop is accompanied by the intrusion of fresh water into the coastal aquifer corresponding to the lowering of the water table (Yechieli 2000). This relatively fresh aquifer receives its water mainly from the rainfall in the west, which feeds both the mountain aquifer and ephemeral streams flowing into the Dead Sea. It has been assumed that much of the water flowing into the coastal aquifer comes from below, under artesian conditions, or by lateral flow from the west. Local annual rainfall is only ~50–100 mm, so direct replenishment from above is small. Air temperature is mild in winter and extremely hot in summer, and annual potential evaporation is >2000 mm.

The south basin of the DS has dried due to the lake level fall in the early 1980s. The basin has been converted into industrial salt ponds, whose level is rising due to salt deposition

3.3.2 Brief tectonic characteristics

The DS depression is a N–S elongated basin that actively subsides. It is formed along the Dead Sea–Jordan transform-plate boundary. It is a pull-apart basin associated with the NW–SE compression and the NE–SW tensile stress (Ben-Avraham 1997, Garfunkel 1997). Numerous faults have been detected throughout the DS area (figure 3.5(b)) mainly trending N–S and NW–SE (Al-Zoubi, and ten Brink 2001, Abelson *et al* 2003, Shamir 2006).

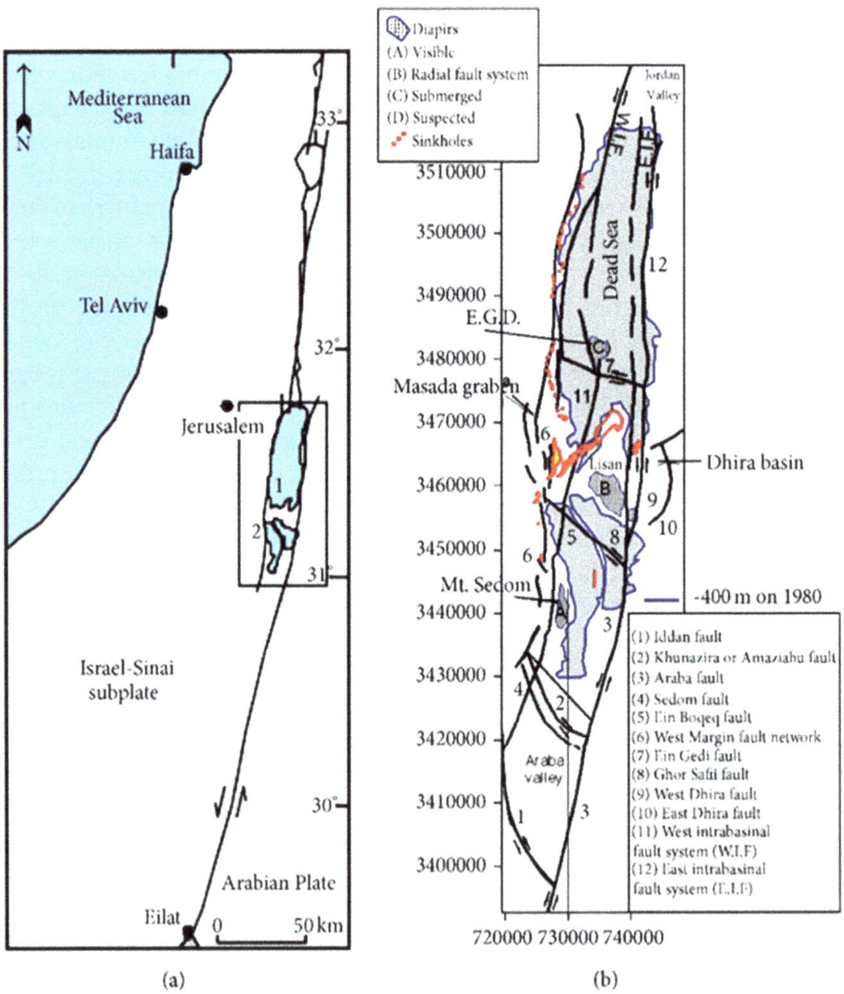

Figure 3.5. Dead Sea tectonic setting: (a) pull-apart basin (based on Garfunkel (1981)) (reprinted from Ezersky and Frumkin (2013), copyright 2013, with permission from Elsevier); (b) faults through the DS area (based on Ben-Avraham (1997) with permission of American Geophysical Union).

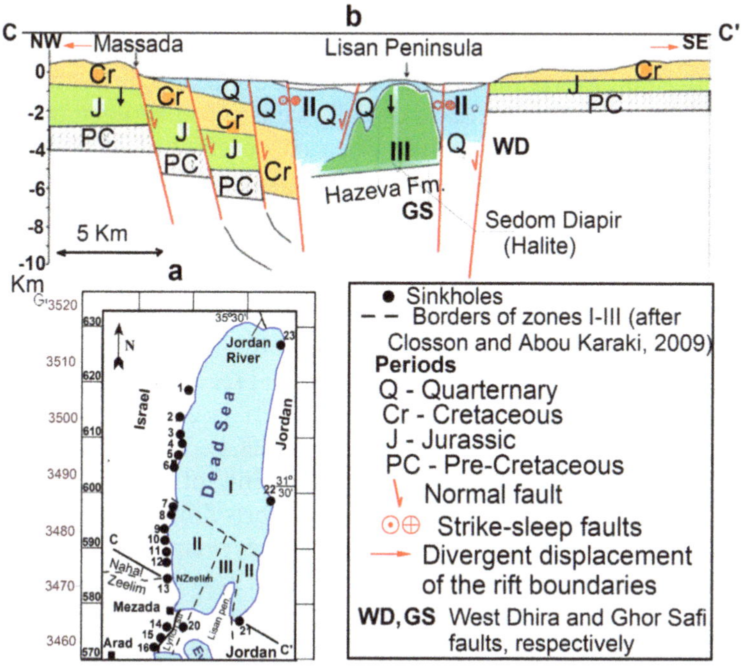

Figure 3.6. Dead Sea basin: a unique region of the world. (a) Plan view; (b) C–C′ (W–E) tectonic section (modified from Garfunkel (1981), copyright 1981 with permission from Elsevier). I–III are areas selected by Closson and Abou Karaki (2009).

The DS basin is tectonically active. Its central part is forming the graben area, subsiding along normal faults (Garfunkel 1997). The subsidence is caused by the pull-apart producing the W–E divergence of the marginal faults (Garfunkel 1997).

At the western shore of the DS basin, several large normal faults exist that are its main structural feature.

A typical deep W–E vertical section through the Lisan Peninsula is shown in figure 3.6(b) (along with line C–C′ in figure 3.6(a)) (Garfunkel 1981, Closson 2005, Ezersky *et al* 2014). In this section, one can see that the 3–6 km space between the bordering faults is composed of clastic quaternary sediments that do not transfer tectonic stresses (see also Shalev and Lyakhovsky 2012, p 166). This is especially understandable taking into account the W–E divergence of the Dead Sea transform boundaries (Wdowinski *et al* 2004).

3.3.3 Geological section

3.3.3.1 Geological section of sinkhole sites

The survey areas are located between the DS shorelines and the western and eastern borders of the DS transform within the graben area. Starting from 1993 a lot of boreholes were drilled along the western DS shore. To detect the salt layer in the subsurface, boreholes were drilled with the saline DS brine or bentonite.

The geological section of the DS western shore (Israel), is based on 80–90 m deep boreholes (Yechieli *et al* 2004). As a rule, it is composed of alluvial fan sediments to a depth of 20–50 m. Lower than the depths of 20–50 m (top) is located a salt layer that has a thickness of 6–30 m. Salt is usually sandwiched between the marl layers occurring below and above, especially when approaching the DS shoreline. The western salt edge is often in contact with the sandy-gravel sediments (Yechieli *et al* 2004). Salt and clayey sediments divide sand and gravel formations into several sub-aquifers (Yechieli *et al* 2006).

3.3.3.2 Generalized geological section

A generalized geological W–E section through the DS (along the C–C′ line in figure 3.6(a)) is shown in figure 3.7. The geological section of the DS western side (Israel), is based on 80–90 m deep boreholes. It is composed of alluvial fan sediments to a depth of 20–50 m (Yechieli *et al* 2004). The coastal area between the fault scarp and the DS shoreline consists mainly of clastic sediments (marl, sand, and gravel) deposited in fan deltas and debris flows. Also found are some intercalations of lacustrine sediments (silt, clay, gypsum, halite, and aragonite) mainly of the Lisan Formation (Begin *et al* 1974) and younger Holocene sediments named the Ze'elim Formation (Yechieli *et al* 1995b). Below these sediments, the salt layer of 10–30 m thickness is located underlined by similar alluvial sediments.

The eastern part of the DS section (Jordan) has been geologically described following an unpublished report of El-Isa *et al* (1995) cited in numerous papers (Taqieddin *et al* 2000, Closson 2005, Frumkin *et al* 2011, Al-Halbouni *et al* 2017, Ezersky *et al* 2021). This section was discovered by boreholes BH-2 (figure 3.7) down to 51 m and did not detect the salt layer, which was proposed by Taqieddin *et al* (2000).

One can conclude from figure 3.7 that the geological conditions in the Ghor Al-Haditha area are similar to Israeli DS shores.

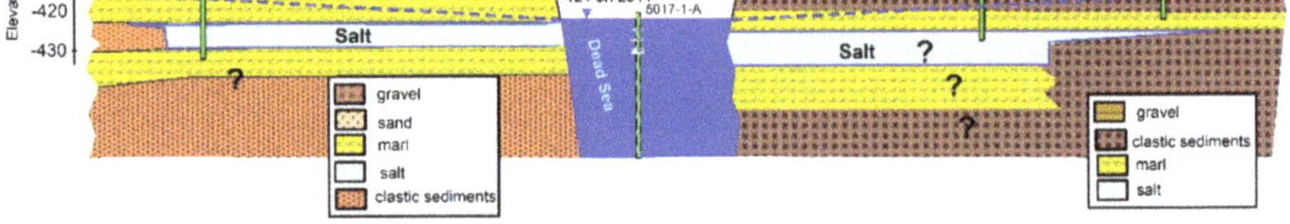

Figure 3.7. W–E generalized geological section (along line C–C′ in figure 3.1) through the southern part of the Dead Sea (after El-Isa *et al* 1995). 5017-1-A is the recent borehole drilled in the water area of the DS (after Ezersky *et al* 2013a, Coianiz *et al* 2019). (The figure from Ezersky *et al* (2013a) is published with the permission of the EEGS. All copyright privileges remain with EEGS. This material cannot be copied or used without the express written permission of EEGS.)

3.3.4 Historical fluctuations of the Dead Sea level and sinkhole problem

Analysis of the Dead Sea evolution indicates that its level has been fluctuating—heightened and lowered (Frumkin and Elitzur 2002). Two different approaches were used to reconstruct the lake-level history. One is the field approach, which uses sediments, geomorphic evidence, and archaeological remains associated with the lake level. The second is the historical approach, which looks for ancient documents about the lake levels. The historical approach includes human records before instrumentally recorded observations. Instrumental measurements of the Dead Sea started recently, less than 100 years ago (Klein 1982), and meteorological observations in the region started only slightly earlier, in the 19th century.

Another method was applied by Stein *et al* (2010). The lake fluctuated between higher and lower levels reflecting the regional hydrological regime that shifted between wet and arid conditions, respectively (Bartov *et al* 2003). During high stands and positive fresh water input, the lake deposited sequences of primary aragonite and silty–detritus, mainly quartz and calcite grains of eolian origin. During low stands, the lake deposited sequences of detrital calcites that reflect the contribution of seasonal floods and occasionally layers of gypsum, which were deposited during episodes of the lake's overturn. The water levels of Lake Lisan (figure 3.8) were reconstructed by identifying sedimentary indicators of the shorelines, which were dated by radiocarbon or U–Th methods (Bartov *et al* 2003).

Recently, the DS level was subjected to human activity. Since the early 1960s, large amounts of fresh water have been diverted from Lake Kinneret and the Jordan River, resulting in a sharp drop in the DS level at the rates of about 0.5 m yr^{-1} since the 1960s, 0.8 m yr^{-1} in 1980, and 1.0 m yr^{-1} recently (figure 3.9). Figure 3.9 shows that until the mid-1960s, the DS level dropped in both the northern basin (DS) and the evaporation ponds (EPs) at the south of the DS coast. Then the level of the EPs

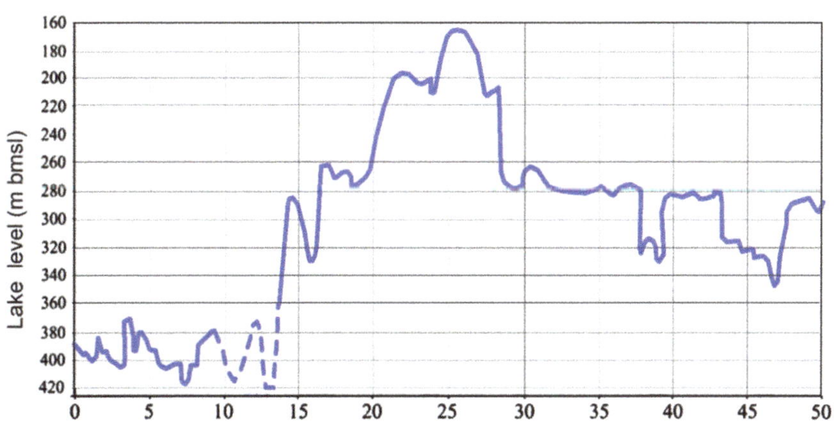

Figure 3.8. Lake level curve: lake levels during the last 50 kyr (Lake Lisan and the Holocene Dead Sea) (Stein *et al* 2010). The high and low levels reflect wet and arid regional hydrological conditions, respectively (modified by the author).

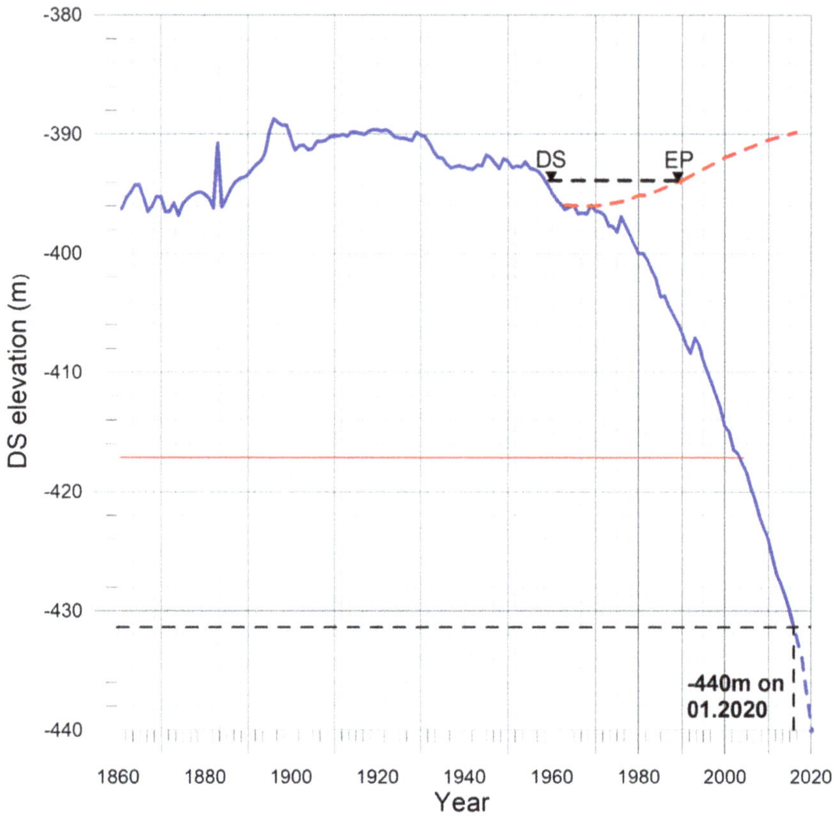

Figure 3.9. DS elevation versus time (years) variations during the last 157 years. The thick blue solid line is for DS northern basin; red dashed line is for EPs southern basin; red thin line—salt top elevation at the Nahal Hever South (NHS) site (has been generated on data of the Geological Survey of Israel).

began to increase because of pumping from the DS into the EPs (Yechieli 2000). In January 2017, the DS level reached ~431 m b.m.s.l. (figure 3.9) and continued to lower.

3.3.5 Hydrology and water use. Hydrogeological peculiarities

According to Yechieli *et al* (2006), three main aquifers draining into the DS can be distinguished in the area to the west of the DS (figure 3.10). Two of them, which extend far beyond the DS area, include the sandstone Kurnub aquifer within Lower Cretaceous rocks, and the limestone and dolomite Judean aquifer within Upper Cretaceous rocks. The third aquifer, in Quaternary alluvium, is confined to the DS transform, bounded to the west by its marginal faults. Alternating silt and gravels within the alluvial sequence divide it into several sub-aquifers with different heads and water compositions. The Quaternary aquifer derives most of its water through lateral flows across the transform faults from the Judea Group aquifer, which is recharged over the mountain terrain, some 20–50 km to the west. Direct recharge to

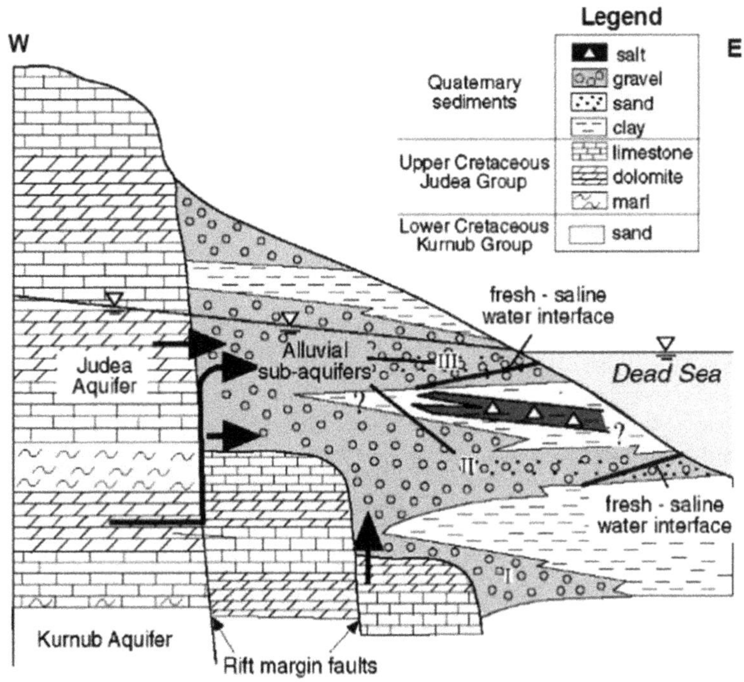

Figure 3.10. Schematic cross-section showing field relations between the different aquifers and the DS. The Kurnub Group aquifer, underlying the Judea Group aquifer, is not shown. Arrows denote possible direction of the groundwater flow (reprinted from Yechieli *et al* (1995a), John Wiley & Sons, with permission of American Geophysical Union).

the aquifer from local rainfall over the alluvial segments is negligible because of the extreme aridity of the region (Yechieli *et al* 2006).

The chemical composition of the DS water and groundwater in boreholes of northern and southern shores is presented in table 3.1 and based on GSI data described in numerous reports.

The density of the DS brine is different in the northern basin (1230 kg m^{-3}) and EP (1240 kg m^{-3}).

GSI has drilled several observation boreholes along the western DS shore that were used for water content determination and of salinity—resistivity calibration (Yechieli *et al* 2004, Ezersky and Frumkin 2017). These boreholes are located in eight sites (figure 4.62 and subsection 4.3.2).

DS water contains also a very high chloride (Cl) concentration (up to 224 g L^{-1} in the northern basin and up to 258 g L^{-1}), which constitutes 98% of the anions, with little sulfate and carbonate. Ezersky and Frumkin (2021) have shown that the dependences of concentrations of sodium chloride (NaCl) versus chloride (Cl) and total dissolved solids (TDS) versus Cl are direct practically functional (figure 3.11).

Note some important features of interrelations presented in figure 3.11:

- Functional, practically correlation between presented concentrations has determination coefficients of 0.997 (figure 3.11(a)) and 0.999 (figure 3.11(b));

Table 3.1. Chemical composition of groundwater from two wells, DS and EP. Reprinted with permission from Yechieli *et al* (2004) with permission from GSI.

Well	Depth	Na (g L^{-1})	K (g L^{-1})	Ca (g L^{-1})	Mg (g L^{-1})	Sr (g L^{-1})	Cl (g L^{-1})	SO_4 (g L^{-1})	Br (g L^{-1})	TDS (g L^{-1})	NaCl (g L^{-1})	Other solids
Mn2*	30	75.7	4.6	11.3	22.8	0.18	202.0	1.10	3.2	320.88	277.7	43.18
Mn4*	30	52.8	7.12	14.75	34.75	0.25	216.7	–	3.85	330.22	269.6	43.18
Mn4	38	53.1	6.45	13.95	34.5	0.235	207.9	–	3.60	319.7	261.0	58.70
EB2*	20	26.0	8.4	19.9	50.8	0.36	229.8	–	5.63	340.89	255.8	85.09
EB2*	28	25.7	8.4	19.8	50.4	0.36	230.7	–	5.65	341.11	256.5	84.61
EB2*	40	30.6	6.7	18.1	47.6	0.32	223.9	–	5.43	332.65	254.5	78.15
DS**		40.1	7.65	17.2	44.0	–	224.9	0.45	5.30	339.6	265.0	74.6
EP**		16.7	11.1	25.2	62.3	–	258.9	0.29	7.47	381.96	275.6	106.36

*Compilation from Yechieli *et al* (2004).

** Dead Sea and evaporation ponds water for comparison.

Comment: Mn denotes the Mineral Beach (northern) site, EB2—Ein Boqeq (southern site), DS—Dead Sea, and EP—evaporation ponds.

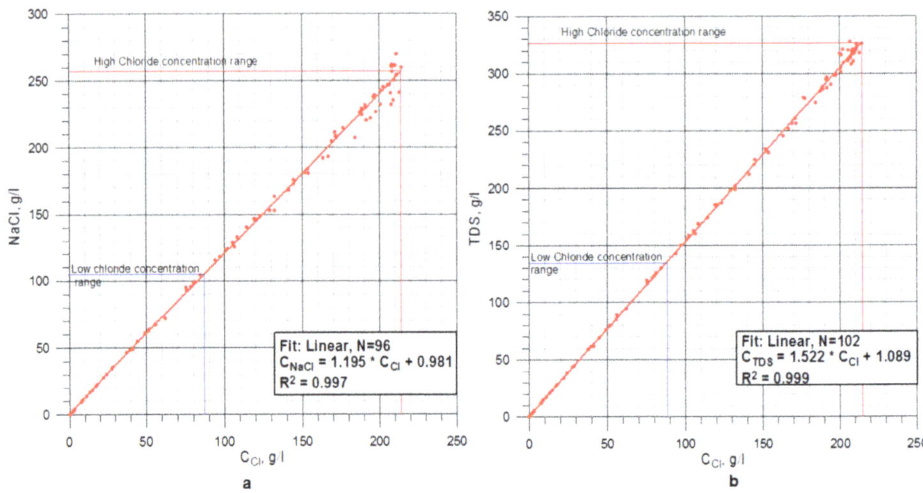

Figure 3.11. Correlation plots between (a) chloride concentration (C_{cl}) and NaCl one and (b) C_{cl} and TDS. Limits of high and low concentrations C_{cl} and C_{NaCl} correspond to the limits of graphs 1a and 1b, respectively, in figure 4 of Ezersky and Frumkin (2021). These limits determining the measurement ranges are characterized as C_{cl}: 87.7 g L^{-1} and 214 g L^{-1}, NaCl: as 106 g L^{-1} and 258 g L^{-1}, respectively, and TDS as 135 g L^{-1} and 328 g L^{-1}. (Reprinted from Ezersky and Frumkin (2021), Copyright 2021, with permission from Elsevier.)

- The limits (maximum concentrations) are determined by a range of parameters where a direct correlation between them exists. Beyond these limits correlations became negative;
- At very high concentrations the scattering of points increases because of the influence of electrolyte density (Serra 1984, p 9, figure 1-3).

Thus, interrelations $C_{Cl} - \rho_x$, $C_{NaCl} - \rho_x$, $C_{TDS} - \rho_x$ are, mutually linked which allows recalculation of available interrelations between chloride concentration and bulk (or groundwater) resistivity (Goldman *et al* 1991, Levi *et al* 2008, Ezersky and Frumkin 2017) to these between bulk resistivity and NaCl or TDS).

Chloride concentration is, therefore, the main parameter used to characterize the groundwater salinity (Yechieli 2000).

3.3.6 The thermal regime in the Dead Sea region and its geological-environment influence

3.3.6.1 Common characteristics of the thermal regime

Undoubtedly, the thermal regime plays a significant role in the intricate complex mechanism of sinkhole generation in the Dead Sea (DS) region. Therefore, in this section, some significant thermal characteristics of this area will be discussed. The generalized thermal conductivities of the most typical rocks and environmental features of this region are presented in table 3.2.

It is interesting to note that the increase of the DS brine salinity (during the last decades) caused its thermal conductivity to decrease (Warren 2020).

Investigations of geothermal fields in the DS basin been carried out by Rosenthal and Eckstein (1968), Ben-Avraham *et al* (1978), Feinstein (1987), Bein and Feinstein (1988), Eppelbaum *et al* (1996), Ivanov *et al* (2002), Eppelbaum and Pilchin (2006), and Nehorai *et al* (2009). These studies indicate that the present-day average heat flows and geothermal gradients within the Dead Sea region are relatively low: 30–35 mW m^{-2} (Ben-Avraham *et al* 1978) and 20 °C km^{-1}–23 °C km^{-1}, respectively (Feinstein 1987, Bein and Feinstein 1988, Eppelbaum *et al* 1996). The latter value increases in separate

Table 3.2. Thermal conductivity of some rocks and environmental features typical for the DS transform zone (revised after Eppelbaum *et al* (2014) with permission of Springer).

Rock	Thermal conductivity (W m^{-1}·°C)	Environmental feature	Thermal conductivity (W m^{-1}·°C)
Dry soil	0.3	Air	0.26 (25 °C)
Sand	0.2–2.9 depending on water presence	Isolated cavity (sinkhole)	0.5
Gravel	0.3–0.5	Water	0.59 (25 °C)
Gypsum	0.5–1.5	Dead Sea brine	0.61 (25 °C)
Limestone	2.5–3.2	Bitumen	0.2
Rock salt	5.3–7.2	Wood	0.09–0.14
sandstone	2.4–2.5	Tuff	0.2–0.5
Marls	1.5–1.54	Oil	0.12–0.15
Shales	0.25–1.08	Gas	0.01–0.015
Siltstone	1.55–1.60	–	–
Clay	0.25–1.08	–	–
Basalt	1.5–2.2	–	–

wells up to 28 °C km^{-1} (e.g., Massada-1 well) (Levitte and Olshina 1985). Low heat flows occur throughout the DS area and are reflected in the small appearance of volcanic activity in this tectonically active zone (Garfunkel *et al* 1981).

For processing and analysis of temperature data, the records observed in more than ten boreholes (figure 3.12) were engaged.

The temperature distribution versus depth for nine boreholes disposed in the southern DS basin (figure 3.13) testifies the specific heterogeneity of these data.

A study of the southern DS Basin using methods of determining paleotemperatures indicates that the geothermal gradient in the Mesozoic interval did not exceed

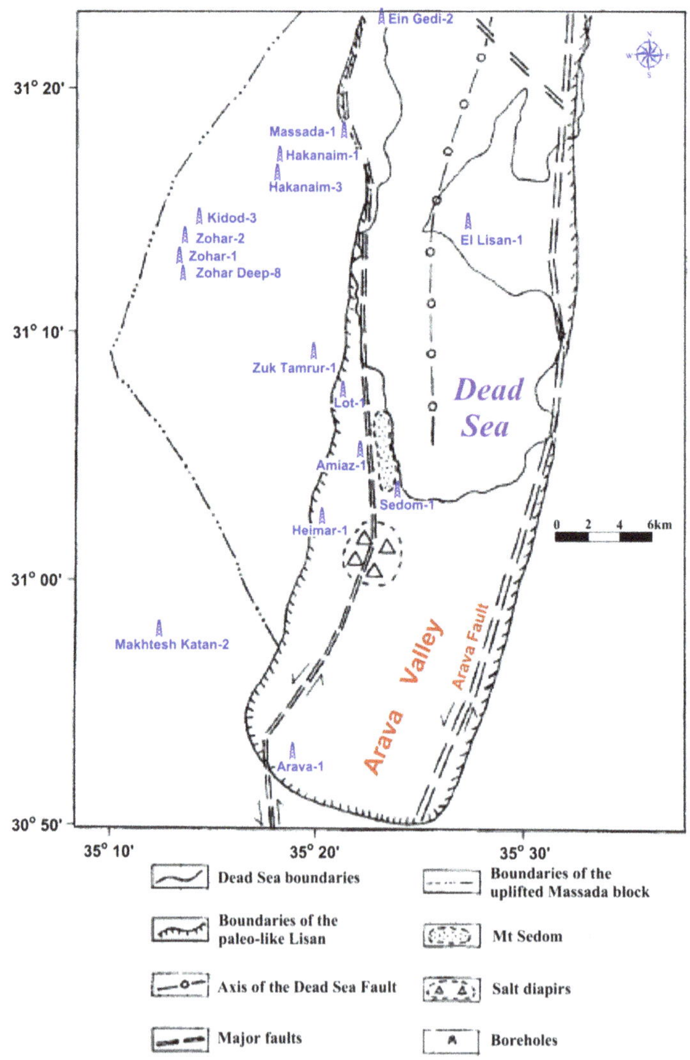

Figure 3.12. Areal map of the study area with location of boreholes where thermal observations were carried out (adapted from Eppelbaum *et al* (1996), John Wiley & Sons).

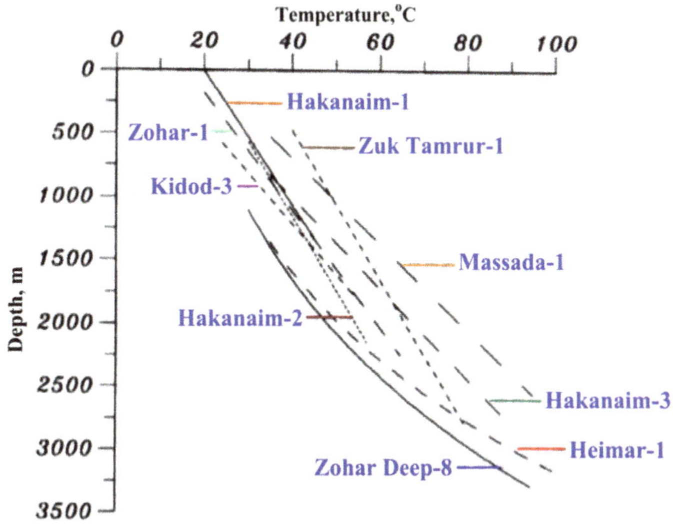

Figure 3.13. Temperatures versus depth for nine boreholes in the southern DS region (modified from Eppelbaum *et al* (1996), John Wiley & Sons).

20 °C km^{-1}–23 °C km^{-1} before the Cretaceous. However, it increased significantly up to 35 °C km^{-1}–38 °C km^{-1} during the Cretaceous and then dropped to 20 °C km^{-1} in the Miocene (Feinstein 1987, Bein and Feinstein 1988).

3.3.6.2 Calculation of horizontal temperature derivative

Calculation of horizontal geothermal gradient (HGG) is realized by the following trivial formula:

$$\frac{T_1 - T_2}{\Delta x} \tag{3.1}$$

where T_1 and T_2 are the temperature observations carried out in the neighboring boreholes (at the same horizontal level), and Δx is the horizontal distance between two of these measurements.

Figure 3.14 displays the calculated HGG between four pairs of boreholes at various depths. The more complete data are presented in table 3.3. The HGGs are generally low in the study area. These values do not usually exceed 1 °C km^{-1} (table 3.3), except for the main DS transform zone. In the southern DS Basin, the highest temperatures at a depth of 1500–2000 m were found in the Lot-1 well, and HGGs increase towards this well. From the north, the HGG increases from 0.21 °C km^{-1} to 1.29 °C km^{-1}; from the south, 0.62 °C km^{-1}–1.3 °C km^{-1}; and from the south-east by an average value of 1.6 °C km^{-1}.

These high HGGs are caused by the numerous inconsequential faults which are associated with the main DS Transform zone. Around the Hakanaim-3, Lot-I, and Massada-1 boreholes, the HGG values vary from 0.69 °C km^{-1} to 1.2 °C km^{-1}. The northern Arava Valley is characterized by low HGG values (0.23 °C km^{-1}–0.46 °C km^{-1}),

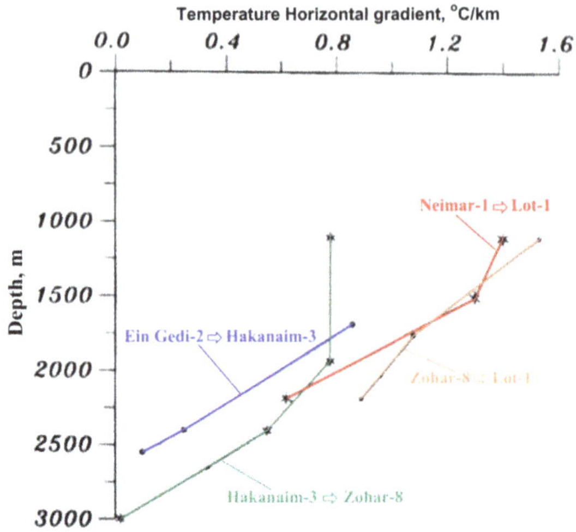

Figure 3.14. HGGs for several boreholes in the southern DS region (modified from Eppelbaum *et al* (1996), John Wiley & Sons).

and low values (0.03 °C km^{-1}–0.78 °C km^{-1}) were also observed some distance from the main graben-bounding fault. At the western flank of the DS S transform zone, HGG values decreased with the depth. Thus, in near-surface layers (bodies) the thermal regime is more active and the thermal homogeneity is greater (figure 3.14). The HGG in the fault zone is usually 10% of the vertical gradient, whereas within the stable blocks it is usually around 4%.

3.2.6.3 Calculation of first and second vertical temperature derivatives

The first vertical derivative of temperature (vertical geothermal gradient (VGG)) is calculated by the well-known formula (e.g., Eppelbaum *et al* (2014)):

$$\Gamma = \frac{dT}{dH} = \frac{T_2 - T_1}{H_2 - H_1} \tag{3.2}$$

where T is the temperature (in °C or K), H is the depth (in m or km), and indexes '1' and '2' designate temperatures and depths in points 1 and 2 taken vertically second below first. The physical dimension of the VGG is °C m^{-1} (or °C km^{-1}).

Some calculated values of the VGG are compiled in table 3.4. Some high VGG values discovered in the upper part of the sedimentary cover of the DS region (see table 3.3), are associated with hydrothermal activity. This parameter is important for a better understanding of sinkholes generation.

Throughout the sedimentary sequence in this area, the VGG values are low, despite the high values in the near-surface intervals. In the Lot-1, Massada-1, Heirnar-1, Arniaz-1, Zohar Deep-8, Hakanairn-1, Arava-l, and Ein Gedi-2 wells, the overall VGG values do not exceed 20 °C km^{-1}, because of the low VGG values in the deeper horizons.

Table 3.3. Horizontal geothermal gradients in the DS region. Reprinted from Eppelbaum et al (2014) with permission of Springer.

Boreholes	Depth (m)	The horizontal geothermal gradient (°C km^{-1})
Massada-1 ⇒ Lot-1	1100	0.47
	2180	0.29
Heimar-1 ⇒ Lot-1	1100	1.30
	1500	1.20
	2180	0.62
Hakanaim-3 ⇒ Massada-1	1700	0.96
	2150	0.69
	2380	0.55
	3000	0.3
Zohar Deep-8 ⇒ Lot-1	1100	1.42
	1785	1.06
	2180	0.89
Hakanaim-3 ⇒ Lot-1	1500	1.27
	2180	0.90
Ammiaz-1 ⇒ Sedom-1	2300	0.01
Arava-1 ⇒ Sedom-1	2300	0.23
Arava-1 ⇒ Ammiaz-1	2650	0.35
Amiaz-1 ⇒ Heimar-1	2370	0.29
Heimar-1 ⇒ Arava-1	2370	0.46
Sedom-1 ⇒ Heimar-1	2370	0.46
Massada-1 ⇒ Ein Gedi-2	1670	0.60
	2370	0.21
Hakanaim-3 ⇒ Ein Gedi-2	1670	0.81
	2370	0.24
	2550	0.10
El Lisan-1 ⇒ Lot-1	2180	0.92
Ammiaz-1 ⇒ El Lisan-1	3580	0.01

In the Lot-1 well, the averaged VGG value in the Triassic sequence (1151–1932 m) is 16.2 °C km^{-1}: 12 °C km^{-1} in the Late Triassic (composed mainly of dolomites and shales); and 23.1 °C km^{-1} in the Middle and Early Triassic (composed mainly of limestones, shales, and sandstones). In the Zohar Deep-8 borehole, the VGG varies from 14.3 °C km^{-1} in the Cretaceous to Upper Jurassic interval (limestones, marls, shales, and sandstones), to 25 °C km^{-1} in Middle-Lower Jurassic carbonates, and 28.5 °C km^{-1} in the Upper Triassic (1850–2000 m; dolomites and shales). At greater depths, the gradient decreases to 14.5 °C km^{-1} in Middle Triassic shales but increases again to 27 °C km^{-1} in the Permian (mainly limestones and shales) to Precambrian (extrusives?) interval. Based on a comparison of the aforementioned data with the paleotemperature determinations (Feinstein 1987, Bein and Feinstein 1988) for the

Table 3.4. Values of the VGG in some DS boreholes. Reprinted from Eppelbaum et al (2014) with permission of Springer.

Boreholes	Depth interval	Vertical geothermal gradient (°C km^{-1})
Massada-1	0–353	52.4
	353–2750	22.0
Lot-1	0–1050	37.6
Heimar-1	0–600	37.3
	600–2370	19.7
Ammiaz-1	2020–4600	21.9
Arava-1	0–417	62.8
	417–2653	12.4
Ein Gedi-2	0–1670	29.0
	1670–2550	17.7
Sedom-1	0–300	56.5
	300–2308	18.5
Zohar Deep-8	1120–3560	26.7

Amiaz-1, Zohar Deep-8, and Zuk Tamrur-1 boreholes, it can be shown that the VHH values in the southern DS Basin have decreased continuously throughout post-Pliocene time (Wakshal and Cohen 1993). The gradient in Pliocene saliferous beds varies from 20.8 °C km^{-1} (El Lisan-1 well) to 13.3 °C km^{-1} (Sedom-1 well). At greater depths (2300 m), the VGG increases to 60 °C km^{-1} in the Sedom-1 well.

The second vertical derivative of temperature is derived from by the following way (Eppelbaum *et al* 2014):

$$\frac{d}{dH}\left(\lambda\frac{dT}{dH}\right) + \kappa = 0, \tag{3.3}$$

where T is the temperature (°C), H is the depth (m), λ is the coefficient of thermal conductivity (W $m^{-1}\cdot$°C), and κ is the total heat generation (mW m^{-3}). The physical dimension of the second vertical derivative is °C m^{-2} (or °C km^{-2}).

For homogeneous (quasi-homogeneous) rocks, where λ is approximately constant, we can present values of the second vertical derivative in the following form:

$$\frac{d^2T}{dH^2} = -\frac{\kappa}{\lambda}, \tag{3.4}$$

Since the λ value of is positive, the total heat generation κ will be positive when $(-\kappa/\lambda)$ is negative (indicating heat-radiating horizon). Obviously, κ will be negative when $(-\kappa/\lambda)$ is positive (indicating heat-absorbing strata). Based upon the $(-\kappa/\lambda)$ relationship, the second vertical temperature derivatives can be determined. Negative or positive values allow the recognition of the geothermal behavior of any stratum under study. This gives the possibility to distinguish whether a considered depth interval is heat-absorbing or radiating.

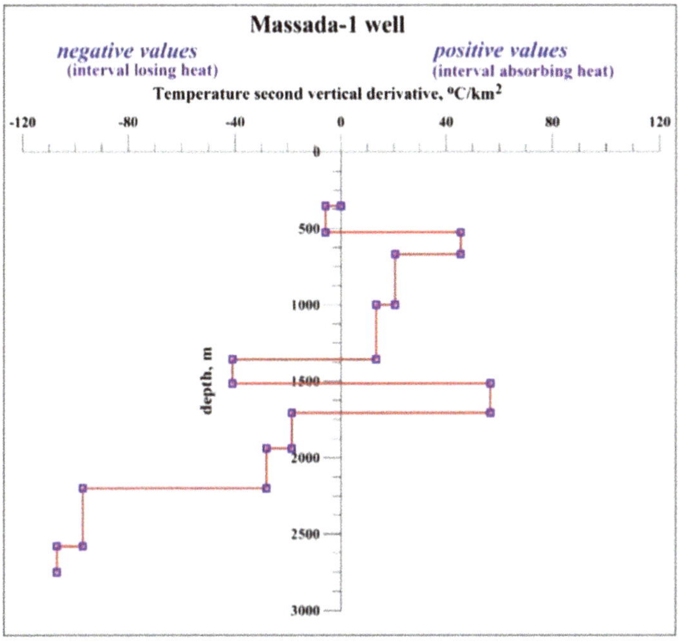

Figure 3.15. The second vertical temperature derivatives for the Massada-1 borehole, southern DS basin. Heat-absorbing (negative) and heat-radiating intervals (positive) are selected.

Calculated second vertical temperature derivatives for the Massada-1 borehole are displayed in figure 3.15. An extended data set (first and second vertical temperature derivatives) for the Lot-1, Heimar-1, Hakanaim-1, and Massada-1 wells are compiled in table 3.5.

3.3.6.4 The influence of thermal regime to the fluid migration

The influence of the thermal regime on the distribution of pressure in hydrothermal fluids can be expressed by the following equation (Eppelbaum *et al* 2014)

$$P = P_0 + \Delta T \frac{\alpha}{\beta} - \frac{1}{\beta} \frac{\Delta T}{V_0} \tag{3.5}$$

where P is the pressure of the fluid (MPa); P_0 is hydrostatic pressure (MPa), a is the fluid expansion ratio (dimensionless), β is the fluid compression ratio (dimensionless), ΔT is the surplus temperature (°C), V_0 is the initial volume of the fluid (m^3), and ΔV is the initial volume of the surplus fluid during the thermodynamic process (m^3). Changes in the $\Delta V/V_0$ ratio are of the most geological interest.

Based on equation (3.5), we can write expressions for the horizontal fluid pressure gradient, dP/dx, of hydrodynamically unified layers (targets):

$$\frac{dP}{dx} = \frac{\alpha}{\beta} \frac{dl}{dx} \tag{3.6}$$

Table 3.5. Calculated average first vertical derivatives (geothermal gradients) and second vertical temperature derivatives in the Dead Sea boreholes.

Borehole	Depth (m)	First vertical derivative (°C km^{-1})	Second vertical derivative (°C km^{-2})
Lot-1	0–1050	37.6	−28.5
	1050–1785	12.2	
			22.4
	1050–2180	23.1	
Heimar-1	0–600	38.9	
			−8.6
	600–1000	32.2	
			0
	1000–1180	32.2	
			−58.7
	1180–1490	6.1	
			33.0
	1490–2370	25.7	
Hakanaim-1	0–250	40.0	
			−64.0
	250–500	8.0	
			26.5
	500–650	13.3	
			44.7
	650–800	20.0	
			0
	800–1000	20.0	
			29.3
	1000–1083	24.1	
Massada-1	0–353	52.4	
			−5.8
	353–527	5.8	
			45.6
	527–670	13.0	
			20.6
	670–1001	18.9	
			13.4
	1001–1356	23.5	
			−40.9
	1356–1516	15.0	
			56.8
	1516–1708	25.0	
			−18.4
	1708–1940	21.1	
			−28.0

(Continued)

Table 3.5. (*Continued*)

Borehole	Depth (m)	First vertical derivative (°C km^{-1})	Second vertical derivative (°C km^{-2})
	1940–2200	14.2	
			96.9
	2200–2580	45.3	
			−106.9
	2580–2750	15.9	

The heat-bearing fluids can bring about shale diapirism in the rocks through which they flow. The transfer of heat to salt bodies (thermal conductivity of salt exceeds this parameter for other geological targets in the DS basin—see table 3.2) is related to the appearance of salt diapirs, which are widely presented in the study area.

The horizontal migration of fluids from the eastern flank of the study area, probably produced by volcanism and the formation of basaltic intrusions, may have contributed to a buildup of surplus temperatures. Basaltic bodies are very extensive on the eastern margin of Lake Lisan (Begin *et al* 1974), and Neogene and Quaternary basaltic eruptions are known in the east and south-east of the DS.

According to equations (3.5) and (3.6), the heat-absorbing process may lead to a sudden increase in the pressure of the interstitial fluids. This factor causes fluids to migrate to the east and west of the north–south trending basalt outcrops on the eastern margin of the DS transform zone. These fluids are composed mostly of overheated waters with a high salt dissolving capacity.

3.2.6.5 Quantitative analysis of temperature anomaly in the southern Dead Sea
Ben-Avraham and Ballard (1984) observed thermal anomaly in the northern DS Basin. The amplitude of the anomaly exceeds 0.2 °C (accuracy of these observations was about 0.01 °C; so the amplitude exceeds the survey accuracy 20 times). The form of this anomaly suggested that the anomalous target (or its upper surface) had enhanced thermal conductivity, and could be approximated by a horizontal circular cylinder (HCC). The application of quantitative methods (description of the methods is described in detail in Eppelbaum *et al* (2014)) enabled the determination of the position of the HCC center (figure 3.16). As can be seen from this figure, the depth is about 60 m. Here both the characteristic point method (parameters d_1 and d_2) and the tangent method (parameters d_3 and d_4) (Eppelbaum *et al* 2014) were employed. Parameter d_3 is the difference in the abscissa of the points of intersection of an inclined tangent with horizontal tangents on one branch; parameter d_4 is the same on the other branch; d_3 is selected from the plot branch with the conjugated extrema).

It was suggested that the anomaly was generated by the upper surface of a small salt dome (numerous salt domes are known in this area). The temperature moment (see section 2.2.15) was calculated to be approximately 9 °C m^{-1}.

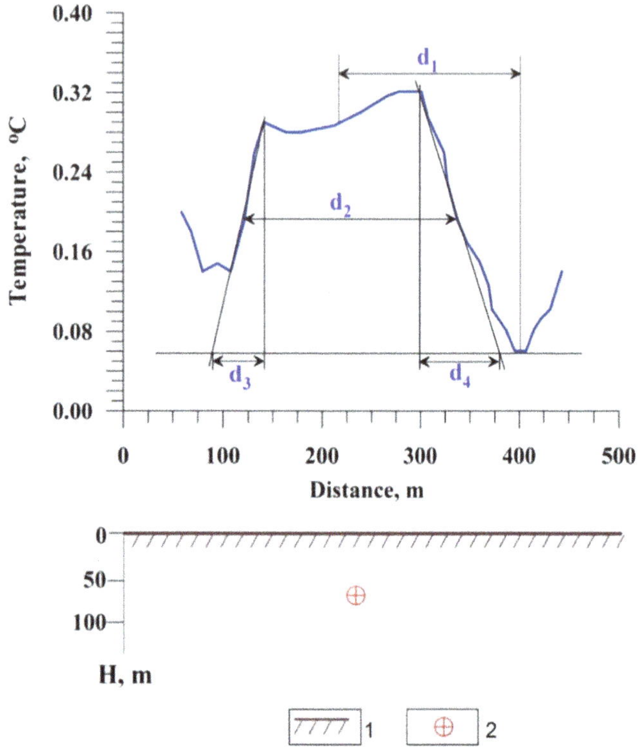

Figure 3.16. Quantitative interpretation of temperature anomaly in the northern DS Basin (observed temperature curve after Ben-Avraham and Ballard (1984), interpretation after Eppelbaum *et al* (1996)). (1) Sea bottom; (2) position of the center of a horizontal circular cylinder (modified from Eppelbaum *et al* (1996), John Wiley & Sons).

3.2.6.6 Estimation of the Dead Sea water flow geodynamics behavior

The Dead Sea (DS) is a unique terminal lake with high concentrated solution (brine) located at the lowest place on the Earth's surface. It has the highest surface temperature and density among the Earth's large water bodies and its level is currently dropping by approximately 1 m yr^{-1} (Ezersky *et al* 2021). The air temperatures in summer can reach 50 °C, and water surface temperature—36 °C (e.g., Nehorai *et al* 2009, Arnon *et al* 2013). Besides the high temperatures of the media, the rate of evaporation is significantly high.

The density of water in the vertical direction varies from 1150 kg m^{-3} at the DS surface up to 1430 kg m^{-3} at a depth of 310–330 m. This density gradient exceeds the normal water density gradient in the Earth's oceans and seas by more than 10^2. Thus, the DS Basin may be considered a distinctive native polygon of the vertically stratified liquid (Eppelbaum *et al* 2010). The most significant source of this phenomenon is temperature and density gradients based on the constant essential inflow of a mineral substance from the great depths (apparently, from the upper mantle). To analyze the water flows geodynamics in the DS Basin, analytical expressions for the motion of stratified liquid with different initial boundary values can be used. Given the very high degree of change of the physical, chemical, and environmental features of this unique basin, the development of a geodynamic predictable model of the area is extremely important. It should be noted that density practically always has a reliable correlation with temperature (Eppelbaum *et al* 2010).

3.2.6.6.1 What awaits the Dead Sea in the future?

At present, the distinctive DS brine is intensively utilized for extracting useful components contained in the liquid. This industrial activity, combined with the extremely arid climate has caused a significant drop in the sea level (Yechieli *et al* 1998, Gavrieli and Oren 2004) and the spontaneous development of hundreds of sinkholes on the DS coast (Ezersky *et al* 2017, 2021). For the DS, the conventional mathematical model describing the sea behavior consists of a set of algebraic and a system of differential equations (e.g., Asmar and Ergenzinger (2002)). The state of the sea is found by solving a system of ordinary differential equations consisting of a system of mass balance equations for the water and eight major chemicals in the sea. The law of conservation is accounted for implicitly through evaporation calculations. However, this approach is unable to take into account all the peculiarities of the unique DS Basin. Interesting results were obtained by Bubnov and Linden (1998) for a homogeneously rotating linear density-stratified fluid. However, this approach demands a significant evaluation.

We suggested applying a more precise physical–mathematical model which has implications for forecasting environmental scenarios for the DS and for estimating the possible ecological impact of excavation of a canal between the Red and Dead seas. The essential differences in the chemical composition of water from these seas shown in table 3.6 indicate that this mixing may be an unpredictable one (figure 3.17).

3.2.6.6.2 Proposed physical–mathematical model

Let us consider the motion of stratified, rotating, and compressible liquid (SRCL) in the Cartesian coordinates (x_1, x_2, x_3) which is rotating together with the SRCL. Let us assume that SRCL rotates around axis Ox_3 and Coriolis vector $\mathbf{f} = (0, 0, f)$, where f is the double angle velocity of rotation. The SRCL is stratified along the axis Ox_3; i.e., its

Table 3.6. Some typical chemical analyses of the Dead Sea (DS) and the Red Sea (RS) (after Abu-Khader (2005), copyright (2005) with permission from Elsevier, Ravizky and Nadav (2007), with supplements).

Element	DS concentrations (mg L^{-1})	RS concentrations (mg L^{-1})	Ratio, DS/RS
Chloride	224 000	23 000	9.74
Magnesium	44 000	1500	29.33
Sodium	40 100	12 500	3.21
Calcium	17 650	500	35.3
Potassium	7650	500	15.3
Bromide	5300	70	75.71

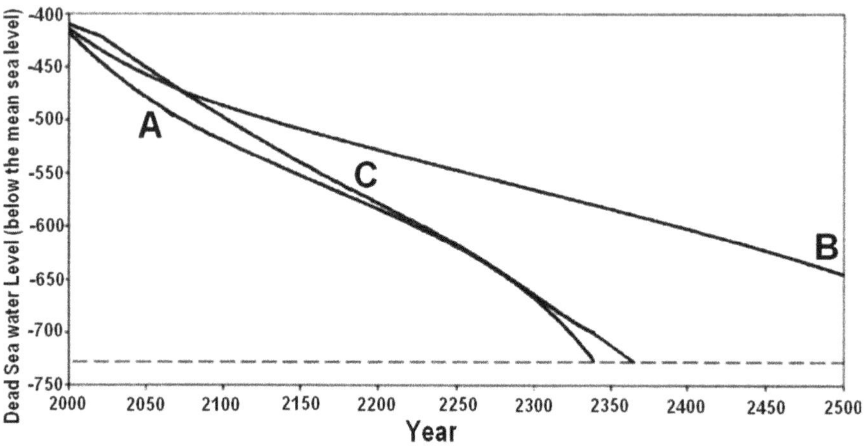

Figure 3.17. Predicted water levels in the DS up to the year 2500. (A) Scenario 1 shows a continuation of the current conditions. (B) Scenario 2 shows cessation of industrial extraction activity. (C) Scenario 3 shows simplified climate change (after Asmar and Ergenzinger (2004)). (Reprinted from Asmar and Ergenzinger (2004), with permission of AGU.)

density in undisturbed conditions depends on x_3, $\rho_0 = \rho_0(x_3)$. We assume below that $\rho_0(x_3) = Ae^{-2\beta x}$, $A > 0$, $\beta > 0$. Small SRCL motions under the gravity field influence without any external forces may be defined as (Yakubov and Eppelbaum 2005):

$$\begin{cases} \rho_0(x_3)\dfrac{\partial \mathbf{v}}{\partial t} + \rho_0(x_3)[\mathbf{f}, \mathbf{v}] + \nabla p + \mathbf{e}_3\rho_1 g = 0 \\ \dfrac{\partial \rho_1}{\partial t} + (\mathbf{e}_3 \cdot \mathbf{v}) + \rho'_0(x_3) + \rho_0(x_3)\operatorname{div}\mathbf{v} = 0 \\ \dfrac{\partial}{\partial t}\rho_1 = \dfrac{1}{c^2}\dfrac{\partial}{\partial t}p + \rho_0(x_3)\omega_0^2(x_3)\dfrac{(\mathbf{e}_3 \cdot \mathbf{v})}{g} \end{cases} \tag{3.7}$$

where g is the gravity acceleration, $\mathbf{v} = (v_1,v_2,v_3)$ is the vector of the SRCL particle motion, ρ_1 is the change in SRCL density caused by its motion, c is the acoustic velocity, p is the dynamic pressure and $\mathbf{e}_3$ is the ort of axis Ox_3 and $[\mathbf{f},\mathbf{v}]$ denoting the vector product between these two vectors. Value $\omega_0^2(x_3) \geqslant 0$ is the quadrate of Vyasyalay–Brent frequency. The condition $\omega_0^2(x_3) \geqslant 0$ indicates an absence of convective motion and stability of the density distribution $\rho_0(x_3)$ in the SRCL. The physical parameters such as the liquid density, acoustic velocity, and pressure depend on the temperature. Thus, the temperature presented in (3.7) is in an implicit form.

After several awkward transformations, equation (3.7) was reduced to the following type:

$$\frac{\partial^2}{\partial t^2}\left[\frac{\partial^2 u}{\partial t^2} - c^2\Delta_3 u + (f^2 + \beta^2 c^2)u\right] - \left[\omega_0^2 c^2 \Delta_2 u + f^2 c^2 \frac{\partial^2 u}{\partial x_3^2} - \beta^2 c^2 f^2 u\right] = 0 \tag{3.8}$$

where u is the scalar function on which the unknown functions $\mathbf{v}$, ρ_1 and p depend.

Let us consider the following problem

$$\begin{cases} L(t, D_t, D_x)u(t, x) = \dfrac{\partial^2}{\partial t^2}\left[\dfrac{\partial^2 u(t, x)}{\partial t^2} - \sum_{i,j=1}^{3}\dfrac{\partial}{\partial x_i}\left(A_{ij}(t, x)\dfrac{\partial u(t, x)}{\partial x_j}\right)\right. \\ \qquad \left. + \sum_{i=1}^{3} B_i(t, x)\dfrac{\partial u(t, x)}{\partial x_j} + C(t, x)u(t, x)\right] \\ \qquad + \sum_{i,j=1}^{3}\dfrac{\partial}{\partial x_i}\left(a_{ij}(t, x)\dfrac{\partial u(t, x)}{\partial x_j}\right) + \sum_{i=1}^{3} b_i(t, x)\dfrac{\partial u(t, x)}{\partial x_j} \\ \qquad + d(t, x)u(t, x)f(t, x), \ (t, x) \in [0, T] \times \Omega = 0 \end{cases} \tag{3.9}$$

where t is the time, A_{ij}, B_i, C, a_{ij}, b_i, d are sufficiently smooth functions (for an application, see equation (3.8)), depending on c, f, β and ω_0.

$$u(t, x)\,|_{x\in\partial\Omega} = 0, \quad t \in [0, T] \tag{3.10}$$

$$\left.\frac{\partial^k u(t, x)}{\partial t^k}\right|_{t=0} = u_k(x), \ k = 0, \ldots, 3, \ x \in \Omega \tag{3.11}$$

Equations (3.10) and (3.11) present the Dirichlet boundary condition and the Cauchy initial conditions, respectively. Thus, equations (3.9)–(3.11) can be utilized as a basis for the development of a comprehensive geodynamical model of the SRCL in the DS Basin (Yakubov and Eppelbaum 2005).

In the DS Basin (at the sea surface and various depths) the following physical parameters were observed: temperature, gravity acceleration, water density, pressure, and acoustic velocity. Modern processing schemes can extend analytically continuation (downward, upward and sideward) of the physical observations to areas not covered by these measurements. Thus, we can create a solvable system of differential equations and exploit this for creating a precise physical–mathematical–geological model of the DS Basin. This model may be used not only for geodynamic predictions but also for the projected canal between the RS and DS. Besides, this theoretical model may be used for retrieving free electrochemical energy released from the unique DS Basin (Eppelbaum *et al* 2010).

3.3.7 Dead Sea salt

The salt of the DS area, its origin, geotechnical and geophysical properties are worldwide described in scientific publications (Ezersky 2006, 2008, Frydman *et al* 2008, 2014, Stein *et al* 2010, Ezersky *et al* 2011, 2017, Ezersky and Livne 2013, Ezersky and Goretsky 2014, 2015, Ezersky and Legchenko 2014).

The halite is usually perceived as rock salt. DS salt is of an ~10 000–11 000 year-old salt layer buried at a depth of 20–50 m below the surface. The unit reaches a thickness of 4–30 m (Yechieli *et al* 2006). The DS halite layers were deposited during the Earlier Holocene, because of the emergence of extremely arid conditions in the shrinking phase between the Lisan Lake and the present DS (Stein *et al* 2010).

The salt was formed within the DS shores area, and its boundary extends west of the existing basin in Israel and east of that in Jordan. Ezersky and Frumkin (2013) have proposed a model of DS salt layer formation (see also sections 3.6.3 and 3.6.4). Stein *et al* (2010) explained varying salt layers elevations in Israel from site to site by heterogeneous descending of layers along faults. Figure 3.18 shows the salt cores extracted from the boreholes. The drilling procedures for extracting salt cores from boreholes in the western DS basin (Israel) are described in Ezersky and Goretsky (2014). In continuation, a brief review of DS salt (halite) characteristics is given.

Weisbrod *et al* (2012) have classified salt types extracted from boreholes. Following salt texture, color, and morphologies two groups of salt were selected on the western shores of the DS. The first is grey coarse 'crystalline transparent salt' (CTS) extracted from a depth of 22 m of the borehole MN-5E (figure 3.18(a)). The second one is fine crystalline 'white massive salt' (WMS) extracted from a depth of 30 m of borehole EB-3E (figure 3.18(b)) (see figure 3.1 for both site locations).

The CTS salt can be found throughout the whole coastal area of the northern DS basin and by places in the EP (Yechieli *et al* 2004). The WMS salt has been found in EP (the southern DS basin) in the Ein Boqeq site and along with the DS dam 5/8 (Charrach *et al* 2007, Frydman *et al* 2008).

Recently, salt layers were discovered in the northern DS basin area during deep drilling. This unit is also located within the Ze'elim Formation at a depth range of 40–100 m (Coianiz *et al* 2019). The DS salt layers are very hydraulically conductive and water-saturated with the DS brine of very high salinity containing the total dissolved solids (TDS) of 340 g L^{-1} in the northern basin up to 380 g L^{-1} in EP (the southern basin) (Yechieli 2000, Frydman *et al* 2008). The geotechnical and geophysical properties of the DS salt will be presented in table 4.14 (section 4.11.4).

a

b

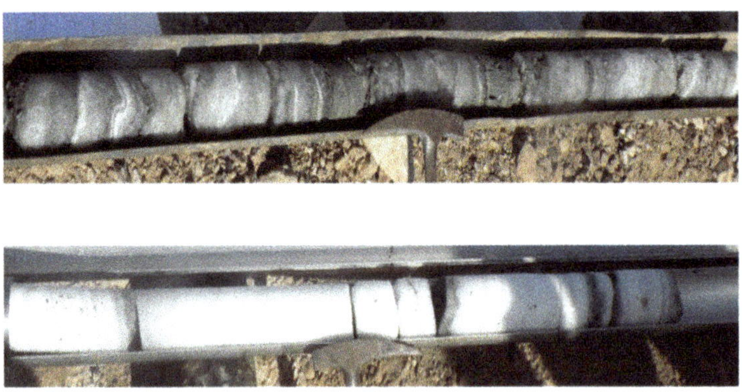

Figure 3.18. Salt columns extracted from the boreholes: (a) CTS (gray coarse) from MN-5E and (b) WMS (fine crystalline) from EB-3 borehole (salt notations are after Weisbrod *et al* (2012)). (Reprinted from Ezersky and Goretsky (2014), copyright 2014, with permission from Elsevier.)

3.4 Dead Sea sinkholes. Dynamic of sinkhole appearance

The DS coastal areas have been dramatically hit by the sinkhole phenomenon occurring since the late 1990s (figures 3.1) (Salem 2020). Most researchers relate the sinkhole phenomenon with the formation of dissolution cavities within the salt layers with the consequential gradual collapse of surface sediments (Ezersky 2006, Yechieli *et al* 2006, Frumkin 2013). The continuous drop of the DS level at a rate of ~1 m yr^{-1} is generally proposed as the main triggering factor (Abelson *et al* 2017, Nof *et al* 2019) (figure 3.19). The lowering water table induces a gradual collapse of shallow unconsolidated sediments overlying buried cavities formed within the salt layer up to 20 m thick, which exist in the shore area at a depth of 25–50 m from the surface and initiated the salt dissolution process (Frumkin and Raz 2001, Yechieli *et al* 2002). The lowering of the DS is accompanied by a corresponding lowering of the groundwater level owing to a good hydraulic connection between the DS and the western aquifer system of the DS shore. It led to the migration of the above-mentioned fresh or unsaturated (concerning Cl^-) groundwater into the coastal area and to their contact with the buried salt layer. For a long period, these salt layers were surrounded by very saline groundwater. It is also possible that additionally ‘piping’, an accessory sinkhole formation process resulting from the washing out of the fines by horizontal and descending water flows via young alluvial fans (Arkin and Gilat 2000, Ezersky *et al* 2009, Ezersky and Frumkin 2020) also takes place here.

a. *Start image of sinkholes*

Sinkholes appear as small holes 0.7–1.0 m in diameter that with time increase and reach 30–50 m in diameter and decades meters in depth (figure 3.20). Sometimes, the inside sinkhole diameter is greater than at entry.

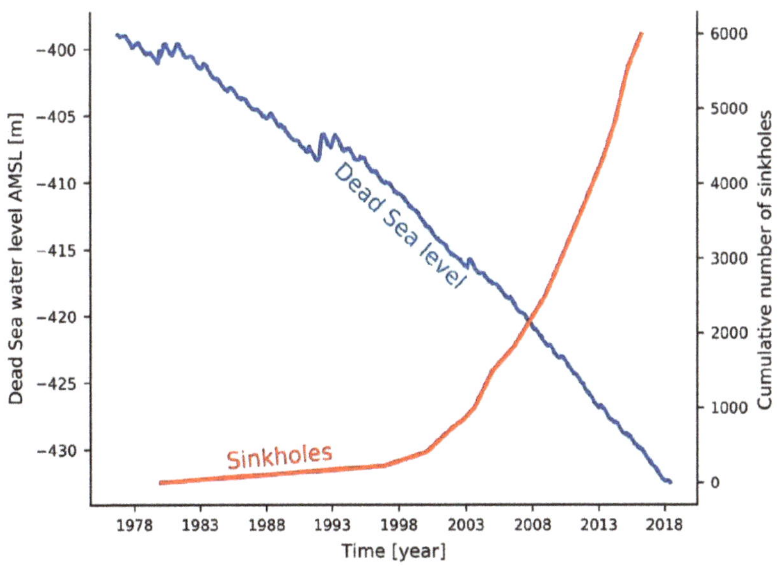

Figure 3.19. A cumulative number of sinkholes from 1980 to 2016 (red line) and the 1976–2018 Dead Sea water levels (blue line). (Reprinted with permission from Nof *et al* (2019), published by MDPI.)

Figure 3.20. Images of small sinkholes. (a) The small tractor fell into the sinkhole, Newe Zohar, 13.10.2002 (photo of Eli Raz). (b) Small sinkhole #23 in the Mineral Beach site (21.12.2010). (c) Small sinkhole at the North of Nahal Hever South (31.03.08). (d) The same sinkhole view inside (b, c, and d—photo of M Ezersky).

b. *DS Sinkhole types*

According to the generic classification of sinkholes proposed by Gutiérrez *et al* (2008), DS sinkholes can be divided into (a) cover sagging and (b) cover collapse holes in mantled karst setting. Cover suffosion sinkholes that result from the downward migration of particles through conduits and the progressive settlement of the cover seem cannot form in the isolated salt layers because of the absence of system fractures typical for carbonate rocks. A cover sagging sinkhole developing during three years is shown in figure 3.21(a). We called it also settling. Cover collapse sinkholes are formed by the upward propagation of soil cavities by progressive roof collapse. Two cover collapse sinkholes are shown in figures 3.21(b) and (c). This subsidence mechanism produces in our study areas scarp-edged sinkholes typically a few meters in diameter (figures 3.21(b) and (c)).

The casing of collapsed borehole HS-2 is seen in figure 3.21(c). In our study area, these are bowl-shaped holes that reach up to ten meters across (figure 3.21(d)). Finally, there is shown sinkhole #15 that has slowed down its development (figure 3.21(e)), and vast vegetation formed at the bottom of the sinkhole.

c. *Temporal development of collapse sinkholes*

Figure 3.22 shows the temporal development of sinkhole collapse in Ein Gedi Holiday Village area.

Figure 3.21. Images of DS sinkholes in the Nahal Hever south: (a) settling #1 developing during 3 years (10.10.2002); (b, c) cover collapse sinkholes #20 (20.05.2007) and #13 (12.2002), respectively; in a sinkhole #13 the casing of collapsed borehole HS-2 is seen; (d) cover sagging sinkhole #5 (05.07.1998); cover sagging sinkhole #15 (10.12.2005) is slowing down, vast vegetation developed at the bottom of the sinkhole (photos of M Ezersky).

Figure 3.22. Temporal development sinkhole collapse through the Ein Gedi Holiday Village area. (a) The first sinkhole near the storage room (July 1998). (b) Holiday village view in 19.03.2006; and (c) in October of 2008 (photo of M Ezersky).

Figure 3.23. Temporal development of cover sagging sinkhole. Sinkhole #5 in Nahal Hever South. (a) 5.07.1998; (b) 29.01.2001; (c) 22.03.2006 (photo of M Ezersky).

d. *Temporal development of cover sagging sinkholes*

The development of the cover-sagging sinkhole in the Nahal Hever South is shown in figure 3.23. As underlined salt was gradually dissolved by groundwater cover soil was subsided (figure 3.23(a)).

The subsidence was washed out also by surface floodwaters during the rain period (figure 3.23(b)). Finally, the cover was subsided when salt was dissolved throughout its entire thickness (figure 3.23(c)).

e. *Lithological section of the Dead Sea sinkholes*

Two types of lithological sections are observed following the available boreholes (figure 3.24). The first type of lithological section is composed of lime carbonates overlain/or no salt layer. This section is characteristic of mudflat areas. In figure 3.24(a), the borehole Mn-5E is shown drilled in the southern part of Mineral Beach (see figure 3.1 for site location). The second type of lithological section consists of alluvial fan sediments like sands, gravels, and their alternations. Such lithology is presented in borehole EB-3E drilled in the Ein Boqeq site in southern DS Basin (figure 3.24(b)). The boreholes HS-3 and HS-2 drilled in the Nahal Hever south site are presented in figures 3.24(c) and (d).

The first of them includes a mud layer that is marl liquefied in the groundwater (lime carbonate) substitutional dissolved salt (figure 3.24(c)). The second borehole, located 40 m apart, includes an 8 m thickness salt layer sandwiched by marl and

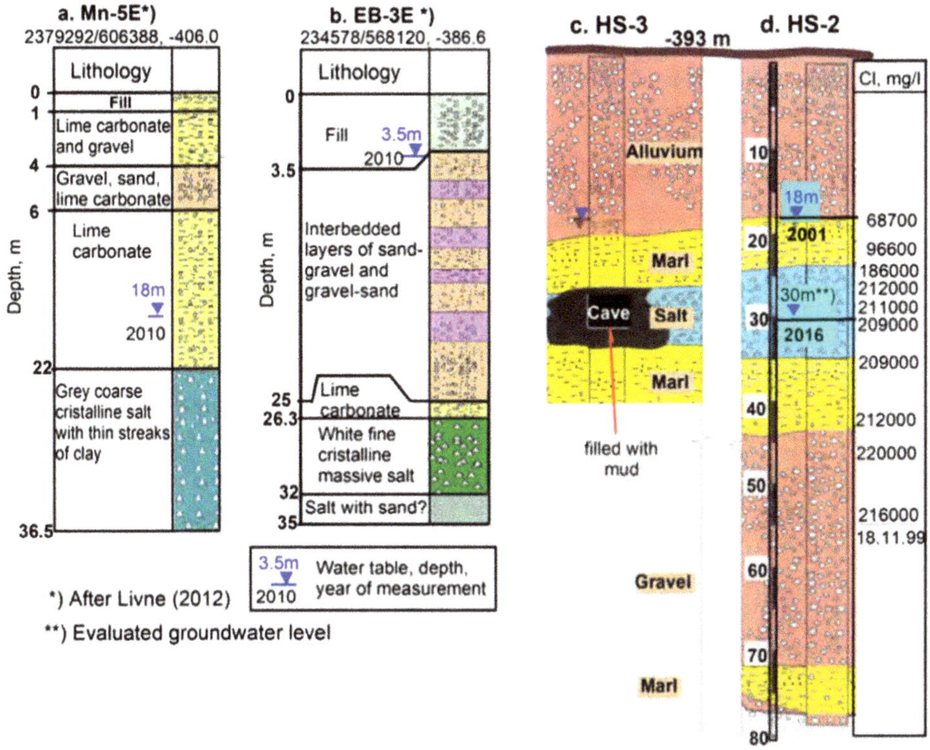

Figure 3.24. Lithological sections of boreholes (a) MN-5E from the Mineral Beach site; (b) EB-3E from the Ein Boqeq site and (c, d) boreholes HS-3 and HS-2 (40 m apart) from the Nahal Hever South area (see figure 3.1 for site location) (a, b). (Reprinted from Ezersky and Goretsky (2014) with permission of Elsevier; (c, d)—Ezersky *et al* (2010) with permission of Schweizerbart Science Publisher www.borntraeger-cramer.de/journals/zfg.)

sandy-gravel sediments (figure 3.24(d)). Later, a typical section for alluvial fansites will be presented.

3.5 Dead Sea shores subsidence

Since the early 1990s, sinkholes have become major problems along the DS shores in Israel and Jordan (Yechieli *et al* 2002, 2016, Abelson *et al* 2017). At the same time, researchers have detected wide, shallow subsidence features (WSSFs) through the large areas of the western shore (Baer *et al* 2002, Abelson *et al* 2003). Sinkholes are readily observed in the field, but their locations and timing were unpredictable. At the same time, a WSSF is often difficult to observe in the field. Nevertheless, being identified, they delineate zones of instability and increasing hazards. Research carried out by Baer *et al* (2002) was aimed to estimate rates of subsidence and reveal if sinkholes and WSSF are associated. Researchers have identified, characterized, and measured rates of subsidence along the DS shores using the interferometric synthetic aperture radar (InSAR) technique. The authors have analyzed data acquired during the years 1992–99 by the European Remote Sensing ERS-1 and

ERS-2 satellites. The interferograms used covered periods of between 2 and 71 months These investigations have shown subsidences observed in the Lisan Peninsula and along the DS shores, in a variety of shapes and sizes, including circular and elongate coastal depressions (a few hundred meters to a few kilometers in length), depressions in ancient alluvial fans, and depressions along salt-diapir margins (Baer *et al* 2002, Closson 2005). Subsidence rates were, as a rule, in the range of 0–20 mm yr^{-1} within the studied period, with exceptionally high rates that exceed 60 mm yr^{-1} in two sites. Authors suggested that subsidence of the marl composed areas is caused by aquifer-system consolidation caused by groundwater lowering and decrease of pore water pressure, resulting in gradual subsidence. Note, that during the study period, the level of the DS and the associated groundwater has dropped by ~6 m.

One-dimensional (vertical) calculations of clay consolidation were assumed according to Terzaghi's theory (Terzaghi 1925, see later edition Terzaghi *et al* 1996), as is common in standard soil mechanics computations. Comparison of InSAR observations with calculations of the expected consolidation have shown that in areas where marl layers are known to compose part of the upper 30 m of the profile, estimated consolidation settlements are agreed well with the order of the measured subsidence. Finally, the observations also showed that despite, in certain locations, subsidence appearing to be structurally controlled by faults, seaward landslides, and salt domes, gradual subsidence is unlikely to be directly related to the sinkholes, excluding the use of WSSFs features as predictable precursors to sinkhole formation

Recent researches by Vey *et al* (2021) based on the innovative Global Navigation Satellite System (GNSS) techniques showed that it behaves synchronously to atmospheric and hydrological changes with a time lag of two months. The authors show that the amplitude and fluctuation period of ground deformation is related to poroelastic hydro-mechanical soil response to the lake level changes. This provides, a first direct link between the shore subsidence, lake-level drop, and evaporation.

3.6 Existing plausible explanations of the sinkhole's appearance. Reported models

3.6.1 Piping model

At the beginning of the DS sinkhole investigations Arkin and Gilat (2000) suggested a piping mechanism connecting with washing out of fine particles from sediments and taking them out to the DS. They have selected three main types of sinkholes: gravel holes occurring in alluvial fans, mud holes occurring in the intervening bays of clay deposits between fans, and a combination of both types at the front of young alluvial fans where they overlap mudflats. Sinkhole development was directly related to the regression of the DS and the corresponding lowering of the regional water table. Thus, the mechanism of sinkhole formation was suggested for two types of sinkholes: the gravel piping sinkhole formation and mud piping sinkholes.

Generally, sinkhole formation is connected with the dissolution of underlying soluble rocks (Benson and Yuhr 2016). Piping sinkholes are caused by turbulent flow in the fine-grain soil with washing out fine particles and their transport to the gully, sea, etc.

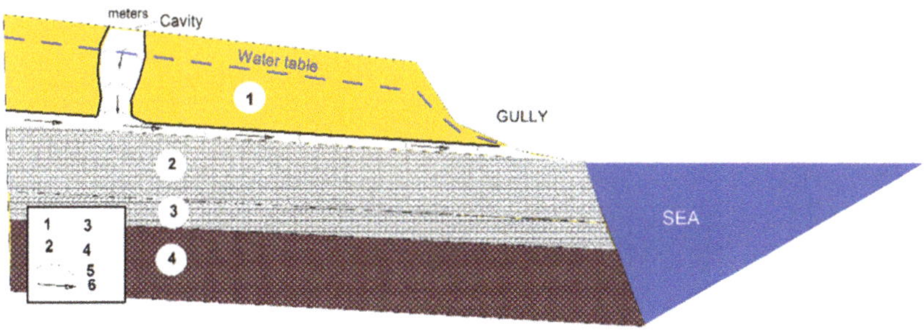

Figure 3.25. Simplified schema of piping sinkhole formation. 1—alluvial fine-grain sediments; 2—gravel; 3—Laminated clay, some sand; 4—bedrock, 5—boundary showing the cavity development, 6—direction of water flow.

The simplest schema of pipe sinkhole formation is shown in figure 3.25. The features of pipe sinkhole formation are as follows. (1) Absence of underlying soluble bedrock; (2) sinkhole development starts from the place of regional lowering of the water table: (3) sinkhole promote inland; (4) there should be fines and thin layers of laminated clays (Higgins 1984, Mendonca *et al* 1993).

Arkin and Gilat (2000) suggested their model when the presence of buried salt in the DS shores was subjected to doubts. Therefore, this model was the most plausible. Following Arkin and Gilat (2000), gravel hole developed in three stages. At the first stage, the subsurface erosion began above turbulent flow in the gravel bed. A hollow and pipe developed upward. Fines were washed out and larger particles were compacted. In the second stage, circular fractures develop at the surface. Finally, the sudden collapse of the upper layer took place (Stage 3). Fine material was washed out at the base of the pipe and was taken out to the DS.

The development of mudholes has slightly differed from gravel sinkholes. In the beginning, subsurface erosion started in gravel or salt crystal layers where the laminated flows became turbulent. A hollow and pipe develop upward; subsidence started at the surface (Stage 1). Then in Stage 2, subsidence continued with appearing concentric fractures at the surface forming a typical funnel shape. Fines were washed out at the base of the pipe. Finally, (Stage 3) walls at the surface collapse and increase the diameter of the hole. As the process continues, several holes may coalesce.

However, on the western DS side, underground water streams bringing washed-out fines to the DS are not energetic enough Frumkin and Raz (2001). Ezersky *et al* (2009), have shown that this mechanism can be reliable in some specific cases as an addition to salt dissolution with the formation of large cavities. Recently, Krawczyk *et al* (2015) and Al-Halbouni *et al* (2017) suggested a similar mechanism for sinkhole formation in Ghor Al-Haditha, presuming that a massive saline clay layer underlies the alluvial sandy-gravel sediments

3.6.2 'Structural' model

'Structural hypothesis' of sinkhole formation was suggested in several studies (Abelson *et al* 2003, 2006, Closson 2005, Closson and Abou Karaki 2009). This

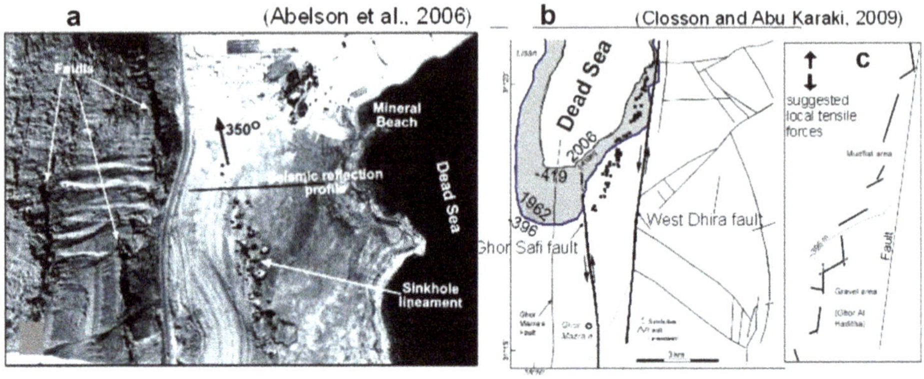

Figure 3.26. Sinkhole lineaments (a) in Mineral Beach mudflat; (b) in Ghor Al-Haditha; (c) tectonic interpretation of Ghor Al-Haditha sinkholes; (a) Abelson *et al* (2006), with permission of AGU; (b and c) Closson and Abou Karaki (2009), with permission of John Wiley & Sons.

hypothesis is based on a visual correspondence between the sinkhole lineaments and the western and eastern fault escarpments. Based on this similarity, structural control of sinkhole formation was assumed and a corresponding numerical model of the fault-induced sinkholes was developed (Shalev *et al* 2006). Abelson *et al* (2003) suggested that sinkholes tend to develop along faults, which serve as preferential hydraulic pathways, bringing confined aggressive water upwards to contact with salt. Closson (2005) considered a similar mechanism for the Jordanian DS coast. Examples of these models are shown in figure 3.26. Abelson *et al* (2006) suggested parallelism of sinkholes and marginal faults through the Mineral Beach area (figure 3.26(a)).

Closson and Abou Karaki (2009) associate the sinkhole lineament in the Ghor Al-Haditha with a hidden fault (figure 3.26(b)). These authors 'suggest that the spatial distribution of most sinkholes (and consequently caverns) is linked to underground fractures caused by regional tectonic movements. One recent example of such geo-tectonic event is the earthquake of April 23, 1979' (p 1420). Closson and Abou Karaki (2009) try to support their theory also by stress analysis (figure 3.26(c)).

The main objection arises that numerous faults are widely distributed through the DS coastal area (e.g. Garfunkel 1981, Garfunkel 1997), whereas sinkholes are usually developed in a single narrow strip 50–100 m wide, roughly parallel to the DS shore.

3.6.3 'Salt edge' model

Another hypothesis explains sinkholes formation by the dissolution of salt layer edge (named also dissolution front), formation of cavities within this edge, with subsequent collapse surface sediments into these cavities. So sinkholes are formed along a dissolution front (buried salt edge) over a tilted salt layer (western edge in Israel and eastern edge in Jordan). The term 'salt edge' was suggested by Jenyon (1986). Visually, DS sinkholes develop along slightly tortuous lines that in aerial photography may look like a series of lineament segments. Ezersky (2006) based on

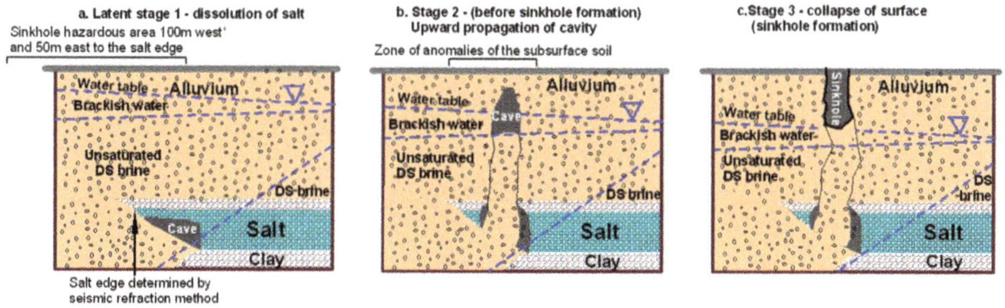

Figure 3.27. Three stages of sinkhole formation at the salt edge: a generalized model. Note that although there is no salt west of the present salt edge, salt has been dissolved there earlier. Therefore, cavities may still propagate towards the surface. It is confirmed by statistics of sinkhole formation. (a) Latent stage, (b) upward, propagation of cavity (stoping); (c) sinkhole formation. (Reprinted from Frumkin *et al* (2011), copyright (2011), with permission of Elsevier.)

seismic refraction mapping shows that this tortuous line repeats the salt border line variations. The spatial pattern of the sinkhole development along the salt edge was observed repeatedly (with the small variations) in all studied sites in Israel and the Ghor Al-Haditha area (Jordan). It allowed the conclusion that the salt edge is a major factor controlling sinkhole formation (Abueladas and Al-Zoubi 2004, Ezersky *et al* 2007, 2010, Legchenko *et al* 2008a). The model of sinkhole development includes three phases (Frumkin *et al* 2011). The salt dissolution may take place a short or long period before any signs are evident on the surface (figure 3.27).

The sinkholes commonly developed along the salt edge which was dissolved by brackish water, unsaturated concerning chloride (Yechieli *et al* 2006, Yechieli 2007, Legchenko *et al* 2009, Ezersky and Frumkin 2017). The first stage is latent (figure 3.27(a)). Instability of the salt cavity ceiling causes upward stoping through the mechanically weak sediments situated between the cavity and the surface (figure 3.27(b)). Ford and Williams (2007, p 385) report cases in different geological settings, where the stoping process propagated through 1000 m of cover rocks. This intermediate stage continues until the surface collapses and the sinkhole forms (figure 3.27(c)).

This hypothesis requires the existence of shallow salt layers on both sides of the DS.

Ezersky *et al* (2013b) have suggested a model (figure 3.28) presuming that the salt belt is girding the northern and southern DS basins. This salt belt comprises all known cases of collapses and accidents including alarming catastrophe at pond 19 in Jordan.

The hypothesis is based on the seismic refraction surveys carried out mainly along the western DS shore (Ezersky 2006, Ezersky and Legchenko 2008, Ezersky *et al* 2010, Frumkin *et al* 2011) and in Ghor Al-Haditha (El-Isa *et al* 1995, Abueladas and Al-Zoubi 2004, Dhemaied 2007). All data available are shown in figure 3.28.

Numerous boreholes drilled in the western DS side have confirmed the presence of a salt layer. Along the Jordanian side, boreholes were drilled using fresh water and there is no direct evidence of salt presence. The salt hypothesis there is based on geophysical investigations only (Ezersky *et al* 2021).

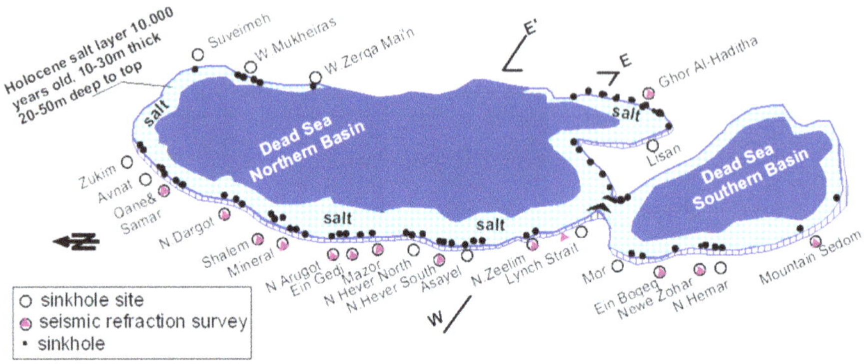

Figure 3.28. Hypothetic salt distribution around the DS at the elevation of −415 to −445 m where the salt layer was formed in the latest Pleistocene. Sinkholes at the north-eastern shore and Ghor Al-Haditha are given after Closson and Abou Karaki (2009) and Frumkin *et al* (2011). (Reprinted from Ezersky *et al* (2013b), copyright (2013), with permission of Springer Nature.)

3.6.4 The salt layer concept model

Let us consider the concept of salt formation and its interaction with faulting (Ezersky and Frumkin 2013). Salt layer deposition and the 'salt edge' concepts are explained in a W–E section presented in figure 3.29.

The first phase of salt deposition corresponding to the latest Pleistocene may relate to a period when the DS water still covered the relatively flat surface topography of the shore (figure 3.29(a)). During the second period, termed by Stein *et al* (2010) 'catastrophic aridity', salt was deposited from the evaporated DS water forming the salt layer (figure 3.29(b)). Finally, around 10 000 years ago the salt layers were covered by fluvial and lacustrine sediments brought by DS wadies (figure 3.29(c)). It is also seen from figure 3.29 that the salt edge may be constrained by vertical (or sub-vertical) up to 30 m-high fault walls articulating often in the uppermost part with thin salt featheredge (figure 3.29(b)). Length of the featheredge and its thickness, depends on the relationships between DS level and intermediate block elevation at the time of salt deposition (figure 3.29(b) left). The right part of the section shows the case where a salt layer was not formed. This section characterizes the central part of the eastern DS shore (figure 3.29 right). This section characterizes the central part of the eastern DS shore. At gently sloping shores the salt feather edge can be thin and long (figure 3.29(b), left). Thus, the genetic connection between the salt layers and the tectonic structure of the DS is seen from our model of salt formation.

3.6.5 Combined 'structural' and 'salt edge' model

The structural model, on the one hand, is based on the visible coupling of sinkholes and faults, the existence of neotectonic faults in the close vicinity of sinkholes. The plausible scenario suggests sinkhole formation via salt dissolution by undersaturated water reaching the salt via faults (Abelson *et al* 2006, Shalev *et al* 2006). Accepting such a mechanism, the structural model ignores any connection of sinkholes with the

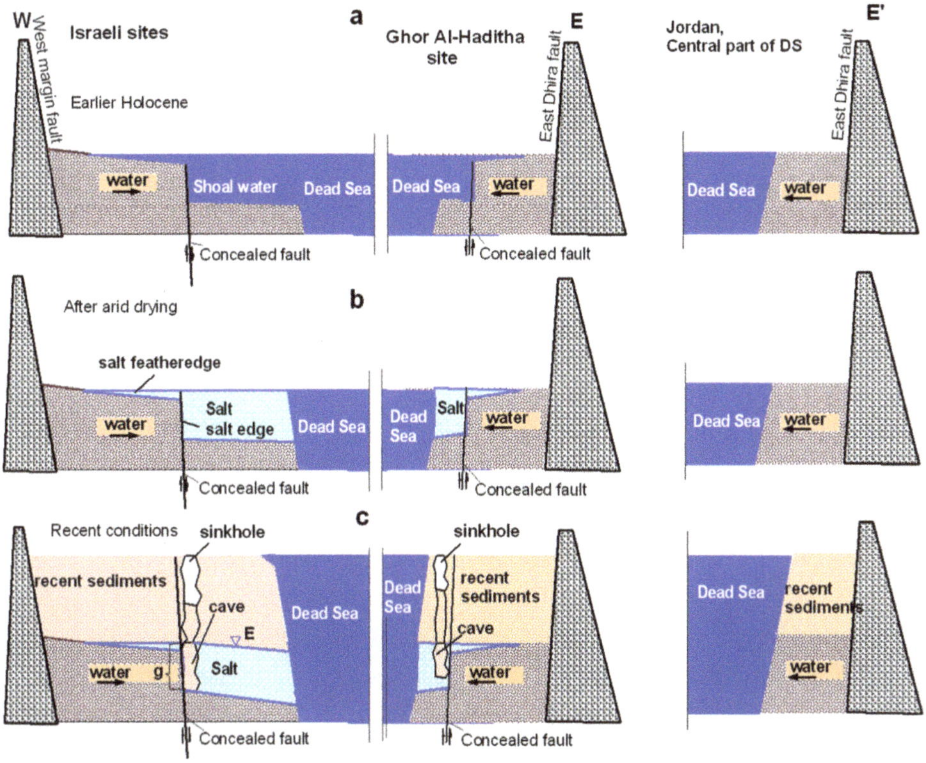

Figure 3.29. A cartoon showing the hypothetic DS sinkhole configuration concept. The left part of the section (W–E) corresponds to gentle shores like the Israeli (western) ones and Ghor Al-Haditha (eastern), the right part of the section (E′) corresponds to steeper shores like the central-eastern shore: (a) latest Pleistocene setting, (b) after salt deposition following arid phase, and (c) recent conditions. g—thickness of the salt, E—salt top. (Reprinted from Ezersky and Frumkin (2013), copyright (2013), with permission of Elsevier.)

ancient or modern DS shores On the other hand, the salt edge model based on geophysical data substantiates sinkhole formation along a dissolution front in the salt zone close to the salt edge. This model ignores faults as an important factor of sinkhole formations, the salt edge model considers the salt edge as forming along the ancient DS shore and connects sinkholes with this shoreline.

Both these hypotheses consider salt karstification as the main mechanism of sinkhole formation. The working hypothesis suggests that brackish groundwater circulation (coming from either fault or wadies) dissolves salt by a gradual increase of salt emptiness and permeability and formation of caverns at the zone adjoining the fault or salt edge. Thus, sinkholes are caused by the collapse of soil to large caverns in a buried salt layer.

To reconcile the mentioned two approaches, we have to consider relations between the salt and faults. Based on both approaches we can suggest a combined model of sinkhole formation comprising elements of both structural and salt edge models. Considering figure 3.29 one would suggest that in some places the walls of sediments constraining the salt edge were originally determined by active normal fault plains (Gardosh *et al* 1990). In places, where large wadies (ephemeral streams) are absent, the salt edge may follow fault plains and, subsequently, sinkhole arrangement along the salt edge also follows faults. Results of the 3D seismic reflection surveys (figure 3.30) show that the salt layers can subside along these faults causing displacements of tens of meters. Faults and fractures located within the dissolution front zone may facilitate the access of aggressive water into the salt, determining the spatial distribution of sinkholes of the salt edge zone.

In contrast, where large wadis flow into the DS, the shoreline is usually forced by alluvial fans. Along with these segments the present and paleo-shoreline deviate from the fault lines, curving according to the fan topography. The dissolution front, defined by the latest Pleistocene shoreline, wriggles accordingly, determining the spatial location of the recent sinkholes.

Thus, Ezersky and Frumkin (2013) analyzing results of seismic reflection surveys suggested that the salt edge was formed either in contact with normal faults or in contact with wadi sediments. At the same time, sinkholes were formed at the dissolution front (salt edge). It explains why sinkhole lineaments are sometimes parallel to faults and sometimes are deflected from them following instead the salt edge (figure 3.31). It

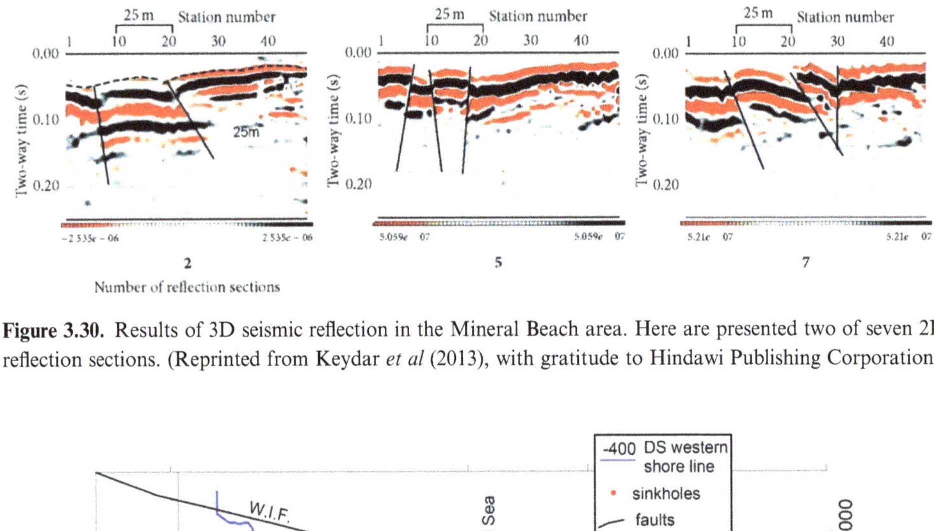

Figure 3.30. Results of 3D seismic reflection in the Mineral Beach area. Here are presented two of seven 2D reflection sections. (Reprinted from Keydar *et al* (2013), with gratitude to Hindawi Publishing Corporation.)

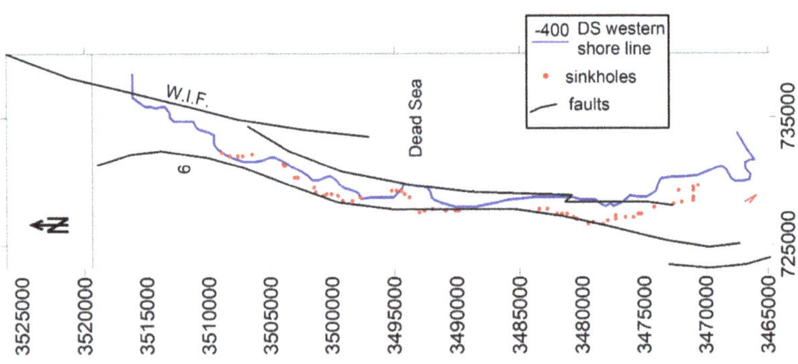

Figure 3.31. Relations between the western faults (gray lines), the DS shoreline at −400 m lake level (blue), and the sinkholes (red) in increased scale. (Reprinted from Ezersky and Frumkin (2013), modified, with permission of Elsevier.)

allows distinguishing sinkholes by kind of dissolving groundwater either coming via fault (like Mineral Beach sinkholes) or from wadies (like Ein Gedi, Arugot).

The sinkhole development in the Mineral Beach study site supports the combined model: before 2006 sinkhole line was followed in the NNW direction along salt edge pre-determined by a fault lineament (Mineral South). The dissolution front at the salt edge was karstified by undersaturated groundwater arriving from deep sources. The recent sinkhole line (Mineral North) develops eastward along the salt edge segment pre-determined by the Nahal (wadi) Hazazon alluvial fan (Ezersky and Frumkin 2013, figure 9d). The dissolution front at the salt edge is karstified by undersaturated groundwater arriving from both deep and shallow sources. Stein *et al* (2010) paid attention to the salt layer top elevation that is differed from site to site varying in the range of 24 m (from −418 to −442 m). It is surprising taking into account that the salt layer formed at approximately the same time by the DS water evaporation (figure 3.29(b)). Therefore, the upper surface level of the salt layer throughout the DS area should have been roughly similar, possibly with some initial dip. Researchers suggest that the salt unit subsides with different rates along the boundary faults, as shown in figure 3.29, whereas salt inclination is caused by tilting of the tectonic blocks (Garfunkel 1997).

Garfunkel and Ben-Avraham (1996, p 155) note that ‘at any time a large part of the basin subsides simultaneously, but the site of fastest subsidence seems to have shifted northward’. The central parts of the DS subside more intensively. Garfunkel (1981) explains it by the pull-apart origin of the DS transform causing a widening of the basin.

It can be presumed that a greater subsidence rate would correspond to a thicker salt layer (‘g’ in figure 3.29(c)). Post-depositional subsidence and basinward tilting would further depress the salt surface elevation (‘E’ in figure 3.29(c)). If our suggestion is right, then salt thickness is a rough measure of the subsidence rate. Also, an inverse correlation would be observed between the salt thickness and its elevation. Such inverse correlation is confirmed by the graph in figure 3.32.

The relationships presented in figure 3.32 thus support our suggestion. An exception is three boreholes located in the southern DS basin. They can be associated with the shallow water depth accelerating salt deposition (Ezersky and Frumkin 2013).

Thus, one can see that there is a close genetic connection between the faults and salt layers. We suggest that superficial coastal basins where salt formed could be generated by the sub-vertical displacements along the faults. The sinkhole lineaments are arranged along the salt layer edge acting as a dynamic dissolution front. This reconciles the two major competing models of sinkhole formation: the structural model, considering control of faults, and the dissolution front model associating sinkhole formation with the salt edge. The combined model suggests that the salt layer edge follows a paleo-shoreline, ~10.5 ka old, constrained by faults and alluvial fans. The sinkholes develop along the dissolution front, controlled by these features. The buried salt layer extends landward, away from the modern DS shoreline, permitting its investigation from the surface.

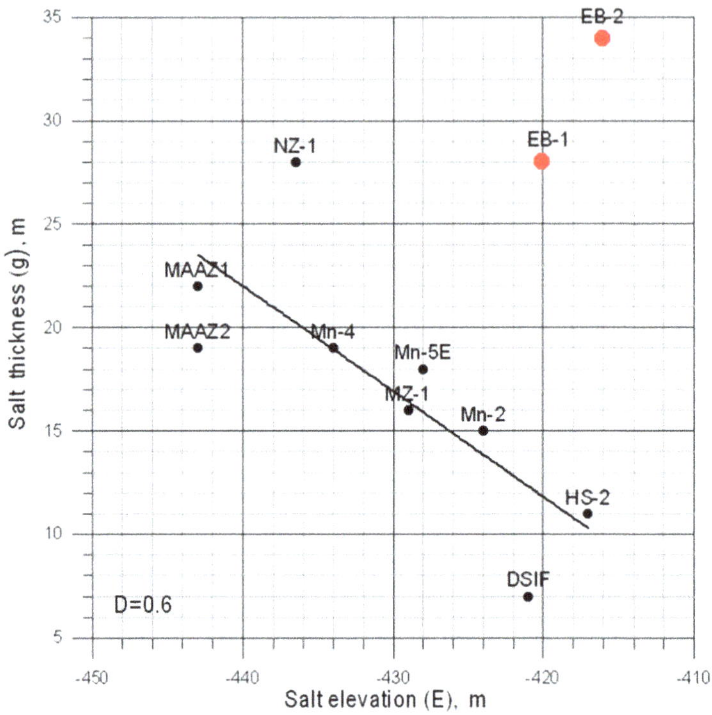

Figure 3.32. Inverse correlation between the salt layer thickness and the salt top elevation in boreholes along the western DS shore. Exceptions are boreholes in the shallow southern DS basin (EB-1, EB-2, NZ-1, see text). (Reprinted from Ezersky and Frumkin (2013), with permission of Elsevier.)

3.6.6 Elements of the piping model

Sometimes, evaporite karst can exist in parallel with some elements of the piping. Such cases were revealed in the Nahal Hever south site (Ezersky *et al* 2009) and in the Newe Zohar site (Ezersky and Frumkin 2020). Experience shows to identify the karst origin it is important to detect if soluble bedrock exists under unconsolidated overburden. For this prospecting, boreholes or geophysical methods should be used. If soluble rock (for instance, salt), exists in the subsurface, salt dissolution seems to be the dominant process causing the cover collapse or cover sagging sinkholes. At the same time, at the start of the process slight dissolution of salt along its border causes overlying sediments to subside, allowing shallow streams to flow through subsurface sediments with increased porosity. If soluble rocks at a depth were not detected, shallow subsurface streams can create an illusion of the piping sinkholes. The case history from the Newe Zohar site indicated that surface observations without boreholes and proper geophysical subsurface methods can be misleading. The example of geophysics application for identification of sinkhole origin will be considered in section 5.2.

3.6.7 Exceptions to the general model/sinkhole sites not investigated yet

Most of the sinkholes have been formed around the northern DS basin figure 3.33(a). These sinkholes range between the depths of 5–50 m and the lateral

Figure 3.33. (a) Map of sinkhole sites distribution, (b) conventional sinkholes in the Nahal Hever South (site I in figure 3.33(a)); (c, d) new sinkhole modifications characteristic for the southern part of the DSa were formed on the dam 5/8 and in the road pavement, respectively (site II in figure 3.33(a)); (e) examples of sinkholes formed in the eastern DS shore, Jordan: near Hotel Marriott, Wadi Mukheira, and Holiday Inn Hotel, Sweimech (site III in figure 3.33(a)); Nahal Mujib with developed sinkholes and tectonic (site IV in figure 3.33(a)). ((f)—photo of D Closson, (g)—photo of N Abou Karaki; Abou Karaki and Closson (2012)).

sizes of 3–100 m, and they are constantly developing in the narrow (50–100 m width) strip along the coastal line, slowly propagating from the south to the north. The example of such sinkholes from the Nahal Hever South site is presented in figure 3.33(b) (site I in figure 3.33(a)). These sinkholes are well studied using boreholes and geophysical methods. The sinkholes recently developing in the southern area of the Israeli shore can be divided into two modes: (a) sinkholes I mode are similar to those of the northern basin although they are smaller in size; (b) sinkholes of II mode (small ones with diameter less than 1 m and depth of several dozens of centimeters) are formed on the DS dam surface figure 3.33(c) (site II in figure 3.33(a)), on the road pavement of route #90 figure 3.33(d) (Mt Sedom area in figure 3.33(a)).

With a lowering of the DS level, various and complicated erosion phenomena began to develop in the Jordanian DS shores. Over time, there began to occur channeling, gullying, landslides, etc. Intensive incision of streams and gullies is propagating upstream towards the infrastructures, causing damage to the roads, bridges, earth dikes, and the hotels/recreation areas (Abou Karaki and Closson 2012).

Sinkhole presented in figure 3.33(e) south of the Marriott Hotel (Wadi Mukheira) was opened in 2004. Two landslides and large sinkholes appeared in 2004 in the Holiday Inn resort figure 3.33(f) (Sweimeh area, site III in figure 3.33(a)). Sinkholes were also formed in the Wadi Mujib (site IV in figure 3.33(a)) and the Arab Potash Company area. The genesis of the northern sinkholes in Jordan and the origin of sinkholes of mode II that appeared on Dam 5, highway#90 on Israeli shores is still not well understood and should be investigated. The same holds for landslides developing along the northern Jordanian shore and Arab Potash Company.

References

Abelson M, Baer G, Stivelman V, Wachs D, Raz E, Crouvi O, Kurzon I and Yechieli Y 2003 Collapse-sinkholes and radar interferometry reveal neotectonics concealed within the Dead Sea basin *Geophys. Res. Lett.* **30** 1545

Abelson M, Yechieli Y, Crouvi O, Baer G, Wacks D, Bein A and Shtivelman V 2006 Evolution of the Dead Sea sinkholes *Geological Society America Special Papers* ed Y Enzel, A Agnon and M Stein vol 401 (Boulder, CO: Geological Society America) pp 241–53

Abelson M, Yechieli Y, Baer B, Lapid G, Behar N, Calvo R and Rosensaft M 2017 Natural versus human control on subsurface salt dissolution and development of thousands of sinkholes along the Dead Sea coast *J. Geophys. Res.: Earth Surf.* **122** 16

Abou Karaki N and Closson D 2012 Field Guidebook, 23 September, 2012, Amman, Jordan. *EAGE Workshop* (www.eage.org)

Abou Karaki N, Fiaschi S and Closson D 2016 Sustainable development and anthropogenic induced geomorphic hazards in subsiding areas *Earth Surf. Process. Landf.* **41** 2282–95

Abueladas A and Al-Zoubi A 2004 The application of a combined geophysical survey (GPR and seismic refraction) for mapping sinkholes in Ghor Al-Haditha Area, Jordan. Fall Meeting Supplement. EOS Transactions, American Geophysical Union, 85, p 47 (Abstract GP11A-0825)

Abu-Khader M M 2005 Viable engineering options to enhance the NaCl quality from the Dead Sea in Jordan *J. Clean. Prod.* **14** 80–6

Al-Halbouni D, Holohan E P, Saberia L, Alrshdan H, Sawarieh A, Closson D, Walter T R and Dahm T 2017 Sinkholes, subsidence and subrosion on the eastern shore of the Dead Sea as revealed by a close-range photogrammetric survey *Geomorphology* **285** 305–24

Al-Zoubi A and ten Brink U S 2001 Salt diapir in the Dead Sea and their relationship to quaternary extensional tectonics *Mar. Pet. Geol.* **18** 779–97

Amit R, Zilberman E, Enzel Y and Porat N 2002 Paleoseismic evidence for time dependency of seismic response on a fault system in the southern Arava Valley, Dead Sea rift, Israel *GSA Bull.* **114** 192–206

Angelakis A N, Valipour M, Choo K-H, Ahmed A T, Baba A, Kumar R, Toor G S and Wang Z 2021 Desalination: from ancient to present and future *Water* **13** 2222

Arandjelović D 1969 A possible way of tracing groundwater flows in karst *Geophys. Prospect.* **12** 404–21

Arandjelović D 1976 Geophysics in the karst. Institute for geological, geophysical, and geotechnical research *Geophys. Inst. Beograd* **17** 1976

Arkin Y and Gilat A 2000 Dead Sea sinkholes—an ever-developing hazard *Environ. Geol.* **39** 711–22

Arnon A, Lensky N G and Selker J S 2013 High-resolution temperature sensing in the Dead Sea using fiber optics *Water Resour. Res.* **50** 1756–72

Asmar B N and Ergenzinger P 2002 Dynamic simulation of the Dead Sea *Adv. Water Res.* **25** 263–77

Asmar B N and Ergenzinger P 2004 The Dead Sea is unlikely to dry up in 50 years *EOS* **85** 69–73

Baer G, Schattner U, Wachs D, Sandwell D, Wdowinski S and Frydman S 2002 The lowest place on Earth is subsiding—an InSAR (interferometric synthetic aperture radar) perspective *Geol. Soc. Am. Bull.* **114** 12–23

Bartov Y, Agnon A, Enzel Y, Reches Z and Stein M 2002 Sequence stratigraphy and reconstruction of Lake Lisan levels *Quat. Res.* **57** 9–21

Bartov Y, Goldstein S L, Enzel Y and Stein M 2003 Catastrophic arid episodes in the Eastern Mediterranean linked with the North Atlantic Heinrich events *Geology* **31** 439–42

Begin B, Ehrlich A and Nathan Y 1974 Lake Lisan the Pleistocene precursor of the Dead Sea *Geol. Surv. Israel, Bull* **63** 30

Beck B F 2003 *Sinkholes and the engineering and environmental impacts of Karst, Geotechnical Special Publication. No. 122* (Reston, VA: American Society of Civil Engineers) p 737

Beck B F and Herring J G 2001 Geotechnical and environmental applications of Karst geology and hydrology *Proceedings of 8th Multidisciplinary Conference on Sinkholes and the Engineering and Environmental Impacts of Karst* (Rotterdam: A A Balkema) p 437

Bein A and Feinstein S 1988 Late Cenozoic thermal gradients in Dead Sea Transform System basins *J. Pet. Geol* **11** 185–92

Ben-Avraham Z, Henel R and Villinger H 1978 Heat flow through the Dead Sea rift *Mar. Geol.* **28** 253–69

Ben-Avraham Z and Ballard D 1984 Near bottom temperature anomalies in the Dead Sea *Earth Planet. Sci. Lett.* **71** 356–60

Ben-Avraham Z 1997 Geophysical framework of the Dead Sea ed T M Niemi, Z Ben-Avraham and J Gat *The Dead Sea: The Lake and its Settings* (New York: Oxford University Press) pp 22–35

Ben Avraham Z 2001 The Dead Sea—a unique global site *Eur. Rev.* **9** 437–44

Benson R C and Yuhr L B 2016 *Site Characterization in Karst and Pseudokarst Terranes. Practical Strategies and Technology for Practicing Engineers, Hydrologists and Geologists* (Berlin: Springer) p 421

Beyth M 2007 The Red Sea and the Mediterranean–Dead Sea canal project *Desalination* **214** 364–70

Bookman (Ken-Tor) R, Enzel Y, Agnon A and Stein M 2004 Late Holocene levels of the Dead Sea *Bull. Geol. Soc. Am.* **116** 555–71

Bubnov B M and Linden P F 1998 Diffusion stratification in a rotating linearly-stratified fluid *Fluids Dyn.* **33** 201–8

Burthens J, Filippi M, Asadi N, Zare M, Slechta S and Churackova Z 2009 Surficial deposits on salt diapirs (Zagros Mountains and Persian Gulf Platform, Iran): characterization, evolution, erosion and the influence on landscape morphology *Geomorphology* **107** 195–209

Charrach J, Goretsky I and Frydman S 2007 Geotechnical studies of soils from an evaporate environment in Israel *Proceedings of 5th International Congress on Environmental Geotechnics, Cardiff, Wales 1* (London: Thomas Telford) 407–14

Closson D 2005 Structural control of sinkholes and subsidence hazards along the Jordanian Dead Sea coast *Environ. Geol.* **47** 290–301

Closson D and Abou Karaki N 2009 Salt karst and tectonics: sinkholes development along tension cracks between parallel strike-slip faults, Dead Sea, Jordan *Earth Surf. Process. Landf.* **34** 1408–21

Coianiz L, Ben-Avraham Z, Stein M and Lazar M 2019 Spatial and temporal reconstruction of the late Quaternary Dead Sea sedimentary facies from geophysical properties *J. Appl. Geophys.* **160** 15–27

De Waele J, Picotti V, Zini L, Cucchi F and Forti P 2009 Karst phenomena in the Cordillera de Sal (Atacama, Chile) ed P L Rossi *Geological Constraints on the Onset and Evolution of an Extreme Environment: the Atacama Area* vol 2 (Geoacta Special Publication) pp 113–27

De Waele J, Guttierez F, Parise M and Plan L 2011 Geomorphology and natural hazards in karst areas: a review *Geomorphology* **134** 1–8

Dhemaied A 2007 *Tomographie seismique d'une zone de subsidence (Sinkhole zone, Ghor Al-Haditha, Jordanie)* Unpublished MS Report (Paris: Paris 6 University) (in French)

DS sinkhole database 1998–2016 Unpublished reports of the Tamar Regional Council (from permission of authors)

Eberhard R and Sharples C 2013 Appropriate terminology for karst-like phenomena: the problem with 'pseudokarst' *Int. J. Speleol.* **42** 109–13

El-Isa Z, Rimawi O, Jarrar G, Abu-Karaki N, Taqieddin S, Atalah M, Abderahman N and Al-Saed A 1995 Assessment of the hazard of subsidence and sinkholes in Ghor Al-Haditha area *Report Unpublished to Jordan Valley Authority* University of Jordan, Amman 141

Eppelbaum L V, Kutasov I M and Pilchin A N 2014 *Applied Geothermics* (Heidelberg: Springer)

Eppelbaum L V, Modelevsky Jr M M and Pilchin, A N 1996 Thermal investigation in petroleum geology: the experience of implication in the Dead Sea Rift zone, Israel *J. Pet. Geol* **19** 425–44

Eppelbaum L V and Pilchin A N 2006 Methodology of Curie discontinuity map development for regions with low thermal characteristics: an example from Israel *Earth Planet. Sci. Lett.* **243** 536–51

Eppelbaum L V, Ezersky M G, Al-Zoubi A S, Goldshmidt V I and Legchenko A 2008 Study of the factors affecting the karst volume assessment in the Dead Sea sinkhole problem using microgravity field analysis and 3D modeling *Adv. GeoSci.* **19** 97–115

Eppelbaum L, Yakubov Y and Ezersky M 2010 Method for comprehensive computing of water flows geodynamics in the Dead Sea basin *Trans. XXXV EGS Meet., Zurich, Switzerland* **P24** 1–4

Ezersky 2006 The geophysical properties of the Dead Sea salt in application to sinkhole problem *J. Appl. Geophys.* **58** 45–58

Ezersky M 2008 Geoelectric structure of the Ein Gedi sinkhole occurrence site at the Dead Sea shore in Israel *J. Appl. Geophys.* **64** 56–69

Ezersky M G, Legchenko A, Camerlynck C and Al-Zoubi A 2007 The salt formation edge as a major indicator of the sinkhole hazard in the Dead Sea Western Coast 13th EAGE Near Surface Meeting and Exhibition 2007 *(Istambul, Turkey, 1–5 September 2007) P67 4 on CD*

Ezersky M G, Legchenko A, Al-Zoubi A, Levi E, Akkawi E and Chalikakis K 2011 TEM study of the geoelectrical structure and groundwater salinity at the Nahal Hever sinkhole site, Dead Sea shore, Israel *J. Appl. Geophys.* **75** 99–112

Ezersky M and Livne E 2013 Geotechnical and geophysical properties of soils in the Dead Sea sinkhole problem 19th EAGE Near Surface Geoscience*(9–12 September, 2013) (Bohum, Mo) P13 4 on CD*

Ezersky M and Frumkin A 2013 Faults—dissolution front relations and the DS sinkholes problem *Geomorphology* **201** 35–44

Ezersky M and Frumkin A 2017 Evaluation and mapping of Dead Sea coastal aquifers salinity using Transient Electromagnetic (TEM) resistivity measurements *C. R. Geosci.* **349** 1–11

Ezersky M G and Frumkin A 2020 Identification of sinkhole origin using surface geophysical methods, Dead Sea, Israel *Geomorphology* **364** 107225

Ezersky M G and Frumkin A 2021 Subaerial morphology affected by groundwater aggressiveness: sinkhole susceptibility above karstified salt, Dead Sea *Geomorphology* **375** 107525

Ezersky M G and Goretsky I 2014 Velocity-resistivity versus porosity-permeability inter-relations in Dead Sea salt samples *Eng. Geol.* **183** 96–115

Ezersky M G and Goretsky I 2015 The scale effect of Dead Sea salt velocities based on seismic-acoustic measurements *M*echanical Behavior of Salt VIII, *Rapid City, SD* ed L Roberts, K Mellegard and F Hansen (Rotterdam: A A Balkema) pp 137–44

Ezersky M G and Legchenko A 2008 Sinkhole hazards. Part I. Mapping of the salt in sinkhole hazardous sites along western Dead Sea shore using Seismic Refraction method *NATO Programme Science for Peace, Project SfP 981128. Report No.* 232/361/08 GII-IRD

Ezersky M G and Legchenko A 2014 Quantitative assessment of in-situ salt karstification using shear wave velocity, Dead Sea *Geomorphology* **221** 150–63

Ezersky M, Legchenko A, Camerlynck C and Al-Zoubi A 2009 Identification of sinkhole development mechanism based on a combined geophysical study in Nahal Hever South area (Dead Sea coast of Israel) *Environ. Geol.* **58** 1123–41

Ezersky M, Legchenko A, Camerlynck C, Al-Zoubi A, Eppelbaum L, Keidar S, Baucher M and Chalikakis K 2010 The Dead Sea sinkhole hazard—new findings based on a multidisciplinary geophysical study *Z. Geomorph. N.F.* **54** 69–90

Ezersky M, Bodet L, Akkawi E, Al-Zoubi A, Camerlynck C, Dhemaied A and Galibert P-Y 2013a Seismic surface-wave prospecting methods for sinkhole hazard assessment along the Dead Sea shoreline *J. Environ. Eng. Geophys.* **18** 233–53

Ezersky M G, Eppelbaum L V, Al-Zoubi A, Keydar S, Abueladas A-R, Akkawi E and Medvedev B 2013b Geophysical prediction and following development sinkholes in two Dead Sea areas, Israel and Jordan *Environ. Earth Sci.* **70** 1463–78

Ezersky M G, Eppelbaum L V, Al-Zoubi A, Keydar S, Abueladas A-R, Akkawi E and Medvedev B 2014 Comments to publication of D Closson and N Abu Karaki 1365 'Sinkhole hazards prediction at Ghor Al Haditha, Dead Sea, Jordan: 'Salt Edge' and 'Tectonic' models contribution—a rebuttal to 'Geophysical prediction and following development *Environ. Earth Sci.* **71** 1989–93

Ezersky M, Legchenko A, Eppelbaum L and Al-Zoubi A 2017 Overview of the geophysical studies in the Dead Sea coastal area related to evaporite karst and recent sinkhole development *Int. J. Speleol.* **46** 277–302

Ezersky M, Eppelbaum L, Legchenko A, Al-Zoubi A and Abueladas A-R A 2021 Salt layer characteristics in the Ghor Al-Haditha area, Jordan: comprehensive combined reprocessing of geophysical data: comprehensive combined reprocessing of geophysical data *Environ. Earth Sci.* **80** 1–20

Feinstein S 1987 Constants on the thermal history of the Dead Sea Graben as revealed by coal ranks in deep boreholes *Tectonophysics* **141** 135–50

Ford D and Williams P 2007 *Karst Hydrogeology and Geomorphology* (New York: Wiley)

Freund R, Garfunkel Z, Zak I, Goldberg M, Weissword T and Derin B 1970 The Shear along the Dead Sea Rift. *Phil. Trans. R. Soc. Lond. A* **267** 107–30

Frumkin A 1994 Morphology and development of salt caves *Natl. Speleol. Soc. Bull.* **56** 82–95

Frumkin A 1995 Rapid entrenchment of stream profiles in the salt caves of Mount Sedom, Israel *Earth Surf. Process. Landf.* **20** 139–52

Frumkin A 2013 Salt karst *Treatise on GeomorphologyKarst Geomorphology* vol 6 ed A Frumkin (San Diego, CA: Academic) pp 407–24

Frumkin A and Raz E 2001 Collapse and subsidence associated with salt karstification along the Dead Sea *Carbonates Evaporites* **16** 117–30

Frumkin A and Elitzur Y 2002 Historic dead sea level fluctuations calibrated with geologicaland archaeological evidence *Quat. Res* **57** 334–42

Frumkin A, Karkanas P, Bar-Matthews M, Barkai R, Gopher A, Shahack-Gross R and Vaks A 2009a Gravitational deformations and fillings of aging caves: the example of Qesem karst system, Israel *Geomorphology* **106** 154–64

Frumkin A, Ezersky M, Al-Zoubi A, Akkawi E and Abueladas A-R 2011 The Dead Sea hazard: geophysical assessment of salt dissolution and collapse *Geomorphology* **134** 102–17

Frumkin A 1997 Liquid crystal cave, Israel ed C Hill and P Forti *Cave Minerals of the World* (Huntsville, AL: National Speleological Society) pp 319–22

Frydman S, Charrash J and Goretsky I 2008 Geotechnical properties of evaporate soils on the Dead Sea area *Eng. Geol.* **101** 236–44

Frydman S, Charrash J and Goretsky I 2014 A geotechnical study of evaporitic, lacustrine sedimentsin the saline environment of the Dead Sea area *Eng. Geol.* **181** 309–22

Galve J, Gutiérrez F, Lucha P, Guerrero J, Remondo J, Bonachea J and Cendrero A 2009a Probabilistic sinkhole modelling for hazard assessment *Earth Surf. Process. Landf.* **34** 437–52

Galve J P, Gutiérrez F, Remondo J, Bonachea J, Lucha P and Cendrero A 2009b Evaluating and comparing methods of sinkhole susceptibility mapping in the Ebro Valley evaporite karst (NE Spain) *Geomorphology* **111** 160–72

Gardosh M, Reshes Z and Garfunkel Z 1990 Holocene tectonic deformation along the western margins of the Dead Sea *Tectonophysics* **180** 123–37

Garfunkel Z 1981 Internal structure of the Dead Sea leaky transform (rift) in relation to plate kinematics *Tectonophysics* **80** 81–108

Garfunkel Z 1997 The history and formation of the Dead Sea basin *The Dead Sea: The Lake and its Setting* ed T M Niemi, Z Ben-Avraham and J Gat (Oxford: Oxford University Press) pp 36–55

Garfunkel Z, Zak I and Freund R 1981 Active faulting in the Dead Sea rift *Tectonophysics* **80** 1–26

Garfunkel Z and Ben-Avraham Z 1996 The structure of the Dead Sea basin *Tectonophysics* **266** 155–76

Goetz P W (ed) 1986 *The New Encyclopedia Britannica 3* 15th edn (Chicago, IL: Encyclopedia Britannica, Inc.) p 937 (http://en.wikipedia.org/wiki/List_of_bodies_of_water_by_salinity)

Goldman M, Gilad D, Ronen A and Melloul A 1991 Mapping of sea water intrusion into the coastal aquifer of Israel by the time domain electromagnetic method *Geoexploration* **28** 153–74

Grimes K G 1975 Pseudokarst: definition and types *Proceedings of 10th Biennial Conference* ed A W Graham (Brisbane: Australian Speleological Federation)

Gutiérrez F, Cooper A and Johnson K 2008 Identification, prediction, and mitigation of sinkhole hazards in evaporite karst areas *Environ. Geol.* **53** 1007–22

Gutiérrez F, Parise M, De Waele J and Jourde H 2014 A review on natural and human induced geohazards and impacts in karst *Earth Sci. Rev.* **138** 61–88

Harza JRV Group 1998 Jordan rift valley—integrated development study: Red Sea–Dead Sea canal project *Prefeasibility Report*

Harza JRV Group 1996 Red Sea–Dead Sea canal project, draft pre-feasibility report, main report *Jordan Rift Valley Steering Committee of the Trilateral Economic Committee*

Higgins C 1984 Piping and sapping ed R LaFleur *Groundwater as a Geomorphic Agent* (Winchester: Allen & Unwin) pp 19–58

Ivanov V A, Lyubartseva S P, Mikhailova E N, Shapiro N B and Gertman I 2002 Model of the Dead Sea. Simulation of the variability of the thermohaline water structure in 1992–2000 *Phys. Oceanogr.* **12** 237–56

Jenyon M K 1986 *Salt Tectonics* (London: Elsevier)

Keydar S, Medvedev B, Al-Zoubi A, Ezersky M and Akkawi E 2013 3D Imaging of Dead Sea area using weighted multipath summation: a case study *Int. J. Geophys.* **2013** 692452

Klein C 1982 Morphological evidence of lake level changes, western shore of the Dead Sea *Isr. J. Earth Sci.* **31** 67–94

Klimchouk A 2005 Subsidence hazards in different types of karst: evolutionary and speleogenetic approach *Environ. Geol.* **48** 287–95

Legchenko A 2013 *Magnetic Resonance Imaging for Groundwater* (London: Wiley-ISTE) p 157

Legchenko A, Ezersky M, Boucher M, Camerlynk C, Al-Zoubi A and Chalikakis K 2008a Pre-existing caverns in salt formations could be the major cause of sinkhole hazards along the coast of the Dead Sea *Geophys. Res. Lett.* **35** L19404

Legchenko A, Ezersky M, Camerlink C, Al-Zoubi A, Chalikakis K and Girard J-F 2008b Locating water-filled karst caverns and estimating their volume using magnetic resonance soundings *Geophysics* **73** 51–61

Legchenko A, Ezersky M, Camerlynck C, Al-Zoubi A and Chalikakis K 2009 Joint use of TEM and MRS method in the complex geological setting *C. R. Geosci.* **341** 908–17

Levi E, Goldman M, Hadad A and Gvirtzman H 2008 Spatial delineation of groundwater salinity using deep time domain electromagnetic geophysical measurements: a feasibility study *Water Resour. Res* **44** W12404

Levitte D and Olshina A 1985 Isotherm and geothermal gradient maps of Israel *Report* GS1/60/84 Geological Survey of Israel, Jerusalem

Lisker S, Vaks A, Bar-Mathews M, Porart R and Frumkin A 2009 Stromatolites in caves of the Dead Sea Fault Escarpment: implications to latest Pleistocene lake levels and tectonic subsidence *Quat. Sci. Rev.* **28** 80–92

Lucha P, Cardona F, Gutiérrez F and Guerrero J 2008 Natural and human-induced dissolution and subsidence processes in the salt outcrop of the Cardona Diapir (NE Spain) *Environ. Geol.* **53** 1023–35

Mendonca A F, Pires A C B and Barros J G C 1993 Pseudo sinkholes in lateritic terrains, Brazilia, Brazil *Applied Karst Geology* ed G Beck (Rotterdam: A A Balkema)

Migowski C, Stein M, Prasad S, Agnon A and Negendank J F W 2006 Dead Sea levels, climate variability and human culture evolution in the Holocene Near East *Quat. Res.* **66** 421–31

Milanović P T 1981 *Karst Hydrogeology* (Littleton, CO: Water Resource Publications)

Milanović P T 2000 *Geological Engineering in Karst: Dams, Reservoirs, Grouting, Groundwater Protection, Water Tapping, Tunneling* (Belgrade: Zebra publishing Ltd) p 347

Milanović P, Maximovich N and Meshcheryakova O 2019 *Dams and Reservoirs in Evaporites* (Cham: Springer)

Nehorai R, Lensky I M, Lensky N G and Shiff S 2009 Remote sensing of the Dead Sea surface temperature *J. Geophys. Res.* **114** 1–8

Nof R N, Abelson M, Raz E, Magen Y, Atzori S, Salvi S and Baer G 2019 SAR interferometry for sinkhole early warning and susceptibility assessment along the Dead Sea, Israel *Remote Sens.* **11** 89

Parise M, De Waele J and Gutiérrez F 2009 Current perspectives on the environmental impacts and hazards in karst *Environ. Geol.* **58** 235–7

Quennell A M 1958 The structural and geomorphic evolution of the Dead Sea rift *Quart. J. Geol. Soc.* **114** 1–24

Ravizky A and Nadav N 2007 Salt production the evaporation of SWRO brine in Eilat: a success story *Desalination* **205** 374–9

Reuter F and Stoyan D 1993 Sinkholes in carbonate, sulphate, and chloride karst regions: principles and problems of engineering geological investigations and predictions, with comments for the construction and mining industries *Applied Karst Geology* ed B F Beck (Rotterdam: A A Balkema) pp 3–25

Rosenthal E and Eckstein Y 1968 Temperature gradients of the subsurface of the Dead Sea area, Israel *Isr. J. Earth Sci.* **17** 131–6

Rybakov M 2005 Microgravity and micromagnetic solution for Dead Sea sinkholes problem Near Surface 2005 *(Palermo, Italy, 4–7 September 2005) B020 4*

Salem H S 2020 Multi- and inter-disciplinary approaches towards understanding the sinkholes' phenomenon in the Dead Sea Basin *SN Appl. Sci.* **2** 667

Serra O 1984 Fundamentals of well-log interpretation *Acquisition of Logging Data* (Amsterdam: Elsevier) p 423

Shalev E, Lyakhovsky V and Yechieli Y 2006 Salt dissolution and sinkhole formation along the Dead Sea shore *J. Geophys. Res.* **111** B03102

Shalev E and Lyakhovsky V 2012 Viscoelastic damage modeling of sinkhole formation *J. Struct. Geol.* **42** 163–70

Shamir G 2006 The active structure of the Dead Sea depression *Geol. Soc. Am. Spec. Pap.* **401** 15–32

Shirav-Schwartz M, Calvo R, Bein A, Burg A, Nof R and Baer G 2006 Red Sea–Dead Sea conduit geo-environmental study along the arava valley *Report* GSI/29/2006 Geological Survey of Israel, Jerusalem

Stein M, Torfstein A, Gavrieli I and Yechieli Y 2010 Abrupt aridities and salt deposition in the post-glacial Dead Sea and their North Atlantic connection *Quat. Sci. Rev.* **29** 567–75

Stiller M, Yechieli Y and Gavrieli I 2016 Rates of halite dissolution in natural brines: Dead Sea solutions as a case study *Chem. Geol.* **447** 161–72

Taqieddin S A, Abderahman N S and Atallah M 2000 Sinkhole hazard along the eastern Dead Sea shoreline area, Jordan: a geological and geotechnical consideration *Environ. Geol.* **39** 1237–53

Terzaghi K 1925 Principles of soil mechanics, IV—settlement and consolidation of clay *Eng. News-Rec.* **95** 874–8

Terzaghi K, Peck R B and Mesri G 1996 *Soil Mechanics in Engineering Practice* 3rd edn (New York: Wiley)

Van Sambeek L L 1996 Dissolution-induced mine subsidence at the Retsof Salt Mine Proc. SMRI Fall Meeting *(Cleveland, OH)* (Solution Min. Res. Inst.) pp 289–309

Vey S *et al* 2021 Delayed subsidence of the Dead Sea shore due to hydro-meteorological changes *Sci. Rep.* **11** 13518

Wakshal E and Cohen A 1993 The geothermal gradient of the nubian sandstone aquifer in the Arava Rift Valley, Central Negev Heights and Sinai *Trans. of the Israel Geol. Society Meeting* 142

Warren J 2020 Brine density and persistence: part 1. Physical properties of a brine *Salty Matters* March 31, 2020 1–6

Weisbrod N, Alon-Mordish C and Yechieli Y 2012 Dynamic dissolution of halite rock during flow of diluted saline solutions *Geophys. Res. Lett.* **39** L09404

Wdowinski S, Bock Y, Baer G, Prawirodirdjo L, Bechor N, Naaman S, Knafo R, Forrai Y and Melzer Y 2004 GPS measurements of current crustal movements along the Dead Sea Fault *J. Geophys. Res.* **109** B05403

Willner S E, Lipchin C, Kronich S, Amiel T, Hartshorne N and Selix S A 2013 A pre-feasibility study on water conveyance routes to the Dead Sea. Arava Institute for Environmental

Studies, Kibbutz Ketura (http://arava.org/wp-content/uploads/2013/05/Dead-Sea-Conduits-Study-English.pdf)

Yakubov Y S and Eppelbaum L V 2005 Geodynamics of water flows in the Dead Sea basin: a proposed model for analysis of stratified, rotating and compressible liquid *Proc. 2nd EUG Meet vol 7 (Vienna, Austria)* pp 1–3

Yechieli Y 2000 Fresh-Saline ground water interface in the western Dead Sea area *Groundwater* **38** 615–23

Yechieli Y 2007 Geochemical monitoring of groundwater in boreholes at the DS coastal area—project of sinkholes (in Hebrew) *Unpublished GSI Report* TR-GSI/18/2007 Geological Survey of Israel

Yechieli Y, Ronen D, Berkovitz B, Dershovitz W S and Hadad A 1995a Aquifer characteristics derived from the interaction between levels of a terminal lake (Dead Sea) and adjacent aquifer *Water Resour. Res.* **31** 893–902

Yechieli Y, Ronen D and Berkowitz B 1995b Are sedimentary salt layers always impermeable? *Geophys. Res. Lett.* **22** 2761–4

Yechieli Y, Gavrieli I, Ronen D and Berkovitz B 1998 Will the Dead Sea die? *Geology* **26** 755–8

Yechieli Y, Wachs D, Shtivelman V, Abelson M, Onn C, Shtivelman V, Raz E and Baer G 2002 Formation of sinkholes along the shore of the Dead Sea—summary of the first stage of investigations *Geol. Surv. Isr. Curr. Res.* **13** 1–6

Yechieli Y, Abelson M, Bain A, Shtivelman V, Onn C, Wacks D, Bar G, Kalbo R and Lyachovski V 2004 Sinkhole phenomena along the Dead Sea shore line (in Hebrew) *Unpublished GSI Report* GSI/21/04

Yechieli Y, Abelson M, Bein M, Crouvi O and Shtivelman V 2006 Sinkhole 'swarms' along the Dead Sea coast: reflection of disturbance of lake and adjacent groundwater systems *Geol. Soc. Am. Bull.* **118** 1075–87

Yechieli Y, Abelson M and Baer G 2016 Sinkhole formation and subsidence along the Dead Sea coast, Israel *Hydrol. J.* **24** 601–12

IOP Publishing

Applied Geophysics for Karst and Sinkhole Investigation
The Dead Sea and other regions
Michael Ezersky, Lev V Eppelbaum and Anatoly Legchenko

Chapter 4

Geophysical methods applied to the sinkhole investigation at the Dead Sea coasts

4.1 An overview of surface geophysical methods applied to sinkhole development at the Dead Sea coasts in Israel and Jordan

Parts of this chapter have been reprinted from Ronen *et al* (2019), Ezersky *et al* (2011), Ezersky and Frumkin (2017), Ezersky and Frumkin (2021), Frumkin *et al* (2011), Ezersky (2008), Ezersky and Goretsky (2014)

More than 6000 sinkholes have been formed during the last 30 years along the Dead Sea (DS) coastal areas in Israel (Abelson *et al* 2006, 2021, Yechieli *et al* 2016, DS sinkhole database 1998–2016) and Jordan (Closson and Abou Karaki 2009, Abueladas 2016—personal communication, Abou Karaki *et al* 2019) (see figure 4.1).

Sinkholes formed in evaporite areas are associated with the dissolution of buried salt layers. Salt is a very soluble material, even in slightly saline (undersaturated) circulating groundwater. Dissolution rates of salt are one to three orders of magnitude higher compared with limestone (Reuter and Stoyan 1993, Gutiérrez *et al* 2008). Therefore, the subsurface dissolution of salt by such groundwater and collapse of the overlying material (sinkhole formation) may also be relatively rapid, from days to some weeks (Frumkin 1995, Van Sambeek 1996, Stiller *et al* 2016).

On the other hand, the DS groundwater contains essentially more salt (~265–275 g L^{-1} NaCl) than normal groundwater. Therefore, DS groundwater slightly slows down the dissolution process (months, years) and allows the observation of changes in geophysical properties of the subsurface on a human life scale.

Sinkholes are surface manifestations of subsurface dissolution and internal erosion and deformation, commonly hidden from the direct observations and most subaerial geomorphological study methods (Gutiérrez *et al* 2014).

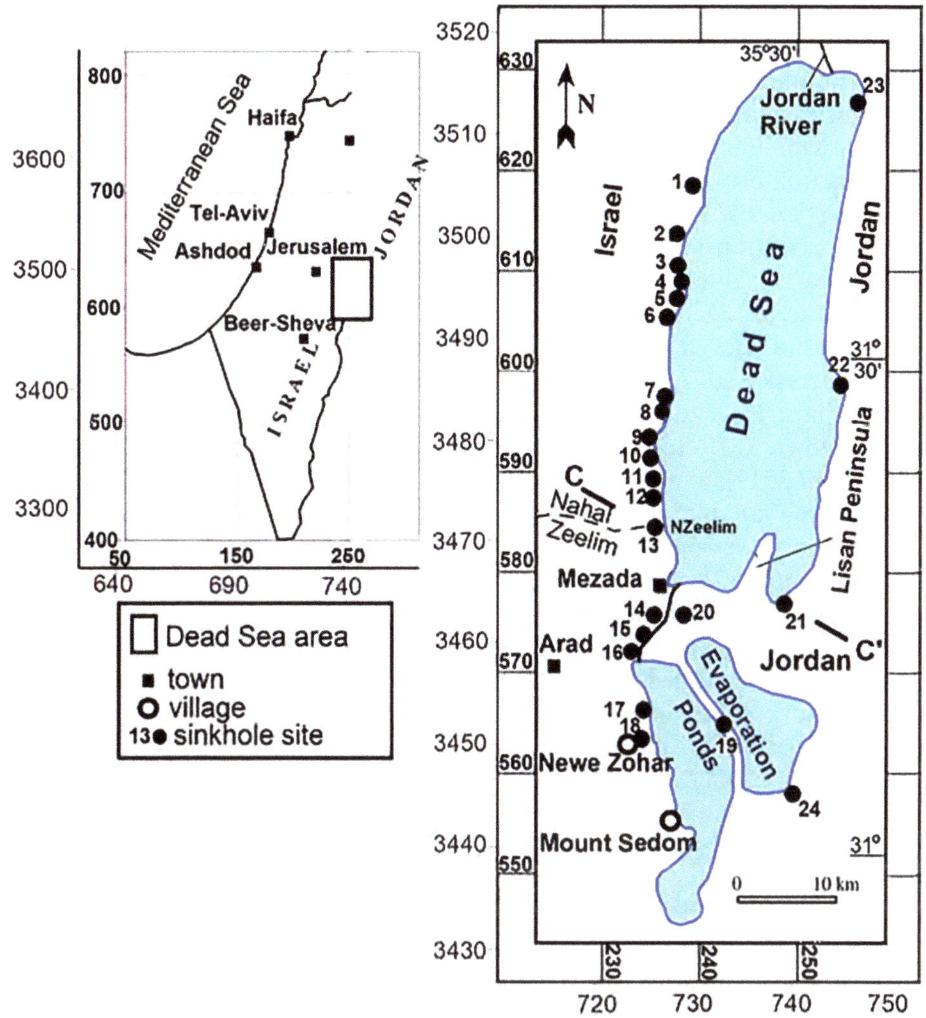

Figure 4.1. Sinkhole sites along the DS shore: (1) Zikim; (2) Avenat; (3) N.Qane; (4) Samar; (5) N. Deragot; (6) Mineral Beach; (7) Ein Gedi and Nahal Arugot; (8) Yesha; (9) Mazor; (10) Nahal Hever north; (11) Nahal Hever south; (12) Asa'el; (13) Nahal Zeelim; (14) Rahaf; (15) Lynch Straight; (16) Mor; (17) Ein Boqeq; (18) Newe Zohar; (19) Dam-2; (20) Lisan; (21) Ghor Al-Haditha; (22) Wadi Mejib; (23) Suweimeh; (24) unnamed; (20–24—after Closson and Abu Karaki (2009)). CC′—the line of the generalized geological section (the coordinates in the frame are new Israel Mercator grid in km, the outer coordinates are Universal Transverse Mercator grid, in km). (Reprinted from Frumkin *et al* (2011), copyright (2011) with permission from Elsevier).)

Different geological formations in the DS coast show strong contrast in their physical properties. That allows the use of surface geophysical methods to study subsurface layers (Ezersky *et al* 2006, Frumkin *et al* 2011). For instance, the contrast in seismic velocities between salt layers (high velocities) and surrounding salt clastic sediments (low velocities) in the DS graben area allows the mapping of salt formations with high reliability (Ezersky 2006). The very low resistivity of the DS aquifers in the coastal area also creates a contrast with the overlying sediments saturated with groundwater. It permits locating of the top of the DS brine as 1 Ωm or 0.5 Ωm interfaces with high accuracy, minimal equivalency, and the distinction between salty and brackish water (Goldman *et al* 1991, Kafri *et al* 1997, Ezersky *et al* 2011, Ezersky and Frumkin 2017, Goldman and Kafri 2021). Moreover, the transient electromagnetic method (TEM) allows one to determine the DS water salinity and aggressivity and to map it through the sinkhole hazardous areas (Ezersky and Frumkin 2017, 2021). The anomaly high resistivity of the sinkhole development sites was discovered in alluvial fans in Israel and Jordan (Ezersky 2008, Al-Zoubi *et al* 2007). It allows effective mapping of the sinkhole hazardous zones and detecting of the signs of sinkhole and subsurface cavities (down to 10–15 m) using the GPR technique (Frumkin *et al* 2009, 2011, Ronen *et al* 2019). Thus, favorable ground conditions for applying geophysical methods are available in the DS coastal area, and the latter can facilitate sinkhole susceptibility assessment in the DS coastal areas both in Israel and Jordan (Ezersky *et al* 2017).

Two aspects of the problem are considered: (1) investigation of the salt and (2) study of the overlying sediments. To resolve these problems different geophysical methods have been tested for the various aspects of sinkhole hazard assessment. In the framework of sinkhole studies old geophysical methodologies were updated and new ones based on seismic and other surface methods were developed (Frumkin *et al* 2011). Detailed considerations on these methods will be given in continuation.

Geophysical investigations were carried out in parallel with the drilling of the boreholes. Numerous 70–90 m depth boreholes were drilled by the Geological Survey of Israel (GSI) (Yechieli *et al* 2004) along with the western DS shores through the sinkhole sites.

In this section, we intend to summarize the available experience and to compare predictions of sinkhole expansion throughout the DS basin made by different authors. We have also tried to compare these predictions with the recent geological–geophysical observations. We hope that such an analysis will produce a better understanding of the phenomenon, which is particularly important at present, when the construction of a projected channel between the Red and Dead seas (Beyth 2007) is being decided.

We anticipate discussing the presence of a salt layer both in the western (Israel) and eastern (Jordan) sides of the DS shores. For instance, such analysis in the Ghor Al-Haditha area (eastern DS side) will be based on an integrated analysis of all available recently observed geophysical data.

A list of geophysical methods used for the study of the DS sinkhole problem by the authors and different researchers has been collected in table 4.1.

In continuation, we will consider all the above-mentioned geophysical methods in detail.

Table 4.1. The geophysical method used for sinkhole studies in the Dead Sea coast (Ezersky *et al* 2017, with permission of IJS).

Method name	Physical properties to be measured	Capability of detection	References
Seismic refraction	The seismic velocity of longitudinal waves (V_p) in solid salt layers (2900–4500 m s^{-1}) is considerably higher than sedimentary overburden (21 090–2700 m s^{-1})	Salt elastic properties and salt delineation	Ezersky (2006), Ezersky *et al* (2010)
Multi-channel analysis of surface waves (MASW)	The seismic velocity of shear (S) waves (V_s)	Salt shear velocities and modulus to estimate salt porosity and karstification	Bodet *et al* (2010), Ezersky *et al* (2013a)
P-wave seismic reflection	Study of the deep (down to 200 m) subsurface structure, detection of sub-horizontal interfaces and sub-vertical faults	Detection of the salt layer within sediments, faults, cavities, etc.	Keydar *et al* (2010, 2012a, 2012b, 2013), Ezersky and Frumkin (2013)
S-wave seismic reflection	Study of the shallow (down to 100 m) subsurface structure, detection of sub-horizontal interfaces and faults	Detection of the salt layer within sediments, faults, cavities, etc.	Krawczyk *et al* (2012, 2015), Polom *et al* (2018)
Transient electromagnetic (TEM, TDEM)	Low resistivity of aquifer formations (1–0.5 Ωm) in contrast to the surrounding sediments (more than 10 Ωm)	Accurate location of the top of the aquifer and mapping water salinity and water aggressiveness variations	Goldman *et al* (1991), Ezersky *et al* (2011), Ezersky and Frumkin (2017, 2021), Goldman and Kafri (2021)
Electric resistivity tomography (ERT)	DC resistivity in subsurface	Search anomalously high resistivity linked to subsurface cavities	Al-Zoubi *et al* (2007), Ezersky (2008)
Magnetic resonance sounding (MRS) or surface nuclear magnetic resonance (SNMR) + 3D-SNMR	Water in the karst voids and water in the surrounding sediments produce nuclear magnetic resonance signals with different relaxation times (less than 150 ms in sediments and more than 600 ms in karst	Detection of water-filled voids, estimate the volume of water and hydraulic conductivity in subsurface layers and karstic caves	Legchenko *et al* (2008a, 2008c), Legchenko (2013)

MRS + TEM	Bulk resistivity, water content (W), hydraulic conductivity (K), and relaxation time (T_1) allows resolving equivalency	Identification of the subsurface lithology	Legchenko *et al* (2008b, 2009)
MRS+MASW	Combining of K and V_s	Estimating of the salt hydraulic conductivity	Ezersky and Legchenko (2014, 2015)
Ground-penetrating radar (GPR)	Structure of most shallow (10–15 m) sediments located over karstified unconsolidated salt	Detection of density deficit caused by cavities moving upward (stoping)	Frumkin *et al* (2009, 2011), Ronen *et al* (2019)
Microgravity and magnetics	Detection of gravity anomalies against a strong noise background	Detection of caverns, estimation of their depth and volume	El-Isa *et al* (1995), Rybakov *et al* (2001, 2005), Eppelbaum (2007a, 2008, 2010, 2011a, 2011b, 2015a, 2015b), Ezersky *et al* (2013b)
Microseismic activity	Detection of microseismic noise	Detection of cavities	Wust-Bloch and Joswig (2006), Abelson *et al* (2018)
Horizontal-to-vertical spectral ratio (HVSR)	Constructing of HSVR	Constructing of the subsurface V_s model	Gorstein and Ezersky (2015)
Borehole measurements	Measurement of V_p and V_s in boreholes	Constructing of the subsurface V_p, V_s model, resistivity–salinity calibration	Ezersky and Goretsky (2014)
Remote sensing methods	Mapping of sinkholes using InSAR, LiDAR	Sinkhole detection, monitoring and inventory	Baer *et al* (2002, 2018), Filin *et al* (2012), Nof *et al* (2019)

4.2 Seismic methods

All seismic methods used in the DS karst investigation have employed the same equipment (apart from geophones) that can be described in the framework of this chapter.

DIGIPULSE seismic source

As a source, this uses an accelerated weight-drop based on a Chevrolet truck (figure 4.2(a)).

The Digipulse seismic source is a vehicle-mounted, high-performance gas-charged accelerated (impact) weight-drop energy source designed for use on seismic exploration surveys such as shallow and deep refraction surveys, 2D and 3D seismic reflection surveys, MASW surveys, and downhole seismic. Vertical stroke is carried out by the 200 kg hammer equipped with a slingshot (figure 4.2(b)). During seismic surveying, the vertical hammer (1(a) in figure 4.2(b)) is lifted and strikes a metallic plate (2a) forced by the slingshot.

For a horizontal stroke, a 30 kg sledgehammer is used (figure 4.2(c)) mounted on the additional tower designed in the Geophysical Institute of Israel (GII). A horizontal stroke is carried out by a 30 kg hammer (1(b) in figure 4.2(c)) on a horizontal beam (2b).

Geophones

Geophones use the following frequency range. For such methods as seismic refraction, reflection, and diffraction, the frequency range is more than 10 Hz, and vertical geophones 10 Hz are used. Horizontal geophones of 8 Hz are used for the S-wave recording. For the MASW method, the frequency range is 1–50 Hz and low-frequency geophones of 2.5 or 4.5 Hz are used (figure 4.3)

Figure 4.2. Seismic data acquisition (Nahal Zeelim site). (a) Digipulse seismic source during work: (1) borehole DS-1, (2) sinkhole, (3) vertical Digipulse hammer tower, (4) the DS; (b) vertical hammer (1a) stroking on the metallic plate (2a) to excite compression and surface waves; (c) horizontal hammer (1b) stroking on the horizontal beam (2b) to excite the shear waves SH. 3—on (b), and (c) is a trigger.

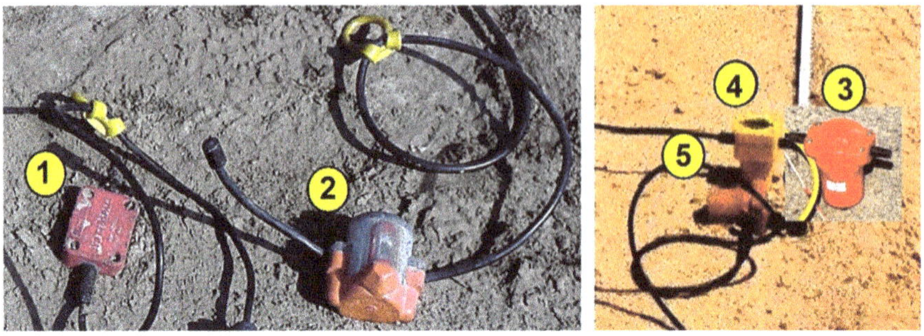

Figure 4.3. Geophones used in the seismic surveys: (1) vertical 10 Hz geophone for P-wave recording, (2) horizontal 8.0 Hz geophone for S-wave recording, (3) vertical 2.5 Hz geophone for surface wave (MASW) survey; (4) and (5) 4.5 Hz geophones for the vertical (V_{SV}) and horizontal (V_{SH}) vibrations, respectively. Note, that vertical geophones of 4.5 Hz were used in the Ghor Al-Haditha (Jordan) and 2.5 Hz vertical geophones were used in the western DS shores (Israel).

Figure 4.4. Data recording in the DS western shore: StrataView seismograph.

Seismographs

Finally, seismic recording systems (seismographs) (figure 4.4) such as Geometrics StrataViewer (48 channel) (Canada); Geode (24 channels) (Canada); Summit-II plus (96 channels) (GMT Germany).

4.2.1 Seismic refraction method

4.2.1.1 The features of the geological conditions in the DS sinkhole sites

The features of the DS geological structure allowing an application of the seismic refraction method in the local specific geological conditions is described in section 3.2.3. From the point of view of seismic methods, it is important to note that (1) there is a salt layer of 6–30 m thickness at 20–50 m depth (top); and (2) salt is located within a graben area filled with unconsolidated sediments. Salt is often sandwiched between the marl layers from below and from above, especially when approaching the DS shoreline.

4.2.1.1.1 Salt and sediments genesis and velocities (Yechieli et al **1995a, 1995b)**

The salt of the coastal DS area is composed of 10 200–10 800 year-old layers of halite (NaCl) (Yechieli *et al* 1995b). These layers were deposited in the Early Holocene during the development of the extremely arid conditions in the shrinking phase between the Lisan Lake and the present Dead Sea (Stein *et al* 2010). The halite appears as rock salt. The salt was deposited within the DS area, and its boundary is extended west of the existing pool (Ezersky *et al* 2013b, Ezersky and Frumkin 2013). The salt unit is located significantly below the groundwater table. The DS salt is characterized by unit weights considerably lower than those usually discussed in the literature, and by the correspondingly different mechanical properties (Frydman *et al* 2014). The unit reaches 4–7 m in thickness at the marginal sites of the lake and thickens up to 30 m toward the deeper lake environment (Stein *et al* 2010). Holocene sandy-gravel sediments and lime carbonates overlie the salt unit. The lime carbonate sediments comprise laminated clay-silt-sized, clastic sediments (calcite), authigenic aragonite, and gypsum (Charrach *et al* 2007).

4.2.1.1.2 Salt layer velocities

Analysis of the salt layer velocities carried out by seismic refraction verified by the borehole drilling at 11 DS sites has shown that longitudinal (compressional) velocities V_p are ranged between 2920 and 4240 m s^{-1} with an average velocity of $V_p = 3468 \pm 402$ (Ezersky 2006). Hence, the most reliable (statistically substantiated) minimum velocity criterion for salt layer presence (based on P-wave velocity only) can be accepted as $Vp_{min} = 2900$ m s^{-1}. During recent years, this criterion was not changed.

On the other hand, how does this criterion interrelate with the P-wave velocities in unconsolidated sediments? Considering the distribution of the compressional velocities in unconsolidated sediments and rock (figure 4.5), it can be seen that $V_p = 2700$ m s^{-1} is an upper limit of the water-saturated gravel (Lyachovitsky *et al* 1989); 2750 m s^{-1} can be a top velocity in the water-saturated clay (Jakosky 1957, Press 1966). As a rule, P-wave velocities of the DS unconsolidated sediments saturated with groundwater are ranged 2200–2500 m s^{-1}.

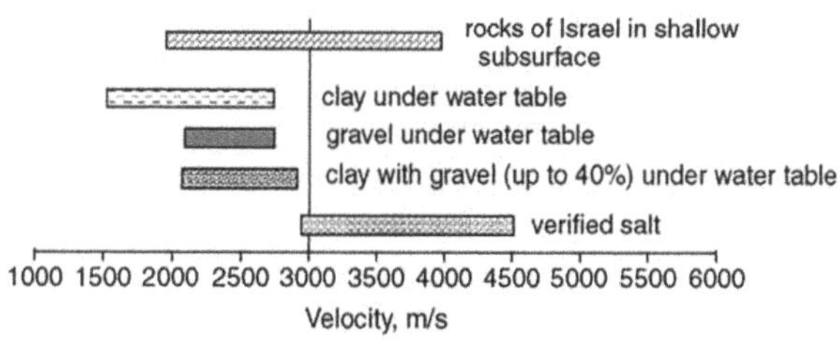

Figure 4.5. Ranges of P-wave velocities in unconsolidated sediments and rocks in Israel (after Ezersky and Sobolevsky (2002)). (Reprinted from Ezersky (2006), with permission from Elsevier.)

In figure 4.5, we note that the compressional seismic velocities of 3000–4000 m s^{-1} are the upper limit for the carbonate rocks in the shallow subsurface in Israel (Ezersky and Sobolevsky 2002). Fortunately, the carbonate rock is absent within the DS graben area or located at considerable depths, whereas the salt unit is mainly extended within the DS graben area. Thus, the above compressional wave velocity criterion Vp_{min} = 2900 m s^{-1} can be applied in the graben area.

4.2.1.1.3 The geological model of the sinkhole sites subsurface

From that considered above it follows that there must be a significant difference in the seismic velocities V_p between the salt and unconsolidated sediments. The *priory* geological model adopted for geophysical mapping is illustrated in figure 4.6. It is based on the previously available seismic refraction data and geological boreholes of the DS area.

The high-velocity unit contacts with the relatively low-velocity unit at depths of approximately 20–50 m (e.g. there are lateral velocity variations along with the refractor). The unconsolidated sediments (alluvium) overlay this unit. The mapping of the salt–alluvium boundary is the main goal of seismic refraction surveys. The sediments coming into contact with the salt layer have higher velocities than the overlying alluvium because of cementation gradation by salt solution.

This analysis has shown that there are two zones with different velocities V_p along with the refractor. The western part of the area is composed of a high-velocity layer with V_p > 2900 m s^{-1}, whereas the eastern part of the area is composed of a relatively low-velocity layer with V_p < 2900 m s^{-1}.

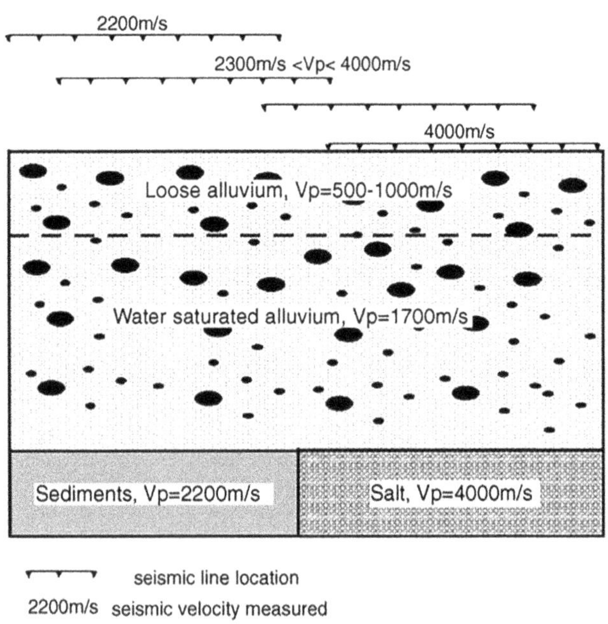

Figure 4.6. Hypothetical geological model of the salt unit environment. (Reprinted from Ezersky (2006), with permission from Elsevier.)

4.2.1.2 Features of the seismic refraction methodology

The seismic refraction method (SRFR) is a geophysical technique intended to determine the elastic characteristics of soil and rock layers and their thicknesses. The travel times of seismic waves refracted at the interfaces between subsurface layers of different velocities are measured.

The general reciprocal method (GRM) uses travel-time data from both forward and reverse shots and provides a graphical solution to resolve the geometry of subsurface refractors using an inversion technique. The method permits the determination of the detailed structure of the refractor and information about any localized lateral variations.

It is based on the principle of refraction migration, which uses the offset (migration) distance which is the horizontal separation between a point on the refractor where a ray is critically refracted and that at the surface where the ray emerges (Palmer 1986, Lankston 1990, Reynolds 2011).

The method was applied to delineate the salt boundary concerning the marine sediments (Ezersky 2006, 2011b, Legchenko *et al* 2008a, Ezersky *et al* 2010, Frumkin *et al* 2011).

For the DS conditions, one should consider the example of data acquisition and processing. Data acquisition was carried out using the 48-channel StrataView seismic recorder by Geometrics Inc (Canada). 48-geophone spreads with 2.5 m separations between geophones were used. Thus, the spread length was 117.5 m.

The DIGIPULSE hydraulic seismic source (figure 4.7) was applied at five points on every refraction line following the GRM technique (Palmer 1986): at both ends of spread (zero-offsets), within the spread (split) and at a certain distance from each end (offsets). A shot offset distance was selected after forwarding modeling. Geophones of 10 Hz (vertical) and 8 Hz (horizontal) were used as seismic sensors.

Figure 4.7. Data acquisition in the Nahal Hever sinkhole site, western Dead Sea (see figure 4.1 for site location). DIGIPULSE seismic source on the Chevrolet track. Two sources are seen as one for the P- wave excitation (grey tower), and S-wave (black tower). The sinkhole is seen in the foreground. (Photo of the author.)

The data processing was carried out using REFRINT software (Shtivelman 1995) implementing two interpretation methods: GRM and slope-intercept method (SIM).

The example of the field records (raw data) showing the seismic signal as a function of time is presented in figure 4.8. There are refracted longitudinal (P) waves and surface waves. No filter was applied. Two zero-offset shots left (a) and right (b) show the same reciprocal time (Rt). Tree layers are seen in the time–distance function. A shot revealed the refracted wave.

The idea of the study is explained for a typical profile intersecting the salt layer boundary. Line EB-7 is 117.5 m long and has 2.5 m intervals between geophones (see figure 4.1 for site location). In this case, noise does not complicate field records and no filter was applied to the raw data. However, in several lines, noise complicates the record, and a high cut filter of 120–250 Hz is usually applied to the record to remove this noise. The strong surface waves are also seen in the raw data. Three separate layers would be distinguished on the records (figure 4.8).

Generally, three lateral blocks with considerably different refractor velocities can be identified from the velocity analysis function (figure 4.9(a)). The northern one is composed of sediments with a velocity of 2590 m s^{-1}, while the second and third ones have a velocity of 2970 m s^{-1} (intermediate), and 3920 m s^{-1} (northern block). Such velocity structure is supported by the refractor velocity graph (figure 4.9(b)) calculated at $XY = 25$ m and a smooth window of 25 m. The depth–velocity section model (figure 4.9(c)) is built using the information on velocities obtained with the GRM method. The section consists of three layers. The uppermost layer from the surface down to approximately 8 m, with a velocity of 690 m s^{-1}, characterizes unsaturated alluvium composed of sand with gravel, clay, or marl; the second layer at the depth interval from 8 to 20–25 m, with a velocity of 2460 m s^{-1}, is interpreted as the alluvial material saturated with groundwater. The lowest layer (refractor) is a velocity heterogeneous structure including three lateral blocks mentioned above. All selected layers are viewed through the profile. One can conclude based on the salt layer velocity criterion (Vp_{min} = 2900 m s^{-1}) that two southern high-velocity blocks of the refractor (V_p = 3930 and 2970 m s^{-1}) are the salt unit, whereas the northern one (V_p = 2590 m s^{-1}) is a low-velocity part of a refractor is the clastic sediments.

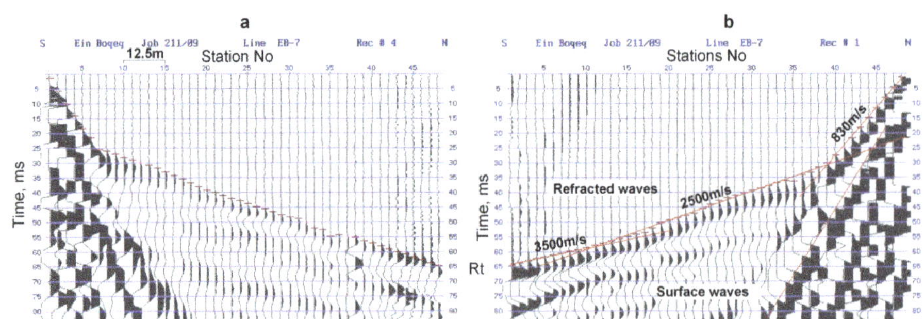

Figure 4.8. The examples of the field record (raw data) showing the seismic signal as a function of time. (a) Forward zero-offset shot and (b) reverse zero-offset shot. The signal was not filtered.

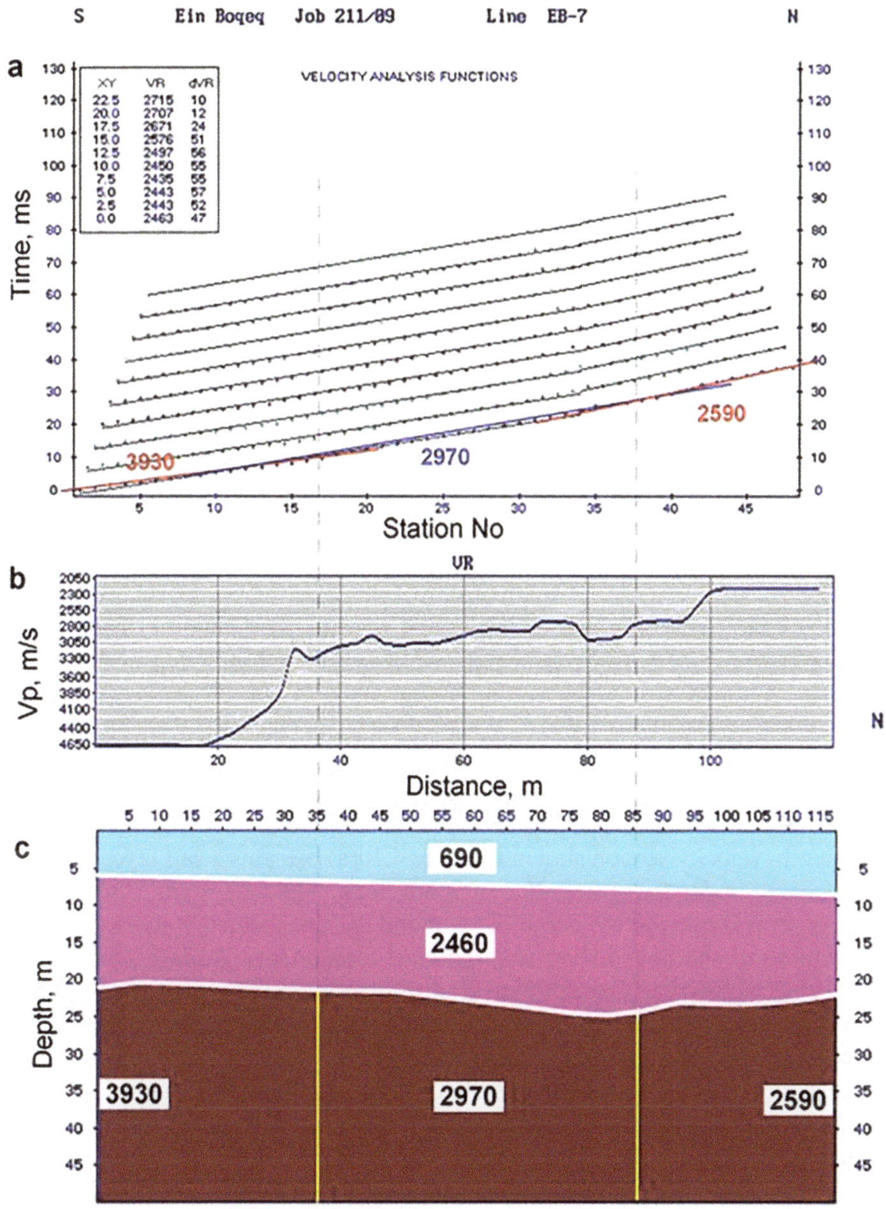

Figure 4.9. Example of data processing of the refraction line EB-7 at the Ein Boqeq area 48-geophone spread, 2.5 m separation between geophones based on GRM method. (a) Velocity analysis function-VAF, (b) refraction velocity graph along the line; and (c) velocity–depth section. There are three velocity layers. The lower layer (refractor) consists of three segments along with the refractor that are selected at the VAF: 2590, 2970, and 3930 m s^{-1}. The box in (a) shows parameters of the seismic segment: various values of the *XY* (parameter of the GRM method), the value of the compressional seismic velocity along the refractor segment (VR), and its average misfit (in m s^{-1}).

This conclusion is strongly supported by both refraction and borehole data: borehole EB-2 some south of the line reveals the salt unit the GRM method), the value of the compressional seismic velocity along the refractor segment (VR), and its average misfit (in m s^{-1}).

The example, presented in figure 4.9 is very stable throughout this area. Shots and sections along the lines crossing the salt sites are repeated from section to section (Ezersky 2006).

4.2.1.2.1 Forward modeling

The forward modeling was aimed to verify the DS subsurface model shown in figure 4.6 and to check the reliability of the GRM technique (Ezersky 2011b). The forward modeling was carried out using Modfar software available in the Refrint packet (Shtivelman 1995). The program computes the arrival times of refracted waves for various depth–velocity models and arbitrary source–receiver geometries using ray tracing. The model is defined using an interactive model generator which allows the user to easily specify the number of layers with the arbitrary interface geometries and with laterally varying velocities. A maximum of six layers can be specified per spread. Nine shots per spread can be used and 20 breakpoints dividing 20 velocity values along the refractor can be defined in the model section.

The procedure of the forward modeling comprises: (1) definition of the velocity versus depth model for various subsurface structures that should be investigated and for different parameters of the acquisition (salt layer within the geological stratified medium, salt layer inclination, etc); (2) calculation of the time–distance graphs for this model; (3) at the next step an inversion of these time–distance graphs is performed; (4) comparison of the inverted and input models to estimate the reliability of the data inversion. Let us consider examples of such modeling (figure 4.10).

The depth–velocity section shown in figure 4.10(a) presents a salt layer (V_p= 3000 m s^{-1}). located under the seismic line of 235 m long. The maximum offset is 120 m. The salt layer is located at the depth of 34 m and its edge located at 47 m (20%) from the beginning of the line. It contacts the clastic sediments saturated with the DS water (V_p= 2500 m s^{-1}). The salt is overlain by an intermediate layer composed of the water-saturated sediments (V_p= 2200 m s^{-1}) and by the loose sediments of the uppermost layer (V_p= 840 m s^{-1}). The calculated time–distance graphs are presented in figure 4.10(b) (forward model). The dip of the graphs is the seismic velocity of the layer. The refractor velocity is calculated for $XY = 25$ m and a smooth window of 25 m (figure 4.10(c)). The velocity decreases gradually from 3000 to 2530 m s^{-1}. If we use 2900 m s^{-1} as salt layer criterion we will locate the salt edge at 55 m (instead of 47 m in the model, i.e. discrepancy is of 20% of the reallocation of the salt edge). The velocity analysis function (figure 4.10(d)) would be approximated by two segments of 2960 and 2530 m s^{-1}. The comparison between the parameters of the model obtained after inversion, and the parameters of the input model show that the velocity of the salt unit (2960 m s^{-1}) differs from the model by 1.3%, velocity of sediments (V_p= 2530 m s^{-1}) is more than model one (2500 m s^{-1}) by 1.2%. The salt layer edge location (at 55 m) differs from the model one by 8 m (15%). Finally, the salt layer depth in the inverted model is 27–34 m (±3–4 m or 10% of the true depth).

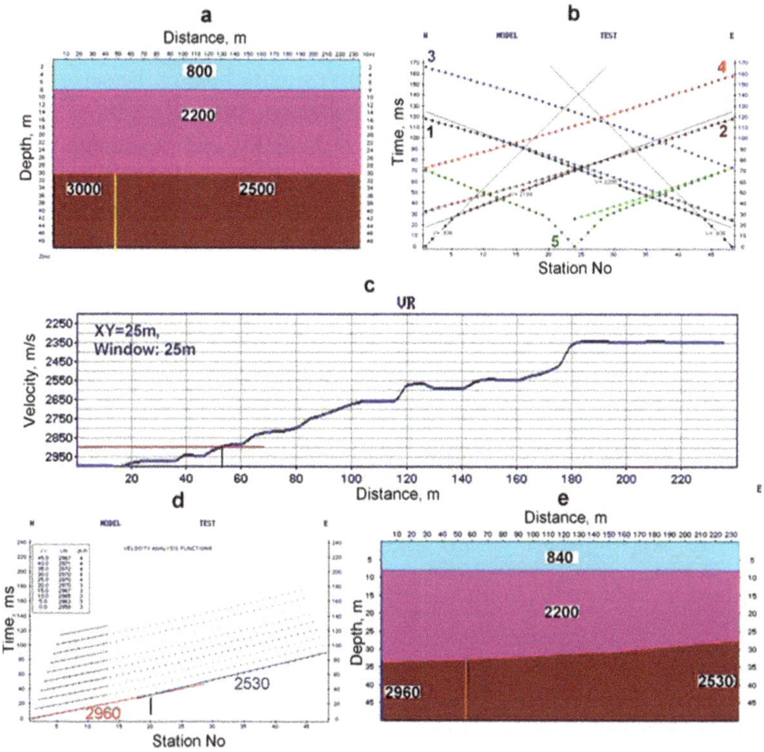

Figure 4.10. Model of the salt layer occupying 20% of the line length (47 m). (a) Input (model) for the velocity–depth section; (b) time–distance graphs calculated by software for the model section; (c) refractor velocity function for $XY = 25$ m; (d) the velocity analysis function for inversion of the time–distance graphs (b); and (d) inverse velocity–depth section revealed from the model section using GRM technique.

By increasing the salt layer length under the seismic line from 20% to 100%, the salt layer edge location varies in the framework of 20–30 m of the true edge location. The velocity of the salt layer is defined with an accuracy of 100–150 m (3%–5%). Depth to the salt top is determined with an accuracy of 10%.

Different variants of the forward modeling have shown that the velocity of the uppermost layer varies in the framework of ±5%. The velocity of the intermediate layer is very stable. The velocity of the refractor does not extend by ±2%.

The salt layer inclined to the surface more than 5° is not detected by the GRM method when its length is significantly less than the length of the seismic line (50% of it, for instance). In contrast, when salt is extended through the entire length of the line, both inclination and velocities are nicely corresponding to the model. Evidently, in this case, the seismic line should be shot along elevation isolines of the salt layer that is presumed to be parallel to the salt edge (generally, parallel to the sea shoreline).

4.2.1.2.2 Salt maps in several sinkhole sites

To check the validity of the salt edge dissolution model in other areas of the DS, several previous seismic refraction data obtained by the Geophysical Institute of

Israel during 1999–2004 were reinterpreted. After confirmation of the validity of the GRM method new seismic refraction surveys were carried out along the western DS shore. Ten sinkhole sites in Israel have been investigated using the seismic refraction method (Ezersky *et al* 2007, Ezersky and Legchenko 2008). Results of salt mapping in the Nahal Ze'elim area were published by Ezersky and Legchenko (2015). All maps are based on the salt velocity criterion $V_p > 2900$ m s^{-1}. The map presented in figure 4.11(a) shows the sinkhole sites studied using the seismic refraction method, the locations of boreholes drilled by GSI, and that of the seismic lines. During this study (Ezersky 2006) it was established that the sinkhole line is being conformed to the buried salt layer edge. Further refraction surveys have confirmed the conclusions of this study. The detailed maps of the salt edge from different sites have been presented in figures 4.11(b)–(d). These sites are distributed along the western DS shores in Israel (figure 4.11) and on the eastern shore in Jordan (figure 4.12).

The Ein Gedi area is situated in the central part of the western DS shoreline on the Nahal Arugot alluvial fan about ~37 m above the DS level in 2017 (see figure 4.1 for site location). The sinkholes at the Ein Gedi district formed in the areas affected by human activity. Among the first sinkholes that appeared from 1995 to 2015, more

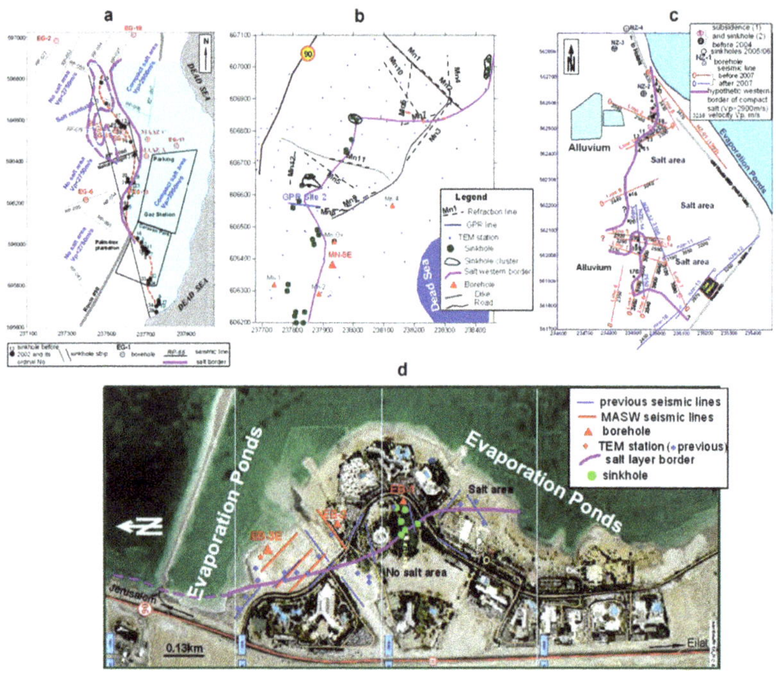

Figure 4.11. Salt location maps based on seismic refraction surveys in several western DS sinkhole formation sites. (a) Ein Gedi–Nahal Arugot area (line length is of 2 km) (Ezersky (2006), copyright (2006) with permission of Elsevier). (b) Mineral Beach (Frumkin *et al* (2011), copyright (2011) with permission of Elsevier). (c) Newe Zohar (Ezersky *et al* (2010), with permission of Schweizer Bart Science Publisher www.borntraeger-cramer.de/journals/zfg). (d) North of Ein Boqeq (DS Hotels area). Salt edge is denoted by a thick lilac line. Sinkholes are arranged (with some small deflections) along this line (Ezersky (2011), unpublished report). (a–d) The salt border is located west of DS. (Compilation from Ezersky *et al* (2017), with gratitude to Tampa, FL (USA).)

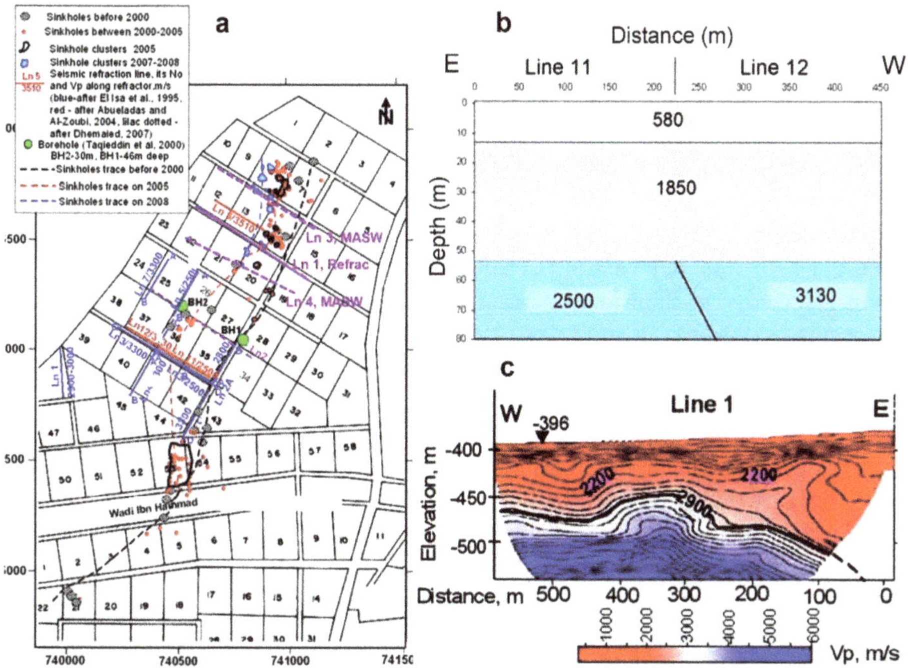

Figure 4.12. Seismic studies were carried out in the Ghor Al-Haditha area. (a) Ghor Al-Haditha area with location of seismic refraction lines carried out by different investigators (El-Isa *et al* 1995, Abueladas and Al-Zoubi 2004, Dhemaied 2007) and salt border map after Frumkin *et al* (2011). (b) Seismic refraction section performed by Abueladas and Al-Zoubi (2004), reprinted with permission of AGU. (c) One of four refraction tomography sections (along with the dashed lilac line #4) carried out by Dhemaied (2007). Note that all seismic sections show high velocity (>2900 m s^{-1}) layer located at a depth of 40–50 m, which in conditions of DS can be salt only (Ezersky 2006). The results acquired by El-Isa *et al* (1995) supporting these findings were presented by Ezersky *et al* (2013b). (Reprinted from (a) Ezersky *et al* (2013b), with permission of Springer Nature. Panels (b) and (c) from Ezersky *et al* (2017).)

than 250 sinkholes have opened up at the Ein Gedi Caravan Park, the Palm-Tree Plantation, and in the Nahal Arugot area (all sites were closed for visitors and agricultural activity). The sinkhole layouts are shown in figure 4.11(a). Sinkholes are arranged along a tortuous line within a narrow, 50–70 m wide zone, about 1750 m long trending south–southeast. Based on the similarity of salt layer and sinkhole arrangement, Ezersky (2006) suggested a model of sinkhole formation along the salt edge. The sinkhole strip crosses route #90, with the sinkholes occurring on both sides of the highway, although, so far no sinkholes have developed immediately under it. However, Nof *et al* (2019) observed the subsidence in the asphalt of Highway #90 using interferometric synthetic aperture radar (InSAR) observations. The sinkholes are continuing to develop today.

The Mineral Beach area. The Mineral Beach area, covering ~1 km^2, is located at the western coast of the central part of the northern DS basin (figure 4.1, for site location and figure 4.11(b)). The study site is situated between the DS shoreline and

Route #90 (the main road along the western DS shore). The area comprises the Mineral Spa, where natural black mud and sulfur-rich hydrothermal water with a temperature of 39 °C, attracts tourists from Israel and abroad. Around the Mineral Spa, sinkholes develop in both mudflat (the south) and alluvial fan (the north) areas. Figure 4.11(b) illustrates that sinkhole development is mainly linear along a salt layer edge or controlled by a fault (Frumkin *et al* 2011). A salt layer with a cavity was intersected by an MN-2 borehole, which crossed DS mud from the surface to 18 m deep. Soon after drilling, the borehole MN-2 collapsed into a newly developed sinkhole, which was then east of the salt layer front. Ezersky (2011a) has delineated the salt layer edge in the Mineral Beach study site (lilac solid line in figure 4.9(b)). The specified salt border is shown in figure 4.11(b). Frumkin *et al* (2011) have shown that the western border of the salt at the north of the area is located in alluvium composed of sandy-gravel sediments.

The Spa was closed in 2015 based on results of the GPR survey because of the intensive development of sinkholes before the great collapse (Ronen *et al* 2019 and section 4.3.1 of this book).

The Newe Zohar (NZ). The Newe Zohar (NZ) site map is shown in figure 4.11(c). (see figure 4.1 for site location). The NZ site is a settlement with several thousand residents. It includes evaporation ponds (EPs) fed by water pumped from the northern DS basin (Ezersky *et al* 2010). Borehole NZ-1 penetrated a 27 m thick salt layer whose top is situated at a depth of 50 m. NZ is the first on the western DS coast where sinkholes affected the main road in 1989. The map of a 1000 × 600 m area using the seismic refraction method was generated to delineate the edge of the salt layer (figure 4.11(c)). The thick lilac line in figure 4.11(c) represents the western edge of the compact salt.

The sinkholes distribution shows a good spatial correlation with the mapped salt edge. Recent sinkholes (white circles) have developed along the salt border. Following this map, the Tamar Regional Council during 2008 carried out 12 excavations along the DS water line bordering the salt edge. No new sinkholes have appeared since. Thus, in the NZ site, the distribution of sinkholes supports the salt edge dissolution model.

The Ein Boqeq area of 1.5–2.0 km long and 250–300 m width extends along the southern DS basin (recently evaporation ponds artificially filled by pumping DS water from the northern basin (see figure 4.1 for location). Numerous hotels are located through Ein Boqeq, also named the Hotel area. The salt layer edge extends in the N–S direction (figure 4.11(d)). The geological profile is composed of pebbles, cobbles, and clayey–sandy sediments. The geological description of the Ein Boqeq site is presented by boreholes EB-1 and EB-2 (Yechieli *et al* 2004). Borehole EB-3E (figure 4.11(c)) was drilled later to collect the salt samples. The boreholes crossed sandy-gravel alluvium overlying the 28 m thick salt unit located at 26 m depth. Embryos of six small shallow sinkholes were detected during the drilling of the EB-1 borehole. No other sinkholes were detected. Different from other sites at Ein Boqeq, the groundwater is located some 3–4 m from the surface, much higher than elsewhere.

Ghor Al-Haditha area (Jordan) is located south-east of the northern DS basin (site 21 in figure 4.1). The lithological structure of the eastern DS side, as the western one, is composed of sediments of Early Holocene overlain in places by alluvium and soil (Bartov *et al* 2006) at approximately the same elevations. Alluvial fan deposits of the Wadi Ibn Hamad cover the southern part of this area. The sinkholes at the eastern coast of the DS started forming in the 1980s (Taqieddin *et al* 2000).

The area of the sinkhole formation is a gently inclined wave eroded platform, about 3 km wide and located east of the Lisan Peninsula. This area formed the bed of the DS during a period of higher DS level (390 m b.m.s.l.).

Considering the generalized geological profile in figure 3.7, one can see that lithological structure from both sides of the geological section through the DS (considering only the Ghor Al-Haditha area) must be similar. This is confirmed by two boreholes, which were drilled by El-Isa *et al* (1995) at the central part of the Ghor Al-Haditha area (figure 4.12(a)) down to 49–51 m.

Following the borehole data (El-Isa *et al* (1995), p 27, figure 2-9) the upper section of the Ghor Al-Haditha is composed of sandy-gravel sediments similar to these located on alluvial fans on the western side (Ezersky *et al* 2017). Boreholes were drilled with fresh water and cannot reveal salt as a whole. Also, the depth of boreholes 49–51 m was not enough to penetrate the salt layer.

To verify the presence of the salt layer in the Ghor Al-Haditha, Frumkin *et al* (2011) and Ezersky *et al* (2013b), have analyzed seismic refraction data acquired by El-Isa *et al* (1995). This analysis has shown that there are two zones with different velocities V_p along with the refractor. The western part of the area (from the DS side) is composed of a high-velocity layer with $V_p > 2900$ m s^{-1}, whereas the eastern part of the area is composed of a low-velocity layer with the refractor velocity of $V_p < 2500$ m s^{-1}. A depth from the surface to the refractor was estimated as 40–50 m. The border between the high and low-velocity layers was interpreted as the salt edge. The high velocities ($V_p > 2900$ m s^{-1}) were interpreted as characterizing a salt layer. We present also data acquired later by other researchers (figures 4.12(b) and (c)). The location of seismic lines is presented in figure 4.12(a). Three seismic refraction profiles were acquired in 2003 by Abueladas and Al-Zoubi (2004). The location of these profiles is shown in figure 4.12(a) (red lines) and one of the refraction sections is presented in figure 4.12(b). The section along the western line 12 shows refractor velocity $V_p = 3300$ m s^{-1} (possibly salt) and eastern line 11 shows refractor velocity $V_p = 2500$ m s^{-1} (obviously not salt). Velocity V_p along the refractor of Line 5 was 3510 m s^{-1}. The entire line is located to the west of sinkholes.

Seismic refraction tomography studies were carried out along four of the MASW lines (Dhemaied 2007). The location of one of the aforementioned four lines is shown in figure 4.12(a) (dashed lilac line #4) and the V_p velocity section is presented in figure 4.12(c). There is a refractor velocity of 2900 m s^{-1} in the lower part of the section at a depth of 40–50 m. All tomography V_p sections have shown a similar structure. Although boreholes in Ghor Al-Haditha did not discover the salt layer recognized from geophysical analysis, the similarity of lithological structure in both western and eastern DS sides, as well as in the upper parts of the boreholes, enables us to suggest that a salt layer would be formed in the eastern DS shore similarly to the western one. Moreover, Taqieddin *et al* (2000) have presumed based on analysis of El-Isa *et al* (1995) surveys that the salt layer occurs at a depth of 25–50 m or more.

4.2.1.3 Seismic refraction method in the sinkhole formation problem

The appearance and development of sinkholes along the DS coast forced the geological community to look for the causes of this phenomenon. The first findings of salt were proposed by Yecheli (1993) in his PhD thesis based on the borehole DS-1 (DSIF) drilled in the Nahal Zeelim area. The drilling was carried out using brine as drilling water (published in Yechieli *et al* (1995b)). However, the salt layer presented in the subsurface caused doubts between geologists for a long time. The state program of the investigation of sinkhole phenomena started in Israel at the beginning of the 2000s. The program included seismic refraction surveys throughout the sinkhole development sites (Yechieli *et al* 2002). Ezersky (2003a, 2003b, 2006) suggested the methodology of salt mapping based on the GRM technique. Using the available seismic field data and carrying new surveys we mapped the western DS shores (Ezersky *et al* 2007, 2010, 2013c, Ezersky and Legchenko 2008). Already the first results showed that sinkholes are formed along the edge of the salt. The model of the sinkhole formation along the 'salt edge' was suggested (Jenyon 1986, Ezersky 2006). Note the GRM method that was subjected to criticism by several geophysicists (Sjögren 2001, Whiteley and Eccleston 2006) has shown very high stability, accuracy, and potential in evaporite karst study. That is explained evidently by the exact correspondence of the method to the real investigated problem and high contrast in the dynamic properties of the salt and unconsolidated sediments.

4.2.2 The multi-channel analysis of surface waves (MASW) method

The drop in the Dead Sea level at a rate of 1 m yr^{-1} and dissolution of the buried salt layers are considered by the geological community as the main factors of the sinkhole formation process. At that, the 10–30 m thick salt layers located at the depths of 25–50 m are dissolved along its edge by undersaturated groundwater with the consequential gradual collapse of surface sediments into the cavities within salt, whereas drop in the DS level is generally considered as a triggering mechanism (El-Isa *et al* 1995, Arkin and Gilat 2000, Taqieddin *et al* 2000, Abueladas and Al-Zoubi 2004, Abelson *et al* 2006, 2017, Ezersky 2006, Yechieli *et al* 2006, 2016, Legchenko *et al* 2008a, 2008c Closson and Abou Karaki 2009, Ezersky *et al* 2010, Frumkin *et al* 2011, Abou Karaki *et al* 2019).

Sinkhole formation at the surface is preceded by: (1) the salt weakening as a result of the intensive circulation of unsaturated (with respect to halite) water, thus enhancing the karstification process (Bernab *et al* 2003, Shalev *et al* 2006); and (2) the decompaction of the sediments above karstified zones. Maimon *et al* (2005, see also section 2.2.1 of this book) have carried out the numerical simulation that supports this presumption. Also, published data and field tests testify to this (Mendonca *et al* 1993). Thus, the location, depth, thickness, and weakening of salt layers along the DS shorelines, as well as mechanical properties of the upper sedimentary deposits, are thus considered as controlling factors of this ongoing process (Ezersky *et al* 2013a).

The shear-wave velocity values (V_s) depend on the porosity, fracturing, and correspondingly, on the mechanical properties of soil and rocks (Stokoe *et al* 1994, Dvorkin 2008). When the medium is water-saturated and target layers are located under the groundwater table the P-wave velocities (V_p) are strongly affected by pore saturation, whereas V_s principally characterizes properties of the medium without fluid filling.

On the other hand, the problem of investigating shallow bedrock using surface-wave methods has been described in several publications (Casto *et al* 2009, Banab and Motazedian 2010, Pasquet and Bodet 2015). All researchers agree that an investigation of the shallow structures (down to 30 m from surface) when the V_s contrast between the overburden (loose soil V_s = 200 m s^{-1}) and firm bedrock (V_s = 2300 m s^{-1}) is not a trivial procedure. In this case, inversion of the fundamental-mode Rayleigh wave could result in V_s of the bedrock much lower than the expected values (Casto *et al* 2009). Another problem is a selection of the number of layers for inversion to avoid the non-uniqueness solution (Cercato 2009, Renalier *et al* 2010). Therefore, to reduce the non-uniqueness of solutions, all available information is required for the parameterization of the initial model.

4.2.2.1 MASW in the Dead Sea sinkholes problem

The main objectives of the MASW application in the DS sinkholes problem were to use V_s to detect decompaction zones (a) in the salt unit (deep target) and (b) in the overlying Layers (shallow target), which could be the sign of forthcoming sinkhole development. Two approaches have been used. The first one was used in Jordan, where roll-along acquisition and dispersion stacking have been performed to achieve the multi-modal dispersion measurements along linear profiles to map decompaction zones in the clastic sediments overlying salt layers (shallow target). The second approach was applied in Israel, where V_s mapping has been performed to discriminate the weak and hard zones within salt layers, after calibration of the inverted V_s near boreholes (deep target) (Bodet *et al* 2010, Ezersky *et al* 2013a). The first issue, in the context of sinkhole hazard, has been recently studied numerically and applied to real data analysis (O'Neill *et al* 2008). The second problem of the salt mapping using surface-wave prospecting methods is carried out for the first time.

The surface-wave prospecting studies have been carried out by French and Jordanian teams in Jordan, and an Israeli team in Israel. Different equipment for data acquisition and data processing techniques was used based on the desired objectives. Both studies were aimed at reducing the sinkhole hazard in the same region of the DS.

4.2.2.2 Methods of V_s measurements—a brief review

The shear-wave velocity measurements are an important part of designing buildings in site-specific conditions such as soil liquefaction, ground-spectral earthquake response, etc. Being mostly independent of soil saturation, shear-wave velocities are more indicative of soil properties and can be used as a diagnostic tool for engineering properties. Seismically, shear-wave velocity (V_s) is the best indicator of shear modulus that is directly linked to a material's stiffness, which is one of the most critical engineering parameters. There are several methods of V_s measurements in laboratory conditions and *in situ* conditions. Each method has its advantages and disadvantages.

Laboratory methods include: (1) the resonant column test; (2) the cyclic torsion tests, and (3) the Bender element tests (ASTM 1996, Terzaghi *et al* 1996, Schneider *et al* 1999, Stokoe and Santamarina 2000). These tests should be carried out on the undisturbed samples, but sometimes they are carried out on compacted or reconstructed (remolded) ones. Laboratory tests allow measurement of shear-wave velocities at controlled conditions and different shear deformations, affecting shear modulus (G) and velocity (V_S). These methods enable taking into account V_s decrease following shear deformations increase (Vucetic and Dobry 1991).

Recently, similar tests were carried out for *in situ* conditions (Stokoe *et al* 1994). Deformations used in calculations are derived for a typical earthquake. However, V_s values measured for the soil samples strongly differ by *in situ* velocities. It should be noted that laboratory testing is relatively expensive.

In situ geotechnical methods. S-wave velocities are being calculated from the correlations between standard penetration tests (SPTs) parameter and V_s velocities (DeJong 2007). The SPT is carried out under ASTM D1586 (2018). Interpreting data is based on Terzaghi *et al* (1996) and DIN 4094-2 (1980). SPT testing is carried out in Israel using standard equipment of 63 kg weight falling from a height of 76 cm. The number of blows required to insert the SPT device 30.5 cm (12 inches) is determined. This value is reported as the raw (uncorrected) SPT blow-count *N*SPT. The correlation between N and V_s is explained by the dependence of SPT blow-count *N*SPT on the relative density of soils (Carter 1983, Iyisan 1996, Terzaghi *et al* 1996). A comprehensive review of *N*SPT—V_s correlations has been carried out by DeJong (2007). The following example shows correlations between the *N*SPT and V_s for DS lime carbonate (figure 4.13(a)).

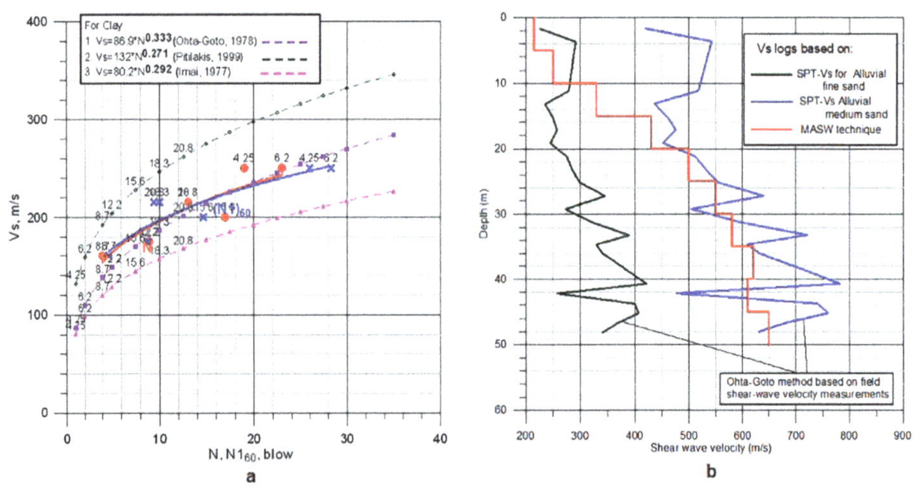

Figure 4.13. Shear wave velocity (V_s) based on *N*SPT blows. (a) V_s versus N relationship for the DS lime carbonate. Relationships of other researchers presented from DeJong (2007) review obtained in similar soils and are presented for comparison (referenced by Sykora (1987)); (b) V_s versus depth graphs calculated for fine and medium sands (Ashkelon area) from SPT log using (*EJGE* 1999, Ohta and Goto (1978)). Graphs are compared with the V_s versus depth curve calculated at the same site using the MASW technique. (Panel (a) reprinted from Ezersky and Livne (2013) with permission from Candemia.)

The analysis shows that the correlation for the lime carbonates is high enough. However, it is significantly higher than that for sandy-gravel sediments (Ezersky and Livne 2013). One can see from figure 4.13(a) that relationships derived by different researches for similar lithology are close to the DS lime carbonate, whereas sandy-gravel sediments show essential scatter. Ohta and Goto (1978) suggested a method to calculate V_s based on the soil properties. It was utilized as a V_s estimator calculating V_s versus the *N*SPT blows and taking into account soil type, the geological epoch of soil, and a combination of other factors (*EJGE* 1999). Examples of V_s distribution calculated for fine and medium sands from SPT log using the Ohta–Goto estimator are shown in figure 4.13(b). One can see that V_s calculated from *N*SPT strongly depends on the lithology, whereas the MASW methods nicely approximate both calculated graphs; showing that upper V_s values are determined by fine sand, whereas the lower part of the graph coincides with medium sand. Another V_s calculation method based on Hardin and Drnevich's (1972) model allows calculation of V_s as a function of void ratio (e), confining pressure (kPa), specific gravity, and soil conditions (dry or saturated).

In situ, geophysical methods for V_s measurements include the borehole and surface techniques. The most accurate is the *cross-hole* method (ASTM 2006). This method requires three or at least two highly parallel boreholes (accuracy of measurement of the distance between two boreholes can be computed to within ±2% to a depth of about 30 m). The velocity is measured in the horizontal direction. The inclinometer is also required to perform accurate measurements of distances between the borehole source and geophone position.

Another method is the *downhole* one (ASTM 2007) allowing V_s measurement of a single borehole along with its casing. The borehole is encased by a PVC pipe or filled with bentonite to stabilize the borehole walls. In the latter case, measurements can be affected by the quality of the borehole walls and fill geometry. Also, soil characteristics in the borehole can differ from those in the site located 100 m away.

Surface methods measurements include seismic refraction SH-measurements. However, from a practical point of view, the measurement of V_s remains a delicate seismic method (Palmer 1986). As an alternative to the S-wave refraction method, we suggested usin MASW (that is in its sense, surface-wave prospecting (SWP) techniques) based on the surface-wave dispersion measurements and inversion (Park *et al* 1999, see, for instance, comprehensive reviews in Socco and Strobbia (2004), Socco *et al* (2010)) to characterize near-surface V_S.

The refraction method is based on direct measurements of the shear wave velocities (V_S), exciting from the surface. The MASW method is based on the excitation of Rayleigh waves at the surface. They are the surface waves. The particle motion of the Rayleigh waves in a half-space is elliptical. Their amplitudes attenuate with a depth. Rayleigh waves are dispersive in a stratified half-space. These waves have a velocity (V_{SV}) slightly less than the shear waves V_s in the homogeneous space by a factor dependent on the elastic constants of the material (Telford *et al* 1993). Recently, the MASW method was added by inversion of Love's waves (V_{SH}) that are transverse and sensitive to horizontal movements—they are recorded only on the seismometers that measure the horizontal ground motion. Another important characteristic of Love's waves is that the amplitude of the ground vibration caused by a Love's wave decreases with a depth since they are surface waves. Love's waves propagate with the phase velocities V_{SH} (Dal Moro 2015, Ivanov *et al* 2017, SurfSeis 2017). Thus, surface waves are of dispersive origin which testifies to the possibility to penetrate to different depths.

In this research only Rayleigh waves propagating with the phase velocity V_{SV} were used. Therefore, here and after we denote these waves as V_s.

4.2.2.2.1 Seismic velocities of soils and rocks in Israel

The ranges of seismic velocities in the soils and rocks of Israel are presented in figures 4.14(a) and (b), respectively. Analysis of figures 4.14(a) and (b) allow us to conclude that ranges of both V_p and V_s seismic velocities in soils and rocks generally conform to those measured by other researchers in shallow subsurface (Jakosky 1957, Press 1966). However, V_p values in rocks reported in the literature are higher (for instance, $V_p = 5800$ m s^{-1} in granite, $V_p = 6100$ m s^{-1} in limestone) than those measured in Israel.

The higher V_p values could be explained by the high stress and better rock quality in deep measurement conditions. Shear-wave velocities measured throughout Israel vary by a wide range: from 100 m s^{-1} in sands and lime carbonates to more than 600 m s^{-1} in gravels, and from 400 m s^{-1} in conglomerates to more than 2400 m s^{-1} in basalts and dolomites. Variability of the V_s and thickness of soft sediments overlying hard rock on the one side, and very limited availability of densely distributed geotechnical information such as V_s at depth call for less expensive and less time-consuming methods to provide the required parameters for site effect assessment. Data presented in figures 4.14(a) and (b) allow consideration of an approximate range of Israeli sediments and rocks for preliminary evaluation of possible site response and modeling. Separately, we consider velocities in the salt constituting firm layers along the DS coastal area.

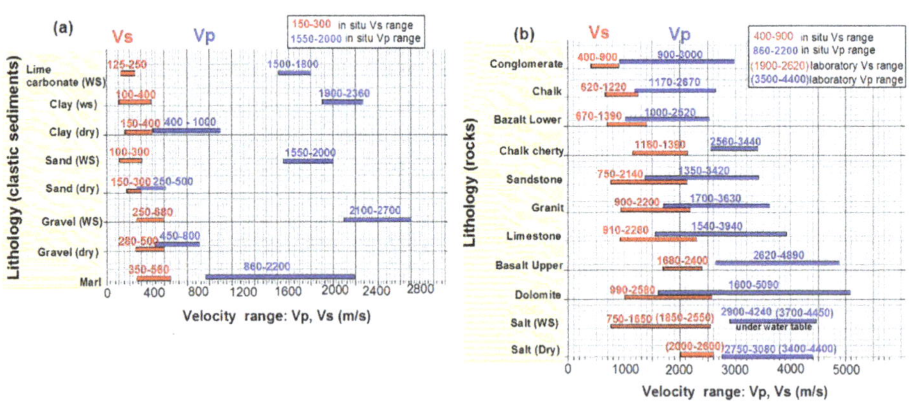

Figure 4.14. Range of V_p and V_s in soils (a) and rocks (b) measured in Israel *in situ* and laboratory (in brackets) conditions. (Reprinted from Gorstein and Ezersky (2015), with permission Camdemia.)

MASW is a standard approach to infer the 1D V_s structure for global earth seismology, near-surface geophysics, geotechnical, and civil engineering applications (Miller *et al* 1999). Usually, surface waves appear on multi-channel seismic data (seismic refraction, reflection, etc) and are considered as a noise component, whereas in the MASW method these waves are regarded as a useful signal. Despite such methods as shear-wave seismic refraction, downhole, and cross-hole surveys can be used for S-wave velocity measurements, they are less economical than the method of surface waves in terms of field operation, data analysis, and overall cost. The great advantage of the surface wave method in comparison with the S-wave refraction one is the quality of the field records. It is seen in figure 4.15, which illustrates field records acquired along the same length line with the employment of different hammers. The data acquisition of S-waves with the refraction method was carried out using the horizontal stroke of a 30 kg sledgehammer (figure 4.2(c)) allowing the excitation of SH waves at the background of other waves (figure 4.15(a)). Raw data of surface wave of R (Rayleigh) type were obtained using vertical stroke by a 200 kg hammer forced by slingshot with vertical geophones 2.5 Hz (figure 4.2(b)). Field records are shown in figure 4.15(b).

It is seen in figure 4.15(a) that field records acquired by SH-wave source (because of its weakness) are complicated by ambient noise allowing interpretation of S-wave, as a rule, to 25–30 channels from 48 ones (e.g., 25–30 m). On the other hand, the record shown in figure 4.15(b) presents surface waves that are characterized by the strongest energy because of strong vertical stroke by the 200 kg hammer (figure 4.2(b)). The quality of the field records is significantly higher and allows interpretation of all records down to depths determined by wavelength (that is usually some tens to hundreds of meters). This advantage is especially important at the excitation of S-waves by a sledgehammer.

Note that in an example presented in figure 4.15 we compare field records of Raleigh (R) and Love (SH) waves. The velocities of these waves are different. This issue can be found by the reader in Dal Moro (2015) and Barton (2006). Generally,

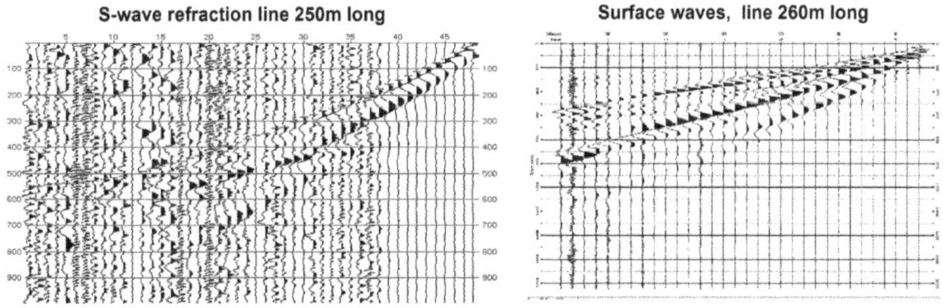

Figure 4.15. Comparison of seismic methods for S-wave acquisition. (a) Raw data were obtained by seismic refraction technique using horizontal stroke on the metallic beam by a 30 kg sledgehammer (figure 4.2(c)). The records are complicated by strong noise. (b) Raw data of surface wave of R (Rayleigh) type obtained using a vertical stroke by a 200 kg hammer forced by slingshot with vertical geophones 2.5 Hz (figure 4.2(b)). (Reprinted from Gorstein and Ezersky (2015), with permission from Camdemia.)

in horizontally stratified rock $V_{SH} > V_{SV}$ potentially by up to 10%. It should be taken into account at data processing.

In general, differences in differently polarized velocity of 5%–10% can be reasonable (Dal Moro 2015).

Active MASW. Active MASW method is based on surface waves excited by an active source like a sledgehammer, Digipulse, whereas the passive method utilizes surface waves generated passively by cultural (e.g. traffic) or natural (e.g. thunder and tidal motion) sources of ambient noise (Park *et al* 2007). Measurement of dispersion of surface waves with active-source is done with typical seismic shot gatherers that are a collection of seismic traces, which share some common geometric features. The wave field is transformed into a frequency-slowness (or a frequency-wave number) domain in which the maxima should correspond to surface-wave signatures. For such dispersion curves can be selected several modes if the propagation mode signatures are well separated. The dispersion curves are then inverted for a 1D V_s profile with a depth. When data are collected in a roll-along mode (e.g., after every shot whole profile, as well as shot point, is shifted by a constant distance equal, for instance, for separation between geophones), each 1D profile is represented at its corresponding midpoint spread, allowing a pseudo-2D V_s section to be drawn. It is assumed that the investigated medium under the spread is 1D and homogeneous.

To record wavelengths large enough for increasing the investigation depth and mitigating near-field effects, long spreads are required (Bodet *et al* 2005, Socco *et al* 2009).

Passive MASW. Passive surface waves are usually of a low-frequency nature with the wavelengths ranging from a few kilometers (natural sources, e.g., tidal motion) to a few tens (or hundreds) of meters (cultural sources, e.g., traffic) (Okada 2003). These waves provide a wide range of penetration depths and therefore a strong motivation to utilize them. For the recording of the ambient kinds of noise different receiver arrays (antennae) are arranged as different geometrical figures (linear, circular, cross-layout, etc) are used. The most accurate estimation is obtained through a survey using a true 2D receiver array (Park and Miller 2006). In built-up urban areas, using the true 2D receiver arrays, such as circular and cross-layout ones, is not practically possible.

Therefore, a method that can be implemented with the conventional 1D linear receiver array can be effective in this case (Louie 2001). The data processing scheme can be found in Park *et al* (2007).

Combination of active and passive MASW measurements. Dispersion images obtained from active and passive data sets should be combined to obtain dispersion curves in widening frequency range The active MASW method generates signals up to about 30 Hz, whereas the passive one allows widening that to the low-frequency range down to geophone eigenfrequency. Moreover, Dal Moro (2015) suggests that a geophone is capable of retrieving signals at frequencies lower than its eigenfrequency.

An actual 'visibility limit' following a rule of thumb proposes half of the formal eigenfrequency. In this case, the visible eigenfrequency of signals acquired, for instance, in Ghor Al-Haditha with geophones 4.5 Hz is of an order of 2–3 Hz. Thus, combining two signals, we can widen the range of frequencies from 2 to 30 Hz, and sometimes even up to 50 Hz. Thus, the penetration depth can be increased to a deep range (low frequencies) and the uppermost depth can be decreased (high frequencies) (Park *et al* 2007).

4.2.2.2.2 The model of the subsurface in the DS sinkhole problem

Considering the MASW investigation in the Ghor Al-Haditha (the Eastern DS shore, Jordan) and the western DS shore (Israel) two approaches of the MASW study can be selected. The model of the subsurface in the sinkhole formation sites can be presented like that in figure 4.16. A dissolution cavity is formed in the salt layer close to its edge.

The model of DS sinkhole formation has been described by Frumkin *et al* (2011). Salt dissolution commonly develops along the salt edge dissolved by brackish water, unsaturated concerning chloride (Ezersky *et al* 2006, Yechieli *et al* 2006). At first, the cavity is formed within the salt close to its edge. Instability of the salt cavity ceiling causes upward stoping of the cavity through the mechanically weak sediments situated between the cavity and the surface.

It may take place a short or long period before any signs are evident on the surface. In such a setting the ascending cavity can be detected by surface measurements at different depths before signs of future sinkhole will appear on the surface. At that, investigation depth will depend on the profile length.

One well-known limitation of the method is its compromise between the lateral resolution and investigation depth (Gabriels *et al* 1987). The inverse problem formulation imposes the investigated medium to be assumed 1D under the profile, e.g., stratified and homogeneous in the layer limit. Also, the line length should be short enough to achieve lateral resolution if profiling is performed. On the other hand, long lines allow the recording wavelengths to be great enough to increase investigation depth and to mitigate near-field effects (O'Neill 2003, Bodet *et al* 2005,

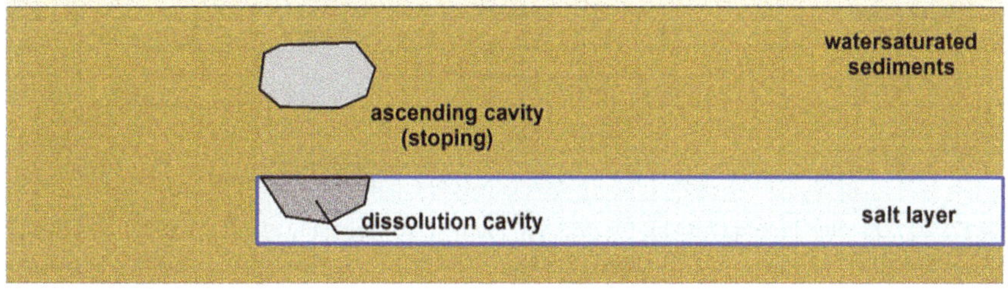

Figure 4.16. Working model for the sinkhole sites investigation using the MASW method.

2009). Several ways exist to overcome such contradictions, particularly when the seismic setup provides redundant data. One of them is specific processing techniques including offset moving windows and dispersion stacking. These can narrow the lateral extent of dispersion measurements (Bohlen *et al* 2004, Grandjean and Bitri 2006). It permits the extraction of more local information about the investigated medium along the line and thus to measure lateral variations of its parameters (Socco *et al* 2009, Boiero and Socco 2010, Strobbia *et al* 2011).

4.2.2.3 Sinkhole sites investigated

The location map of sites investigated by the MASW method in the DS coastal areas is shown in figure 4.17(a). A location map of the MASW lines in site I (Ghor Al-Haditha) is shown in figure 4.17(b). Location maps of the MASW lines in Israeli sites will be shown later in the V_S maps through sites II–IV.

Let us consider separately MASW studies carried out to resolve the aforementioned problems.

(1) *Shallow targets*: detection of reduced V_S values testifying to decompaction zones in the unconsolidated sediments overlying salt layers. It is reached by using the roll-along acquisition and dispersion stacking that allow achieving multi-modal dispersion measurements along linear profiles (Jordan) (Bodet *et al* 2010); and (2) *Deep targets*: V_S mapping of salt layers to discriminate weak and hard zones within the salt layers, after calibration of inverted V_S near boreholes (Israel) (Ezersky *et al*

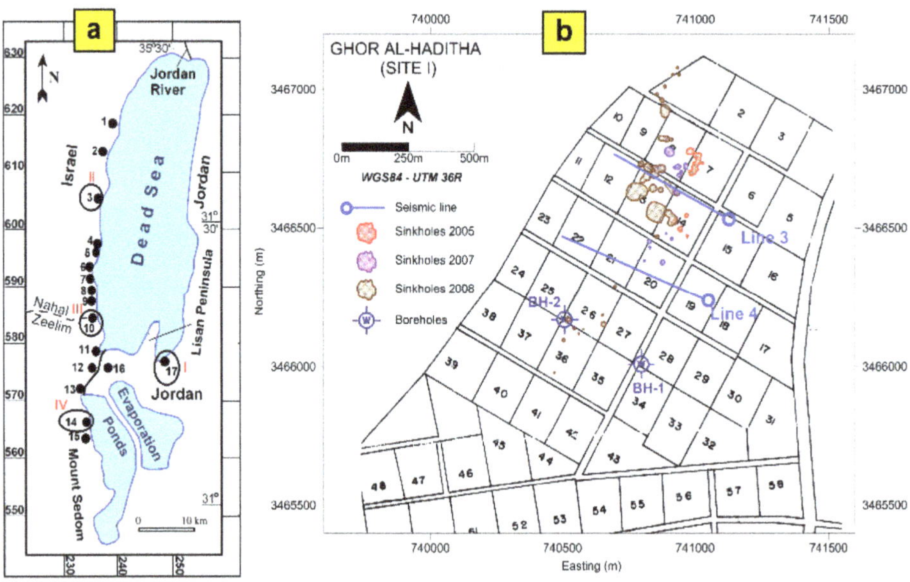

Figure 4.17. (a) DS sites investigated by the MASW method, (b) location of the MASW lines in the site I (Ghor Al-Haditha). (The figure of L Bodet in Ezersky *et al* (2013a) is published with the permission of the EEGS. All copyright privileges remain with EEGS. This material cannot be copied or used without the express written permission of EEGS.)

2013a). To better understand the methods of data processing we refer the reader to the above-mentioned papers.

4.2.2.3.1 Data acquisition

In Ghor Al-Haditha (Jordanian site I in figures 4.17(a) and (b)) French researchers (Bodet *et al* 2010) on one hand anticipated strong lateral heterogeneities caused by cavities in the shallow subsurface but on the other hand, try to study great investigation depths to reach the salt layer (see figure 4.16). The aforementioned investigations applied dense multifold acquisition geometries to enable stacking strategies to extract the dispersion data.

In the Ghor Al-Haditha (Site I in figure 4.17(a)) data acquisition was carried out with a 24-geophone spread (4.5 Hz, vertical component). A 72-channel geometrics seismic recorder was used. The separation between geophones was 5 m. Shot location was 5 m away from the nearest trace (off-end shooting), and move up between the shots was one receiver interval. The seismic source was a sledgehammer. Two lines (#3 and #4, see figure 4.17(b) for location) were shot (Bodet *et al* 2010, Ezersky *et al* 2013a).

*In Israeli Sites (*II–IV *in* figure 4.17(a)), Israeli researchers used conventional seismic data (i.e., the vertical component of the wave field from common shot records obtained in the shallow refraction surveys) (Ezersky *et al* 2013a). The data excitation was carried out using a Digipulse hydraulic source mounted on a pickup truck (see figure 4.2(b)). A Summit II plus seismic recorder was used with a 24-geophone spread (vertical 2.5 Hz geophones). Receiver spacing was 5 m, shot location was 5–10 m away from the nearest trace (off-end shooting), and the move up between shots was a one to two-receiver interval.

4.2.2.3.2 Results

Shallow target problem: detection of decompaction zones in the clastic sediments above the salt layers (Jordan) (Bodet *et al* 2010).

4.2.2.4 Surface-wave data processing

Ghor Al-Haditha (Site I, figures 4.17(a) and (b)). All recorded seismograms display coherent wavefields, sometimes with strong surface scattering and lateral velocity variations. The automatic dispersion data analysis is not possible because of the wide variability in the surface-wave data quality along both lines. An example of the recorded seismograms is presented for line 3 in figure 4.18(a) (as more complicated).

A part of dispersion images is hardly usable, as shown by the example presented in figure 4.18(b). Such poor quality can be partly explained by either the presence of lateral heterogeneities below the seismic spread or a not powerful enough seismic source that is a sledgehammer and confirms the need to narrow the lateral extent of the dispersion measurements. Roll-along stacking of dispersion images was thus applied to the multifold data. This processing technique involves an offset window that moves along as it records to extract local wavefields with common midpoints. The stacking is achieved by summing the frequency-slowness spectra of the windowed data. For this survey, the best compromise between the wave number resolution and stacking efficiency was found to have an aperture of 12 traces (55 m)

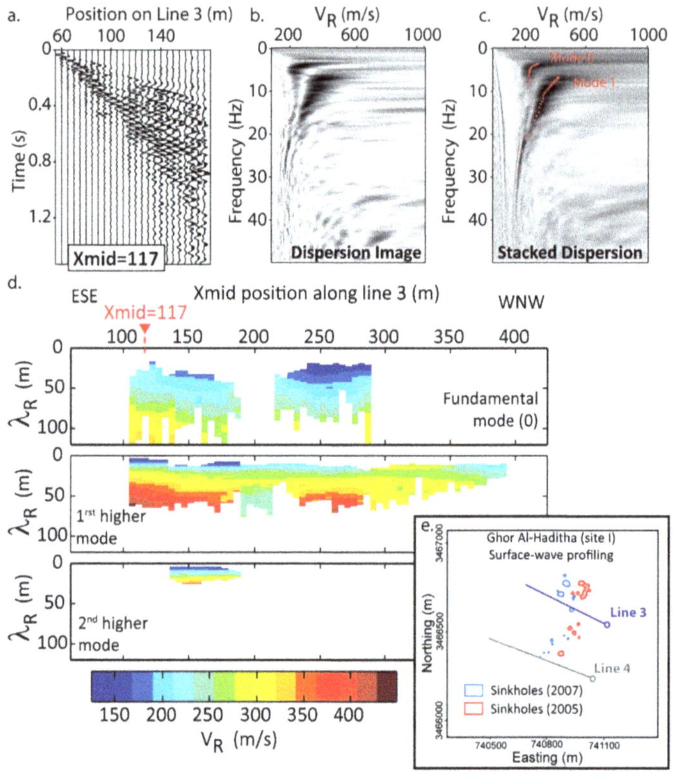

Figure 4.18. Data acquisition in Ghor Al-Haditha (Jordan). The effect of dispersion stacking along Line 3. Xmid is the spread midpoint position on the line. (a) Surface waves field records for midpoint Xmid = 117. (b) The corresponding normalized seismogram is given with its dispersion image. (c) The result provided by the roll-along dispersion stacking centered on Xmid is presented for comparison. (d) Pseudo-section of Rayleigh-wave phase velocity (V_R) dispersion curves picked along Line 3 after dispersion stacking. λ_R is the wavelength. (e) Line 3 location map. (The figure from *Journal of Environmental and Engineeding Geophysics* in Ezersky *et al* (2013a) is published with the permission of the EEGS. All copyright privileges remain with EEGS. This material cannot be copied or used without the express written permission of EEGS.)

(see figure 4.16 for explanation), a minimum offset of 5 m, and a maximum offset of 65 m. A stack of 13 normalized dispersion images at each spread midpoint location (Xmid). The effect of dispersion stacking is illustrated by one example along Line 3 (figure 4.18(c)). Xmid is the spread midpoint position on the lines. The corresponding normalized seismogram (figure 4.18(a)) is given with its dispersion image computed with all 24 traces (figure 4.18(b)) (Bodet *et al* 2010).

The result provided by the roll-along dispersion stacking centered on Xmid is presented for comparison (figure 4.18(c)). Such a stacking process improved dispersion image quality. The signal-to-noise (S/N) ratio is highly increased so that mode identification is improved, which enables semiautomatic picking.

The collections of 57 dispersion images for Line 3 and 79 dispersion images for Line 4 were obtained. On each stacked dispersion image, coherent maxima were picked and extracted with an estimated standard error in-phase velocity (the dispersion measurement uncertainties were defined according to the workflow that can be found in O'Neill (2003) and Pasquet and Bodet (2015)). Up to three modes were observed and identified as fundamental (0), first or second modes, using visual browsing of dispersion images along the line. The resulting dispersion curves are presented as Rayleigh-wave phase velocity pseudo-sections (figure 4.18(d)), which allows the coherence in mode identification to be checked. Indeed, mode identification is a delicate task, particularly in the presence of low-velocity layers, which tend to make the energy 'jump from one mode to another' (O'Neill 2003). The forward modeling was performed using the final models obtained at some particular locations to assess the relative importance of the different modes (Socco and Strobbia 2004).

1D inversion of dispersion data was performed at each spread midpoint (Xmid) position using the neighborhood algorithm (NA), implemented for near-surface applications by Wathelet *et al* (2004) and Wathelet (2008). Assuming a vertically heterogeneous layered model, theoretical dispersion curves were computed from the elastic parameters using the Thomson–Haskell matrix propagator technique (Thomson 1950, Haskell 1953).

A stack of three layers overlaying the half-space was selected to keep the number of parameters as low as possible. A maximum half-space depth (HSD) of 60 m was evaluated based on the maximum observed wavelengths (figure 4.18(d)). Frequency-phase velocity curves of each mode, with associated uncertainties, were then inverted with five independent runs, generating a total of 25 500 models (figure 4.19).

Surface-wave profiling was carried out at the site along lines 3 and 4 (450 m and 550 m long, respectively). Line layouts are presented in figure 4.17(b). Line 4 crosses the sinkhole course to the south of the main subsidence area, whereas line 3 intersects the currently active area. During the survey (May 2007), a sinkhole opened suddenly along line 3. From that location, accelerated development of sinkholes occurred towards the north in 2008, where several farms and an industrial site were

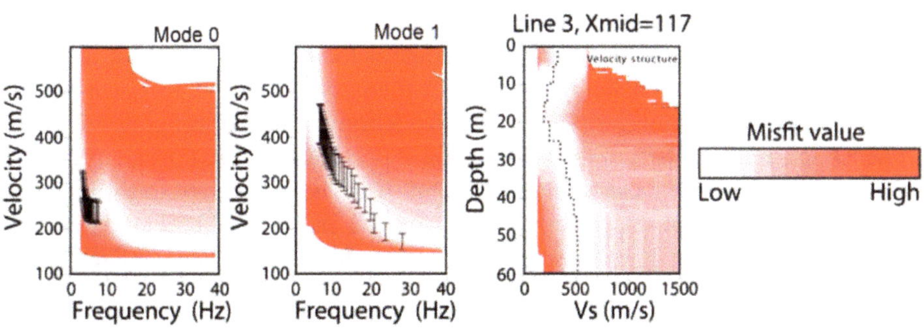

Figure 4.19. An example of 1D inversion of dispersion data (error bars) extracted at the spread midpoint position 117 m along Line 3, using the neighborhood algorithm (NA) as implemented by Wathelet *et al* (2004). (The figure from *Journal of Environmental and Engineeding Geophysics* in Ezersky *et al* (2013a) is published with the permission of the EEGS. All copyright privileges remain with EEGS. This material cannot be copied or used without the express written permission of EEGS.)

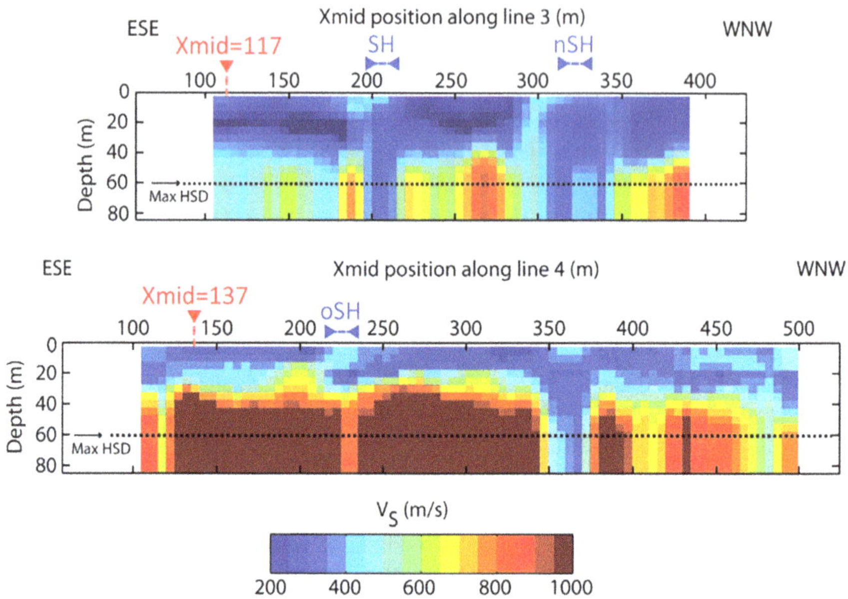

Figure 4.20. Pseudo-2D V_s sections along line 3 (top) and line 4 (bottom), constructed from all 1D profiles. The dotted lines correspond to the maximum half-space depth (max HSD). The Xmid positions of the example are shown in figure 4.18(d) are marked by a red arrow. An existing sinkhole (SH) in the vicinity of line 3 (at 200 m), and intensive sinkhole development that occurred during four years following the survey (nSH) are indicated. Along line 4, the position of an old refilled sinkhole (oSH) is shown between 200 and 250 m. (The figure from *Journal of Environmental and Engineeding Geophysics* in Ezersky *et al* (2013a) is published with the permission of the EEGS. All copyright privileges remain with EEGS. This material cannot be copied or used without the express written permission of EEGS.)

abandoned. At the same time, new sinkholes also opened in a southerly direction. The dispersion stacking (see example in figure 4.18(c)) and multi-modal inversion were performed as described above (see example in figure 4.19). Final sampled V_s profiles with the lowest misfits were extracted and resampled in-depth, every 5 m down to the half-space, to estimate an average velocity structure, as shown in the example of figure 4.19. Each 1D profile was then represented at its corresponding position (Xmid) to obtain the pseudo-2D V_s section presented in figure 4.20.

The 'O' marks the origins of the lines. Sinkhole occurrences were tracked in 2005 and 2007, during the survey. accelerated development of sinkholes occurred toward the north in 2008. The background image corresponds to the former agricultural land division.

Deep target problem: V_s mapping of salt layers (Jordan and Israel) (Ezersky *et al* 2013a).

4.2.2.4.1 Ghor Al-Haditha (salt map and vertical E–W section)

A seismic refraction survey was initially carried out by El-Isa *et al* (1995) enabling us to estimate the depth (35–40 m) and velocity (V_p = 3000–3500 m s^{-1}) of the salt layer. The salt top map was generated based on these data by Frumkin *et al* (2011) and Ezersky *et al* (2013b) (figure 4.21).

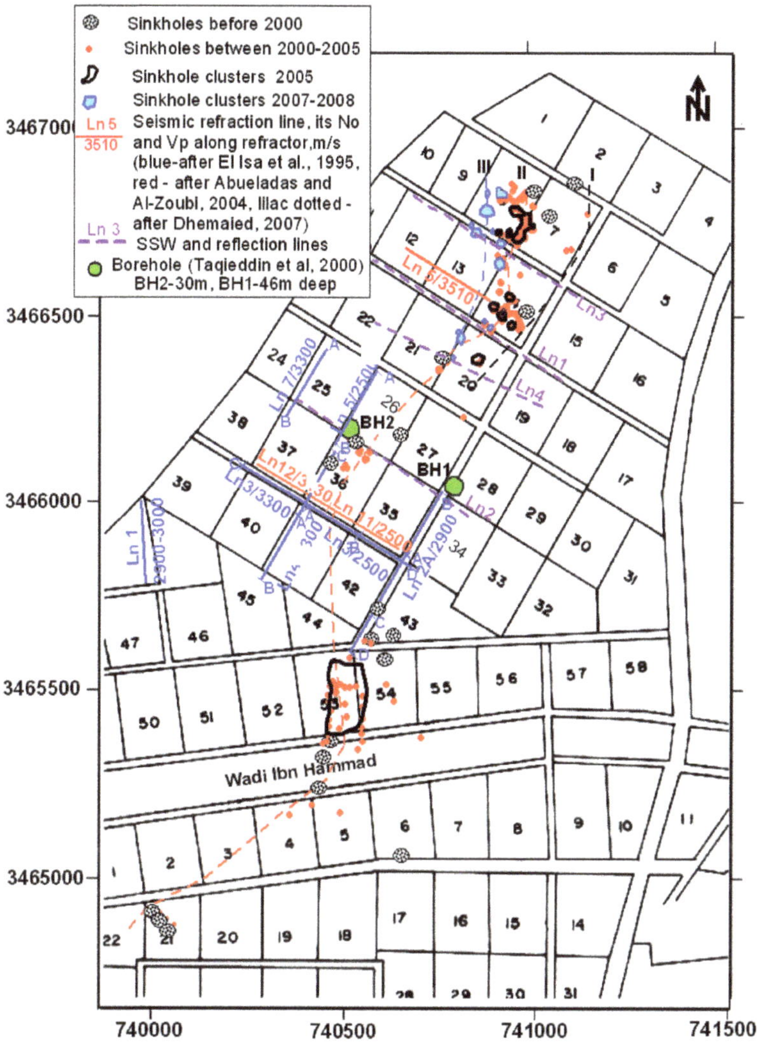

Figure 4.21. A Ghor Al-Haditha layout map. Seismic refraction lines (after El-Isa *et al* (1995), Abueladas and Al-Zoubi (2004)), boreholes and sinkholes before 2000 (after Taqieddin *et al* (2000)), sinkholes between 2000 and 2008 (after Abueladas, personal communication), sinkhole clusters in 2005 and 2008 (after Bodet *et al* 2010) are shown. I, II, and III are sinkhole lines, respectively, in 2000, 2005, and 2008. Coordinates are in m, Universal Transverse Mercator grid. (Reprinted from Ezersky *et al* (2013b) with permission from Springer Nature.)

The red dashed tortuous line in figure 4.21 is the eastern border of the salt layer through the Ghor Al-Haditha area. The shear wave velocity (V_s) structure with $V_s >$ 1000 m s^{-1} in figure 4.20 was identified by Bodet *et al* in Ezersky *et al* (2013a) as the salt layer top of which is located at the 40 m depth.

Later, we re-processed the available data to reveal some additional features of the structure and properties of the subsurface. Because the DS salt layer investigations were carried out for the first time, some special procedures were applied.

1. Forward modeling.
2. Calibration near boreholes.
3. Estimation of MASW method resolution.

1. The forward modeling enables us to select a model and compare dispersion images calculated theoretically (by the forward modeling) and measured in the field.
2. Calibration near boreholes allows (1) checking properties of the subsurface (e.g., V_p and V_s, etc) and identifying properties of the soils and salt, (2) selecting model parameters for inversion, and (3) estimating the resolution of the surface-wave method.
3. Resolution of the MASW method determines the depth range of the reliable values of measured parameters (e.g., V_s).

We have re-processed the MASW data acquired along Line 4 by Bodet *et al* (2010) (with their permission) from Ghor Al-Hadith (see figure 4.17(b) for location) to validate if the geometry of the salt layer is possible (e.g. depth and thickness of salt) in the Ghor Al-Haditha area (Ezersky *et al* 2021). The procedure of the data re-processing is explained in figure 4.22. WinMASW software by Eliosoft was used (with permission of Geotec Ltd).

One can see that there is a similarity between the dispersion image presented in figure 4.22(a) measured in the field and that calculated by the forward modeling (figure 4.22(b)) for the five-layered model shown in figure 4.22(c). The experimentally obtained dispersion image and the extracted dispersion curve are shown in figure 4.22(d). In the dispersion image, one can see only the fundamental mode, higher modes being obscured in this case. The model for inverting comprises three low V_s velocity layers above a high-velocity salt layer and a fifth low-velocity layer below that. The resolution of the uppermost layer limited by the minimum wavelength is $\lambda_{\min} = 11.7$ m ($V_{\text{ph}_{\min}}$ =281 m s^{-1}; $f_{\max} = 24$ Hz), where $V_{\text{ph}_{\min}}$ is the minimum phase velocity. Thus, V_s in a layer shallower than $H_1 = 5.8$ m thick is unreliable following the half-wavelength criterion. The maximum wavelength is 142 m ($V_{ph_{\max}} = 415$ m s^{-1}, $f_{\min} = 2.93$ Hz), defining a maximum penetration depth of 71 m concerning wavelength. Taking into account that line length is 120 m, the penetration depth formally is defined by half of that and is 60 m.

Then we applied the inversion based on a genetic algorithm (Dal Moro 2015) to investigate variations of the salt V_s and thickness at different parameterization and start models. The solution converges after 80 iterations.

Results of the data inversion are summarized in figure 4.22(e) where a thin red solid line presents an experimental mean V_s model (with ± bars) and the blue line is the best V_s model. 'Mean' is the average value of the parameter from several iterations (in our case 80) during the inversion process. A convergence of the solution is estimated by the closeness of the extracted dispersion curve and inverted

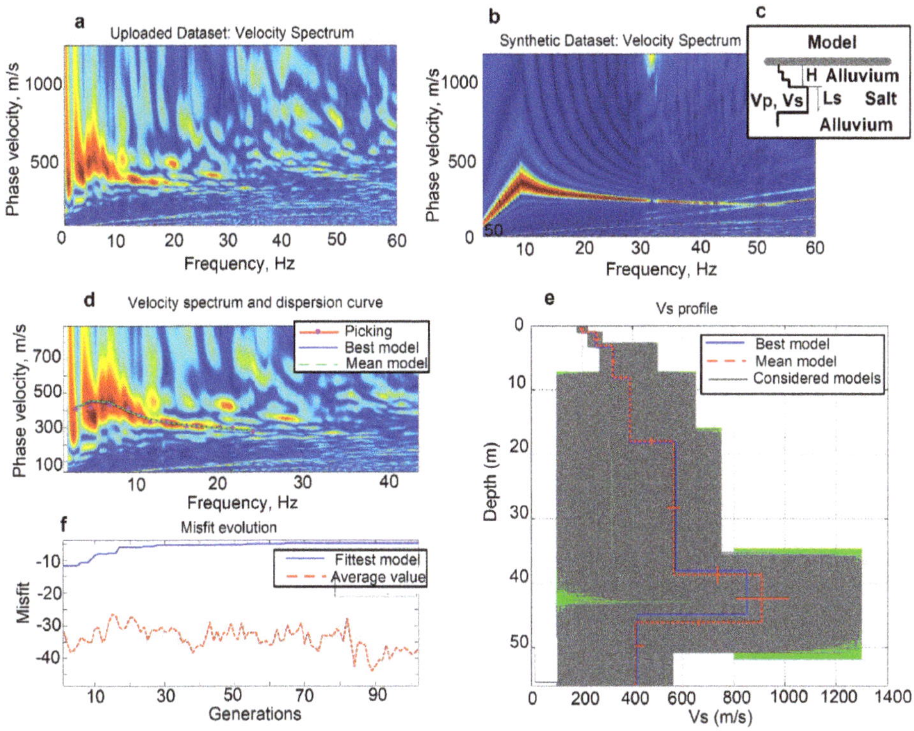

Figure 4.22. Results of the re-processing of data acquired in the Ghor Al-Haditha area (line 4, record 1249, Xmid = 320 m). (a) Dispersion image. (b) The dispersion image derived by the forward modeling for the model shown in figure 4.22(c). (c) V_s versus depth five-layer inversion model used for the forward calculations. (d) Dispersion image and extracted dispersion curve with the mean and the best models fitted. (e) V_s versus depth profile of the dispersion image shown in figure 4.22(a) derived by the inversion: (f) misfit evolution during inversion process versus the number of generations (blue curve)) and the average (mean) value of models calculated for every layer (red curve). (Reprinted from Ezersky *et al* (2021), with permission of the Springer Nature. Field data of L Bodet *et al* with permission.)

ones (see figure 4.22(d)). The value of the parameter corresponding closest inversion result is named 'Best'. Grey background is a combination of considered models limited by the search space (green area). Poisson's ratio is 0.35. The mean misfit of the best dispersion curve along the line is 5.03%, misfit of the mean dispersion along the line is 39.3%

4.2.2.5 Ghor Al-Haditha (vertical E–W section)—inversion

Using modeling results, we have processed data to reveal parameters of the salt layer along line 4. The corresponding section has been shown in figure 4.23.

Table 4.2 presents the mean parameters of the salt layer obtained by the totalization of alternative models during the inversion process (red line in figure 4.22(e)). The best value is the model giving the dispersion curve closest to that derived from the field data (figure 4.22(d)). This graph is shown in figure 4.22(e) by the blue line.

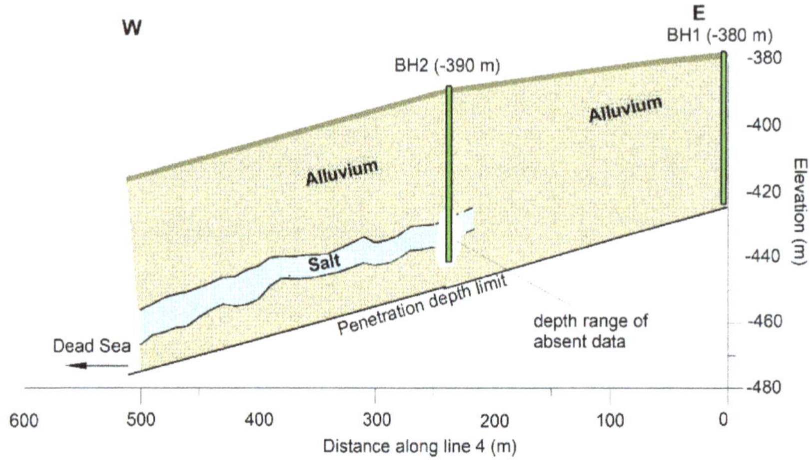

Figure 4.23. The geological section along line 4 with the calculated salt layer parameters using the MASW technique. Two boreholes are shown (BH1 and BH2). At the crossing of the salt layer with the borehole BH2, lithological data are absent. The parameters of the salt layer are presented in table 4.2. (Reprinted from Ezersky *et al* (2021), with permission of the Springer Nature. Field data of L Bodet *et al* with permission.)

Table 4.2. Calculated parameters of the salt layer along line 4. (Reprinted from Ezersky *et al* (2021) with permission of Springer Nature.)

	Mean	Standard deviation	Best	Standard deviation
Depth to salt top, m	39.4	±1.44	39.1	±2.01
Salt thickness, m	8.4	±1.71	8.35	±1.87
Salt V_S, m s^{-1}	990	±97.4	1023	±154

Thus, the depth of the salt layer varies between 37 and 41 m and it slightly increases with an approach to the DS. The salt thickness in 2007 varies between ~6.5 and 10 m, and velocity V_s is ~ 900–1090 m s^{-1}. Both parameters are slightly increased at the end of the interpreting line.

4.2.2.6 Israeli sites (II–IV in figure 4.17(a))

At the Israeli sites, inversion of the dispersion data has been carried out using the linearized inversion with a gradient-based iterative method (Miller *et al* 1999, Park *et al* 1999, Xia *et al* 1999, 2003) with commercially available SurfSeis v.3 software (Park *et al* 2008). The root mean square error (RMSE) between the theoretical dispersion curve and the measured one is usually used as an indicator of the closeness between measured and iteratively calculated dispersion curves. Usually, in the linearized inversion methods, constraints are applied to the solution to reduce the degree of non-uniqueness (Cercato 2009). The dispersion equation depends mainly on the V_s and thickness of the layers. An appropriate choice of these parameters (the initial model) is considered a fundamental issue for the successful application of inversion (Socco and Strobia 2004, Renalier *et al* 2010). Some parameters can be selected using the well-known rules of thumb (Xia *et al* 2003): (a) the ratio of V_p versus V_s can be considered bounded for near-surface materials, assuming Poisson's ratio values ranging from 0.2 to 0.48; and (b) the Rayleigh wave velocity in a uniform half-space is very close to the V_s of the layer. As a reference model for inversion, V_s can be approximated by the phase velocity multiplied by a correction factor (less than unity for fundamental-mode data). Also, starting models were constrained by existing external geological and geophysical data (see below).

We have processed the MASW data acquired in the Israeli sites to validate the geometry of the salt layer (e.g. depth and thickness of salt) and its velocity V_s (V_{SV}) (Ezersky *et al* 2013a). An example of data processing is shown in figure 4.24 for the

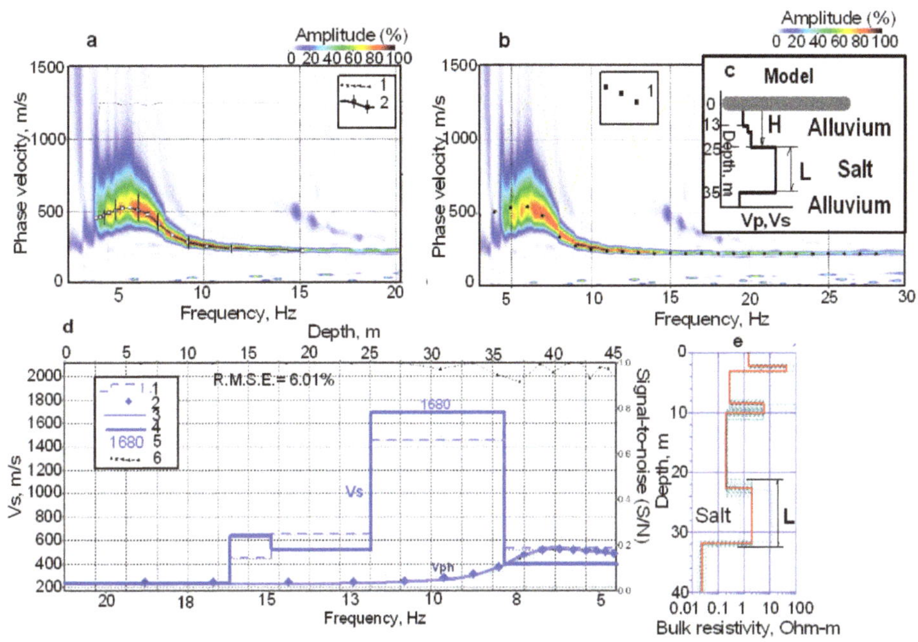

Figure 4.24. Results of the processing of data acquired in the Nahal Ze'elim (line 4). (a) Dispersion image and extracted dispersion curve; (b) dispersion curve matching the dispersion image derived by forward modeling for the five-layered model V_s versus depth shown in figure 4.24(c); (d) V_s (e.g. V_{SV}) versus depth profile of the dispersion image shown in figure 4.24(a) derived by the least-square inversion (RMSE = 6.01%); and (e) inverted TEM resistivity versus depth graph (dotted lines are equivalent solutions). Denoted in figure 4.24(a): 1—S/N curve, 2—extracted points and dispersion curve with the error bar; in figure 4.24(b): 1—dotted line matching the dispersion image derived by forward modeling; in figure 4.24(d): 1—V_s versus depth initial model, 2—experimental extracted points (see in figures 4.24(a)), 3—dispersion curve matching the experimental points, 4—final inverted V_s versus depth layered model, 5—inverted V_{SV} of the salt layer (m s^{-1}) with RMSE = 6.01%, 6—S/N ratio versus frequency graph. (The figure from *Journal of Environmental and Engineeding Geophysics* in Ezersky *et al* (2013a) is published with the permission of the EEGS. All copyright privileges remain with EEGS. This material cannot be copied or used without the express written permission of EEGS.)

Nahal Ze'elim (Site III in figure 4.17(a)). Data were acquired deploying 24 2.5 Hz geophones placed 5 m apart. A seismic refraction survey was initially carried out along the same line, enabling us to estimate the depth (25 m) and velocity ($V_p = 3500$ m s^{-1}) of the salt layer. The layer thickness (~10 m) was estimated from the TEM resistivity–depth function (figure 4.24(e)). The seismic signal is comprised of four stacks. The S/N ratio is high (0.9–1.0) (curve 6 in figure 4.24(d)).

The dispersion image and the extracted dispersion curve are depicted in figure 4.24(a). Only the fundamental mode is seen in the dispersion image, higher modes being absent or obscured in this case. In figure 4.24(b), the similar dispersion image derived by the forward modeling for the five-layered model V_s versus depth is shown in figure 4.24(c). The similarity of dispersion curves testifies to the rightness of the selected model to invert, that comprises three low V_s (e.g., V_{SV}) velocity layers above a high-velocity one (salt) and a fifth low-velocity layer for considering the existing geological information.

The resolution of the uppermost layer is limited by $\lambda_{min} = 10.8$ m: ($V_{ph_{min}} = 226$ m s^{-1}, $f_{max} = 21$ Hz in this case). Thus, V_s in a layer shallower than $H_1 = 5.4$ m thick is unreliable following the half-wavelength criterion (e.g., Rix and Leipski 1991). The maximum wavelength is 108 m ($V_{ph_{max}}$ = 484 m s^{-1}, f_{min} = 4.48 Hz), defining a maximum penetration depth of 54 m following the half-wavelength criterion. Next, we applied the least-squares inversion to investigate variations of the salt V_s for different parameterization and to develop initial models. An inversion was carried out at a fixed V_p value of 3500 m s^{-1}. A solution converges after 20–25 iterations. Results of the data inversion are summarized in figure 4.24(d); a rhombus (2) represents the experimental extracted points and a thin solid line (3) is the matching dispersion curve. For an optimum five-layer model (starting $V_s = 1450$ m s^{-1}), the inverted V_s value is 1680 m s^{-1} at RMSE = 6.01%, the Poisson's ratio is 0.35 (4 in figure 4.24(d)). Varying the initial velocity model slightly changes the salt layer velocity by approximately 8%. Likewise, changing the salt thickness from 6 to 11 m (uncertainty in the TEM determined L in figure 4.24(e)) results in varying the misfit error in the realm of 6%–10%. Thus, variations of V_s of the salt layer induced by different initial models, do not exceed 8%.

4.2.2.7 Calibration measurements

As mentioned above, the objectives of calibration were: (1) to compare V_p and V_s velocities measured by different methods with those from surface-wave prospecting, and to distinguish V_s under the different lithologies; (2) to select model parameters for inversion; and (3) to estimate the resolution of the surface-wave method.

Two boreholes were drilled for the project at the Ein Boqeq (Site IV) and Mineral Beach (Site II) (see figure 4.17(a) for location). The borehole EB-3E crossed sandy-gravel sediments and encountered the salt layer at a depth of 26 m (figure 4.25(b)). The borehole MN-5E, located at the southern margin of the Mineral Beach site on a mudflat found the DS carbonate clay (mud) overlaying and underlying a 20 m thick salt layer located at a depth of 22 m (figure 4.25(e)). The lithological description and properties of both lithology types are presented in section 4.12 and figures 4.163 and 4.164. The depth–velocity graphs are presented in figures 4.25(a) and (d)), along with two vertical electrical resistivity profiles derived from the TEM soundings

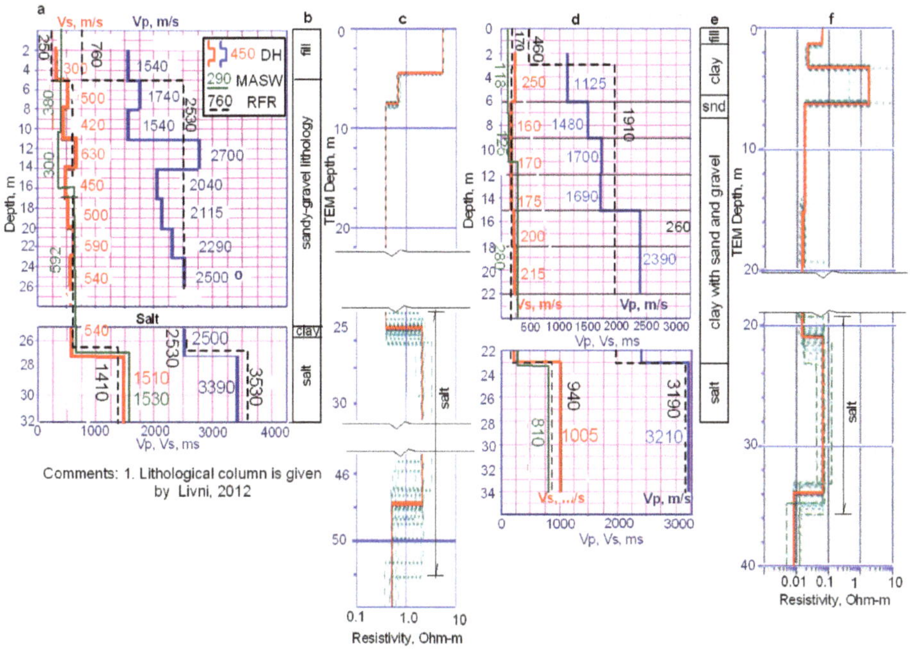

Figure 4.25. Results of the calibration measurements in the EB-3E and MN-5E boreholes: downhole data are combined with surface-wave and seismic refraction graphs (a) and (d), respectively; corresponding lithological columns of the EB-3E and MN-5E boreholes (b) and (e), respectively; and TEM resistivity–depth functions of the EB-3E and MN-5E boreholes (c) and (f), respectively, the scale of the TEM depth is distorted. (The figure from *Journal of Environmental and Engineeding Geophysics* in Ezersky *et al* (2013a) is published with the permission of the EEGS. All copyright privileges remain with EEGS. This material cannot be copied or used without the express written permission of EEGS.)

(figures 4.25(c) and (f)). Downhole (DH) surveys within the boreholes and close located seismic refraction surveys were used for calibration measurements.

As follows from calibration measurements, seismic results obtained by different methods (downhole, MASW, refraction) are consistent one to another, showing very close values of S-wave velocities and corresponding with the TEM sections by structure.

Also, the calibration measurements allow determining the V_s of the lithological units penetrated by the boreholes: clayey sediments are characterized by V_s = 150–280 m s^{-1}; sandy-gravel sediments have V_s = 300–600 m s^{-1}, and the salt layer exhibits a V_s of 810–1650 m s^{-1}.

4.2.2.8 Salt layer mapping (the western DS shore)

In the Mineral Beach area (site II in figure 4.17(a)), the V_s map of the salt, which is based on surface-wave measurements along with ten profiles (figure 4.27(a)), shows a gradual eastward increase in V_s from 810 m s^{-1} near the salt edge to more than 1600 m s^{-1} near the DS bottom. The geological cross-section A–B (figure 4.27(d)) shows a similar W–E variation in V_s with distance from the salt edge to the DS. The large

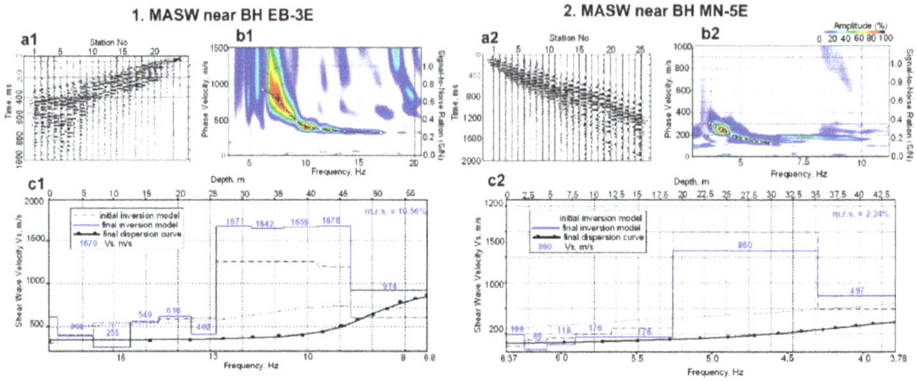

Figure 4.26. Results of surface wave prospecting along the line near EB-3E borehole in the Ein Boqeq (site IV in Figure 4.17(a), sandy-gravel) (**14-1**), and line MN-5E in the Mineral Beach site (site II in figure 4.17(a), mudflat) (**14-2**). **14-1**: (a1) Signal, (b1) dispersion image, and (c1) V_s inversion model (five layers above salt). 1—initial V_S–depth model, 2—experimental extracted points (see (b1)), 3—dispersion curve matching the experimental points, 4—final inverted V_s versus depth layered model, 5—inverted V_s of the salt layer (m s^{-1}), 6—S/N ratio graph. **14-2**: (a2) Signal, (b2) dispersion image, and (c2) V_s inversion model (five layers above salt). 1—initial V_S–depth model, 2—experimental extracted points (see (b2)), 3—dispersion curve matching the experimental points, 4—final inverted V_s versus depth layered model, 5—inverted V_s of the salt layer (m s^{-1}). (The figure from *Journal of Environmental and Engineeding Geophysics* in Ezersky *et al* (2013a) is published with the permission of the EEGS. All copyright privileges remain with EEGS. This material cannot be copied or used without the express written permission of EEGS.)

sinkholes (about 30-m diameter) are located in the low-velocity zone $V_S = 850–950$ m s^{-1}, whereas the latent karstic cavities (revealed by borehole Mn-O, figure 4.27(a)) are located in the low-velocity zone $V_S = 950–1050$ m s^{-1}. A similar seismic structure is observed in the Nahal Ze'elim area (site III in figure 4.17(a)) where the V_S map (figure 4.27(b)) was drawn from six seismic lines with a SE–NW direction. As above, V_S gradually increases from 760–950 m s^{-1} close to the salt edge to more than 1600 m s^{-1} near the eastern line nzl6ss, located 300 m west of the DS. Note the presence of superficial, relatively small sinkholes (3–5 m diameter) arranged along the salt edge mapped using the seismic refraction method (Ezersky *et al* 2010). In the Ein Boqeq area (site IV in figure 4.17(a)) the V_S map is less detailed as it is based on surface-wave dispersion measurements along with four profiles (figure 4.27(c)). Again, the *Vs* velocities quickly increase from 850 m s^{-1} near the salt edge to more than 1600 m s^{-1} at the DS shoreline (some 150 m east of the salt edge). This area is close to the evaporation pond where the salt layer was found at a depth of 26 m below sandy-gravel sediments. According to the borehole data, highly mineralized water (~230 g L^{-1} chloride) envelops the salt layer (Yechieli *et al* 2004). Sinkholes have not been detected in this area.

We used all available information for the parameterization of the initial model before inversion to reduce the non-uniqueness of solutions. Every seismic line was accompanied by seismic refraction and TEM surveys, allowing the determination of the seismic line on the western DS shore, where linearized inversion was used. We

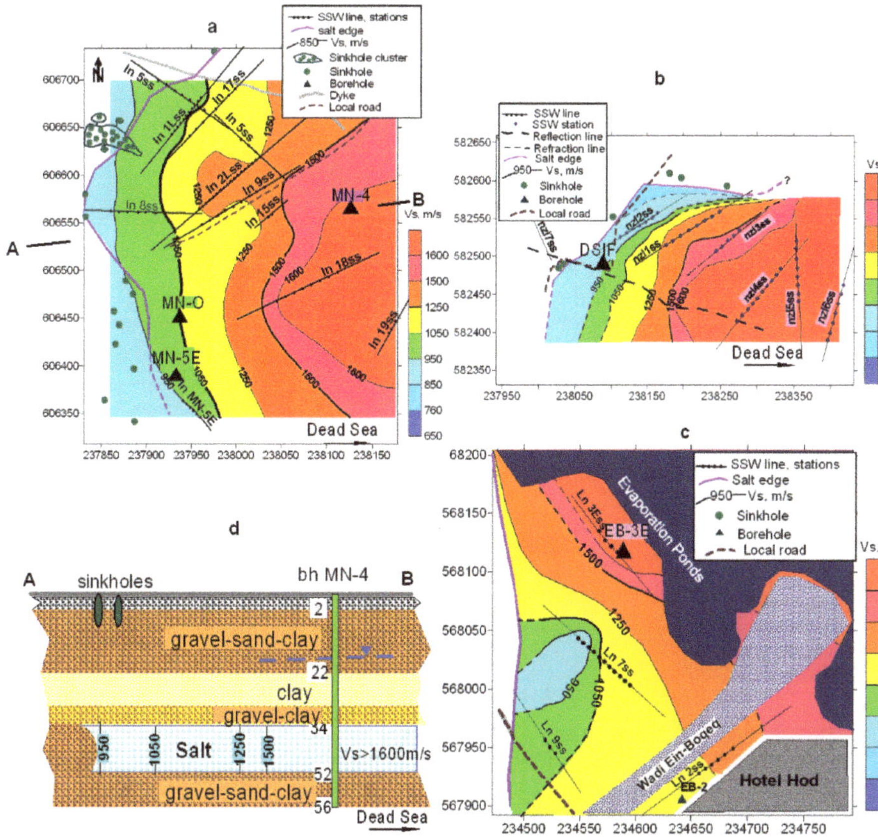

Figure 4.27. V_s maps in the salt unit at (a) Mineral Beach (site II in figure 4.17(a)), (b) Nahal Ze'elim (site III in figure 4.17(a)), and (c) Ein Boqeq (site IV in figure 4.17(a)). (d) Mineral Beach W–E geological section along A–B line via the borehole MN-4. (The figure from *Journal of Environmental and Engineeding Geophysics* in Ezersky *et al* (2013a) is published with the permission of the EEGS. All copyright privileges remain with EEGS. This material cannot be copied or used without the express written permission of EEGS.)

obtained V_s values consistent with borehole measurements, which provided confidence in the inversion results, using the calibration of the method near two boreholes where other methods were applied to reveal the V_s structure of the subsurface. As a result, we have mapped the salt layers through three Israeli sites.

The V_s (V_{SV}) map of Israeli sites (II–IV) shows the increase of V_s from the salt edge (that is coincident with sinkhole area) to the DS shoreline (figure 4.27). The V_s variation with distance for the three investigated sites in Israel is displayed in figure 4.28.

Note that principally all dependencies between V_s and distance from the salt edge are similar, although they slightly differ for different sites. Data were updated using a linear function for sites III and IV, and a power of that for site II. The r-squared coefficient of determination varies from $D = 0.65$ (in site IV) to $D = 0.93$–0.95 (in two other sites). The latter show well the relative relationships between the measured

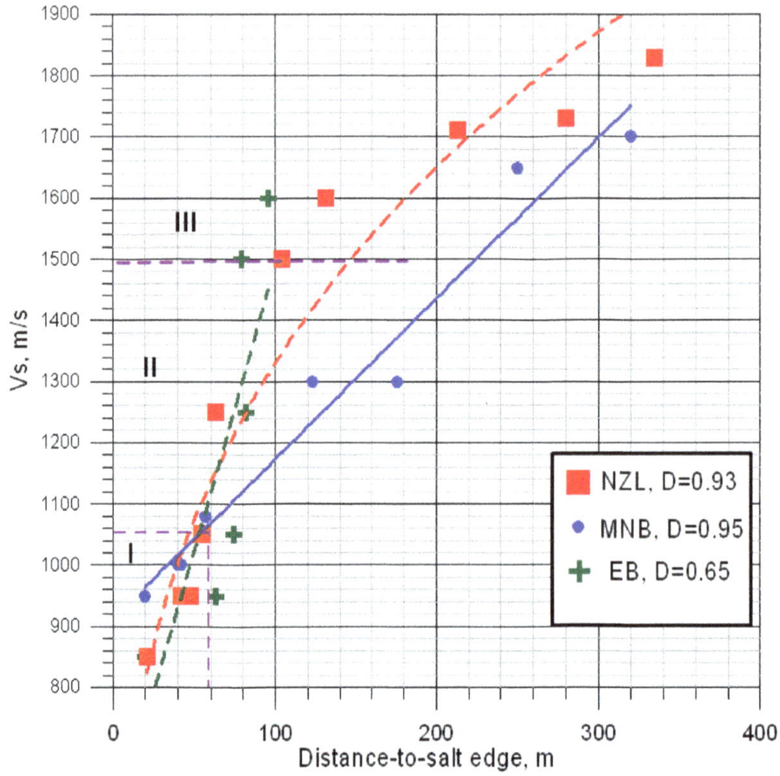

Figure 4.28. Variations of V_s versus distance to salt edge for three sites of the western DS shore (based on table 4.3). *D* is the coefficient of determination (*r*-squared) in a separate site (see figure 4.17(a) for site location). (The figure from *Journal of Environmental and Engineeding Geophysics* in Ezersky *et al* (2013a) is published with the permission of the EEGS. All copyright privileges remain with EEGS. This material cannot be copied or used without the express written permission of EEGS.)

values. All graphs converged in the area nearest the sinkholes (salt edge). V_s values vary from 760 m s^{-1} at the salt edge and increasing to over 1500 m s^{-1} with the distance from the salt edge to the direction of the DS. Such variation could be caused by a change of either salt voidness or lithological changes, such as in clay content. Analysis of the borehole logs has shown that clay content in the salt layer does not exceed 3%–5%, which could lead to a V_s decrease of 10%–15% (Dvorkin 2008). Since V_s variations of 200% are observed, they probably result from variations in the salt porosity. The circulation of undersaturated water concerning chloride causes salt dissolution. This water circulation is more intensive near the salt edge and attenuates with distance from it. Correspondingly, karstification (voidness) of salt is active near the layer edge and gradually decreases towards the DS. This process fits the sinkhole formation along the salt edge (Legchenko *et al* 2008a, Ezersky *et al* 2010, Frumkin *et al* 2011) and is supported by the salt dissolution model proposed by Shalev *et al* (2006). The dissolution belt along the salt edge is wide at the western side and in the northern and southern parts of the Jordanian side of the DS. It can

Table 4.3. Shear wave velocities were measured in the DS shore area in Israel using the MASW technique. (Ezersky *et al* (2013a), 18 (4), with the permission of the EEGS. All copyright privileges remain with EEGS.)

Site[1]	Line no	Salt depth (m)[2]	Phase velocity, $V_{ph_{max}}/V_{ph_{min}}$ ($m\ s^{-1}$)	Frequency, f_{min}/f_{max}, (Hz)	Z_{max} (m)	RMSE[3] %	V_s in salt[4] ($m\ s^{-1}$)	Z_{lim}, (m)	V_p in salt, ($m\ s^{-1}$)	Distance from MASW line to salt edge (DS), m
II	1SSL	35–40	503/353	4.8/14.16	54±3	6±1.9	975±75	60	2900	30 (350)
II	MN5E	22.3	290/130	3.0/6.4	48.5	6.9±3.1	810±80	60	3200	20(360)
II	17SS	35–40	528/363	4.8/10.9	55±2	4.2±1.3	1215±103	60	3000	50 (330)
II	8SS	35–40	526/348	4.6/13.1	58±13	7.9±4.5	990±260	40	2940	30 (350)
II	2LSS	35–40	585/409	4.4/14.7	66.2±4.4	5.4±1.5	1350±230	60	3420	110 (270)
II	5SS	35–40	541/362	4.9/14.0	56±26	9.6±10.7	1070±160	60	3080	50 (330)
II	18SS	34	588/176	3.6/17.1	88±30	24±10.5	1650±40	60	3470	250 (130)
II	19SS	20	578/259	3.7/17.3	78±13	29±15.5	1670±150	40	3780	340 (40)
Salt edge-DS										0(380)
III	2SS	22–25	555/271	4.89/17.1	57±5.1	3.1±0.54	904±65	60	2930	40 (600)
III	1SS	22–25	601/246	5.8/16.9	51±2.6	6.2±5.1	1430±45	60	3100	100 (540)
III	3SS	22–25	634/257	5.77/15.1	55.3±5.9	4.9±0.80	1492±60	60	3200	160 (480)
III	4SS	22–25	615/221	5.59/24.6	55.8 ±10.4	6.1±3.6	1710±40	60	3500	220 (420)
III	5SS	22–25	631/214	4.97/12.1	63.5±6.0	9.9±7.5	1730±43	60	3600	280 (360)
III	6SS	22–25	615/208	4.56/10.7	67.4 ±10.1	15.4±9.4	1830±120	60	3600	320 (320)
Salt edge-DS										0 (640)
IV	9ss	26–30	640/381	5.12/18.2	61.7±7.2	7.9±0.99	880±107	60	3400	25 (145)
IV	7ss	26–30	718/335	5.57/19.8	65±6.5	6.06±2.5	984±130	60	3570	70 (90)
IV	EB3E	26	930/320	7/17	57.5	36	1540	60	3530	95 (30)
IV	2ss	26–30	848/357	6.08/20.6	69.8±5.3	9.03±7.3	1655	60	3430	180 (80)
Salt edge-DS										0(~125–160)

The salt layer in the DS area is characterized at three sites. The measured parameters (table 4.3) are the maximum and minimum frequencies, phase velocities, maximum penetration depths (Z), RMSE, V_s, V_p, depths limited by the line length, and the distances of each line to the salt edge and the DS shoreline.

[1] Site II—Mineral Beach; Site III—Nahal (Wadi) Ze'elim; Site IV—Ein Boqeq (see figure 4.17(a) for location).

[2] Depth to the salt top.

[3] RMSE±δ variations along with the SW spread.

[4] V_s±δ variation along with the SW spread. Comment: lines located in the sinkhole area are highlighted in grey. To investigate the spatial variations in the salt V_s maps from the surface wave measurements (figure 4.27) have been constructed for three sites on the western DS shore.

become narrow and even completely disappear in the central part of the Jordanian side. As a result of the heterogeneous dissolution, the seismic properties in salt should strongly vary along (e.g., parallel to DS shoreline) and gradually change (increase) perpendicular to the shoreline (i.e., towards the DS direction).

The MASW method has shown high potential in the decision of the key problems of the DS sinkhole phenomena: (shallow problem) search of decompacted sediments and (deep problem) validation of salt layer properties and geometrical features (depth, thickness, shear wave velocities V_S and their distribution from salt edge to the DS direction). That in turn shows the location of possible sinkhole formation zones.

Moreover, a combination of the MASW method with other ones (SNMR, HVSR, for instance) enables us to determine the porosity, hydraulic conductivity, degree of karstification, etc (Ezersky and Legchenko 2014, 2015, Gorstein and Ezersky 2015). These problems are considered in chapter 5 of the book.

4.2.3 Seismic reflection method

A new methodology for imaging shallow subsurface is presented. This methodology combines several, recently developed imaging techniques. The first one is zero-offset common shot point (CSP) stacking (Keydar *et al* 1996). The heart of this technique is a new new normal moveout (NMO) travel-time correction formula as a function of reference velocity and wavefront parameters namely radius of curvature of the common shot wavefront and angle of entry of the arriving wavefront. The other one is a diffraction imaging technique (Landa and Keydar 1998, Keydar *et al* 2010) that utilizes diffracted waves to detect inhomogeneous objects such as faults, sinkholes, cavities, caves, tunnels. The diffraction travel-time correction curve is defined as a function of the velocity of diffracted waves (the diffraction imaging is considered separately in section 4.2.4).

The CSP stacking and diffraction imaging formulae can be extracted provided the model of the medium is known. However, in reality, the model of the medium is unknown; therefore, some kind of inverse problem must be solved to construct these parameters. This is done, like in the conventional CMP stacking procedure, by finding a maximum of a chosen coherence measure. However, unlike in conventional CMP stacking one optimal parameter (Vnmo) has to be estimated, in the case of CSP stacking there are three unknown optimal parameters. These parameters are estimated by a time-consuming procedure of optimization problem involving visual examination.

In our new approach, we use an alternative, a more formal recently proposed multipath summation with proper weights (Keydar 2004, Keydar and Shtivelman 2005, Shtivelman and Keydar 2005, Shtivelman *et al* 2009). The multipath summation is performed by stacking the target waves along all possible time surfaces having a common apex at the given point. This approach does not require any explicit information on parameters since the involved multipath summation is performed for all possible parameter values within a wide specified range. We will show how using the CSP stacking using weighted multipath summation makes it possible to obtain a better image than using the conventional CMP method.

Here we give a short description of CSP stacking and multipath weighted summation. For a more comprehensive discussion of the methods, one can read the relevant papers.

4.2.3.1 The common shot point (CSP) stacking method (Keydar *et al* 2012b)
The CSP stacking of reflection waves (Keydar *et al* 1996) is a special case of a homeomorphic imaging approach (Gelchinsky 1988, Cruz *et al* 2000). This approach is based on the fundamental principle of topological equivalence of a reflector and its image as constructed from the stack parameters. The basis of the zero-offset CSP stacking method is a new NMO time correction formula. To get the time correction formula to let us assume a source located at point A_o and receivers located at point Aκ (figure 4.29). A wavefront, with a center of curvature denoted by C_o*, is generated by this source and reaches the receivers A_K. The ray $A_oC_oA_o$ is normal.]

For an arbitrary set of receivers A_K and sources A_o at the surface, from simple geometrical consideration, the following expression of common shot time curve is obtained

$$T_{sr}^{csp} = \frac{(x^2 + 2 \cdot \sin\beta \cdot x \cdot R + R^2)^{0.5}}{V_0} \quad (4.1)$$

where R is the radius of curvature of the common shot wavefront, β is the angle of entry of the arriving wavefront, x-distance between shot A_o and receiver A_K and V_o is reference velocity at the subsurface. CSP imaging of subsurface usually can be performed by a spatial summation of the wave field along time curves defined by equation (4.1) as a function of optimal parameters, namely radius of the curvature R, angle of entry of the arriving wavefront, and the reference velocity of the subsurface V_O. Note that equation (4.1) is, unlike conventional CMP stacking, stretch-free. This property makes it possible to get a better reflection image of the

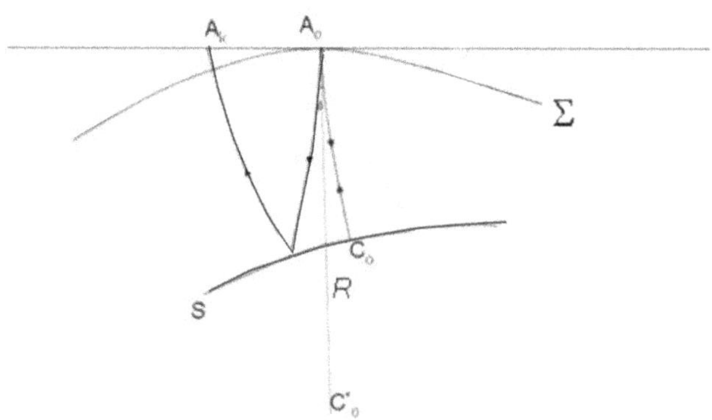

Figure 4.29. Schematic description of the ray path from reflector S. (Reprinted from Keydar *et al* (2012a), copyright World Academic Publishing.)

shallow subsurface than by using the conventional CMP stacking. The optimal parameters β, R, and V_0 are estimated in (Keydar *et al* 1996) using an optimization problem that consists of finding parameters that maximize some correlation functional. Still, now the estimation of these parameters is a very challenging problem. In our new methodology, we use an alternative approach to the so-called weighted multipath summation. This approach replaces the complex optimization problem of estimating NMO parameters by stacking along all possible trajectories that are created from small variations of those parameters. In the following, we give a short description of weighted multipath summation.

4.2.3.2 Multi-pass summation

The summation of CSP imaging and diffraction imaging performed along time curves is defined by formulas (1) and (2), respectively. The summation can be implemented using either of the two approaches. In the first, conventional, approach, that was implemented in (Keydar *et al* 1996, 2010, Landa and Keydar 1998) the target waves are stacked along the time surfaces defined by some optimal parameters such as velocity in the case of diffraction imaging or wavefront parameters and velocity in the case of CSP imaging. These optimal parameters usually are estimated using an optimization problem that consists of finding parameters that maximize some correlation functional. We use an alternative, a more formal recently proposed multipath summation with proper weighting (Keydar 2004, Keydar and Shtivelman 2005, Shtivelman and Keydar 2005, Landa *et al* 2006). The weighted multipath summation allows us to replace the complex optimization problem of estimating the optimal parameters, by summation along all possible trajectories that are created from small variations of those parameters. The weighted multipath summation (WMPS) can be described by the following expressions (Keydar *et al* 2012):

$$I_{\mathrm{WMPS}} = \sum_{P} I_P \mathrm{W}(I_P) / \sum_{P} W(I_P) \tag{4.2}$$

$$I_P = \sum_{P} U_{SR}(T_{SR}(P)), \tag{4.3}$$

where: U_{SR} is the amplitude of receiver R and shot S, $W(Ip) = \exp(\lambda I_P)$ is an optimal weighting function, λ is a unidimensional large number. Ip is the image obtained for a fixed radius of curvature, velocity in the case of CSP imaging, and fixed velocity value in the case of diffraction imaging using expression (4.1). The summation (4.3) with the proper weight $W(Ip)$ is performed for all possible parameter values within a specified range. The imaging defined by Ip means that, for every point to be imaged, seismic amplitudes are stacked together along all possible time curves defined by equation (4.1). The constructive and destructive interference of the amplitudes contributed by each time curve produces an image close to that obtained by stacking with the 'correct' parameter (Shtivelman and Keydar 2005).

The real example has been taken from Mineral Beach (site II in figure 4.30(a)). which is located at the DS shoreline (Keydar *et al* 2011). To specify the structural

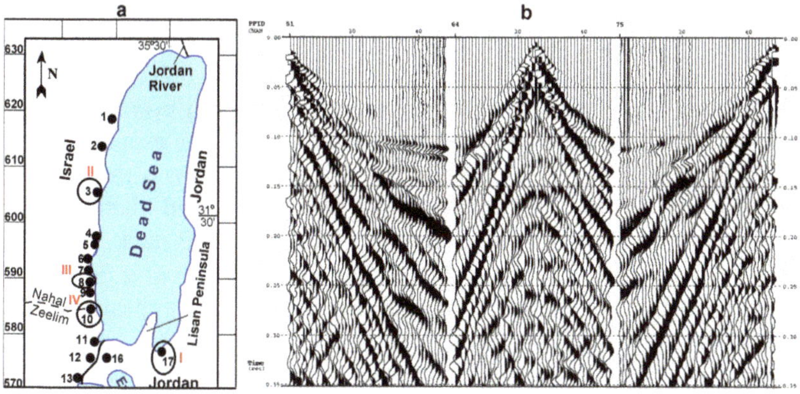

Figure 4.30. (a) Location map of the DS reflection sites. (b) Example of three common shot gathers. Black points are sinkholes, I–IV -sites of reflection surveys. I—Ghor Al-Haditha, II—Mineral Beach site, III—Nahal Hever South site; IV—Nahal Zeelim site. (Reprinted from Keydar *et al* (2012a), copyright World Academic Publishing.)

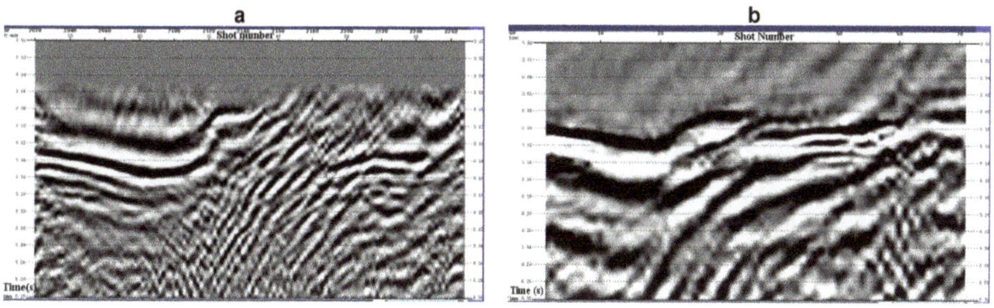

Figure 4.31. (a) CMP stacked section obtained using conventional processing. (b) CSP time section obtained using the weighted multipath summation. (Reprinted from Keydar *et al* (2012a), copyright World Academic Publishing.)

geological conditions of the site, a 345 m multi-channel reflection survey has been acquired. Field parameters of this survey included 71 shot points with a 5 m interval and 48 receivers for each shot with a 2.5 m interval between the receivers. As a source, an accelerated weight-drop ('DigiPulse') was used. The data was recorded using a 0.5 ms sample rate and 0.5 s record length. Typical common shot gathers seen in figure 4.30(b). The time sections, obtained using conventional CMP (figure 4.31(a)) and CSP methods (figure 4.31(b)), reveals a complex structure at subsurface that can be interpreted as two blocks separated by a normal fault at ~ sp 30. Both sections were processed using the same flow including band-pass filter, gain, and static correction. The CMP stacked section includes 269 traces with 1.25 m intervals and can be divided into three zones, till sp 30, sp 30–45, and 45 to the end (sp 71). The first zone includes a clear coherent reflector at the time 125–155 msec with a dominant frequency of 40Hz. In the second zone, data quality is poor and the reflector disappears and can be seen again in the third zone till sp 65. The CSP section includes 71 traces with a 5 m interval.

In general, this section is similar to the CMP stacked section but it is different in detail. A reflector is seen and unlike in the CMP stacked section, this reflector is continuous along all the lines (sp-1–65). The dominant reflector has a similar character but with a lower frequency contained. The main event appears on the conventional CMP stacked section at time ~140 ms, the same event appears on the CSP section at time ~110 ms. One of the possible explanations is that in the case of CSP stacking the NMO time correction is stretch-free and therefore it allows imaging the first arrivals of the event. In the case of the CMP stack, the NMO time correction is not stretch-free and the stacking along the NMO curve destroys the first arrivals of the event and only the later phase of the event appears on the CMP stacked section.

Thus, the new approach includes recently developed imaging techniques, namely CSP stacking and multipath summation. The resulting time section, obtained using CSP free stretch stacking, is superior to the time section obtained by the CMP method. The diffraction image produces a clear image of subsurface heterogeneities such as faults, sinkholes, cavities, caves, tunnels, or voids. This approach does not require any information about the subsurface, since the multipath summation is performed for all possible parameter values within a wide specified range.

4.2.3.3 Example of 2D reflection sections from the Nahal Ze'elim (Israel) and Ghor Al-Haditha

Nahal Ze'elim (site IV in figure 4.30(a)) is located south of the northern DS basin. A 340 m-long W–E 2D seismic reflection profile was shot across the sinkhole line (figure 4.32(a)). The edge of a 7–10 m-thick salt layer has been delineated by seismic refraction study and verified by borehole DS-1 (figure 4.32(a)). The dataset was originally collected using a 96-channel Summit II seismic recorder (GMT, Germany) with a 60-geophone spread (10 Hz, vertical component). Receiver spacing was 2.5 m, shot location was 5 m away from the nearest trace (off-end shooting west side), and the move up between shots was a two-receiver interval.

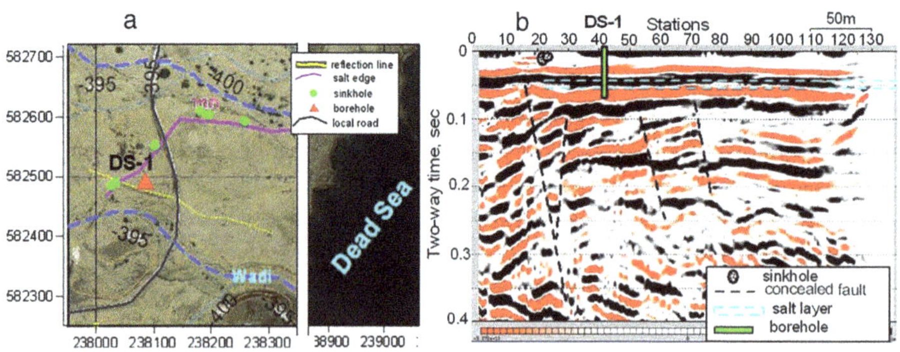

Figure 4.32. Nahal Ze'elim site. (a) Location map: reflection line (yellow) crosses the salt layer edge and a small sinkhole which is a part of a sinkhole line. (b)W–E 2D seismic reflection section through the site. (Reprinted from Ezersky and Frumkin (2013), copyright (2013) with permission from Elsevier.)

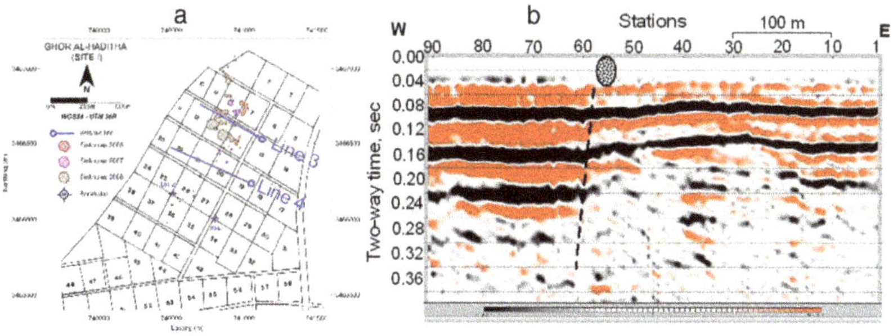

Figure 4.33. Ghor Al-Haditha (*Site I* in figure 4.30(a), Jordan). (a) Location map. (b) E–W 2D reflection section along line GH-4. The gray ellipse indicates the sinkhole cluster. (Gratitude to L Bodet for permission to reuse data, processing by Keydar *et al* (2011), the figure has been reprinted from Ezersky and Frumkin (2013), copyright (2013) with the permission of Elsevier.)

The seismic reflection data were processed using the weighted multipath summation (Keydar *et al* 2011, 2013). A distortion zone is evident where the reflection profile intersects the sinkhole line (figure 4.32(b)). The section shows the presence of faults close to the sinkhole zone. (4.4) The upper reflector is smooth but distorted between stations 10 and 30. Two sinkholes and the salt edge coincide with the fault site.

Ghor Al-Haditha (*Site I in* figure 4.30(a)) is located south of the northern DS basin. An E–W 2D seismic reflection section along line 4 through the Ghor Al-Haditha site is shown in figure 4.33(b) (see figure 4.33(a) for line location). The dataset was originally collected by Bodet *et al* (2010) to analyze surface-wave dispersion. Field records were acquired using a 72-channel geometrics seismic recorder with a 24-geophone spread (4.5 Hz, vertical component). Receiver spacing was 5 m, shot location was 5 m away from the nearest trace (off-end shooting east side), and the move up between shots was one receiver interval. The recorded seismograms were also compiled and processed by Keydar *et al* (2011) to extract CSP stacking and diffraction images along the lines.

The black dashed line divides the section into two parts differing in their reflection pattern. The line separates zones of different lithology. We suggest that the western reflector might be a salt layer characterized by strong reflections. The eastern part of the section might represent the clastic sediments. The separation between the two parts of the section could also be a fault although the evidence seen in the section is not conclusive.

4.2.3.4 3D imaging using weighted multi-pass summation: a case history

To understand the relationship between a developing sinkhole and its tectonic environment, a high-resolution (HR) 3D seismic reflection survey was carried out at the western shoreline of the DS (Keydar *et al* 2013).

Numerous HR seismic reflection surveys were carried out using a CMP technique (Abelson *et al* 2003, 2006, Yechieli *et al* 2006). The HR seismic reflection method is intended for the study of faults and conduits in the shallow subsurface (down to 100–200 m deep), which could bring the fresh water to the salt layer presumably from below. The conventional processing technique, the so-called CMP method, essentially consists of a 'stack' (summation) of properly corrected traces thus increasing the S/N ratio. However, in the shallow subsurface, the conventional CMP method causes a loss of information because of the problem of 'stretching' of data caused by the NMO time correction formula. That is why a new 2D and 3D free stretch imaging approach had been applied in the framework of the MERC project M27-050 to examine the tectonic hypothesis connecting the sinkholes with tectonic faults (Keydar *et al* 2011, 2012a, 2012b). This approach involves a zero-offset CSP stacking and diffraction imaging method (Keydar *et al* 2012b).

Based on those methods in combination with a new weighted multipath summation technique (Keydar 2004, Keydar and Mikenberg 2008), a package of programs was written. The CSP stack and diffraction method (see section 4.2.4) are complementary to each other and reveal useful information about the subsurface. The CSP stacking contains information about the structure of the subsurface, whereas the diffraction method serves as a tool for the detection of faults and voids. These methods were applied along the DS's shorelines.

4.2.3.5 The 3D zero-offset stacking method

The basis of the 3D zero-offset stacking method is a new NMO time correction formula for 3D media as a function on wavefront parameters (Keydar and Mikenberg 2008, 2010). One of these parameters on which the proposed time correction formula is based is the emergence angle β, defined as the acute angle between the wavefront normal and the normal to the acquisition plane at the CSP. Let us consider a central ray and its associated wavefront arriving at a CSP (figure 4.34(a)).

An ellipsoidal wavefront emerges at the CSP location O. It has two independent vectors **e**1 and **e**2 tangent to the wavefront and the normal **n**. The emergence angle β is the angle between **n** and the normal to the acquisition of the plane, **k** (Keydar *et al*

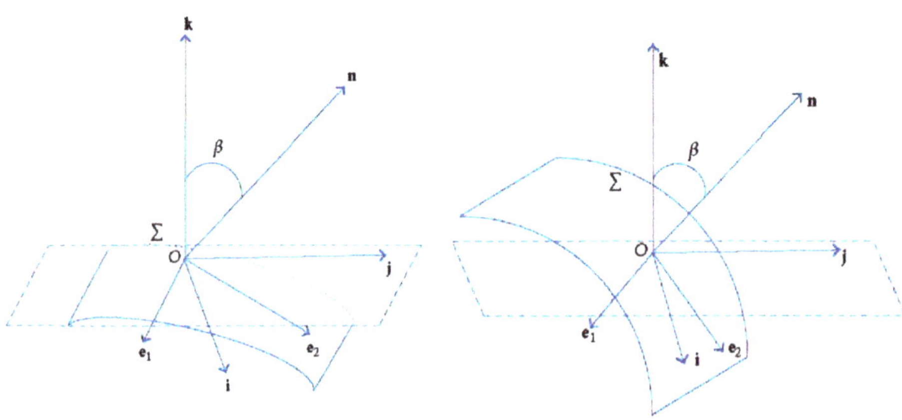

Figure 4.34. (a) Scheme of a wavefront arriving at a CSP. (b) The same as (a) in the case of one direction is normal to vector **k**. (Reprinted from Keydar *et al* (2013), copyright (2013) with permission of Hindawi Publishing Corporation.)

2013). Additional fundamental parameters on which the time correction formula is based are the curvatures of the wavefront. It is well known from the differential geometry of surfaces that at each point of the wavefront two principal curvatures can be defined, being the eigenvalues of the 2×2 matrix of second derivatives. These curvatures are the minimal and maximal curvatures and are related to the mean and Gaussian curvatures. The principal curvatures are associated with two perpendicular principal directions, which together with the wavefront normal form a rectangular system of axes. The curvature of the wavefront is completely described by the two principal curvatures and the angle the principal directions make with the general coordinate system. This is denoted in figure 4.34 as **i**, **j**, and **k**. The principal curvatures can be expressed in terms of principal radii of curvature, which play an important role in the travel-time correction expressions derived in this paper. Figure 4.34(b) illustrates the case where one of the principal directions is normal to **k**. Based on those principles and eikonal equation only, the following 3D travel-time correction formula for an arbitrary system of coordinates was obtained:

$$t = \frac{1}{V}(a_{31}x + a_{32}y) + \frac{1}{2VR_1}(a_{11}x + a_{12}y)^2 + \frac{1}{2VR_2}(a_{21}x + a_{22}y)^2 + \dots \quad (4.4)$$

$$a_{31} = \sin\beta \sin\varnothing;\ a_{32} = -\sin\beta\cos\varnothing;$$

$$a_{11} = \cos\varnothing;\ a_{12} = \sin\varnothing;$$

$$a_{21} = -\cos\beta\sin\varnothing;\ a_{22} = \cos\beta \cos\varnothing,$$

where: β is the emergence angle of the wavefront at point O, ϕ is the azimuth angle, defined as the angle between the axis $O\widetilde{X}$ and the principal direction of the wavefront at point O, $\tilde{x}$ and $\tilde{y}$ are the offsets along with the axis $O\widetilde{X}$ and $O\widetilde{Y}$. For a given source–receiver, we gather the moveout equations that express the moveout correction concerning a zero-offset trace by five parameters measured at the central point. The parameters are principal radii of the wavefront R_1, R_2, the azimuth angle ϕ, the emergence angle β, and the reference velocity V.

In the following section, we give a short description of the weighted multipath summation.

4.2.3.6 Multipath summation

The summation of 3D stacking is performed along time surfaces defined by equation (4.4). The summation can be implemented using either of the two approaches. In the first conventional approach, the target waves are stacked along the time surfaces defined by wavefront parameters and velocity. These optimal parameters usually are estimated using an optimization problem that consists of finding parameters that maximize some correlation functional. We use an alternative, more formal recently proposed multipath summation with proper weighting (Keydar 2004, Keydar and Shtivelman 2005, Landa *et al* 2006, Schleicher and Costa 2009). The weighted multipath summation allows us to replace the complex optimization problem of estimating the optimal parameters, by summation along all possible surfaces that are created from small variations of those parameters. The weighted multipath summation (WMPS) can be described by the following expressions:

$$I_{\text{WMPS}} = \frac{\sum_P I_P W(I_P)}{\sum_P W(I_P)}, \tag{4.5}$$

$$I_P = \sum_P \sum_S \sum_R U_{SR}(T_{SR}(P)),$$

where $W(I_P) = \exp(\lambda I_P)$ is the 'optimal' weighting function and λ is the unidimensional large number. I_P is the image obtained for fixed parameters P, namely, for radii of curvatures azimuth and emergence angles. U_{SR} (T_{SR}) is the seismic trace for a given source–receiver pair. T_{SR} is the travel time from shot S to receiver R. The summation (4.4) with the proper weight (I_P) is performed for all possible parameter values within a specified range. The image defined by I_P means that for every point to be imaged, seismic amplitudes are stacked together along all possible time surfaces defined by (4.4). The constructive and destructive interference of the amplitudes contributed by each time surface produces an image close to that obtained by stacking with the 'correct' parameter (Shtivelman *et al* 2009).

4.2.3.7 Case study

Study site. The Mineral Beach study site (site II in figure 4.30(a)) is located between the DS shoreline at the east and route #90 (the main road along the western DS shore) at the west. The area is characterized by N–S normal faults (Abelson *et al* 2006). Around Mineral Beach, sinkholes develop in both mudflat (southern boreholes Mn-2, Mn-5) and alluvial fan (northern borehole Mn-4) areas. Boreholes show that the salt layer at the northern part of the area is located at 10 m deeper (−434 m elevation) than that at the southern one (−424 m). At the northern part of the area, salt is overlain by sandy-gravel sediments, whereas the southern part is composed of the DS mud overlying the salt.

The field acquisition covers a 120 by 60 m area and consists of seven receiver lines in a 10 m interval between them (figure 4.35(a)). The shooting was carried out using six shot lines (288 shots in total) and each shot includes 96 channels in 2.5 m intervals between them. The distance between shot lines is 10 m. The data was recorded using a 0.5 ms sample rate and 0.5 s record length. We used truck-mounted accelerated weight (Digipulse) as an energy source (see section 4.2 figure 4.2(a)) and a single 10 Hz geophone per station. Typical common shot gathers are seen in figure 4.35(b).

To image the new developing fault in detail, the survey was designed with a full azimuth cover for offsets less than 30 m (figure 4.36).

The data were processed using the new 3D stacking algorithm. The range of variation of parameters in (4) was as follows: radii of curvature from 3 to 600 m, emergence angle from −5° to +5°, and velocity from 300 m s^{-1} to 1000 m s^{-1}. The value of λ was 900. The results from processed 3D imaging are presented in figure 4.37(a) as the time sections along the reflection lines and in figure 4.37(b) as a 3D cube time section.

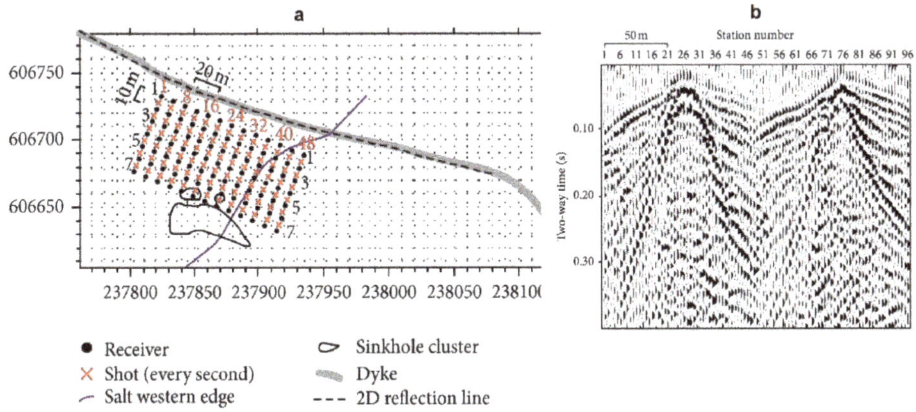

Figure 4.35. (a) The geometry of 3D seismic survey (b) Typical common shot gathering including 96 channels. (Reprinted from Keydar *et al* (2013), copyright (2013) with the permission of Hindawi Publishing Corporation.)

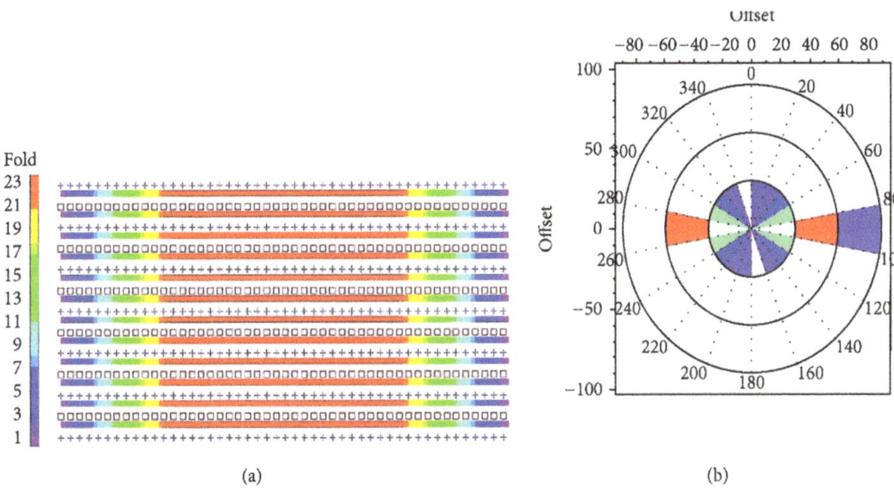

Figure 4.36. Fold scheme. (a) The shot locations are marked by '+' and the locations of the receivers are marked by squares, (b) fold as a function of azimuth. (Reprinted from Keydar *et al* (2013), copyright (2013) with the permission of Hindawi Publishing Corporation.)

A time map of the first reflector is presented in figure 4.37(c). On all the figures are seen sub-horizontal coherent reflectors at an approximate depth of 50–80 m, which incline on a closer location to the exposed sinkhole. Also, faults are seen in all sections (figure 4.37(a) sections denoted by 2, 5, and 7). These faults are seen in the reflector presented in figure 4.37(c). The results are consistent with the results that were obtained from a previous 2D study (Yechieli *et al* 2006) at the same site. This work provides the first 3D HR imaging on the edge of a sinkhole and nearby fault.

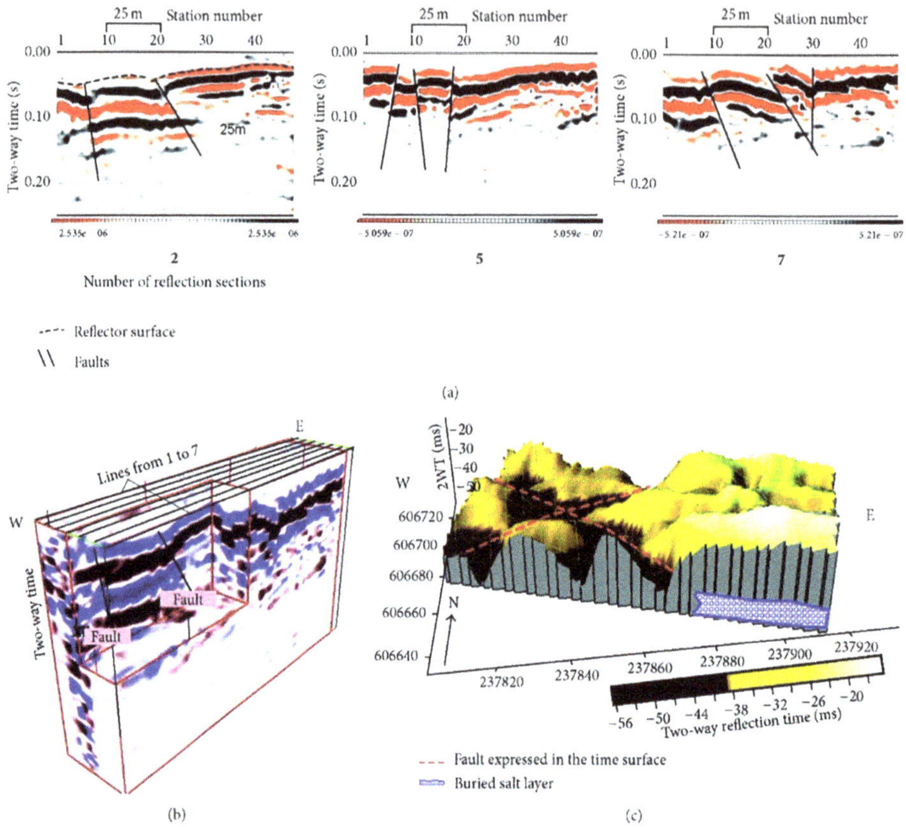

Figure 4.37. Results of 3D seismic reflection in Mineral Beach area. (a) Typical two-way time sections constructed from 3D time cube obtained using 3D imaging approach (b). (c) Two-way time map of reflector in the 3D presentation. Two faults seen in all sections are expressed in the time surface of the reflector (Reprinted from Keydar *et al* (2013), copyright (2013) with the permission of Hindawi Publishing Corporation.)

The results of the seismic interpretation of 3D images suggest a possible linkage between the revealed fault, sinkholes, and field observations.

4.2.3.8 Conclusions

We have presented the 2D and 3D imaging techniques that are a new approach for imaging shallow subsurface (Keydar *et al* 2012a, 2013). The proposed approach includes the recently developed CSP stacking and multipath summation. The resulting time section, obtained using CSP free stretch stacking, is superior to the time section obtained by the CMP method.

The core of this approach is a new 3D NMO time correction surface formula. Imaging of subsurface is performed by a spatial summation of seismic waves along these time surfaces using the proposed multipath summation with proper weights. The multipath summation is performed by stacking the target waves along all possible time surfaces having a common apex at the given point. This approach does not require any explicit information on parameters since the involved multipath summation and is performed for all possible parameter values within a wide specified range.

New techniques were used to understand the relationship between a developing sinkhole and its tectonic environment. Several new seismic surveys along a line crossing the sinkhole lineaments and several previous reflection sections reinterpreted using new techniques were carried out.

The results from the processed 3D time volume show sub-horizontal coherent reflectors at an approximate depth of 50–80 m which incline on a closer location to the exposed sinkhole and suggest a possible linkage between the revealed fault and the sinkholes.

4.2.4 Diffraction method

The method of diffraction imaging is developed aimed to detect underground heterogeneities (Landa *et al* 1987, Klem-Musatov 1994). Scattered (diffracted) waves are used in a developed approach called 'diffraction imaging' (Landa *et al* 1987, Kanasevich and Phadke 1988, Gelchinsky *et al* 1993, Belfer *et al* 1998, Landa and Keydar 1998, Keydar and Shtivelman 2005, Shtivelman and Keydar 2005). This approach has been originally proposed for the detection of geological fault location. In the framework of this work, we have improved the existent diffraction imaging technology for detecting and implemented it on the data acquired in the DS.

The diffraction imaging technique (Landa and Keydar 1998, Keydar *et al* 2010), utilizes diffracted waves to detect inhomogeneous objects such as faults, sinkholes, cavities, caves, tunnels, is one of the new methodologies for imaging shallow subsurface. The diffraction travel-time correction curve is defined as a function of the velocity of diffracted waves. The CSP stacking discussed in the previous section, and diffraction imaging formulae can be extracted provided the model of the medium is known. However, in reality, the model of the medium is unknown; therefore, some kinds of the inverse problem must be solved to construct these parameters.

A new methodology for imaging shallow subsurface combines several recently developed imaging techniques. The first one is zero-offset CSP stacking described in the previous section (Keydar *et al* 1996), diffraction imaging, and multipath weighted summation. For a more comprehensive discussion of the methods, one can read the relevant papers.

4.2.4.1 Diffraction imaging technique

The imaging algorithm for the detection of hidden anomalies should emphasize scattering objects such as voids and faults and not be too sensitive to the presence of continuous and smooth interfaces, which generate reflected/or refracted waves. For example, it is clear that the so-called stacked section, conventionally used in seismic reflection imaging, cannot be employed for faults and facies changes detection, since the stacking procedure emphasizes reflected waves and tends to eliminate diffraction/scattering events generated by faults and facies changes. For the detection of scattering objects, it was proposed to use a so-called diffraction imaging algorithm (Landa and Keydar 1998) which is based on the phase correlation of the diffraction signals of the observations. The data are analyzed along the different diffraction curves to find the curve closest to the travel-time curve of the signal.

Assume a diffractor located at a point (X_D, Z_D) (figure 4.38). Then for a set of shot points (X_0, 0) and receiver points (X_k, 0) located at the surface, the kinematic response of the diffractor (diffraction time curve) can be defined as follows:

$$T_{SR}^{DIFF} = \sqrt{(\xi_{SD}/V)^2 + T_D^2} + \sqrt{(\xi_{DR}/V)^2 + T_D^2}, \tag{4.6}$$

where $\xi_{SD} = \sqrt{(X_D - X_0)^2}$ is the horizontal source-to-diffraction distance, $\xi_{DR} = \sqrt{(X_K - X_D)^2}$ is the horizontal diffractor-to-receiver distance, V is the velocity in the medium and $T_D = Z_D/V$ is the vertical time above the diffractor. The velocity term V defines the shape of the diffraction time curve, whereas the time T_D defines its apex. For various values of V and fixed values of T_D, equation (4.6) defines a family of diffraction curves with different shapes and with a common apex at time $2T_D$. The unknown parameter V is estimated (Landa and Keydar 1998, Keydar *et al* 2010) using the wave correlation procedure which consists of finding optimal parameter V which maximizes the coherency functional.

A new 3D approach to imaging subsurface inhomogeneities was presented by Shtivelman and Keydar (2005). The proposed method is based on the multipath summation of the wave field along all possible diffraction trajectories in 3D space, where every subsurface point is assumed to be a point diffractor. The above results were obtained under the assumption that the subsurface inhomogeneity to be imaged (a cavity) is a localized object that can be represented as a point diffractor. For other types of objects, such as tunnels or fault planes, the method should be modified to improve the spatial resolution and the S/R ratio of the images. This will be the subject of further investigations.

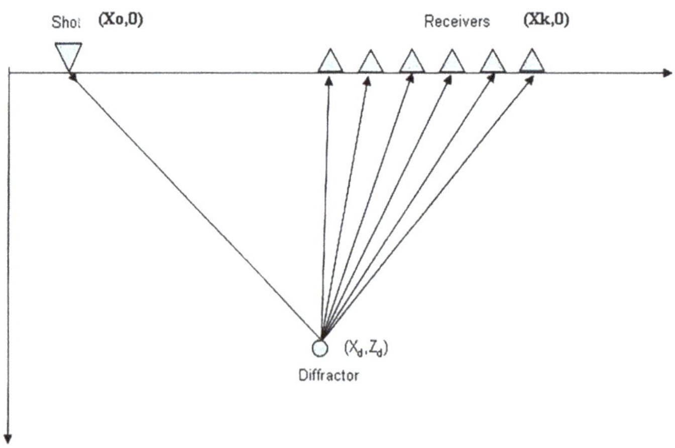

Figure 4.38. Schematic description of the ray path from diffractor X_D, Z_D. (Reprinted from Keydar *et al* (2012a), copyright World Academic Publishing.)

4.2.4.1.1 Real example

The Mineral Beach (study site II in figure 4.30(a) in the previous section) is located at the western DS shoreline. The area, covering ~1 km^2, is situated in the central part of the northern DSa basin (figure 4.39(a)). Around Mineral Beach, sinkholes develop in both mudflat (south) and alluvial fan (north) areas.

To specify the structural geological conditions of the site, a 345 m multi-channel reflection survey has been acquired. The field parameters of this survey included 71 shot points with a 5 m interval and 48 receivers for each shot with a 2.5 m interval between the receivers. As a source, an accelerated weight-drop ('DigiPulse') was used. The data was recorded using a 0.5 ms sample rate and 0.5 s record length.

As mentioned in section 4.2.3, the CSP reflection study was carried out along the MERC1 line (see figure 4.39(a) for location). The diffraction imaging was applied in addition to CSP to detect the fault using the same field records (figure 4.39(b)).

Both CSP and diffraction sections are shown in the section (figure 4.40). One can see that the strong anomalies in the diffraction image are located between shots 22 and 30 (figure 4.40(a)) are consistent with the discontinuities of the reflector on the CSP time sections (figure 4.40(b)).

From the above example, one can see that the diffraction image produces a clear image of the subsurface heterogeneities such as faults, especially, in their uppermost part.

Another problem is associated with subsurface cavities that can be the embryo of future sinkholes. Nowadays there are thousands of sinkholes that represent a danger to life and property and negatively affect any developments in the area (Abelson *et al* 2017).

4.2.4.2 Nahal Hever South (site III in figure 4.30(a))

The diffraction imaging of sinkholes has been constructed in the Nahal Hever South site using several seismic reflection profiles that were shot there (figure 4.41(a)). The typical common shot gazer from Line SV128 is shown in figure 4.41(b).

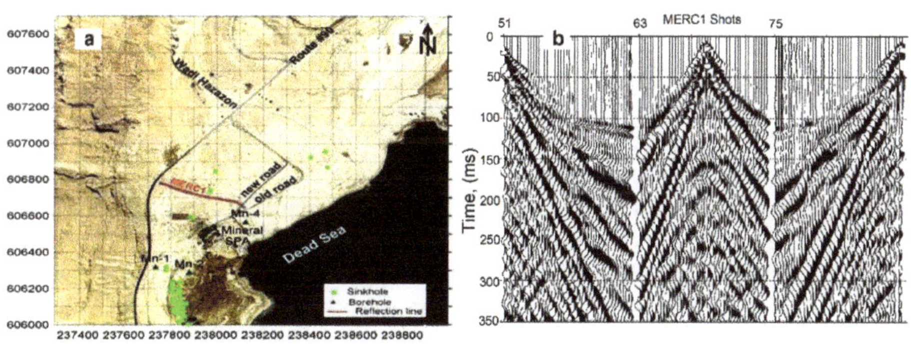

Figure 4.39. (a) The Mineral Beach sinkhole development area with the location of (a) the MERC1 seismic reflection line (red line). (b) Example of three common shot gathers. (Reprinted from Keydar *et al* (2012a), copyright World Academic Publishing.)

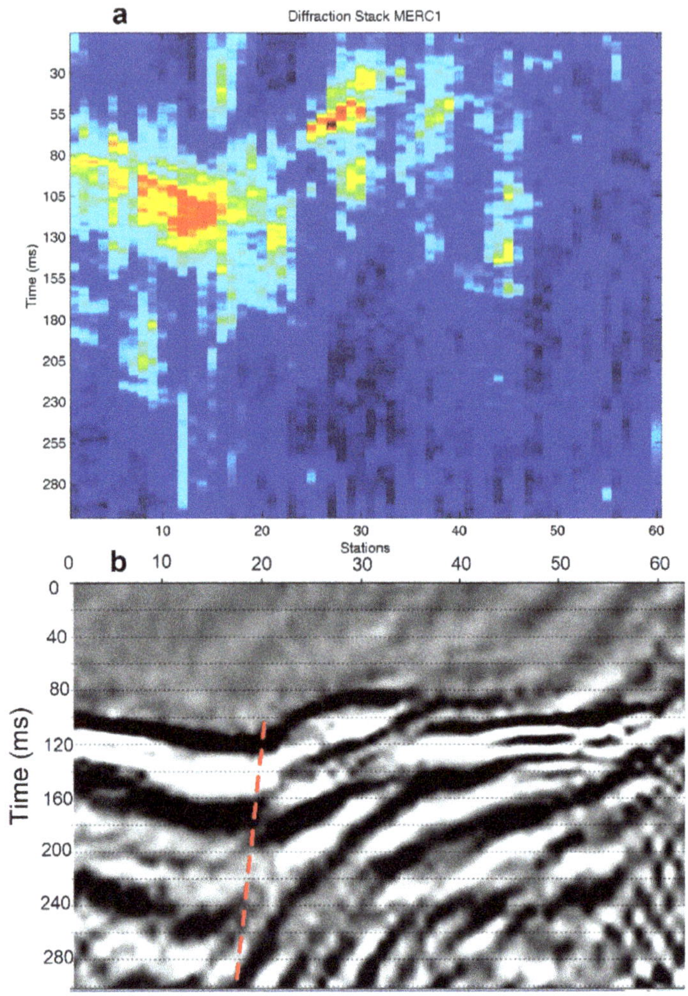

Figure 4.40. (a) Diffraction image obtained by diffraction method using weighted multipath summation, and (b) CSP time section obtained using the same method. The red dashed line is a concealed fault. (Reprinted from Keydar *et al* (2012a), copyright World Academic Publishing.)

A schematic presentation of the geological structure of the site is shown in figure 4.41(a) (Yechieli *et al* 2004). The section (borehole SH-2) is composed of alluvial fan sediments down to 18 m deep, a 5 m thick marl layer, and an 11 m thick salt layer. The salt layer is underlain by clay and gravel. The elevation of the area is approximately −393 m. The water table in the area was measured at a depth of 18.5 m in April 2002. The top of the salt layer is at a depth of 24 m from the surface and 5–6 m below the water table. In borehole HS-3, a mud-filled cave was detected at the depth of 23–29 m, which is assumed to be the result of the salt layer dissolution. The visible section of sinkholes consists of sand–gravel intercalating layers. A group of sinkholes about 20–50 m apart appeared in July 1998. These sinkholes (2.0–3.0 m

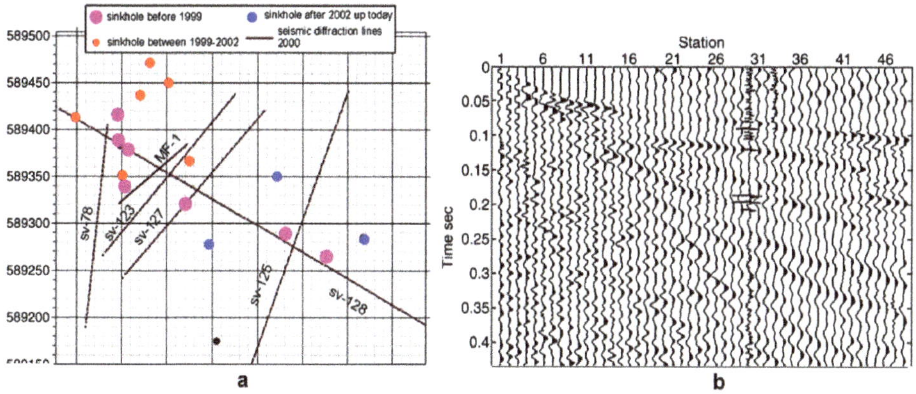

Figure 4.41. (a) Sinkholes and seismic line scheme location in the Nahal Hever south area. (b) Typical common shot gather from line SV 128. (Reprinted from Keydar *et al* (2010), copyright (2010) with the permission of Elsevier.)

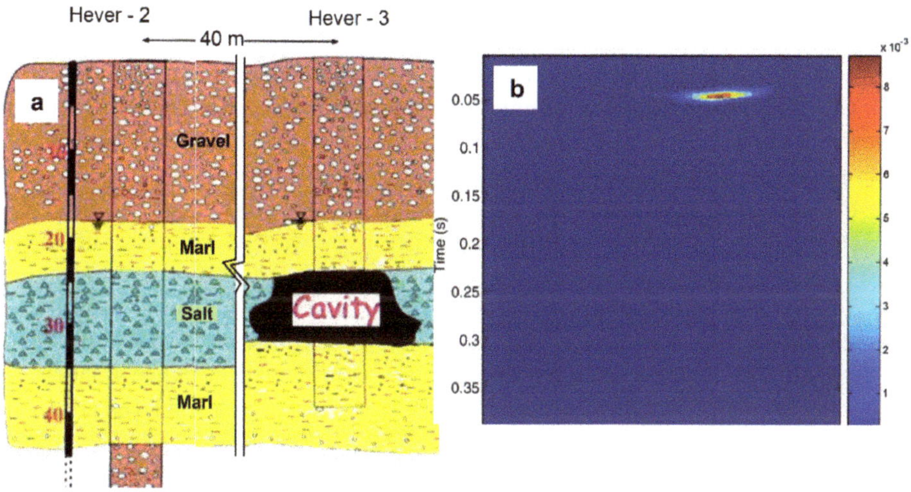

Figure 4.42. (a) The geological section crossing the HS-2 and HS-3 boreholes in the South Nahal Hever South area. (b) D-section above cavity shown in (a). (Reprinted from Keydar *et al* (2010), copyright (2010) with the permission of Elsevier.)

deep and up to 3 m in diameter) varied over some time and, at present, there are sinkhole clusters up to 50–100 m in diameter. To test the feasibility of the diffraction imaging to detect cavities, the data were acquired in the vicinity of the borehole Hever-3 shown in figure 4.42(a) where the cavity was encountered in a salt layer within the depth range of 21–27 m.

The cavity can be assumed to serve as a diffractor (Keydar *et al* 2010). The diffraction imaging is based on the results of the seismic reflection survey that was carried out at this site. The length of the receiver line was 117.5 m and was fixed;

receiver spacing was 2.5 m. The number of source points was 48 with 2.5 m spacing between them. The source of seismic energy was the DigiPulse (truck-mounted accelerated weight drop). The data on the records are dominated by surface waves; no indications of diffracted events can be found. Figure 4.42(b) shows the D-section obtained using the diffraction imaging algorithm. The regions of high amplitude (displayed in red color) are related to the diffractor location. Maximum amplitude is achieved at the location of $X = 77$ m and $Z = 27$ m (1200 m s^{-1} × 0.023 s). The estimated location of the diffractor is in good agreement with the true location of the cavity, as revealed in the well.

The diffraction imaging was applied to other real data from the Nahal Hever South site that was acquired in the year 2000. The survey included five seismic lines. The location of the lines is shown in figure 4.41(a), a typical common shot gather is shown in figure 4.41(b) (the trace 30 is a 'dead trace').

All diffraction images that were obtained in the Nahal Hever South using a diffraction imaging algorithm are shown in Keydar *et al* (2010, figures 7–11). Here we present examples of two diffraction images from the lines SV 123 and SV 128.

The strong anomaly on line SV 123 (figure 4.43(a)) is associated with the borehole.

The most interesting results were obtained along the longest line SV 128 (figure 4.43(b)) that intersects the borehole and some sinkholes (see figure 4.41(a) for location). The leftmost anomaly is associated with the borehole, another anomaly in the center of the line that is associated with the subsurface karst was observed on the surface only 1 year after the acquisition.

At the same time, we cannot see any strong anomalies along the lines of SV 125. The absence of the anomalies indicates that there is no diffraction in the subsurface.

4.2.4.3 The Ghor Al-Haditha (site I in figure 4.30(a))

Two MASW profiles data along lines 3 and 4 were used to construct the diffraction imaging and CSP stacked sections (line location is presented in figure 4.33(a)).

These CSP and diffraction sections are presented in figure 4.44 (Keydar *et al* 2011).

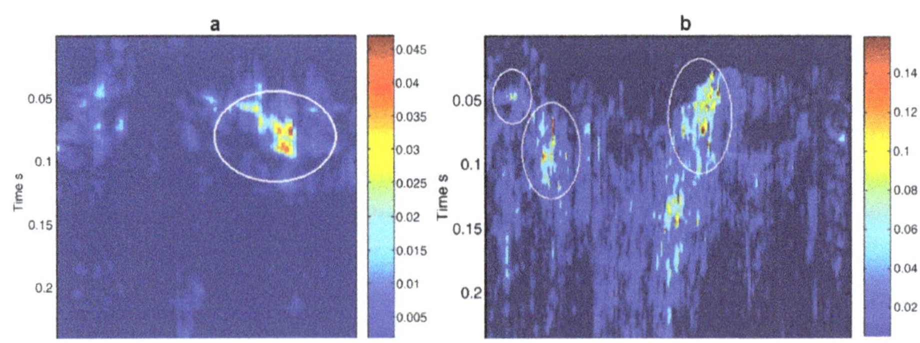

Figure 4.43. (a) D-section was obtained using diffraction imaging of the line SV 123, and (b) using diffraction imaging of line SV 128. (Reprinted from Keydar *et al* (2010), copyright (2010) with the permission of Elsevier.)

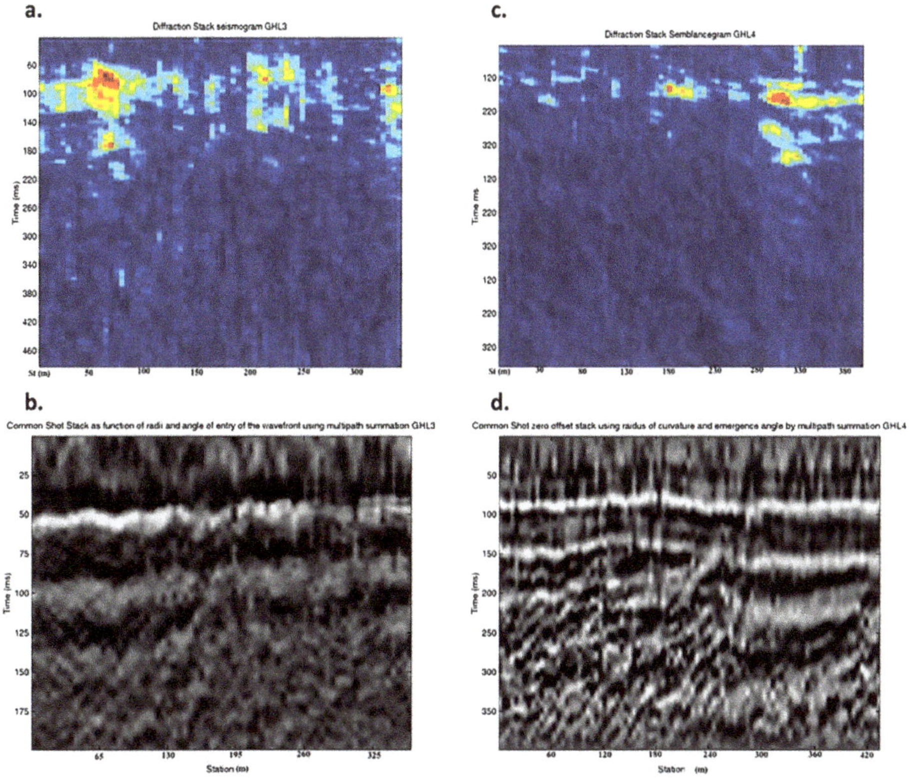

Figure 4.44. (a) Diffraction image and, (b) CSP stacked section along Line 3. (c) Diffraction image and, (d) CSP stacked section along Line 4. (Reprinted from Keydar *et al* (2011), with permission of EAGE.)

Both diffraction images (figures 4.44(a) and (c)) express the strong lateral heterogeneities possibly correlated with some of the low-velocity anomalies depicted by the dispersion inversion results (around position 200 m along Line 3 and position 220 and 350 m along Line 4). The CSP section along line 3 (figure 4.44(b)) indicates a deep reflector possibly related to compacted sediments and complicated by a fault (at the position interval 165–190 m). Along line 4, the CSP stack section (figure 4.44(d)) shows a comprehensive anomaly at position interval 270–290 m clearly expressed on the diffraction image (figure 4.44(c)).

From this position, the CSP stack also shows a deep reflector, correlated with the lateral velocity change in the dispersion inversion results.

4.2.4.4 Conclusions

The diffraction image produces a clear image of subsurface heterogeneities such as faults, sinkholes, cavities, caves, tunnels, or voids. This approach does not require any information about the subsurface, since the multipath summation is performed for all possible parameter values within a wide specified range. The CSP stack and diffraction method are complementary to each other and reveal useful information about the subsurface. The diffraction method serves as a tool for the detection of faults and voids, while the CSP stacking contains information about the structure of the subsurface.

4.3 Electromagnetic methods

4.3.1 The GPR method

4.3.1.1 Features of GPR in the Dead Sea area

Thousands of sinkholes have opened along with western DS shores from south to north and in the Ghor Al-Haditha (the Eastern DS) shore (figure 4.45) (Abelson *et al* 2017, Abou Karaki *et al* 2019, Nof *et al* 2019).

The DS karst has an evaporite origin associated with the formation of dissolution voids within buried salt layers (Abelson *et al* 2006, Ezersky 2006, Legchenko *et al* 2008a, 2008c, Closson and Abou Karaki 2009, Yechieli *et al* 2016, Ezersky *et al* 2017).

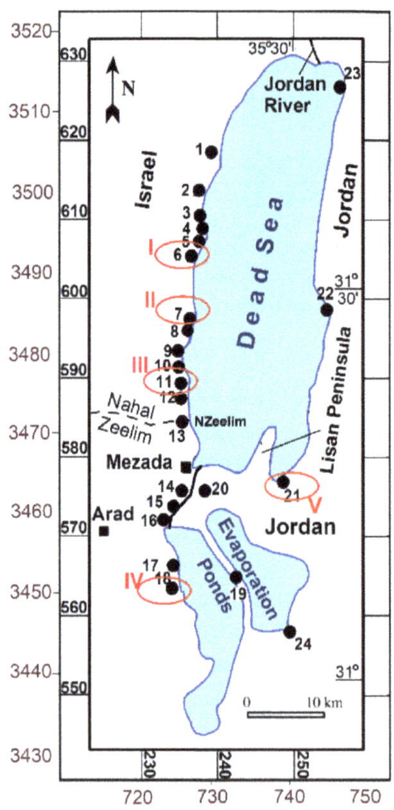

Figure 4.45. Sites of sinkhole development along DS shores, where GPR surveys were carried out. I—Mineral Beach; II—Ein Gedi–Nahal Arugot; III—Nahal Hever South; IV—Newe Zohar; V—Ghor Al-Haditha (the coordinates in the frame are new Israel Mercator grid in km, the outer coordinates are Universal Transverse Mercator grid, in km). The complete list of sinkhole sites is shown in figure 4.1 in section 4.1.

In the DS environment, subsurface cavities are formed by the dissolution of salt by unsaturated halite water flows through the subsurface (Yechieli *et al* 2006, Frumkin *et al* 2011). Thus, sinkholes are associated with the subsurface karst, e.g. system of cavities within salt layers. These cavities collapse the overlain sediments and gradually ascend to the surface and there form sinkholes. This process is named as stoping (Ford and Williams 2007, p 385, Frumkin *et al* 2011).

There are different geophysical methods to detect subsurface cavities such as microgravity (Rybakov *et al* 2001, Eppelbaum *et al* 2008), Magnetic resonance sounding (MRS) (Legchenko *et al* 2008b); Electrical resistivity tomography (ERT) (Gibson *et al* 2004, Cardarelli *et al* 2006) and other methods (see comprehensive review in Chalikakis *et al* (2011), section 2 of that manuscript). However, generally, the GPR is the most utilizable method for cavity detection in the shallow subsurface. Also, the method has the highest resolution within other geophysical methods (Greenfield 1979, Benson and Yuhr 1993, Benson 1995, Mellet 1995, Stevens *et al* 1995, Casas *et al* 1996, Reppert *et al* 1999, Beres *et al* 2001, Bristow and Jol 2003, Hoover 2003, Rodriguez *et al* 2014, Billi *et al* 2016, Gutiérrez *et al* 2018). The high resolution of the GPR method is caused by the high frequencies of the scanning (50 MHz–1 GHz). The same factor is the reason for the depth limitations. The penetration depth depends also on the properties of the subsurface and its dielectric constant. As a rule, the penetration depth is ranged from the first meters to the first tens of meters.

It deserves to be mentioned GPR's ability to produce 2D and 3D images of soils and sediments at shallow depths means that combined images of geological materials associated with cavities and archaeological objects are possible (Conyers 2016).

The problem of cavities detection in the DS shores is caused by two factors: (1) the evaporite karst of the DS area is located at a great depth (20–70 m), unreachable for the GPR method penetration depth of which is 5–15 m; (2) the cavities are located under the very saline groundwater table resulting in the screening of the electromagnetic waves radiated by the GPR systems.

Despite the above-mentioned problems subsurface void detection using the GPR technique has been repeatedly carried out along the DS shore area (Shtivelman *et al* 1999, Arkin *et al* 2000, Batayneh *et al* 2002, Frumkin *et al* 2008, 2011, Goldshleger *et al* 2018).

Our experience has shown that the GPR technique in its 2D and 3D modifications can be a very efficient tool to predict sinkhole susceptibility in the DS coastal areas.

4.3.1.1.1 Features of the GPR methodology

The GPR method is used here to detect the shallow structural deformations in sediments above the salt layer. GPR is based on the transmission of electromagnetic waves into the subsurface and the recording of reflected waves from objects or interfaces with the different dielectric constants. A radar system comprises a signal generator, antennae (for transmission and reception), and receiver. The system has an onboard computer that facilitates data processing both while acquiring data in the field and post-recording (Reynolds 2011). As the antenna is moved along the ground surface, the signals of the reflected waves are displayed as a function of their two-way travel time, i.e. the time elapsed between the transmission and detection by the receiver, in the form of a radargram (also named the radar section) that is used for the interpretation of the GPR study. The horizontal axis of the radargram is the distance along with the GPR profile. The vertical axis is usually two-way time from surface to target and back. The time can be converted to the depth if the dielectric constant ε_r is known or has been determined.

The GPR main parameter determining the propagation and reflections of electromagnetic waves is the relative dielectric constant ε_r, which varies from 1 for air to 81 for water (examples of the ε_r values, velocity, and electrical conductivity σ for different rocks are represented in table 2.1 in section 2.2.1.5). The basic formulas are also presented there. In this section, we repeat the main properties related to the DS area. The dielectric constant of the DS deposits varies in a wide range from $\varepsilon_r = 3$–6 (alluvial fans) to $\varepsilon_r = 25$–45 (mudflats) (see table 4.4). Such a wide range is caused by the variable soil salinity, from very high at the mudflat sites (where sediments are saturated with the DS brine) to very low at the alluvial fan sites (where sediments have a high porosity).

4.3.1.1.2 The GPR equipment

In the DS area, two GPR systems have been used. The first one is the SIR-20 systems (Production GSSI, USA) with antennae (figures 4.46(a) and (b), respectively) 100 MHz, 200 (or 250), 300 MHz, 400 (or 500) MHz, 900 MHz (Frumkin *et al* 2008, 2011). Another GPR system is the CU-II RAMAC GPR system (Production MALA, Sweden) (figure 4.46(c)), which includes the CU-II control unit, a modern and upgraded unit known for its high versatility. The system enables us to carry out the ground-penetrating radar surveys, as well as surveys in the boreholes and cross boreholes (tomography and reflection) (Ronen *et al* 2019). The system has several electronic control units, suitable for use with a variety of antennas—50, 100, 250 and 500 MHz antennas, both shielded and unshielded (Mala 2010).

4.3.1.1.3 The GPR data processing

Data acquired with the SIR-20 system were processed using WINRAD A/VA 5 software intended for post-processing of the radargrams. The acquisition range was 50 ns (for 500 MHz antennae) and 200 ns (for 100 MHz antennae).

With the CU-II RAMAC GPR system, the signal processing was performed by the complete, state-of-the-art REFLEXW software package produced by the German company K J Sandmeier (Daniels 2007, Booth *et al* 2010, Sandmeier 2020) used for processing and interpretation of reflection and transmission data. The software allows 2D and 3D interpretation of the GPR data, as well as a 2D simulation of the electromagnetic wave propagation based on a finite difference approximation. Another software, GPR-SLICE (2012), was used for 3D data processing and creating images of ground-penetrating radar reflection data. GPR-SLICE integrates radargram profiles taken over a survey grid to produce horizontal time slice maps of radar anomalies in the ground (Conyers 2013).

Table 4.4. GPR parameters in the DS sites of different geology.

Conductivity, σ, mmho m^{-1} (mho m^{-1})	Resistivity, ρ_x, Ωm	Dielectric permittivity, ε_r	Penetration depth, D, m* (100 MHz Antenna, 200 ns range)	DS site geology	DS cite name
1	1000*	5.5	10–20	Alluvial fan	Mineral Beach (N), Ein Gedi, Newe Zohar
10	100**	8	7–8	Alluvial fan	Nahal Hever (S); Ghor Al-Haditha
100	10***	12–15	3–5	–400 m elevation and lower	Mineral Beach (S)
1000	1****	30 and more	0.2	Mudflat	Mineral Beach (S), Ghor Al-Haditha near DS
10 000	0.1	>50	0.1	Mudflat	

* Ronen *et al* (2019),
** Ezersky *et al* (2006), Al-Zoubi *et al* (2007),
*** Frumkin *et al* (2008),
**** Ezersky (2008), Khlaifata *et al* (2010).

Figure 4.46. Two types of GPR systems. (a) SIR-20 radar and (b) set of antennas (from left-to-right): 100, 250, 300, 400 MHz, and (SIR-10) 500 MHz; (c) CU-II RAMAC GPR system with the 100 MHz antennae in the field.

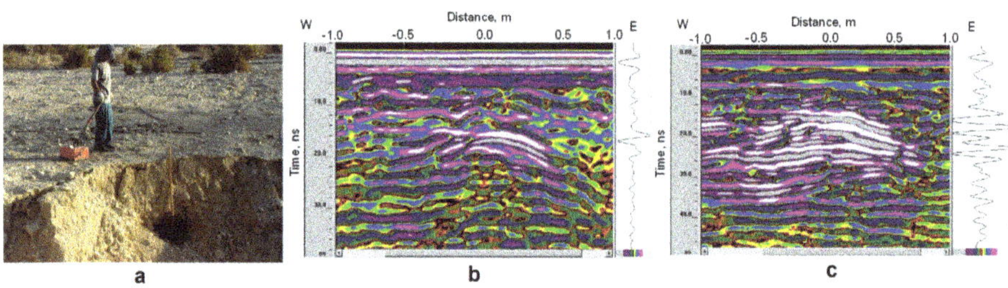

Figure 4.47. GPR field studies in the Newe Zohar site (Site IV in figure 4.45): radargrams obtained over the man-made void (tunnel) within the southern vertical wall of the sinkhole 19S. (a) Void of 0.3 m diameter and 1.0 m length is located in a depth of 1.0 m from the surface. Towing the antenna was performed in a traditional mode along the E–W line. (b) 500 MHz antenna, 0.5 m from the edge of the sinkhole (c) 400 MHz antenna, 0.5 m from the edge. GPR signal is shown at the right sides of both radargrams (Frumkin *et al* 2008).

4.3.1.1.4 Interpretation of complicated GPR images

Conventionally, a GPR signal above a void within homogenous material will appear as a strict hyperbolic image (Daniels 2007). Therefore, a GPR technique successfully detects voids, caverns, and tunnels. In real geological conditions, the signal can be complicated by geological features such as different interbedding lithological layers, their depth, different antennas, geometric shape of the void, etc. Examples of the conventional radargram are shown in figure 4.47. The radargram was obtained over a man-made void (horizontal tunnel) within a vertical wall of the sinkhole, the void's dimensions are 0.3 m diameter, 1.0 m in length, and its depth is 1.0 m from the surface (figure 4.3.1-4(a)). Two GPR anomalies are shown in figures 4.47(b) and (c). The GPR signal is shown on the right-hand side of the radargrams (Frumkin *et al* 2008). Time translation to depth shows that the value of the dielectric constant $\varepsilon_r = 4$.

4.3.1.1.5 The GPR images above non-deep cavities

The images obtained in the Mineral Beach site indicate a shallow cavity connected to the opened sinkhole (figure 4.48). The cavity is located 0.5 m out of the sinkhole.

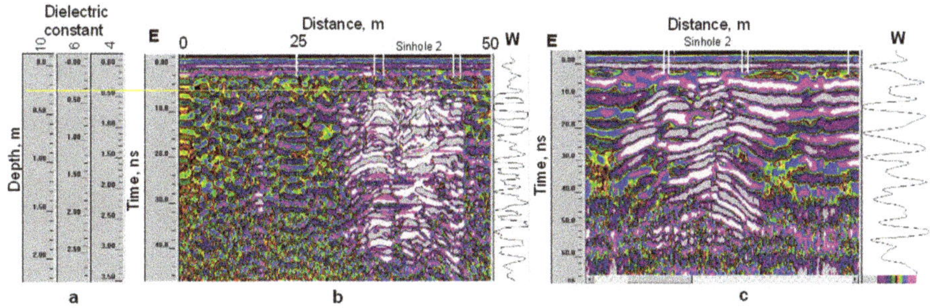

Figure 4.48. GPR field studies Mineral Beach (the site I in figure 4.45). Radargrams over the natural voids were recorded towing the antenna in traditional mode along the E–W direction. (a) Scale depth for different dielectric constant; (b) 400 MHz antenna, 0.5 m from sinkhole; (c) 200 MHz antenna, 0.5 m from sinkhole; 2 marks denote sinkhole location. The yellow line serves for depth calibration. It corresponds to the depth of 35 cm from the surface to the void ceiling. GPR signal is shown at the right sides of both radargrams (Frumkin *et al* 2008).

Radargrams have been recorded with a 400 MHz antenna (figure 4.48(b)), and with 200 MHz antennae (figure 4.48(c)). The two-way time is shown along the vertical axis (figure 4.48(a)) has been translated to penetration depth using different values of dielectric constant 4, 6, and 10.

Knowing the depth to void ceiling (0.35 m) we can estimate the dielectric constant of the subsurface by comparing the measured depth with the calculated one. In this case $\varepsilon_r = 6$. One can see in figures 4.48(b) and (c) (right sides) that signals in both cases are of good quality. Note that images are more complicated than those recorded above a man-made cavity (figure 4.47). However, anomalies look very impressive testifying to the cavity in the subsurface.

Reverberations. Another type of GPR image has been suggested by Kofman *et al* (2006). Carrying out GPR scanning above subsurface voids, the authors revealed the resonant oscillations above the existing voids. According to Glover (1992), the reverberations generated by void targets may be considered to arise from reflections of the appropriate traveling waves from the walls of the enclosure and other GPR responses. The results of the experiments have provided different radar signatures, which could be useful in understanding the void's reflections (Kofman *et al* 2006, Frumkin *et al* 2008, 2009). Based on past practical GPR experience of the void detection and presently obtained experimental data, two rules of thumb may be adopted for the prediction of the appearance of resonant radar pictures:

1. The void diameter is larger than the wavelength in the air of the antenna used.
2. The vertical size of the empty void is not significantly smaller than its horizontal dimension.

It follows from above that to obtain reverberations, the vertical size of the empty voids should not be significantly smaller than their horizontal dimension. However, it appears that if the vertical dimensions are significantly larger than the horizontal dimensions, the creation of reverberation phenomena is not diminished.

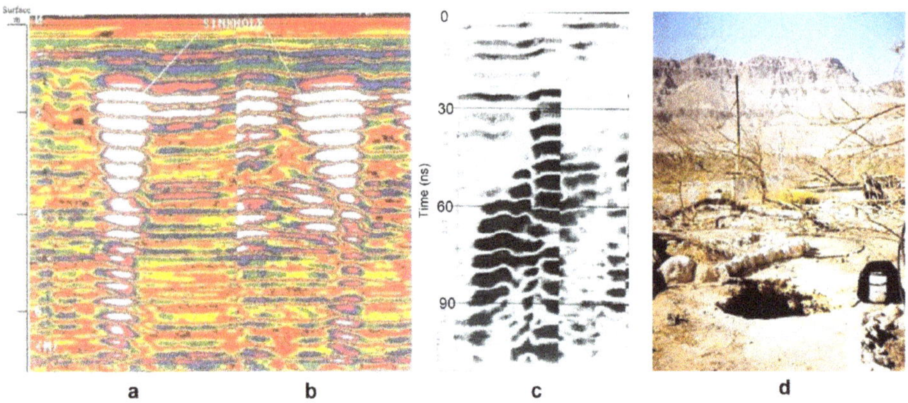

Figure 4.49. Anomalies recorded above existing voids in the Ein Gedi sinkhole site. (a and b) Radar image of air-filled sinkhole revealed by towing of 300 MHz bistatic antennae in N–S and W–E one (b) directions, respectively, 120 ns range (Kofman 1994); (c) GPR image of a developing sinkhole (real-time thermal printouts): void detected—February 1998, (d) sinkhole collapsed—2006 (Kofman 1996, Arkin *et al* 2000, Frumkin *et al* 2008).

For example, figures 4.49(a) and (b) indicate the clear resonant oscillations generated by the inner surface of the sinkhole at a roof depth of about 1.1 m revealed in July 1997 during a practical GPR survey by 300 MHz antennae in an alluvial fan at the DS shore near Ein Gedi, Israel (Arkin *et al* 2000). According to the frequency spectrum analysis, the dominant frequency is 107.8 MHz. The GPR data were collected in both orthogonal north–south (figure 4.49(a)) and west–east (figure 4.49(b)) directions. This cylindrically shaped void collapsed due to a passing truck, 2 months after its subsurface detection; its diameter was 1.5 m and its height was about 8 m.

Another example is presented in figure 4.49(c) and (d). Sinkhole embryo (void) was detected in the Ein Gedi site of the DS area in February 1998. The surface collapsed eight years later in 2006 (Arkin *et al* 2000). The image of the voids is seen as black areas on a white background under the real-time thermal printouts. They are prominently standing out against clay, silt, marl, salt, and gypsum lenses and layers. The roof of the voids is shown by the first high amplitude oscillation (1st phase) and can be measured fairly precisely. The depth shown is converted from travel time using a single average velocity of 0.13 m ns^{-1}.

4.3.1.1.6 Classification of GPR images

Conventional GPR images have a simple shape and signs considered above that testify to the presence of voids in the subsurface. However, images recorded in the real field conditions are often more complicated. Extensive experience of GPR measurements in the DS area enables us to suggest several GPR anomalous images showing a potential collapse hazard based on a probabilistic approach (figure 4.50). Suggested GPR image classification corresponds to that used in the geological hazard map (Abelson *et al* 2004, Yechieli *et al* 2016).

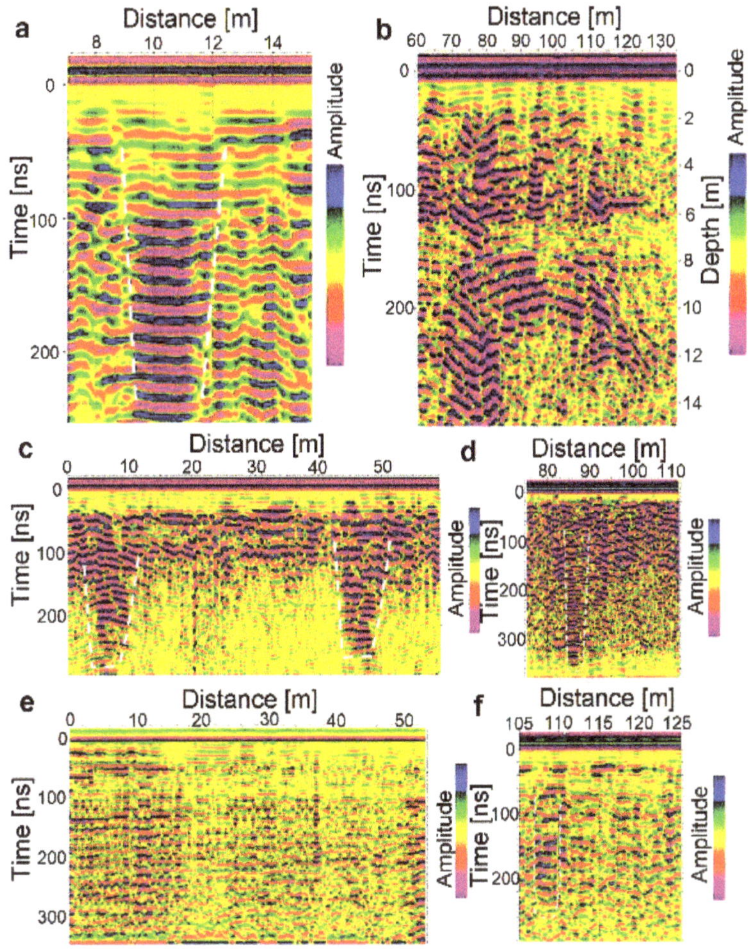

Figure 4.50. Examples of *very high probability anomalies* relate to the underground cavity. (a) Ein Gedi area; (b) Mineral Beach area; (c and d) *high probability anomaly* Nahal David site; (e) *no anomaly* (Mineral Beach); (f) *low probability anomalies* related to the underground cavity (Ein Gedi). (Reprinted from Ronen *et al* (2019), copyright (2019) with permission from Elsevier.)

Abelson *et al* (2004) suggested the classification of sinkhole hazardous areas along the DS shores based on the appearance of sinkholes (see Yechieli *et al* (2016) for explanation).

The potential levels depicted on the enclosed maps are as follows:

- *Sinkhole sites zone* (the yellow zone)—Areas around sinkhole clusters marking the highest potential level for sinkhole development.
- *High potential zone*—Zones of high potential for sinkhole appearance (red color in maps), comprising the major conditions for sinkhole formation, namely: (i) presence of a salt layer in the sub-surface; (ii) decline in the DS water level and the associated groundwater.

- *Low potential zone*—Zones in which the possibility of sinkhole formation is low but cannot be ruled out (green color in maps). These include areas where the probability of finding a salt layer in the subsurface cannot be ruled out. In the southern basin, where the freshwater flow in the subsurface is limited due to the rising level of the brines in the evaporation ponds, areas suspected of the presence of salt layer are also marked in green rather than red, as in the northern basin.

It deserves mentioning that geological criteria of sinkhole susceptibility are based on the appearance of sinkholes only. The salt layer presence is only suspected. At the same time, geophysical methods allowed mapping of the salt and suggested the sinkhole formation model based on knowledge of geological conditions, including salt mapping, groundwater salinity, and corrosiveness evaluation (Ezersky *et al* 2007, Ezersky *et al* 2017, Ezersky and Frumkin 2017, 2021). Moreover, following the salt edge hypothesis, sinkhole susceptibility zones are located at the boundary of the salt layer in Israel and Jordan. The width of the sinkhole-prone zone is 100–150 m from the salt edge inland. To the east of the salt edge in Israel, the salt will dissolve slowly over time and the sinkhole belt will promote to the east. West of the salt edge, the non-dissolved salt can be found. Therefore, the 50 m areas west of the salt edge are accepted as a dangerous zone, where sinkholes can also form (see red high likelihood zone).

Therefore, the criteria of subsurface void detection are more sure. These criteria are considered in terms of *very high probability* anomalies characterizing the subsurface cavities, *high probability*; *low probability* anomalies, and *no anomalies* (figure 4.50).

4.3.1.1.7 Examples of cavity detection in the DS site subsurface using GPR

In this section, we consider examples of cavity detection in different sites of the DS shores.

Mineral Beach area (Site 1 in figure 4.45), covering ~1 km^2, is located on the western coast of the central part of the northern DS basin. The study site is located between the DS shoreline and Route #90 (the main road along the western DS shore) (figure 4.51(a)).

Figure 4.51(b) illustrates that sinkhole development at the beginning of the process is mainly linear, along a salt layer edge, or controlled by a fault (Ezersky 2010, Frumkin *et al* 2011). A salt layer including cavity was discovered by the Mn-2 and Mn-5E boreholes (see figure 4.51(b) for location). The borehole MN-2 crossed the DS mud from the surface to 18 m deep. The borehole MN-1 crossed a sandy-gravel section. Soon after drilling, the borehole MN-2 collapsed into a newly developed sinkhole.

The GPR study was aimed to detect susceptibility of sinkhole formation at two sites: (A) a dirt road connecting the Mineral Spa to Route #90 and (B) a parking lot near the spa (figure 4.51(a)). The geological structure of both sites is characterized by sandy-gravel sediments. The groundwater level was 18.5 m deep in 2008 and ~27 m in 2018. The subsurface is characterized by the relatively high resistivity allowing a GPR penetration depth of 8–10 m at a dielectric constant of 5–6 (V = 0.13–0.12 m ns^{-1}, respectively).

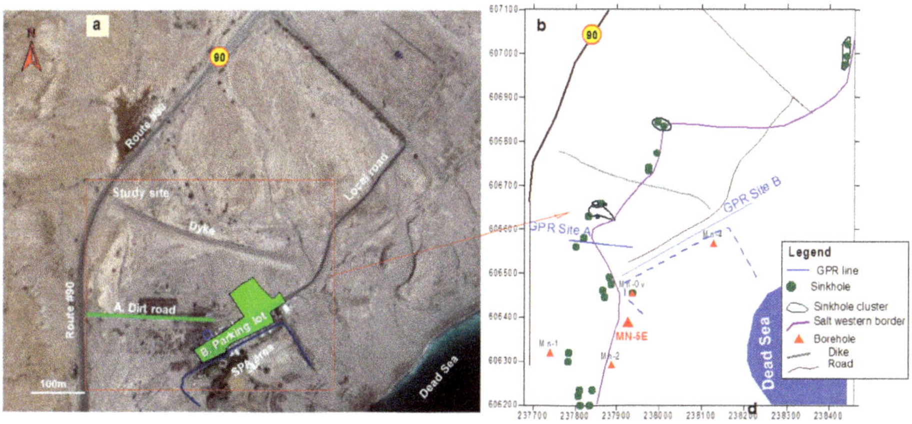

Figure 4.51. Mineral Beach area: (a) air map with two GPR study sites (A and B) location; and (b) geopysical measurements layouts, recent sinkholes, and salt western border. (Reprinted from Ronen *et al* (2019), copyright (2019) with permission from Elsevier.)

4.3.1.1.8 Site A. Dirt road

The GPR measurements were carried out in 2008 as a 2.5D study (Frumkin *et al* 2011). The dirt road runs perpendicular to the salt edge, crossing it at the south-western part of site A (figure 4.51(a)). Five GPR lines were carried out with 500 MHz antennae along the dirt road and two additional oblique lines (figure 4.52(a)). The radargram presented in figure 4.52(b) shows three anomalies which correspond: (1) to a vertical structure, apparently, a fault or fracture; (2) a disturbing medium interpreted as fractures and voids filled with air (figure 4.52(b)), and (3) a strong reflector whose pores can be filled with air or water (figure 4.52(b)). Anomalies 2 and 3 reveal an inclined and sagged layer.

No signs of sinkholes were seen on the Earth's surface in November 2008. The anomalous reflector is shown in the 3D presentation (figure 4.52(c)) generated with the profiles acquired along five lines (see the location of five lines in figure 4.52(a)). Two weeks later a scarp formed at the place of anomaly 1. By 17 December 2008, the scarp reached 0.8–1.0 min height, and a new collapse sinkhole was formed (figure 4.53(a)).

A 5 m diameter sinkhole, 2 m deep, formed above the anomaly 2 (1 in figure 4.53(b)), and concentric fractures 5–10 mm wide were seen at the surface above anomaly 3 (2 in figure 4.53(b)). The surface further subsided at the location of anomalies 1 and 2.

Finally, on 21 December 2010, a new 3 m-deep sinkhole was observed above anomaly 3 (figure 4.53(b)). Thus, sinkholes have appeared above all anomalies detected earlier by GPR studies.

Mineral Beach (site B), Parking area near SPA. (Ronen *et al* 2019). Beginning from December 2008 and after, a great number of sinkholes developed through all territory of Mineral Beach. In site B the GPR survey was carried out because the collapse of a sinkhole in the spa parking lot caused suspicion of further sinkhole development in this area (Ronen *et al* 2019). Sixty-two GPR lines maximum 180 m

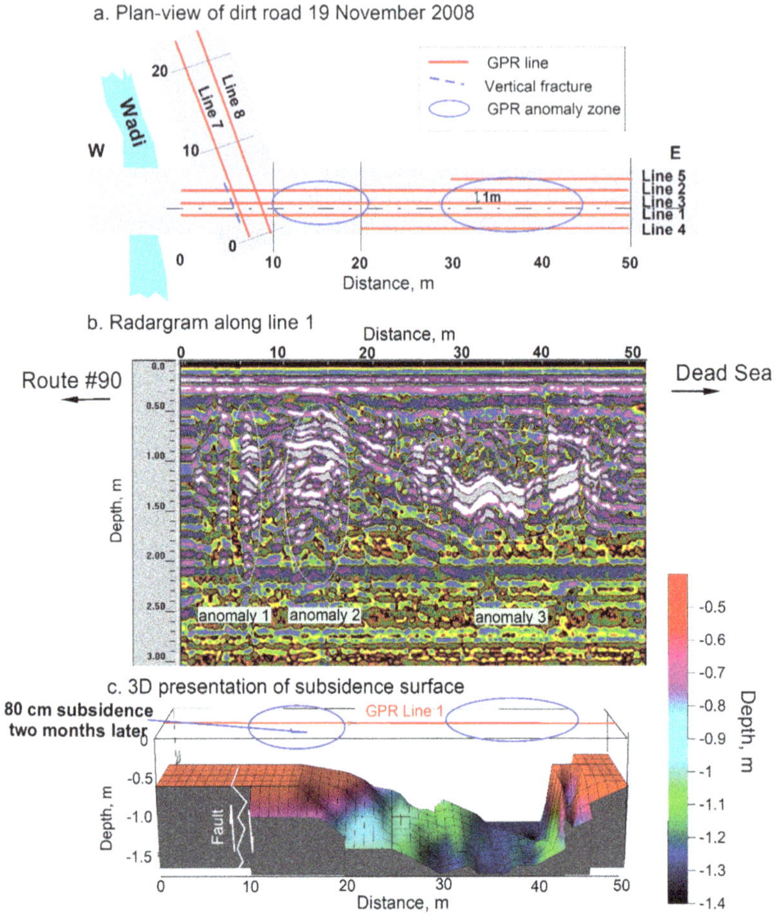

Figure 4.52. GPR study at the Mineral Beach (western Dead Shore). (a) GPR line layout. (b) Radargram along the central line 1 obtained with a 500 MHz antenna. (c) 3D presentation of the anomalous reflector detected by the GPR scanning along the dirt road. (Reprinted from Frumkin *et al* (2011), and from Ronen *et al* (2019), copyright (2019) with permission from Elsevier.)

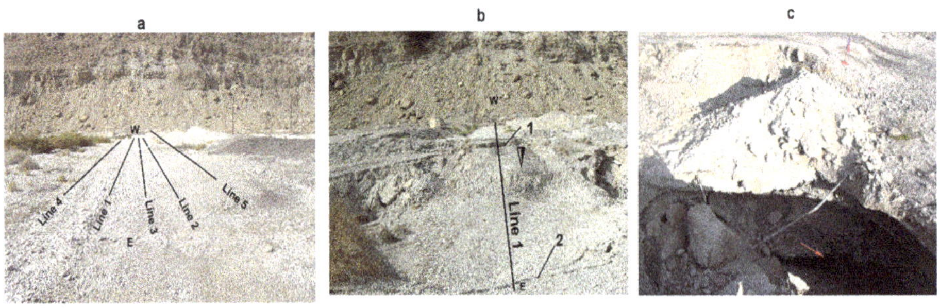

Figure 4.53. GPR study at the Mineral Beach, site A. (a) Surface of the site before subsidence in November 2008; (b) the surface after subsidence 21 December 2008. (c) Excavation south to GPR line. The flags in (b) and (c) show a scarp formed on 17 December 2008 (photo by Eli Raz). Line 1 (W–E) shows the deformation of the study site (photo by Amos Frumkin).

long were spaced approximately 1.5 m apart. Scan lines made in this area are depicted in figures 4.53(a) and (b). A CU-II RAMAC GPR system (Production MALA, Sweden) with a shielded 100 MHZ antenna was used.

Strong GPR anomalies of a technogenic origin and anomalies induced by the hidden cavities can be seen in radargrams obtained along the lines 009 to 027. An example of such a radargram along the 012 lines is presented in figure 4.55 (see figure 4.54 for location). At first, one can see hyperbolic distortions of a right-hand shape caused by the

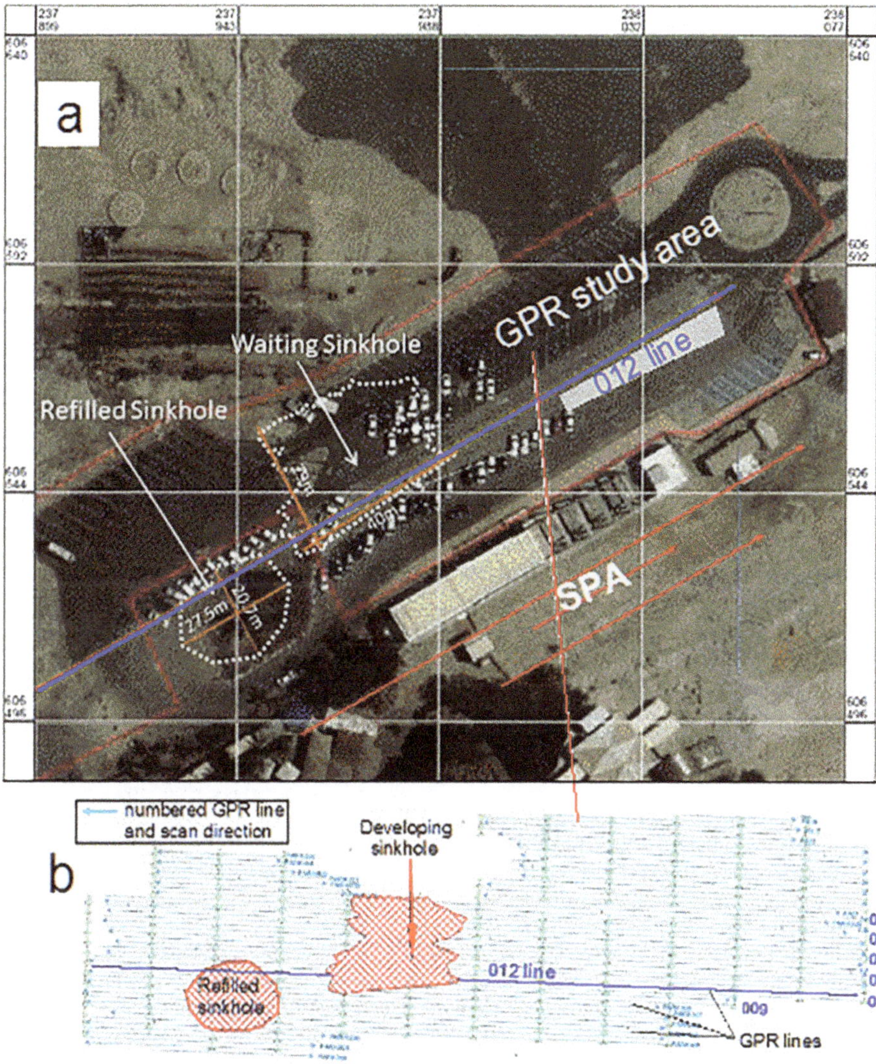

Figure 4.54. Mineral Beach (site B). GPR layout at the parking lot area: (a) air map of the study area. (b) Scheme of the GPR scanning. The location of the two-cavity-induced anomalies is shown on the scheme. GPR line 012 is marked by dark blue in both figures. The example of radargram along this line will be later presented. (Reprinted from Ronen *et al* (2019), copyright (2019) with permission from Elsevier.)

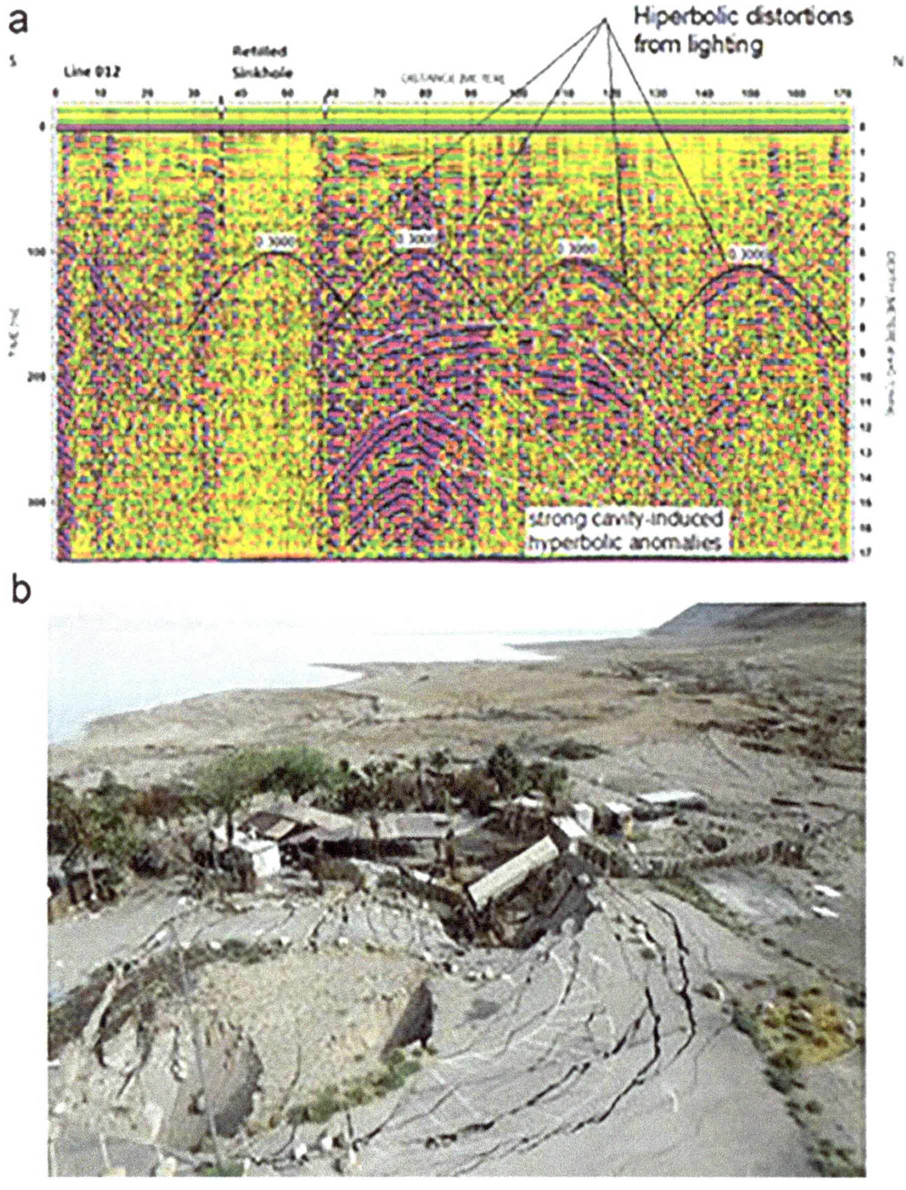

Figure 4.55. Mineral Beach (site B), (a) GPR radargram along the line 012 in January 2015 (see panels a and b in figure 4.54 for location). Digits are light velocity in a vacuum—0.3 m ns^{-1}. (b) View of SPA area on June 2016 after the collapse of the surface with sinkhole formation. ((a) Reprinted from Ronen *et al* (2019), copyright (2019) with permission from Elsevier; (b) with permission of BBC from 1 November 2016.)

interference from electric street lights. These anomalies are located 30 m apart at a depth of 5 m. Digits shown at the top of hyperboles denote the velocity of light in vacuum $C = 0.30$ m ns^{-1}. Another two anomalies located in the center of radargram at a depth of 9–13 m are those induced by a large subsurface cavity.

These anomalies also have a hyperbolic shape but with a more complex form. Detected anomalies allowed the prediction of a possible collapse that followed several months later. The spa area after the collapse is presented in figure 4.55(b). This GPR study resulted in the advance closer to the beach area thus avoiding potential harm to visitors.

4.3.1.1.9 The Ein Gedi–Nahal Arugot area (Site II in figure 4.45)

The Ein Gedi–Nahal Arugot area is situated in the central part of the western DS shoreline on the Nahal Arugot alluvial fan, about 20 m above the DS level. The sinkholes at the Ein Gedi district have appeared in areas affected by human activity. The first sinkholes appeared in 1995, and to the present N100 sinkholes have opened up at the Ein Gedi Caravan Park, the Palm-Tree Plantation, and in the Nahal Arugot area. The site was studied extensively by boreholes (Yechieli *et al* 2004) and seismic refraction lines (Ezersky 2006, Ezersky *et al* 2013b). As revealed by the analysis of aerial photographs, no sinkholes were detected in the area before 1995. Sinkholes are arranged along a tortuous line within a 50–70 m, wide zone, about 1700 m—long, trending south-south-east (figure 4.56).

The sinkhole strip crosses route #90, with sinkholes occurring on both sides of the highway although, so far, no sinkholes have developed immediately under it. The sinkholes are continuing to develop until the present time. The typical geological section is composed of cobbles, pebbles, and boulders of varying sizes, as well as sand, silt, marl, and clay. A 20 m thick salt layer was detected at a depth of approximately 50 m (top)–70 m (bottom) in several boreholes. The western limit of the salt layer is shown in figure 4.56 by a black dashed line. The Nahal David study site is located north of the Nahal Arugot area (figure 4.56).

Generally, Nahal David (David's Stream) is a major attraction in the Ein Gedi Reserve. A lot of tourists from Israel and abroad visit this site. That is why the visitors' safety in this place is the focus of Tamar Regional Council. A sinkhole hazard was raised in the parking lot (figure 4.57(a)) where subsidence of 3–5 m in diameter and 10–20 cm deep began to develop in the asphalt (Ronen *et al* 2019).

GPR survey at the site was carried out along seven GPR lines (2D study) and around the subsidence (3D study) across 8×12 m^2 of ground (figure 4.57(b)). Data were acquired using the CU-II RAMAC GPR system with a 100 MHz antenna (figure 4.56(a)). Data processing was carried out with REFLEXW software (2D radargrams) and GPR-SLICE v.7 software for 3D cube generation. GPR radargrams are shown in figure 4.58 as a 3D cube.

In figure 4.58(a) anomaly A2 is seen from the west at 8–20 m depth and a distance of 8 m along the x-axis. An amplitude anomaly marked as A3 (red area in figure 4.58(b)) is viewed clearly from the south at a depth between 11 and 21 m and $X = 5$ m. The latter anomaly is located just under subsidence at asphalt (figure 4.57(a)).

GPR anomalies detected in the Nahal David site testify to sinkhole hazard (figure 4.56). Although they were not still formed there, however, this place is potentially hazardous. It can be seen because it is located on the continuation of the salt layer edge mapped by the seismic refraction method.

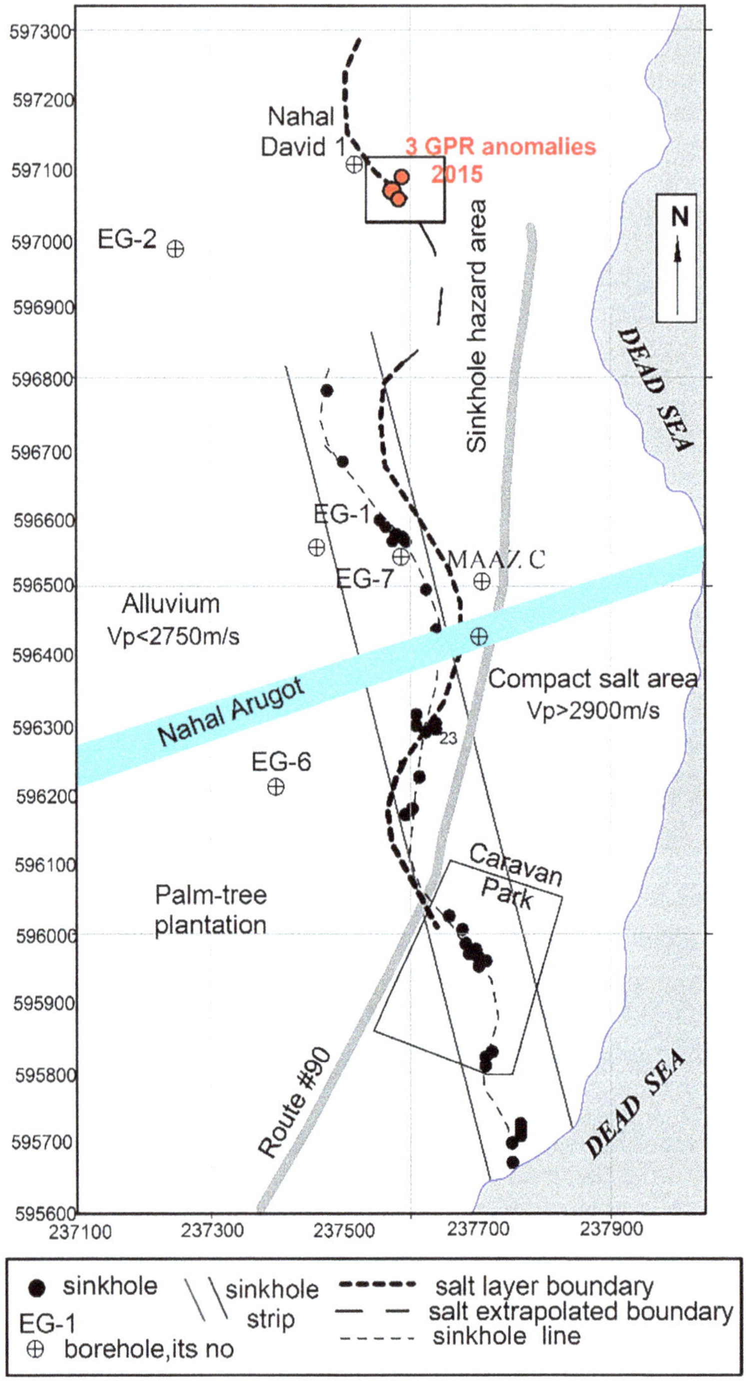

Figure 4.56. Ein Gedi–Nahal Arugot—Nahal David area with salt border. GPR study was carried out at Nahal David site, located north of Nahal Arugot area (study site has been marked by the black square). Coordinates are in m, new Israel Mercator grid. (Reprinted from Ezersky (2006), copyright (2006) with permission from Elsevier.)

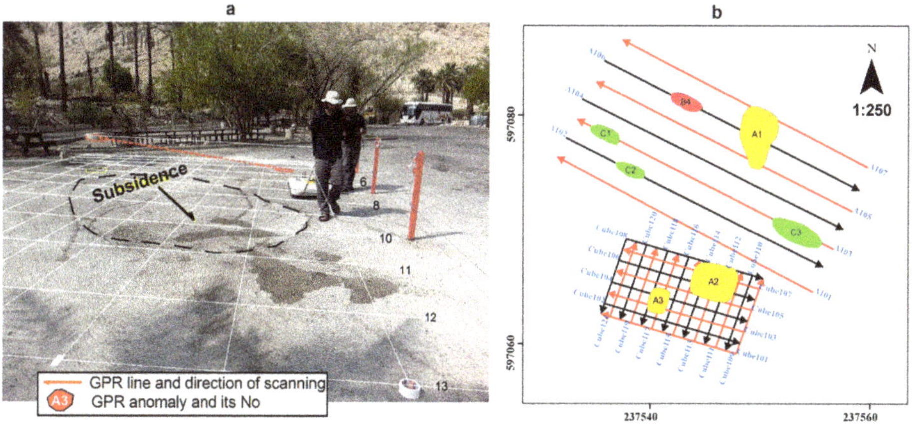

Figure 4.57. GPR study in the Nahal David site. (a) photo of 3D data acquisition with GPR anomalies; (b) layouts of 2D and 3D GPR study with notifications of anomaly locations. Color of anomalies designates yellow—sinkhole in very high probability; red—high probability, green—moderate probability (after classification of Geological Survey of Israel) (figure has been taken from Ronen *et al* (2015)). (Reprinted from Ronen *et al* (2019), copyright (2019) with permission from Elsevier.)

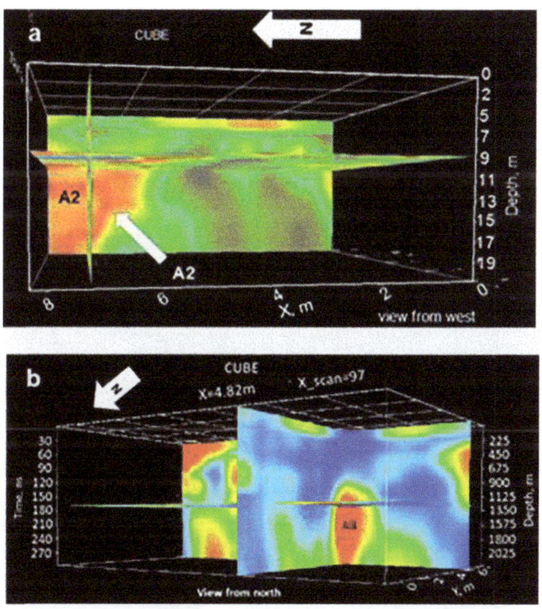

Figure 4.58. Radargrams obtained for the 3D GPR survey presented in panel b. (a) Anomaly A2 view from the west is located at 8–20 m deep and $X = 8$ m; (b) anomaly A3 view from south to north at a depth between 11 and 21 m and at $X = 5$ m. Both anomalies are seen in the 3D cube (figure has been taken from Ronen *et al* (2015)). (Reprinted from Ronen *et al* (2019), copyright (2019) with permission from Elsevier.)

4.3.1.1.10 Nahal Hever South **(Site III in figure 4.45) in (Ezersky *et al* 2006)**
As mentioned above (chapter 3) a barely noticeable fracture-surrounded subsidence S3 was firstly noted at the beginning of April 2002 (figure 4.59). On April 16, 2002, a GPR survey was carried out at the settling 3 site. A 100 MHz antenna was used. No anomalies were detected from the surface down to 5 m depth. The next GPR survey was carried out in May 2002 along line MF-2 with 500 MHz antennae. A 30 m long GPR line parallel to the line MF-2 crossed settling 3 (figure 4.59(a)). Figure 4.59(b) (left) shows the radar line location concerning the settling 3 one. In figure 4.59(c) (upper) we can see the radargram with a strong anomaly, 7–10 m long, which was detected in the northern part of settling S3.

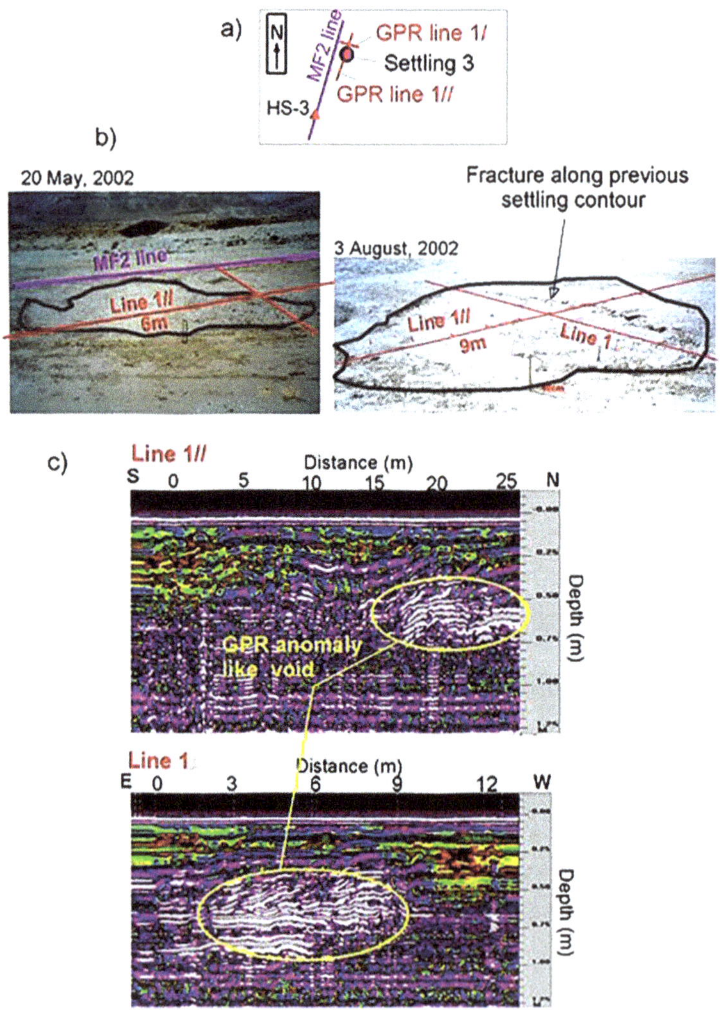

Figure 4.59. GPR study of the settling 3 site: (a) layouts, (b) photos of settling 3 before (left) and after (right) measurement, (c) radargram along line 1 (upper) and perpendicularly it (lower). (Reprinted from Ezersky *et al* (2006), copyright (2006) with the permission of EAGE publications.)

Figure 4.60. Sinkholes formed in the GPR anomaly site in September 2003 (photo of M Ezersky). (Reprinted from Ezersky *et al* (2006), copyright (2006) with the permission of EAGE publications.)

In the perpendicular section, the same anomaly is 6 m long and located at its eastern side (figure 4.59(c), lower). This anomaly was interpreted as a void, presumably, a sub-lateral fracture at 0.5–1.0 m deep. In August 2002 the significant fracture crosses settling 3 at the surface (figure 4.59(b), right). During 2002–03 this settling transformed to a great sinkhole 30 m in diameter and some 1–1.5 m deep (figure 4.60).

Ghor Al-Haditha (Jordan) (Site-V figure 4.45) (Batayneh *et al* 2002). It has to be noticed, that Jordanian geophysicists (Camerlynck *et al* 2006) conducted successful GPR outlining of the buried sinkholes at depths up to 12 m at the Ghor Al-Haditha area, on the east side of the DS shore. The recent studies performed in the Ghor Al-Haditha in the framework of the NATO SfP 981128 Project using the SIR-10B GSSI system are presented in figure 4.61.

4.3.1.1.11 Discussion

The examples considered above show that underground voids—embryos of future sinkholes can be detected using the GPR technique from 2–3 to 10–15 m deep depending on the soil dielectric properties.

1. The relatively deep location of the formed caverns within the salt layer (significantly more than the penetration depth of the electromagnetic waves (EWs)) in combination with the saline (but not saturated concerning halite) groundwater covering the salt layer from above, makes it seems to be an application of EW impossible. However, our experience indicates that the GPR technique in its 2D and 3D modifications can be a very efficient tool for predicting sinkhole susceptibility.
2. It is associated with the fact that the GPR potential to detect subsurface cavities is well introduced into the model suggested by Frumkin *et al* (2011). The model considers three stages as follows:

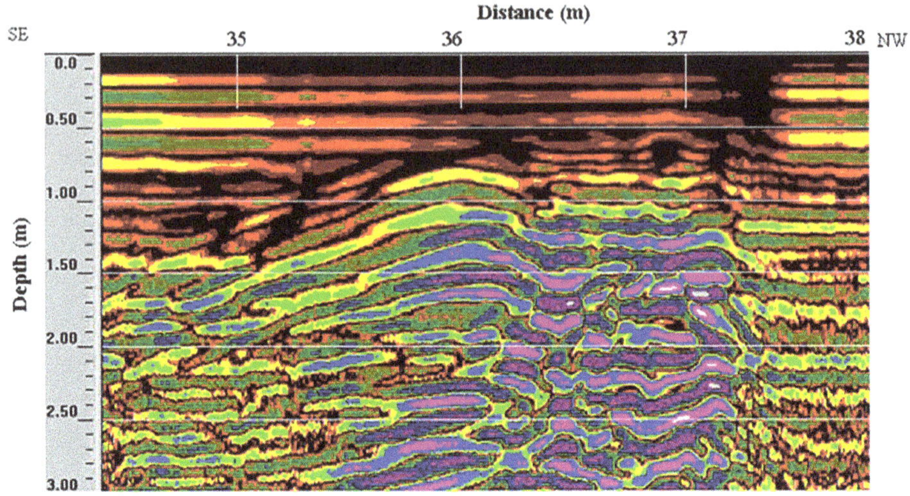

Figure 4.61. Part of radar cross-section along line 2 in the Ghor Al-Haditha sinkhole site (Jordan). The main anomaly along this line at the distance between 35 and 37 m is located at approximately 1 m depth may represent a buried sinkhole. (Reprinted with permission from Camerlynck *et al* (2006)).

Stage I (latent). Salt dissolution may take place a short or long period before any signs are evident at the surface. The sinkholes commonly developed along the salt edge which was dissolved by brackish water, unsaturated concerning chloride (Yechieli *et al* 2006, Yechieli 2007, Legchenko *et al* 2009, Ezersky *et al* 2010).

Stage II (destroying of the salt cavity ceiling). Instability of the salt cavity ceiling causes upward stoping (upward propagation of cavities by progressive roof breakdown) through the mechanically weak sediments situated between the salt cavity and the surface. Ford and Williams (2007, p 385) report cases in different geological settings, where the stoping process propagated through >1000 m of cover rocks. This intermediate stage continues until the surface collapses and the sinkhole forms. (Stage III. (sinkhole formation)).

It is important to underline that during the first stage the salt layer (and consequently, dissolution cavern) is located under very saline seawater brine that acts as a screen for EWs and at a depth beyond GPR penetration depth. During the second stage the dissolution cavity slowly ascends to the surface, *comes out from brine, and hereafter can be detected by the surface geophysical methods.*

3. Is there enough time after destroying the roof ceiling to predict sinkhole appearance using geophysical methods? Numerical calculations for the plastic behavior of a medium have shown that the gradual lifting of the cavity (soil damage) from below up to the surface is developed with time (Roche *et al* 2000, Maimon *et al* 2005, section 2.2.1 of this manuscript). Stoping speed depends on many factors such as geological structure, mechanical properties of soil, drop in the DS level, concurrent phenomena

like an earthquake, precipitation, etc. It is understood that the stoping within the mud sediments and sandy-gravel ones is different because of differences in their mechanical properties (Shalev and Lyakhovsky 2012). by the HS-3 borehole at the 24–29 m depth interval in the Nahal Hever south site Frumkin *et al* (2011) have shown that due to the 'stoping' process, we have enough time to predict and possibly even prevent a collapse of the surface before it occurs. Some evidence suggests that the void in the subsurface exists over a long duration. One example is the cavity discovered in 2001. The surface is still stable at the HS-3 site. The GPR anomaly detected in April 2002 transformed during 2003 to a large sinkhole 30 m in diameter and 1.5 m deep (figure 4.59, Ezersky *et al* 2006). There are also other examples like the boreholes Mn-2, Mn-O in the Mineral Beach area (Ronen *et al* 2019).

4. The anomalous high resistivity sites were discovered in alluvial fans in Israel and Jordan, where sinkhole hazard is most important because of the human activity throughout these areas (Taqieddin *et al* 2000, Yechieli *et al* 2002, Al-Zoubi *et al* 2007, Ezersky 2008). The high resistivity of these sites is caused by the decompaction of subsurface due to subsidence. Resistive media creates promising conditions for utilizing the GPR technique for the detection of the subsurface signs of sinkhole hazard through alluvial fan areas. Thus, GPR is a valuable instrument for sinkhole prediction in the coastal DS areas.
5. The GPR method is effective in all urban sites and at most of the sites of human activities of the DS coast (farms, dwellings, routes, agricultural areas, etc), but it fails in areas composed of saline mud in the subsurface (some areas lower than −400 m elevation). Through the fan areas composed of sandy-gravel sediments, the penetration depth of the EWs varies and is usually limited to the first meters down to 8–10 m (maximum to 13–15 m), but not deeper than the water table (depending on the dielectric constant, which as a rule is 4–25). In areas covered by DS, the mud penetration depth is limited to a depth of 1 m.
6. The selection of 2D/3D surveys: advantages and disadvantages. In most cases, a 2D survey provides a good solution. However, greater reliability and improved results can be achieved with true 3D approaches, eliminating ambiguities caused by spatial aliasing and out-of-plane reflections and diffractions in 2D GPR sections (Grasmuck and Green 1996). Moreover, only a 3D survey can provide reliable information regarding the depth and areal extent of small-scale features (Buker *et al* 1996). On the other hand, the 3D technique has limited applicability due to an increase in cost that stems from greater data acquisition and hence an increase in the processing effort required. Therefore, method selection should take into account the size of the possible target, the urgency, and the allocated budget. A suitable compromise can be reached by scanning parallel profiles where the distance between them equals or is less than the expected size of the anomaly (see figures 4.52(a) and 4.54(b)). Measurements along parallel or non-parallel disjoint profiles are also related to the 3D observations, similar to GPR measurements along a grid of the intersecting mutually parallel lines (figure 4.57(b)) (Sandmeier 2020).

4.3.2 The transient electromagnetic TEM method

4.3.2.1 Features of TEM in the Dead Sea area

We have preliminarily considered the TEM method in section 2.2.1.6. As previously mentioned, the DS is a hypersaline lake. The main feature of the DS aquifers is a very low resistivity in the range of 0.2–2.0 Ωm (Yechieli *et al* 2000, Goldman and Kafri 2021). Such low resistivity is measured down to hundreds of meters (Levi *et al* 2008). It determines using the TEM (TDEM) method sensitive to the very low resistivity of the subsurface (Kaufman 1978). The TEM method was routinely used in Israel for locating the fresh-saline water interface in coastal areas (Goldman *et al* 1991, Kafri *et al* 1997, Yechieli 2000, Goldman and Kafri 2021). In the DS sinkhole problem, the TEM method was also used as: (a) a quasi-3D technique based on numerous 1D soundings to map variations in resistivity throughout the areas (Ezersky *et al* 2011); (b) to determine the salt layer geometry (location of top depth, thickness) and properties (Frumkin *et al* 2011, Ezersky and Legchenko 2014, 2015); (c) to map the salinity and aggressiveness of DS aquifers (Ezersky and Frumkin 2017, 2021); (d) to interpret data of the SNMR method (Legchenko *et al* 2008c) and to distinguish lithology of the aquifer (Legchenko *et al* 2009). It was shown also that the TEM method enables understanding the sinkhole origin (Ezersky and Frumkin 2020).

The map in figure 4.62 displays the sites studied using the TEM method.

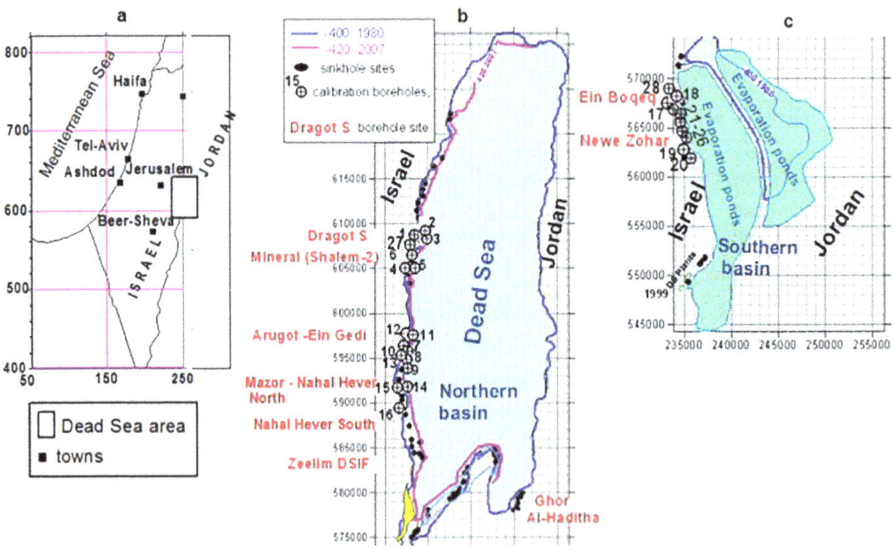

Figure 4.62. DS sites were studied using the TEM method. (a) Study area; (b) northern DS basin; (c) southern DS basin (evaporation ponds). The location of the calibration boreholes in the western DS shore is shown. (Reprinted from Ezersky *et al* (2011), copyright (2011) with permission from Elsevier.)

4.3.2.2 Features of the TEM FAST methodology

The TEM FAST methodology (based on the TEM FAST 48 system) is limited by the maximum transient time of 16 ms (as against 100 ms of geonics systems). Correspondingly, the penetration depth depending on resistivity will be significantly less. That is why the TEM FAST methodology is 'little in-depth research' (Kamenetsky 1997).

Physical and mathematical bases of the TEM are well and fully described in the literature (Kamenetsky 1997, Albouy *et al* 2001, Barsukov and Fainberg 2001, 2016, Kamenetsky *et al* 2010, Barsukov *et al* 2015). One of the few models of media for which were developed the simple and accessible formulas, for analysis of the TEM signals, is the model of homogeneous half-space.

Three features of the method should be denoted here: (1) stage of the measurement; (2) efficient radius of the 'current ring'; (3) penetration depth (AEMR 2018, *TEM manual from the permission of authors*).

1. Stage of measurements. To understand factors influencing the TEM measurements let us consider expressions for asymptotic estimation of the TEM signals received on the homogeneous half-space at late and earlier stages. Parameter characterizing the stages is as follows

$$t_0 = t\rho/\left(\mu_0 L^2\right), \tag{4.7}$$

where $\mu_0 = 4\pi 10^{-7}$ [H m^{-1}] is the magnetic permeability in a vacuum, ρ [Ωm] is the specific electrical resistivity of the half-space, t [s] is the transient time, and L [m] is the loop side size. Theoretically, the late stages of transient $t_0 \gg 1$ for one-turn round antennas R (transmitter) and r (receiver) lying above homogeneous half-space with resistivity ρ and magnetic permeability of vacuum μ_0, are described by the formula:

$$E(t)/I = 0.05 \times \left(\pi^{1/2}\mu_0^{5/2}\right)/\rho^{3/2}\left(r^2R^2\right)t^{-5/2} \tag{4.8}$$

At the *late stage*, the registered signal is proportional to the conductivity $\sigma^{3/2} = \rho^{-3/2}$ and to the product r^2R^2. Thus, in TEM the amplitude of a signal at the late stages is rather sensitive to the changes of conductivity of section in comparison, for example, with the methods of direct current. Besides, the signals $E(t)/I$ at $t_0 \gg 1$ do not depend on a site of receiving loop $r < R$. The same would be said for the square antennas of L and l side sizes.

For the early stage of transient, the generalized time must be $t_0 \ll 1$. Then for coincided loop configuration $l = L$:

$$E(t)/I = \mu_0 L\pi^{-1/2}/2t. \tag{4.9}$$

parameters of signal do not depend on the resistivity of the media.

The above-mentioned testifies that the size of the antennas is very important for an earlier stage of a transient. The finger law says at a choice of the optimum size of antennas it is necessary to remember that the large size of antennas increases limiting the depth of researches, however, does not allow stratifying subsurface horizons. A depth range h, for which reliable interpretation of the received results is possible for coincided loop configuration, with $TR = REC = L$, is usually laid within the range

$h_{min} > L/10$ and $h_{max} < 3L$. It means for a 25 × 25 m^2 loop minimum investigation depth can be 2.5–3 m, whereas the maximum depth can be 75 m. For a 50 × 50 m^2 loop, the same depth range can be between 5 and 150 m.

In our study, we used only coincident transmitter/receiver square loops with the side sizes of L = 25 × 25 m^2, 50 × 50 m^2, and 100 × 100 m^2 taking into account the above-mentioned limitations.

For antennae 25 × 25 m^2, for homogeneous half-space with 100 Ωm resistivity and transient time varying from 0.01 to 20 ms generalized time, t_0 is varied from 1.3 to 2500, practically completely in the late stage of the process. In the case of 1 Ωm resistivity of half-space, characteristic for the DS area, in the same transient range, the generalized response t_0 will vary from 0.013 to 25, passing all three stages from earlier to late. It should be noted that when measuring the transient response of the medium from microseconds to milliseconds, we usually go through all three stages: from early to late. At each stage, the resolution is a different one.

2. *Efficient radius of the current ring.* The signal, registered in the receiving loop at the late stage of transient, is caused by currents, induced in the current ring inside section with the effective radius R_{eff} and 'attitude' depth $H_{eff} \sim R_{eff}$ (figure 4.63).

Effective radius $R_{eff} = (t\rho/\mu_0)^{0.5}$, is considerably more than the radius of transmission loop R ($R_{eff} \gg R$). The vertical magnetic field created by the contour is practically homogeneous within the limits of its area at $r < R_{eff}$. Therefore, registered signals proportional to the derivative of the magnetic field over time do not depend on the site of their receiving. This property of the late TEM stages determines, on the one hand, a high investigation depth in combination with good resolution of the method ($E \sim \sigma^{3/2}$). On the other hand, resistivity measured at the depth H_{eff} characterizes the average properties of an area within the 'current ring'. It means that R_{eff} determines the 'measurement scale' or the 'scale of testing'. The radius R_{eff} versus transient time graphs are presented in figure 4.64.

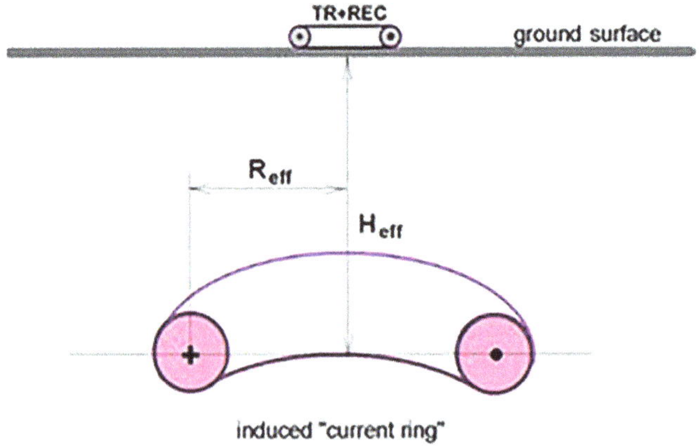

Figure 4.63. Efficient 'current ring' in TEM measurements (AEMR 2018). Reprinted with permission from P Baruskov.

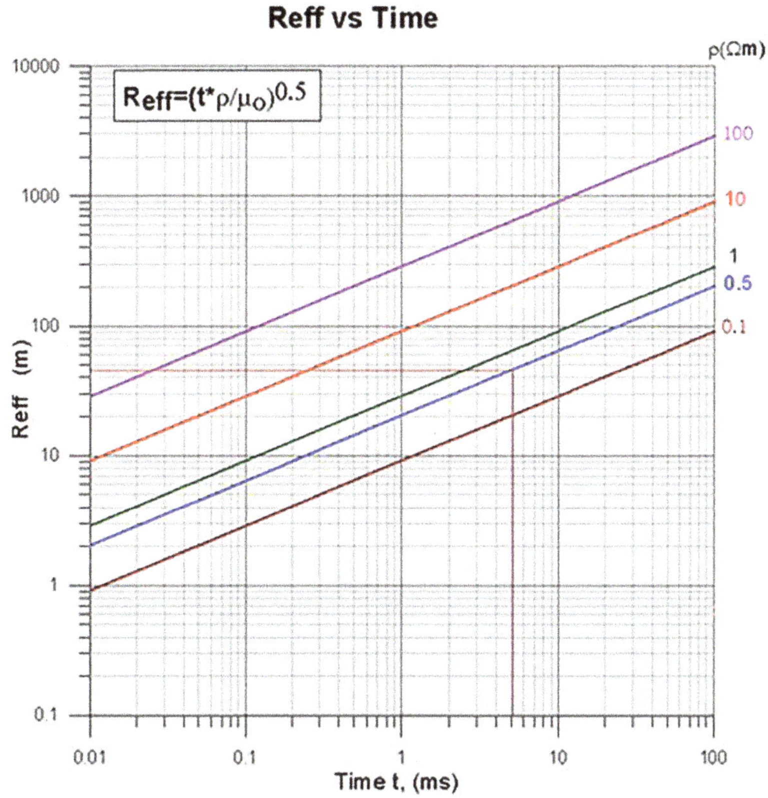

Figure 4.64. The effective radius of the induced 'current ring' versus transient time. The parameter of curves is the half-space resistivity. Reprinted with permission from P Baruskov.

Table 4.5. Diffusion depth of TEM transient at different resistivity of half-space. Reprinted with permission from P Baruskov.

Bulk resistivity ρ_x, Ωm	50	10	2	1	0.5	0.2
Diffusion depth h_t, m	1128	504	226	159	112	71

It must be denoted that R_{eff} does not depend on the loop size but only on the depth and resistivity of the subsurface.

3. *Penetration depth.* Penetration (diffusion) depth is determined by the formula $h_t = (2t\rho/\mu_0)^{0.5}$, where ρ [Ωm] is the resistivity of the half-space, t [s] is the transient time, and μ_0 is the $\mu_0 = 4\pi 10^{-7}$ [H m^{-1}] is a magnetic permeability in a vacuum.

Let us evaluate the diffusion depth h_t at different resistivities of half-space at a transient time of 16 ms (that is the maximum time of the TEM FAST 48). The results are presented in table 4.5.

The last three columns of this table characterize the resistivities of the DS subsurface. The calculated depths will be limited by the relationship between the signal strength and noise conditions as well as by the subsurface structure (resistivities and thickness of layers). The signal strength in turn depends on the loop size. Average site transient signal $E_{av}(t)$ is decreased with a depth and reaches the average site noise level E_{averr} ($E_{av}(t) = E_{averr}$) at a depth h_m that is real diffusion depth.

Ezersky *et al* (2008) have evaluated the diffusion depth in two sites of the DS—Nahal Hever South and Ein Gedi—with 25 × 25 and 50 × 50 m^2 antennas based on 20 measurements in each site. The studied sites are characterized by different structures and noise levels. It was shown that in the Nahal Hever South noise allows measuring up to 10–16 ms with the 25 × 25 m^2 loop and 16 ms with the antenna of 50 × 50 m^2. In the Ein Gedi site measurements are limited by 8–10 ms with the antenna 25 × 25 m^2 and up to 10–16 ms with the 50 × 50 m^2 loop.

These data are preliminarily presented to illustrate the possible time ranges for the DS sites. The real-time range is defined in the course of performing this study.

4.3.2.3 TEM fast equipment

The TEM method can be used for vertical depth sounding, profiling, and mapping. A procedure commonly utilized for ground exploration using transient techniques is to lay a circle or square loop in the vicinity of the area to be examined. The high productivity of the TEM FAST equipment in a light, portable unit that uses a single Rx/Tx loop is an important advantage. Such a configuration Rx/Tx is termed a coincident, with the same loop acting both as a transmitter (Tx) and receiver (Rx) (Barsukov *et al* 2015). It is known that TEM results with a coincident loop may be less reliable than those made with a central loop configuration. Additional errors in TEM results may be caused by the induced polarization (IP) effect and the magnetic viscosity influence or the superparamagnetic effect (SPM). The rocks in the DS coastal area are characterized by low magnetic susceptibility ($5 - 8 \times 10^{-5}$ SIU, Rybakov *et al* 2005). A comparison of results obtained using the GEONICS and TEM FAST systems revealed a very good correspondence suggesting that IP and magnetic viscosity effects do not affect measurements (Ezersky *et al* 2011).

In TEM-FAST 48 the opportunity to use several power batteries is foreseen:

- Standard internal battery 12 V 2000 mAh (B0).
- External battery 12 V for the supply of the receiver and generator (B1).
- External battery 12–24 V only for the supply of the generator (B2).

The battery B1 is used usually for monitoring long-term supervision. At the moment of turning on the B1, the battery B0 is automatically disconnected. When turning on the B2, the internal battery B0 automatically is used only for the receiver. The inventor provides current 1–4 A. In the very low resistive environment of the DS coast, only a current of 1–2 A was reached.

The fieldset of the TEM FAST system is shown in figure 4.65.

The TEM specification is as follows:

Loop size 25 × 25 m^2, 50 × 50 m^2, and 100 × 100 m^2;

Loop configuration: coincided loop or loop-in-loop;

Current: 1 A, 4 A;

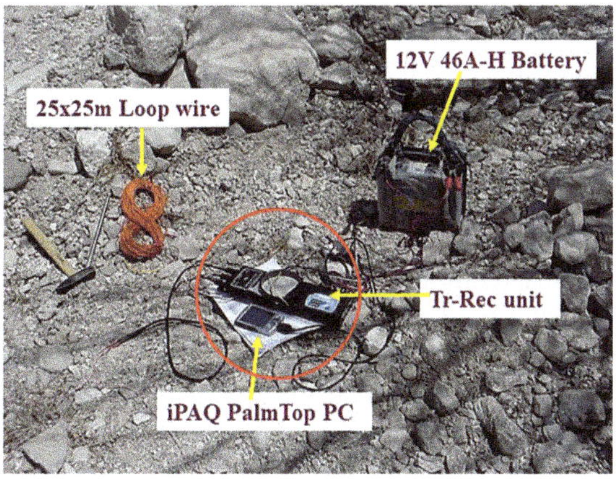

Figure 4.65. TEM FAST 48 equipment in the field. 1—Transmitter–Receiver–Controller unit, 2—hp IPAQ pocket PC h2200 series, 3—0.5 mm^2 wire for 25 × 25 m^2. Reprinted with permission from P Baruskov.

Dynamic range (max/min)—10 V/1 μV;
Transient time 4–16 000 μs;
Internal battery life is not less than 120 soundings (I = 1 A, stacking 3 min).

Depth measurements at the TEM Fast methodology can consist of about 1–300 m. Shallow measurements can be made with a small transmitter loop of 1 × 1 – 5 × 5 m^2. In this antenna size, the TEM FAST is used as a metallic detector. Deeper measurements can require a large transmitter loop of 18–500 m.

With the increase of antenna size, the more powerful signal is radiated to the Earth and more depth penetration is reached. However, increasing the antenna size decreases the resolution of the uppermost depths. The sure interpretation of measurements is possible in the range of $H_{min} = 0.1L$ to $H_{max} = 3\ L$. Where H_{min} and H_{max} are min and max of resolution depth, and L is the size of the antenna side.

The weight of the TEM FAST equipment set is 2–3 kg packed in the notebook bag. The whole set with 1–2 employees is transported by car. 25–30 TEM stations are carried out during the working day. In a desert area not complicated by trees and bushes, productivity for two people with an antenna of 25 × 25 m^2 is 80 soundings per day, they just carry the antenna on poles, stop, measure, and go further—this is what they do even in the tundra.

4.3.2.4 The TEM FAST results are interpreted with two 1D inversion commercially available software programs

The first program, TEM-RESEARCHER (AEMR 2018), is intended for the modeling and inversion of soundings by determination of field (transient) data, obtained with the usage of the ungrounded horizontal magnetic antennas (loops) in two configurations: coincided and central loop configurations. It includes an option for taking into account IP and SPM effects. Two ways of generating a section include transformation and inversion. The transformation of the transient curves $\rho_a(t) = \mathrm{Re}\, s(t)$ in pseudo-section $\rho(h) = \mathrm{Re}\, s(h)$. The layered model is selected based on the transformation curve and layered inversion is performed (Barsukov *et al* 2015, TEM-researcher 2017). The software permits the generating of both transformation and inversion sections and maps. The software is designed for fast performing 'quasi' 2D and 3D pseudo- and inverse-resistivity sections and maps from the multiple 1D sounding (tens-hundreds soundings). It should be performed often in field conditions. Therefore, the option for estimation of the equivalency and misfit error was not foreseen in the software. For more accurate estimations of the inversion results, this disadvantage was filled up by the software IX1D by Interpex Ltd (Interpex Ltd 2019).

The IX1D software (Interpex Ltd 2019) is considered to be the standard software for interpreting Geonics TEM data allowing forward and inverse modeling for layered and smoothed (1D) models, equivalence analysis, and calculation of the RMS misfit error for every iteration. The RMS misfit error is used as an indication of the fit between the synthetic data generated from the model and the measured data. For TEM data (in-loop), RMS is estimated in the log domain. The error is estimated according to the following formula:

$$E_{\log} = \sqrt{\left\{\sum_{i=1}^{i=n}[\log(V_s/V_m)]^2\right\}/n} \tag{4.10}$$

where V_s/V_m is the synthetic field value (voltage) and is the measured value, n is the number of samples (measurement points). The RMS log error can be converted to linear space and expressed in percent as follows:

$$E\% = (10^{E_{\log}} - 1) \times 100 \tag{4.11}$$

The equivalence of the inversion parameters of selected layers is estimated based on the resolution matrix, which is a byproduct of the inversion process (Stoyer 2010). The matrix is used in the equivalence analysis process. When the inverse process reaches the best fit, the parameters are incrementally adjusted until the fitting error reaches 1.2 times the best-fit error.

Despite the above-mentioned differences, both types of software give very similar inversion results (Ezersky *et al* 2011).

TEM-3D_Wizard program is the tool for 3D TEM direct and inverse problem solution. The program gives to the user a lot of possibilities, however, the user needs to have some experience to work with it. A demo version and description is available at www.aemr.net.

3D Problem The TEM FAST equipment permits 1D soundings. It was shown that such measurements should be very careful, especially in the areas with disturbed geometry of the interfaces (Rabinovich 1995). To avoid the wrong results the consequent TEM measurements along the traverses should be performed. Performing soundings along traverses and throughout areas allows the generating of pseudo-2D sections and 'quasi' 3D maps at the different levels. Because the TEM 3D inversion is presently only in progress, 3D forward modeling is recommended to be performed (Rabinovich 1995, Barsukov *et al* 2015). Nevertheless, 3D effects complicate results when the scale of measurement is more than the size of the heterogeneity (or equal to that). When the scale of measurement is less than the size of heterogeneity at least by factor 2, the 3D effect would be negligible. The measurements with loops of different sizes would characterize the stability of decisions in the places where the 3D problem is suspected.

4.3.2.4.1 Equivalency problem

The equivalency problem (named also non-uniqueness) complicates often the TEM results themselves. Fitterman *et al* (1988) have shown that all types of the resistivity section (A-ascending, H-bowl, K-bell, and Q-descending) are less or more affected by the equivalency. At the same, the resistivity log decreasing with the depth (Q-type) is minimally affected by this factor. However, for solving the equivalency problem, any additional information like geological data available at the area under investigation or data obtained from other methods, for instance, seismic refraction, must be used (Flathe 1976, Fitterman *et al* 1988). The joint inversion of both TEM and CVES data permitting considerably reducing equivalency in the layer interpretation would also be efficient (Raiche *et al* 1985). On the other hand, the constrained inversion would also decrease the equivalency (Auken and Christiansen 2004). Goldman *et al* (1994) suggest starting measurements at least at 10 μs. That is possible with the TEM FAST HPC 48 equipment. To avoid (or at least, estimate) equivalency we have employed the IX1D software (Interpex Ltd 2019). We have applied in our study a combination of seismic refraction and borehole data, as well as the MRS method, most efficient because of its possibility to detect water in the subsurface (Ezersky *et al* 2008).

4.3.2.5 Comparison of TEM results with borehole log

In figures 4.66(a) and (b) TEM results are compared with the geological section of the borehole HS-2 located close to the TEM station 4 in the Nahal Hever site (see figure 4.62 for site location). The geological column in figure 4.66(c) is composed of five lithological units. *A priori*, we can expect different levels of resistivity of the selected units. Adding the factor of high salinity groundwater, we can expect considerable differentiation of the section concerning resistivity.

Indeed, uppermost alluvium has a resistivity of more than 10 Ωm, whereas underlying soil saturated with salty water (if it exists), and especially with the DS brine, should have a very low resistivity. In water-saturated sediments, we select two layers of water-saturated clay enveloping the salt layer that, based on our preliminary knowledge derived from the laboratory measurements, has a resistivity of up to 5–9 Ωm. Finally, the layer underlying the lower clay unit comprises clay with gravel saturated with the DS brine. This unit is also expected to have a resistivity somewhat higher than that of the saturated clay. Taking into account that layers of higher and lower resistivity are interbedded, we can expect *a priori* at least six layers: (1) unsaturated alluvium; (2) saturated alluvium; (3) saturated clay; (4) salt; (5) saturated clay; and (6) saturated sandy-gravel–clayey lowermost unit.

4.3.2.6 Selection of the number of layers in the resistivity model

Although more layers provide for a better inversion (Zhdanov and Keller 1994), we do not want to add extra layers. We have to select the optimal model that, on one

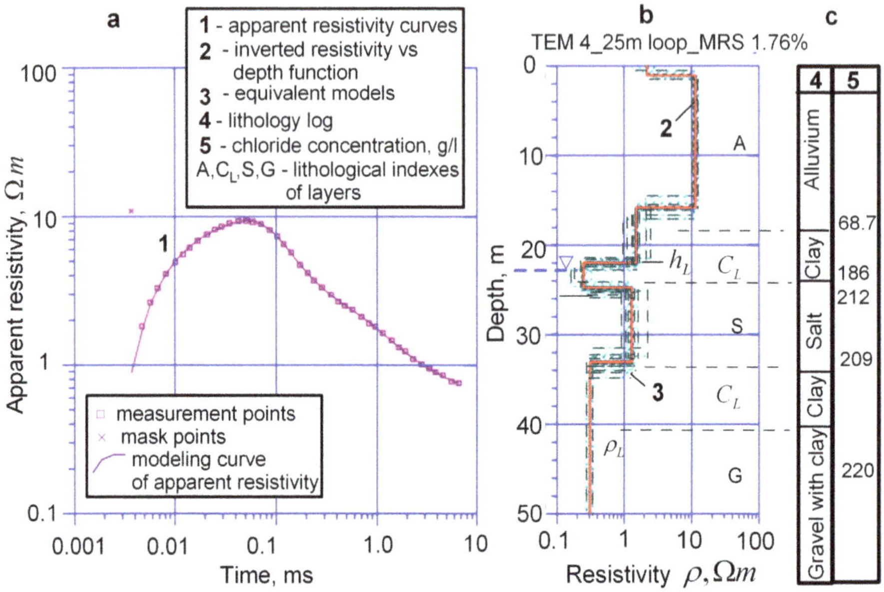

Figure 4.66. Results of TEM| sounding near HS-2 borehole in the Nahal Hever South site: (a) TEM apparent resistivity versus time, and (b) inverted resistivity–depth function, (c) lithological (d) and hydrogeological (e) columns from the borehole HS-2. Column e shows chloride concentration [g L^{-1}] (see figure 4.62 for location). Lithological indexes of the layers are designated by letters as A—alluvium above the water table, C—clay saturated with the DS brine, S—salt, G—clay with gravel, saturated with the DS brine. Also noted is h_L depth from the surface to 1 Ωm interface, and ρ_L bulk resistivity of sediments saturated with the DS brine. (Reprinted from Ezersky *et al* (2011), copyright (2011) with permission from Elsevier.)

hand, correlates well with the geological section, and on the other hand, provides the highest resolution of the layer parameters (resistivity ρ_i and layer thickness h_i). The option to select a multi-layered model is a matter of equivalency and resolution. This issue has been discussed extensively in the literature (Raiche *et al* 1985, Fitterman *et al* 1988, Sandberg 1993, Goldman *et al* 1994). Inverse resistivity versus depth functions are presented in figures 4.67(b)–(f).

Generally, as the number of layers increases from 4 to 7 from the left (b) to right (f), the RMS value decreases. For the inversion procedure with all parameters free, an RMS error of 1.76 is minimal for the six-layered model. It increases to 2.52% and 2.94% for the five-layered and four-layered models presented in figures 4.67(c) and (b), respectively. For the three-layered model, the RMS error achieves 7.07% (not shown in figure 4.67). It increases slightly to 1.80% for the seven-layered model as well (figure 4.67(f)). For the six-layered model data were inverted also with a fixed fifth layer thickness (figure 4.67(e)) that slightly increased RMS to 1.86%. One can see that the six-layered model is most suitable with the geological section (figures 4.66(b) and (c)).

The origin of the 1 Ωm interface has been discussed by Kafri *et al* (1997). In sandy aquifers, the 1 Ωm interface allows locating the upper boundary of sediments

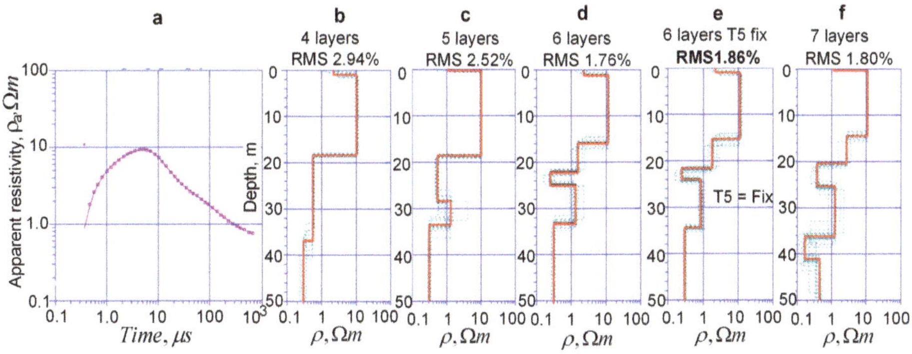

Figure 4.67. (a) Experimental data (squares) and model transient curve (solid line) and (f) inverse resistivity versus depth functions calculated for different layered models: four-layered (b), five-layered (c), six-layered (d and e), seven-layered (f). For models (b–d) and (f) all parameters are free, for model (e) thickness of the fifth layer (salt) was fixed, a three-layered model is not shown. (Reprinted from Ezersky *et al* (2011), copyright (2011) with permission from Elsevier.)

saturated with the DS brine (Ezersky *et al* 2011), and we expect that the 1 Ωm interface should coincide with the water table (Goldman and Kafri 2021). However, in clay saturated with the DS brine, this interface either does not follow the water table or follows it at a rate essentially lower than that in sands depending on the intensity of the circulation of meteoric water (Yechieli *et al* 2001). It seems to be the same case. In all, the 1 Ωm interface is determined most certainly with practically no equivalence problem.

4.3.2.7 Resolution of the TEM method

Stoyer (2010) has suggested a method for estimation of equivalency and resolution based on the resolution matrix parameters (RMP) and implemented it in the IX1D software (Stoyer 2010, Interpex Ltd 2019). The optimal number of selected layers is based on the minimal misfit error (RMS), calculated according to equation (4.4). For instance, the resolution of the layer parameters for TEM station 4 (closest to borehole HS-2) is estimated using the resolution matrix. The experimental measurement points and the model transient curve are shown in continuation.

4.3.2.8 Application of the TEM| measurements for DS water salinity evaluation

Following a hydrogeological model (Yechieli *et al* 1995a) in the DS region (figure 4.68), the recent drop in the sea level and the associated drop of fresh-saline water interface allow the incursion of fresh groundwater coming from the west or deep layers via faults into the coastal aquifer. The start model of sinkhole formation suggested that lowering of the DS level and migration of chemically undersaturated (concerning halite) groundwater into the coastal aquifer has caused diverse environmental changes (such as landslides, land subsidence, disturbance of coastal springs, etc) via the variation of the groundwater salinity, resulting in the dissolution of the buried salt layers and, most significantly, in wide-scale collapse and sinkholes

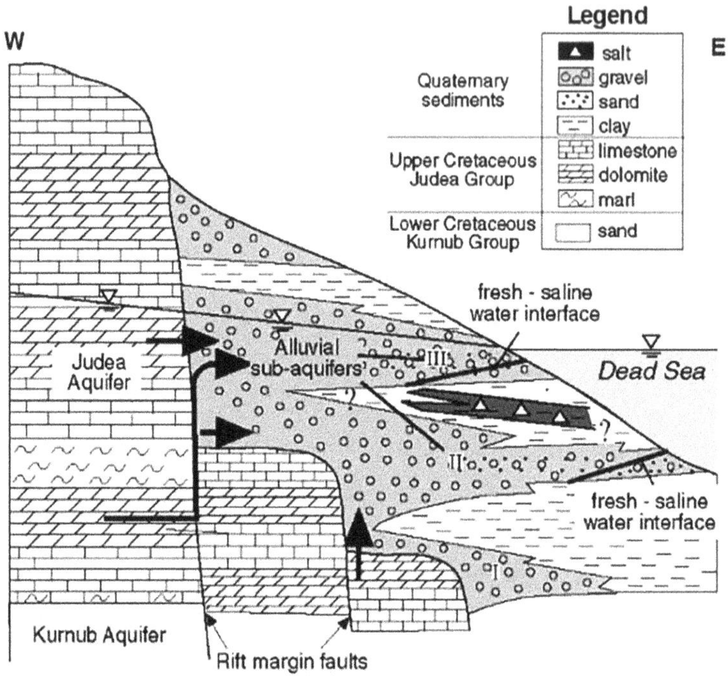

Figure 4.68. Schematic cross-section showing field relations between the different aquifers and DS. The Kurnub Group aquifer, underlying the Judea Group aquifer, is not shown. Arrows denote the possible direction of groundwater flow. (Reprinted from Yechieli *et al* (1995a), John Wiley & Sons, with the permission of the American Geophysical Union.)

development above the salt (Salameh and El-Naser 2000, Closson *et al* 2005, Yechieli *et al* 2006).

It was presumed that the coastal groundwater intervention to the DS (as a result of lowering its level) caused the DS brine to be considerably diluted to be enabled to dissolve buried salt (Goldman *et al* 2000, Ezersky *et al* 2009). Geophysicists as a rule looked for groundwater with a resistivity of more than 1 Ωm (Goldman and Ezersky 1999).

Yechieli *et al* (2006) have presumed that really salt is dissolved by very saline groundwater. Moreover, the monitoring of groundwater salinity in some hydrological boreholes confirmed that groundwater salinity does not decrease noticeably with time (Yechieli 2007). Further TEM investigations enabled us to understand the phenomenon of DS salt dissolution. This study included: (1) calibration of TEM resistivity–water salinity inter-relations using hydrological boreholes drilled by Geological Survey of Israel (GSI) and TEM measurements near those boreholes carried out by Geophysica Institute of Israel (GII); (2) generating of inter-relations between corresponding parameters to using TEM resistivity measurements to calculate groundwater salinity and aggressivity; and (3) determine the spatial distribution of these parameters to understand the mechanism of salt dissolution —sinkhole formation in the DS coastal area.

4.3.2.9 Principle of resistivity–salinity calibration

Twenty-eight observation boreholes have been drilled by the GSI along the DS and at the evaporation ponds' (EPs) coastal area (Yechieli *et al* 2004). These boreholes are located in eight sites, most of which are near sinkhole clusters (figure 4.62). In the northern DS basin, the chloride concentration of chemically saturated groundwater is 224 g L^{-1}; it is 259 g L^{-1} in the southern EPs. At the depth range corresponding to the salt layer (20–60 m), the majority of the boreholes revealed a groundwater salinity value between 54% and 93% of chloride saturation. Only three boreholes located in the Ein Gedi–Arugot area displayed groundwater with relatively low chloride concentration, 15–70 g L^{-1} (7%–33% of saturation). Near every borehole TEM sounding was carried out with loops 25 × 25 m^2 and, more rarely, 50 × 50 m^2.

The sinkhole sites along the DS shores include two parts with different structures: salt area and no salt one divided by the salt border (figure 4.69(a)). The western to the salt boundary part of the site ('no salt area') is composed of loose sediments, the upper part of which is not saturated with the groundwater (figure 4.69(b)). This part of the section is composed of 2–4 layers characterized by bulk resistivity ρ_x = 1–20 Ωm. The lower part of the section saturated with the DS brine is characterized by resistivity ρ_x = 0.1–1.0 Ωm.

The eastern part ('salt area') adjacent to the DS is characterized by a more complicated resistivity structure caused by the salt layer (figure 4.69(c)). Figure 4.69(d) shows the map of the Mineral Beach site taken as an example. Examples of interpretation of TEM sounding (resistivity–depth plots) performed near boreholes Mn-1 and Mn-4 are shown in figure 4.70.

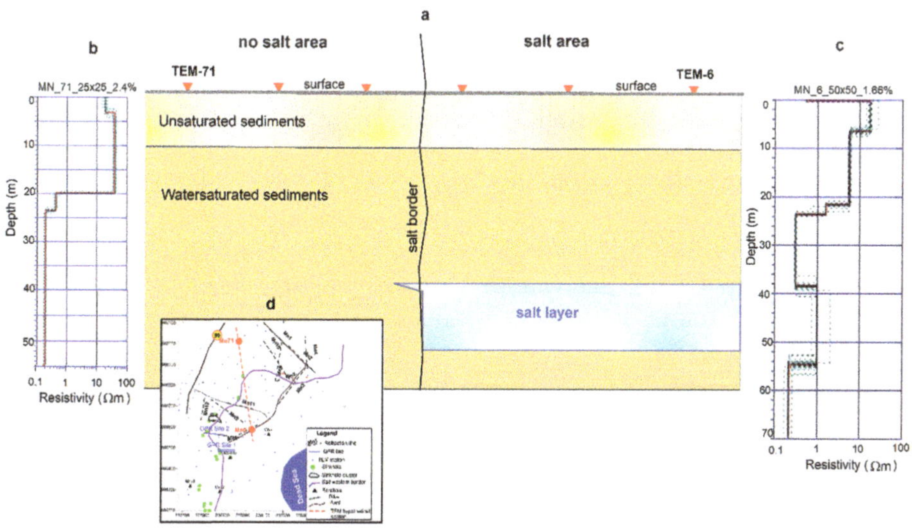

Figure 4.69. Scheme of TEM measurements for mapping of the DS coastal area. (a) Hypothetical section of the site along the red dashed line in figure 4.69(d); (b) TEM resistivity versus depth graph (MN-71 station) in the 'no salt area'; (c) TEM resistivity versus depth graph (MN-6 station) in the 'salt' area; (d) salt map through Mineral Beach area, red dashed line.

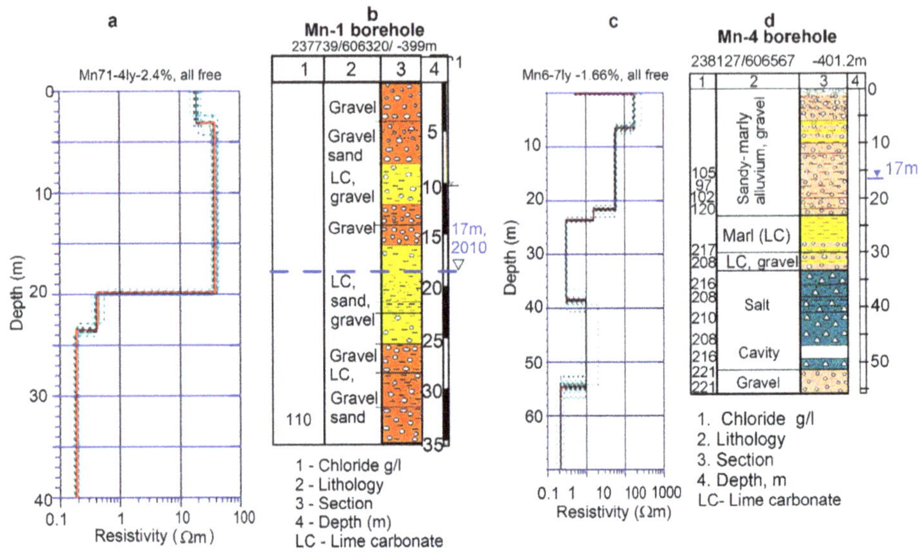

Figure 4.70. Borehole Mn-1: (a) inverse TEM resistivity versus depth graph and (b) results of its testing, in respect to depth: (1) chloride concentrations in the groundwater (g L^{-1}) (after Yechieli *et al* (2004) with permission from GSI); (2) lithology; (3) borehole lithological section; (4) depth. Thick line in figure 4.70(a) denotes the inversed resistivity, whereas thin dashed lines are equivalent solutions. Borehole Mn-4: (c) and (d) the same as (a) and (b), respectively.

Both boreholes Mn-1 and Mn-46 were drilled in the Mineral Beach area (the northern DS basin) near highway #90 and near the Mineral Beach Spa, respectively (see figure 4.62 for location). The lithology consists mainly of marl interbedded with gravels and sands. Marl consists of fine grain carbonate named by Frydman *et al* (2008) *lime carbonates* (LCs). Brine-saturated LCs (named also DS mud) do not exhibit cohesion. This feature allows considering saturated LCs as cohesionless sediments, like fine sands (Ezersky and Livne 2013). Note that the resolution of the resistivity determination in calibration boreholes (without salt layer, like Mn-1) is high enough, which is seen from the RMPs presented in table 4.6.

Considering the resolution of the most important layers, which are fourth (aquifer) in figure 4.70(a), one can see that for this layer RMP = 0.88, e.g., the resistivity of the fourth layer is constrained well (relate to Stoyer (2010) for interpretation).

The second TEM resistivity graph in figure 4.70(c) comprises the salt layer in the depth range between 33 m and 52 m. The RMP values are compiled in table 4.7.

4.3.2.10 Water resistivity–salinity relationship based on borehole calibration

Earlier, water resistivity–salinity relationships of the DS aquifers were first obtained by Yechieli (2000). A good qualitative correlation was established between the salinity and the electric conductivity (EC) of diluted water from both the DS (TDS = 340 g L^{-1}) and the groundwater from the well DSIF (TDS = 320 g L^{-1}). EC measurements were carried out *in situ* and under laboratory conditions. No

Table 4.6. The RMPs of Mn-71 station: four layers, RMS = 2.4%.

$\rho 1$	**0.75**						
$\rho 2$	0.13	**0.75**					
$\rho 3$	0.07	−0.14	0.59				
$\rho 4$	0.00	−0.01	−0.06	**0.88**			
T1	−0.27	−0.14	−0.03	0.0	0.18		
T2	0.04	0.04	0.04	0.00	0.16	**0.96**	
T3	0.00	0.00	0.11	0.09	−0.01	0.00	0.01
	$\rho 1$	$\rho 2$	$\rho 3$	$\rho 4$	T1	T2	T3

ρ_N is the resolution of resistivity of layer N.
T_N is the resolution of the thickness of layer N.
RMP of well- and fairly well-constrained layers are bold.

Table 4.7. The resolution matrix of MN-6 station: seven layers RMS 2.62%.

$\rho 1$	0.59												
$\rho 2$	0.04	0.02											
$\rho 3$	0.02	−0.03	**0.91**										
$\rho 4$	0.00	−0.01	0.01	0.01									
$\rho 5$	0.01	0.01	−0.01	−0.01	**0.98**								
$\rho 6$	−0.02	0.00	0.00	0.00	0.01	0.49							
$\rho 7$	0.02	−0.01	0.00	0.00	0.00	0.03	0.51						
T1	−0.42	−0.19	0.03	0.00	0.00	0.00	0.04	0.52					
T2	−0.02	0.23	0.16	0.00	0.01	0.00	0.00	0.00	**0.81**				
T3	0.02	−0.11	−0.07	0.02	0.01	0.00	−0.01	0.01	0.12	**0.97**			
T4	−0.01	−0.15	0.00	0.00	0.00	0.00	−0.01	−0.03	0.02	−0.05	**0.85**		
T5	0.03	0.09	0.03	0.01	−0.01	0.03	0.00	−0.01	0.00	0.00	0.06	0.29	
T6	0.02	0.01	−0.05	0.01	−0.19	0.11	0.00	0.01	0.01	0.05	0.01	0.01	**0.76**
	$\rho 1$	$\rho 2$	$\rho 3$	$\rho 4$	$\rho 5$	$\rho 6$	$\rho 7$	T1	T2	T3	T4	T5	T6

ρ_N is the resolution of resistivity of layer N.
T_N is the resolution of the thickness of layer N.
RMP of well- and fairly well-constrained layers are bold.
RMP of the thickness of the sixth layer (salt) is 0.76. The layer is a good constraint. The resistivity of the sixth layer (salt) RMP = 0.49 is not resolved well. Similarly, the seventh layer (aquifer) with RMP = 0.51 is not resolved well.

quantitative relationships were found. However, some conclusions were inferred from this study. During a consequent number of years, many boreholes were drilled along the DS shores. Laboratory determination of the salinity of water samples along with the EC measurements in boreholes was carried out (Yechieli *et al* 2004). Other investigators measured similar relationships in the resistivity range of 2 Ωm and over (Goldman *et al* 1991, Kafri *et al* 1997, Kafri and Goldman 2005, Levi *et al* 2008). Ezersky and Frumkin (2017) have widened the resistivity–salinity calibration to the range of <2 Ωm which is very important for the evaluation of the potential of groundwater to dissolve the salt.

In figure 4.71(a), there are two segments of the data with different slopes. We approximate the data plot by two linear segments in log–log axes: curve 1a (in the range of high chloride concentration) and curve 1b (in the range of low concentration). The data plot for high chloride concentration is separately displayed in figure 4.71(b). We observe a generally good correlation ($R^2 > 0.95$) between the water resistivity and chloride concentration in both segments (empty circles, figures 4.71(a) and (b)). One segment 1a (figures 4.71(a) and (b)) corresponds to high chloride concentration (C_{cl} = 88–214 g L^{-1}) and another one segment 1b (figure 4.71(a)) to low concentration (C_{cl} < 88 g L^{-1})] (we relate the reader to the article of Ezersky and Frumkin (2017) for details).

Now we can write Archie's equation. The relationship between the bulk resistivity ρ_x and the water resistivity ρ_w is shown in log–log coordinates in figure 4.72.

The graph for sandy-gravel sediments allows fitting data to an equation for sediments with an effective porosity $\varnothing_e = 0.32$, and LCs with effective porosity $\varnothing_e = 0.44$. The corresponding equation is seen in figure 4.72.

We compare the resistivity–salinity relationships obtained by various researchers in Israel (figure 4.73).

One can see that the data of this study (triangles) agree well with the relationships obtained within similar quaternary sandy-gravel sediments along the DS shore (solid circles in figure 4.73) and the Mediterranean coastal areas in Israel (Kafri and Goldman 2006). At the same time, the relationship obtained by Levi *et al* (2008) shows resistivity values that are approximately 2.5–3.0 times higher than our results

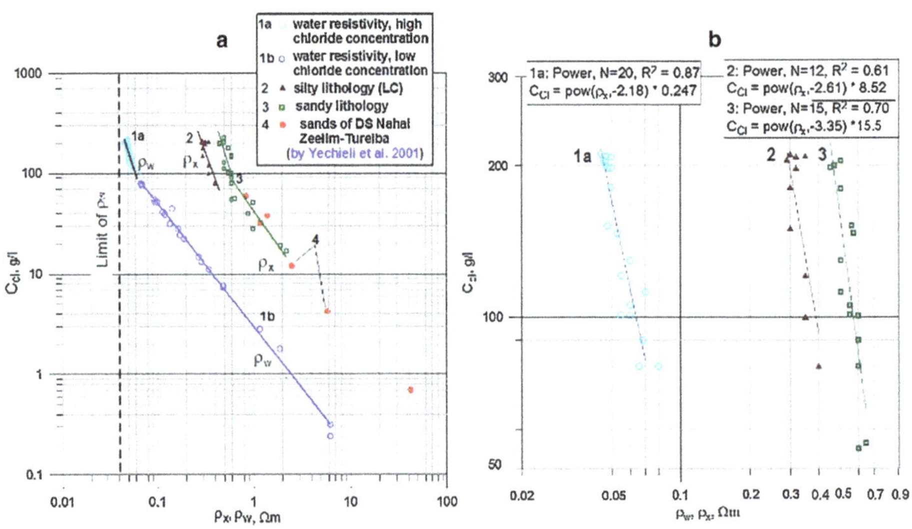

Figure 4.71. Various inter-relations between the chloride concentration and resistivity: (a) all data; (b) high chloride concentration (new) data. Curves 1a and b represent the high and low chloride concentrations in pore water (fluid) respectively versus the water resistivity based on GSI data, Yechieli *et al* 2004, (2) chloride concentration versus bulk resistivity in LC sediments; (3) the same in saturated sandy sediments; (4) the same in sandy sediments of Nahal Zeelim-Tureiba, after Yechieli *et al* (2001). (Reproduced from Ezersky and Frumkin (2017), copyright 2017, published by Elsevier Masson SAS. All rights reserved.)

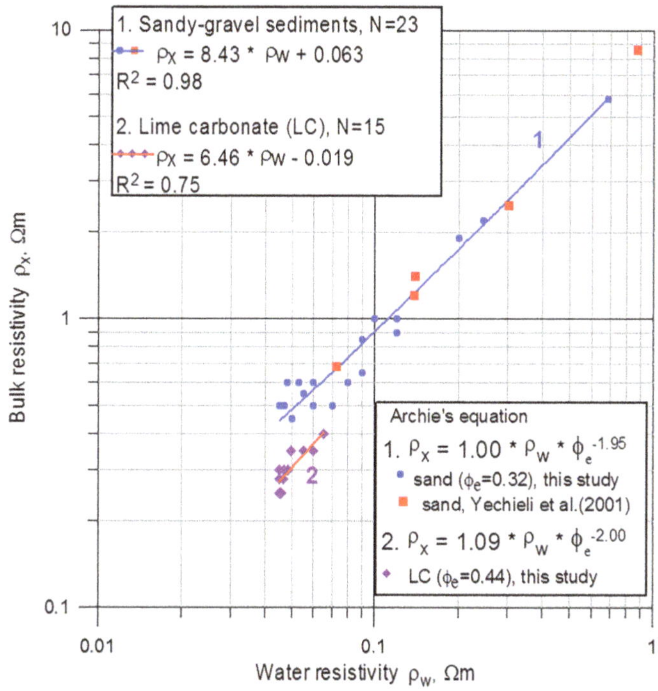

Figure 4.72. Relationships between the water resistivity and bulk resistivity for sandy-gravel lithology and LC. Archie's equation shown in the lower window is fitted to data for effective porosity $\varnothing_e = 0.32$ (sandy-gravel) and $\varnothing_e = 0.44$ (LC). (Reproduced from Ezersky and Frumkin (2017), copyright 2017, published by Elsevier Masson SAS. All rights reserved.)

at the same salinity (crosses in figure 4.73). The relationships presented in figure 4.72 at first confirm the conclusions of Kafri and Goldman (2006) that the correlation between TEM bulk resistivity and groundwater salinity is excellent for high-salinity brines exceeding 10 g L^{-1} of chloride but greatly deteriorates with decreasing salinity associated with the brackish and freshwaters. It corresponds well to the conclusion of Yechieli *et al* (2001, p 373) that in cases of high groundwater salinity measured within boreholes (decades–hundreds of g L^{-1} chloride), salinity is the main factor governing TEM resistivity, regardless of differences in other factors such as lithology (i.e. composition, texture, grain size, etc) and porosity.

Our study supports this conclusion for a range of very low resistivity values (lower than 1 Ωm). It can be seen from figure 4.73 (triangles) that there is practically no scatter in this resistivity range.

4.3.2.11 Water aggressiveness and its inter-relations with resistivity

The aggressiveness is defined (in simple form) as a measure of the relative capacity of water to dissolve rock material (Lowe and Waltham 1995). It relates also to the DS groundwater that is capable of dissolving the salt. If the Cl concentration in salt-saturated groundwater is, C_{Clmax}, then water with the lower Cl concentration C_{Cl} has

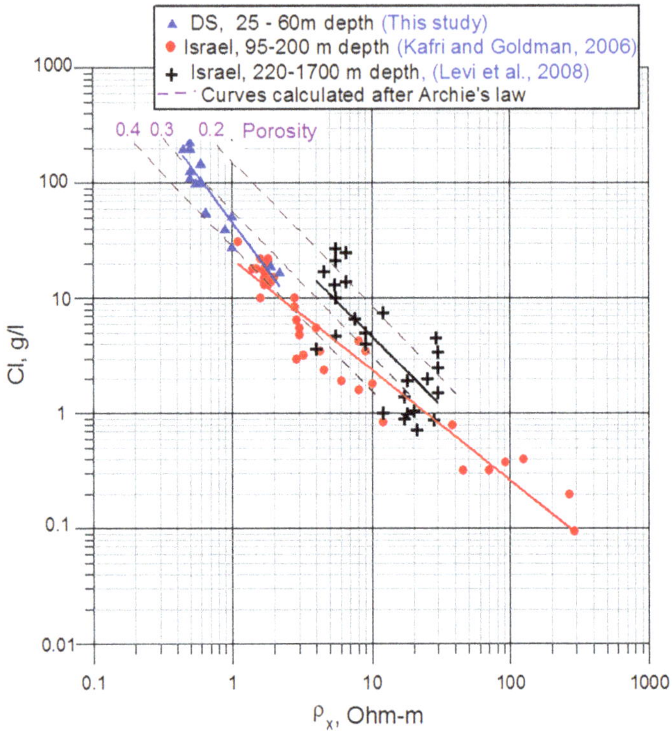

Figure 4.73. Bulk resistivity–salinity relationships obtained in the DS area by various researchers. The dashed lines are the relationships calculated for different porosities using Archie's Law (equation (2.2) in section 2.2.1.6). (Reproduced from Ezersky and Frumkin (2017)), copyright 2017, published by Elsevier Masson SAS. All rights reserved.)

the potential to dissolve halite $C_{\mathrm{ClPOT}} = (C_{\mathrm{Clmax}} - C_{\mathrm{Cl}})$. Then, the aggressiveness of the groundwater can be expressed as:

$$A = \frac{C_{\mathrm{ClPOT}}}{C_{\mathrm{Clmax}}} = \frac{C_{\mathrm{Clmax}} - C_{\mathrm{Cl}}}{C_{\mathrm{Clmax}}} = 1 - \frac{C_{\mathrm{Cl}}}{C_{\mathrm{Clmax}}}. \tag{4.12}$$

When groundwater comprises C_{Clmax} chloride, $A = 0$, e.g. groundwater is not aggressive (we presume the temperature of groundwater is constant). Without any dissolved chloride $C_{\mathrm{Cl}} = 0$ and $A = 1$. In this case, the aggressivity concerning salt is maximal. As mentioned above, the chloride concentration C_{Cl} can be evaluated by measuring bulk resistivity of the aquifer using TEM.

The maximum chloride concentration of the DS water (C_{Clmax}) is 224 g L^{-1} in the northern basin and 259 g L^{-1} in the southern one. These chloride concentrations correspond to the total dissolved solids (TDS) values of 340 and 380 g L^{-1}, respectively. Thus, knowing the resistivity of the aquifer and using equation (4.12), one can evaluate the groundwater aggressiveness A and the associated sinkhole susceptibility. Note also that water aggressiveness is an important index allowing a distinction between the resistivity of sandy and lime sediments that can

Table 4.8(a). Groundwater classification of aggressiveness concerning DS aquifers salinity within *Sandy-gravel sediments*. (Reproduced from Ezersky and Frumkin (2021), copyright (2021) with permission from Elsevier.)

ρ_x, Ωm LC*	ρ_x, Ωm SG**	ρ_w, Ωm	A*** equation (4)	C_{ClPOT}	C_{Cl}, g L^{-1}	Water aggressiveness
>0.54	>0.80	>0.1	>0.75	>166	<58	Very aggressive
0.40–0.54	0.55–0.8	0.07–0.1	0.50–0.75	109–166	115–58	Aggressive
0.30–0.40	0.45–0.55	0.045–0.07	0.10–0.50	0–109	224–115	Slightly-to-medium aggressive
<0.30	<0.45	<0.045	0–0.1	0	224	Conventionally non-aggressive

* LC—lime carbonate;
** SG-sandy-gravel sediments;
*** $C_{Clmax} = 224$ g L^{-1}.

Table 4.8(b). Groundwater classification of aggressiveness concerning DS aquifers salinity within. (Reproduced from Ezersky and Frumkin (2021), copyright (2021) with the permission of Elsevier.)

ρ_x, Ωm LC*	ρ_x, Ωm SG**	ρ_w, Ωm	A*** equation (4)	C_{ClPOT}	C_{Cl}, g L^{-1}	Water aggressiveness
>0.54	>0.80	>0.09	>0.75	>176	<48	Very aggressive
0.40–0.54	0.55–0.8	0.065–0.09	0.50–0.75	176–119	105–48	Aggressive
0.30–0.40	0.45–0.55	0.05–0.065	0–0.50	0–119	224–105	Slightly-to-medium aggressive
<0.30	<0.45	<0.05	0	<0	224	Conventionally Non-aggressive

* LC—lime carbonate;
** SG—sandy-gravel sediments;
*** $C_{Cl}^{max} = 224$ g L^{-1}.

have the same chloride concentration (e.g. aggressiveness) at different resistivity values (see table 4.8(b) in continuation). It can be used for the comparison of evaporite environments worldwide.

Aggressiveness *A* versus bulk resistivity plot is shown in figure 4.74. The groundwater aggressiveness in sandy-gravel sediments is shown in figure 4.74(a). Similar inter-relations for the LCs are presented in figure 4.74(b). These inter-relations are based on the equations presented in Ezersky and Frumkin (2017).

Because there is no established classification, we suggest a novel groundwater aggressiveness classification for sandy-gravel sediments and LCs using plots presented in figure 4.74. We divided resistivity values into four ranges corresponding to: (1) conventionally non-aggressive groundwater ($A = 0$); (2) slightly-to-medium aggressive ($A = 0\%$–50%); (3) aggressive groundwater ($A = 50\%$–75%); (4) very

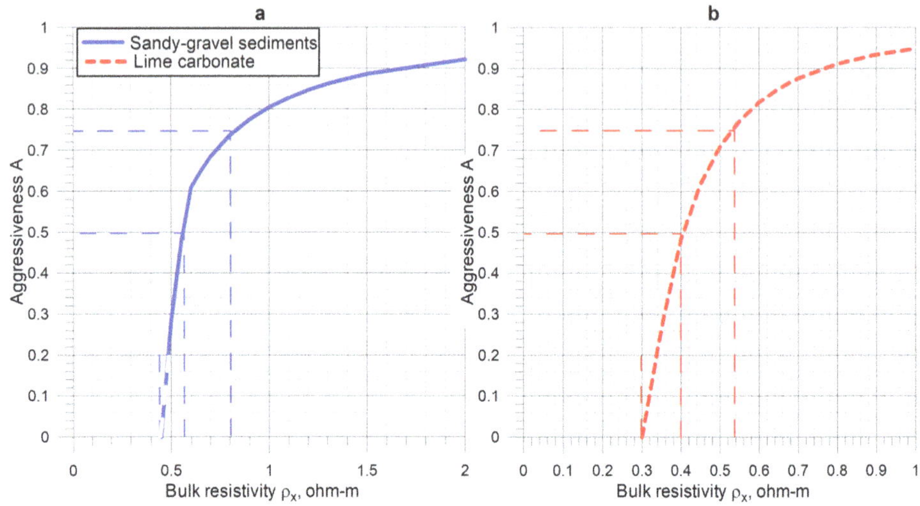

Figure 4.74. Aggressiveness A versus bulk resistivity plots for (a) sandy-gravel sediments and (b) for LCs. (Reproduced from Ezersky and Frumkin (2021), copyright (2021) with permission from Elsevier).

aggressive one ($A > 75\%$). Parameters of the above classification are presented in tables 4.8(a) and 4.8(b).

In tables 4.8(a) and 4.8(b), the aggressiveness (A) and its categories are the same for both sandy-gravel sediments and lime carbonates. The resistivity ρ_x of the same aggressiveness in the LC is approximately 30%–35% lower than in the sandy-gravel sediments.

ρ_w is very close to the two types of sediments. C_{Cl} concentrations differ slightly. Resistivity ranges in the LCs are much narrower than those in sandy-gravel sediments. The difference in bulk resistivity between the sandy-gravel and LC sediments at the same salinity level is explained by the different porosity of these lithologies; 0.5 and 0.4 (total), and 0.44 and 0.32 (effective) for LCs and sandy-gravel sediments, respectively. Moreover, in the DS coastal area, LCs are presented as water-saturated marshy mud having higher porosity than laboratory samples.

The interesting conclusions can be formulated as a result of the above classifications.

(1) The aggressiveness of LCs at the same ρ_x is higher than that of sandy-gravel sediments. For instance, the bulk resistivity of slightly-to-medium aggressive sandy-gravel sediments ranges from 0.45–0.55 Ωm, whereas LCs with approximately the same resistivity range of 0.40–0.54 Ωm are aggressive. It is explained by the higher effective porosity of LCs discussed above. The difference in the same aggressiveness category is approximately 30%–35%.

(2) The DS sandy-gravel aquifer with bulk resistivity $\rho_x < 1.0$ Ωm, whose water was classified by Kafri *et al* (1997) as DS brine, contains only ~50 g L^{-1} NaCl, thus having the potential to dissolve up to ~ 170–180 g L^{-1} salt.

The first conclusion explains the very low resistivities $\rho_x = 0.3–0.5$ Ωm measured in the Nahal Hever South site (western DS coast) sinkhole area (figure 4.75(a)).

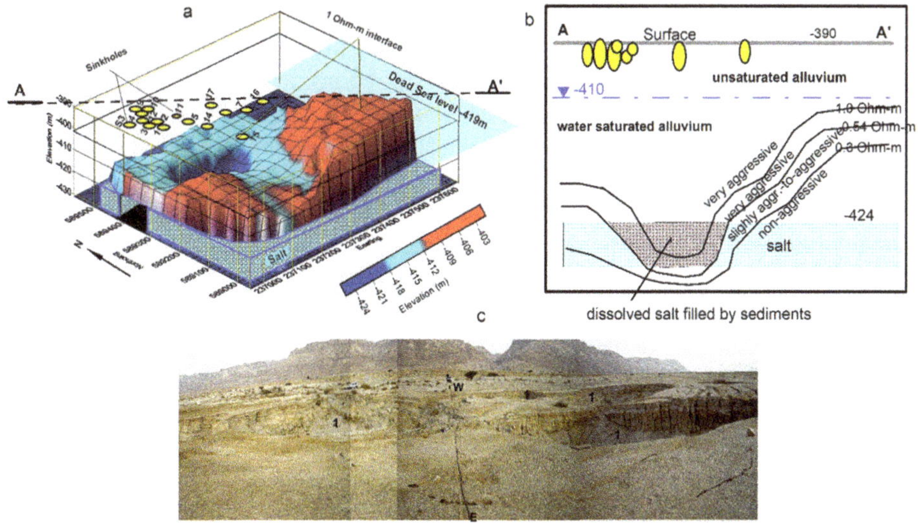

Figure 4.75. (a) 3D presentation of the lower 1.0 Ωm interface generated from 1D TEM resistivity sections in the central-southern parts of the Nahal Hever site. Numbered figures are sinkholes visible on the surface. (b) The section along line A–A′ (see (a) for location) explaining the sense of 1 Ωm funnel; (c) a large sinkhole cluster formed in 2006 at the funnel place. By digit 1 are denoted sinkholes. ((a) Reprinted from Ezersky *et al* (2009), with permission of Springer.)

Figure 4.75(b) presents the topography of the lower 0.5–1 Ωm interface in a 3D presentation showing its spatial geometry. This interface descends from the southeast, where it is located close to the surface, toward the northwest. The prominent feature of this map is the large funnel-shaped hole in the central part of the flat. The hole is approximately 200 m in diameter. Most of the sinkholes are located in the vicinity of this hole. The configuration of the 0.5 Ωm interface is shown also in the resistivity section crossing the area in the northwest–southeast direction. This section, in combination with the MRS data, enables the interpretation of the Nahal Hever hydrogeological situation and the mechanism of sinkhole formation

Note that in the northern part of the Nahal Hever South site sinkholes have intensively developed 1–2 years later.

4.3.2.12 Groundwater salinity mapping

A new approach based on the separate processing of TEM data through salt and the aquifer was suggested by Levi *et al* (2011). The first stage is the delineation of the salt layer, which is an essential part of the DS shore conditions. As was mentioned above, the salt layer (~10–40 m thickness) inserts a significant heterogeneity of the seismic velocities and electrical resistivity values to the investigated subsurface. Therefore, after locating the salt layer edge using the seismic refraction method (Ezersky 2006), we can separate the area into two quasi-homogeneous sites of investigation, relatively salt boundary (figure 4.69): (1) 'no salt' area located east of the salt edge, where the measured resistivity is influenced by the groundwater salinity; and (2) 'salt' area east of the salt edge, where the measured resistivity is affected by salt porosity at a constant water resistivity value. Besides, it allows avoiding the 3D effects possible in the vicinity of the karstified salt. Note that in areas where the subsurface is composed of quasi-homogeneous sediments, seismic surveys will not be needed.

Through the study area, 50–70 TEM soundings are carried out with distances between the sounding points (stations) of 40–50 m. The resistivity readings correspond to the average resistivity through an area of an effective ring which is equal to the depth of interest (figures 4.63 and 4.64). Thus, laterally, the resistivity characterizes an area of a circle with a radius equal to the depth of interest. It is presumed that the electrical resistivities of unconsolidated sediments in the highly saline brine are not changed sharply. Geophysical mapping allows the generating of averaged lateral maps.

The vertical resolution of the resistivity inverted section is determined by the resolution matrix that is calculated for the thickness and resistivity of every layer (Stoyer 2010). An example of the resolution estimation is presented in Ezersky *et al* (2011).

An example of the separate mapping of aquifer and salt layer resistivity is shown in figures 4.76(a) and (b), respectively.

In limits of the 'salt area' resistivity (after Archie's law) is a measure of the salt porosity (Frumkin *et al* 2011), whereas, in the 'no salt area' resistivity is interpreted in terms of groundwater salinity based on the resistivity–salinity calibrations generated for the DS region (Ezersky and Frumkin 2017). The new methodology permits also 2D and 3D presentation of the acquired results despite the fact that the methodology itself of the data acquisition is of 1D in its origin. In such an interpretation, we have applied the TEM FAST methodology to investigate salt conditions through the Mineral Beach site. The methodology comprises a fast-

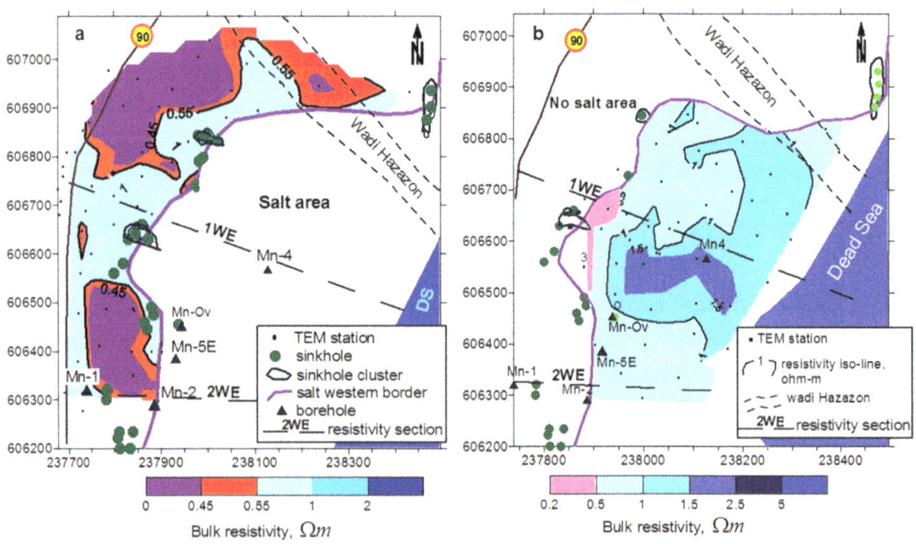

Figure 4.76. Two resistivity maps were generated separately for aquifer (a) and salt layer (b), respectively, west and east of the salt layer border (thick solid line). (Reproduced from Ezersky and Frumkin (2017), copyright 2017, published by Elsevier Masson SAS. All rights reserved.)

spacious mapping of the bulk resistivity in both vertical and horizontal directions through the great area covering ~1 km.

The 'salt area' survey reveals the clear zonation of the area following the distance from the DS. The zone nearest to the salt edge is characterized by resistivity less than 0.5 Ωm. These zones are not regular and they are associated with the salt karst. Sinkhole formation in these zones is most probable. The next zone is characterized by the resistivity of 0.5–1.0 Ωm and increased porosity. Finally, a zone characterized by the resistivity of more than 1 Ωm and, especially more than 1.5 Ωm, is a zone of the background porosity. Estimation of the salt porosity can be based on a calibration of the resistivity that has been carried out by Ezersky and Goretsky (2014). The map of the salt top surface shows the deepest position in the central part of the area (−434 m elevation) and the shallower one at the margins of the studied area (−424 m). Mapping the 'no-salt area' shows the high relative resistivity of the clastic sediments (0.6–0.75 Ωm) along the salt edge. Such resistivity can characterize the brackish water with a chloride concentration of 60–100 g L^{-1} (at a maximal concentration of C_{Cl} = 224 g L^{-1}). To avoid equivalent solutions, one more additional method (seismic refraction) or drilled boreholes is suggested.

Two sections along the lines 1WE and 2WE (figure 4.77) show the deep resistivity structure of the Mineral Beach area (see figure 4.76(a) for location).

Section 1WE (figure 4.77(a)) is completely composed of sandy-gravel sediments (Frumkin *et al* 2011). The section shows inclined resistivity layers ascending seaward. This characterizes undersaturated groundwater rising from deep parts at the west of the section. The salt layer edge is located within the more resistive zone (0.55–1.0). At the 2WE section unlike 1WE, one-two low resistivity zones (<0.45 and even 0.35) envelop the salt layer from above and below (figure 4.77(b)). However, the salt edge is

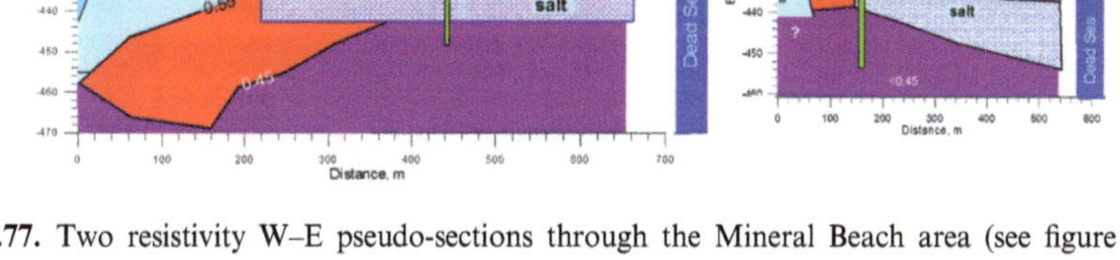

Figure 4.77. Two resistivity W–E pseudo-sections through the Mineral Beach area (see figure 4.76(a) for location). (a) The 1WE section crossing the borehole MN-4 presents the central part of the area composed of sandy-gravel sediments; (b) the 2WE section crossing the boreholes MN-1 and MN-2 in the southern part of the area composed in its eastern part by lime carbonates. (Reproduced from Ezersky and Frumkin (2017), copyright 2017, published by Elsevier Masson SAS. All rights reserved.)

about the resistivity zone of 0.55. Such resistivity structure of the 2WE section is explained by the lithology of the southern part of the Mineral Beach area: sand at the west part of the section, and LCs at the east (Frumkin *et al* 2011). The low porosity of LCs limits groundwater circulation across above and below the salt layer, resulting in stagnant, high-salinity water. Besides, fresh groundwater flows from the west of the area. A second difference in the 2WE section is the salt layer geometry. The borehole Mn-4 shows that the salt layer top is at −434 m b.m.s.l., whereas the borehole Mn-1 crossed the salt edge at −424 m, e.g. 10 m higher than in Mn-4. Such a distortion of the salt layer geometry caused by fault activity was discussed in our previous papers (Frumkin *et al* 2011, Ezersky and Frumkin 2013).

As an example, we constructed a map of the calculated water salinity C_{cl} throughout the aquifer of the Mineral Beach area west of the salt edge (figure 4.78). The map is inversed to the resistivity map of the aquifer (figure 4.76(a)), e.g., zones of

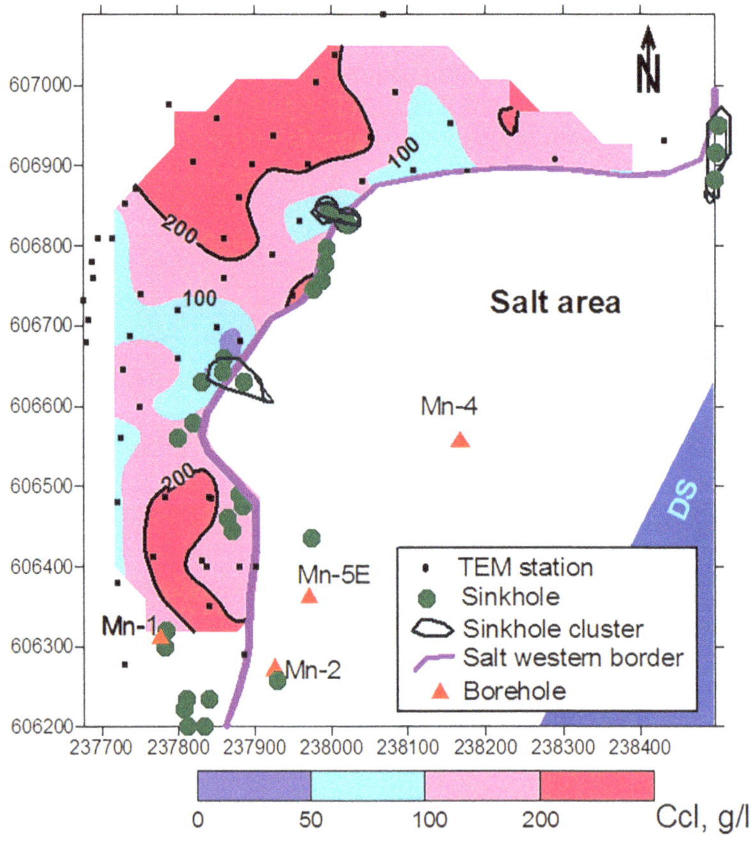

Figure 4.78. Evaluation of groundwater salinity through the Mineral Beach area using resistivity distribution. The map of water salinity at the depth of the salt layer is based on the measured ρ_x values and calculated values of chloride concentration (after figure 4.71). (Reproduced from Ezersky and Frumkin (2017), copyright 2017, published by Elsevier Masson SAS. All rights reserved.)

high chloride concentration in figure 4.78 correspond to the zones of low bulk resistivity in figure 4.76(a). Note, zones of the low chloride concentration correlate with the sinkhole clusters line along the salt edge.

4.3.2.13 Underground water streams mapping

It is a very useful application of TEM resistivity mapping from the point of view of sinkhole formation to reveal the salinity of the underground streams that would dissolve the salt layer at its edge. Such mapping was carried out in the Ein Gedi site (Ezersky *et al* 2008).

The Ein Gedi–Arugot area of 1000 × 1800 m has been mapped in 2005–6 to determine the resistivity distribution. A total of 90 TEM stations of 25 × 25 m and 50 × 50 m were set up. As a rule, transient curves are of good quality.

The map presented in figure 4.79 shows the resistivity ρ_L distribution throughout the lower low-resistivity layer located under the 1 Ωm interface at a depth range of 45–55 m (−435–445 m elevation corresponding to the top of the salt layer). The TEM stations, sinkholes, and boreholes are also shown in the resistivity map.

A common characteristic of inverse resistivity sections is the presence of a low resistivity layer with resistivity lower than 1 Ωm. Because TEM stations are located at different elevations and distances from the DS, the depth of the low resistivity layer in the examples presented here varies between 16 and 58 m. The depth of the water table in the Ein Gedi–Arugot area is 25–30 m. The penetration depth is 60–80 m.

It can be seen in figure 4.79 that generally resistivity varies in the range 0.25–1 Ωm. However, within this area, we can select two resistivity zones (marked by red and pink colors): the bigger parts of the area (red) with a resistivity lower than 0.6 Ωm (down to 0.3 Ωm), and the pink zone characterized by a resistivity of $0.6\ \Omega m < \rho_L < 1\ \Omega m$. This zone shaped like lineament is extended in the SW–NE direction reaching the DS. It crosses the sinkhole strip and boreholes eg1, eg19, eg7, MAAZ 1C, MAAZ2A, and eg13. It can be said that this lineament merges with the local branch coming from the eg6 borehole, and with another one coming from the west where Nahal David is located.

We suggest this zone can be associated with the underground stream. The suggested stream flows from the west where wells show freshwater to the DS where groundwater is characterized by high salinity. Following graph 3 in figure 4.71(a), the sandy-gravel aquifer with resistivity $\rho_L > 0.6\ \Omega m$ contains chloride concentration less than 100 g L^{-1}. Thus, following table 4.8(a) this aquifer is aggressive concerning salt.

That can explain sinkhole distribution along the salt edge.

4.3.2.14 DS water aggressiveness mapping

The geological section of the Mineral Beach area (see figure 4.69 for location) is composed of sandy-gravel sediments with a salt layer at the east of the site and LC lithology at its southern part, south of the dot-dashed line in figure 4.80.

The resistivity–aggressiveness map is shown in figure 4.70. The map presents a coastal aquifer that is in contact with the edge of the buried and faulted salt layer. Then, using table 4.8(a) one can observe zones of different aggressiveness using the

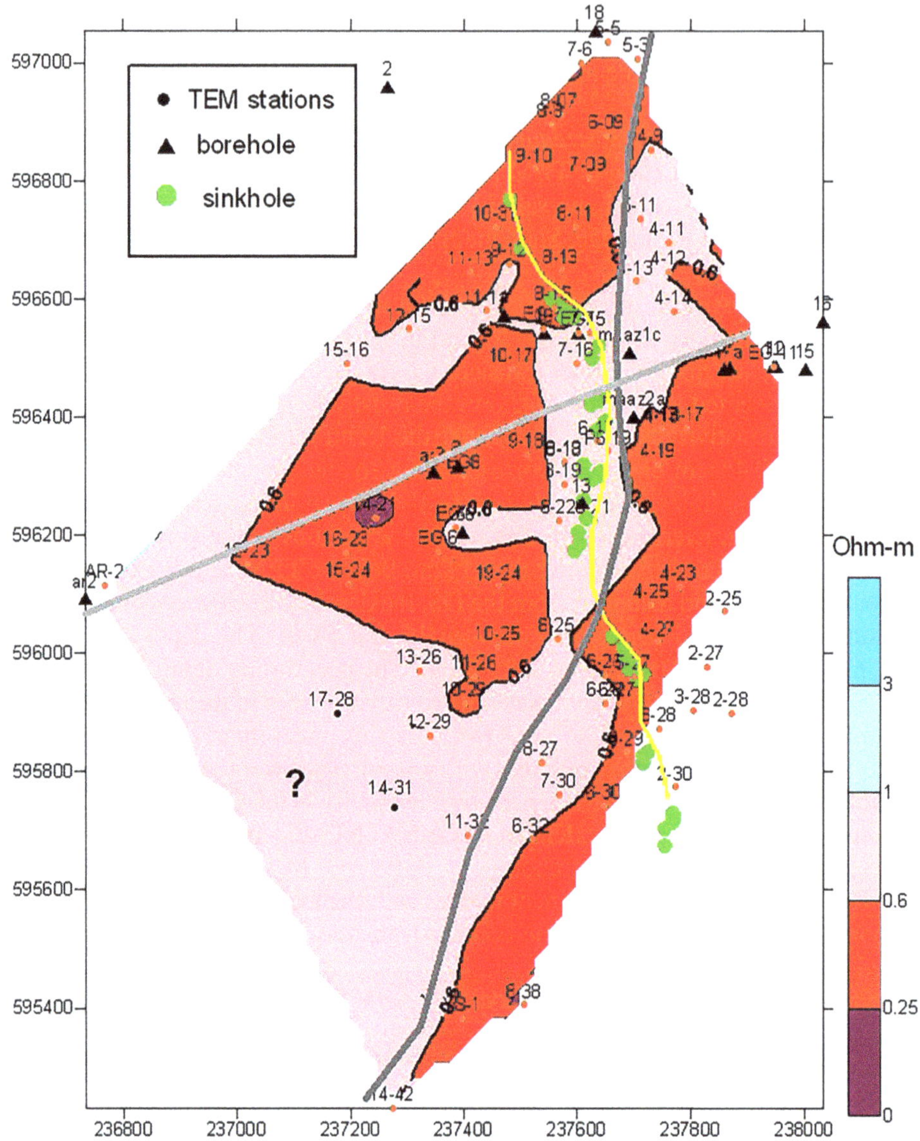

Figure 4.79. Inverted TEM resistivity map (ρ_L) at a depth of 45–55 m (−435–445 m elevations) throughout the Ein Gedi–Arugot site. The salt edge is arranged along the line of sinkholes (green circles). Grey WE line is Nahal Arugot, grey SN line is Route #90 (Ezersky *et al* 2008).

color-coded legend of figure 4.80 (Ezersky and Frumkin 2021). Most sinkholes in the area were formed within the medium to the very aggressive aquifer (table 4.8(a)) characterized by a resistivity zone between 0.45 Ωm and >1.0 Ωm. In the high aggressive zones (low chloride concentration) that are characterized by resistivity higher than 0.80 Ωm, the major sinkhole clusters have formed. They are Snk. 1 and Snk. 2, smaller sinkholes are located within the less aggressive zones of 0.45–0.80

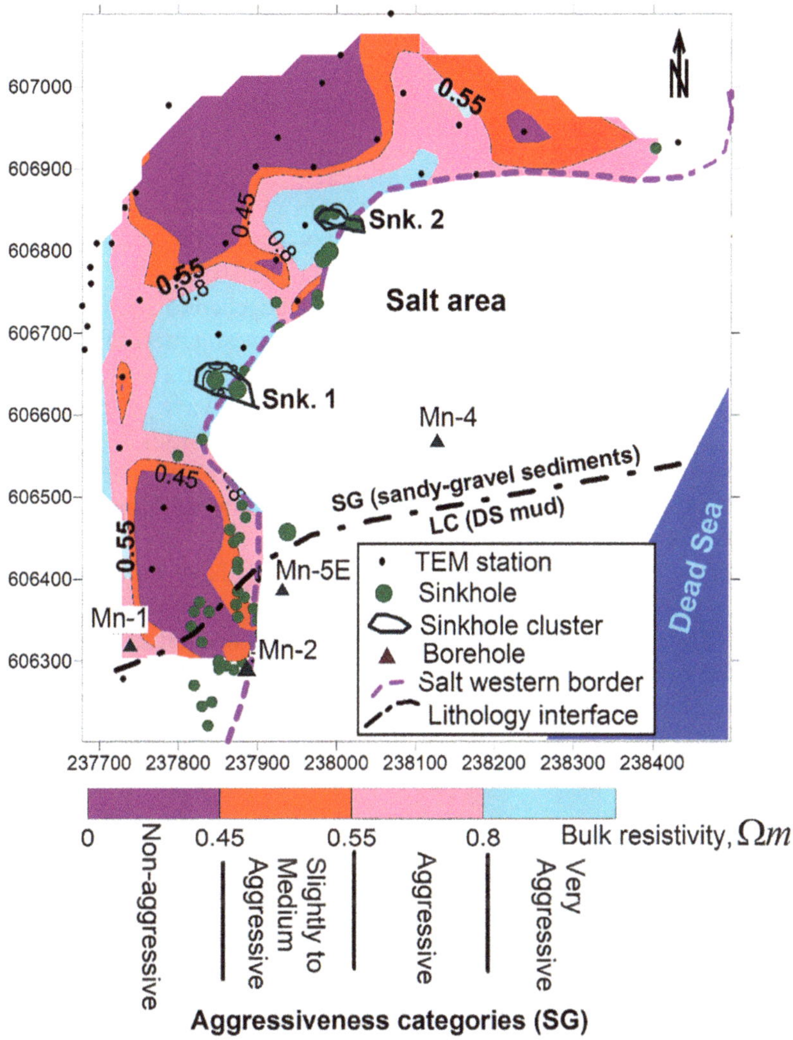

Figure 4.80. The Mineral Beach site (western DS shore). Resistivity map (Ωm) generated for sandy-gravel aquifer west of the salt layer border (thick dashed lilac line). At the color scale, the groundwater aggressiveness is denoted following table 4.8(a) for sandy-gravel sediments. (Reprinted from Ezersky and Frumkin (2021), copyright (2021) with the permission of Elsevier.)

Ωm resistivity (figure 4.80). Zones of high chloride concentration (low aggressive with $\rho_x < 0.45$ Ωm) in figure 4.80 do not contain sinkholes. Note, the southern zone (below the dot-dash line) is composed of LCs and its aggressiveness concerning salt follows table 4.8(b). Sinkholes in this zone were intensively developed on the mudflat sufficiently earlier (during 2000–11) (Abelson *et al* 2017, figure 4b).

Now it is clear from the resistivity–salinity and resistivity–aggressiveness interrelations that sinkhole formation is associated with a very low resistive (highly saline), but aggressive aquifer. In the generated maps and resistivity sections one can estimate the aquifer salinity and aggressiveness based on corresponding interrelations (Ezersky and Frumkin 2017, 2021) and understand the potential of the aquifer to dissolve the salt and to form sinkholes. Nowadays, knowing aggressiveness, we can explain sinkhole development in the very low resistivity zones (Ezersky and Frumkin 2021).

4.3.2.15 Features of salt layer investigation

We have discussed above the TEM resistivity–depth graphs, their structure, equivalency, and resolution. These graphs reflect the structure of the subsurface (Frumkin *et al* 2011, Levi *et al* 2011, Ezersky and Legchenko 2015).

On the other hand, TEM is not still used for the investigation of some salt layer characteristics: its depth, thickness, and porosity. These data (e.g. salt depth and thickness) can be used for constraining the surface wave inversion, e.g. for constructing the initial model for inversion (Ezersky *et al* 2013c). In this subsection, we present an example of the application of our approach to the investigation of salt layer and coastal aquifer in the sinkhole development sites in Israel.

1. *The buried salt layer depth and thickness*

   Ezersky *et al* (2011) and Frumkin *et al* (2011) have shown that the TEM inverse resistivity versus depth graph reflects the structure of the salt layer depth from the surface and the salt layer thickness. It is explained by the location of high resistive salt ($\rho_x > 1.5$–$2.0\ \Omega$m) at the background of the very low resistive aquifer ($\rho_x < 0.1$–$0.5\ \Omega$m) It is seen in figure 4.81(b) that both the depth and thickness of the high resistivity unit at a depth of 22 m

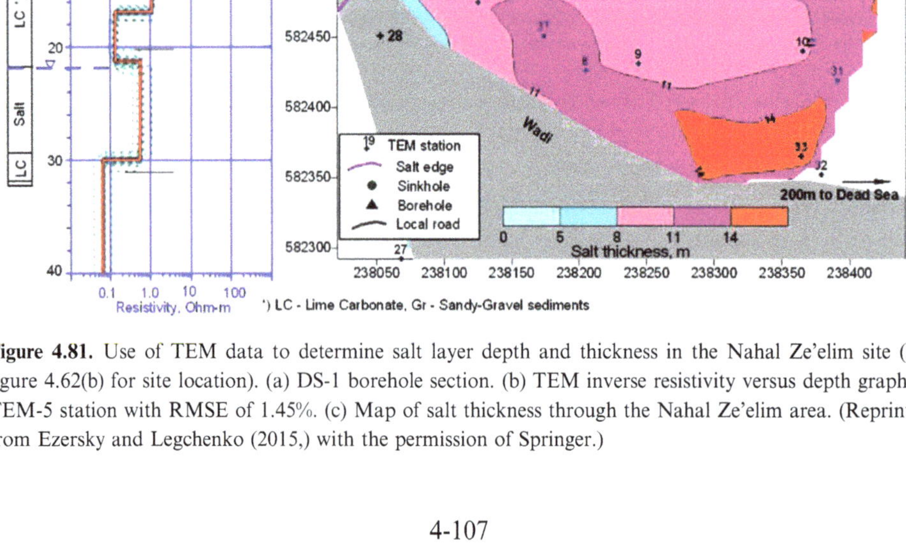

Figure 4.81. Use of TEM data to determine salt layer depth and thickness in the Nahal Ze'elim site (see figure 4.62(b) for site location). (a) DS-1 borehole section. (b) TEM inverse resistivity versus depth graph at TEM-5 station with RMSE of 1.45%. (c) Map of salt thickness through the Nahal Ze'elim area. (Reprinted from Ezersky and Legchenko (2015,) with the permission of Springer.)

corresponds well to the salt layer in the borehole log in figure 4.81(a) (Ezersky and Legchenko 2015).

The map of the salt layer thickness through the Nahal Zeelim area is shown in figure 4.81(c). The Nahal Ze'elim survey area is located at the southern part of the northern DS basin some 2–3 km to the east of the Judea Mountains. The geological section is composed of pebbles, cobbles, and clayey–sandy sediments. The TEM survey was carried out in the sounding mode using loops of size 25 × 25 m (rarely, 50 × 50 m). The salt thickness map shows that the thickness of salt is increased towards the DS.

Another example of the salt layer conditions mapping was carried out in the Mineral Beach site based on a new mapping methodology (Levi *et al* 2011). The salt top elevation map is shown in figure 4.82(a). The salt top surface is downward towards the DS. Its western part is located at an elevation of −424 m, and the eastern part of the salt layer top is located at the elevation of −440 m. Thus, the difference in the salt layer elevations is 16 m. Similarly, the salt layer is thickened towards the DS. Its thickness varies from 1–5 m at the west to 15–19 m at the east of the area (figure 4.82(b)).

2. *Salt layer porosity.* The aforementioned interpretation of the resistivities is based on Archie's Law (Archie 1942), which establishes that in a saturated medium with ionic pore water conductivity, the bulk electrical resistivity depends on the porosity, pore volume occupied by the fluid, and resistivity of the fluid filling the pore space. In this case, the bulk resistivity is expressed as:

$$\rho_x = a\rho_\mathrm{w}\phi^{-m}, \tag{4.13}$$

where ρ_w is the resistivity of the solution filling the pores, ϕ is the porosity, parameters a and m appear to depend on the cross-section, geometry of the pores along the flow path. The values of parameters a and m vary mostly within the range of 0.6 to 1.4 and 1.37 to 1.95, respectively (Frumkin *et al* 2011).

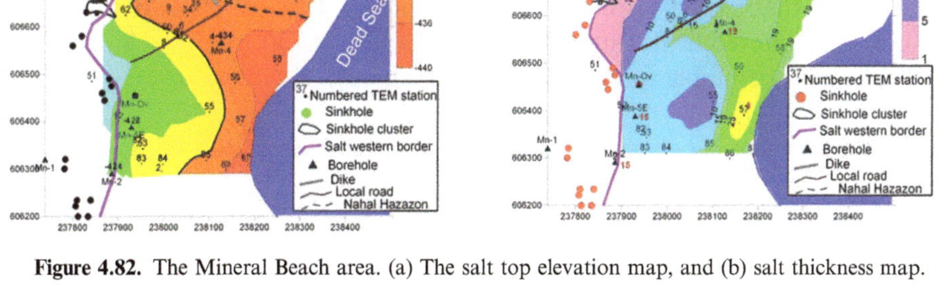

Figure 4.82. The Mineral Beach area. (a) The salt top elevation map, and (b) salt thickness map.

We presume based on the model proposed by Shalev *et al* (2006) that the porosity ϕ of the salt is laterally decreasing with a distance from the salt edge towards the DS. This is because the intensity of water circulation in pores presumably decreases in this direction. It was confirmed also by the MASW method using shear wave velocities V_s (Ezersky *et al* 2013a, and section 4.1.2 in this publication). Undersaturated water coming from outside (either from the Judean Mts aquifer or from the deeper aquifer through faults) circulating through salt pores becomes saturated concerning halite in a few days (Frumkin 1994). Consequently, fluid resistivity ρ_w will be practically constant and equal to that of the DS brine, 0.04 Ωm (Kafri and Goldman 2005). Then bulk resistivity ρ_x in equation (4.13) would therefore be a measure of porosity; higher porosity is associated with the lower bulk resistivity. Matching empirical parameters *a* and *m* to experimental data, Archie's law for DS salt can be used to estimate a porosity from the measured resistivity. Spangenberg *et al* (1998) propose the following Archie's Law for salt artificial specimens:

$$\rho_x = \rho_w \times 0.62 \times \phi^{-2.33} \tag{4.14}$$

This expression gives us a rough estimation of the salt porosity from bulk resistivity.

Ezersky and Goretsky (2014) have obtained the following ρ_x–ϕ inter-relation for the DS salt cores:

$$\rho_x = \rho_w \times 2.9 \times \phi^{-1.98} \tag{4.15}$$

TEM resistivity of the salt layer measured in the Mineral Beach site (figure 4.76(b)) varies in the range of 0.5–2.5 Ωm (Frumkin *et al* 2011). Such a porosity based on equation (4.15) would vary from 40% to 18%. However, the scale of salt cores does not correspond to the scale of salt layers. This evaluation requires additional investigation and verification.

4.3.3 Very low frequency (VLF) method

The VLF-EM technique is *a passive* method that uses radiation from ground-based military radio transmitters (used for navigation, of which there are about 42 worldwide) operating in the VLF band (15–30 kHz) as the primary EM field.

The Mottled zone is situated in the vicinity of the DS transform (DST) and exhibits intriguing rock complexes (e.g., Vapnik *et al* 2006). Within this zone, VLF-EM investigations along five profiles have been conducted (Rajab 2021). The obtained data were processed by Karous and Hjelt (1983) and other filters, and interpretation was performed using a 2D inversion of the VLF-EM data (Huang *et al* 1998).

One of the interpreted profiles is shown in figure 4.83. Analysis of this profile indicates that sub-vertical sinkholes characterized, as a rule, by high electric resistivity, will be suitable targets for the VLF survey.

4.3.4 Frequency domain electromagnetic (FDEM) method

A frequency domain electromagnetic (FDEM) method is related to the electromagnetic active remote-sensing method. The FDEM method provides a geophysical tool for measuring the apparent values of electric conductivity (EC) and magnetic

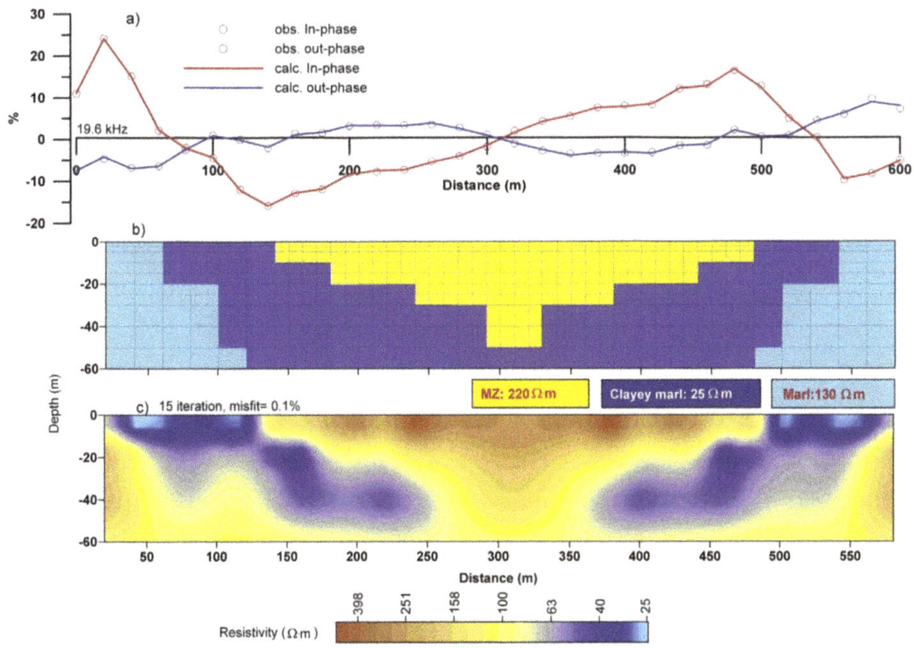

Figure 4.83. VLF-profiling and resistivity models in the Mottled zone (DS area): (a) VLF-EM data, (b) resistivity model under flat topography, (c) 2D inverted resistivity model of created synthetic VLF-EM data. (Reprinted from Rajab (2021), copyright (2021) with the permission of Elsevier.)

susceptibility (MS) of subsurface rocks, soils, and minerals (McNeill 1980a). Goldshleger *et al* (2018) have used a GEM-2 system working in the multi-frequency mode (625 Hz, 1975 Hz, 6125 Hz, 19 175 Hz, and 60 625 Hz). In the DS sinkhole problem, the FDEM limited application is known (Ronen and Beck 1997, Goldshleger *et al* 2018). The FDEM instrument provides two measurement values simultaneously—the quadrature component which is strongly dependent on the EC, and the in-phase component, which responds strongly to MS. The study was aimed to monitor the process of sinkhole development by observing the variations of EC parameters. Two EC maps in the Ein Gedi south site, which were conducted at 625 Hz are shown in figures 4.84(a) and (b), and EC distribution along the NE–SW line (c).

Sinkhole development is characterized by higher EC values. Correspondingly, the areas in maps occupied by the sinkhole development are marked by the red-pink color. Similarly, graph 2 in figure 4.84(c) indicates that in April 2012 in the northeastern part of the site the subsurface voids were developed more actively than in November 2011 (graph 1). In the center and the southwestern part of the site, the sinkhole processes were not developed.

4.4 Surface nuclear magnetic resonance (SNMR) method

Observed since 1980, the rapid development of sinkholes in the DS coastal area is linked up with the formation of karst voids in the salt rock. Results of geological and

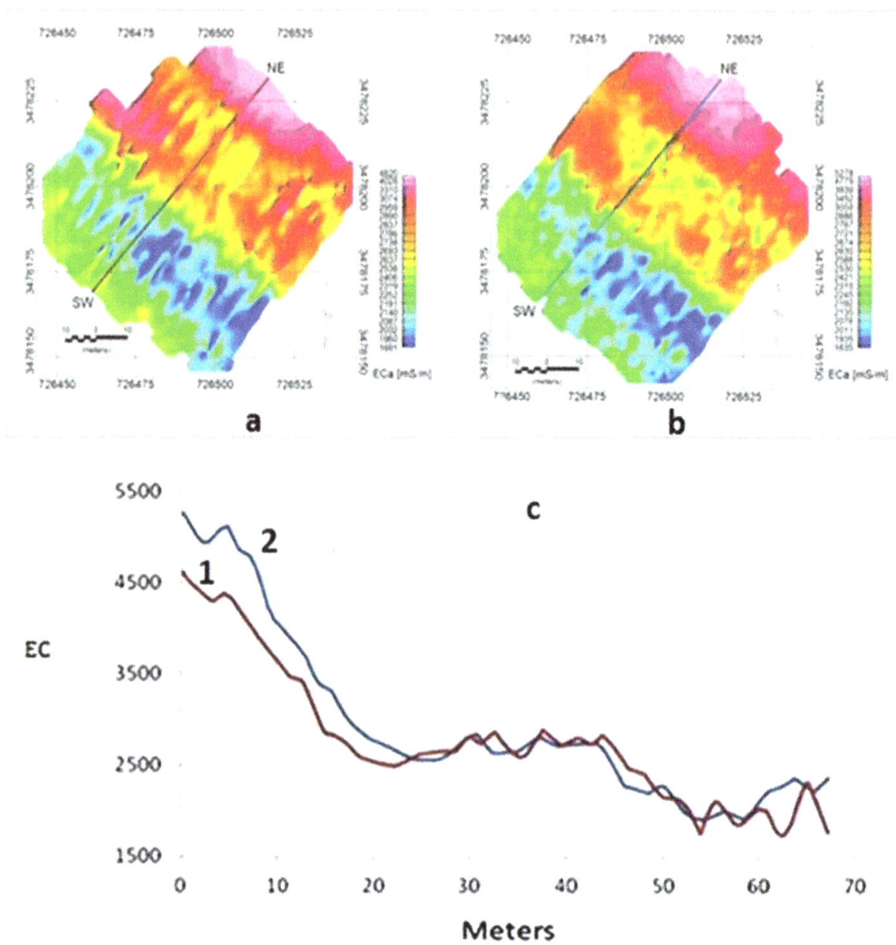

Figure 4.84. FDEM maps of the spatial variations of the EC values in the 'Ein Gedi South' site, for two measurement dates: 11 November 2011 (a), 12 April 2012 (b); (c) distribution of EC values along the NE–SW line. 1—November 2011, 2—April 2012. (Reprinted from Goldshleger *et al* (2018) with the permission of U Basson, GeoSence Ltd, with gratitude to *Journal of Remote Sensing & GIS*, an open access journal.)

geophysical investigations carried out by the Geological Survey of Israel and the Geophysical Institute of Israel revealed that the sinkhole development correlates with the drop of the DS level, which lowers the groundwater level. In the framework of the NATO project SfP N° 981128 ('Application of a new integrated geophysical approach to the prediction of the natural hazards related to the development of sinkholes caused by the drop of the Dead Sea level and human activity'), the salt karst in the Nahal Hever area (figure 4.85) was investigated with the SNMR method. In October–November 2005 and Match 2007, two SNMR surveys have been carried out.

The shallow part of the subsurface in Nahal Hever is composed of alluvial fan sediments down to a depth of 25 m and an 11 m thick salt layer below 25 m. Three boreholes were drilled in the area (figure 4.86) (Yechieli *et al* 2004).

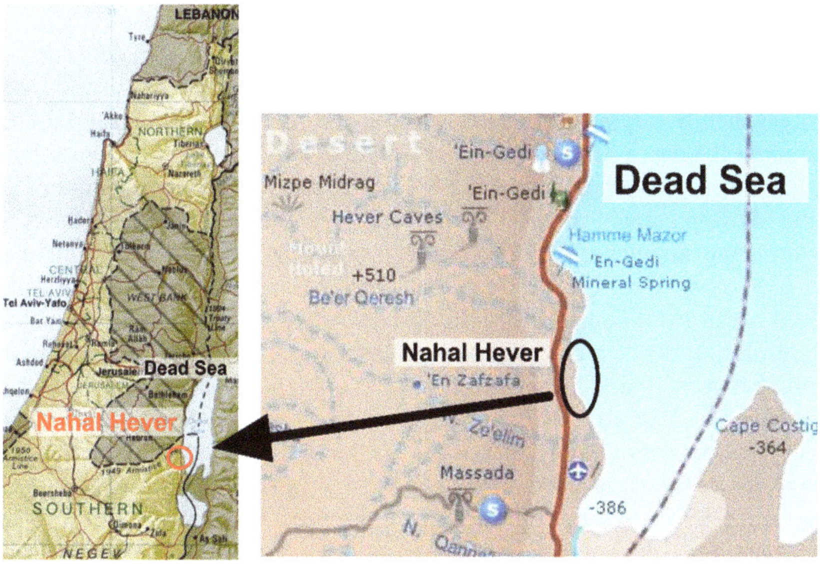

Figure 4.85. Location of the investigated area in Israel.

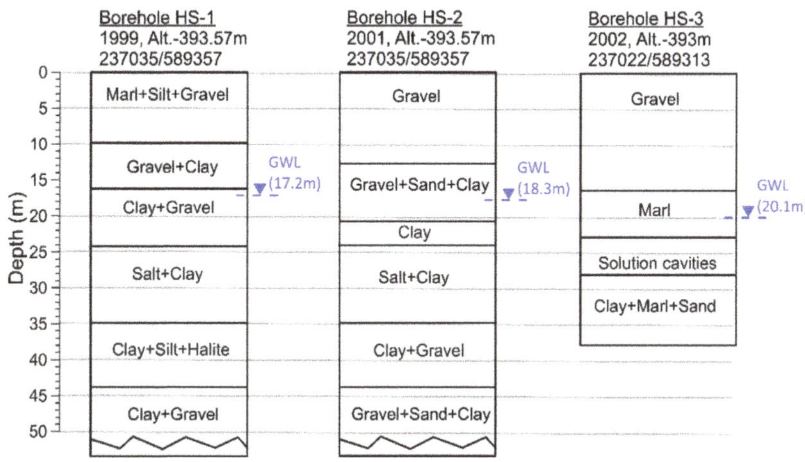

Figure 4.86. Boreholes drilled in the Nahal Hever site (reprinted from Yechieli *et al* (2004) with permission from GSI).

The boreholes HS-1 and HS-2 detected the salt layer at the depth interval of 24–35 m. In the borehole HS-3, a cavity filled with mud was detected at a depth of 24–28 m, which is possibly connected with the dissolved salt layer. The water table in the area was measured at a depth of 17.2 m in 1999 and 18.5 m in borehole HS-3.

The particularity of the DS coastal area comprises a very high electrical conductivity of the subsurface caused by the very saline groundwater. SNMR is an EM method and electrically conductive layers make an EM screen that attenuates the EM field. Under these conditions, the depth of investigation with the SNMR is reduced (Legchenko *et al* 2008c). For estimating the depth of water detection with SNMR, we use a two-layer geoelectrical model typical for the Nahal Hever site. The first layer is 24 m thick and its resistivity is variable. The resistivity may vary between 1 and 20 Ω.m depending on the water and clay content. The second layer is saturated with the saline water and the resistivity is slightly varying around 0.6 Ω.m. We assume a 1 m thick layer of bulk water and compute the depth at which this layer can be detected with SNMR. Figure 4.87 shows the normalized amplitude of the SNMR signal versus depth computed considering the above-presented model. At the depth between 40 and 50 m, the screening causes a strong attenuation of the signal, which suggests the maximum depth of investigation of 40–45 m. Thus, the depth of investigation under the DS coast conditions is reduced about twice comparing with the expected 80–100 m (Legchenko 2013).

The resolution of the method in Nahal Hever was estimated assuming an aquifer occurring between 20 and 80 m with the water content of 10% and the resistivity of the first layer of 1, 5, and 20 Ω.m. Figure 4.88(a) shows the model of the water content (gray rectangle) and results of the inversion. Inversion was carried out using the Tikhonov regularization method (Legchenko and Shushakov 1998) and taking into account the resistivity of the subsurface. Figure 4.88(b) shows the amplitude of the SNMR signal (circles) and the inversion fit (solid lines) versus pulse moment. Despite a strong attenuation of the signal caused by the conductivity of the first layer, the model was well resolved down to 30 m. The bottom of the aquifer was not detected because of the limited depth of investigation.

Under the DS coast conditions, knowledge of the electrical resistivity of the subsurface is necessary for the interpretation of SNMR measurements (Legchenko *et al* 2008c, Lehmann-Horn *et al* 2011). Figure 4.89 shows an example of erroneous

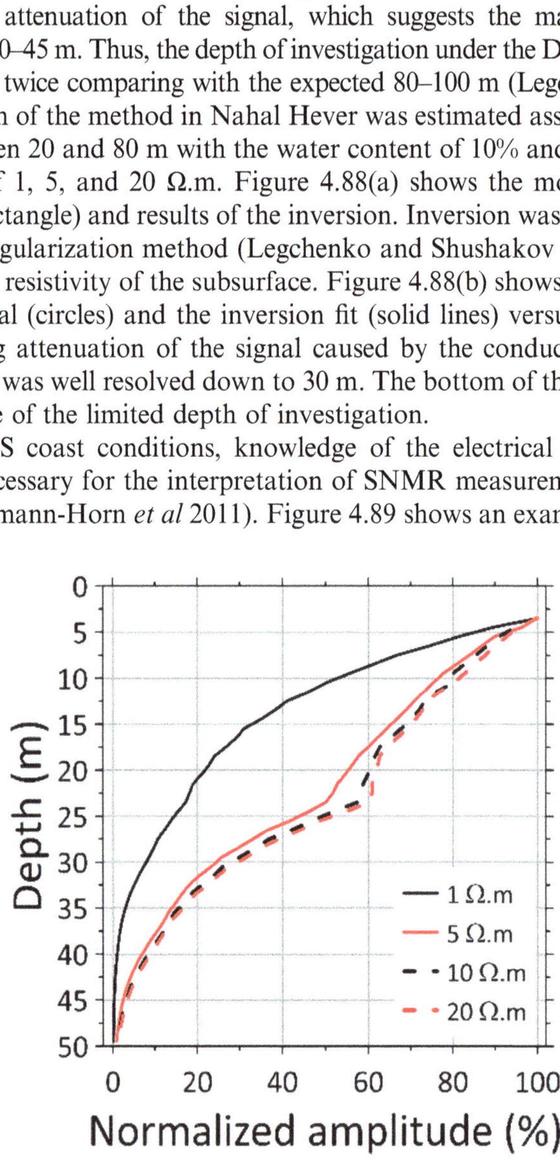

Figure 4.87. Normalized amplitude of the SNMR signal computed considering a 1-m-thick layer of the bulk water versus depth computed assuming a variable electrical resistivity of the unsaturated sediments (0–24 m) overlying a deeper layer (24–80 m) with the resistivity of 0.6 Ωm.

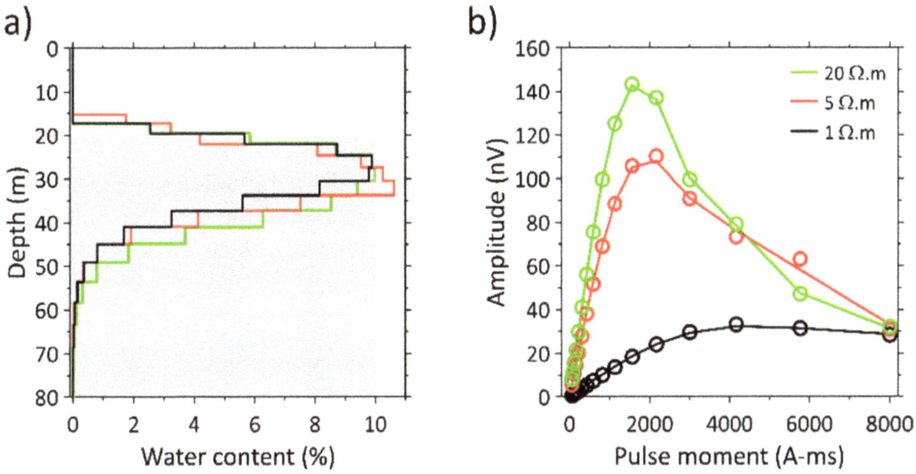

Figure 4.88. A 1D inversion of the synthetic data sets computed considering a water-saturated layer between 20 and 80 m (w = 10%) with the variable electrical resistivity of the unsaturated sediments between 0 and 24 m and a very conductive deep layer of 0.6 Ωm between 24 and 80 m: (a) the water content versus depth and the water content model (gray rectangle); (b) synthetic amplitudes (circles) and the inversion fit (solid lines) versus pulse moment.

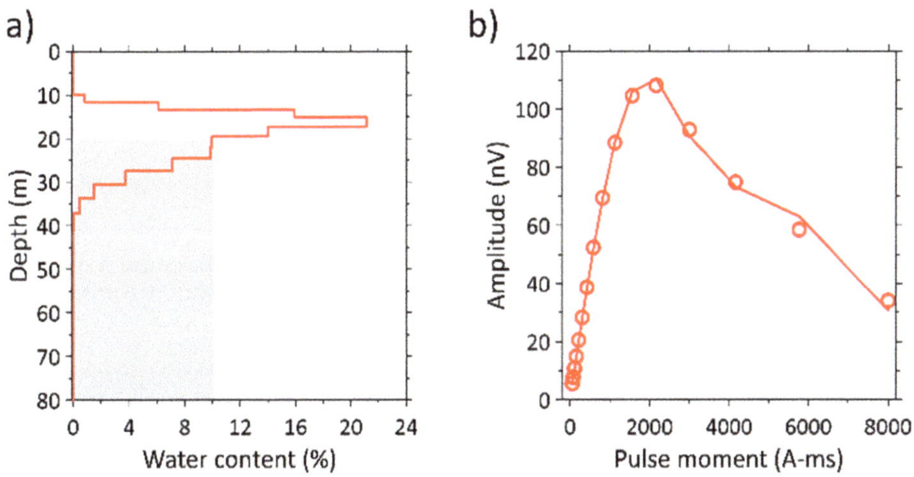

Figure 4.89. A 1D inversion of the synthetic data sets computed considering a water-saturated layer between 20 and 80 m (w = 10%). The electrical resistivity of the unsaturated sediments was assumed at 5 Ωm (0–24 m) and a very conductive deep layer of 0.6 Ωm (24–80 m). (a) The water content versus depth shown by the inversion and the water content model (gray rectangle); (b) synthetic amplitudes (circles) and the inversion fit (solid lines) versus pulse moment. Inversion was performed using the intentionally erroneous geoelectrical model of the subsurface, which comprises the resistivity of 1 Ωm instead of the true value of 5 Ωm.

inversion caused by an erroneous electrical resistivity used for the inversion. The water content profile provided by the inversion (red line) and the model (gray rectangle) shown in figure 4.89(a) is not consistent when the resistivity was erroneous. Note that the synthetic signal (figure 4.89(b)) was fitted well by the theoretical signal computed after the inverse model.

Figure 4.90 displays the position of the SNMR loops and boreholes in the Nahal Hever south area. We used a NUMIS^{plus} system developed by IRIS Instruments (France) with a 100 × 100 m square loop. At each SNMR station, a TEM measurement was done. TEM soundings were carried out using the TEM FAST 48 HPC system manufactured by AEMR (Netherlands).

Around the area investigated by the use of SNMR methodology, we observed a strong variation of both the signal amplitude and the relaxation time T_1. Figure 4.91 illustrates two examples of the aforementioned investigations. Parameters of the

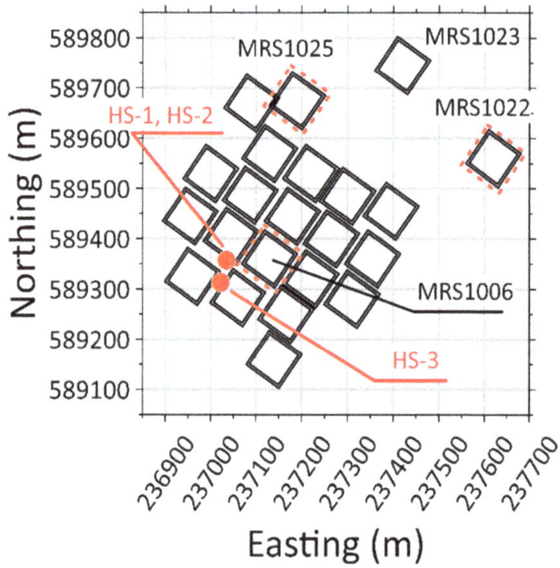

Figure 4.90. Position of the SNMR loops (squares) and the boreholes in Nahal Hever (red dots). The red dashed lines show two stations with the maximum (MRS1025) and the minimum (MRS1022) SNMR signals, and the MRS1006 station investigated in 2005 and 2006.

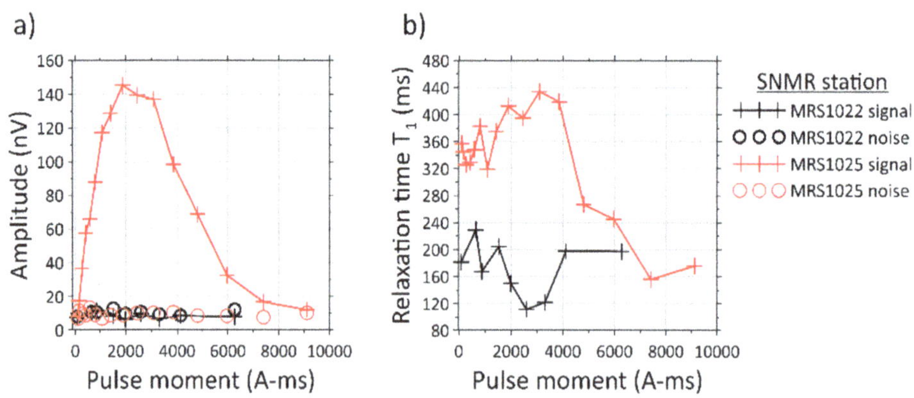

Figure 4.91. The amplitude and the relaxation time versus pulse moment were measured at two SNMR stations with the largest (MRS1025) and the smallest (MRS1022) SNMR signals.

SNMR signal measured with the MRS1025 sounding point to an aquifer. The relaxation time T_1 is long and may correspond to a coarse sand or gravel formation (Seevers 1966, Dunn *et al* 2002) but also to karst voids in the subsurface (Legchenko *et al* 2008b). Measurements corresponding to the MRS1022 station reveal a low permeable subsurface, for example, a clay layer or a compact salt rock. SNMR does not show water-filled karst at this location.

The amplitude of the SNMR signal versus pulse moment (see for example figure 4.91(a)) has a half sinus-wave shape and we use the maximum amplitude as a parameter for characterizing the signal amplitude. Analysis of raw data not affected by heavy processing is a preliminary step before performing an inversion. Figure 4.92(a) shows variations of the signal amplitude throughout the investigated area. The relaxation time T_1 corresponding to the aquifer is shown in figure 4.92(b). Both the amplitude and the relaxation time point to an aquifer in the southern part of the area.

Figure 4.93 illustrates a comparison of the SNMR inversion with the lithological log in the borehole HS-2. An aquifer location observed with the SNMR corresponds to the salt layer revealed by the borehole. The relaxation time is long and hence salt is hydraulically permeable. The water content is low for a continuous aquifer formation and may correspond to water-filled voids in salt. This observation corroborates well with a dissolved salt found in the borehole HS-3 (figure 4.86).

The subsurface in Nahal Hever is heterogeneous and for delineating aquifer formations we use a 3D inversion of the SNMR data (Legchenko *et al* 2011, Chevalier *et al* 2014). Figure 4.90 shows the position of SNMR stations used for the 3D inversion. We exclude the data to set two distant soundings (MRS1022 and MRS1023) because of their remote location. Figure 4.94 indicates the amplitude of the SNMR signal versus the pulse moment recorded at each SNMR station. The normalized pulse moment of each sounding is a set between the station numbers. For example, the amplitude measured at station number five is shown by the red dots between the sounding numbers of five and six. The black line shows the

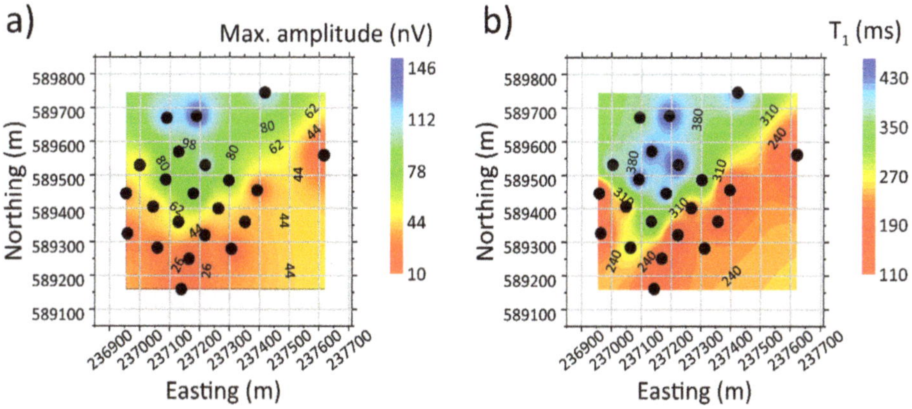

Figure 4.92. Parameters of the SNMR signal measured in the Nahal Hever site: (a) map of the maximum amplitude; (b) map of the relaxation time T_1. Black dots show the positions of the SNMR loops.

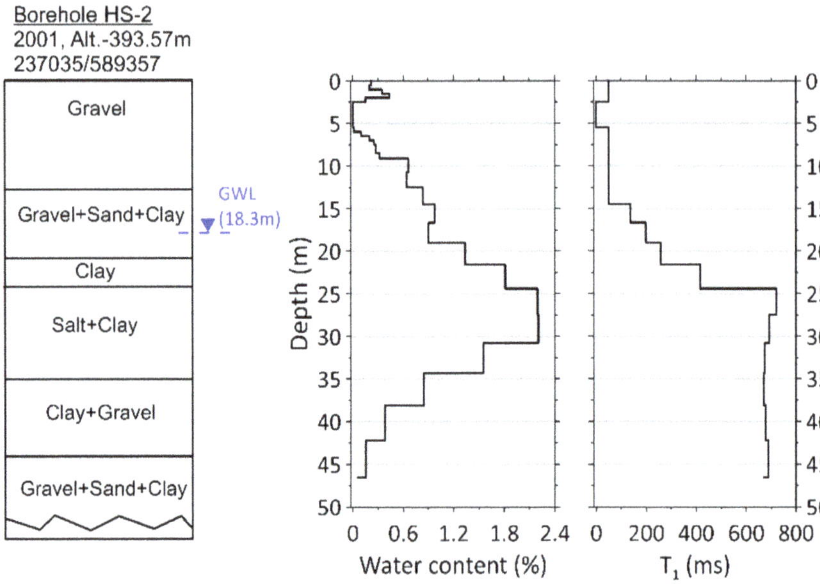

Figure 4.93. A 1D inversion of SNMR sounding made in 2005 around the boreholes HS-1 and HS-2: (a) the borehole log; (b) SNMR water content; (c) SNMR relaxation time T_1. Note that the groundwater level in 2005 (SNMR measurement time) is about 4 m lower than that measured in 2001 (drilling time).

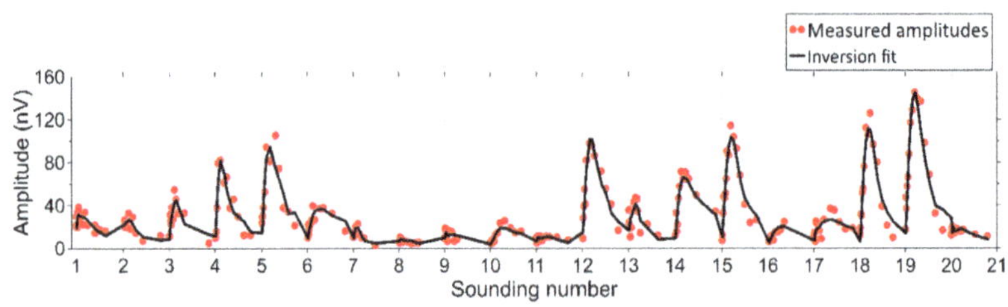

Figure 4.94. Measured amplitude of the SNMR signal versus normalized pulse moment (red dots) and the inversion fit corresponding to the 3D inverse model (black line). The amplitudes corresponding to each sounding are plotted between the sounding numbers along with the profile. For example, data corresponding to the sounding No. 1 is plotted between 1 and 2.

theoretical signal computed after the 3D inverse model. A good correspondence between the experimental and theoretical amplitudes makes sure that the inverse model is the correct one.

Figure 4.95 displays a horizontal slice derived from the 3D inverse model. The altitude of −420 m is close to the static water level in the Nahal Hever site. The uncertainty of the inversion is 40–50 m (Legchenko *et al* 2011). The aquifer may be composed of coarse sand or gravel with intercalations of less permeable sediments like silt or clay. Water in the karst channels and voids developed in a salt formation also may produce a similar SNMR response. The SNMR cannot distinguish between these two interpretations. SNMR analyses indicate that the boreholes

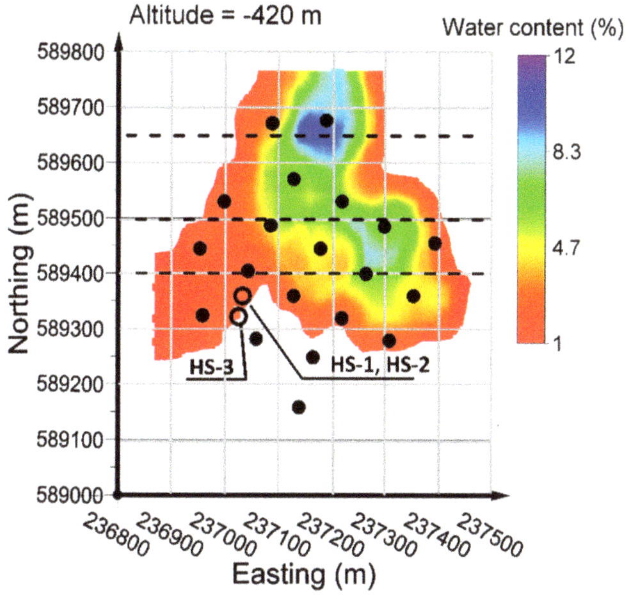

Figure 4.95. The water content in a horizontal slice at the altitude of −420 m. The black dots show SNMR stations, black circles the boreholes and the black dashed lines show the position of three vertical cross-sections.

drilled in 1999–2001 are outside of the aquifer formation. In the southern part of the area, the 3D inversion shows a narrow band corresponding to the aquifer (figure 4.95). However, after figure 4.92(a), the amplitude and the relaxation time measured with the MRS1023 sounding (S–E part of the area) correspond to the aquifer formation. An insufficient sampling along the eastern border of the area and the consequent exclusion from the MRS1023 data of the 3D data set to cause this underestimation of the aquifer extension. The black dashed lines show the position of three vertical cross-sections.

Figure 4.96 shows the vertical cross-sections of the water content oriented in the E–W direction. The water level was not measured, but estimated relative to the known DS level. SNMR shows a high water content above the static water level being −418 m in 2007. This happens because SNMR detects water both in the saturated and unsaturated sediments and does not show the static water level. For example, one can see the water content profile presented in figure 4.93. In the Nahal Hever site, the subsurface is very heterogeneous, and approximate measurements of the resistivity using a 100 × 100 m square loop, and 1D TEM inversion may cause errors in the SNMR inversion results. For example, figure 4.89 displays that an inaccurate resistivity used for the inversion leads to errors in the aquifer geometry. The depth of investigation with the SNMR in the Nahal Hever site is limited by the first 35–40 m, which limits the capacity of SNMR to accurately detect the bottom of the aquifer.

SNMR measurements were done in 2007 in the areas affected by ground instability. Several sinkholes were visible at the surface. More sinkholes appeared

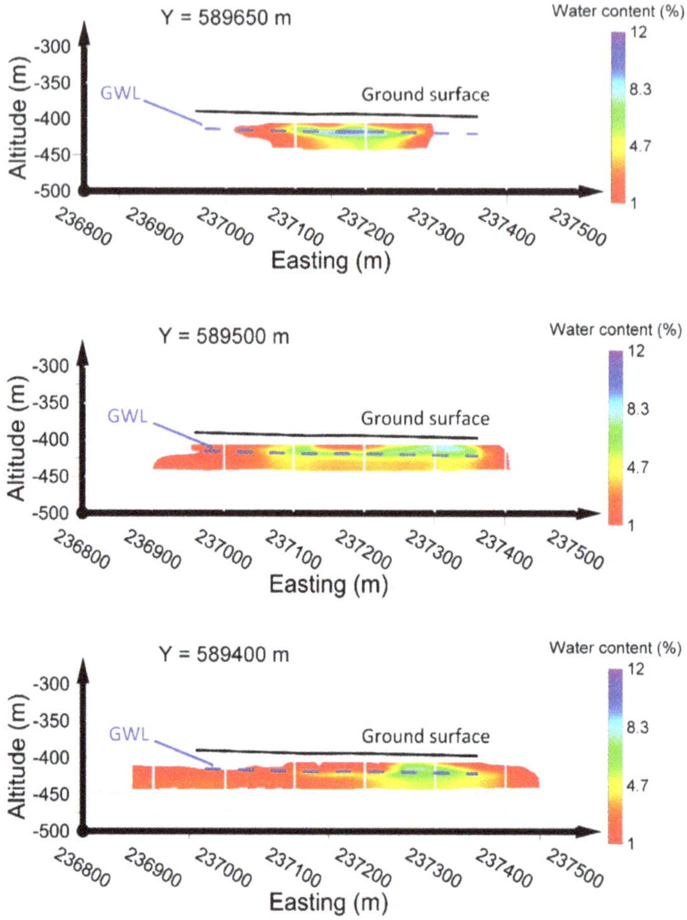

Figure 4.96. Three vertical cross-sections of the water content in the E–W direction.

a few years later. Figure 4.97 shows the location of sinkholes observed in 2008 (blue dots). The red squares show sinkholes that appeared between 2008 and 2015. Figure 4.98 summarizes the number of sinkholes that appeared before 2008 and between 2008 and 2015 (Ezersky *et al* 2017). In figure 4.97, all these sinkholes are shown over the aquifer observed with the SNMR.

A large number of sinkholes occupy the southern and western borders of the aquifer formation. In figure 4.97, the old sinkholes lie over the low-permeable zone (blue dots) outside the aquifer. However, one should take into consideration that the ground collapse causes refilling voids in the subsurface with fine sediments (silt, clay) causes a lowering the hydraulic conductivity (Legchenko *et al* 2008b). This fact suggests that some ten or twenty years before the SNMR survey, the southern part of the aquifer moved north. Sinkholes in the southern part of the area appeared in a very permeable formation (blue color) that may correspond to the presence of karst voids in salt. During the SNMR survey in 2007, sinkholes were not visible in this zone.

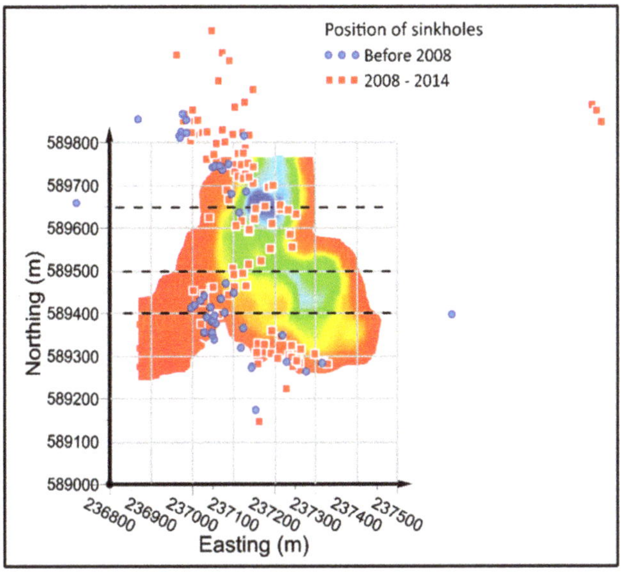

Figure 4.97. Position of sinkholes observed in Nahal Hever before 2008 (blue dots) and between 2008 and 2015 (red squares) overlying the aquifer location shown by SNMR.

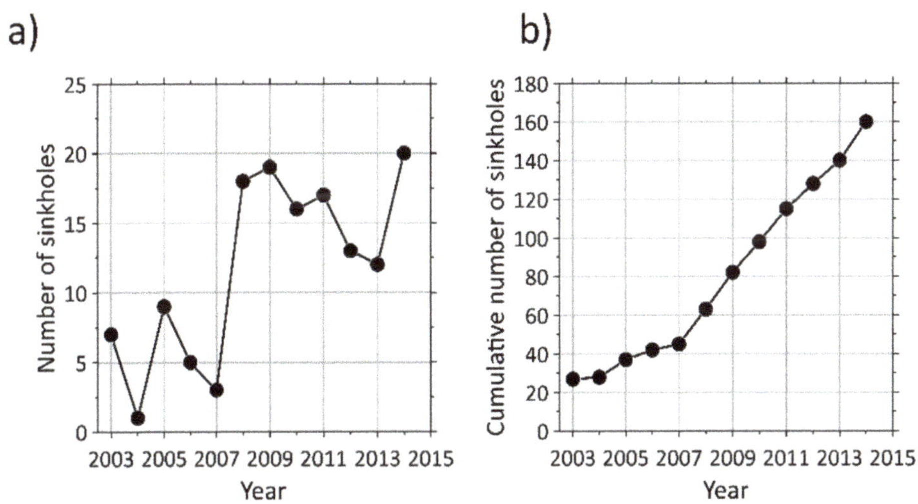

Figure 4.98. A number of sinkholes versus the year of appearance: (a) the annual records; (b) the cumulative number of sinkholes.

The sinkhole development was triggered by the drop of the DS level, but the mechanism itself has different explanations. One of the popular models assumes a rapid dissolution of the salt rock by the intrusion of freshwater caused by the lowering of the water table (Yechieli *et al* 2006). However, the salt layer observed in boreholes in 1999 was protected by 7 m of saltwater, and the dissolution should be a slow process. Sinkholes appeared before the water table become at the level of the salt formation and hence, and the salt was already affected by the karst development. Another model assumes that the cavities in salt have existed for a long time and just the lowering of the groundwater level caused subsidence (Legchenko *et al* 2008a). The SNMR survey in the Nahal Hever site indicates that the sinkholes phenomenon is linked up with the water circulation in the subsurface. SNMR results shown in figure 4.97 suggest that the sinkholes developed along the border of the aquifer. Then, a horizontal movement of the sediments and surface erosion expanded the damage in a larger area. This process was sped up by surface water that dissolves salt particles mixed up with the sediments. Salt cemented sand, clay, and gravel, but water can dissolve such an unconsolidated material much faster than salt rock.

4.5 Microgravity

4.5.1 Field microgravity surveys

The field microgravity surveys were carried out in three sites of the DS coastal areas (figure 4.99).

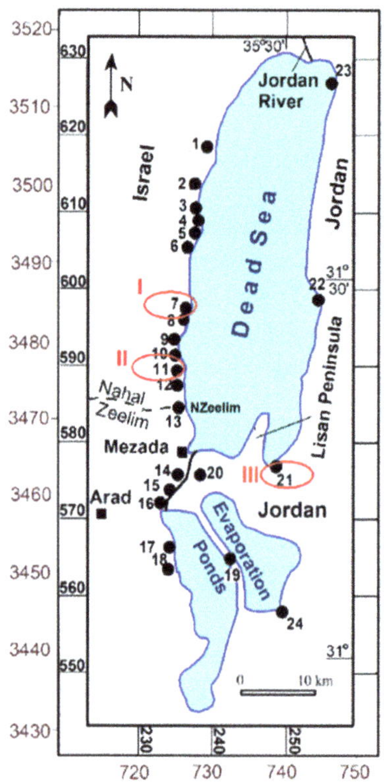

Figure 4.99. Location map of the microgravity study sites. Black points show the sinkhole sites. Red ellipses display the study sites: 7—Ein Geddy (Site I); 11—Nahal Hever south (Site II); 21—Ghor Al-Haditha (Site III). The complete list of sinkhole sites is shown in figure 4.1 in section 4.1 (the coordinates in the frame are the new Israel Mercator grid (km), the outer coordinates are the Universal Transverse Mercator grid (km)). *Nahal Hever South* (figure 4.99, Site II).

To reveal the potential of the microgravity method, the feasibility study was carried out throughout the Nahal Hever South sinkhole development area (Rybakov *et al* 2001, Eppelbaum *et al* 2008). About 5000 gravity stations were observed in this area by the use of a Scintrex CG-3M AutoGrav gravity meter. Microgravity mapping of 200×200 m^2 area (with the observation grid of 3×3 m^2) was carried out in 1999 after the first sinkholes appeared. The location of the gravity observation points in the Nahal Hever South area is displayed in figure 4.100(a). All necessary corrections (including complete terrain correction) were inserted (Rybakov *et al* 2001). The residual microgravity anomaly of –(0.8–0.14) mGal, extending in the north–south direction was delineated in the area. The anomaly consists of three sections with sizes from about 20×20 m^2 to 50×50 m^2.

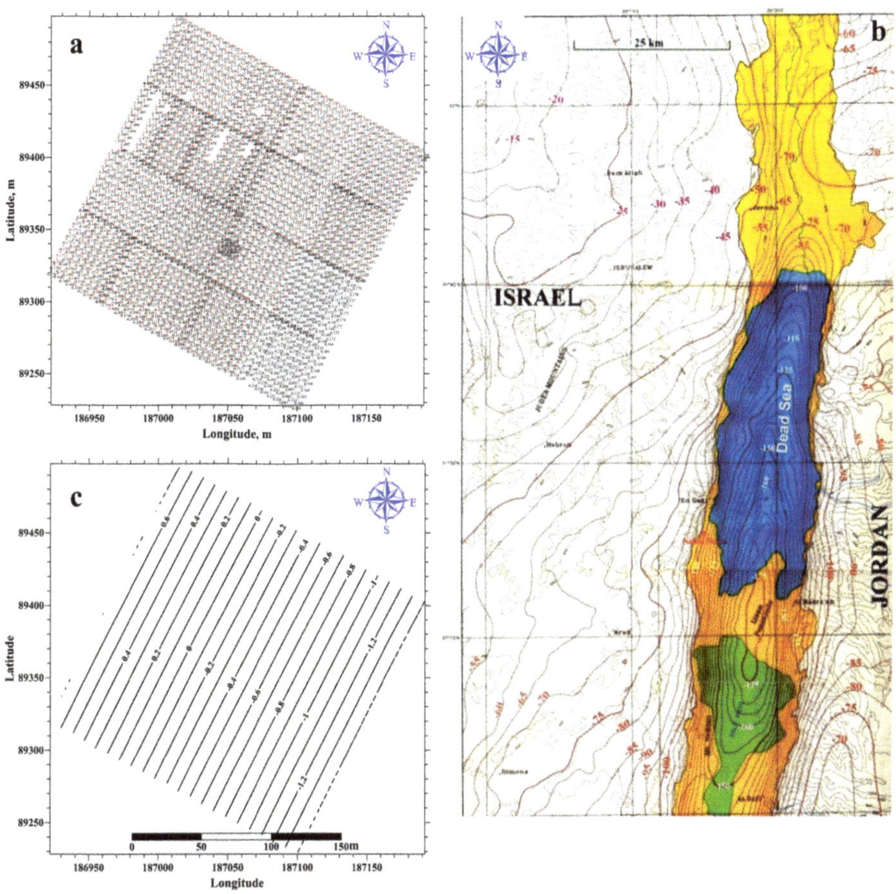

Figure 4.100. Nahal Hever South area. (a) Location map of gravity points. (b) Regional gravity map in the Bouguer reduction with a topographic (bathymetric) map of the studied area. (Reprinted from ten Brink *et al* (1993), with the permission of the American Geophysical Union.) (c) Scheme of the residual regional horizontal gradient of Δg_B in the studied area (compiled using Rybakov (2007) data; reprinted from Eppelbaum et al (2008) with permission from *Advances in Geosciences*).

These investigations were carried out in the zone of the great regional horizontal gradient ($\Delta g_{B(RHG)}$) (up to 10 mGal km^{-1}) caused by the influence of the regional DS gravity minimum (figure 4.100(b)). The values of the $\Delta g_{B(RHG)}$ were polynomial computed and subtracted from the observed Δg_B map (Rybakov 2007, Eppelbaum *et al* 2008). The residual scheme of the horizontal Δg_B gradient for the Nahal Hever South area is presented in figure 4.100(c). As shown in this figure, $\Delta g_{B(RHG)}$ reaches high values for such a small area.

The residual gravity anomaly map of 1999 is shown in figure 4.101. The combination of the residual gravity with the site topography (white contours in figure 4.101) suggests that the main gravity minimums are superimposed by the regional trend (third-order polynomial). Because the gravity data are terrain corrected, both subsidence and the gravity minimum can be caused by the existence of the underground cavities.

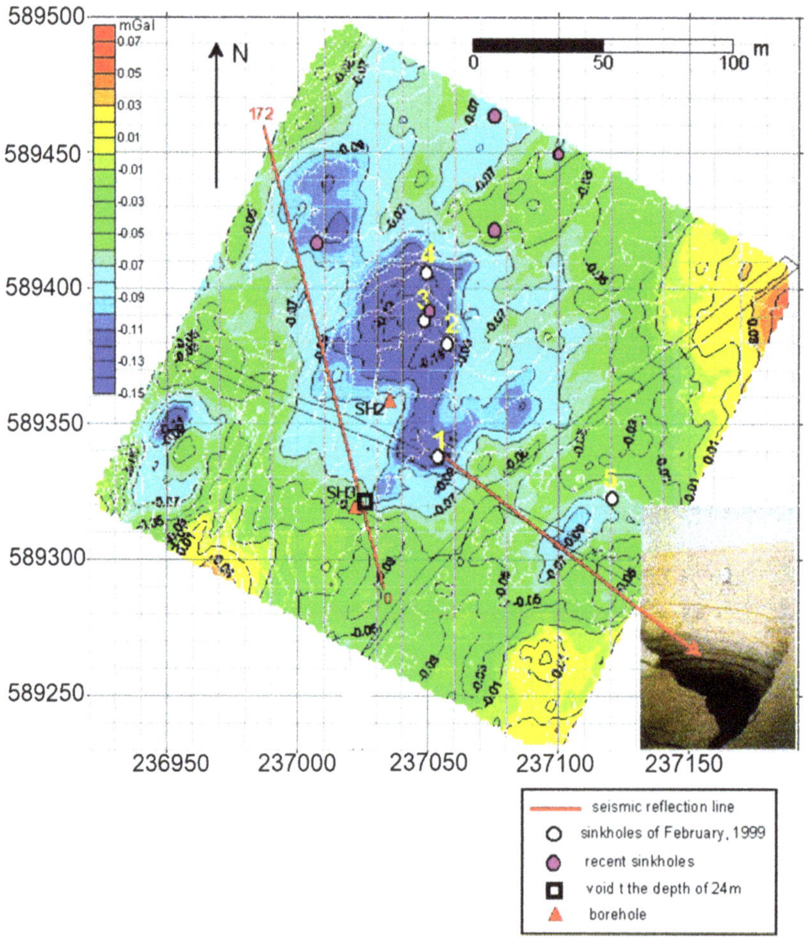

Figure 4.101. Residual gravity anomaly map of the Nahal Hever South (western DS coast) superimposed by detailed topography (white contours −0.25 m interval). Photo shows sinkhole No. 1. (Reprinted from Rybakov *et al* (2001), with the permission of SEG) *Ein Gedi (EG)* (figure 4.99, Site I.)

Ein Gedi–Nahal Arugot is one of the earlier sinkhole sites mostly studied using different geophysical methods (Ezersky 2002b, 2003a, 2003b). The microgravity study was carried out there by Rybakov (2003, 2005). The data acquisition and processing were described above. The microgravity Bouguer map shows the regional trend decreasing significantly eastward (figure 4.100(c)). Removing the trend allowed generating the residual gravity map of the Ein Gedi site presented in figure 4.102. The elongated negative anomaly extended in the north direction occupies most of the western part of the site, and the noticeable negative anomaly is located in the SE corner (figure 4.102). This anomaly seems to include almost all the sinkholes which have opened up in the study area. The microgravity shows a positive anomaly along with highway #90 caused by the fact that microgravity surveys were performed during and after industrial reconstruction (dynamic compaction). Rybakov (2005) notes that along this part of the road, the gravity measurements were repeated at monthly

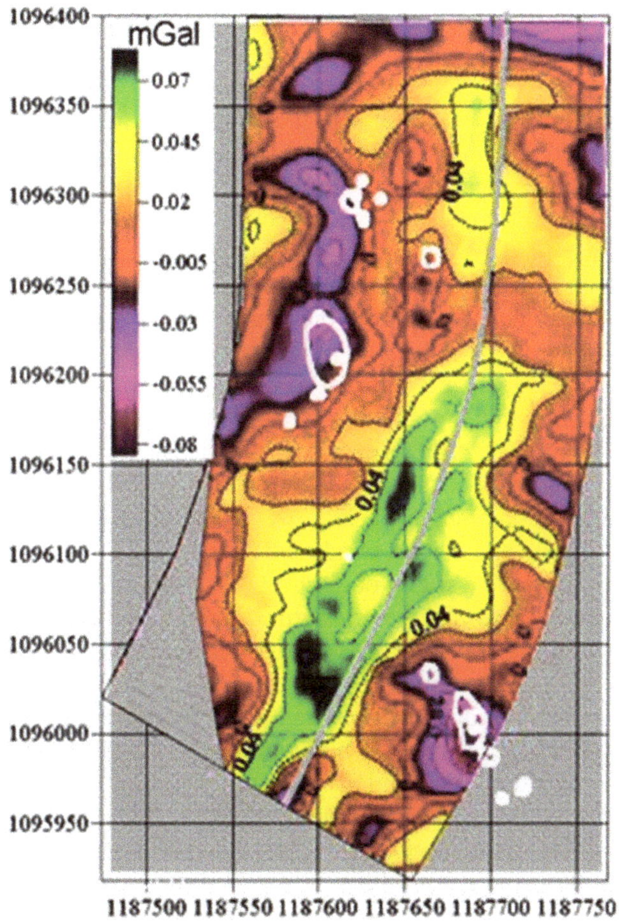

Figure 4.102. Residual microgravity map of the Ein Gedi site. The contour interval is 0.02 mGal. White polygons denote the open and filled sinkholes. (Reprinted from Rybakov (2005), with the permission of EAGE.)

intervals during 1999. The monitoring suggested a decrease in the gravity values, which took place between April and May of 1999 and was interpreted as being caused by the development of a concealed cavern below this section of the main road. Indeed, the ground was collapsed here in 2002.

4.5.2 Features of the microgravity modeling in the salt karst areas

In the areas with developed salt tectonics, the karst targets are often associated with the salt bodies. The model example (figure 4.103) illustrates a well-known (e.g., Veselov and Sagitov 1968) very simple, but effective calculation of the second horizontal derivative of gravity potential over the sub-horizontal body with the contrast density properties. The graph of the second horizontal derivative of gravity potential may be interpreted by the use of methods developed in magnetic prospecting (Eppelbaum 2011b).

A simple PGM presented in figure 4.103 indicates that from analysis of the gravity field behavior (figure 4.103(A)) it is difficult to recognize where the left end of the salt layer occurs (under conditions of gravity anomaly superposition revealing this point will be more complicated). At the same time, the calculation of the second horizontal derivative of gravity potential W_{xz} (figure 4.103(B)) allows determining exactly this point as a projection of W_{xz} max value (figure 4.103(C)). Besides this, W_{xz} anomaly could be quantitatively interpreted by the use of methodologies developed for complex environments in magnetic prospecting (Eppelbaum 2011b).

Figure 4.104 shows a more complex example of 3D gravity field computations.

Determination of the right end of salt body occurrence is no less important a problem. Five variants of gravity field were computed here (figure 4.104(A)): (1) at infinity, (2) at 300 m (within the limits of the considered model), (3) at 350 m, (4) at 400 m, and (5) at 700 m (as in figure 4.103). The difference between the behavior of the curve in the right part of the section is obvious. Similarly, it is interesting that a significant difference between Δg_B (infinity) and Δg_B (700 m) over the left end of the anomalous body was found (figure 4.104(B)). Some small disagreements were discovered and for the differences anomalies between Δg_B (700 m) and Δg_B (300, 350, and 400 m).

Analytical continuation of gravity field to the lower semi-space (underground) should underline the anomalous effects from small geological targets. However, it is necessary to take into account the possible arising of disturbances of two kinds: (1) noise components in the observed gravity field may be transformed to significant fictitious anomalies, (2) possible erroneous calculation of the downward continuation below the upper edge of the anomalous body may initiate an effect of the 'field destruction' (Strakhov 1976).

Among the majority of downward continuation methods for potential fields, we will shortly consider two: (1) method based on the Gauss' theorem of the averaged value of the harmonic function, and (2) Strakhov's (1976) method.

(1) Procedure of downward continuation of the gravity field, based on the known Gauss theorem of the averaged value of the harmonic function (e.g., Blakely 1995):

$$U(0) = \frac{1}{4\pi R^2} \int_s U(M)\, dS, \quad M \in S, \tag{4.16}$$

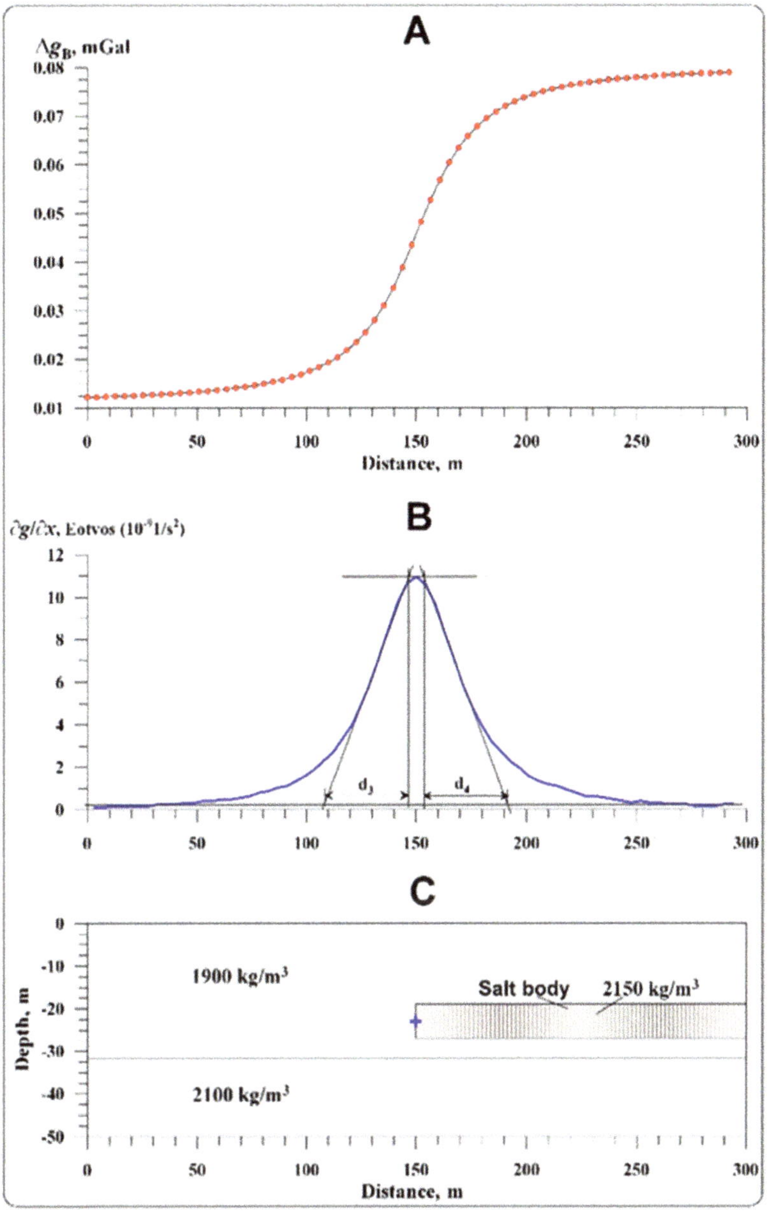

Figure 4.103. Computation of second horizontal derivative of gravity potential (W_{xz}, or ∂_g/∂_x). (A) Computed gravity curve, (B) calculated horizontal derivative of gravity field, (C) physical–geological model (PGM) (right end of the salt body was computed as occurring at 700 m). Symbol '+' indicates the determined position of the half of the vertical thickness amplitude of the salt body (modified after Eppelbaum *et al* (2008)). (Reprinted from Eppelbaum *et al* (2008) with permission of EGU, copyright the Authors 2008. This work is distributed under the Creative Commons Attribution 3.0 License.)

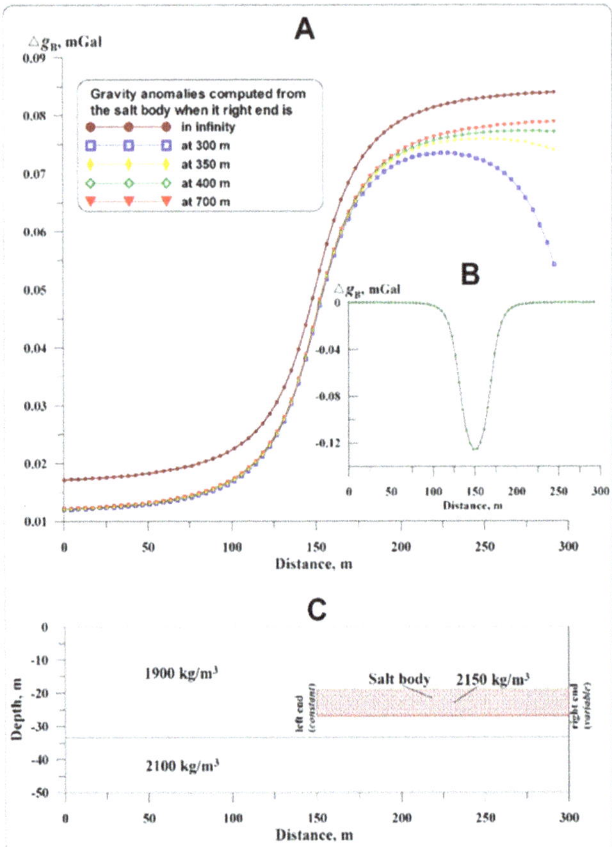

Figure 4.104. The computing of gravity effects from the model of the salt body by different positions of the right end (buttoned) of the body. (A) Computed gravity anomalies for different positions of the salt body right end. (B) Computed difference between Δg_B (infinity) and Δg_B (700 m). (C) PGM. (Reprinted from Eppelbaum *et al* (2008) with permission of EGU, modified. Copyright the Authors 2008. This work is distributed under the Creative Commons Attribution 3.0 License.)

where S is the sphere, R is the radius of the sphere, point 0 is the sphere's center and M is the point located at a sphere.

If we select six points at the sphere with radius $R = h$, we can write the following equation for the downward continuation of the gravity field at some level $-h$:

$$\Delta g(0,0,-h) = \left\{ \begin{array}{c} 6\Delta g(0,0,0) - \Delta g(0,h,0) - \Delta g(h,0,0) \\ -\Delta g(-h,0,0) - \Delta g(0,0,h) - \Delta g(0,-h,0) \end{array} \right\}, \tag{4.17}$$

where $\Delta g(0,0,h)$ is the observation at the level h over the Earth's surface. We can receive this value, for instance, by (1) the direct measurements, or by (2) the upward continuation of gravity data.

The equality is more exacting with the sphere radius decreasing (thus, this procedure is preferable for precise gravity investigations). An example of such downward continuation to the levels of 0.7 and 1.0 m below the Earth's surface is shown in figure 4.105. It is obvious that the gravity field intensity increases by 30% and 50%, respectively, comparing with the surface (0.3 m over the Earth's surface) computations.

(2) Strakhov's scheme of downward continuation (2D) consists of applying the following expression:

$$\Delta g(0, -h) = \left\{ \begin{array}{c} 7.303\Delta g(0,0) - 2.326\,[\Delta g(0.5h, 0) + \Delta g(-0.5h, 0)] \\ -\,0.568\,[\Delta g(h, 0) + \Delta g(-h, 0)] - 0.193\,[\Delta g(2h, 0) + \Delta g(-2h, 0)] \\ +\,0.018\,[g(3h, 0) + \Delta g(-3h, 0)] - 0.041\,[\Delta g(6h, 0) + \Delta g(-6h, 0)] \\ -\,0.003\,[\Delta g(9h, 0) + \Delta g(-9h, 0)] \end{array} \right\} \quad (4.18)$$

,

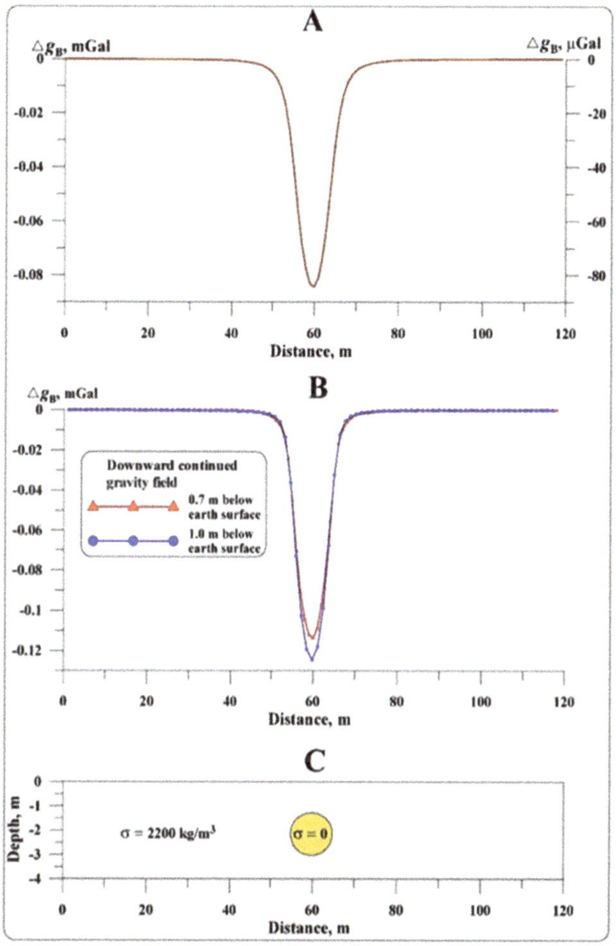

Figure 4.105. Downward continuation of gravity field using the theorem of the averaged value of the harmonic function (3D): (A) gravity field observed at the level of 0.3 m over the Earth's surface, (B) gravity field downward continued to levels of 0.7 and 1.0 m below the Earth's surface, (C) PGM. (Reprinted from Eppelbaum *et al* (2008) with permission of EGU, copyright the Authors 2008. This work is distributed under the Creative Commons Attribution 3.0 License.)

where h is the depth of the downward continuation (and simultaneously length of horizontal intervals where Δg values are selecting for calculation of expression (4.18)).

The preference of Strakhov's method is that the employment of this procedure does not demand any additional data besides the horizontal gravity observations along with the studied profile. However, it is a 2D procedure with known imperfections. Figure 4.106 shows a comparison of Strakhov's and 'averaged value' methods computed for more deeply occurring targets (comparing with the previous example). It is not hard to see that the amplitudes of the computed curves (figure 4.106(B)) differ for some values. At the same time, it is necessary to underline that the application of these procedures gives only localization and some qualitative estimation of karst terranes and their classification.

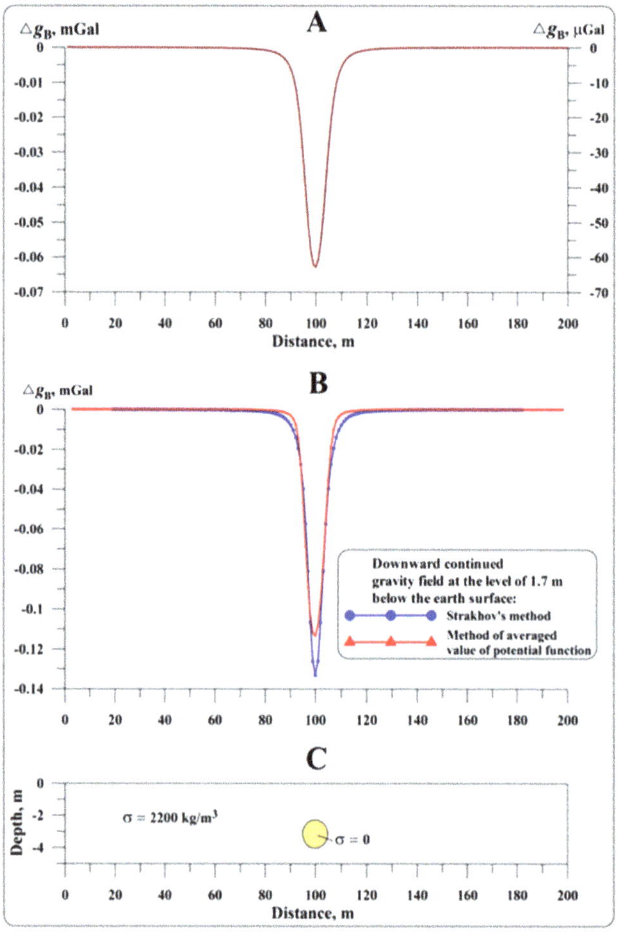

Figure 4.106. Comparison of two procedures of gravity field downward continuation: (A) Gravity field computed at the level of 0.3 m over the Earth's surface, (B) Gravity field continued downward to the level of −1.7 m below Earth surface using (1) the Strakhov's method (2D), and (2) averaged value of the harmonic function (3D), (C) PGM. (Reprinted from Eppelbaum *et al* (2008) with permission of EGU, copyright the Authors 2008. This work is distributed under the Creative Commons Attribution 3.0 License.)

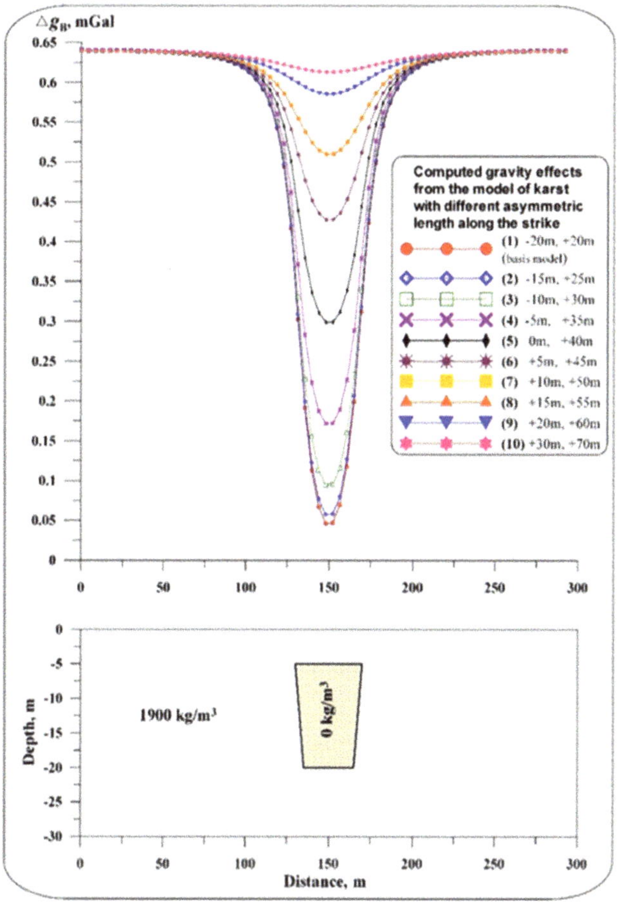

Figure 4.107. Computation of gravity anomalies from the karstic body with different asymmetric positions of the body ends along the strike. (Reprinted from Eppelbaum *et al* (2008) with permission of EGU, copyright the Authors 2008. This work is distributed under the Creative Commons Attribution 3.0 License.)

Figure 4.107 shows gravity effects from the underground cave by different asymmetric locations of the body ends (buttends) along the strike. Analysis of these effects indicates that gravity anomalies computed from the various models range in the wide diapason and the application of 3D modeling for investigating such hidden objects is highly desirable.

4.5.3 Examples of the gravity field modeling in the Dead Sea area

Gravity data examinations were successfully applied for delineation of salt layers in several areas in complex conditions of the DS western coast (Rybakov *et al* 2001, Eppelbaum *et al* 2008, Ezersky *et al* 2010, 2013b). Several prospective methodologies for gravity field transformations in complex physical–geological environments were tested on the eastern DS coast (Al-Zoubi *et al* 2013).

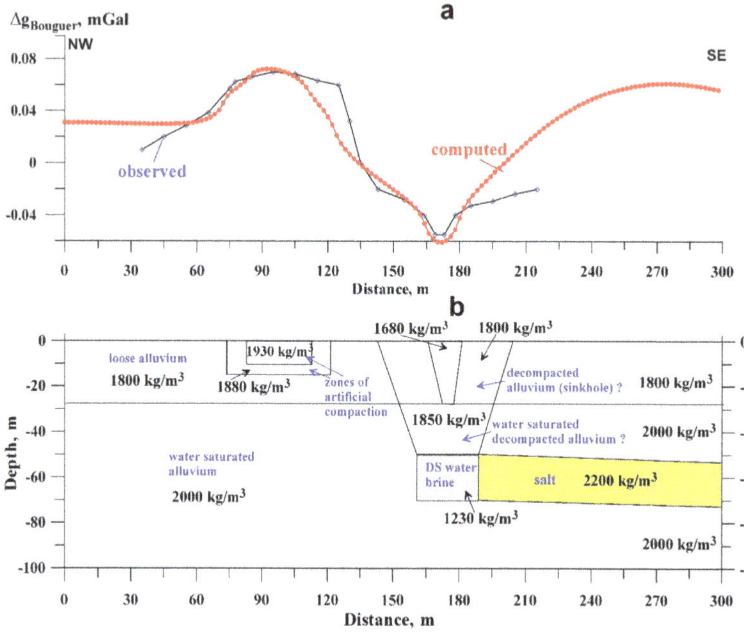

Figure 4.108. 3D gravity modeling over the complex geological section in the area of Ein Gedi, (a) observed and computed gravity fields, (b) PGM. (Reprinted from Ezersky *et al* (2013b) with permission of Springer, modified.)

Without hesitation, 3D modeling of real *PGM* is the most powerful tool of the gravity field examination (Eppelbaum and Khesin 2012). Results of 3D gravity field modeling along a profile crossing the Ein Gedi area (western Dead Sea coast, Israel), is shown in figure 4.108.

Location of Profile C′–C in the Nahal Never south site (Israeli coast) selected for gravity field modeling is shown in figure 4.109.

The PGM developed based on 3D interactive modeling of gravity field along the almost all length of profile C–C′ in the Nahal Hever south site are displayed in figure 4.110 (modified after Eppelbaum *et al* (2008)). The dissolution karst volume was approximately estimated as 35 000 m^3 (its length along strike was selected as $y_1 = 50$ m and $y_2 = -45$ m, which does not contradict the available geological data).

4.5.4 Revealing 3D gravity effects from sinkholes occurring in the Dead Sea area: model and field cases

The area under study—Ghor Al-Haditha—is situated in the eastern coastal plain of the DS (Jordan) in conditions of a very complex regional gravity pattern (see figure 4.99 (Site III) and figure 4.111 for location) (Al-Zoubi *et al* 2013). Satellite gravity observations displayed in figure 4.111 were made with the regular global 1-minute grids that can differentiate these data from the previous odd surface and airborne gravity measurements. This complex gravity field distribution in the

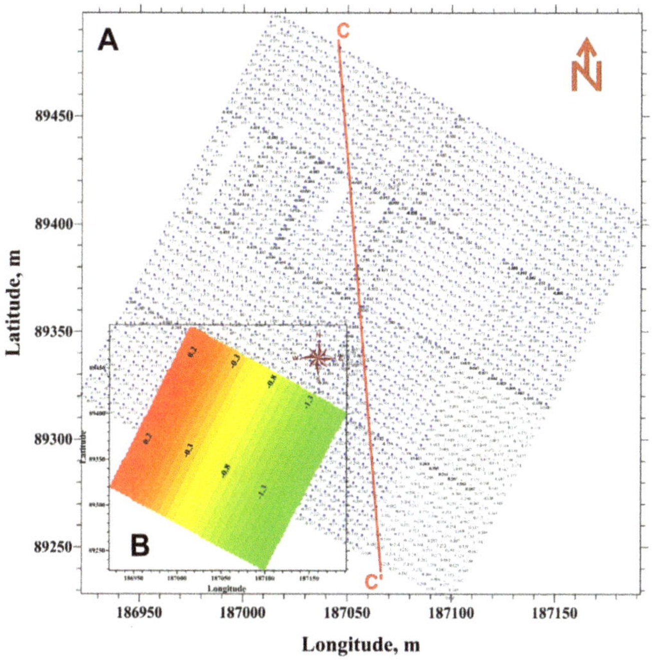

Figure 4.109. A): Gravity data observed in the Nahal Never south (after Rybakov (2007), with permission) (B): polynomial calculated horizontal gravity gradient. The red bold line indicates the location of the interpretation profile C′–C.

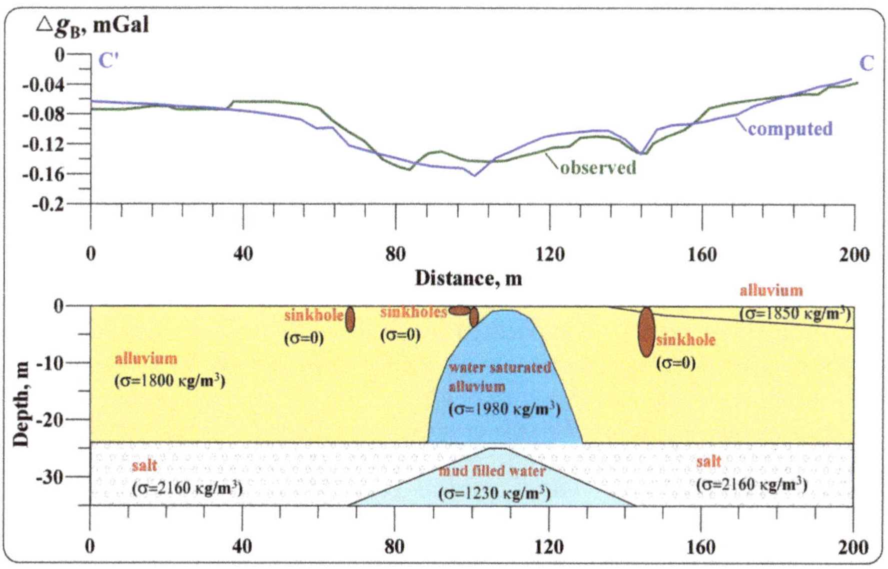

Figure 4.110. Results of 3D modeling of gravity field along the almost all length of profile C′–C (location of this profile is shown in figure 4.109(A)). (Reprinted from Eppelbaum *et al* (2008) with permission of EGU, copyright Authors 2008. This work is distributed under the Creative Commons Attribution 3.0 License.)

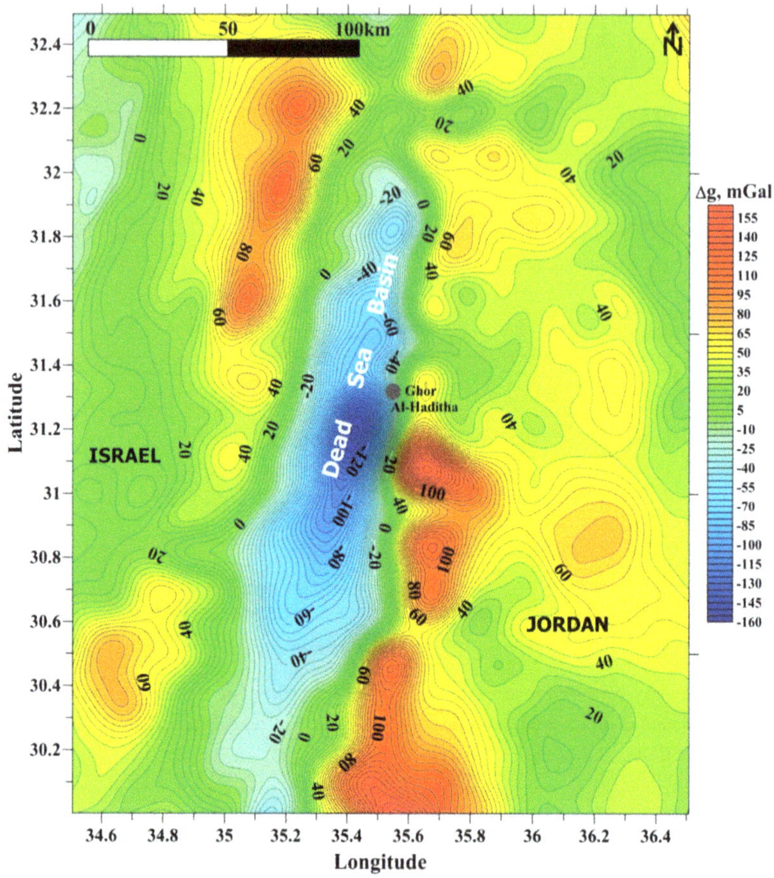

Figure 4.111. Areal map of the region under study superimposed to the satellite-derived gravity map retracted to the Earth's surface. (Reprinted from Al-Zoubi *et al* (2013), copyright Hindawi Publishing Corporation.)

vicinity of the area under study is caused mainly by the strong negative effect of the low-density sedimentary rocks and the salt layers accumulated in the DST.

To test the methods of regional trend elimination, two theoretical PGMs for (1) sinkholes, and (2) DST were developed. The computed gravity effects from these PGMs were also artificially complicated by randomly distributed noise.

4.5.4.1 Computation of the 3D gravity effect from the physical–geological model of sinkholes against the strong noise background

4.5.4.1.1 Construction of the initial physical–geological model

To calculate the 3D gravity field, a scheme of 12 parallel profiles with a distance between them of 5 m was applied (figure 4.112). A PGM with two layers (σ_1 = 2000 kg m^{-3} and σ_2 = 2100 kg m^{-3}) and with two types of ellipsoidal sinkholes was prepared (figure 4.113). The center of the first (large) sinkhole is located at a depth of −60 m below the Earth's surface in the second subsurface layer, with a contrasting density of −900 kg m^{-3}. The

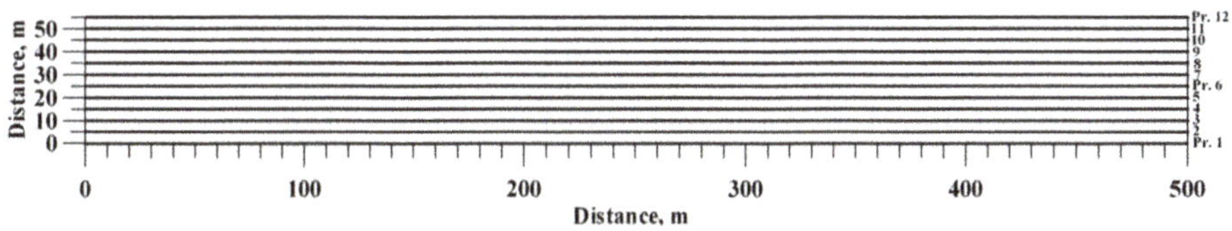

Figure 4.112. Scheme of gravity field 3D computation for the developed model example. (Reprinted from Al-Zoubi *et al* (2013), copyright Hindawi Publishing Corporation.)

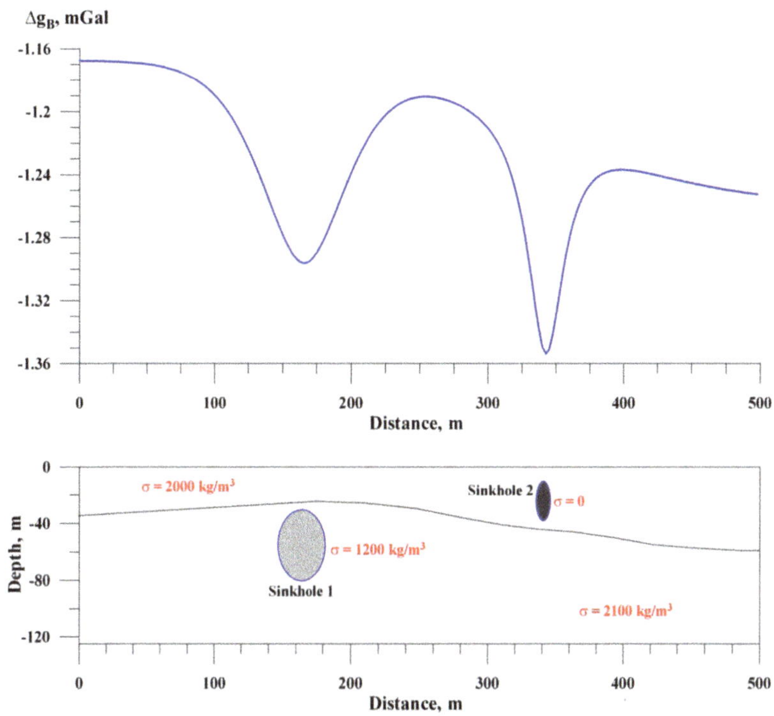

Figure 4.113. Gravity field anomalies along profile 6 from the models of sinkholes. (Reprinted from Al-Zoubi *et al* (2013), copyright Hindawi Publishing Corporation.)

center of the second (small) sinkhole is located at a depth of −20 m below the Earth's surface in the first subsurface layer, with a contrasting density of −2000 kg m^{-3}. Profile 6 is selected as the central one, and the left and right ends of sinkhole 1 were defined as −30 and +30 m, and for sinkhole 2 as −12 and +12 m, respectively. For the 3D gravity field modeling for this and the following examples, mainly the GSFC program (Eppelbaum and Khesin 2004) was employed. The number of computation points along the sinkholes PGM is chosen to be 200; i.e. every 2.0 m.

The compiled gravity map for all 12 profiles from the sinkhole PGM is shown in figure 4.114. As can be seen from this map, the gravity anomaly from sinkhole 2 is narrower than sinkhole 1, but the latter is characterized by comparatively high gravity amplitude.

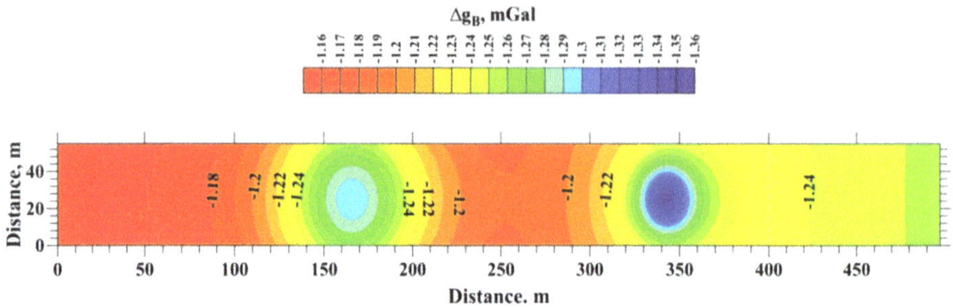

Figure 4.114. Compiled gravity map (in milliGal) for 12 profiles (see scheme in Figure 4.5.10). (Reprinted from Al-Zoubi *et al* (2013), copyright Hindawi Publishing Corporation.)

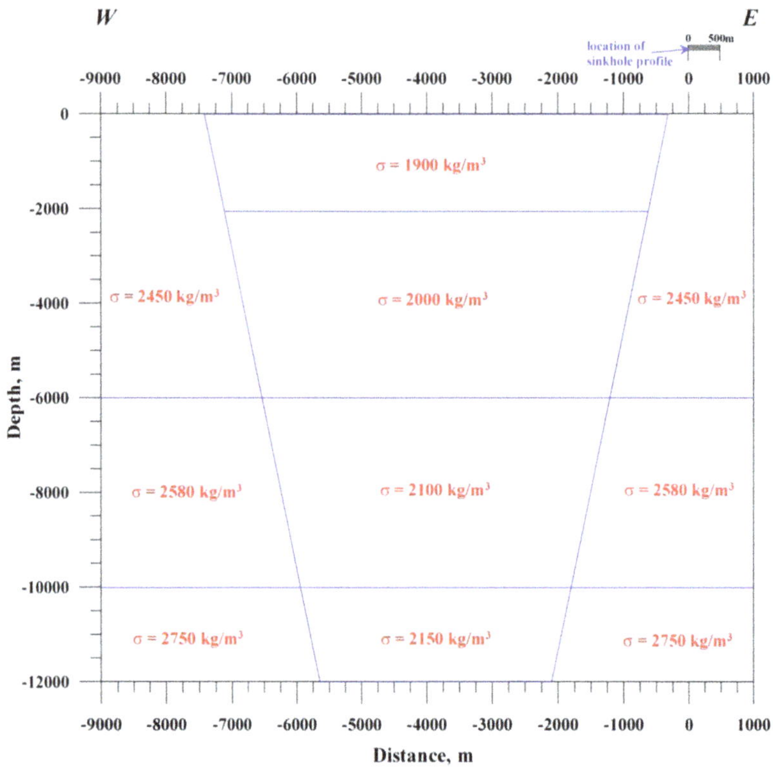

Figure 4.115. Simplified density-geological model of the DST (modified after Al-Zoubi *et al* 2013). (Reprinted from Al-Zoubi *et al* (2013) copyright Hindawi Publishing Corporation.)

4.5.4.1.2 Computation of the 3D gravity effect from the Dead Sea transform

The simplified PGM of the DST for its deepest part (figure 4.115) was constructed from data presented in Ginzburg and Ben-Avraham (1997), Weber *et al* (2021), and the author's experience (Eppelbaum and Pilchin 2006, Eppelbaum 2019). The location of the 500 m sinkhole profile is shown in the upper right section of the model (figure 4.115). The PGM of the DST is computed as the same for all 12 profiles. The computed gravity effect from the DST is added to the gravity field obtained from the sinkhole PGM (figure 4.115). As can be seen from this figure, an anomaly from sinkhole 2 can be visually detected, but the anomaly from sinkhole 1 is practically undetectable against the regional gravity trend produced by the DST.

Given that the real geological medium is usually more complex than in the models presented in figures 4.113 and 4.115, a random number generator is used to introduce a noise factor into the calculations. For random noise generation, the algorithms developed by Bichara *et al* (1981) and Wichura (1988) were employed. The parameters of the randomly distributed noise—the mean values and the standard deviations along 12 profiles—are displayed in table 4.9. In other words, the randomly distributed non-recurrent noise was added to 200 computation points for each of the 12 profiles (figure 4.116).

Figure 4.117 shows a gravity map compiled after adding the randomly distributed noise (see table 4.9). The combined gravity effects from (1) the sinkhole PGM, (2) the DST PGM, and (3) randomly distributed noise is used to compute the integrated gravity map integrating effects of these three factors (figure 4.118). As can be seen from the map (figure 4.118) there are no visual signatures of the negative anomalies from sinkholes 1 and 2.

4.5.4.1.3 Results of the different methodologies applied to eliminate the regional gravity trend

To remove the regional trend, different algorithms and methods were applied: the first and second derivatives, self-adjusting and adaptive filtering, Fourier series, wavelet decomposition, principal component analysis, inverse probability, and some others (altogether more than 30 different procedures).

Examples of applications of (1) the entropy parameter using a moving window with self-adapting size, (2) gradient sounding, and (3) power estimation by the Morlet transformation are presented in figures 4.119(A), (B), and (C), respectively. Computing the entropy with the moving window (figure 4.119(A)) revealed a clear

Table 4.9. Inserted randomly distributed noise.

Profile #	Mean value	Standard deviation
1	0.150	0.040
2	0.160	0.030
3	0.140	0.035
4	0.130	0.038
5	0.170	0.029
6	0.120	0.033
7	0.150	0.038
8	0.140	0.032
9	0.110	0.024
10	0.160	0.031
11	0.125	0.025
12	0.15	0.028

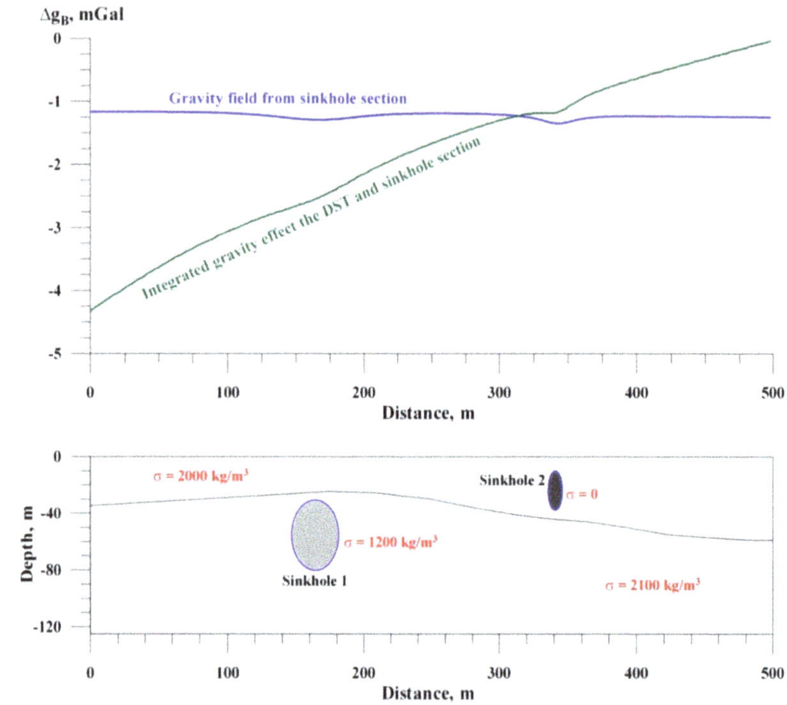

Figure 4.116. Combined gravity field along profile 6 from the models of sinkholes and effect of the DST. (Reprinted from Al-Zoubi *et al* (2013), copyright Hindawi Publishing Corporation.)

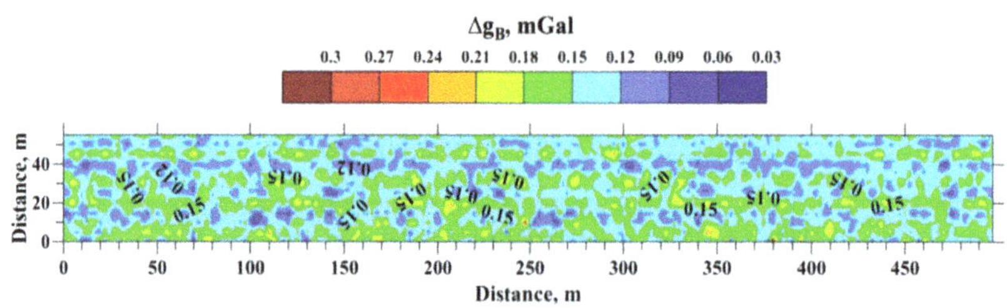

Figure 4.117. Compiled gravity map of the random noise for 12 profiles. (Reprinted from Al-Zoubi *et al* (2013), copyright Hindawi Publishing Corporation.)

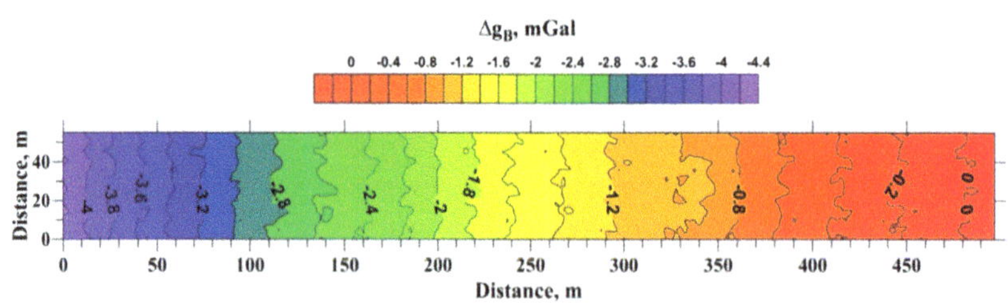

Figure 4.118. Compiled gravity map for 12 profiles with the combined effect from (1) DST, (2) sinkholes, and (3) random noise. (Reprinted from Al-Zoubi *et al* (2013), copyright Hindawi Publishing Corporation.)

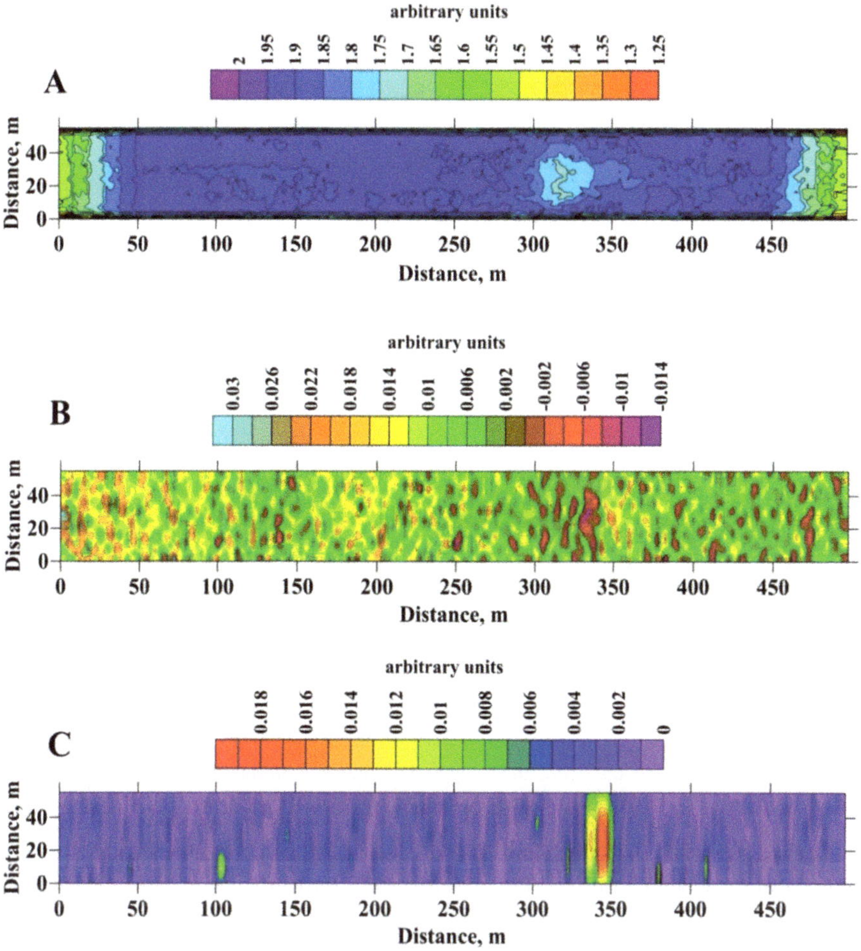

Figure 4.119. Results of three different methodologies application: (A) entropy computation using a moving window with self-adapting size, (B) gradient sounding, (C) power estimation by Morlet transformation (after Al-Zoubi *et al* 2013). (Reprinted from Al-Zoubi *et al* (2013), copyright Hindawi Publishing Corporation.)

ring anomaly from sinkhole 2; the anomaly from sinkhole 1 was difficult to recognize. At the same time, the boundary effect at the map edges (figure 4.119(A)) complicates the obtained image visualization. The results of gradient sounding (figure 4.119(B)) suggested the presence of an anomaly from sinkhole 2. A power estimation based on a Morlet transformation (figure 4.119(C)) very clearly indicates the location of sinkhole 2. However, a superposition of computed gravity anomalies and noise effects gives a false weak anomaly (located at 105–108 m) for sinkhole 1.

Regression analysis is now considered one of the most powerful methods for removing trends of different kinds (e.g., Shatterjee and Sadi 1996, Rawlings *et al* 1998, Bingham and Fry 2010). Two regression methods were selected. Figure 4.120 shows the residual gravity map after subtracting a bi-linear saddle ($F(x,y) = a + bx +$

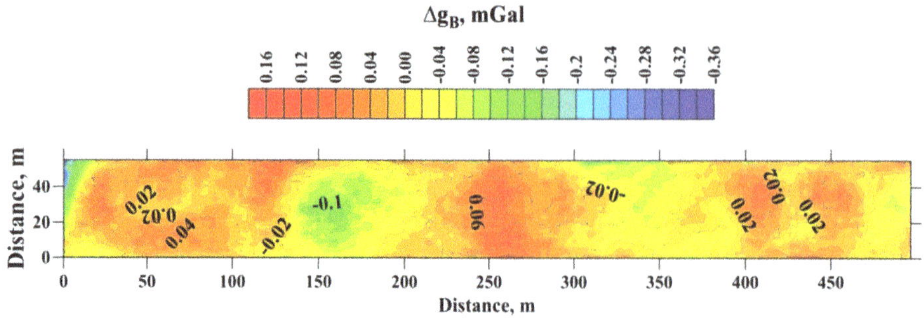

Figure 4.120. Residual gravity map after subtracting the bi-linear saddle regression (modified after Al-Zoubi *et al* (2013)). (Reprinted from Al-Zoubi *et al* (2013), copyright Hindawi Publishing Corporation.)

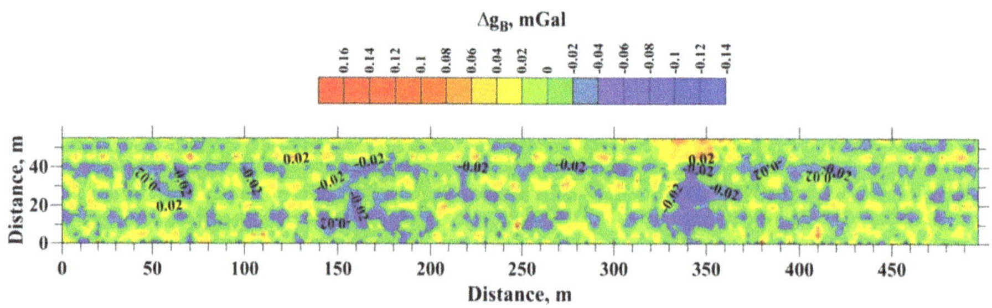

Figure 4.121. Residual gravity map after subtracting the local polynomial (modified after Al-Zoubi *et al* (2013)). (Reprinted from Al-Zoubi *et al* (2013), copyright Hindawi Publishing Corporation.)

$cy + dxy$) regression. The negative gravity anomaly from sinkhole 1 in the area of 160 m (see figures 4.113 and 4.116) is detected, whereas the negative anomaly from sinkhole 2 in the area of 340 m is small and could not be reliably recognized.

Figure 4.120 shows the residual gravity map after subtracting a bi-linear saddle ($(x, y) = a + bx + cy + dxy$) regression. The negative gravity anomaly from sinkhole 1 in the area of 160 m (see figures 4.113 and 4.117) is detected, whereas the negative anomaly from sinkhole 2 in the area of 340 m is small and could not be reliably recognized.

The gravity map after subtracting a local polynomial regression ($F(x,y) = a + bx + cy + dxy + ex^2 + fy^2$) is presented in figure 4.121. Here the negative anomaly from sinkhole 1 is weak and is difficult to detect, but the anomaly from sinkhole 2 was unmistakable. These findings suggest that there are advantages to using a combination of transformation methods.

4.5.4.1.4 Removing regional gravity trend in the area of Ghor Al-Haditha, the eastern coastal plain of the Dead Sea (Jordan)

The Ghor Al-Haditha area is located southeast of the northern DS basin (see figure 4.111). Alluvial fan deposits from the Wadi Ibn Hammad cover the southern part of this area. Borehole sections indicate that the geological material of the shallow subsurface consists of laminated sand interbedded with layers of calcareous silts and possibly clay or marl. The sinkholes at the eastern coast of the DS can be dated to the mid-1980s (Taqieddin *et al* 2000).

The compiled Bouguer gravity map (figure 4.122) testifies to the strong influence of the negative gravity effect due to the DST (and possibly, other geological factors). Performed analysis indicated that computing the first and second derivatives, self-adjusting filtering, gradient directional filtering, Fourier series, principal component analysis, and other methods were less successful than the application of the bi-linear saddle and local polynomial regressions.

Figure 4.123 displays the results of the residual gravity map after subtracting the local polynomial. After regional trend removal, two local anomalies were found: one complex in the center of the area and the other one—near the western border. However, this type of analysis is valid only for target qualitative delineation.

The gravity profile was constructed along the line A–B in figure 4.123. Performed 3D modeling indicates that such a gravity anomaly can be produced by a sinkhole (similar to model 2 in figure 4.116, but enlarged roughly twice) with its upper edge occurring at a depth of 4 m below the Earth's surface (figure 4.124). The location of this sinkhole and its size are consistent with the available geological data (Al-Zoubi

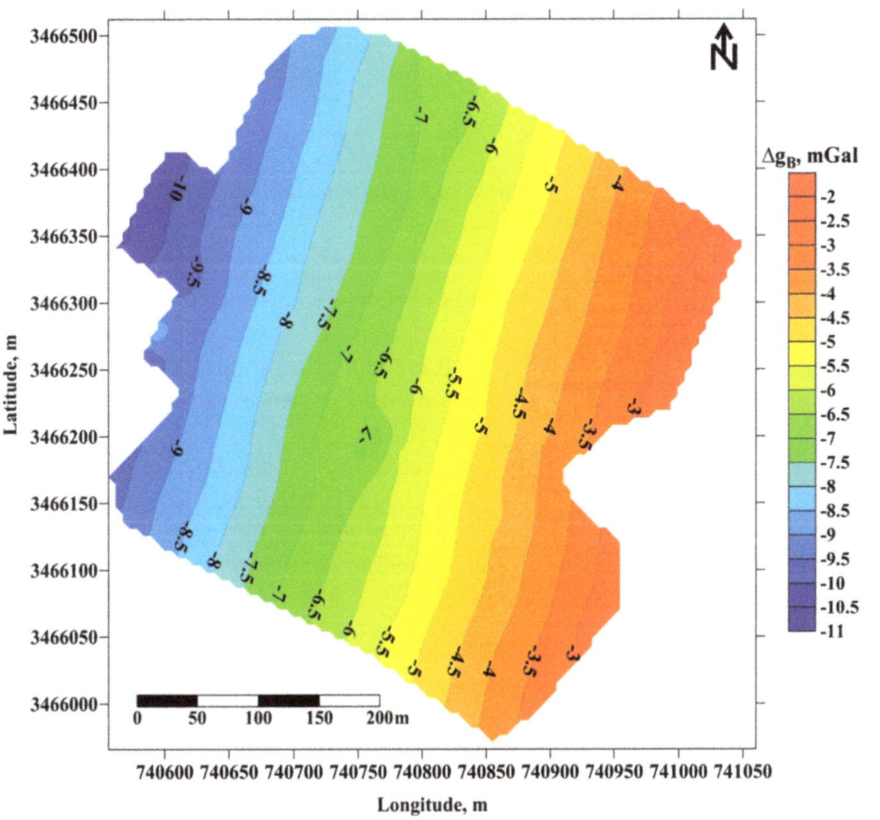

Figure 4.122. Bouguer gravity map of the Ghor Al-Haditha area (Jordan). (Reprinted from Al-Zoubi *et al* (2013), copyright Hindawi Publishing Corporation.)

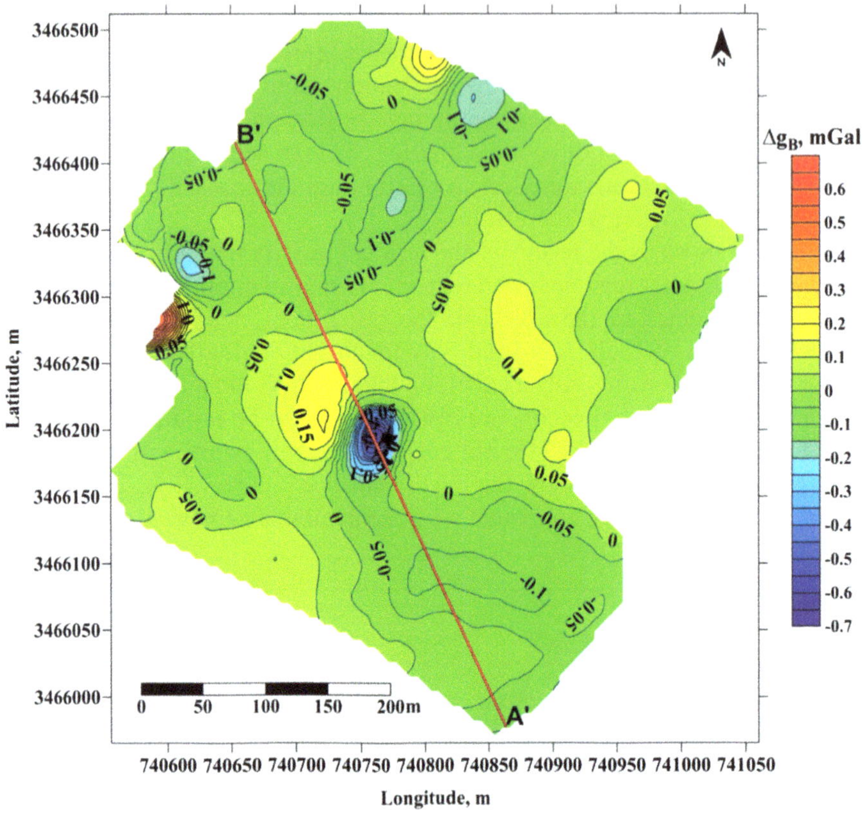

Figure 4.123. Residual gravity map of the Ghor Al-Haditha area after subtracting local polynomial (modified after Al-Zoubi *et al* (2013)). (Reprinted from Al-Zoubi *et al* (2013), copyright Hindawi Publishing Corporation.)

et al 2013). The disparity between the observed and computed Δg_B in the right part of the profile may have been caused by the presence of an additional small underground cavity with an irregular shape.

Thus, detailed gravity studies are an effective tool for sinkhole localization even in significantly complex geological environments. Nevertheless, it should be taken into account that the density-thermal dependence of sedimentary associations (especially in the case of thick strata) calls to reinterpreting detailed gravity surveys (Gadirov and Eppelbaum 2015).

4.6 Magnetometry

A magnetic map observed at 1 km level over the Earth's surface for the western DS area (International Geomagnetic Reference Field has been removed) is presented in figure 4.125. This map indicates that the regional magnetic field increases from the south to the north (it is mainly caused by the crystalline magnetic basement structure). However, within the DS western coastal line, the magnetic field has not designated any significant anomalies.

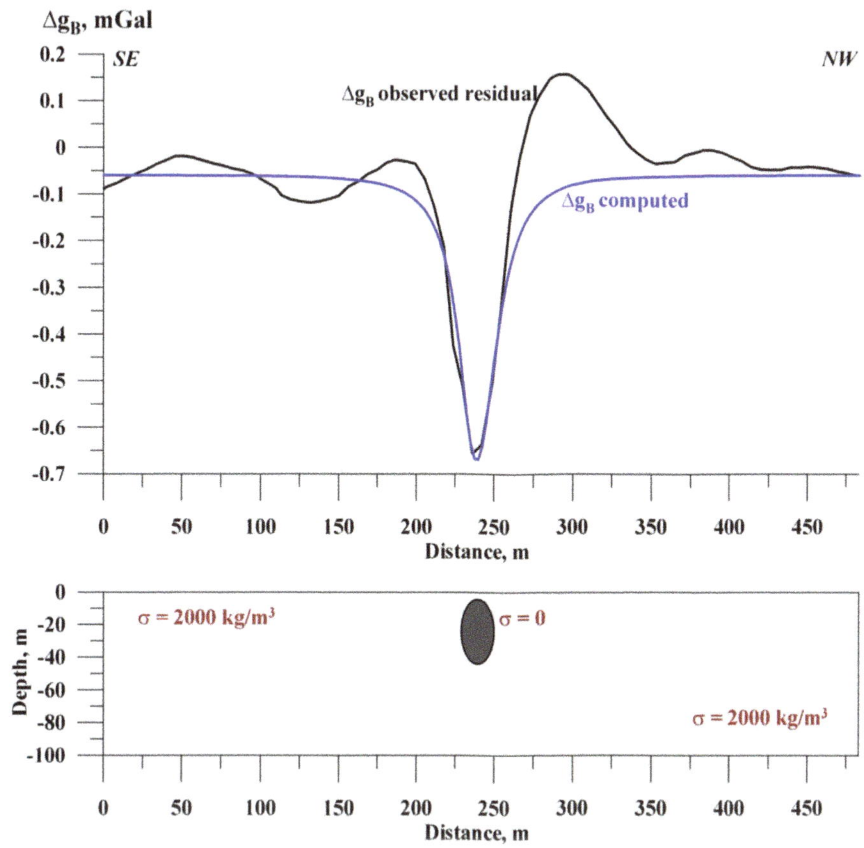

Figure 4.124. An initial PGM along profile A′–B′ based on 3D gravity field modeling. (Reprinted from Al-Zoubi *et al* (2013), copyright Hindawi Publishing Corporation.)

Geological media in the DS region are characterized by low magnetization that complicates the direct localization of buried sinkholes (Eppelbaum 2015a). Nevertheless, sometimes comparatively significant magnetic anomalies were observed in this region (Rybakov 2005, Rybakov *et al* 2005).

The precursors of micromagnetic use for sinkhole detection are based on the low magnetization of shallow sediments in the coastal DS region. Measurements of the magnetic susceptibility on outcrops have shown the value of 16–40 mA m^{-1}.

In the DS area, the known sinkholes cause negative anomalies of few nanoTesla (nT) (figure 4.126(b)).

Rybakov *et al* (2005) reported about the feasibility micromagnetic studies carried out in five sinkhole sites of the DS western shore located in alluvial fans and mudflats.

The anomalies from sinkholes were recognized from the magnetic data analysis. Magnetic studies also indicated that most collapse features correlated with the known tectonic lineaments, suggesting conduits for groundwater and thus the

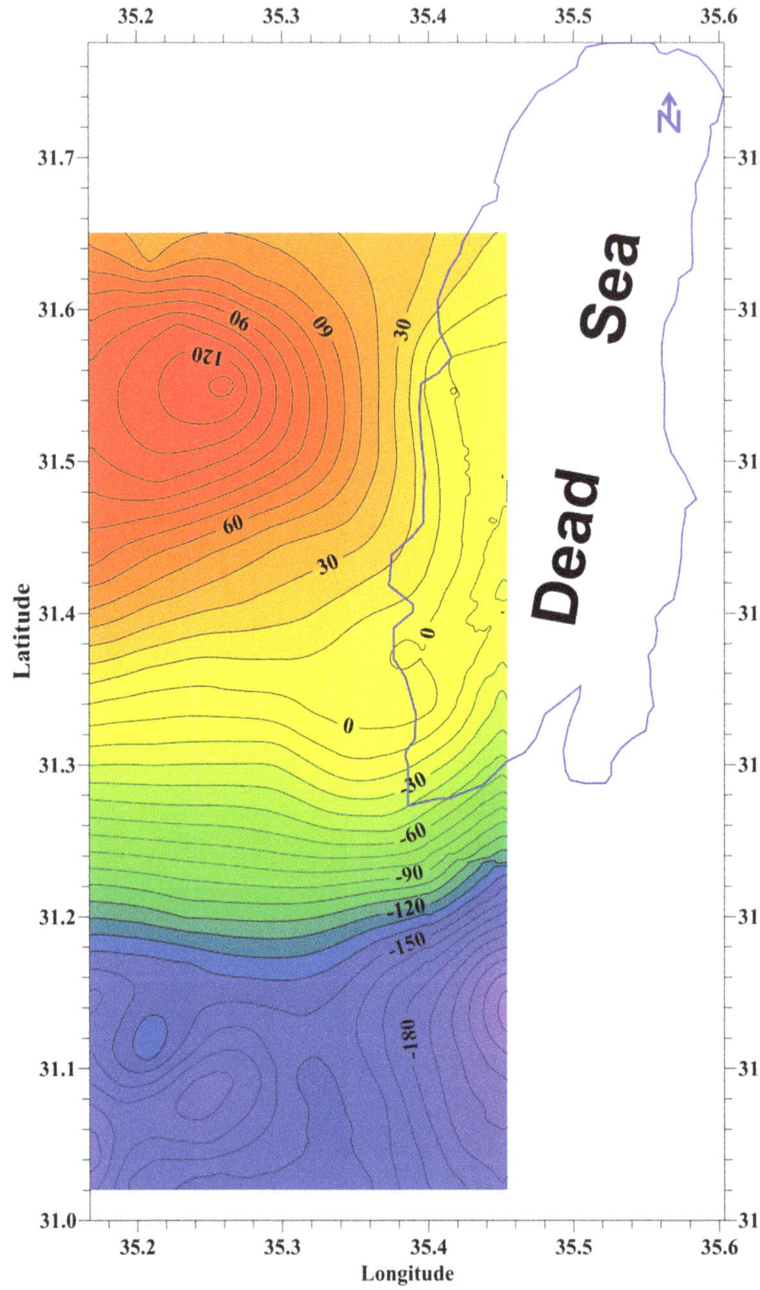

Figure 4.125. Airborne observed magnetic map over the Israeli south and central parts of the DS region.

dissolution of evaporites in the sediments. An example of the micromagnetic survey in the Newe Zohar site is presented in figure 4.127(a).

Zohar Alluvial fan. The first ground collapse in the DS area occurred suddenly at the Zohar site in 1980. Recently, two small sinkholes opened up in the Zohar alluvial

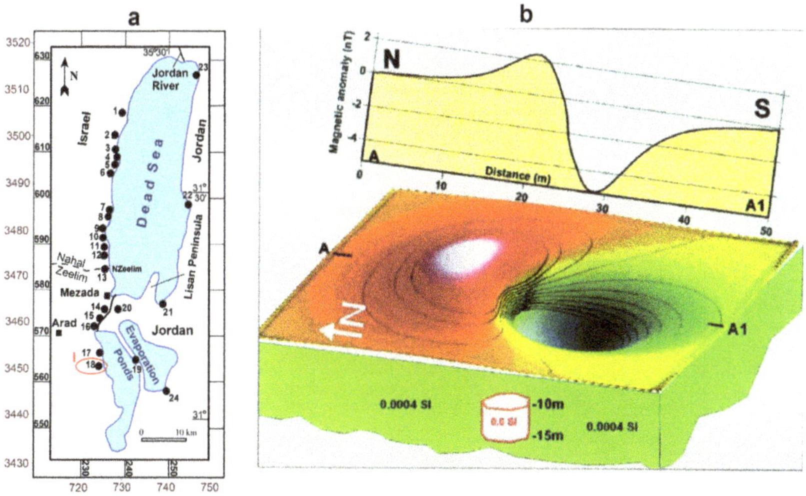

Figure 4.126. Micromagnetic study. (a) Newe Zohar site location map. (b) Synthetic magnetic anomaly caused by a concealed cavity in low magnetic media. (a) 18-Newe Zohar (Site I). The coordinates in the frame are the new Israel Mercator grid in km, the outer coordinates are the Universal Transverse Mercator grid, in km. (Reprinted from Rybakov *et al* (2005), with permission of SEG.)

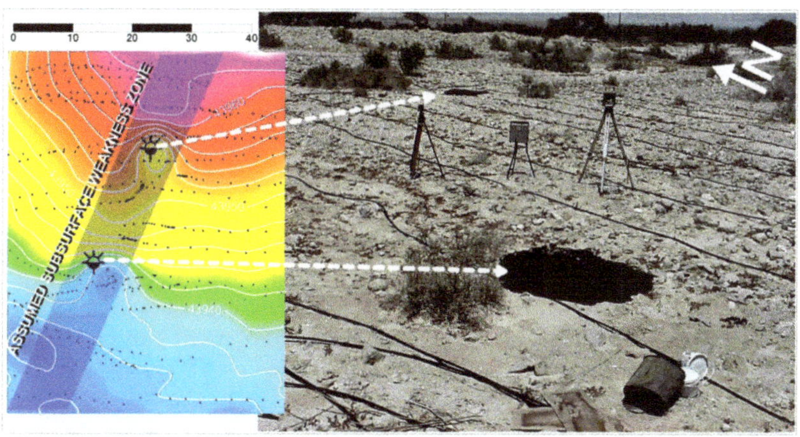

Figure 4.127. Micromagnetic study in the Newe Zohar south site. Two small sinkholes have recently opened (photo on right). The high-resolution magnetic map (at left) shows magnetic anomalies caused by cavities. (contour interval is 2 nT). (Reprinted from Rybakov *et al* (2005), with permission of SEG.)

fan (photo in figure 4.128). The slightly inclined cone-shaped sinkholes are small holes, about 1.5 m in diameter at the surface and about 5 m in diameter at a depth of 7 m. Since the sinkhole volume can be reliably measured and magnetic measurements can be taken directly above the cavern, this site was a perfect natural laboratory to examine the magnetic technique. The 90 × 90 m site was covered by a dense micromagnetic survey.

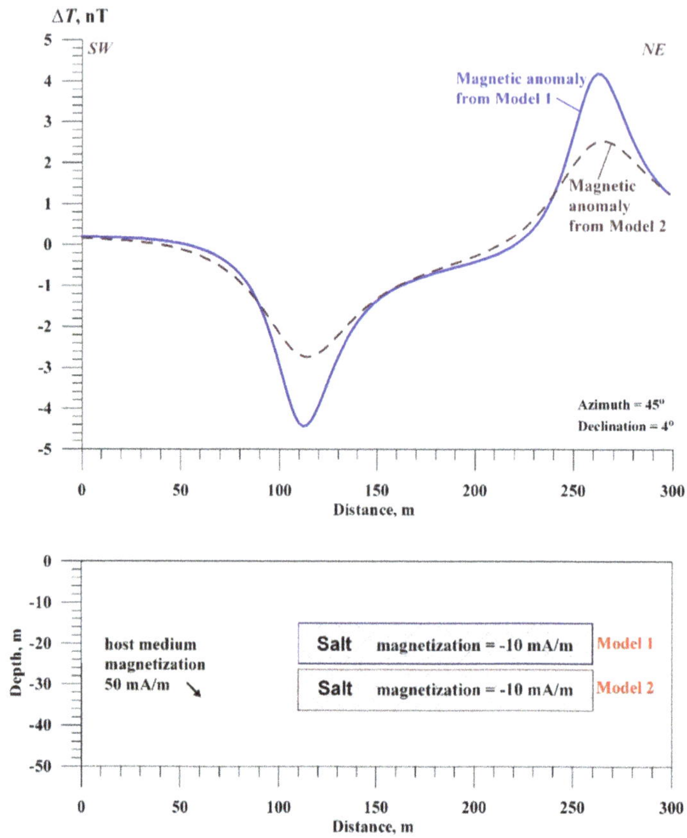

Figure 4.128. Magnetic field modeling over two salt bodies of intermediate configuration.

The magnetic map (figure 4.128, left) shows two typical magnetic anomalies associated with the sinkholes.

However, these magnetic anomalies can be explained by the observed holes only partially. This suggests that additional sinkhole development can be expected.

4.6.1 Localization of salt bodies

3D magnetic field computation indicates that the salt bodies of intermediate form occurring at different depths create differentiable anomalies (figure 4.128). Commonly, these anomalies could be interpreted by the methodology developed for the analysis of anomalies produced by thick bed bodies (Eppelbaum 2013, 2015b). However, it is a subject of a separate investigation.

Let us consider the magnetic effect from the 'classic' thin horizontal plate characterized by large horizontal thickness $2b$, small vertical thickness ($2b \gg h1$ and $h2$; h_1 and h_2 are the depths to upper and lower surfaces of the horizontal plate, respectively) and near-surface occurrence (figure 4.129).

Of course, this model cannot be interpreted as a 'quasi-thick' model. If the parameter $2b$ is sufficiently large, we observe two independent anomalies, which we

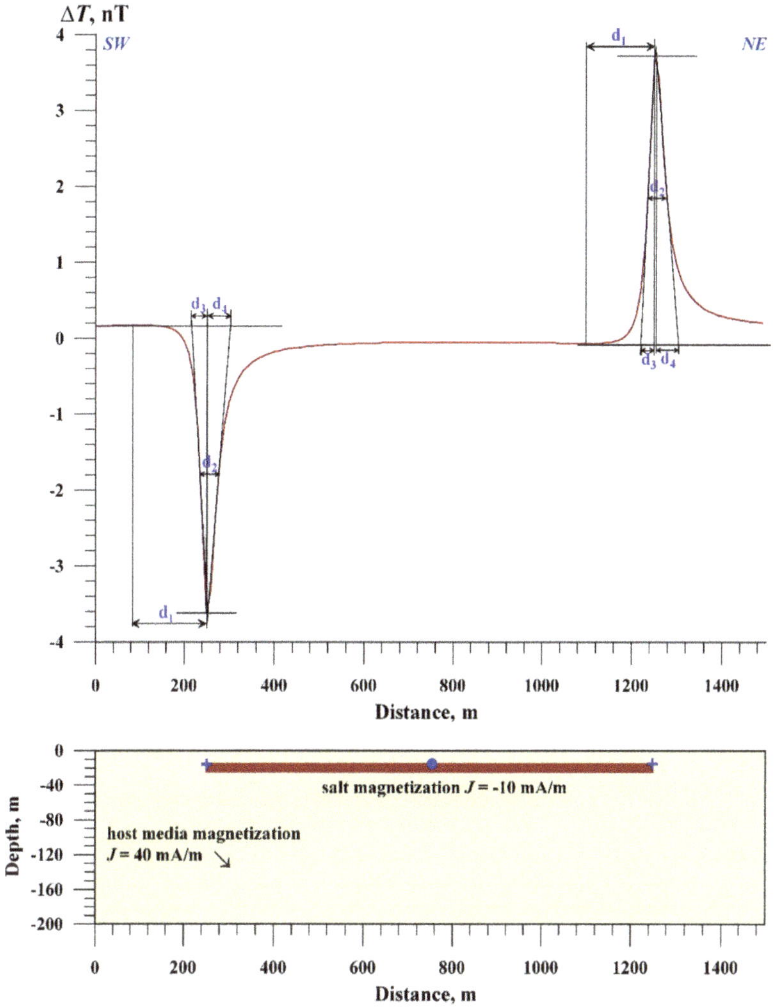

Figure 4.129. Quantitative analysis of magnetic anomalies produced by the model of the salt bed. Here blue crossings show determined positions of the left and right ends of the body, and blue filled circle—position of the center of the body. (Reprinted from Eppelbaum (2015b), with permission of Springer.)

can interpret by the use of methodologies developed for complex physical–geological conditions for the thin-bed models (Eppelbaum *et al* 2001). In this case, we assume that the magnetization of the *L.H.* bed is positive, and the *R.H.* bed is a negative one (in this case, a determination of magnetization is not possible). As is shown in figure 4.129, the results of interpretation indicate the position of the center of the upper edges of two 'fictitious' thin beds in the left (positive anomaly) and right (negative anomaly) of the considered thin horizontal plate model.

Analysis of borehole sections and surface geology enabled the calculation of the following PGM in the Shalem area in the western (Israeli) DS coast (figure 4.130).

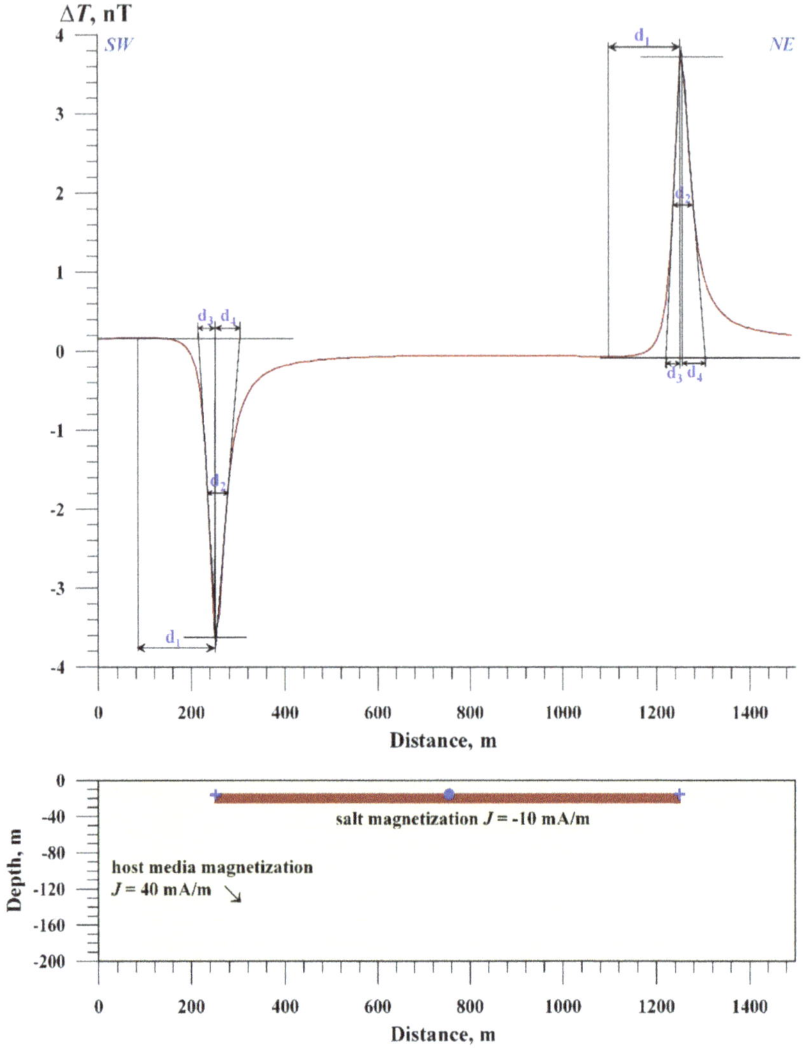

Figure 4.130. Model magnetic curve ΔT over the salt body having a form of the thick bed along with profile C′–D′ in the Shalem area in the western DS coast.

4.6.2 Qualitative analysis of magnetic anomalies

Field magnetic investigations have been carried out in the eastern DS coast (Jordan) (Al-Zoubi *et al* 2011) (figure 4.131). Two magnetic sensors were applied with a vertical distance of 1 m at the levels of 2 and 1 m, respectively. Thus, both maps of the vertical magnetic gradient $\boldsymbol{T_z}$ and total magnetic $\boldsymbol{\Delta T}$ maps could be compiled. Altogether 1122 points of an irregularly displaced grid of the SW–NE (Al-Zoubi *et al* 2011) trend were observed (figure 4.132). After removing the diurnal magnetic variations, a map of the total magnetic field (compiled for the top magnetic sensor—

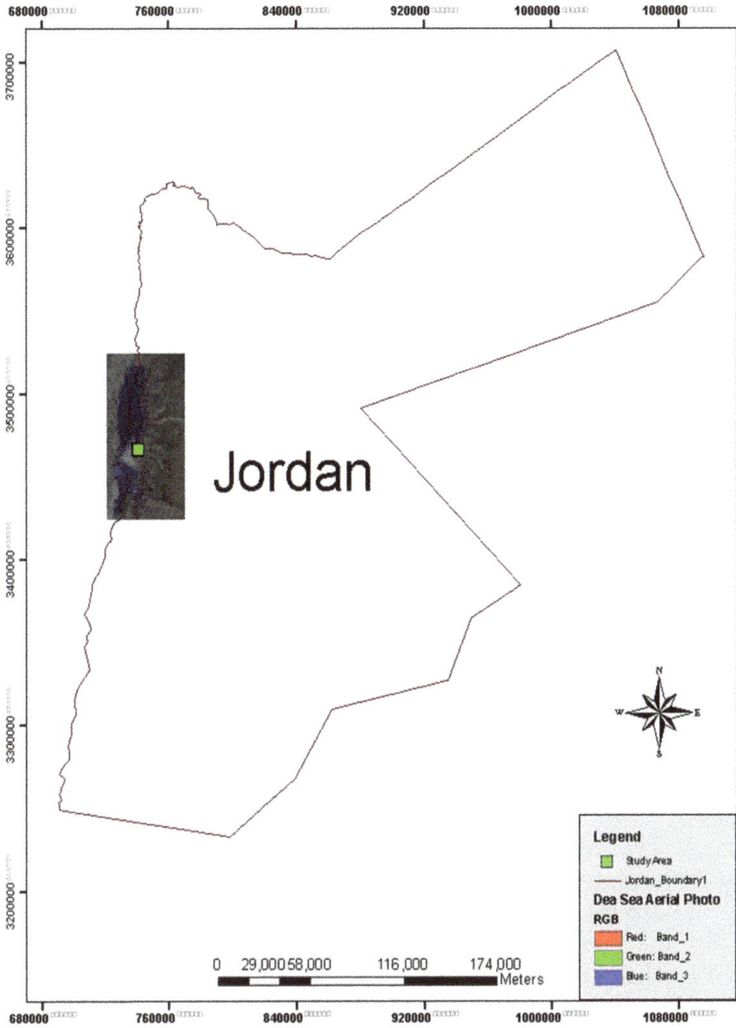

Figure 4.131. Areal map of the magnetic observations at the eastern DS coast (Jordan). (Reprinted from Al-Zoubi *et al* (2011), with permission of EAGE.)

2 m over the Earth's surface) was compiled (figure 4.133). For increasing the visibility, from the total magnetic field, some averaged level of 44 350 nanoTesla (nT) has been removed.

Taking into account that the significant negative magnetic anomaly in the lower part essentially distorts the common normal level, the same magnetic map was reconstructed without several disturbing points caused obviously by the modern magnetic noise (figure 4.133). The elliptical zone '3' of local anomalies is determined more clearly. A linear fault zone 6 of the SE–NW trend is delineated, and contact zone 7 (northern part) is revealed.

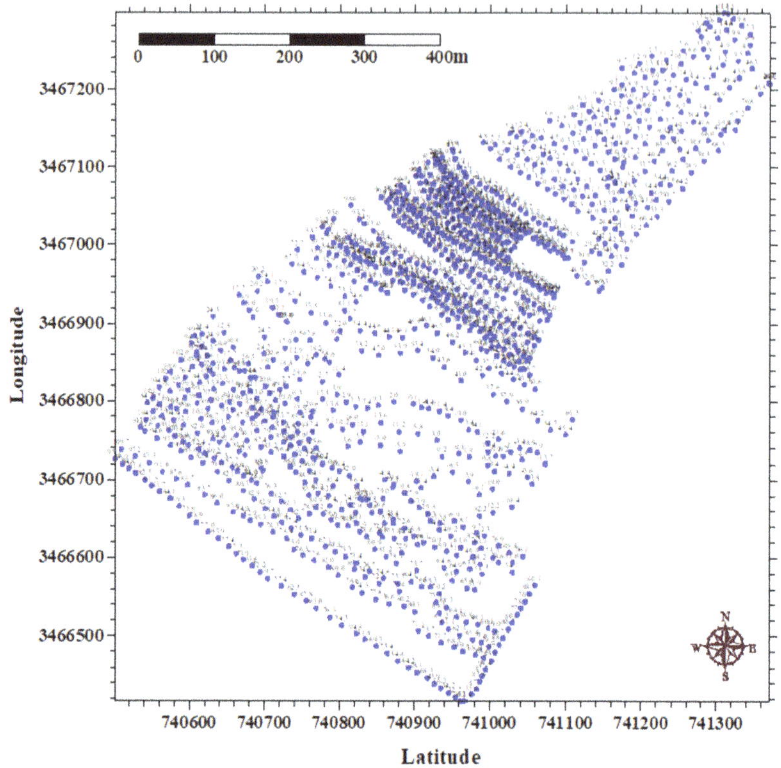

Figure 4.132. Location of magnetic field observation points. (Reprinted from Al-Zoubi *et al* (2011), with permission of EAGE.)

4.6.2.1 Preliminary magnetic data analysis

In figure 4.133, zones 1A, 1B, 2A, and 2B are distinctly delineated. The most interesting for the present investigation is the elliptical zone '3', which includes several local positive and negative magnetic anomalies. In the southern part of the area a zone of linear anomalies '4' was selected. Finally, a very intensive local negative magnetic anomaly between the longitudes 740 900 and 741 000 is caused, apparently by some modern metallic noise.

A topographic map of the area is shown in figure 4.134. Figure 4.135 displays the map of the vertical gradient of the magnetic field (four noisy points were removed). We could propose that the magnetic anomalies compiled in this map reflect mainly near-surface anomalies. Most parts of these anomalies are associated with the elliptical zone '3' and linear-gradient zone '4'.

4.6.2.2 Some magnetic data transformations

It is well known that various transformations play an important role in the potential geophysical field analysis (e.g., Blakely 1995). It is especially significant when a superposition of anomalies of different order exists. Application of gradient operator (calculation of the gradient magnitude) allowed us to delineate some buried geological objects (figure 4.136). It is necessary to note that the linear fault zone

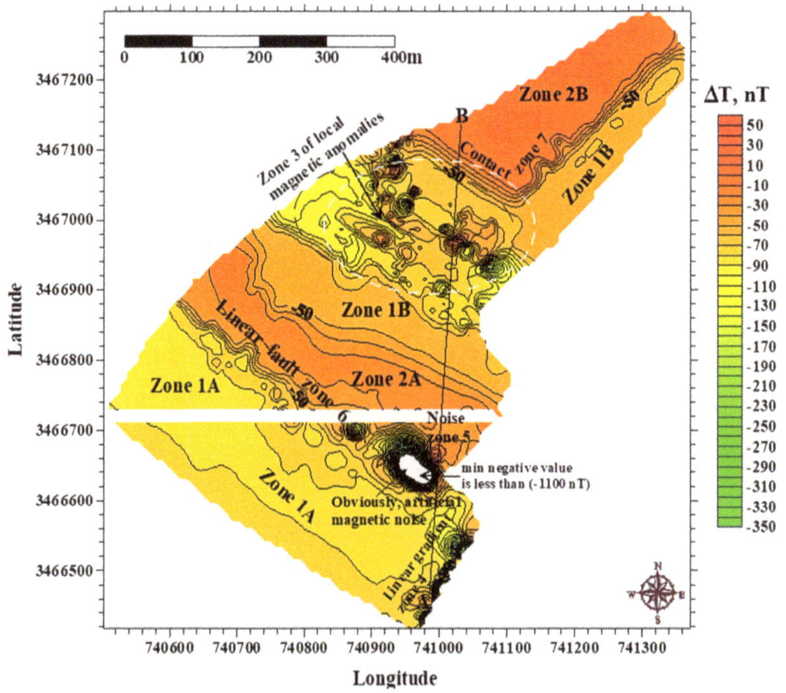

Figure 4.133. Total magnetic map for the area under study with some initial zonation. (Reprinted from Al-Zoubi *et al* (2011), with permission of EAGE.)

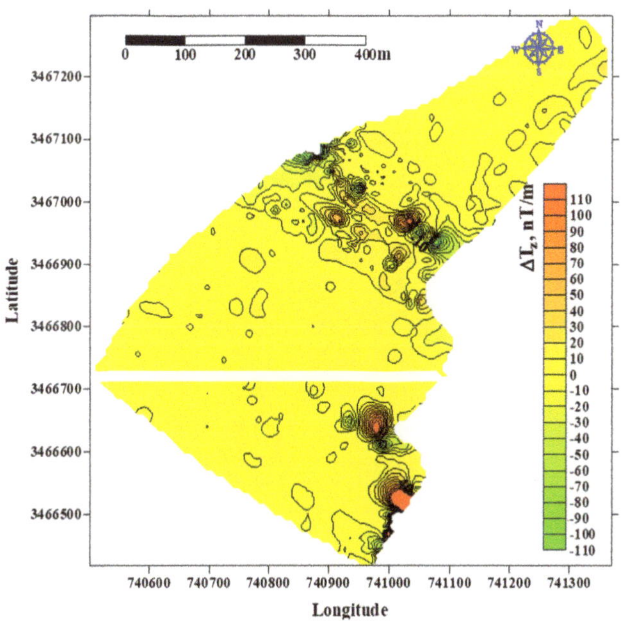

Figure 4.134. Topographic map of the area under study. (Reprinted from Al-Zoubi *et al* (2011), with permission of EAGE.)

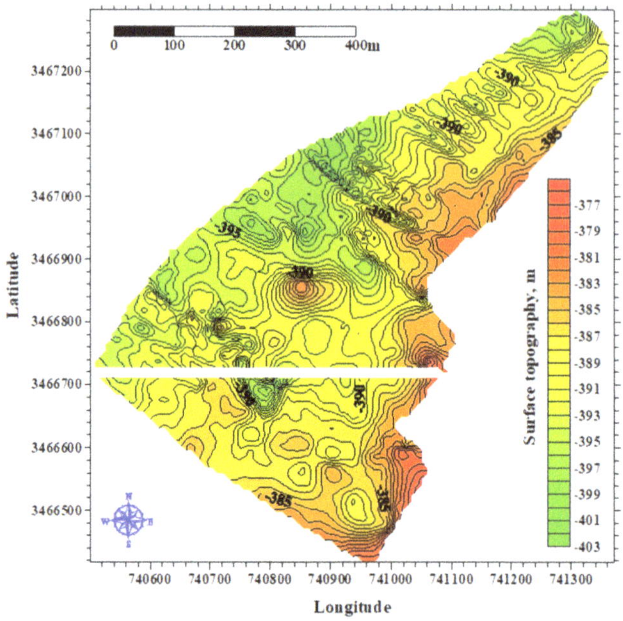

Figure 4.135. Map of the vertical magnetic gradient for the area under study (revised after Al-Zoubi *et al* 2011). (Reprinted from Al-Zoubi *et al* (2011), with permission of EAGE.)

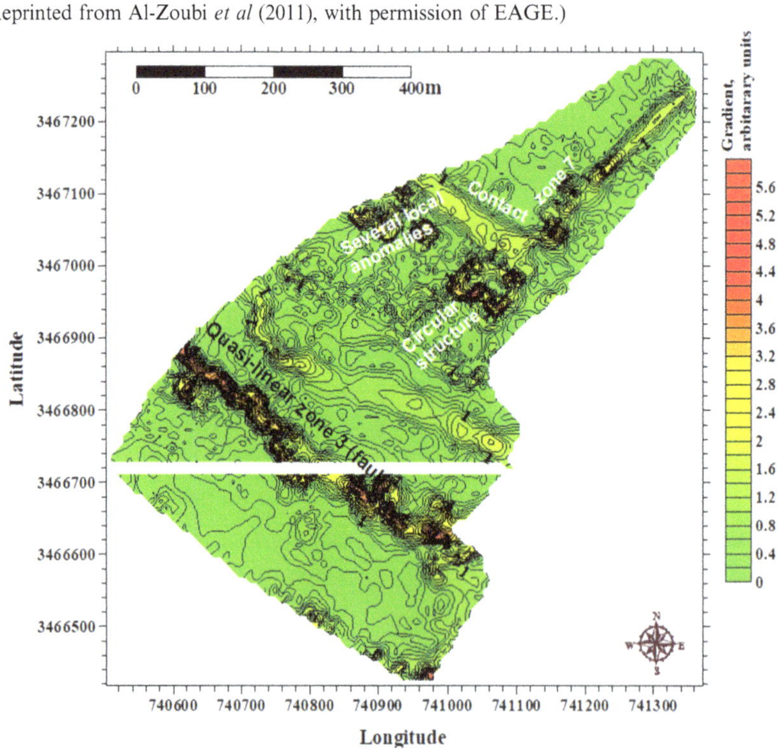

Figure 4.136. Total magnetic field map transformed by the use of gradient operator. (Reprinted from Al-Zoubi *et al* (2011), with permission of EAGE.)

'6' is detected in the map of the total magnetic field (figure 4.133), in the total magnetic field map transformed by gradient operator (figure 4.136) and partially—in the topographic map (figure 4.134). The contact zone 7 has been distinctly expressed in all three above-mentioned maps.

Note that application of gradient operator permitted to localize a quasi-circular structure (a general methodology of ring structures identification is presented in Eppelbaum (2007b)) in the upper part of the studied area at the longitudes 741 000–741 100.

4.6.3 Quantitative analysis of magnetic anomalies

Quantitative analysis of magnetic anomalies has been conducted by the use of improved methodologies presented in section 2.2.1.6 (see also Eppelbaum *et al* (2001), Eppelbaum (2011a, 2015b, 2015c), Eppelbaum and Mishne (1988, 2011)). Visual analysis of the magnetic curve presented in figure 4.137 indicates that probably here are two hidden sinkholes occurring one below the other. Anomalies from these targets were interpreted separately. Red and green symbols '◉' show positions of the centers of two buried sinkholes.

Figure 4.138 shows quantitative analysis of a magnetic anomaly produced by a salt layer having a form of thick bed. Since salt is a paramagnetic matter with a small

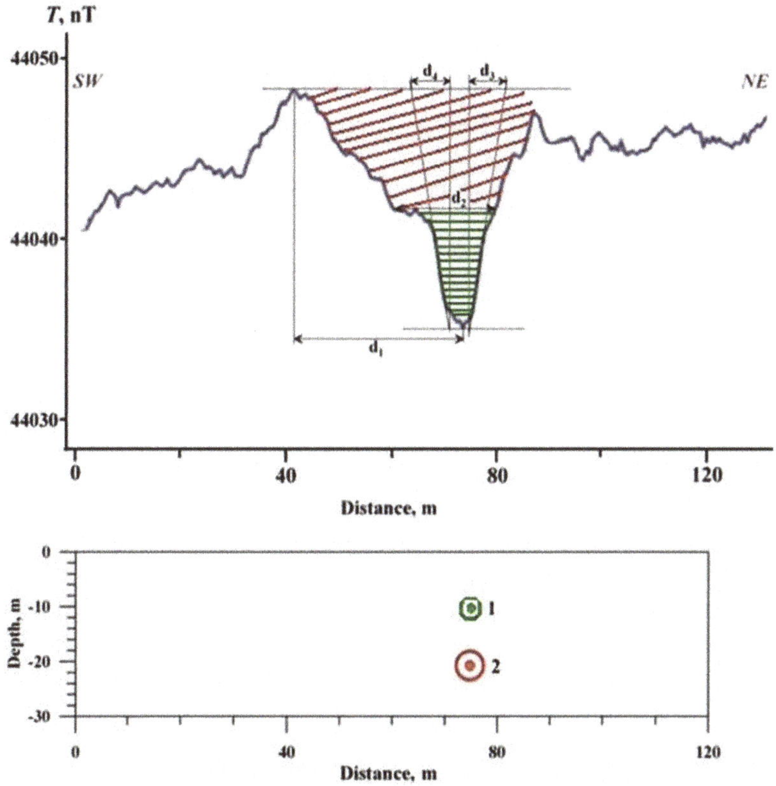

Figure 4.137. Quantitative analysis of magnetic anomalies from the sinkholes in the Never south site (western DS shore, Israel) (observed magnetic curve after Rybakov *et al* (2005), with permission from SEG). Symbol ◉ designates the position of the center of sinkholes.

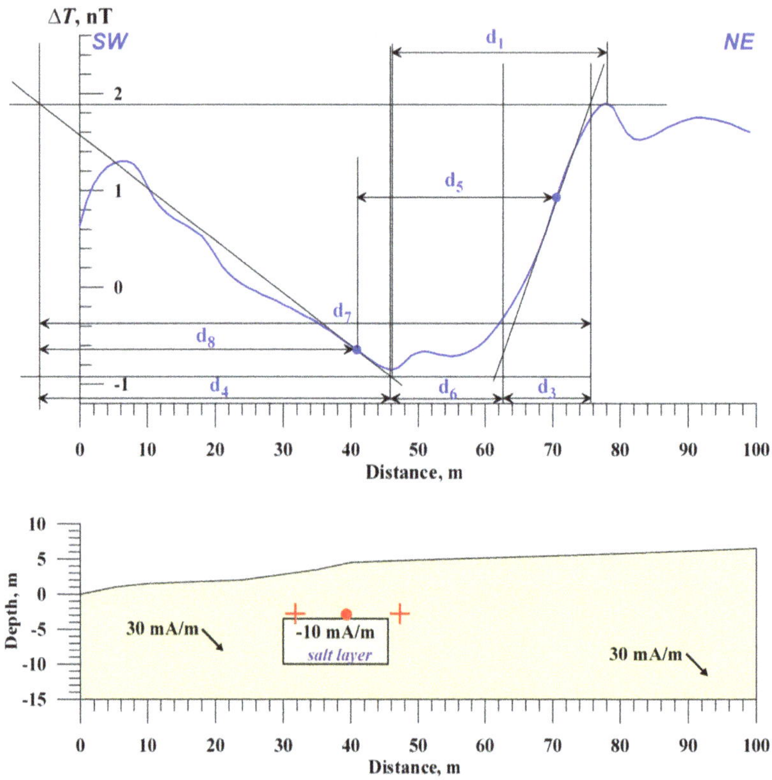

Figure 4.138. Quantitative analysis of magnetic anomaly over the model of salt thick bed computed for the conditions of rugged terrain relief. Here red crossings show determined positions of left and right ends of the body, and blue filled circle—position of the center of the body. (Reprinted from Eppelbaum (2015a), with permission from SAGEEP, copyright 2015. All rights reserved.)

magnetization (–10 mA m^{-1}) usually occurring in host media with a very low magnetization (here a magnetization of 30 mA m^{-1} was selected), the amplitude of the modeled anomaly is only about 3 nT. The characteristic peculiarity of this model is the presence of rugged terrain relief disturbing the magnetic anomaly (Eppelbaum 2015a). Results of quantitative interpretation (the interpretation methodology for such bodies was presented in detail in Eppelbaum (2013, 2015b)) do not ideal coincide with the available PGM, but demonstrate sufficient accuracy (figure 4.136). The calculated *Je* value is about (–7) mA m^{-1} which is the admissible discrepancy with the initial data. The future of comprehensive gravity and magnetic investigations in almost inaccessible areas is associated with applying different kinds of remotely operated vehicles (e.g., Bashirov *et al* 1992, Eppelbaum and Mishne 2011, Eppelbaum 2016).

4.7 Electrical resistivity tomography (ERT), IP, SP

The electrical resistivity methods have the advantage of low cost and ease of operation. It is assumed that electrical parameters (such as resistivity or conductivity) are related to the formation properties and may vary with time. For example, high porosity materials containing voids manifest themselves as high resistivity zones. In contrast, the water (of differing degrees of salinity) as well as zones containing clay appears in the resistivity section as low resistivity zones (Keller and Frischknecht 1966, Nelson and Haigh 1990). That is why geoelectric methods are used widely for solving the sinkhole problem (Arandjelovic 1969, Werner 1984, Smith and Randazzo 1993, Roth *et al* 1999, Zhou *et al* 1999, Milanović 2000b, van Shoor 2002). Based on this knowledge we applied the ERT method (named also 2D resistivity imaging, electrical imaging, continuous vertical electrical sounding—CVES, etc) to achieve a better understanding of the subsurface geoelectrical structure at sinkhole development sites. ERT is in its origin the resistivity method working on direct (or slowly changing with the frequency of 1 Hz) current and it relates to DC methods (Barker 1992). Being also the resistivity method ERT, has all its advantages and disadvantages. For instance, it has more resolution concerning the resistive medium than EM methods (Kaufman 1978). On the other hand, it has considerably lower penetration and resolution in the high conductive medium than the TEM method.

The ERT method has been used in Israel from the end of the 1990s to the middle 2000s and has yielded good results (Ezersky 2002a, Levi *et al* 2011, Ezersky *et al* 2013c).

The question of resistivity of the subsurface in the sinkhole occurrence sites is very important because: (a) resistivity can be an indicator of the underground caverns and (b) because it can answer if GPR would be used in the sinkhole hazardous areas.

The resistivity distribution at the shallow subsurface was recorded using the 'Quasi-3D' technique permitting the studying of the vertical and lateral resistivity variations (Ezersky 2002b, 2003a, 2003b, 2008). We added our data by information acquired during previous studies. Let us consider resistivity distribution in the sites.

4.7.1 The method

2D resistivity prospecting yields information about both the lateral and vertical distribution of the Earth's resistivity and can therefore be used in both qualitative and quantitative ways for the identification of structure and features at shallow depths. The ERT method has been used for this purpose (Barker 1992). The subsurface resistivity distribution is determined by making measurements on the ground surface. From these measurements, the true resistivity distribution of the subsurface can be estimated. The ground resistivity is related to various geological parameters, such as mineral and fluid contents, porosity, and degree of water saturation in the rock (McNeill 1980c). The geoelectric method is adopted in geological engineering and hydrogeological applications, because of its high efficiency for detecting the presence of heterogeneities (cavities, ore bodies, etc). Also, it can differentiate between the rock layers such as clay, marl, sand, sandstone, limestone, and water-saturated soils and rocks.

ERT method is in part a development of the traditional vertical electrical sounding (VES) method, which in its simplest form may be described as four electrodes laid in a line and inserted a few centimeters into the ground (figure 4.139(a)). If the ground resistivity is homogeneous (the half-space model), then it could be calculated exactly using the simple equations dependently on electrode configuration (see section 2.2.1.9).

The most useable among the different electrode installations are linear ones. In these installations, MN electrodes are placed usually in the middle third part of the AB segment, where the normal field is close to the homogeneous one.

For the *Wenner* configuration (figure 4.139(a)) in which the distances between the electrodes are equal, e.g., AM = MN = NB = a, increase in AB separation causes an increase in MN allowing ΔU constantly to increase and be at a measurable level. The geometrical factor for the Wenner configuration is $K = 2\pi a$, where K is a geometrical factor.

In the DS area resistivity decreases quickly with the depth and it results in a decrease of measured potential between the surface MN electrodes as the electrode spacing increases. The Wenner array maintains the potential between the MN electrodes at measurable levels for larger spacing.

An electric current is sent through one pair of electrodes (current or transmitting electrodes A and B: the potential difference established in the Earth by this current is measured using a second pair of electrodes (potential or measuring electrodes, M and N).

While studying the variation of resistivity with a depth, the spacing between the current and potential electrodes is gradually increased. The greater the spacing, the more pronounced the effect of material at the depth. Depending on the geometric configuration of both current and potential electrodes, various geoelectric arrays are

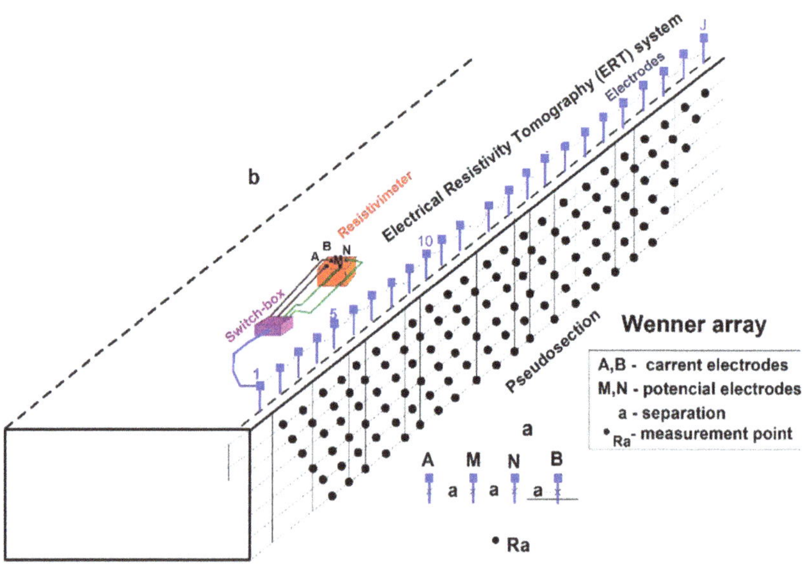

Figure 4.139. Scheme of ERT measurements using Wenner array. (a) Four-electrode setup. (b) Schema of field measurements. (Reprinted from Ezersky (2002a), with permission from EAGE.)

defined. Since the ground is not generally homogeneous, the apparent resistivity of the Earth is determined. The behavior of the apparent resistivity with separation AB roughly reflects the behavior of the true resistivity with depth. To obtain a quantitative distribution of the resistivity with the depth, a so-called inversion of the measured data is required. ERT is based on the earlier developed VES method. Several electrodes are laid out at similar intervals along the line (figure 4.139(b))) and the VES procedure is repeated successively on each pair of electrodes (Loke and Barker 1996).

The first measurements were carried out with a Bison 2390 resistivity meter, updated for multi-electrode measurements (Ezersky 2002b, 2003a, 2003b, 2008) or later an IRIS SYSCAL R2 system. The data have been inverted using RES2DINV (Loke 2004), an inversion routine based on the smoothness-constrained least-squares method (deGroot-Hedlin and Constable 1990). For fast inversion, quasi-Newton optimization is used (Loke and Barker 1996).

At sites of complicated topography and geometry, we performed simple 'quasi' 3D mapping derived from multiple 2D lines. The data from each 2D survey line was initially inverted independently to provide a 2D cross-section. Then a resistivity section in different directions could be constructed. The quality of such a 3D survey is expected to be poorer than that produced with a complete 3D survey but may reveal major resistivity variations across the survey lines like the geological section derived from separate boreholes. The resolution of these 3D measurements corresponds to the approximate spacing between the lines. It means the resistivity anomalies traced between two and more ERT lines have high reliability, whereas anomalies of the smaller size that appeared either at only one line or between lines would be considered as either of low reliability or even as artifacts. At that term 'anomaly' means zone at the resistivity maps and sections with the resistivity values differing by factor 5 and more from the background resistivity (really it is a prominent anomaly). A similar approach was applied by Kidanu *et al* (2016, 2020) using Res3D inversion software.

4.7.1.1 ERT measurements in the DS shores

The ERT studies were carried out in some sites of the DS shores in Israel and Jordan (figure 4.140)

4.7.1.2 Parametric investigation (2D resistivity modeling)

The porosity of the background zone (located out of sinkhole sites) may be calculated using the technique suggested by Kafri and Goldman (2005) based on the TEM method. The conventional expression to evaluate the porosity in sands is:

$$F = 0.81/\phi^2 \tag{4.19}$$

where $F = \rho_x/\rho_w$ is the formation factor. The expression (1) permits calculating the soil porosity in the saturation zone (below the water table) if bulk resistivity ρ_x and resistivity of the pore fluid ρ_w is known. Using data of the EG-6 borehole located in the Ein Gedi (Site II in figure 4.140) west of the sinkhole zone where water salinity varies in the range of 50–100 g l^{-1} or ρ_w = 0.13–0.07 Ωm (Yechieli et al 2004) and

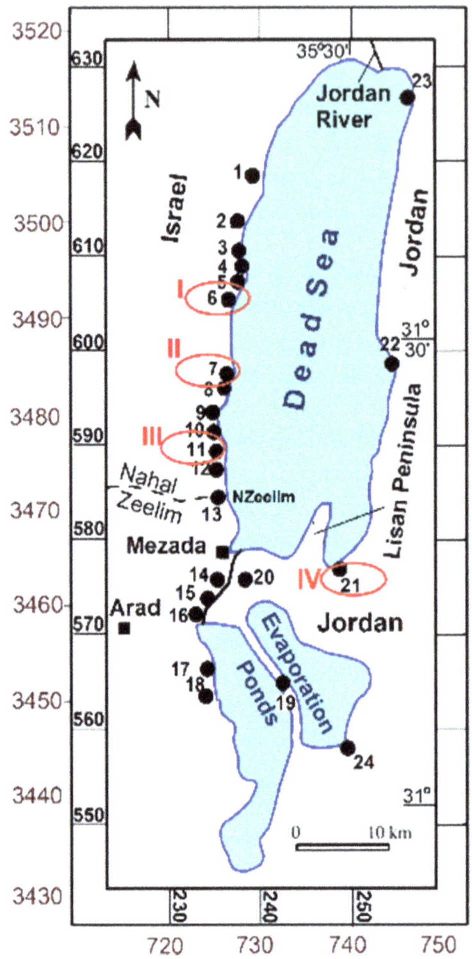

Figure 4.140. Location map of the ERT study sites. Black points—sinkhole sites. Red ellipses—study sites, 6—Mineral Beach (Site I); 7—Ein Gedy (Site II); 11—Nahal Hever south (Site III); 21—Ghor Al-Haditha (Site IV). The complete list of sinkhole sites is shown in figure 4.1 in section 4.1. (The coordinates in the frame are the new Israel Mercator grid in km, the outer coordinates are the Universal Transverse Mercator grid, in km.)

ρ_x = 0.86–0.46 Ωm—(unpublished TEM data) we can estimate the porosity out of the sinkhole zone as $\phi = 0.35$. These values are in good agreement with the laboratory porosity estimates (0.30–0.41) for alluvial fans in Jordan for similar geological conditions (after Taqieddin *et al* 2000). Then we will fit the model resistivity section to the measured one varying the porosity in the anomaly zone.

The resistivity model is represented in figure 4.141(a).

There are two layers in the section. The uppermost layer from the surface to 20 m deep with a resistivity of 20 Ωm simulates the soil matrix of the aerial zone. The upper layer is underlain by the low resistivity one (of 1 Ωm resistivity) representing the sediments saturated with the DS brine. Let us assume the section is of 1 m thickness and its properties are varied through the area (depth–distance plane) and

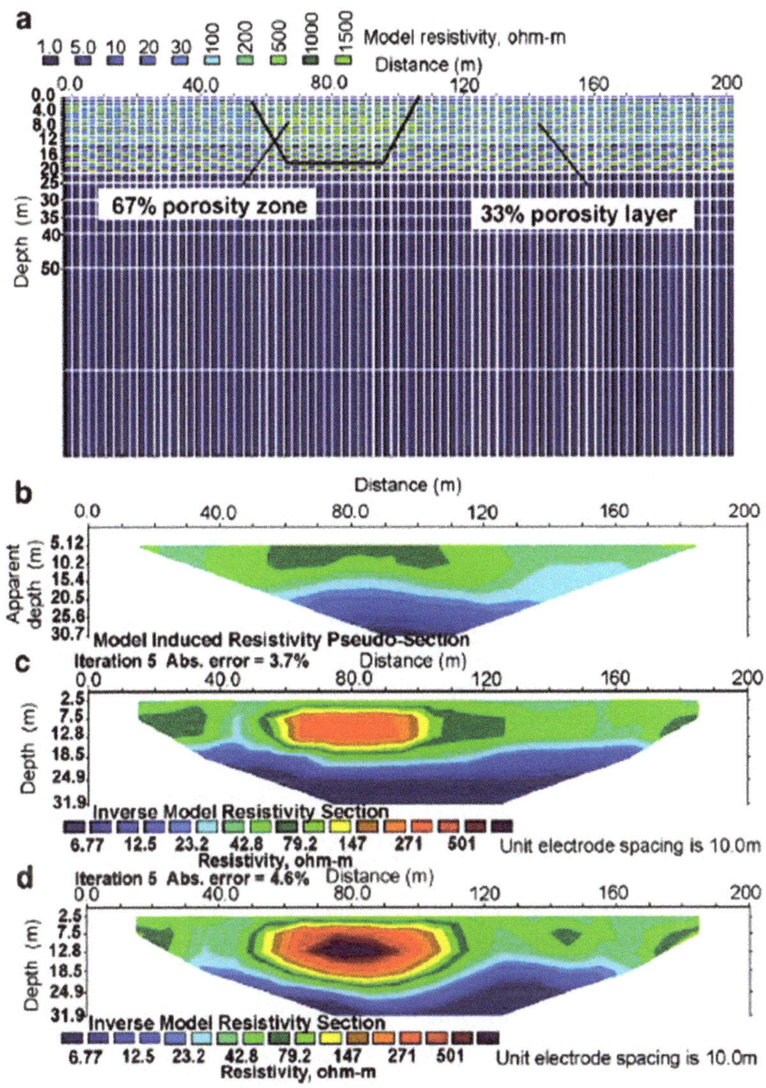

Figure 4.141. Ein Gedi site. (a) Start resistivity model section; (b) model induced apparent resistivity pseudo-section; (c, d) and 2 inverse resistivity sections: modeling the air-filled elements by the resistivity $\rho_x = 1500\ \Omega\text{m}$ (c) and $\rho_x = 10\ 000\ \Omega\text{m}$ (d) for the Ein Gedi sinkhole development area. (Reprinted from Ezersky (2008), copyright (2008) with permission from Elsevier.)

they are constant along with its thickness (in normal to figure direction). Then we can consider the volumetric characteristics of the layer (like porosity, for instance). The uppermost layer is divided by small ($1 \times 2\ \text{m}^2$) elements. The size of elements is less than electrode separation by factor 4 and it is less than the thickness of the layer by factor 10. Every third element has a resistivity of 1500 Ωm modeling the air-filled void. High resistivity elements are homogeneously distributed throughout the layer. Thus 33% of the layer volume is occupied by the air-filled elements and the other

67% are the soil matrix. Taking into account that size of the elements is considerably less than the characteristic size of the measurement array and geometric size of the section we suggest that our model roughly simulates the 33% porosity soil. Then we select a zone in the central part of the section (between distance 50 and 100 m) and add there more 33% high resistivity elements. The central part of the section simulates now the zone of 67% porosity.

The forward calculation of the above model using the finite element method results in the apparent resistivity pseudo-section shown in figure 4.141(b). Finally, the inversion procedure of this section results in the inverse model resistivity section represented in figure 4.141(c). 5% noise was added to the apparent resistivity pseudo-section.

It can be seen from figure 4.141(c) that the decompaction zone in the central part of the upper layer composed of porous soil with air-filled pores (porosity $\phi = 67\%$) generates the high resistivity anomaly of 200–500 Ωm, whereas the soil with a normal porosity of 33% has a resistivity of 40–80 Ωm. Altering the resistivity of the high resistivity elements by factor 10 (up to 10 000) we obtained the high resistivity anomaly of the same approximate area in the range of 200–900 Ωm (figure 4.141(d)).

An important conclusion may be made from the results of modeling. Despite the non-uniqueness of the performed calculation, modeling indicates that the soil decompaction in the range porosity of 33%–67% would be the plausible reason to explain the shallow resistivity anomaly of 200–900 Ωm and beyond (Ezersky 2008). This result gives the basis for the performance of ERT and GPR studies in the fan areas of the DS for cavity detection. On the other hand, the joint interpretation of the ERT and GPR data would narrow down the non-uniqueness of results (Ezersky 2008). The same can be said concerning microgravity and ERT combining that will be shown in continuation (Ezersky 2003a).

4.7.1.3 The electrical structure of the subsurface in the sinkhole occurrence sites based on ERT studies Mineral Beach (Site I in figure 4.140)

Five ERT lines were surveyed at the Mineral Beach area (Site I in figure 4.140) during a period of active sinkhole development (2011–12) over an area of 300 × 550 m^2 (figure 4.142(a)). All ERT data have a high noise level caused by the high contact resistance of the electrodes with the dry alluvium. Despite the noise, the 2D inversion sections (figure 4.142(b) Lines 1–5) show a high resistivity anomaly ($\rho_x > 1000$ Ωm) at depth of ~13 m and width of 10–30 m extended perpendicularly to all the lines. This anomaly matches well with the salt edge delineated using the seismic refraction method (Frumkin *et al* 2011) that is presented in figure 4.142(a) (lilac thick line).

The spatial resistivity distribution is well seen on the resistivity maps constructed through the Mineral Beach area presented in figure 4.143. Such presentation is similar to the quasi-3D ERT method described by Ezersky (2008). The method is based on extrapolation resistivities between the separate lines and allows detection of heterogeneities with size more than the separation between the ERT lines.

One can see in figure 4.143(b) that there are high resistivity anomalies presumably associated with a decompaction of the subsurface. Zones of higher resistivities are

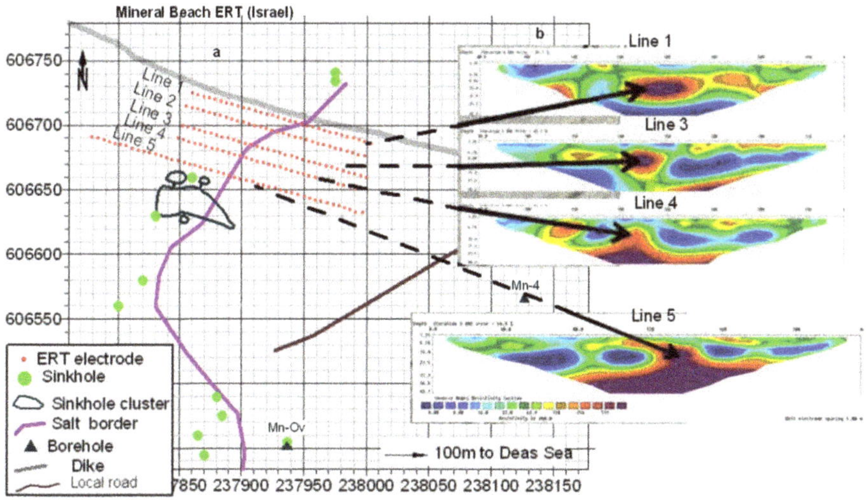

Figure 4.142. ERT results in Israel and Jordan. (a) ERT location map at the Mineral Beach area; (b1–b4) inverse resistivity sections along ERT lines; (c) location map of two ERT sections at the Ghor Al-Haditha area; (d1 and d2) inverse resistivity sections along respective ERT lines.

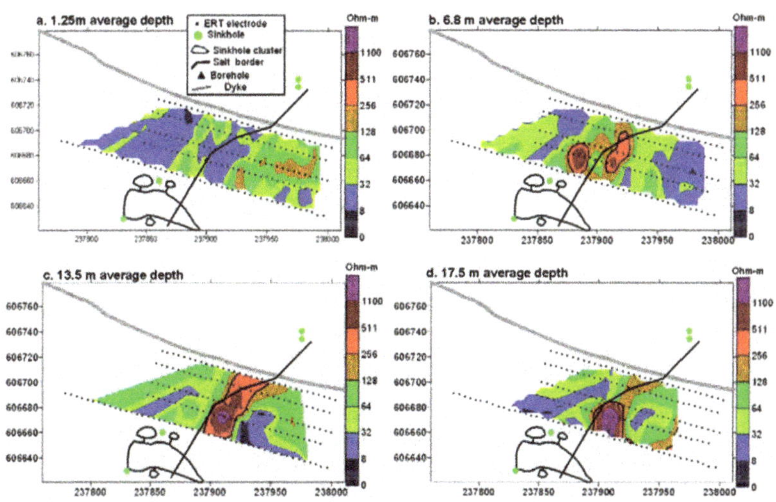

Figure 4.143. Inverse resistivity maps through the Mineral Beach area constructed for different depths using 2D resistivity sections (quasi-3D survey after Ezersky (2008)). (a)–(d) Maps for different average depths denoted at maps. A large sinkhole cluster is shown south to resistivity maps.

drawn towards the salt edge (shown in all figures 4.143(a)–(d) by the solid black line). Note that the shallowest map is composed of resistivity values ranged mainly between 8 and 128 Ωm. High resistivity anomaly ($\rho_x > 260$ Ωm) appears on maps at 6.8 m (figure 4.143(b)) and deeper depths (figures 4.143(c) and (d)). Sections 4 and 5 in figure 4.142(b) show the high resistivity zone at 24–26 m deep that is directed in the SE direction where the large sinkhole is located.

Following the results obtained at Mineral Beach, we hypothesize that the high resistivity anomalies detected by ERT lines are related to soil decompaction caused by the gradual collapsing of soil into dissolution cavities at 40 m deep (Frumkin *et al* 2011). The decompaction gradually ascends with time to the surface resulting in sinkhole formation. Perhaps, cavities were existing for a long time like that detected by the borehole Mn-Ov (figure 4.142(a)). Availability of the pre-existing sinkholes was suggested by Legchenko *et al* (2008a). Abelson *et al* (2018) reported the heightening microseismic activity in the vicinity of the salt edge detected during nanoseismic monitoring in Mineral Beach. Thus, we suggest that the high resistivity of the subsurface is caused by dissolution cavities within the salt layer along the dissolution front. Falling of the soil and stones into these cavities causes micro-seismic events recorded by Abelson *et al* (2018). They suggested also that the dissolution of cavities within salt was formed 15 years before their appearance at the surface. Note that a great sinkhole collapsed at the location of ERT lines in 2015 (see photos in figure 4.55(b)section 4.3.1 this manuscript).

4.7.1.4 Ein Gedi area (figure 4.140, site II)

The Ein Gedi area is situated in the central part of the DS shoreline in the Nahal Arugot alluvial fan about 20 m above the DS level (figure 4.140, Site II). The typical geological section 1 consists of cobbles, pebbles, and boulders of varying sizes, as well as sand, silt, and marl. A 4–20 m thick salt layer was detected at a depth interval of 50 to 70 m at several boreholes. The western limit of the salt layer is located between the boreholes EG-7 and EG-1/EG-6. The groundwater is characterized by a very high salinity similar to the DS water. However, hydrogeologists distinguish some conventional interfaces separating zones of different salinity, based on electrical resistivities, such as a brine (with electrical resistivity $\rho_w < 1\ \Omega$m) in the eastern part of the area towards the DS, and a fresh (or brackish) water in the west part of the area ($\rho_w > 3\ \Omega$m). The intermediate zone is interpreted as diluted brine (with resistivity of $1\ \Omega\text{m} < \rho_w < 3\ \Omega$m) (Kafri *et al* 1997).

Fifteen image lines were surveyed at the Ein Gedi area during a period of active sinkhole development (in 2001–2) over an area of 300×550 m^2 (figure 4.144) (Ezersky 2008).

The sinkholes in the surveyed area are formed in a strip 50–70 m wide and 300–500 m long, extending approximately in a north–south direction. The sinkholes are arranged along a tortuous line within this strip.

Typical inverse resistivity sections along lines are shown in figure 4.145.

As a rule relatively high resistivity anomalies ($\rho_x > 200\ \Omega$m) are surrounded by a low resistivity background ($\rho_x = 50$–$100\ \Omega$m) in all sections. Resistivity maps for different layers were constructed from 2D linear surveys (figure 4.146).

Two inverse resistivity–depth sections were compiled along lines A–A′ and B–B′ (figure 4.146(a)).

From the aforementioned materials follow some interesting and important conclusions.

1. The resistivity maps of lower layers (figures 4.146(a) and (b)) testify that large high resistivity anomalies are penetrating under Highway #90

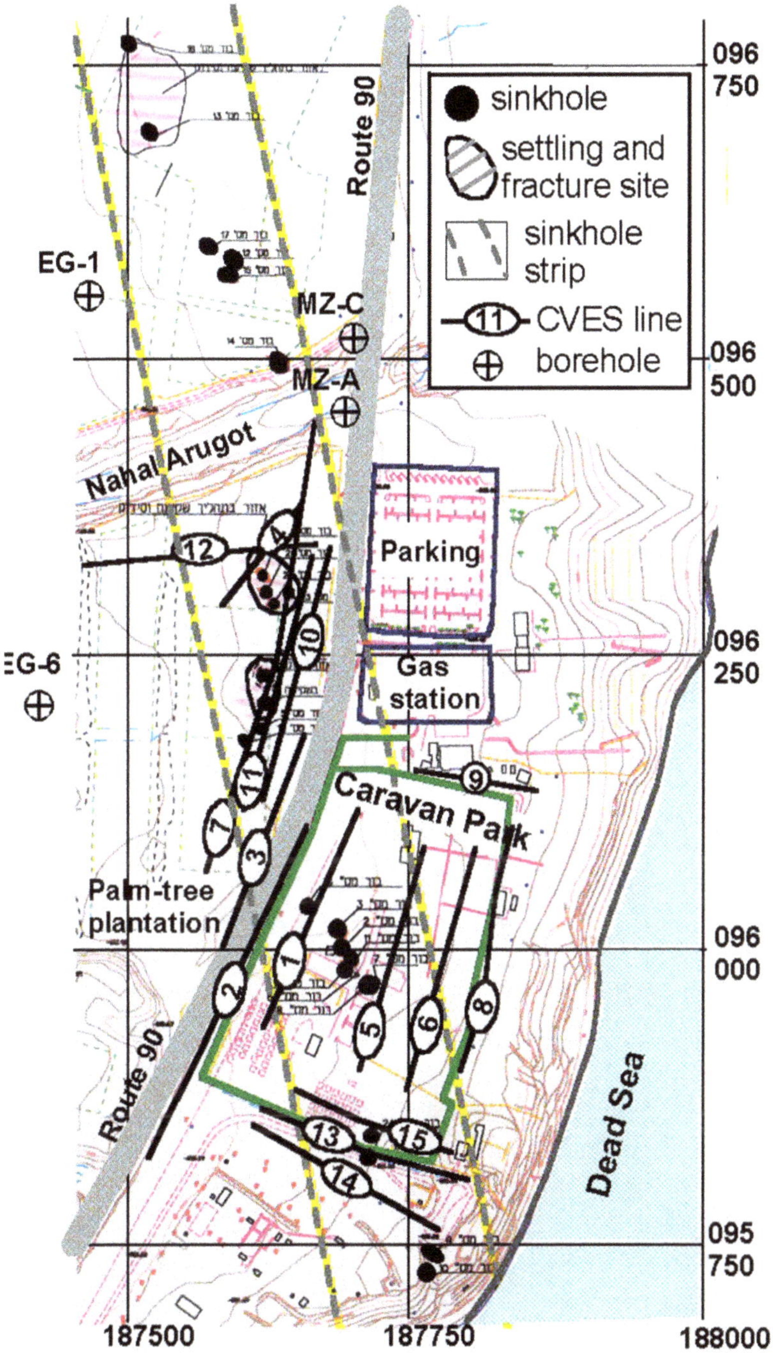

Figure 4.144. ERT survey layouts in the Ein Gedi area. (Reprinted from Ezersky (2008), copyright (2008) with permission from Elsevier.)

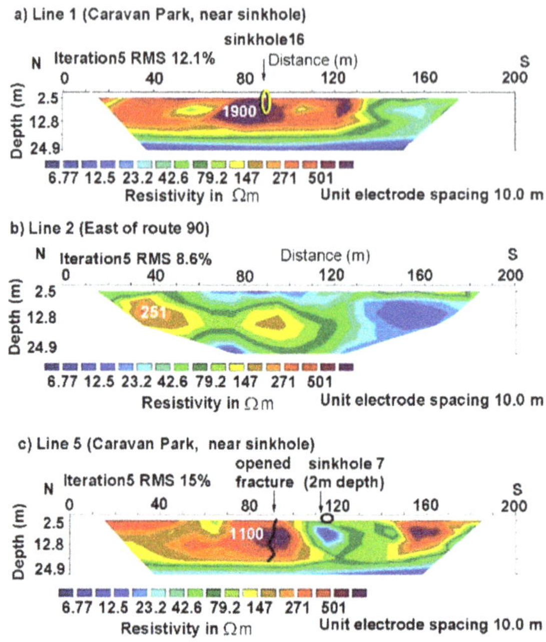

Figure 4.145. Three typical inverse resistivity sections from the EG area are shown (numbers within anomalies are maximum electric resistivities in Ωm). (Reprinted from Ezersky (2008), copyright (2008) with permission from Elsevier.)

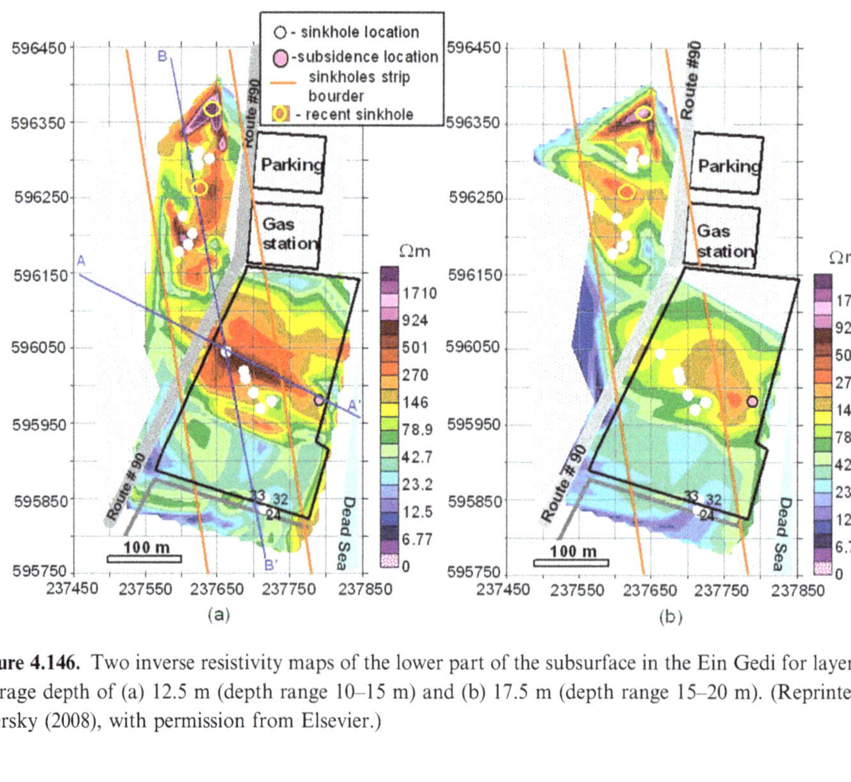

Figure 4.146. Two inverse resistivity maps of the lower part of the subsurface in the Ein Gedi for layers at an average depth of (a) 12.5 m (depth range 10–15 m) and (b) 17.5 m (depth range 15–20 m). (Reprinted from Ezersky (2008), with permission from Elsevier.)

(measurements of 2000–1 years). These anomalies predict sinkhole susceptibility in the highway pavement. The highway was repaired and fortified in 2008–9. A strong membrane was installed under the asphalt. However, in 2011 accelerating subsidence linked to sinkhole was measured by InSAR (Nof *et al* 2019). We have discussed it in section 4.10 (figure 4.193).
2. Within sandy-gravel sediments, sinkholes induce high resistivity anomalies.
3. The large sizes of resistivity anomalies (250 × 300 m^2), which are considerably greater than those of the sinkholes, form one of the features of the sinkhole sites in the Ein Gedi area. The anomalies continue down to the water table or even deeper (maximum of 25–35 m depth). A low resistivity layer of 1–8 Ωm underlies high resistive layers. The combined analysis of the image results and other geophysical data shows that the high resistivity anomalies are associated with the soil mass decompaction at the sinkhole development sites and surrounding areas.

4.7.1.5 Nahal Hever south (Site III in figure 4.140)

The Nahal Hever South site is located in the northern DS basin some south of the Ein Gedi (figure 4.140 for location). The geological structure of the site according to

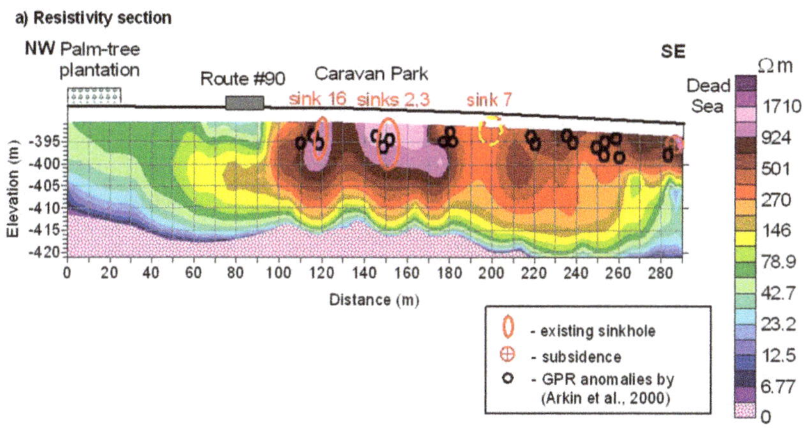

Figure 4.147. 2D inverse resistivity cross-section along line A–A′ (see figure 4.146(a) for location) scale ratio $X:Y = 2:1$. (Reprinted from Ezersky(2008), copyright (2008) with the permission of Elsevier.)

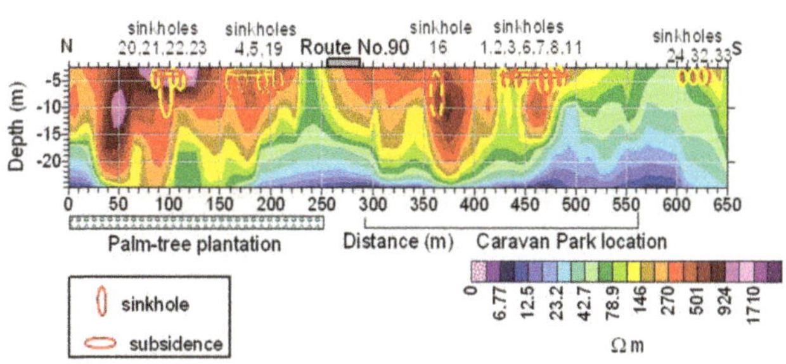

Figure 4.148. 2D inverse resistivity cross-section along section B–B′ (see figure 4.146(a) for location). Scale ratio $X:Y = 1:5$. (Reprinted from Ezersky (2008), copyright (2008) with the permission of Elsevier.)

the boreholes HS-1, HS-2, and HS-3 (Yechieli *et al* 2004), is schematically presented in figure 4.149. The section is composed of alluvial fan sediments down to a depth of 18 m, a 5 m thick marl layer, and an 11 m thick salt layer. The salt layer is underlain by marl and gravel. The DS marl was named by Frydman *et al* (2008) as 'lime carbonate' (see section 4.8 for details). The elevation of the area is approximately −393 m. The water table in the area was measured at a depth of 17.2 m in October 2000 and 18.5 m in April 2002. The top of the salt layer lies at a depth of 24 m. In borehole SH-3, a cavity (filled with mud) was detected at the depth of 23–29 m, which is assumed to be the result of the salt layer dissolution The visible section of sinkholes consists of sand–gravel intercalating clay–sand layers; iron oxide mineralization is also visible.

Thirteen sinkholes were formed in the Nahal Hever South area (Ezersky *et al* 2009). A few sinkholes that were identified in this area in 1999 also appear in aerial photographs as far back as 1990.

Different geophysical measurements were carried out in the Nahal Hever South area during the international project NATO SfP 981128. The results of these studies are presented in numerous articles (Ezersky *et al* 2009, 2010, Legchenko *et al* 2008a, 2008b, 2008c, 2009).

ERT study was carried out in the framework of the Ministry of National Infrastructure of Israel (Ezersky 2003a, 2003b). Examples of four Nahal Hever South ERT sections are shown in figure 4.150.

The Nahal Hever South inverse resistivity sections differ from those in the Ein Gedi by a lower resistivity. The resistivity range of the Nahal Hever South site is 1–200 Ωm, whereas that of the Ein Gedi sections varies between 5 and 800 Ωm (figure 4.145). The background resistivity in the Nahal Hever area is 20–40 Ωm, whereas in Ein Gedy that is 40–60 Ωm. The subsurface section is complicated by a relatively high resistivity of 50–200 Ωm which appears as a separate anomaly. There it seems a good agreement between these high resistivity anomalies and the sinkholes shown in figure 4.150. The lower part of the sections has a very low resistivity of 1–2 Ωm.

Resistivity distribution in the Nahal Hever area represented in figure 4.151 is heterogeneous (Shtivelman *et al* 1999, Ezersky 2003a, 2003b). The eastern part of the area composed of marl (LC) is characterized in its upper part (0–8 m, deep) by

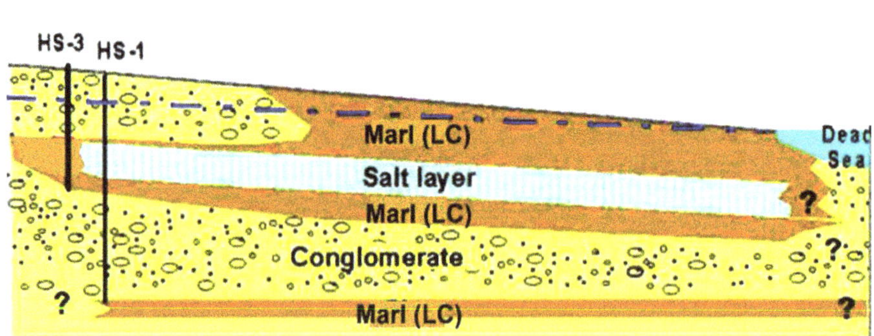

Figure 4.149. Geological section of the Nahal Hever southern area (after Yechieli *et al* (2002) with permission from GSI).

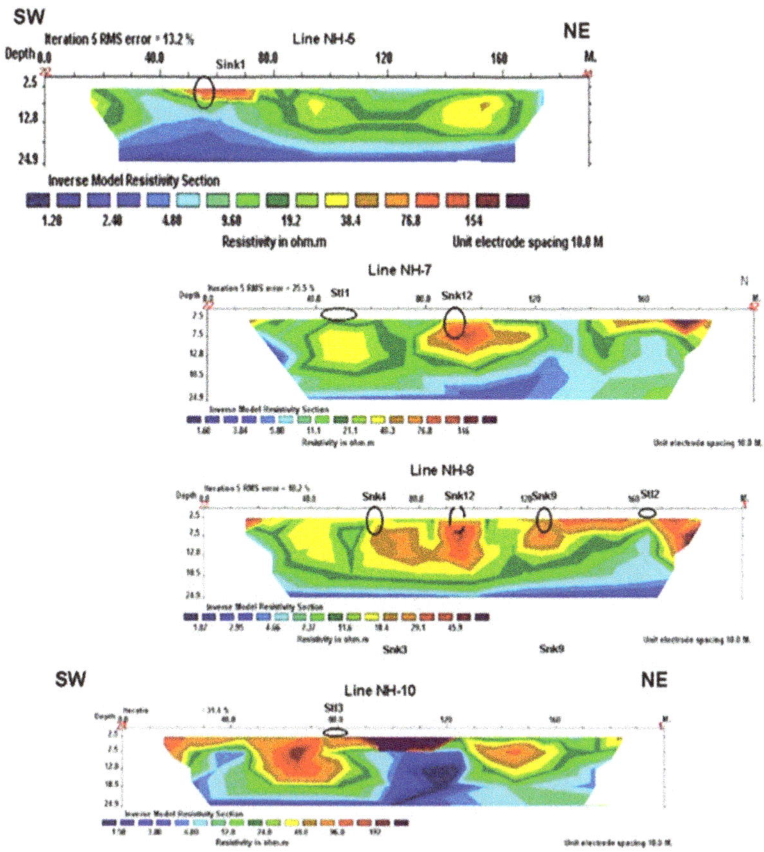

Figure 4.150. Examples of the resistivity sections obtained at the Nahal Hever site. Black ellipses are the sinkhole locations on the corresponding section.

the resistivities of 8–10 Ωm, whereas the area west of the sinkholes is characterized by resistivities of 25 Ωm and beyond (figures 4.151(a) and (b)). It is caused by the features of the geological structure of the site composed of the carbonate clays to the east and sediments of alluvial fans to the west. At the resistivity section (figure 4.151(c)) one can see characteristic low resistivity anomalies shaped like a bell. These anomalies are associated with runoff water flows descending from the surface during the rainfalls and floods, interfering with the salt dissolution (Ezersky *et al* 2007, Legchenko *et al* 2007).

Two more compiled inverse resistivity sections along the sinkhole lines are presented in figure 4.152. The boreholes, sinkholes, and ERT line location maps are also shown in these figures. As a rule, the sinkholes coincide with the relatively high resistivity anomalies at the uppermost subsurface. If 20 Ωm is the background resistivity of the upper part of the sections, then the *sinkhole site resistivity-to-background ratio* (later referred to as *resistivity ratio*) will be 2–6. For recently formed sinkholes (after 2000) this ratio is usually higher.

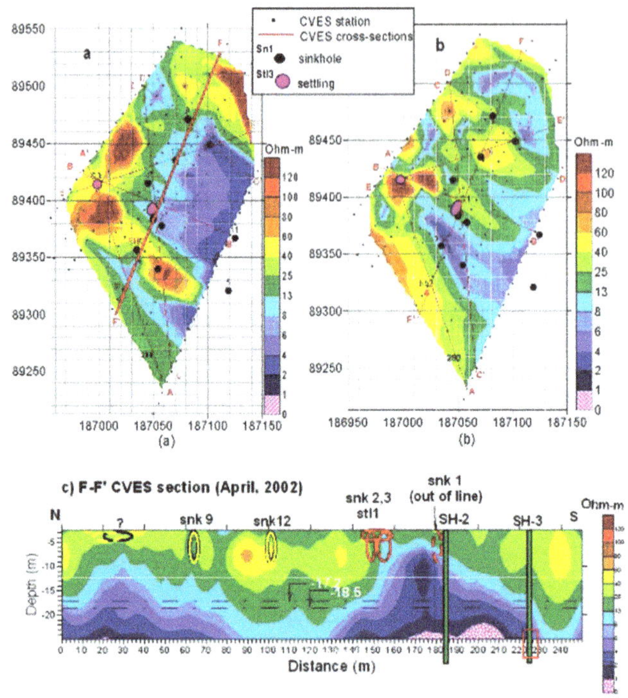

Figure 4.151. The Nahal Hever south area. Inverse resistivity maps at depths range (a) 0–5 m; (b) 5–10 m; and (c) resistivity section along with F′–F line.

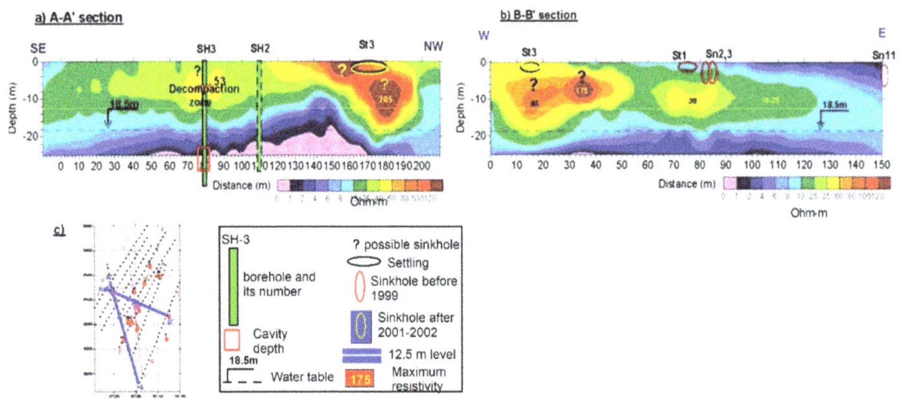

Figure 4.152. A′–A (a) and B–B′ (b) sections and their layouts (c) at the Nahal Hever area.

Section A–A′ (NW–SE) is presented in figure 4.152(a). The section intersects settling (subsidence) 3 and borehole SH-3, which revealed the void within the presumed salt layer (Yechiely *et al* 2004). The resistivity anomaly of 40–60 Ωm appears at the SH-3 location at a depth of 6–12 m (resistivity ratio is 2–3). The drilling report notes that the soil decompaction (rapid drill passage) occurs precisely at this depth interval. The large resistivity anomaly of 100–200 Ωm is seen at the settling 3 location at the subsurface of the northwestern part of the section. The anomaly descends downward to a depth of 18–20 m. The very low resistivity horst anomaly (of 8.1 Ωm) is located close to the high resistivity anomaly. The water table level (18.5 m) is shown by a dashed line.

Section B–B′ (W–E) intersects settling 3 and a group of old sinkholes 2 and 3 and settling 1 (figure 4.152(b)). A similar resistivity anomaly (as in figure 4.152(a)) is seen at the site of settling 3 (maximum at depth 5–12 m) and a relatively weak anomaly (resistivity ratio of 1.5–2.0) is located just under the old sinkholes group (with a maximum at depth 6–12 m). The old sinkhole group is located at the marl boundary (resistivities are 13 Ωm).

Discussion. The resistivity distribution throughout the Nahal Hever southern area is characterized by a more complicated resistivity structure than the Ein Gedi area. The distinguishing futures of this area are:

- Low resistivity of the uppermost near-surface layer (10–20 Ωm) composed of carbonate clay (marl).
- Slight cavity-related resistivity anomaly. The maximum resistivity, as a rule, is located at the depth interval of 5–12 m. The most resistive (expressive) anomaly is located in the northwestern part of the area where marl passes to the sandy-gravel sediments (Ezersky *et al* 2009).
- Presence of the very low resistivity anomalies under or very close to sinkhole locations. These anomalies have a horst-like (bell) shape with the base located at the bottom of the section and the top situated at a depth of 5–10 m. Increased resistivity anomalies are connected with the soil decompaction in the subsurface. This is partially confirmed by the drilling log of the borehole HS-3, where the fast promotion of the drill was noted in the depth interval 6–12 m.

Generally, considering the section in figure 4.151(c) one can see that in a number of the cases there is some correlation between the low resistivity horst-anomalies and the sinkhole locations, whereas in other ones such correlation is absent.

4.7.1.6 Ghor Al-Haditha (Jordan) (Site IV in figure 4.145)

Numerous ERT resistivity studies performed in the sinkhole development sites in Jordan have shown a relatively high resistivity structure of the uppermost section (Al-Zoubi *et al* 2007, Frumkin *et al* 2011, Ezersky *et al* 2013c). For instance, profile 3, is presented in figure 4.153(b). The location is shown in figure 4.153(a). The profile is 590 m long, the resistivity varies vertically along the first 400 m (figure 4.153(b)). The moderate resistive layer (95–250 Ωm) at the southeast part of the profile has the thickness of this layer gradually increased toward the northwest. The continuous high resistive layer with a resistivity of more than 500 Ωm is extended between the distances '0' and '390 m'. The low resistance values less than 15 Ωm at the bottom of the section characterizes the water-saturated part of the section. The western part of the section where the resistivity changes horizontally and vertically represents the area of the active sinkhole, and the two high resistive zones at distances 400 and 450 m represent the known buried sinkholes.

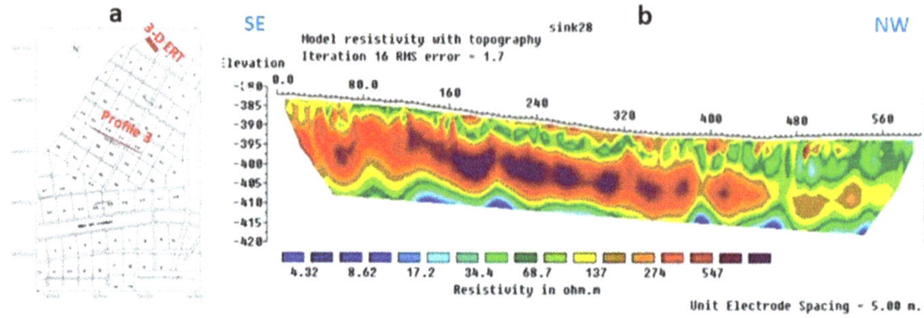

Figure 4.153. ERT measurements in the Ghor Al-Haditha (Jordan). (a) Location map, (b) the line 3 inverse resistivity cross-section. The southeast part of the cross-section where the resistivity is varied both laterally and vertically represents the active sinkholes area and the two high resistive zones at distances of 400 and 450 m represent known buried sinkholes. (Reprinted from Al-Zoubi *et al* (2007), with permission from Science Publication.)

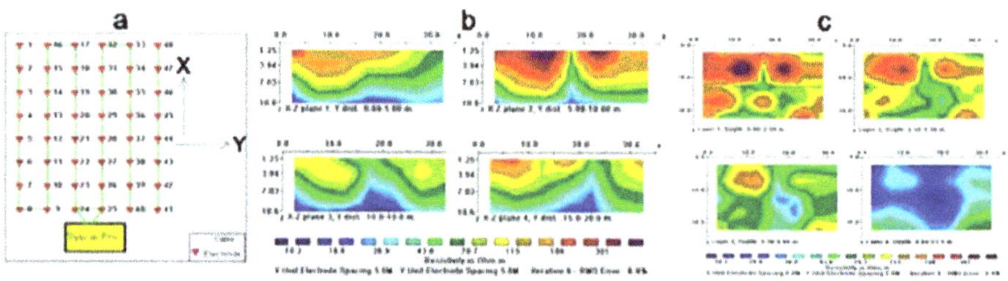

Figure 4.154. (a) 3D ERT study in Jordan. (a) The layouts of potential electrodes for the 3D Wenner–Schlumberger electrical array. (b) Vertical *X–Z* inverse resistivity cross-sections along *Y*-axis at distances = 0–5, 5–10, 10–15 and 15–20 m. (c) Horizontal *X–Y* inverse resistivity maps along *Z*-axis at layer depths of 0–2.50, 2.5–5.0, 5.0–7.5, and 7.5–10 m. (Materials Al-Zoubi *et al* in Ezersky *et al* (2013c), reprinted with permission from Dr Abdelrahman Aqel Abueladas.)

Figure 4.154 shows an example of a 3D multi-electrode imaging survey that was carried out at the north of the Ghor Al-Haditha area (see figure 4.153(a) for location) (materials of Al-Zoubi in Ezersky *et al* (2013c)). The Wenner–Schlumberger configuration was used for the survey. The 48 multi electrodes cables were oriented in the *X*-direction, which is approximately east to west with a 5 m spacing for *X* and *Y* directions (figure 4.154(a)). The outputs of the 3D modeling are a series of vertical sections (35 m length and 11 m deep) along with one of the horizontal axis (figure 4.154(b)) and horizontal slices (resistivity maps) 25 × 35 m^2 at different depths ranges across the areas of the survey grid (figure 4.154(c)). In our case, the vertical resistivity cross-sections are presented by *X–Z* planes located along the *Y*-axis with distances of 5 m (figure 4.154(b)). Because of the limited sizes of the resistivity array (here, 35 m along *X* and 25 m along with *Y* directions) investigation depth is also limited by 11 m (compare with 30–40 m investigation depth along line 3). Figure 4.154(c) shows the resistivity changes for the slices with depth through

X–Y horizontal plane from the surfaces to 10 m deep. The resistivity of the layers decreased as the horizontal distance increased from the surface.

Conclusions. In the DS coastal areas, the resistivity of the uppermost subsurface depends on the site elevation, water table location, and lithology. The water table defines the limit of the penetration depth of direct (low frequency) current. In the northern DS basin area depth from the surface to the water table varies in the range of 10–27 m. With the drop in the DS level, this depth increases. In the southern basin (evaporation ponds) where the water level is heightened at a rate of 20 cm yr^{-1}, depth to the water table is decreased, and as a rule that is of 3–5 m. In the southern part of the Newe Zohar area located in an alluvial fan, the depth to the water table is exceptionally larger, 8–10 m. In sites located close to −400 m elevation and lower, subsurface resistivity varies in the range of 1–20 Ωm.

The geoelectric studies carried out in the sinkhole development sites indicate resistivity of the subsurface of tens to hundreds of Ωm and reach thousands of Ωm in the sinkhole development sites located in alluvial fans (Ein Gedi, central and northern Mineral Beach, Newe Zohar). It enables using the GPR method for void detection in the areas of the sinkhole development along the DS shore. However, low resistivity areas also occur (as a rule at below −400 m elevation). Analysis of geoelectric studies has shown that the resistivities in these sites are directly associated with the subsurface structure and are the most important for human activity (dwellings, routes, agricultural areas, etc). The subsurface structure is characterized by tens to hundreds of Ωm that allow penetration of the electromagnetic waves down to 5–15 m. Our conclusions are based on the real studies and measurements carried out in the sinkhole development sites.

It should be denoted that a 3D resistivity survey requires great efforts in the organization of the field setup, but the obtained results are limited by the number of electrodes and multi-electrode boxes. We consider that the technique of quasi-3D with the organization of parallel 2D lines and data processing of pseudo-3D sections will be more effective (Ezersky 2008, Kidanu *et al* 2016, 2020)

Separately, we have to consider the reliability of resistivity measurement results in comparison with results of other methods such as microgravity, GPR, etc. This issue will be considered in chapter 5.

4.8 Borehole and laboratory geophysical testing of soil and salt

(The method and results are presented in detail in Report M6/M27-050 by Ezersky and Livne (2013).)

To investigate geophysical and geotechnical properties of the lithology of sinkhole sites and relationships between the geophysical and geotechnical rock and soil properties *in situ* boreholes drilling was carried out in the DS sites composed of typical rocks.

4.8.1 Boreholes drilled for the project

Two types of lithology were studied *in situ* and laboratory conditions to reveal the geotechnical and geophysical properties of these soils and salt layers. Studied sinkhole sites are shown in figure 4.155. Two boreholes MN-5E and EB-3E, of

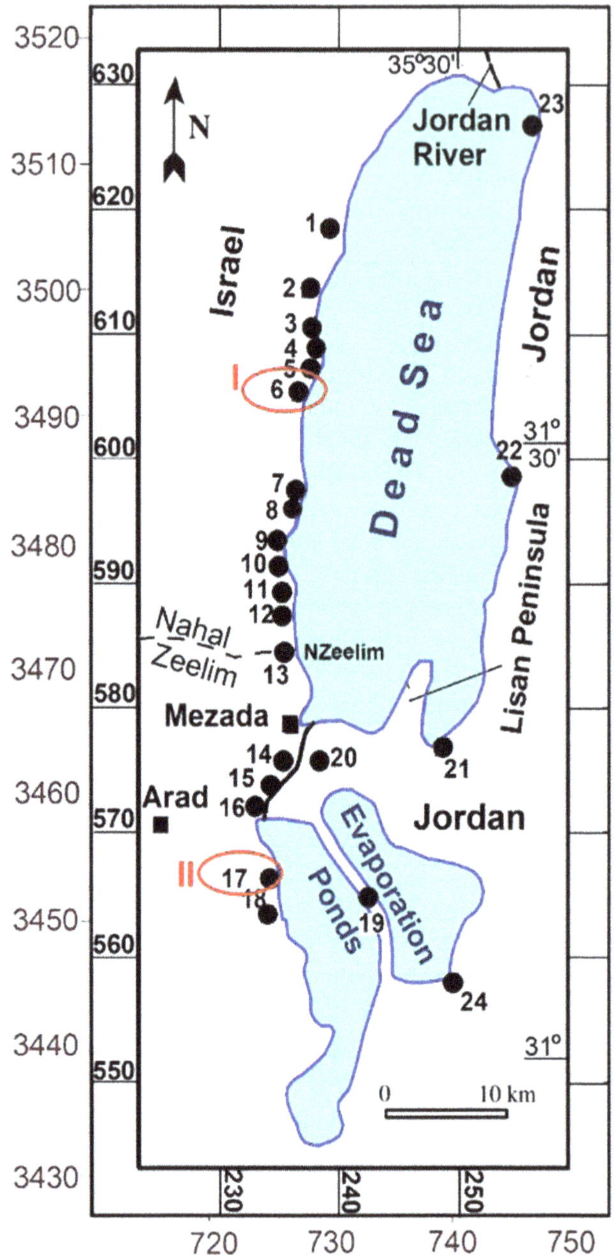

Figure 4.155. Location map of the survey sites. Black points—sinkhole sites. Red ellipses—drilling sites, 6—Mineral Beach (Site I), 17—Ein Boqeq (Site II). The complete list of sinkhole sites is shown in figure 4.1 in section 4.1 (the coordinates in the frame are the new Israel Mercator grid in km, the outer coordinates are the Universal Transverse Mercator grid, in km).

Figure 4.156. (a) Drilling of the Mineral borehole Mn-5E, and (b) its location on the Mineral Beach map. (c) Drilling of the Ein Boqeq borehole EB-3E, and (d) its location on the Mineral Beach map.

clayey and sandy-gravel, respectively lithology, were drilled in the Mineral Beach (figure 4.156(b)) and Ein Boqeq (figure 4.156(c)) sites jointly by Geophysical Institute of Israel (GII), Geological Survey of Israel (GSI) and Ben Gurion University of the Negev (BGU).

Figures 4.156(a) and (c) display, respectively, the Mineral and Ein Boqeq drilling sites photographs, and figures 4.156(b) and (d) show, respectively, the MN-5E and EB-3E drilling site locations. The Mineral area, south of the spa (figure 4.156(b)) is composed of the very plastic carbonate clays characterizing the mudflat areas of the DS (Frydman *et al* 2008). The Ein Boqeq site is composed of sandy-gravel sediments characterizing the young alluvial fan (Arkin and Gilat 2000). In both sites, clastic sediments are underlain by a massive salt layer.

One can see in figure 4.156(d) that the Ein Boqeq site borehole is located some 10–20 m from the DS shoreline, whereas in the Mineral Beach, site distance from DS water to the borehole was ~200 m. The drilling water in the Ein Boqeq was pumped from the DS directly. For the Mn-5E borehole, a 1 m^3 tank was filled by the DS water brought from the sea.

4.8.1.1 Goal and objectives of drilling

Two drilling sites have presented typical geology of the western DS shore (sandy-gravel sediments and carbonate clays (marl) overlaying salt layer) (Arkin and Gilat 2000).

Therefore, the borehole drilling was aimed to reveal the geotechnical and geophysical properties of the above-mentioned soils and salt and relationships between the geophysical parameters (V_p, V_s, ρ_x) and geotechnical ones (porosity, strength, etc) for use of these factors in calculations of the Earth stability in the sinkhole development areas along the DS shores.

The objectives of the study are as follows:

1. To determine the distribution of the geophysical parameters (V_p, V_s, and ρ_x) along the borehole and in different lithologies.
2. To investigate the geotechnical properties of DS soils and salt *in situ* conditions (boreholes, SPT and V tests).
3. To study the laboratory properties of soils and salt (physical properties, undrained shear strength, s*u*, unconfined uniaxial strength, *qu*).
4. To search the inter-relationships between the laboratory and *in situ* properties and between geotechnical and geophysical parameters.

4.8.1.2 Methods of borehole drilling and geophysical investigations

The borehole drilling technique.

Drilling was carried out in two different stages.

Stage 1. The drilling of clastic sediments

The boreholes were drilled from the surface down to the salt top with the 5 inches instrument using a protective drill pipe. The bentonite mixture was used as a drilling fluid. The SPT testing (in some cases Vane test) were carried out with the interval of 1.5–2.0 m following the ASTM D 1586-08 requests. The cores were packed into the air isolated tube and transferred to the physical laboratory. Two–three undisturbed samples 0.5 m long were collected with the 4 inches tube.

Stage 2 of drilling -salt

After the top of the salt was reached, drilling was continued using the 4 inches instrument within salt without a protective pipe to allow measurements to be in contact with the salt walls. The drilling was carried out by the use of the DS brine as a drilling fluid. Salt cores were collected, packed, and transferred to the laboratory.

When drilling of the salt interval was finalized, some measurements were carried out, such as: (a) downhole; (b) ultrasonic logging; and (c) apparent resistivity logging.

Then, the protective pipe was replaced by the PVC pipe of 3 inches diameter, and the gap between the PVS pipe and soil was filled with the bentonite mixture following the ASTM D 4428 (ASTM 2006) request. Further downhole measurements were carried out within the PVC pipe to study clastic sediments following D 7400-07 standard (ASTM 2007).

I. *Geophysical investigations of boreholes*

Different geophysical measurements were carried out in boreholes just after their drilling: (1) downhole measurements; (2) ultrasonic logging and (3) resistivity logging by the method of apparent resistivity. Note: ultrasonic logging was carried out in the frequency range of 100–200 kHz, whereas downhole measurements are ranged from 100–300 Hz.

The following geophysical methods of the borehole investigation were used.

*Downhole (*ASTM D-7400-07)

The downhole survey is aimed to determine both the compressional and shear wave velocities of the soil mass. For the elastic wave excitation, the Digipulse source or the handle sledgehammer are used. The measurement scheme procedure for the *downhole study* is shown in figure 4.157. The measurement procedure following D 7400-07 ASTM Standard (ASTM 2007) for the downhole seismic testing is as follows: The generation of compression waves is carried out by the vertical stroke on a metallic plate (figure 4.157(a)).

The generation of shear waves involves setting a metal plate, such as a railroad tie, oriented horizontally and parallel to the transversal *t*-component and perpendicular to the *l*- and *v*-axes of the geophone array and striking with a sledgehammer in the horizontal direction.

In the present survey, we have used one borehole geophone intended for the velocity measurements and located in the borehole. The borehole geophone was moved (lowered) within the borehole. For the P-wave surveying, this geophone at the test start was situated at a depth of 1.0 m and coupled to the borehole wall (figure 4.157(a)). The vertical strokes were made on the surface close to the borehole (figures 4.157(a) and (b)). The vertical, longitudinal, and transversal oscillation components were received by

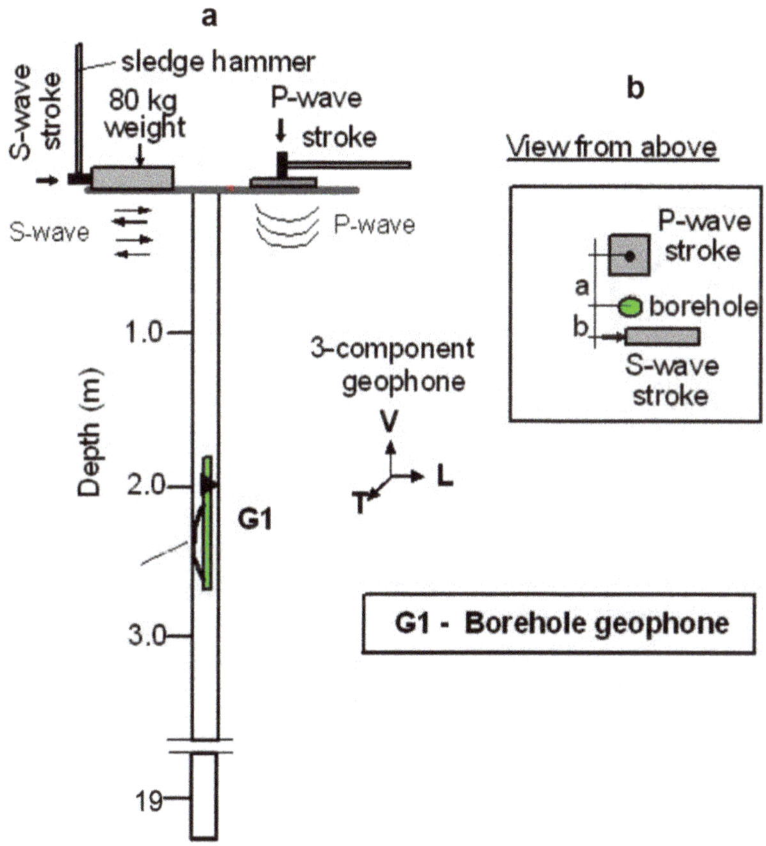

Figure 4.157. (a) Scheme of the downhole measurements. (b) A plan view of seismic source disposition at the site.

the geophone and recorded separately on the three channels of the *Geode* seismograph. The geophone is then moved down 1 m and the procedure is repeated. The S-wave surveying differs only by the strike direction as mentioned above. At each level, we applied the compression and shear energy using a sledgehammer.

Data acquisition parameters for the downhole measurements are as follows:

- Record length: 250 ms for both P and S waves;
- Sampling rate: 0.125 ms;
- Distance between the boreholes and sledgehammer during the downhole survey—2.80–3.30 m;
- Distance between the geophone levels within the borehole during the downhole survey—1.0 m.

In situ data acquisition in the borehole is represented in figure 4.158, source of seismic energy has been mounted on the Chevrolet truck (c1) (only the Digipulse frame has been shown).

There are two types of hammers mounted on the track. The first one (named Digipulse) is the hydraulic hammer for the P-wave excitation (figures 4.158(a2)). It strikes vertically on a metallic plate (a3) along the arrow. The second one is intended for S-waves excitation (figures 4.158(b4) and (c4)). It is a 30 kg weight sledgehammer striking horizontally on beam (c5) placed under the Digipulse weel. In some cases, the

Figure 4.158. Data acquisition in the downhole: (a) seismic source for P-wave excitation; (b) sledgehammer for shear wave (SH) excitation. (c) Digipulse: the seismic source for SH waves; (d) geophone. 1—Digipulse source based on a Chevrolet truck (the only frame that has been shown); 2—P-wave hammer; 3—metallic plate; 4—sledgehammer for horizontal stroke (SH wave excitation); 5—beam for horizontal stroke; 6—borehole, 7—geophone, 8—compass.

handle sledgehammer 9 kg weight was used (figure 4.158(b4)). The horizontal beam, in this case, was loaded by the wheel of a 4×4 tender as shown in b4. A three-component geophone BG-3 (manufactured by Geostaff Ltd, USA) is used (figure 4.158(d7)) for the elastic wave receiving within the borehole (d6). This geophone comprises three sensors: one is vertical and two others are mutually perpendicular horizontal sensors. The compass (d8), which is connected to the BG-3 geophone, allows orienting one of the sensors (longitudinal) in the borehole in the northern direction. Then the second sensor (transversal) is oriented in the west–east direction. Thus, orienting seismic horizontal source along either northern or western direction we reach the S-wave excitation in direction of most sensitivity of borehole geophone and measure SH wave velocity. In this survey, the S-wave source was directed along the east–west axis (in both boreholes).

Ultrasonic testing (D2845–05) (Ezersky-Goretsky 2014, Ezersky and Eppelbaum 2017.)

Method and equipment. [(4.12) In ultrasonic logging (USL) the elastic waves are spreading from a high frequency (50–100 kHz) transmitter to receivers along the borehole wall under investigation (Beck 1981). The 'dotty' (detailed) ultrasonic logging enables measuring elastic velocities within 0.1–0.2 m along the borehole (Savich *et al* 1969). Wave penetrates the rock approximately 0.1–0.2 m following the wavelength. The scheme of the dotted ultrasonic logging is presented in figures 4.159(a)–(c). A seven-transducer sonde (developed by Hydroproject Institute, Moscow, Russia), with 0.2 m separation between the sensors (figure 4.159(c)), is

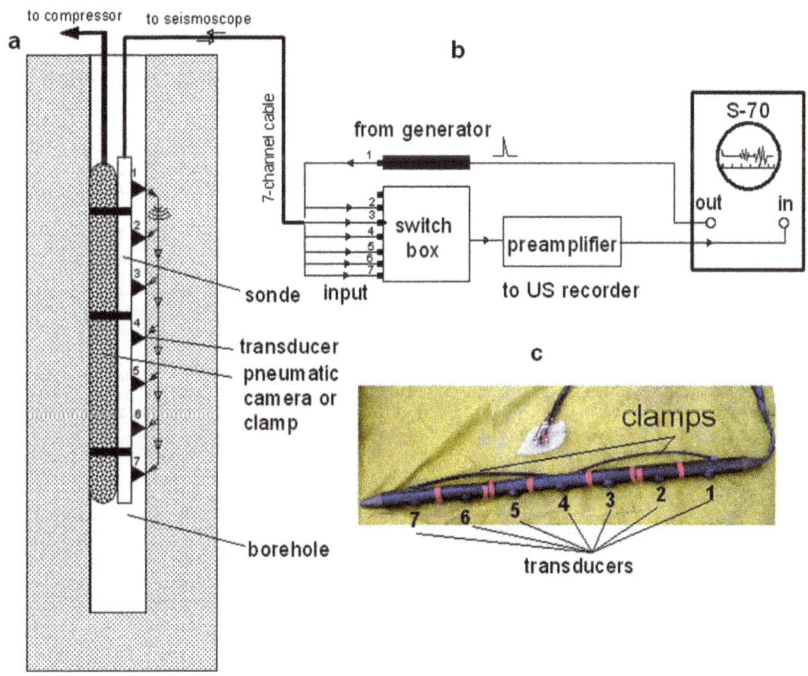

Figure 4.159. Borehole logging layouts. (a) Ultrasonic logging: seven-transducer sonde within the boreholes; (b) communication scheme; (c) and sonde photography. (Reprinted from Ezersky and Goretsky (2014), copyright (2014) with permission of Elsevier.)

placed into the borehole (figure 4.159(a)) and clamped onto the wall using both metallic clamps and a pneumatic camera pumped by air from the surface. It provided excellent contact between the sonde and the borehole wall. Every transducer (made of 70–100 kHz frequency piezoelectric ceramic) can act both as a transmitter and as a receiver when connected to a generator or oscilloscope, respectively. Switching is carried out with a switch-box (figure 4.159(b)). Elastic wave spreading along the borehole wall from transmitter arrives consequently to the receiving sensors where it is converted, gained, and visualized on the oscilloscope's electronic tube of the US-meter S-70 that was described above.

Considering ultrasonic oscillograms one can see that there are also the moments of wave triggering (start of transmission) on the tube (see figure 4.160). The time

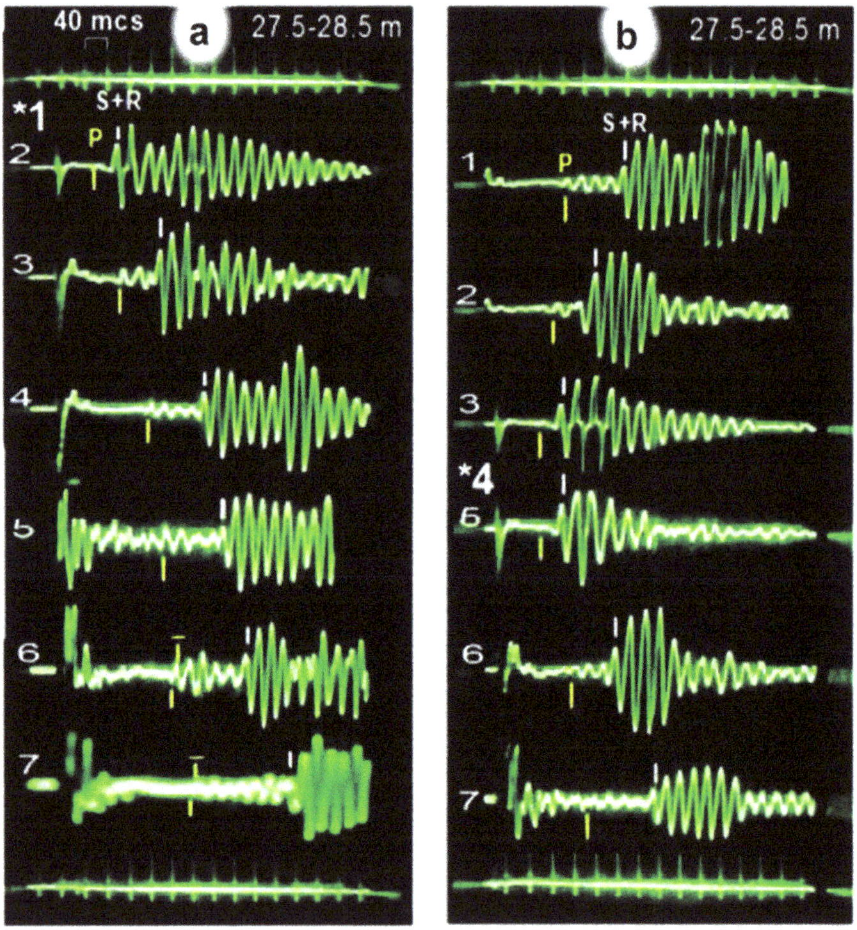

Figure 4.160. Oscillograms of ultrasonic profiling in boreholes. (a) First transmitter; second to seventh transducers are receivers (direct shot); (b) fourth the transducer is a transmitter, the first to third and fifth to seventh are receivers (split shot). P and S + R arrivals are marked by yellow and white, respectively. (For interpretation of the references to color in this figure legend, the reader is referred to the web version of this article.) (Reprinted from Ezersky and Goretsky (2014), copyright (2014) with permission of Elsevier.)

difference (t) between the time triggering and signal arrival at the nth receiver is the spread time between the transmitter and receiver along with a fixed distance equal to the $L = (n - 1) \times 0.2$ m (that is, separation) between the neighboring sensors. The wave velocity is then defined by simply dividing: $V = L/t$. Examples of the field records are presented in figures 4.160(a) and (b).

Transmitting transducer in figure 4.160(a) is that No. 1 marked on oscillograms by an asterisk (see ultrasonic sonde in figure 4.159(c), for explanation). Figure 4.160(a) displays oscillograms of ultrasonic pulses received at receivers 2–6. Arrival times of the P-waves are denoted by yellow vertical markers.

The character of the wave pattern. Generally, two types of waves are observed in the oscillograms (figure 4.160): (1) P-wave that contains different types of longitudinal waves (direct, head waves, and screw waves); and (2) S+R wave, that is, interfering vibrations of shear and surface waves that do not usually separate in records. The S+R waves are more intensive than P-waves. Arrival times of S+R waves are marked by green vertical markers. In fractured rocks, the intensity of the S +R wave is decreased in comparison with P waves. At that, the visible amplitude of the S+R wave is 10–20 folds more than the P-wave amplitude. In strongly fractured rocks only P-wave arrivals would be selected.

Detailed analysis of kinematic and dynamic characteristics P- and S+R waves shows:

1. P-wave is interference vibration. Mainly, its part is the longitudinal waves propagating along the borehole wall (zero modes). When there is no radial velocity gradient or an elastic stress distribution around the hole, it is a direct longitudinal wave (P-wave). If there is a zone of low velocities near the hole wall (tunnel, for instance) or elastic–plastic stress distribution, it is a head wave that at the beginning of the way propagates as a refracted wave and then it propagates along with a zone interface between a zone of low velocity (near the wall) and a zone of preserved undamaged rock. In our case (surface borehole, diamond drilling) the zone of low velocities is absent and only P waves spread along the borehole wall.
2. S+R wave is more complicated for even the zero-wave mode. In first arrivals, the S-wave is recorded (either direct or head one). The following phases are interfered with by the surface R wave described by Biot (1952) and Stilke (1959). This wave is excited by the propagation of elastic waves along a hole in the solids. Its phase and group velocity depend on the ratio between the wavelength (λ_R) and the circumference of the cylindrical hole ($\lambda_R/2\pi r_0$), where r_0 is the diameter of the hole.

The repeatability of measurements was numerously checked during different studies performed on objects at hydropower stations (Savich *et al* 1983). Repeatable ultrasonic logging was carried out in the same drill hole. Measurements were carried out 4–5 times during the same day. Variations of the ultrasonic velocities V_p and V_s measured at a base of 0.6 m during the same day do not differ more than 1%–2% (Savich *et al* 1979). This value includes differences in the position of the transducer by repeatable measurements and, perhaps, some short-time variations in the rock properties. For details of the method see Ezersky and Goretsky (2014).

4.8.1.3 Measurements by the method of apparent resistivity

The method is based on the differentiation of rocks around a borehole following their specific electrical resistivity (Khmelevskoy 1979). A three-electrode potential sonde was used (figure 4.161).

In the potential sonde, the separation between the potential electrodes MN extends separation between the current electrodes A and M. Electrode current *B* is earthed at the surface. Resistivity relates to the point O named the 'measurement point'.

Apparent resistivity is calculated using the expression $\rho_a = k_a \times (\Delta U/I)$, where geometric coefficient $k_a = (4\pi \times AM \times AN)/MN$, ΔU—is the potential difference between *MN*, *I*—current in the measuring system, *AM*, *AN*, and *MN* are the separations between electrodes. In our survey, $MN = 0.8$, $AM = 0.4$ m (in standard denotation, A04M08N) (Serra 1984).

The borehole diameter is 96 mm. It is suggested that there is no zone of changed rock adjoining the borehole wall because of the shallow (20–40 m deep) location of the salt layer and, respectively, absence of high destructive stress. Also, a diamond drill provides a good quality borehole wall.

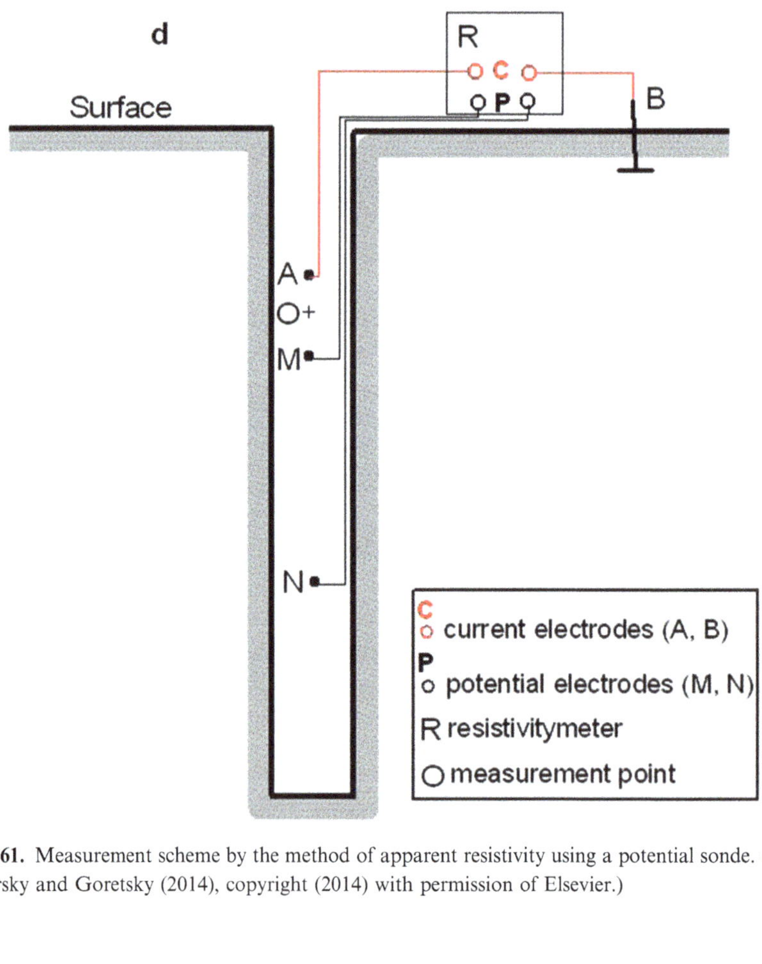

Figure 4.161. Measurement scheme by the method of apparent resistivity using a potential sonde. (Reprinted from Ezersky and Goretsky (2014), copyright (2014) with permission of Elsevier.)

The main objective of the potential sonde is an estimation of the apparent resistivity of the rock massif. The potential sonde enables judging the character of variations of rock resistivity and its variation along the borehole wall. The zone of investigation of a potential sonde is ~0.5 m.

The Syscal R2 (IRIS, Orlean, France) digital device was used as a resistivity meter (*R* in figure 4.161) to generate electric current and to measure potentials between the *MN* electrodes.

4.8.1.4 N*SPT measurements*

SPT is a widely used soil exploration tool that involves using an SPT hammer to drive a split barrel sampler at the bottom of a drill string to obtain soil samples. The split barrel sampler is introduced into the ground to 45 cm (18 inches) and the number of blows is counted for every 15 cm (6 inches). Then, the blows of the first 6 inches are neglected and the next 12 inches are taken into account. The number of blows required to penetrate the last 30 cm (12 inches) is the '*N* value' or 'SPT blow count value', which following Terzaghi *et al*'s (1996) suggestion is related to soil strength and is correlated with that.

The procedures for determining a standardized blow count were presented (Skempton 1986), which allow for hammers of varying efficiency to be accounted for. This corrected blow count is referred to as 'N_{60}' because the original SPT (Mohr) hammer has about 60% efficiency, and this is the 'standard' to which other blow count values are compared. N_{60} is given as:

$$N_{60} = \frac{E_m\, C_B C_S C_R N}{0.60} \tag{4.20}$$

where N_{60} is the SPT *N*-value corrected for field procedures and apparatus; E_m is the hammer efficiency; C_B is the borehole diameter correction; C_S is the sample barrel correction; C_R is the rod length correction, and *N* is the raw SPT *N*-value recorded in the field. Skempton (1986) provides charts for estimating the appropriate values of C_B, C_S and C_R.

This equation generally can be written:

$$N_{60} = (\mathrm{ETR}/60) \times N \tag{4.21}$$

N_{60} = SPT blow counts corrected for hammer efficiency;
ETR = hammer energy transfer ratio;
N = uncorrected (raw) SPT blow count.

4.8.1.5 laboratory measurements of geophysical parameters (Ezersky *et al* 2013c)

4.8.1.5.1 The bender element measurement

The bender element (BE) test is the laboratory measurement. However, we present it in this chapter to compare the shear wave velocity measured in the laboratory with those measured in the borehole with the same soil.

BE is a non-destructive test performed in soil specimens to determine the small-strain shear modulus (G_{max}) of the soil (figure 4.162(a)). G_{max} is an important soil

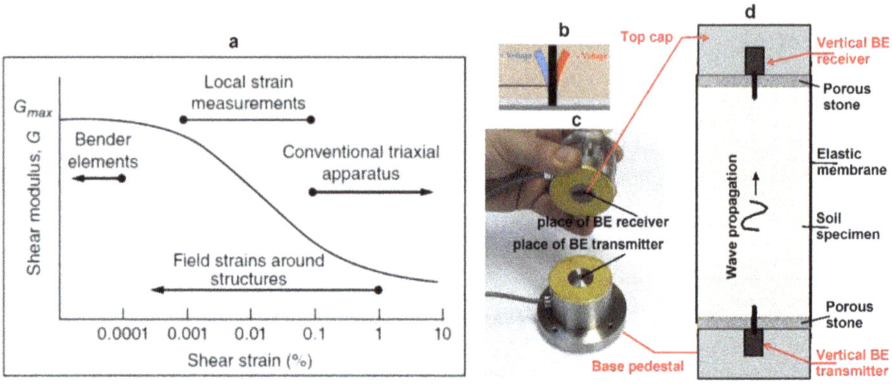

Figure 4.162. BE measurements. (a) G_{max} versus shear strain graph. (b) Vertical S-wave BE; (c) BE holders. (d) Scheme of measurements using vertical BE. (Compilation from manual of GDS Bender Element system (UK).)

property that helps us to understand the elastic behavior of the soil and to evaluate its response to dynamic loadings, such as earthquakes, passing vehicles, and vibrations (Kumar and Foufoula-Georgiou 1997, Kumar *et al* 2013).

The GDS BE system (UK) enables easy measurement of the maximum shear modulus of soil at small strains in a triaxial cell. The BE measurements of LC sediments were carried out in the Building and Infrastructure Testing Laboratory, Haifa, Israel following ASTM D 8295-19 (ASTM 2019). The undisturbed specimens from the Mn-5E borehole were subjected to water saturation and consolidation and tested in the three axial cells at the confining stress from 50 to 500 ka.

The BEs (figure 4.162(b)) were installed along the vertical axis of triaxial specimens, and the shear modulus (*G*) at the vertical direction was determined. These are called *vertical bender elements* (figure 4.162(b)) and consist of the transmitting element (transmitter) at the base, which produces the shear wave, and the receiving element (receiver) at the top which captures the shear wave (figure 4.162(c)). When there is a need to determine the shear modulus in different planes, mostly to determine soil's anisotropy, BEs can also be installed at the mid-height of the sample, at opposite sides with the elements facing each other. A scheme of BE testing is shown in figure 4.162(d).

Ultrasonic laboratory measurements on samples

During the DS project, the ultrasonic laboratory measurements were carried out on the salt and soil samples (Ezersky and Goretsky 2014). The ultrasonic method is, generally, intended for laboratory measurements of the pulse velocities of compression waves (V_p) and shear waves (V_s) in rocks and the determination of ultrasonic elastic constants of anisotropic rock, or one exhibiting slight anisotropy (Silaeva and Shamina 1958, Crampin *et al* 1980). The velocity measurement procedure is carried out following ASTM D 2845 (ASTM 2008). The primary advantages of ultrasonic testing are that it yields compression and shear wave velocities in the ultrasonic frequency range (decades to hundreds of kHz) to measure the elastic constants of intact homogeneous isotropic rock specimens (Cannaday 1964). As a rule, the measured wave velocities differ from the seismic ones measured by *in situ* conditions in the same rock but offer a good approximation. This is explained by the heterogeneity of rock mass comprising fractures, weakness zones, etc, causing, the so-called, anomaly dispersion of elastic wave velocities (Savich *et al* 1969). The latter is determined as a dependence of the elastic velocities upon the frequency range of waves used for the rock investigation.

Apparatus. For the laboratory measurements, a SonicViewer SX Model 5251C digital device based on an XP operating system with two coupled piezoelectric transducers was used: 200 kHz P-wave sensors (Model-5223) and 100 kHz S-wave ones (model-5224), (OYO Corporation, Japan). Another device, the S-70 (Terentiev 1993), was the compact field ultrasonic meter (hereafter, US-meter) with the analogous oscilloscopic signal registration (Institute of the Physics of Earth, Russian Academy, Moscow). The US-meter comprises the time-mark quartz generator that provides highly accurate arrival time readings of not less than 0.1–0.2 μs. The absolute accuracy of V_p determination, concerning intact salt ($Vp = 4460$ m s^{-1}), at a sample length of 70 mm is 0.8%–1.2%.]

The US-meter is intended mainly for the ultrasonic measurements in boreholes drilled in rock mass using a special multi-channel probe (see section US logging above). However, at slight signals of SonicViewer, we used this oscilloscope for sample measurements. For this, joint calibration of both devices was carried out to determine delay time in every system. Shear-wave records are identified following Rao and Lakshmi (2003) as a sharp arrival on a background of forerunner P-wave light oscillations (Ezersky and Goretsky 2014). In several cases, verification of shear waves was carried out by rotating the shear wave transducer (changing the polarization direction from horizontal to perpendicular) and checking the corresponding variation in a wave phase. After ASTM D 2845 (ASTM 2008) for P-waves, the diameter of the sample must be more than five wavelengths λp. In this case, velocity V_p characterizes the infinite massif and λp can be determined from the V_p/V_s ratio (see section 9.2 of ASTM D 2845 (ASTM 2008). The frequency of the OYO transducers is 200 kHz, whereas the highest frequencies forming the front of the pulse are considerably higher (250–400 kHz). The frequency of the first phase is 300–400 kHz. Thus, at $Vp = 4000$ m s^{-1}, $\lambda p = 4000{:}300\,000$ Hz $= 0.013$ m, where the diameter of the sample is 0.068 m and $D/\lambda p = 5.23$.

4.8.1.5.2 Resistivity laboratory measurements

The salt and soil resistivities were also measured (Ezersky and Goretsky 2014). Salt samples were investigated using the resistivity method at 25 °C after saturation in the DS brine and air after thorough drying. At first, the sample's resistivity was calculated from the resistance value by applying the following formula: $\rho = R\bullet(A/L)$, where ρ is the resistivity in Ωm, R is the resistance in Ω, A is the cross-sectional area of the current electrodes in m^2, and L is the sample length in meters. Consequently, the ratio A/L represents a multiplication factor that needs to be applied to the resistance reading to obtain the sample's resistivity value. The bulk resistance of salt samples R was measured using two-electrode configurations without a guard ring (Duba *et al* 1978, ASTM 2012) with both electrode disks made of brass. In the brine-saturated conditions, the electrodes were placed at the ends of the samples. A cellulose sponge soaked with brine solution was used to hold the core sample between the electrodes in dry conditions. This enabled the core sample to make full contact with the current source. Before the measuring process after drying, it was important to ensure that the surface of the sample was fully dried so that there was no surface conductivity of current through the sample side. To avoid surface current, the side surface of cores was isolated by hygroscopic paper.

Apparatus. A Nilsson 400 resistance meter was used. It is an alternating current (AC) meter or a 12 V direct current (DC) meter utilizing a Wien Bridge (AC bridge) with a phase-sensitive detector and a square-wave inverter that produces a nominal alternating signal at 97 Hz. Because this signal is at a very irregular electrical frequency, the meter is insensitive to AC or DC ground currents, which can distort resistance measurements. The meter is a four-terminal, null-balancing ohm-meter that measures resistances from 0.01 Ω to 1.1 MΩ.

4.8.2 Results of drilling (Ezersky *et al* 2013c)

MN-5E borehole (see figure 4.156(c) for location) was drilled in the northern DS basin in the margin of an alluvial fan near the Mineral Beach SPA. The water table at the moment of drilling was 18.5 m. The section presented in figure 4.163(c) consists mainly of fine grain sediments, silts with a little clay (named after Frydman *et al* (2008) *lime carbonates* (LCs)) with a few inter-layers of gravels. The LC sediments are typically comprised of laminated, clay to silt size clastic sediments (predominantly calcite), and authigenic aragonite and gypsum. The clastic detrital material was brought into the basin by seasonal floods. The DS LC sediment is made up of about 95% calcite, with the other 5% being aragonite, quartz, and gypsum (Ezersky and Livne 2013). The 16 m thickness salt layer is located at the depth of 22 m. Note the section is typical for DS mudflats.

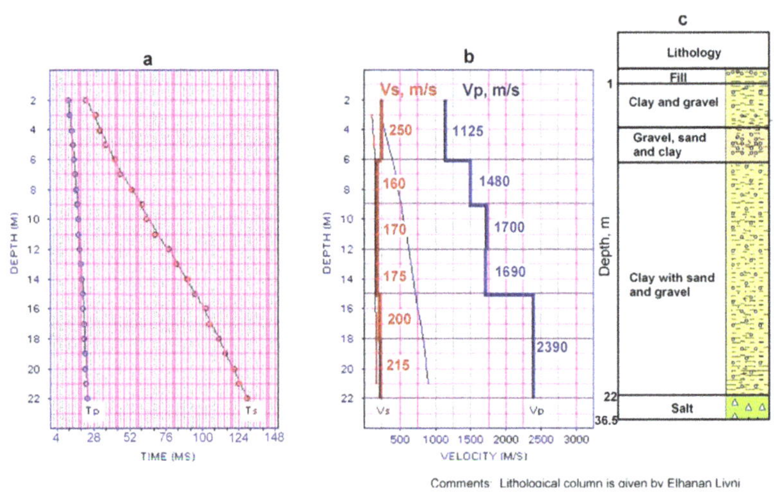

Figure 4.163. (a) Arrival time versus depth graphs and (b) P-wave velocity (V_p) and S-wave velocity (V_s) versus depth graphs for downhole survey in MN-5E borehole. (c) Lithology column.

The EB-3E borehole was drilled in the alluvial fan of the Ein Boqeq site (see figure 4.156(d) for location). The water table is located 3.5 m deep. The section consists mainly of the coarse clastic sandy-gravel sediments (Livne 2012).

The salt layer is located at 26.3 m depth. Note the section is typical for the DS fans. Similar coarse sandy-gravel lithology is reported by Taqieddin *et al* (2000) on the Jordanian shore.

4.8.2.1 Downhole results

Examples of raw data obtained in the boreholes Mn-5E and EB-3E are shown in figure 4.165.

The P-wave seismograms are presented in figures 4.165(a) and (c) for the Mn-5E borehole only. Seismograms of SH-waves are presented in figures 4.165(b) and (e) for clastic sediments, and in figures 4.165(d) and (e) for the salt layer. The seismograms are of a good quality allowing the calculation of P-wave and SH-wave velocity logs.

Results of the downhole measurements in clastic sediments of boreholes are shown in figures 4.163(a) and (b) (for borehole Mn-5E) and figures 4.164(a) and (b) (for borehole EB 3E).

Figures 4.163(a) and 4.164(a) present variations of S- and P- wave arrival times in the boreholes Mn-5E and EB-3E, respectively. Corresponding velocity logs are presented in figures 4.163(b) and 4.164(b) for the same boreholes. Sediments in the borehole Mn-5E (figure 4.163(c)) are characterized by velocities of compression waves (V_p) varying in the range of 1100–2400 m s^{-1}. Shear wave velocities (V_s) are varied from 160 to 250 m s^{-1} (figure 4.163(b)).

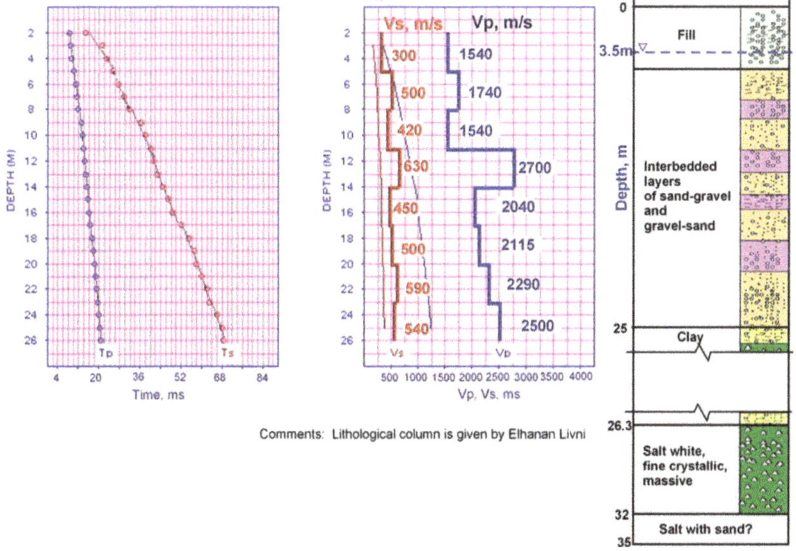

Figure 4.164. (a) Arrival time versus depth graphs and (b) P-wave velocity (V_P) and S-wave velocity (V_s) versus depth graphs for downhole survey in EB-3E borehole. (c) Lithology column.

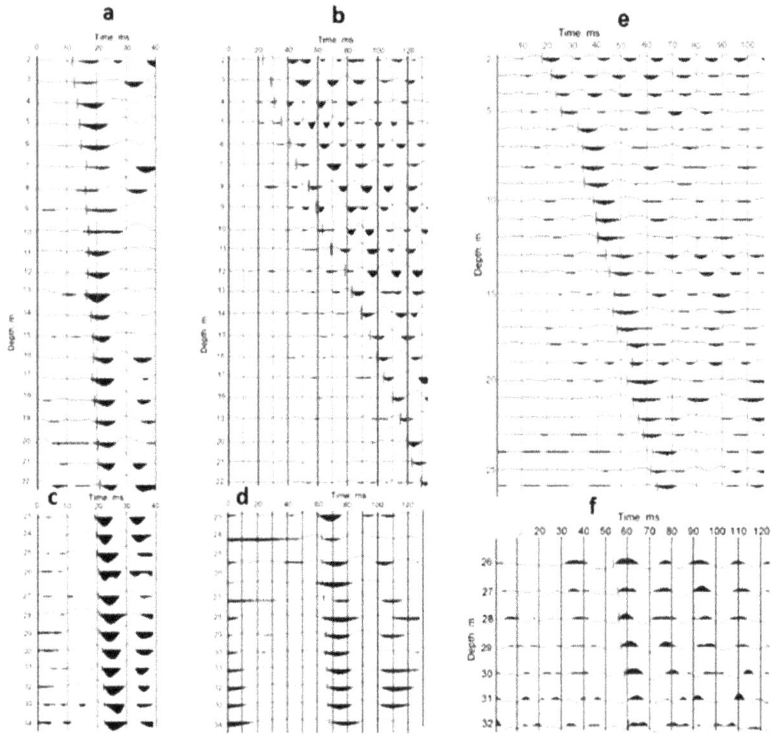

Figure 4.165. Raw data of the downhole measurements in boreholes Mn-5E (a, d) and EB-3E (e, f). (a) seismograms of the P-waves measured in clastic sediments; (b) seismograms of the P-waves measured in salt; (b, e) seismograms of shear waves measured in clastic sediments; (d, f) seismograms of shear waves measured in salt.

Sediments in the borehole EB-3E (figure 4.164(c)) are characterized by velocities of the compression waves (V_p) varying in the range of 1500–2500 m s^{-1}. The shear wave velocities (V_s) vary from 300 to 630 m s^{-1} (figure 4.164(b)).

Comparing velocities measured in different lithologies one can conclude that P-wave velocities measured in both boreholes are of the same range. It is caused by the water filling the pores of sediments in the boreholes Mn-5E and EB-3E. At the same time, in the borehole Mn-5E water table is located at 18.5 m where the LC retains water.

The S-wave velocities measured in boreholes are essentially different. LCs (borehole Mn-5E) have V_s 2–3 fold lower than sandy-gravel sediments of borehole EB-3E.

4.8.2.2 The velocity of the salt layer

Velocities measured in the salt layer underlying the clastic sediments were defined as $Vp = 3580$ m s^{-1}, $V_s = 1000$ m s^{-1} in the borehole Mn-5E and $Vp = 4020$ m s^{-1} and $Vs = 1510$ m s^{-1} in the borehole EB-3E.

Apart from the downhole measurements, SPTs were carried out in the borehole; also performed were surface surveys near the boreholes: such as seismic refraction method, and TEM.

Table 4.10. Parameters measured in the boreholes Mn-5E (LCs) and EB-3E (sandy/gravel).

Borehole	Mn-5E	EB-3E	Mn-5E	EB-3E
Soil type	**LC**	**SG**	**Salt**	**Salt**
Range V_p, m s^{-1} (downhole)	1100–2400	1500–2700	3210	3390
Range V_s, m s^{-1} (downhole)	150–250	300–600	1005	1510
ρ_x, Ωm (TEM)	0.2	0.3	1	2
NSPT, blows	3–26	5–60		

Downhole surveys within the boreholes and close surface methods (seismic refraction, MASW, and TEM) were used for the calibration measurements.

Parameters measured during these tests are presented in table 4.10 and are compared in figure 4.25.

Analysis of downhole graphs in boreholes Mn-5E and EB-3E is described in section 4.2.2, MASW (figure 4.25). However, the velocity logs obtained by various methods characterized by different resolutions, are very similar and vary within limits of +10%. Thus, the calibration measurements allow determining the V_s of the lithological units penetrated by the boreholes: clayey sediments are characterized by V_s = 120–250 m s^{-1}; sandy-gravel sediments have V_s = 300–600 m s^{-1}, and the salt layer exhibits a V_s of 810–1650 m s^{-1} (Ezersky and Livne 2013).

4.8.2.3 Ultrasonic and electrical measurements

The ultrasonic and resistivity logs (V_p, V_s, and $\rho_a$versus depth graphs) are shown in figures 4.166(a) and (b), respectively. Taking into account the more complicated origin of S-wave velocities (discussed in section 4.2.2) and the apparent origin of the resistivity graph, we compare the V_p log measured in boreholes with that in salt samples. A V_s graph has been calculated from V_p using the inter-relation derived from the ultrasonic measurements in salt samples in our tests (Ezersky and Goretsky 2014):

$V_s = 0.945 \bullet V_p - 1626.2$ (R-squared = 0.78). (4.8-4)

In this case, we can also compare the V_s log with V_s values measured on samples. Note at first that both V_p and V_s velocities (figure 4.166(a)) and apparent resistivity (figure 4.166(b)) show a trend increase with depth (thick dashed blue lines). Trend curves were calculated by smoothing V_p and V_s interval values using a running average of 13 points. Resistivity was measured between the MN electrodes spaced 0.8 m and the resistivity graph is properly smoothed. The V_p and V_s velocities measured in salt samples are shown in figure 4.166(a) by the red circles and green rhombs, respectively. Similarly, resistivity values ρ_x measured in samples are shown in figure 4.166(b) by the lilac squares. One can see from comparison data that the velocity values obtained from salt samples are higher than those measured in the borehole at the corresponding depths. The difference is greater in the upper part of the salt layer. Sharp variations of the velocities in the borehole logs correlate with the fractured zones (7 and 8 in figure 4.166(a)).

Both velocities of salt samples and ultrasonic logs converge with a depth. The resistivity data indicate similar regularities. It has to be taken into account that the

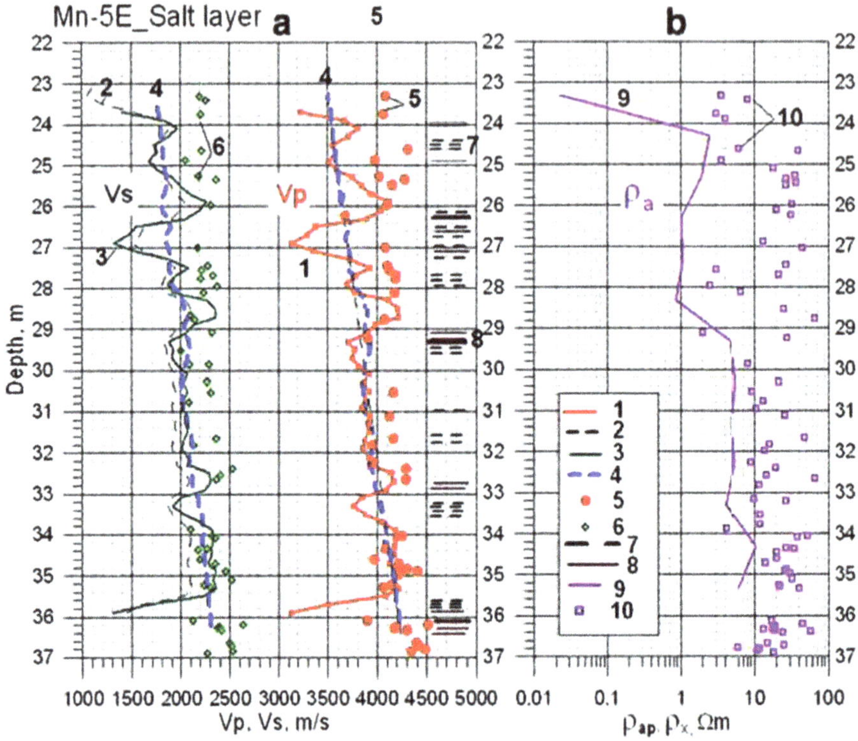

Figure 4.166. Combined interpretation of data acquired in the borehole Mn-5E. (a) Velocity versus depth graph. Denoted: 1—V_p log, 2—VS+R log, 3—V_s calculated from (4.8-1), 4—V_p and V_s trends, 5—V_p, ultrasonic measurements (USM) on samples, 6—V_s, USM on samples, 7, 8—brine and LC, respectively, filled fractures; (b) resistivity versus depth graph: 9—apparent resistivity log in borehole and 10—bulk resistivity of salt samples (Ezersky). (Reprinted from Ezersky and Goretsky (2014), copyright (2014) with permission of Elsevier.)

in situ properties can be much more affected by dissolution processes and it explains relations in figure 4.166.

4.8.2.4 Geotechnical investigation in boreholes and laboratory

4.8.2.4.1 SPT results

The *N*SPT in lime carbonates sediments is ranged from 5 to 25 blows (30 cm^{-1}) that allow classifying of clayey sediments as a firm to a stiff consistency. *N*SPT in sandy-gravel sediments varies from 16 to 50 blows that characterize these sediments as from medium to dense ones (Ezersky and Livne 2013).

Vane test results. Results of borehole testing using the Vane test were calculated in the geotechnical laboratory of the Ben Gurion University (Livne 2012).

Two shear strength values are presented in table 4.11. The first one is s_{up}-peak value measured at the moment of the start of the rotation, and the second one s_{ur} is resistance at the stable rotation of the vane instrument.

The BE test results. Examples of the typical oscillograms obtained in the BE tests are presented in figure 4.167 that shows the signal measured at 150 kPa confining

Table 4.11. Results of undrained shear strength calculation using field Vane test.

Number	Depth (m)	Peak resistance (N*m)	Residual resistance (N×m)	Peak shear strength S_{up} (kPa)	Residual shear strength S_{ur} (kPa)
1	3.1	47.45	20.34	202.42	86.75
2	10.3	6.78	0	28.92	0

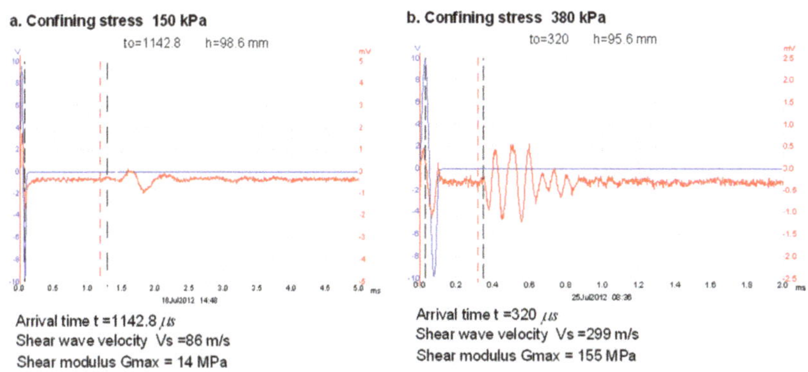

Figure 4.167. Examples of oscillograms obtained in the BE tests: (a) at confining pressure of 150 kPa; (b) at confining pressure of 380 kPa. The vertical red dashed line is the arrival time of the signal, the vertical black dashed line is the time of the first phase of the signal.

Table 4.12. Results of bender element tests* (Goretsky 2012).

Spec. No	Confining pressure, kPa	Height, mm	Wet density, kg m^{-3}	Dry density kg m^{-3}	V_s, m s^{-1}	G, MPa	V_p m s^{-1}	ρ_x, Ωm	q_u,** kPa	s_u,*** kPa
1	50	102.1	1806	1301	119.1	26	1771	1.87	53.2	26.6
2	150	98.6	1878	1416	86	14				
3	270	97.6	1896	1446	132	33	1980	0.59	98	49
4	380	95.6	1900	1185	299	170				
5	500	94.6	1920	1259	350	235	1780	1.28	243.5	121.8

Notes:
*Tests were carried out after the water saturation and consolidation.
**Unconfined compression strength (preliminary consolidated specimens to 50–500 kPa).
***Undrained shear strength at zero confining pressure.

pressure (a) and the signal recorded at 380 kPa confining pressure (b). One can see that with the increase of confining pressure signal is improved and velocity V_s increases too.

Results of five V_s measurements carried out at different confining pressures using BE are presented in table 4.12. There are shown also other parameters determined during these tests as the geometry of specimens, wet and dry density, dynamic shear moduli, UCS parameters q_u, and undrained shear strength s_u.

It follows from table 4.12 that V_s varies in the range of 86–350 m s^{-1} increasing with the increase of confining pressure.

4.8.2.5 Analysis of borehole geophysical and geotechnical results (Ezersky *et al* 2013c) *correlations between SPT and* V_s

Two SPT parameters were calculated: *N*SPT (uncorrected values) and *N*SPT$_{60}$ subjected to different corrections (Terzaghi *et al* 1996).

Two inter-relationships were looked for: V_s versus *N*SPT (red graphs in figures 4.168(a) and 4.168(b)) and V_s versus *N*SPT$_{60}$ (blue graphs in the same figures).

Analysis of both graphs shows that there is no significant difference between the two above correlations that can be written as follows:

(a) for clayey sediments of the Mn-5E borehole

$$V_S = N\text{SPT}^{0.261} \times 107.20\ (D = 0.82) \quad \text{and} \quad V_S = N\text{SPT}_{60}^{0.238} \times 113.86 \qquad (4.22)$$

$(D = 0.81)$ (D—is coefficient of correlation)

(b) for sandy-gravel sediments of the EB-3E borehole

$$V_S = N\text{SPT}^{0.347} \times 151.75\ (D = 0.61) \quad \text{and} \quad V_S = N\text{SPT}_{60}^{0.281} \times 177.82 \qquad (4.23)$$

$(D = 0.41)$

One can see comparing expressions (4.22) and (4.23) that the coefficient of correlation for the LCs is significantly higher ($D = 0.82$ for *N*SPT and $D = 0.81$ for *N*SPT$_{60}$) than that for sandy-gravel sediments ($D = 0.61$ for *N*SPT and $D = 0.41$ for *N*SPT$_{60}$).

At the same time coefficients of correlation for both *N*SPT and *N*SPT$_{60}$ are equal in clayey sediments ($D = 0.81$–0.82) and are significantly different in sandy-gravel

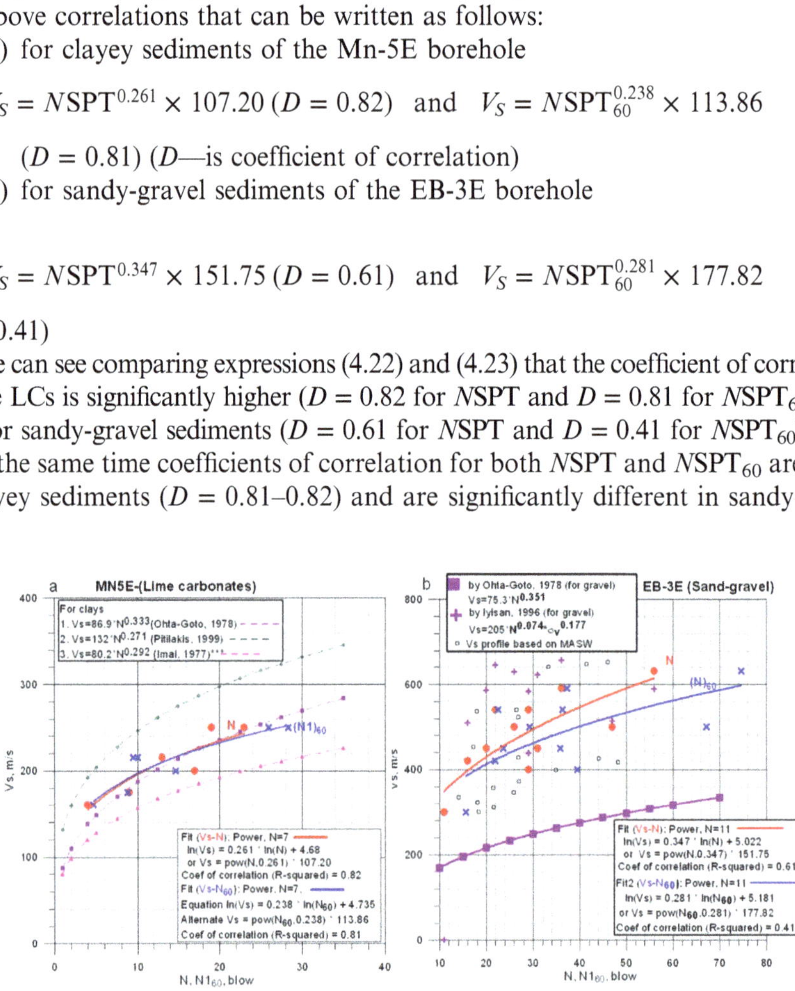

Figure 4.168. *N*SPT versus V_s inter-relationships obtained in two boreholes: (a) Mn-5E, and (b) EB-3E. Relation *N*SPT × 151.75 ($D = 0.61$) and $V_S = N\text{SPT}_{60}{}^{0.281} \times 177.82$ ($D = 0.41$). (Reprinted from Ezersky and Livne (2013), with the permission of EAGE.)

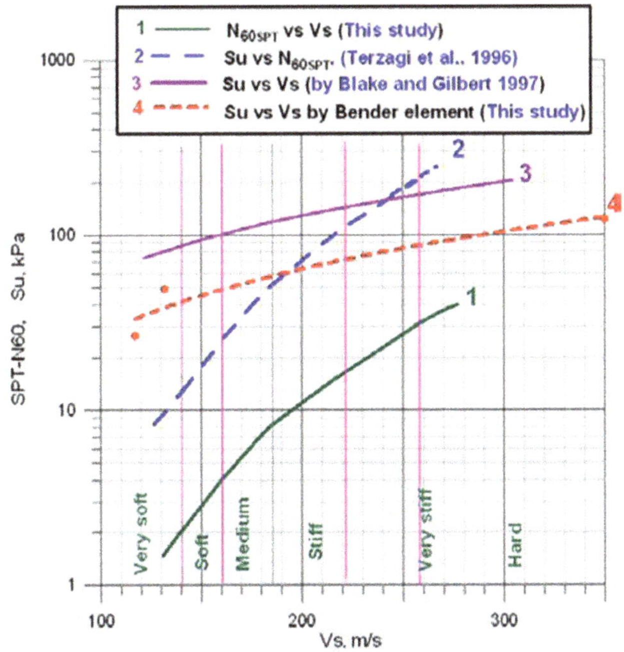

Figure 4.169. Inter-relations between the V_s and undrained shear strength. 1—*N*SPT versus V_s from borehole data (this study); 2—S_u derived from SPT based on Terzaghi *et al* (1996); 3—clay by Blake and Gilbert (1997), 4—LC (BE, this study). (Reprinted from Ezersky and Livne (2013), with the permission of EAGE.)

ones so that $D = 0.61$ for *N*SPT is higher than $D = 0.41$ for *N*SPT$_{60}$. The regularities noted above are confirmed at the comparison of different relationships derived by different researches for similar lithology. Clayey sediments are characterized by closer relationships (figure 4.168(a)), whereas sandy-gravel sediments show essentially different correlations (figure 4.168(b)). A similar difference in correlation for these soils was obtained by Ezersky and Frumkin (2021).

S_u versus Vs relationships. LCs demonstrate the correlation between the undrained shear strength S_u and V_s (4 in figure 4.169). Similar relationships were revealed in the laboratory tests by Blake and Gilbert (1997) (3 in figure 4.169). Curve 1 presents *N*SPT$_{60}$ versus V_s from the borehole data (this study). Curve 2 in figure 4.169 based on SPT testing suggested by Terzaghi *et al* (1996) has been presented for comparison.

4.8.2.6 Comparison of V$_s$ *measured using downhole and BE methods*

Curve 4 presented in figure 4.169 shows that there is a dependence between S_u and V_s.

To compare V_s measured in the borehole and those by BE (figure 4.170) we have recalculated confining pressure of the BE test to depth. Depth of soil specimens was calculated from consolidation pressure in the presumption that:

(1) Consolidation pressure is equal to the stress σ_H at the depth H;
(2) At the depth range between 0 and 23m and water table at the depth of 18 m we can neglect the pore pressure and accept σ_H as: $\sigma_H = (\sigma_1 + \sigma_3)/2 = (\sigma_1 + \sigma_1 K_o)/2$ where K_o =1—sinϕ is the coefficient of Earth pressure (horizontal-to-vertical stress

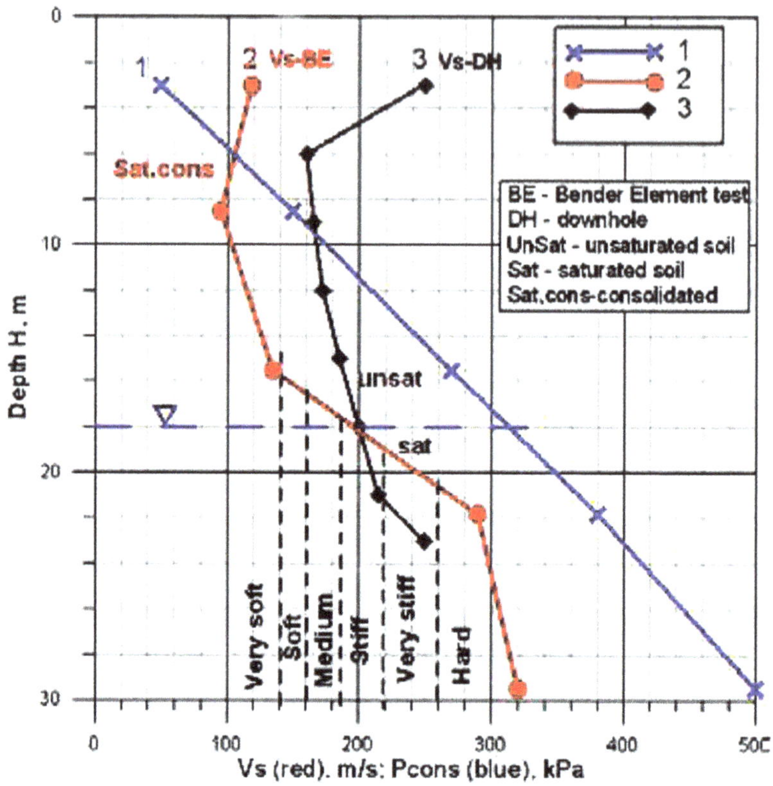

Figure 4.170. (a) Variation of different parameters with the calculated depth (curve 1). Depth of soil specimens was calculated from consolidation pressure at an average naturally wet unit weight of 17.5 kN m^{-3} (above the water table) and saturated unit weight 18.8 kN m^{-3} (below that). V_s versus depth graphs are derived from the BE test of the undisturbed saturated and consolidated LC specimens (curve 2) and downhole data in the MN-5E borehole (curve 3).

ratio) (Terzaghi *et al* 1996, section 16.5, p 104) and ϕ is the friction angle that is 2.4° above the water table and 34° under that.

(3) Then $H = \sigma_H/\gamma$ where γ is the unit weight of soil that is 17.5 kN m^{-3} above the water table and 18.8 kN m^{-3} under that.

A calculated depth versus consolidation pressure graph is presented in figure 4.170 (curve 1). The V_s versus calculated depth graph (derived from the BE tests) is presented in figure 4.170 (curve 2). One can see that difference of V_s between the DH and BE test velocities in the upper unsaturated part of the subsurface is 50%–70% and gradually converged in the deeply saturated soils. Generally, this discrepancy is known. Terzaghy *et al* (1996) note that invariably when field and laboratory V_s values are compared, values of Vs_{lab} range from slightly less to considerably less than the *in situ* values, Vs_{fld}. Stokoe and Santamarina (2000) suggest several factors that can affect the (Vs_{lab}/Vs_{fld}) ratio. Particularly, based on testing of 47 specimens they concluded that the loss in the stiffness in laboratory samples results from *irrecoverable damage at interparticle contacts.*

On the other hand, several researchers (Schneider *et al* 1999, Szczepański 2008) report that *in situ* measurement of the low-amplitude shear moduli from seismic piezocones, seismic flat dilatometers, SASW, and cross-hole tests were found to be in good agreement with laboratory values measured by the resonant column or the BE tests.

In our tests, undisturbed specimens of good quality were used and other reasons would be responsible for the difference in measured V_s values (Ezersky *et al* 2013c). Recall that V_s values in the laboratory BE tests were obtained from the water-saturated samples, whereas the downhole measurements were carried out in the desaturated DS mud. It means that the difference in the saturation state would be responsible for the difference in the measured V_s values.

Theoretically, Dvorkin (2008) suggests an explanation based on the following presumed relationship between the V_s values in dry and saturated soils:

$$V_s = V_{\mathrm{sDry}}\sqrt{\frac{\rho_{\mathrm{bDry}}}{\rho_b}} = (1-\varphi)^2 V_{\mathrm{sS}}\sqrt{\frac{(1-\varphi)\rho_b}{(1-\varphi)\rho_s + \varphi\rho_f}} \tag{4.24}$$

Where ρ_{bDry} and ρ_b are the bulk densities of dry and wet sediments, respectively; and ρ_s ρ_f are the densities of the solid and fluid phases, respectively. Values V_{sS} as well as V_{pS} can be calculated from the elastic moduli and density of the appropriate mineral mix as the square root of the modulus divided by the density. Using (4.24) we can estimate $V_{\mathrm{sDry}}/V_s = \sqrt{\rho_b/\rho_{\mathrm{bDry}}}$ using data of laboratory testing as 1858/1321 = 1.40 = 1.18. This theoretical evaluation is less than the experimental ratio, that is, of 1.5–1.7 folds. This discrepancy can be explained by the experimental study of Sharma and Bakkapatham (2008) that has shown that at soil (sand) saturation between 80% and 100% G_{max} decreases from 15 to 6 MPa, e.g., 2.5-fold (respectively, V_s should decrease as 2.5 = 1.6 fold) (figure 4.171).

Thus, V_s values measured by the BE are in good accordance with the field data.

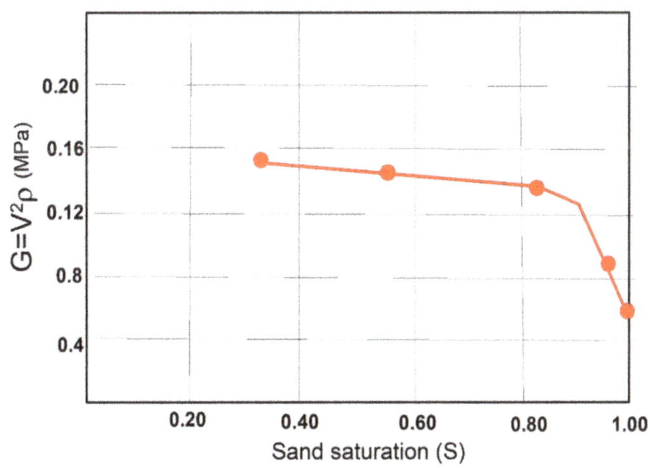

Figure 4.171. Approximate dependence of stiffness $G_{max} = V^2\rho$ versus degree of sand saturation S.

4.9 Other geophysical methods in the DS sinkhole problem

4.9.1 HVSR method

We have considered the HVSR method in section 2.2.1.17. Recall that the method is based on Nakamura's hypothesis (1989) that site response could be estimated from the spectral ratio of the horizontal versus vertical components of noise observed at the same site (site of interest). The method is intended to identify the main characteristics of site responses for soft deposits (i.e. resonance frequencies and amplification factor). Recall also that the H/V technique alone is not sufficient to characterize the complexity of site effects and in particular the absolute values of seismic amplification (SESAME 2006, Dal Moro 2015). The method required the use of an additional method that can decrease the number of unknown parameters, such as the thickness of layers and their V_s values. In such a setting the method was used in recent decades in Israel, where the determination of H/V spectral ratios from ambient noise (HVSR) was supplemented with on-site borehole and geological information, as well as with the seismic refraction, downhole, or cross-hole S-wave surveys to derive the required models of the subsurface (Zaslavsky *et al* 2005, 2008, 2009). Recently, Gorstein and Ezersky (2015) have shown that a successful addition to the HVSR is MASW. Several studies were carried out in the DS area to combine HVSR with the MASW method to map the subsurface (Ezersky *et al* 2012, 2013c). Other studies were aimed to combine the active and passive MASW with HVSR methods for increasing the depth and improving the accuracy and reliability of V_s model in site response assessment (Ezersky *et al* 2014, 2015).

This issue is considered here for two southern sites of the DS western shore.

Estimating possible site effect is an integral part of the evaluation of the seismic hazard and reduction of earthquake damage. In regions with low or moderate seismicity (as in Israel), the site response should be determined by analytical tools. These computations require the knowledge of the subsurface geological structure in terms of V_s profile down to seismic bedrock. Conventionally, this problem is resolved by the joint implementation of the HVSR (named also Nakamura's) technique, which is based on the ambient noise measurements and seismic methods such as S-wave refraction or MASW method. The first one is significantly limited in resolution because of the weak source of S-waves. The MASW method conventionally using 4.5 Hz geophones is restricted in penetration depth of surface waves because of frequency (wavelength) limitations. In this study, geophones of 2.5 Hz and special data processing to provide constructing V_s section down to bedrock located at 100 m and even more than 200 m deep were used. The suggested methodology has been tested last year at several sites on the DS shore, where the subsurface model is represented by two compact salt layers serving seismic reflectors that occur inside the unconsolidated sediments. The depth of the upper salt layer is in the range of 20–70 m, while the depth of the deeper salt layer was estimated greater than 200 m.

HVSR data acquisition: The methodology of HVSR data acquisition and processing is considered in detail in Zaslavsky *et al* (2009). The ambient noise measurements are conducted using portable instruments (Shapira and Avirav 1995) consisting of a multi-channel amplifier, a GPS for timing, and a laptop computer to digitize and store the data. (figure 4.172).

Each seismograph station consists of three (one vertical and two horizontal) L4C velocity transducers (Mark Products) with a natural frequency of 1.0 Hz. The sample rate is 100 samples per second and the filter band-pass is between 0.2 and 25 Hz. All the equipment: sensors, power supply, amplifiers, personal computer, and connectors are portable allowing the performing of measurements in the autonomous mode (figure 4.172(b)). Examples of three-component noise records are shown in figure 4.172(c). The presented pattern of measurements is typical and repeated from site to site.

HVSR data processing. Data processing is explained in figure 4.173. For each site, the average H/V spectral ratios and their corresponding standard deviations are determined by applying the following process: (1) time windows, each of 30–60 s long depending on fundamental frequency, are selected; (2) a Fourier transform is applied on the time windows, using cosine-tapering (1 s at each end) before transformation and then smoothed with a triangular moving Hanning window;

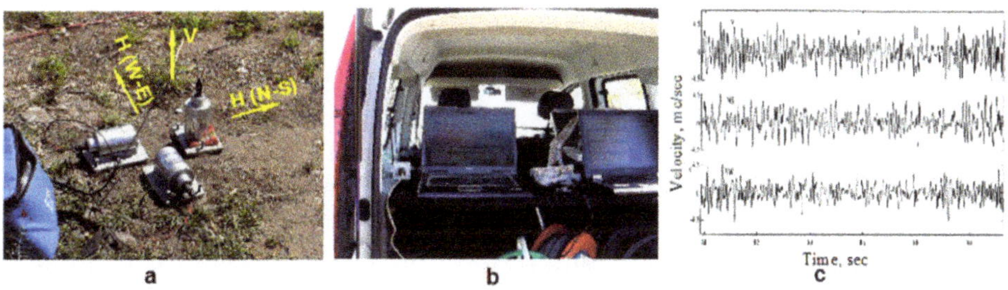

Figure 4.172. Photographs of noise measurements: (a) three-component 1 Hz seismological station; (b) recording of seismic noise; (c) field records.

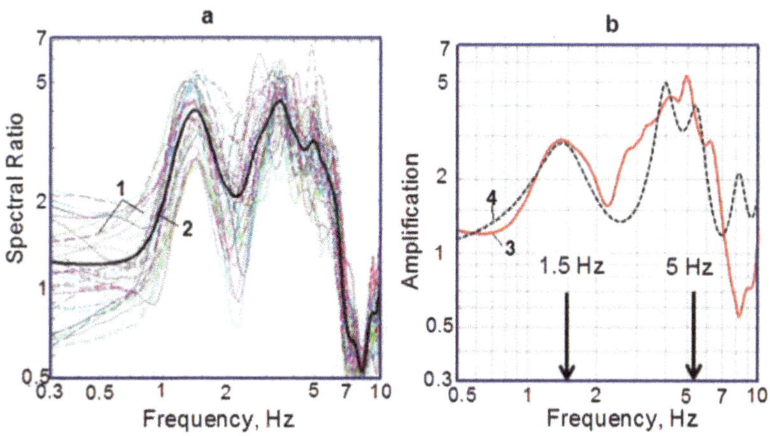

Figure 4.173. (a) Example of an individual (1) and average (2) spectral ratios obtained on the DS shore. (b) Analytical transfer function (4) in comparison with observed H/V spectral ratio (3). Arrows denote the resonance frequencies.

(3) for each site a set of up to 50 time windows is selected, records within these time windows are compiled, each window provides an H/V spectral function.

Data processing is carried out using 'SEISPECT' software developed in the Geophysical Institute of Israel (Perelman and Zaslavsky 2001). SEISPECT is a MATLAB application for the spectral analysis and processing of the ground motions including seismograms recorded by the short-period and broadband seismic stations, as well as strong-motion accelerometers. The main modules realized in the program are: visualizing and editing of the input data; selecting time window and computing fast Fourier transform (FFT) and H/V spectral ratios; saving and displaying results. The average spectral ratio for each of two horizontal components is computed; if the curves of average spectral ratios of the two components are similar then the average of the two H/V ratios is defined as:

$$A(f) = \left[\frac{1}{2n} \sum_{i=1}^{n} \frac{S_{\mathrm{NS}}(f)_i}{S_{\mathrm{V}}(f)_i} + \sum_{i=1}^{n} \frac{S_{\mathrm{WE}}(f)_i}{S_{\mathrm{V}}(f)_i} \right. \tag{4.25}$$

] where $SNS(f)i$ and $SEW(f)i$ are individual spectra of the horizontal components and $SV(f)i$ is an individual spectrum of the vertical component.

4.9.2 The MASW

Active and Passive techniques. The active and passive data acquisition are presented in section 2.2.1.1 (see also Ezersky *et al* (2014), Gorstein and Ezersky (2015)). Here we consider data processing.

Active MASW. Active-source surface-wave dispersion measurements are done with the typical seismic shot gatherers that are a collection of seismic traces, which share some common geometric features. The wave field is transformed into a frequency-wave number (or frequency-slowness) domain in which the maxima should correspond to surface-wave signatures (figure 4.174(a)). Several modes can be picked out for such dispersion curves if the propagation mode signatures are well separated. The dispersion curves are then inverted for a 1D V_s profile with depth. When data are collected in a roll-along mode, each 1D profile is represented at its corresponding midpoint spread, allowing a pseudo-2D V_s section to be drawn.

The inverse problem formulation imposes that the investigated medium is assumed dimensional under the spread. Long spreads are required to record the

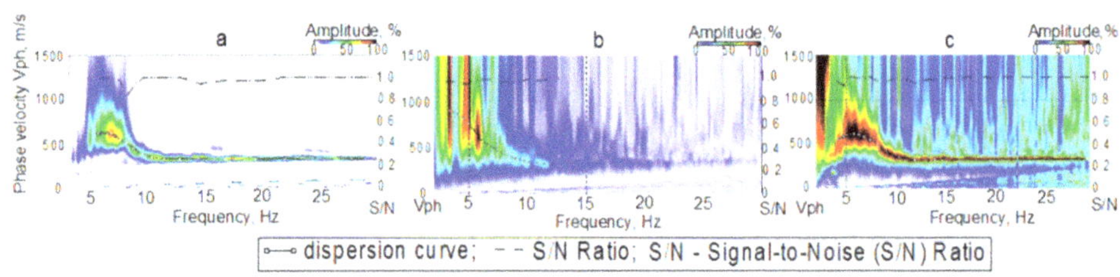

Figure 4.174. Examples of dispersion images (DI) and extracted curves from the Ein Boqeq site (western DS). (a) Active MASW; (b) passive MASW; and (c) combined (active + passive) image.

wavelengths large enough for increasing the depth of investigation and for mitigating the near-field effects (Bodet *et al* 2005, Socco *et al* 2009).

Passive MASW. The passive surface waves generated from natural (e.g., tidal motion) or cultural (e.g., traffic) sources are usually of a low-frequency nature with the wavelengths ranging from a few kilometers (natural sources) to a few tens (or hundreds) of meters (cultural) (Okada 2003) providing a wide range of penetration depths and therefore a strong motivation to utilize them. The ambient noises are recorded using the receiver arrays (antennae) arranged as different geometrical figures (linear, circular, cross-layout, etc arrays). The dispersion image and extracted dispersion curve are extracted (figure 4.174(b)). The most accurate estimation is obtained through a survey using a true 2D receiver array (Park and Miller 2006). However, because the true 2D receiver array, such as circular and cross-layout ones are not a practical or possible mode of the survey in built-up urban areas, a method that can be implemented with the conventional 1D linear receiver array can be effective in this case (Louie 2001). The data processing scheme can be found in Park *et al* (2004).

Combination of active and passive MASW measurements. Dispersion images processed from the active and passive data sets should be combined to obtain the improved dispersion curves (figure 4.174(c)). The active MASW method generates signals in the range of 5–30 Hz, whereas a passive one allows for widening that to the low-frequency range down to the geophone frequency. Combining two signals, we widen the range of frequencies from 2.5 to 30 Hz, and sometimes even up to 50 Hz. Thus, the penetration depth can be increased to a deep range (low frequencies) and the uppermost depth can be decreased (high frequencies) (Park *et al* 2004).

Then we will consider two examples of constructing the subsurface V_s models in the south of the DS: Ein Boqeq North and Dam 5/8 separating evaporation ponds of Israel and Jordan.

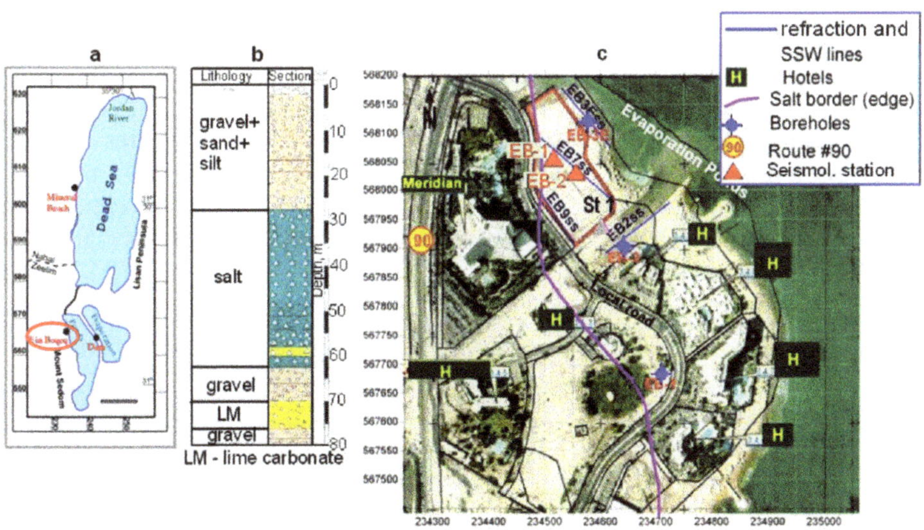

Figure 4.175. Ein Boqeq north site. (a) Site location map; (b) EB-2 borehole section (after Yechieli *et al* (2004), with permission from GSI); and (c) geophysical measurements layout.

4.9.2.1 Ein Boqeq north site (Ezersky *et al* 2013d)
The Ein Boqeq area of 1.5–2.0 km long and 250–300 m width is extended along the southern DS basin (recently evaporation ponds were artificially filled by pumping the DS water from the northern basin (figure 4.175(a)).

Several hotels are located in Ein Boqeq which is also named the Hotel area. The study area is situated in the northern part of Ein Boqeq, about 2–3 km east of the Judea Mts (see figures 4.175(a) and (c)). According to data from the boreholes EB-1 and EB-2 (Yechieli *et al* 2004), a geological section of this area is represented by pebbles, cobbles, and clayey–sandy sediments (figure 4.175(b)). The borehole EB-3E crossed the sandy-gravel alluvium layer overlying the salt unit at 26 m deep. The thickness of the salt unit is 28 m. The western salt border (lilac line in figure 4.175(c)) was mapped by the seismic refraction surveys.

4.9.3 MASW measurement results

Active MASW measurements. Dispersion image of the active MASW measurements along line 9ss (see figure 4.175(c) for location) is shown in figure 4.174(a).

Passive MASW measurements. Passive MASW measurements were carried out using roadside technique along a 24-channel line located south to the Hod Hotel, 35–45 m from a local road (see figure 4.175(c) for location).

The geophone array is shown in figure 4.176(a). An example of an ambient noise record from traffic is shown in figure 4.176(b). The noise was recorded for 1 min with a sample rate of 4 ms. A dispersion image with the extracted dispersion curve is depicted in figure 4.176(c).

Combined MASW calculations.

The combined image shown in figure 4.174(c) integrates the dispersion images of passive and active MASW shown in figures 4.174(a) and 4.176(c). Combined inversion of the dispersion curve with the five-layered model is presented in figure 4.177 (RMSE = 12%).

Note that every type of MASW would be used separately, whereas combined use allows improvement of resolution at both lower and high frequency. Passive MASW shows better resolution at low frequency (Z max = 115 m), whereas at high frequency Z min = 21.2 m. The active method shows Z max = 58.5 m and Z min = 5.0 m. Finally, combined MASW shows resolution of the passive method at low frequency and resolution of the active MASW at high frequency.

HVSR measurement results

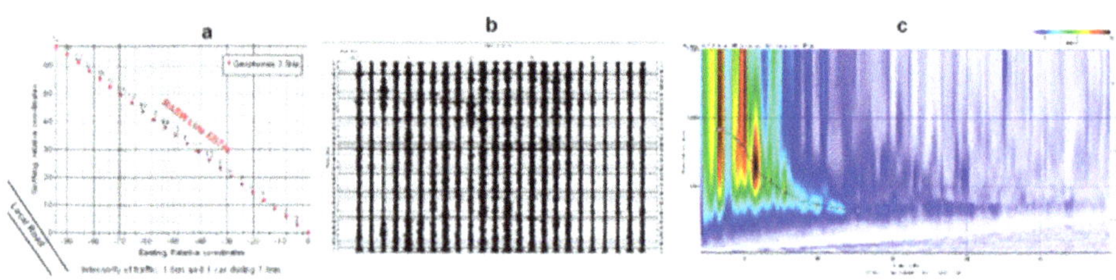

Figure 4.176. Ein Boqeq site. (a) Passive MASW roadside array, 5 m separation, (b) field records, record length 1 min, sample rate 4 ms, and (c) dispersion image with extracted dispersion curve.

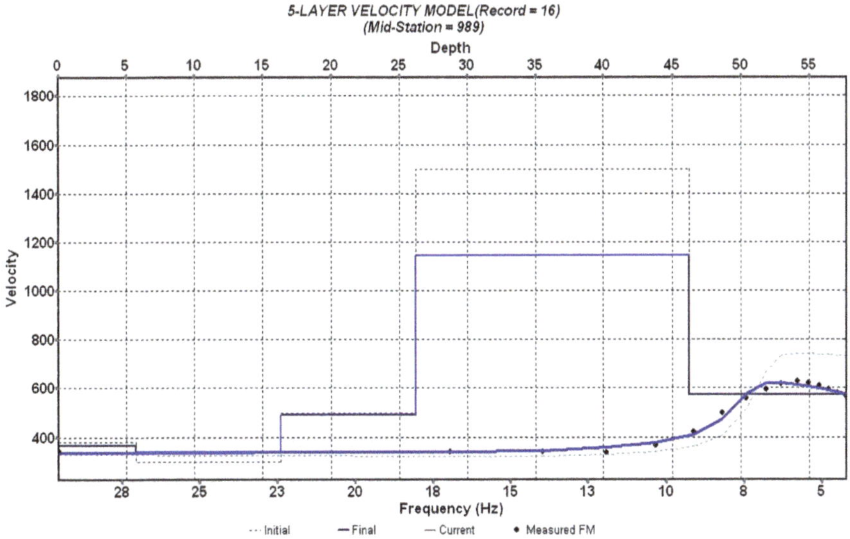

Figure 4.177. Ein Boqeq north site, Combined inversion line EB7SS: inversed depth–velocity (V_s) graph for the five-layered model. Denoted: initial model (thin blue dotted line), corresponding initial dispersion curve (thin green dotted line), experimental points (blue rhombs) and fitted dispersion curve (blue solid line), inverse depth–velocity model (blue layered graph). The lower the horizontal axis is the frequency (Hz) and the top axis is depth (m).

Table 4.13. Optimal 1D subsurface model for determination of analytical site response function at the Ein Boqeq north site.

Layer	Thickness, m	Depth, m	V_s, m s^{-1}	Density, kg m^{-3}	Damping, %	Possible lithology
1	5	5	200	1600	5	Soft soil
2	5	10	300	1600	5	Sand, gravel
3	17	27	410	1800	4	Sand, gravel
4	28	55	1200	2100	1	Salt, gravel
5	135	190	600	1800	2	Gravel, loam
Half-space	∞		1500	2100		Massive salt

Ambient noise measurements were carried out close to the borehole MB-2 (for location see figure 4.175(c)). Individual and average H/V spectral ratios obtained at Station 1 are shown in figure 4.175(a)). Two prominent resonance peaks with the amplification of 2.5–3.5 are observed at the frequency ranges 0.6–0.7 Hz and 3.5–6 Hz. The optimal analytical function superimposed on the average spectral ratio is shown in table 4.13.

Figure 4.178(b) The best fit of the analytical and empirical estimations is reached by assuming the thickness of the gravel, clay, and limestone layer equal to 140 m and V_S of the deep reflector (probably another salt layer) is 1500 m s^{-1}. Geotechnical parameters of the upper part of the subsurface model are kept as they are in the MASW section or with minor changes. The optimal subsurface model is given in table 4.13.

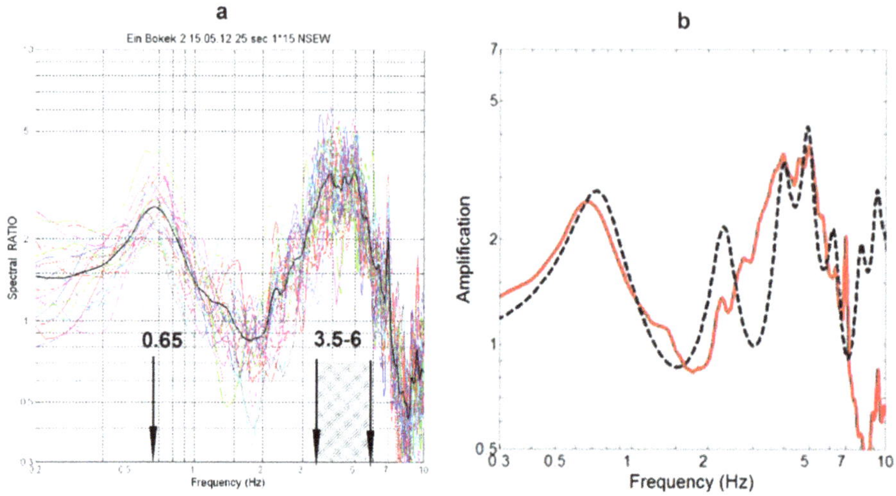

Figure 4.178. HVSR results in the Ein Boqeq north site. (a) Individual and average H/V spectral ratios obtained at the point EB-2 (figure 4.175(c) for location). (b) Analytical transfer function (black dashed line) in comparison with H/V spectral ratio of the ambient noise obtained at the Ein Boqeq north site (red line). (Reprinted from Gorstein and Ezersky (2015), with permission from Camdemia.)

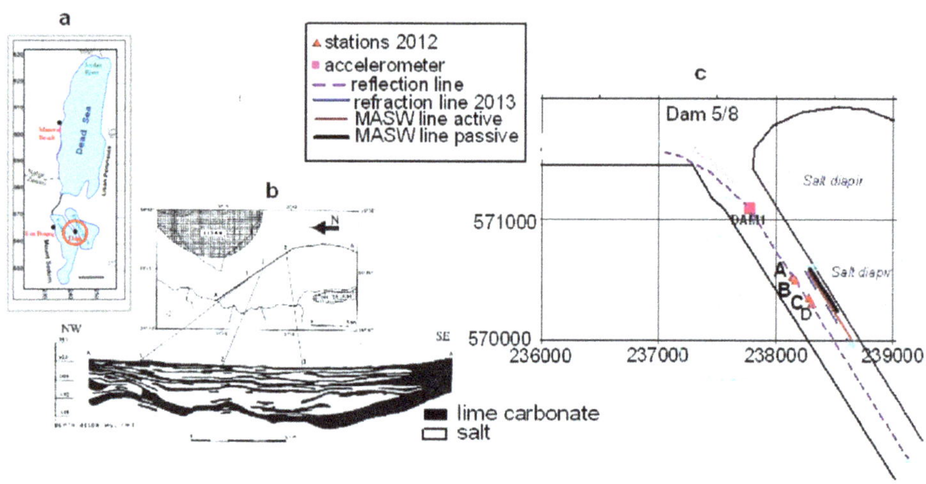

Figure 4.179. Dam 5/8 general layout of measurements. (a) Site location map. (b) Geological section after Bruner *et al* (2003) geophysical and seismological measurements layouts (A–D recording stations). (Reprinted from Ezersky *et al* (2013d), with permission from Ministry of Energy and Water Resources (MEWR) of Israel.)

4.9.3.1 Dam 5/8 example (Ezersky *et al* 2013d)

The ambient noise measurements using the MASW and HVSR methods were carried out at the Dam 5/8 (located in the southern DS basin) used nowadays as evaporation ponds (figure 4.53(a)). Interpretation of the drilling data conducted before the dike was built (figure 4.179 (b)) testifies that the area is covered by young

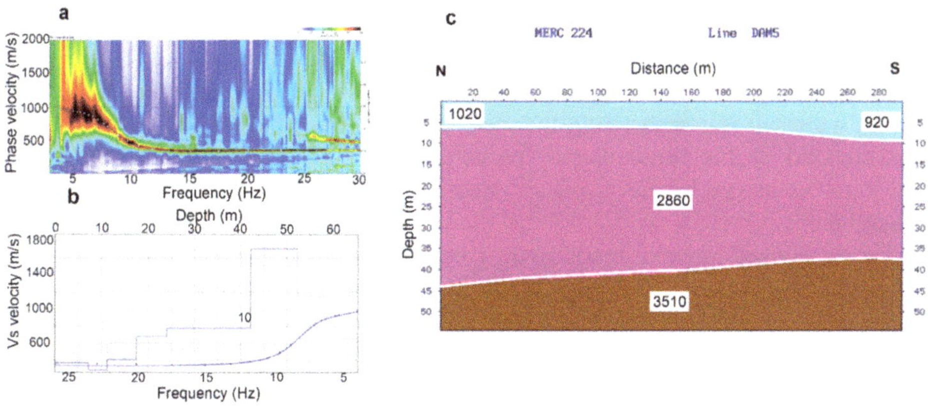

Figure 4.180. Results of the geophysical surveys on the Dam 5/8. (a) Combined MASW dispersion image with the extracted curve (31 geophones passive roadside array 10 m separation plus active MASW line), (b) MASW V_s profile (RMSE = 8.9%, (c) seismic refraction V_p velocity–depth section. (Reprinted from Ezersky *et al* (2013d), with permission from Ministry of Energy and Water Resources (MEWR) of Israel.)

sediments of halite intercalated with soft clay to silt-sized limestone (previously described as clay or marl) (Charrach 2019).

The properties of these LCs have been described by Frydman *et al* (2008). These sediments show signs of the recent active tectonic faulting and salt diapirism. In the southern part of the DS basin, the uppermost part of the sequence is dominated by 60–80 m of salt layers alternating with some lime mud deposited over the last 15 000 years. These sediments belong to the Ze'elim Formation. The sequence below to a depth of hundreds of meters consists of lime mud, clay, and possibly some sand and gravel of the Lisan and Samra Fms. The area is intersected by salt diapirs of different dimensions and depths below the surface and is disturbed by complex fault systems and hallokinesis (Shalev *et al* 2005, Charrach 2019).

Geophysical measurements similar to that aforementioned in the Ein Boqeq north site were carried out on Dam 5/8. The measurements that compare the MASW passive and active measurements, the seismic refraction line of 260 m long, and the HVSR recording at four points (A–D) are presented in figure 4.179(c).

Results of the geophysical survey on Dam 5/8 are shown in figure 4.180.

The combined image is shown in figure 4.180(a) and consists of dispersion images of the passive and active MASW. Combined inversion of the dispersion curve with the seven-layered model is presented in figure 4.180(b) (RMSE = 8.9%). There is also seismic refraction V_p-depth section (figure 4.180(c)) (Ezersky *et al* 2013d).

4.9.4 HVSR measurement results

Until now, the ambient noise measurements were carried out at four points (see locations in figure 4.179(c)). A common feature of all the H/V ratios is an interval of amplification at the frequencies of 3–7 Hz, and only at point D can one see an additional peak at the frequency of 1.5 Hz. Note that this frequency is observed in the H/V spectral ratios of two earthquakes (receiver functions), which were recorded by the accelerometer DAM1 located 50 m to the north (see figure 4.179(c)). As seen from the H/V spectral ratios, a frequency of 1.5 Hz is fundamental. Additional peaks which are observed in the frequency range of 3–7 Hz have lower amplitudes. For this reason, we use it in constructing the analytical model of the HVSR at the point D characterized by the fundamental frequency of 1.5 Hz. We note that these data testify to some preliminary results and need to be confirmed by additional measurements.

The analytical transfer function has been calculated based on the V_S section retrieved from the MASW results. We note that according to the MASW, the upper salt layer is marked at a depth of 42–43 m (figure 4.180(a)) and that is in complete agreement with the results of the P-wave refraction survey (figure 4.180(b)). This upper salt layer is a shallow reflector and corresponds to the resonance peak of an analytical transfer function which is positioned at the frequencies 4–6 Hz. To approximate both the fundamental resonance peak at a frequency of 1.5 Hz and the second high-frequency peak, the thickness of the intermediate between the two reflectors layer and V_S of deep reflector were adjusted considering the MASW data. The analytical function superimposed on the HVSR obtained at the DAM-D point is shown in figure 4.181(b). The optimal subsurface model providing the best fit with the experimental function is given in table 4.14.

Studies carried out using the HVSR and MASW methods have shown the potential of the method to investigate deep (more than 250 m) parts of the subsurface. This study directly relates to the sinkhole problem associated with fact that the southern DS basin is complicated by diapirism. The Mt Sedom (west of the southern basin, see figure 4.179(a) for location) is a salt diapir, the southern part of which is raised above the Earth's surface. After different analyses, a rate of diapir

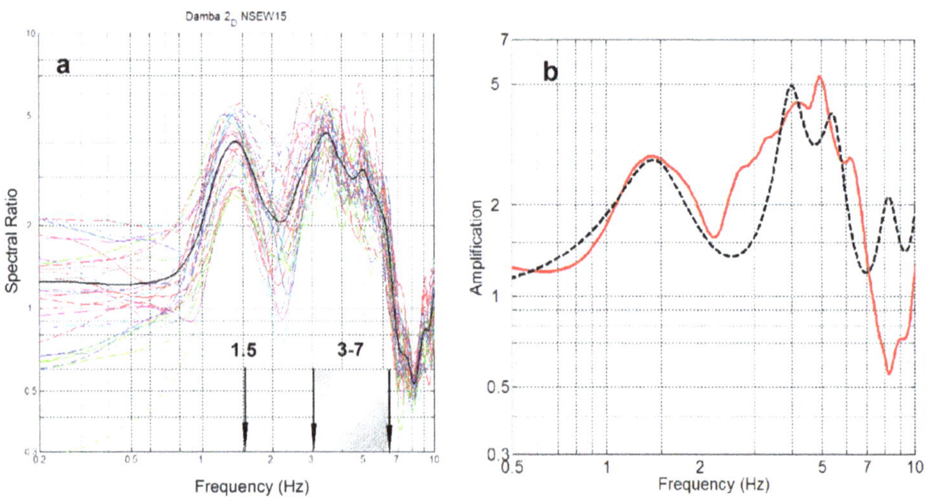

Figure 4.181. Dam 5/8 HVSR results. (a) H/V spectral ratios that were obtained at point D located at the Dam 5/8 site. (b) Comparison between the analytical response function calculated for the Dam 5/8 site and the HVSR obtained at point DAM-D. (Reprinted from Ezersky *et al* (2013d), with permission from Ministry of Energy and Water Resources (MEWR) of Israel.)

Table 4.14. Optimal 1D subsurface model for calculation of the analytical site response function at the Dam 5/8 site. (Reprinted from Ezersky *et al* (2013d), with permission from Ministry of Energy and Water Resources of Israel (MEWR).)

Layer	Thickness, m	Depth, m	V_S, m s^{-1}	Density, kg m^{-3}	Damping, %	Possible lithology
1	7	7	378	1800	4	Soft soil
2	6	13	290	1700	5	Salt, LC
3	4	17	400	1900	4	Salt, LC
4	6	23	700	1700	2	Salt, LC
5	18	41	790	1900	2	Salt, LC
6	10	51	1700	2300	1	Salt
7	145	196	1110	2200	1	Salt, LC
Half-space	∞		2240	2500		Massive salt

arising is estimated at 6–9 mm per year (reference!). Some underground salt diapir bodies were revealed by the seismic reflection study. They are arranged along the SW–NE line and located at different depths. Most of these bodies occur south of the Lisan Peninsula at about 100 m depth (Al-Zoubi and Ten Brink 2001). Another shallow diapir is suspected to be under engineering dams near Mt Sedom that cause deformations of subsurface and induces a new type of sinkhole in Dam 5/8 and Highway #90. That is a subject of further investigation.

4.9.4.1 Nanoseismicity, 'nanoseismic monitoring'

DS sinkholes formation is associated with the collapse of near-surface material into dissolution cavities in salt layers (e.g., Abelson *et al* 2003, 2006, Ezersky *et al* 2006, Frumkin *et al* 2011, Yechieli *et al* 2016). The sinkhole evolution is controlled primarily by the rock solubility and secondarily by the competence of the sediment cover above the dissolution cavity related to its thickness and rheological properties (Maimon *et al* 2009, Parise and Lollino 2011, Shalev and Lyakhovsky 2012). The high solubility of evaporitic rocks, e.g., gypsum and salt, enhances sinkhole formation when exposed to undersaturated groundwater (e.g., Yechieli *et al* 2006). During the past 20 yrs, more than 6000 sinkholes have occurred along the western shores of the DS (Abelson *et al* 2018). According to Abelson *et al* (2006) and Yechieli *et al* (2006), this extremely rapid evolution results from the dissolution of a salt layer buried within the young sediments along the DS shores. The onset of the salt dissolution is caused by the intrusion of fresh groundwater from the west, triggered by the lowering of the DS level at the rate of ~1.0 m yr^{-1}. Another reason can be decrease in the groundwater head accompanying this drop in the DS level. Ezersky and Frumkin (2017) have shown that the very saline water (named also the DS brine) has the potential to dissolve a large quantity of the salt, forming the underground cavities. In addition, Legchenko *et al* (2008a) suggested the availability of the pre-existing cavities in the DS subsurface. However, in both cases development of the covered cavities can be accompanied by acoustic noise at the surface.

Two studies using the recording of noise produced by the falling of stones within the covered cavities (named 'nanoseismic monitoring') were carried out in the DS area in the Nahal Hever site by Wust-Bloch and Joswig (2006) and Abelson *et al* (2018) in the Mineral Beach (see figure 4.182(a) for location).

The working model of the study is presented in figure 4.182(b). The cavity is formed by the gradual falling of stones consisting of clay with gravel from the roof of the cavity to its floor or groundwater. Seismic pulses arising can be recorded on the surface above the cavity and in its vicinity at some distance. This model is similar to that suggested by Hardy and Belesky (1987) and discussed in section 2.2.1.17. The study of such a process should include the presence of a salt layer, a multi-channel system for noise recording, calibration of recorded pulses to distinguish between the real pulses from falling stones and the ambient noise of another origin. Wust-Bloch and Joswig (2006) have acquired and processed data using the approach of the 'nanoseismic monitoring' (Joswig 2005) based on the Seismic Navigation Systems (SNS) and the SparseNet software suite. The six-channel SNS consists of four short-period sensors: one single three-component sensor at the center of the configuration and three vertical component sensors at the periphery.

Wust-Bloch and Joswig (2006) have calibrated the events arising from the falling of stones. The waveform analysis by sonograms recognizes two main groups of seismic events: impacts on dry material and impacts in liquid (figure 4.180(b)). Performed analysis has demonstrated that the discrimination between both types of source functions is robust despite the extreme nature of the scattering media. Finally,

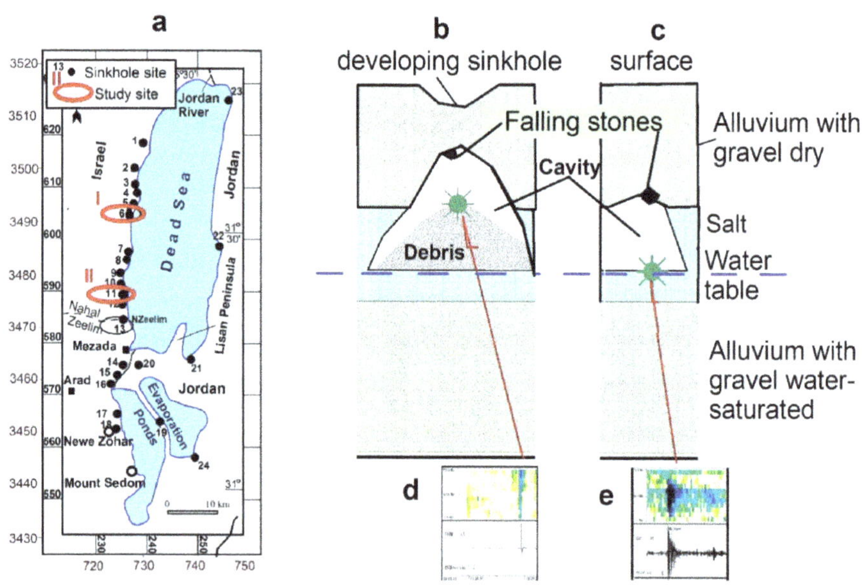

Figure 4.182. (a) The DS sites investigated by the nanoseismic monitoring I—Mineral Beach II—Nahal Hever South where groundwater depth is 18–20 m from the surface, salt is found between the 24 and 32 m depth (reprinted from Yechieli et al (2002), with permission from GSI). (b and c) Schematic diagram of a sinkhole and covered cavity in the subsurface according to the local borehole section (b) noise is generated by stone falling on debris); (c) noise is formed by stone falling to groundwater. Two sonograms are shown in the lower part of this section: the left one (d) is connected with falling of stones on dry material (debris) and the right one (e) is associated with the falling on liquid (groundwater).

the presence of the high-frequency sonograms can be employed to indicate that cavitation is taking place above the groundwater level and, therefore, progressing towards the surface. Another result of sonogram analysis was crucial for successfully suppressing a multitude of noise bursts. The distribution of the nanoseismic sinkhole-generated events that arose during 12 h is shown in figure 4.183.

It seems the SNS system provided optimal event coverage for the most sinkhole clusters 200–300 m away. Further development of the method was found in the study of Abelson *et al* (2018). The study was carried out in Mineral Beach (see figure 4.182(a), Site I). Abelson *et al* (2018) used seismic monitoring system that detects the rock mass stability overlying subsurface cavities, similar to those used in the underground mines (Mendecki 1997). In contrast to the previous seismic monitoring of the DS sinkholes (Wust-Bloch and Joswig 2006), geophones were placed in boreholes at various depths to reduce background noise and estimate hypocenter depth. Five triaxial geophones were buried in five boreholes in the alluvial fan around the Mineral Beach sinkhole cluster, at spacings of several hundred meters. The distribution of the recording stations is shown in figure 4.183(a). The locations of the seismic events were calculated by fitting the observed arrival times and a velocity model based on seismic refraction profiles. To estimate the event location which occurred within a radius of 350 m from the array

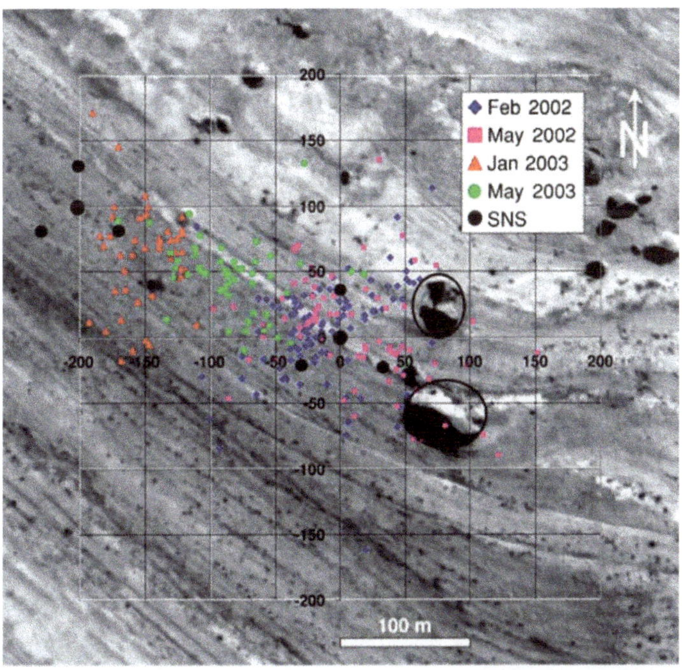

Figure 4.183. Location plot of the sinkhole-generated events recorded at the Nahal Hever site (DS) during four 12 h periods. An aerial photograph (1999.02.12) of the same location with two fully developed sinkholes (black circles) is provided for comparative purposes. Black circles indicate the location of the SNS and its sensors. (Reprinted from Wust-Bloch and Joswig (2006), with permission of Oxford Academic-Journal Compilation, copyright 2006 RAS.)

center the following velocities were used: $Vp = 750\ \text{m s}^{-1}$ and $Vs = 350\ \text{m s}^{-1}$. The seismograms recorded during 25.08.2012 are presented in figure 4.183(b).

Seismic monitoring through the Mineral Beach area was continued from 28 June to 7 September 2012. Here 82 seismic events were recorded. Excluding seven events located several kilometers away from the test site, most of the recorded events (with local magnitude ML = −2 to −3) were distributed from the sinkhole lineament eastward (figure 4.184(a)). This distribution of the nanoseismic events is in agreement with the general pattern of eastward propagation of the sinkhole occurrence. As shown in figure 4.184(a), the mudflat sinkholes initiated along a narrow belt at the westernmost part of the present lineament. Since 1996, the sinkhole lineament widened only eastward. Taking into account that the sinkhole lineament is distributed along the western edge of the salt layer (Ezersky 2006, Frumkin *et al* 2011, Ezersky and Frumkin 2013), this eastward propagation is a surface expression of the eastward migration of the salt dissolution front (figure 4.184(a)) (Ezersky and Frumkin 2013). This study testified that the source of the nanoseismic events is from cracks and faults rather than rockfall impact within a subsurface cavity, as previously suggested by Wust-Bloch and Joswig (2006). That is caused, evidently, by the observed sinkhole-induced tensile cracks at that location.

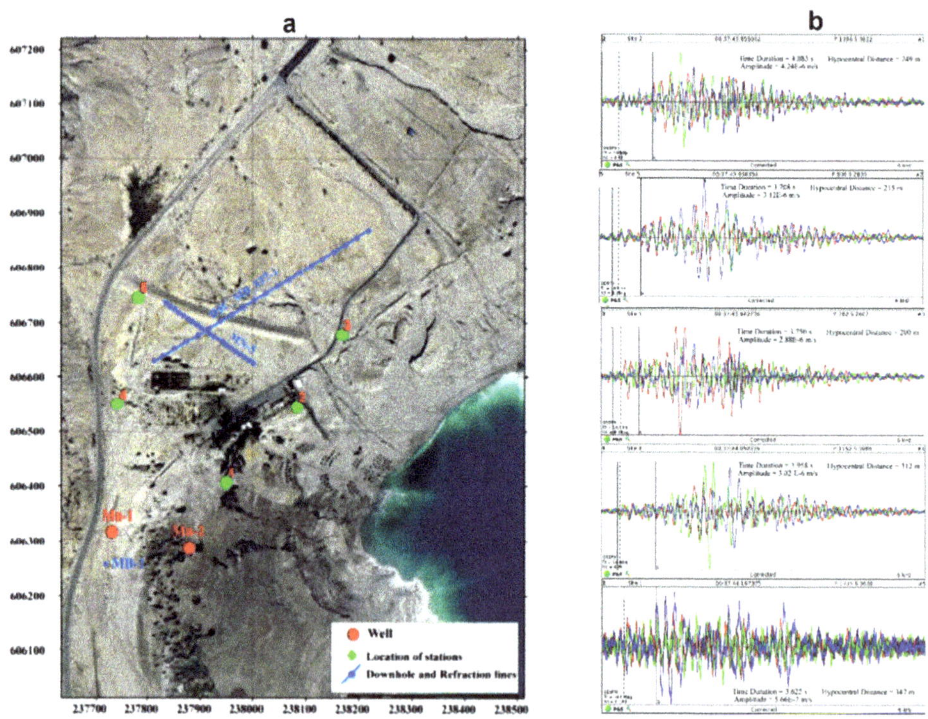

Figure 4.184. The Mineral Beach site (western DS shore). (a) Location map of the nanoseismic recording system. (b) Waveforms example of event 25.08.12 (00:37:43). (Reprinted from Aksinenko *et al* (2012), copyright (2012) with permission from Geophysical Institute of Israel.)

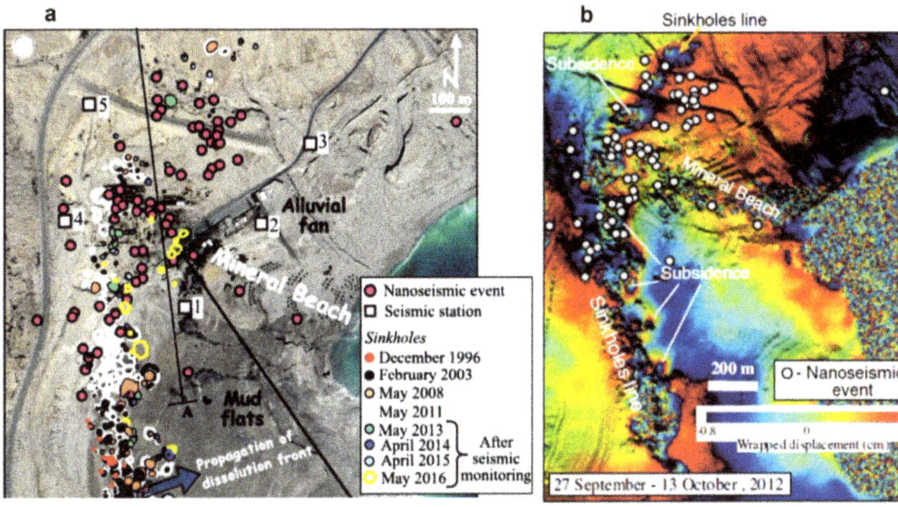

Figure 4.185. The Mineral Beach study site. (a) Distribution of the recorded nanoseismic events during the study compain between 28 June and 7 September 2012. (b) Nanoseismicity plotted upon interferometric synthetic aperture radar detected subsidence. The epicenters are projected on an interferogram showing land subsidence ~1 month after recording the nanoseismic events. Note the absence of subsidence above the seismic events east of the sinkhole line. (Reprinted from Abelson *et al* (2018), with the Permission of Geological Society of America. Modified by M Ezersky.)

Abelson *et al* (2018) compare results of the nanoseismicity monitoring with InSAR observations (Figure 4.184(b)). Along the sinkhole lineament, the nanoseismicity was accompanied by subtle subsidence, recorded by InSAR. However, further to the east, no subsidence was detected above most of the nanoseismic events during 2012–16. That allows suggesting that the stiff sediments of the alluvial fan hinder subtle land subsidence above the concealed activities, extending the period of nanoseismic precursory signals by several years relative to the short-term InSAR precursors (generally months; Nof *et al* 2013) (figure 4.185).

This two-month-long monitoring does not foreshadow a single cavern, as does the InSAR, because it is significantly shorter than the several-year period of upward propagation of a cavity. For precise detection of the cavity, much longer monitoring with more nanoseismic events is required.

4.10 Nongeophysical methods

4.10.1 Boreholes: chemistry (Yechieli *et al* 2004)

The boreholes in the DS sinkhole sites were drilled by the Geological Survey of Israel (GSI) in the framework of the State program starting from the beginning of 1993 (Yechieli *et al* 2004). In the start, the drilling was carried out using a spiral. Later the drilling was carried out with a rotary machine without drilling fluids (no water or mud), using casing following the rock-bit. Such drilling enabled the recording of the hydraulic head distribution over the different sub-aquifers encountered and sampling undisturbed groundwater at any required water horizon below the water table (Yechieli *et al* 2016).

For a collection of salt samples boreholes drilled with the DS brine as drilling fluid using casing following down to the salt top (Ezersky *et al* (2013c), see also section 4.8 of this manuscript). In some boreholes, SPT was carried out.

Water samples collected either through pumps or by a bailer were analyzed for their chemical composition. Analyses were conducted according to the standard methods and procedures adopted at the laboratories of the GSI (Yechieli *et al* 2016). An example of such results can be found in table 3.1 (chapter 3 of this book) presenting the chemical composition of groundwater from two wells, the DS and evaporation ponds.

Salt samples were prepared in the laboratory of GII where elastic velocities and resistivities were measured, geotechnical parameters (porosity and permeability were measured in the Building and Infrastructure Testing Lab (BITL) Ltd, (Haifa) (Ezersky and Goretsky 2014).

Boreholes were distributed along the entire DS from north to south (see figure 4.62(a) and (b) in section 4.3.2)).

A demonstrative example of the borehole report is shown in figure 4.186. The typical geological section of the DS western side (Israel) based on 80–90 m deep

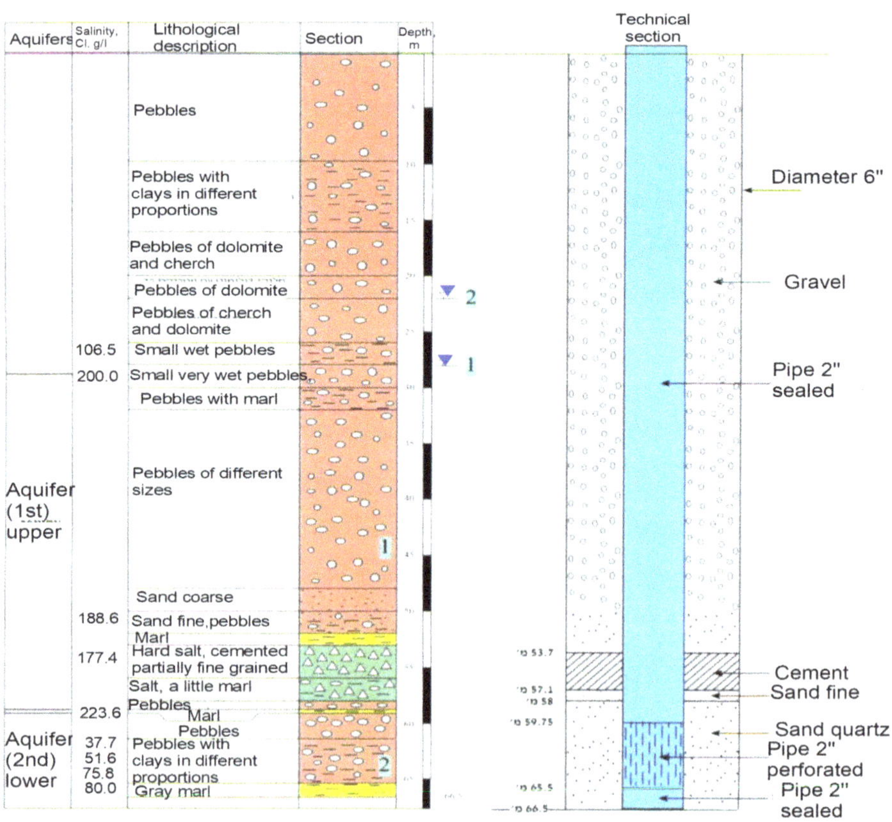

Figure 4.186. An example of the borehole EG-13 report from the Ein Gedi site. (Reprinted from Yechieli *et al* (2004), with permission from Geological Survey of Israel.)

boreholes (Yechieli *et al* 2004) is composed of alluvial fan sediments to a depth of 20–50 m. The salt layers have a thickness of 6–30 m and are located at depths of 20–50 m (top). Salt is usually sandwiched between the marl layers from below and from above, especially when approaching the DS shoreline. The salt edge to the west of the DS is often in contact with the sandy-gravel sediments (Yechieli *et al* 2004).

Boreholes were used: (a) to monitor the groundwater level; (b) to measure the groundwater electrical conductivity. Monitoring of the groundwater level was carried out by GSI in 20 boreholes and was aimed to explore the response of the groundwater level to the shrinkage of the DS (Yechieli *et al* 1995b, 2010). Figure 4.187 presents variations of the groundwater level with a time in two sites: Nahal Arugot (a) and Nahal Zeelim (b), and the DS level.

One can see that variation of groundwater in the Nahal Arugot is composed of sandy-gravel sediments (located ~700 m from DS shoreline) (figure 4.187(a)) are conformed to variations of the DS level testifying the good hydrogeological connection. Variations of the head in the Zeelim boreholes located at a much larger distance from the shoreline (~4 km), the drop of the water level is significantly lower than the drop of the DS levels (figure 4.187(b)).

The monitoring of electrical conductivity was aimed to explore the response of the fresh-saline groundwater interface to the shrinkage of the DS. In all cases, the EC was measured by *in situ* devices from the groundwater table to a few meters below the fresh-saline interface (Yechieli *et al* 2010). Following Yechieli (2000), the EC values below 160 mmho cm^{-1} increase with salinities, while the EC values higher than 160 mmho cm^{-1} do not correlate with the salinity. The value of 150 mmho cm^{-1} is approximately equivalent to the salinity of 200 g L^{-1}.

The monitoring has shown that generally, salinity does not vary stably in different boreholes and sites. However, there are some regularities in the spatial salinity (EC) distribution. Below some of the boreholes, EC values are exhibited by the decrease at given depths, for instance, in the borehole EG-11. This process is probably controlled by the hydraulic properties of the sediments near the DS, whereby the

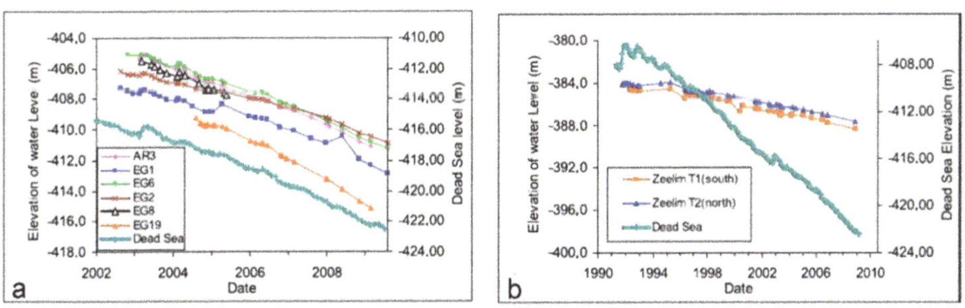

Figure 4.187. Groundwater levels in monitoring boreholes in the DS coastal aquifer at different locations and distances from the shoreline along with the DS level data from the Hydrological Service of Israel). (a) AR3, EG1, EG2, EG6, EG8, and EG19, located at Wadi Arugot ~500–700 m from the DS shoreline; (b) Zeelim T_1 and T_2, located in Wadi Zeelim (~4 km) (see figure 4.62 for site location). (Reprinted from Yechieli *et al* (2010), with the permission of American Geophysical Union.)

gravel in the Arugot alluvial fan allows more rapid flushing, while in other locations slower flushing is expected.

(Yechieli 2007, Kiro *et al* 2008, Yechieli *et al* 2010). In borehole EG-19 EC increases with a time above the fresh-saline interface and it does not vary noticeably below that (Abelson *et al* 2017).

4.10.2 Sinkhole database

More than 6000 sinkholes along the western DS shores forming starting from the end of the 80th were monitored and processed by GSI (DS sinkholes, Abelson *et al* 2006).

Sinkholes were detected from the aerial photographs (Raz 2000, Abelson *et al* 2002) and surface examinations. The DS sinkhole database includes measurement of the coordinates at the time of appearance, diameter, depth, lithological description, and monitoring of the sinkhole development. The monitoring was carried out by Eli Raz geologist of the Ein Gedi kibbutz (DC sinkhole database, Raz 2000). GSI has created a sinkhole database in the ArcGIS framework (egozi.gsi.gov.il). Data were processed in the GSI. An example of sinkhole distribution along the DS western shore is shown in figure 4.188.

4.10.3 Monitoring of sinkholes (database Raz 2000, Abelson *et al* 2006)

In figure 4.189, two graphs are presented (a) analyzing the temporal sinkhole development in the DS sites (Abelson *et al* 2006, 2009) and (b) showing a correlation between the time of sinkhole appearance and easting of the salt layer defined by the seismic refraction method (Ezersky *et al* 2010). The first graph (figure 4.189(a)) shows the history of the decline of the DS level (red line) and the altitude of the sinkhole sites (horizontal bars). The DS decline has accelerated significantly since the mid-1970s. The appearance of sinkhole sites began in 1980 when the DS level was at ~400 m below sea level. Most sinkhole sites are found below −400 m while accompanying the DS decline, except for those in the alluvial fans. Furthermore, the relative timing of the DS decline as well as sinkhole appearance and acceleration (figure 4.189(a)) indicate that the inception of sinkhole formation occurred as the DS level declined below the −400 m contour. Exceptions are found at seven sites (out of 34) located on the alluvial fans of major streams above the −400 contour (figure 4.189(a)). General regularities of the timing of sinkhole appearance are: (a) the northward propagation of sinkhole activity through time; (b) a link between the occurrence of sinkholes and the lowering of the DS level (Abelson *et al* 2006).

On the other hand, the seismic refraction study has revealed a spatial correlation between the salt edge and the sinkhole distribution in the studied sites (Ezersky *et al* 2010). The map in figure 4.62(b) shows the western shore of the DS shifts to the east towards the north. We suggest that the sinkhole formation is controlled by the regression in the DS and the variable position of the salt edge that is most eastward at the north of the DS than south of that. This interpretation is confirmed by the graph presented in figure 4.189(b) showing the direct correlation between the positions of the salt edge (based on the seismic refraction mapping) and the time

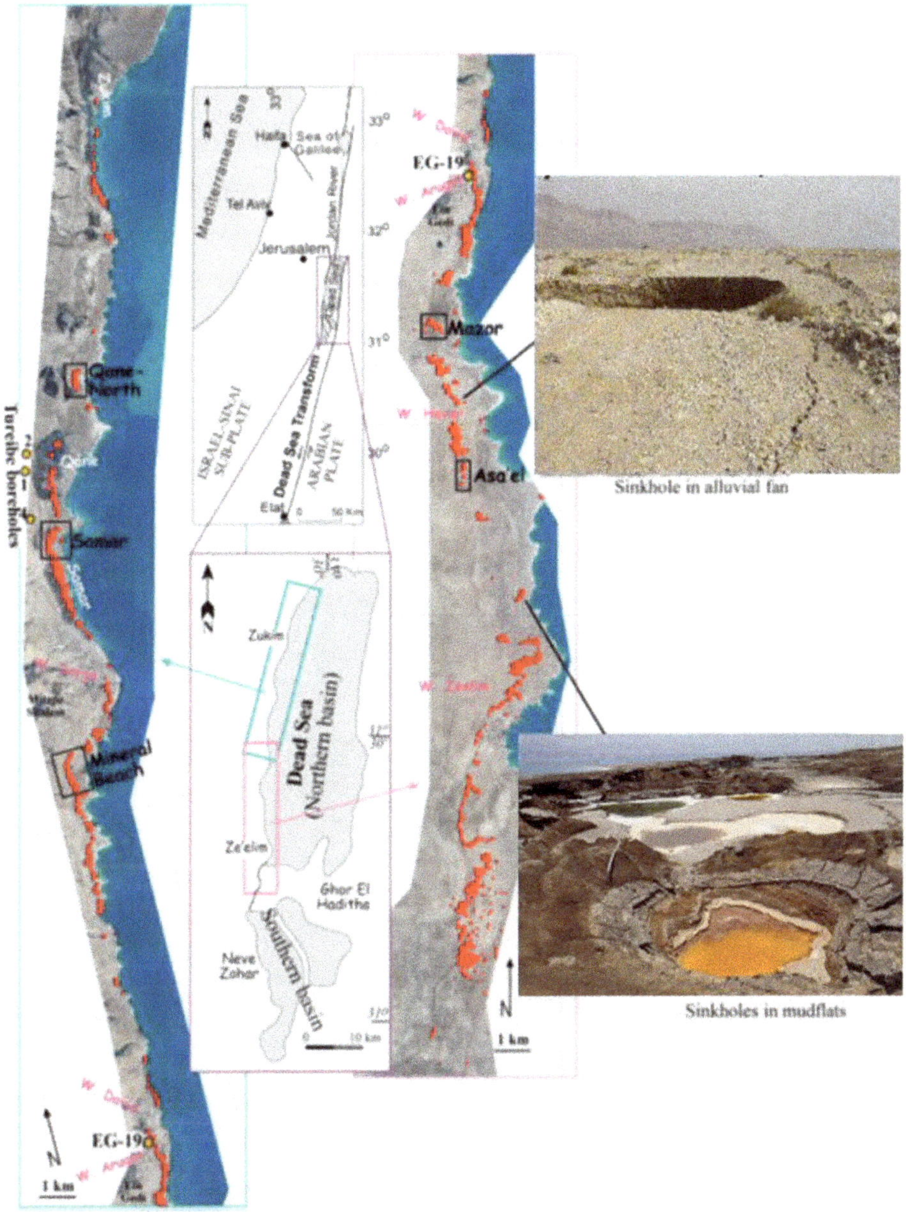

Figure 4.188. Distribution of sinkholes along the west coast of the northern basin of the DS on a rectified collage of the aerial photographs. Red dots designate the sinkholes enrolled by a buffer zone of 35 m, mapped on lidar images from May 2015. The two panels are the (left) northwestern and (right) southwestern coasts. The inset shows the location of the two panels. Photographs on the right side show examples of sinkholes in (top) alluvial fans (photo of M Ezersky) and (bottom) mudflats (photo of Shmuel Brown, used from permission http://israel-tourguide.info). (Reprinted from Abelson *et al* (2017), with the permission of John Wiley & Sons.)

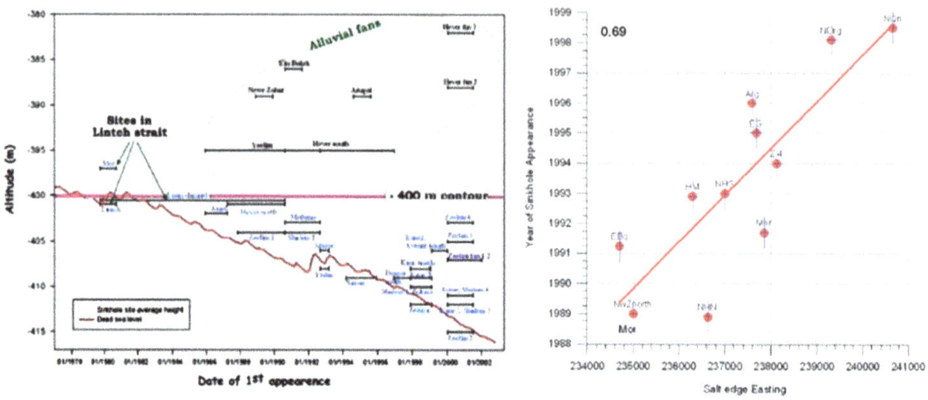

Figure 4.189. (a) History of the decline of the DS level (red line) and altitude of sinkhole sites (horizontal bars). The bars describe the error in the time of the first appearance of sinkhole sites. Most sinkhole sites are found below −400 m, while accompanying the DS decline, except for those in the alluvial fans. (b) Time of the first appearance of sinkholes versus the shift to the east of the salt edge revealed by the seismic refraction survey at 10 DS sites. The error bar reflects the uncertainty about the time of sinkhole appearance and the range of variations of the salt edge at various sites. ((a) Reprinted from Abelson *et al* (2009), with permisison from the GSI; (b) reprinted from Ezersky *et al* (2010), with the permission of Schweizerbart Science Publisher, available at www.borntraeger-cramer.de/journals/zfg.)

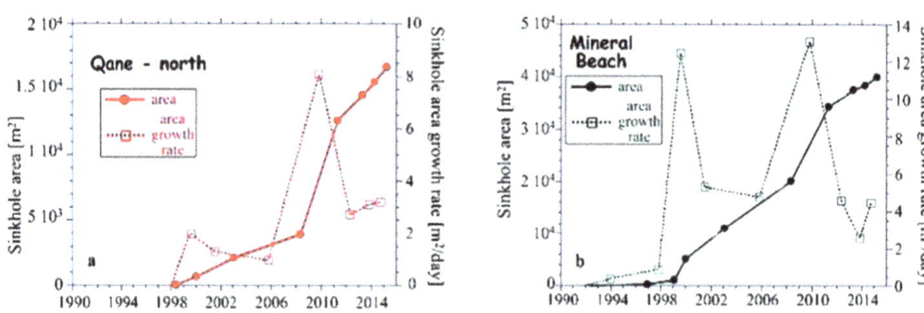

Figure 4.190. Evolution of two sinkhole clusters (a) Qane-north and (b) Mineral Beach. The analysis is carried out in terms of the increase in the sinkhole area (m^2) (solid line) and the growth rate of the sinkhole area (m^2 d^{-1}) (pointed line). The selected clusters are within the mudflats, to avoid delay bias in the more competent alluvial fans. (Reprinted from Abelson *et al* (2017), with the permission of American Geophysical Union.)

of sinkhole appearance, e.g., the development of sinkholes in the western DS shore is associated with the salt edge, which is situated in progressively more eastern positions towards the north.

Numerous studies are devoted to the observation of the temporal sinkhole development (Ezersky *et al* 2010, 2013b, Abelson *et al* 2009, 2017). An example of the temporal development of sinkhole clusters in two sites is shown in figure 4.190(a) and (b).

One can see in figure 4.190 the area growth measured in seven to nine dates of two sinkhole sites (see figure 4.188 for locations). Rates of the areal growth were calculated as the area change between two dates divided by the number of elapsed days. Both sites display similar patterns of the areal growth rates showing two growth peaks, the first around 1999–2000 and the second about 2010.

4.10.4 Aerial photographs

As we considered in section 2.2, the passive remote sensing methods found wide use in the sinkhole mapping and susceptibility problem (Galve *et al* 2009b, Gutiérrez *et al* 2011, 2019, Benson and Yuhr 2016, Orhan *et al* 2020). These methods were widely applied in the DS sinkhole problem (Abelson *et al* 2002, 2003, 2017, Baer *et al* 2002, 2018, Closson 2005, Closson *et al* 2005, Closson and Abou Karaki 2009, Nof *et al* 2013, 2019, Yechieli *et al* 2016, Al-Halbouni *et al* 2017, Arav *et al* 2020).

Methods used in the DS sinkhole areas included remote sensing—interferometric synthetic aperture radar (InSAR), airborne LiDAR, and aerial photographs.

Airborne LiDAR, also known as airborne laser swath mapping, uses a laser ranging device, together with information on the position and orientation of the aircraft platform. This is an active remote sensing system that records very subtle topographic features. LiDAR data were acquired at an average flight height of 700 m. The densities of the ground measuring points were set to be at least 4 points/m^2. The digital elevation models (DEMs) were produced from the ground targets ('point clouds') at 0.5 m/pixel with the vertical precisions of 10–20 cm and were used for the annual to decadal elevation change detection and sinkhole mapping and catalog creation (Abelson *et al* 2017). Filtering for vegetation was not carried out because of the practical absence of vegetation (Filin *et al* 2011, Avni *et al* 2016).

The mapping and monitoring of the sinkholes were conducted through the field surveys, analysis of the repeated aerial photographs, and comparison between the LiDAR DEMs (Yechieli *et al* 2016). The directions of the sinkhole lineaments were measured and compared to those of the main faults in the DS Rift (Abelson *et al* 2003).

The spatio-temporal sinkhole cluster development in the mud area of the Mineral Beach (south) area is presented in figures 4.191(a) and (b).

Figure 4.191(b) displays the same area based on the rectified aerial photographs until 2011. The color marks the time of sinkhole formation. It is the well-seen promotion of sinkholes with a time to the eastern direction earlier denoted by Frumkin *et al* (2011) in the Mineral Beach (Israel) and Ezersky *et al* (2013b) in the Ghor Al-Haditha area (Jordan).

4.10.5 Maps of sinkhole susceptibility

Based on the model of the DS sinkhole formation, geologists of the GSI constructed maps of the potential sinkhole hazards in different parts of the DS area. The principles of the construction of such maps are explained in figure 4.192. The most important governing factor for the sinkhole formation is the existence of a salt layer, without which there is no sinkhole potential (no color in the map (figure 4.192)). The location of the sub-surface salt layer was determined by the observation boreholes (Yechieli *et al* 2004), seismic refraction surveys (Ezersky 2006, 2011a, 2011b,

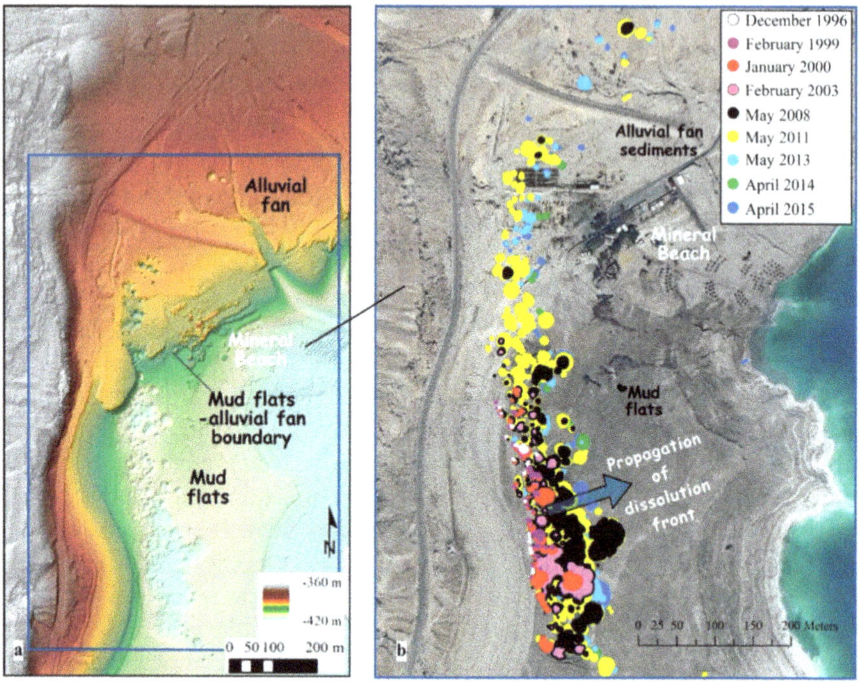

Figure 4.191. Evolution of the sinkhole cluster near the resort site of Mineral Beach presented (a) an image taken by *airborne LiDAR* from April 2013 and (b) on a corrected *aerial photograph* from May 2011. Sinkholes were digitized on the corrected aerial photographs until 2008 and in the airborne LiDAR images from 2011 to May 2015. (Reprinted from Abelson *et al* (2017), with the permission of American Geophysical Union.)

Ezersky *et al* 2007, 2010, 2013b, Frumkin *et al* 2011), and some geological considerations. An additional requirement for sinkhole occurrence is the contact between the relatively fresh groundwater and the salt layer. Other sinkhole potential zones in figure 4.192 are: (1) sinkhole area (yellow on maps); (2) high probability (red) and (3) low probability zone (green).

The map of sinkhole probability based on such principles has been presented as an example in Yechieli *et al* (2016) and completely in GSI (2012).

Note, seismic refraction mapping of salt layer edge throughout the entire area of the Newe Zohar site has been carried out (Ezersky *et al* 2010). The geophysical mapping shows a more accurate pattern of salt distribution and allows mapping when no sinkholes are visible on the surface.

4.10.6 InSAR in the Dead Sea shores

InSAR is used to measure the elevations of the Earth's surface from a satellite. The principle of interferometry exploits carefully engineered phase differences between two or more radar images. The phases of images with a difference of position (e.g., two antennae onboard one vector acquire images simultaneously) (single-pass interferometry such as SRTM mission) or with a difference of time (e.g., one

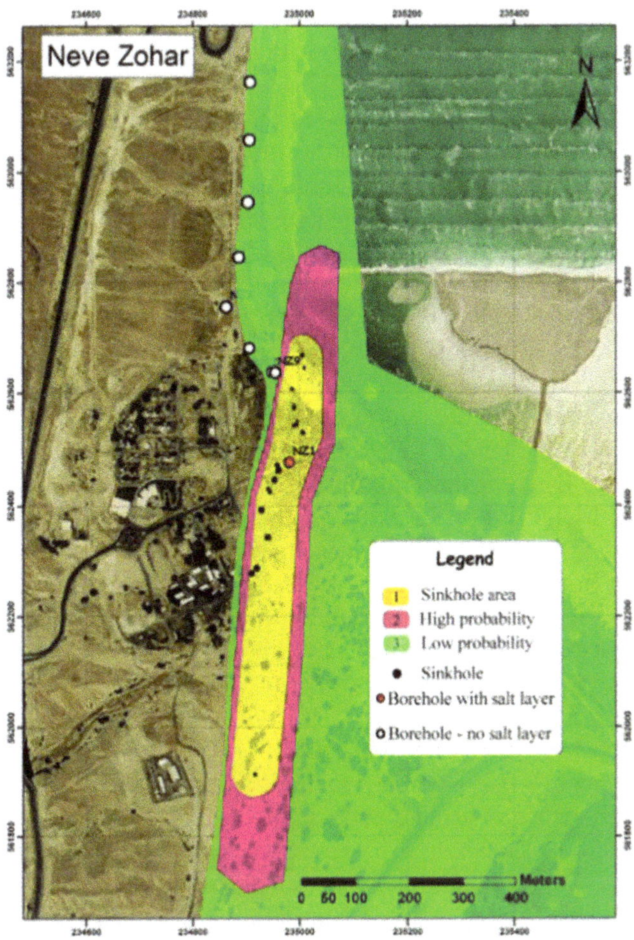

Figure 4.192. An explanation of the construction of a map of potential levels of sinkholes formation around the Newe Zohar site. (Reprinted from Abelson *et al* (2009, with permission from GSI.)

antenna acquires images at two distinct times or multi-pass interferometry) can be compared after proper image coregistration (Massonet and Feigle 1998). The InSAR observations in the DS area are aimed to evaluate the ground subsidence and map the sinkhole development (Baer *et al* 2002, 2018, Abelson *et al* 2003, 2018, Closson 2005, Nof *et al* 2013, Shviro *et al* 2017).

Baer *et al* (2002) identify, characterize, and measure rates of subsidence along the DS shores using the InSAR technique. 16 SAR scenes acquired during the years 1992 to 1999 by the European Remote Sensing ERS-1 and ERS-2 satellites were analyzed. It was the first publication that has analyzed the wide, shallow subsidence features (WSSFs) through the DS shores in Israel and Jordan. Special attempts were done to connect the WSSF with the DS sinkholes. The subsidence was explained by dehydration of the massive marl layers as a result of a drop in the DS level when the thickness of the marl layers is 15–20 m and over. The phase differences measured in calculated interferograms correspond to subsidence rates generally in the range of 0–20 mm yr^{-1} within the studied period, with exceptionally high rates that exceed 60 mm yr^{-1} in two specific regions. The authors concluded that the gradual subsidence is unlikely to be directly related to the sinkhole formation.

Closson (2005) and Closson *et al* (2005) studied subsidence and sinkhole hazard assessment in the southern DS area in Jordan. Attempts were made to explain the leak from evaporation pond 18 (SEP 18) and dyke collapse of evaporation pond 19 (SEP-19) in the Arab Potash Company in March 2000. The InSAR data were added by the microgravity measurements. The authors analyzed an interferometric digital terrain model of the recently emerged platform of the Lisan peninsula and interpreted the radar differential interferograms contemporary with gravity measurements for the peninsula. Closson *et al* (2005) have discussed the possibilities to detect, assess and monitor areas prone to collapse on the Jordanian side of the southern DS coast. The field of negative gravity anomalies helped to complete the missing part of the DinSAR deformation field for the subsidence areas. The authors have shown the usefulness of the integrated combination of gravimetry and SAR interferometry data to monitor the subsidence zones. We believe that following up and intensifying further interferometric analysis, complemented by other appropriate investigations in specific areas will be of great importance for insuring or later monitoring the stability of the particularly exposed segments of the whole existing set of dykes as well as future ones.

Nof *et al* (2013, 2019) to detect sinkholes have used systematic high temporal and spatial resolution InSAR observations, incorporated with and refined by detailed LiDAR measurements. It is possible because the sinkhole formation takes place at gradual local subsidence of ground surface, preceding their collapse during periods ranging from a few days to about five years. Some case studies illustrate the timelines and effectiveness of such methodology as well as its limitations and complementary methodologies used for the sinkhole monitoring and hazard assessment. The InSAR-generated subsidence maps have become fundamental for the sinkhole early warning and mitigation along the DS coast in Israel and are incorporated into all sinkhole potential maps which are mandatory for the planning and licensing of new infrastructure.

Road 90 is the main eastern route that connects the southern desert districts of Israel to the more populated and the humid northern districts. This is also the road that connects the DS area to the rest of the country. The road has been severely threatened by sinkholes in the area of En Gedi. An elongated cluster of sinkholes and their associated gradual subsidence crosses Road 90 obliquely from NNW to SSE (figure 4.193(a) (Nof *et al* 2019)). The sinkhole site has been active since the mid-1990s.

To protect the road from sinkhole collapse, geosynthetic sheets were placed in 2002 ~1 m under its surface at its intersection with the sinkhole lineament. Between 2002 and 2012, subsidence and sinkhole collapse continued along the lineament, while the road at its intersection with the subsidence lineament showed no subsidence, most likely due to its reinforcement.

Different geophysical methods were carried out in the EG sinkhole area. The seismic refraction method has delineated the salt dissolution front (Ezersky 2006).

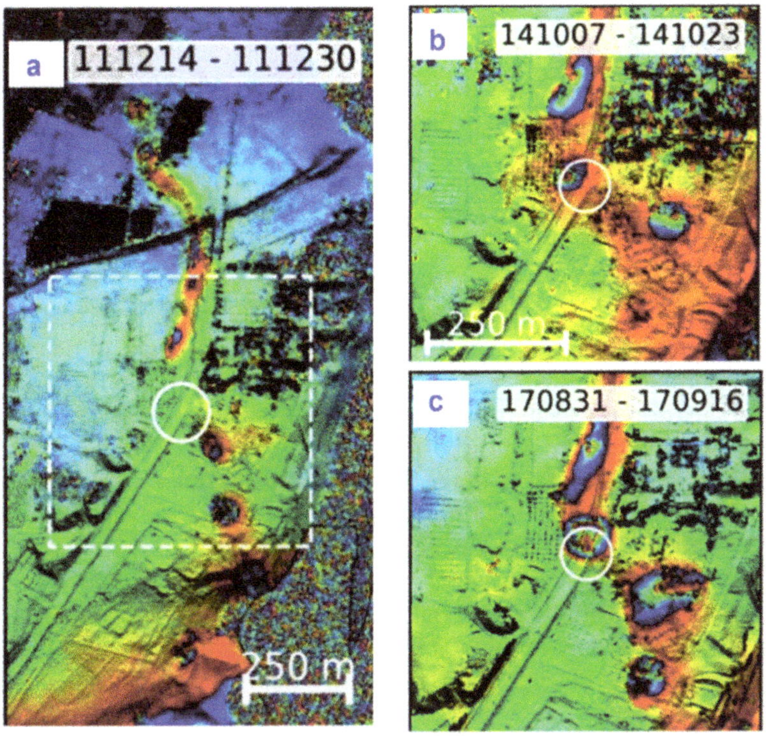

Figure 4.193. (a) An interferogram in December 2011 showed no subsidence at the intersection between the road and the subsiding lineament (solid white circle). The rectangle marks the area covered in panels (b) and (c). (b) Interferogram in October 2014 and (c) that in August 2017 showing the beginning and acceleration of subsidence on the road. Interferograms are projected over the hill-shaded relief of LiDAR DSM. (Reprinted from Nof *et al* (2019), with permission of R Nof, published by MDPI. Modified by M Ezersky.)

ERT study allowed detection of the zone of sinkhole susceptibility based on the heightened resistivity anomaly in the sinkhole area (Ezersky 2008). This study had shown in 2002 that the high resistivity zone has penetrated Highway #90 and testified to collapse of the road. The InSAR measurements began later. The example of InSAR monitoring on Highway #90 at the Ein Gedi site is shown in figure 4.193 (Nof *et al* 2019).

One more example was presented in section 4.9.1 (figure 4.184(b)) (Abelson *et al* 2018).

4.10.7 Photogrammetry

Close-range photogrammetry is used to generate the 3D surface models from measurements made in the photographic images taken from different known positions. The comparison of the surface models produced with the data collected at different dates may be used to resolve the sinkhole-related surface changes. Al-Halbouni *et al* (2017) performed a detailed characterization of a highly active and vegetation-free sinkhole site in the eastern DS shores (Jordan), through a photogrammetric survey with aerial images of the area that were acquired by a Ricoh GRII camera with 16.2 MP resolution. They mapped surface changes related to the sinkhole activity and estimated volume loss related to subsidence as well as an average subsidence rate.

4.10.8 Lazer (Filin *et al* 2011, 2012, Arav *et al* 2020)

Filin *et al* (2012) have suggested the application of high-resolution airborne laser scanning data for the detection and characterization of geomorphic processes. The authors have shown the ability of airborne laser scanning data to detect the submetric geomorphic features, such as thin and small gullies, small head cuts, and embryonic sinkholes on a sub-meter scale. This combined with the accurate location of these features given by the airborne laser scanning is of prime importance in describing the sinkhole hazards in the DS. Furthermore, the ability to calculate the 3D dimensions of geomorphic features such as gullies and sinkholes is an important tool for constructing a sinkhole susceptibility map.

Arav *et al* (2020) have studied the evolution of sinkhole fields along the DS coastal plains, from initiation to stabilization. The prevailing hypothesis attributes their formation to the dissolution of a buried halite layer by groundwater but fails to explain their evolution in some places and their stability in others. Based on thirteen years of high-resolution laser scanning and field surveys, the mentioned authors trace a complete evolutionary cycle of sinkholes and show that the prevailing theory explains only the first out of a four-stage cycle. Following the model, an acceleration phase begins when the sinkholes interact with the drainage system.

4.11 Properties of rocks soils and salt

4.11.1 Some features of rock mass properties

An analysis of the rock mass nature features is based on the works of Stagg and Zienkevich (1968) and Palmstrom (1995). The main terms will be entered for a better understanding of the present investigation.

A rock mass is a heterogeneous material and quite often discontinuous, but is one of the materials in the Earth's crust, which is mostly used in construction.

The first feature is that a *rock mass* is composed of *rock material* and *joints* (discontinuities) of different properties, sizes, and directions (figure 4.194). The discontinuities ranging in length from less than a decimeter to several kilometers, divide the bedrocks into units, volumes, or blocks of different scales forming a material in which all elements behave in mutual dependence as a unit (Palmstrom 1995).

The second important feature in a rock mass and its utilization in contrast to other construction materials is the size or volume of material involved in measurement (in other words, *scale effect*). Generally, a rock mass comprising defects and inhomogeneities of different sizes (from microfractures in the rock sample to extended joints in the rock mass) is characterized by the scale effect; e.g., the

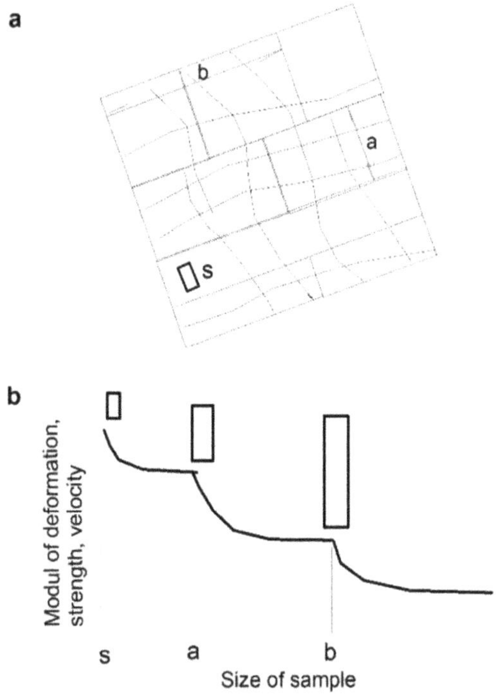

Figure 4.194. (a) Simplified model rock consisting of various dimensions units or blocks formed by discontinuities of different sizes families of fissures. (b) Different characteristics of rock mass (deformation modulus, strength, elastic velocities etc) depends on the size of the sample tested. Reprinted from Jaeger (2009), with permission from Cambridge University Press.

dependence of rock properties (strength, deformability, etc) on the scale of rock volume involved in measurements (Brace 1981, Janelid and Kvapil 1965).

Jaeger (2009) concerning Bernaix's (1967) private report on the rock at Malpasset, shows schematically the structure of the rock mass explaining also scale effect (figure 4.194). According to Bernaix (1967), the crushing strength of the rock samples depends on the spacing of the fissures. This pattern of fissures also explains why the test results depend on the size of the sample (e.g., scale effect). The pattern of microfissures and microfractures is, on a very small scale, an image of the macrofractures which cross the rock mass. This complexity justifies the remark of Bernaix (1967) that microfractures and macrofractures should be classified separately.

The complicated structure of the rock mass with its defects and inhomogeneities and the wide range of its applications cause challenges and problems in rock engineering and construction which often involve considerations that are of relatively little or no concern in most other branches of engineering.

Deer (1968), denotes that also the *hydraulic conductivity* of the rock samples determined by water flow through the porosity channels (primary hydraulic conductivity of rock material defined by the pore throats) essentially differs from the permeability of rock mass (secondary hydraulic conductivity) because the latter may be markedly affected by the flow through fissures. Consequently, estimation of the hydraulic conductivity experimentally obtained at different scales can be different (Schon 2011, section 1.2.5). *Porosity* that is closely related to permeability (Schon 2011, section 1.2.5) is also a scale-dependent parameter.

The magnitude of scale effect is determined by the degree of rock heterogeneity and the volume of the heterogeneities involved (denoted as W_i). Generally, different properties of the rock mass (deformability, permeability, porosity, strength, elastic characteristics, etc) decrease with the increase of W_i: the larger is volume, the fewer of the above properties. At that, properties of rock matrix as determined in the laboratory are only a maximal evaluation of the rock's properties that is decreased by geological engineers relative to the properties of the matrix by using different ratings (Bieniawski 1974).

On the other hand, the scale of testing by dynamic methods is defined by the half-wavelength $\lambda/2$, so that the volume of investigation is $Wi \cong 0.25\Delta l_i\lambda_i^2$, where Δl_i is the base of measurement (length of the sample, for instance), $\lambda_i = V_i/f_i$ is wavelength.

Ultrasonic measurements correspond to $W_i = 10^{-6}$ m^3, whereas seismic measurements correspond to a measured volume $W_i = 10^{-4}$–10^3 m^3. Thus, the measured geophysical parameters (V_p, V_s) are also scale-dependent parameters (Savich *et al* 1969). Note, that the scale effect induces the anomaly dispersal of velocities the lower is the length wave λ.

The less the volume of measurement, the higher is the velocity. An example of a variation of the elastic P-wave velocities (V_p) versus the measured volume W_i for different rocks is shown in figure 4.195 by red graphs (Savich 1984). It can be understandable from this figure that the elastic velocities of rocks decrease with an

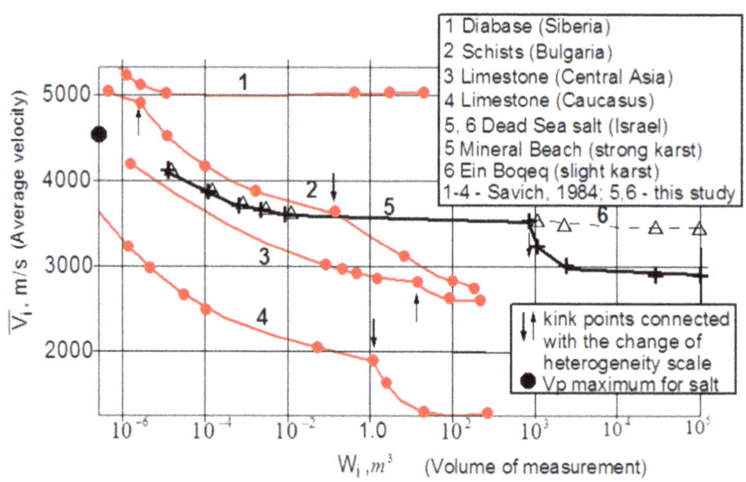

Figure 4.195. Curves showing the variations of average velocity (V_i) versus measured volume (W_i) for different rocks: 1—diabase (Siberia); 2—schists (Bulgaria); 3—limestone (Central Asia); limestone (Caucasus); 5, 6—DS salt: 5 in the sinkhole development site Mn-5E, 6 in the no sinkholes site EB-3E. (1–4 after Savich (1984), 5, 6—Ezersky and Goretsky (2014)). The arrows designate kink points connected with the change of the heterogeneity scale. (Reprinted from Ezersky and Goretsky (2015), with permission of CRC Press/Balkema, London.)

increase in volume W_i. Sometimes this increase is quite sufficient. Ezersky and Goretsky (2015) had added these graphs (P-wave velocities of the DS salt) based on the ultrasonic and seismic measurements (the black graph in figure 4.195).

Considering V_p graphs in figure 4.195 one can see characteristic kink points. These points following Savich (1984) correspond to a change of the heterogeneity scale. In these points the small increase in W_i causes the sharp decrease of V_p (it is similar to figure 4.194(b)). The salt velocity graph (6 in figure 4.192) smoothly changes with an increase in the volume of sampling. It corresponds with the properties of salt plasticity. Otherwise, the graph of salt velocity (curve 5 in figure 4.195) behaves in the sinkhole sites. One can see the kink point at $W_i = 10^3$ m^3. In this case, a sharp decrease in the P-wave velocity is caused by the cavity in the salt layer.

4.11.2 Seismic velocities of soils and rocks in Israel

The compressional (longitudinal) (V_p) and shear (V_s) wave velocities are important parameters characterizing the elastic dynamic properties of soil and rocks. The velocities of seismic wave propagation through elastic media are determined by the elastic moduli and densities of the rocks through which they pass. As a rule, the velocities increase with increasing density.

The P-waves velocity is:

$$Vp = (E_d/d)^{0.5}$$

The S-waves velocity is:

$$Vs = (G_d/d)^{0.5}$$

The ratio V_p/V_s is defined in the terms of Poisson's ratio (ν_d) that is given by:

$$\nu_d = \frac{\left(\frac{Vp}{Vs}\right)^2 - 2}{2\left(\frac{Vp}{Vs}\right)^2 - 2}$$

$$E_d = 2G_d(1 + \nu_d),$$

where: E_d is the dynamic Young's modulus (MPa), G_d is the dynamic modulus of rigidity (shear modulus) (MPa), d is the density (kg m^{-3}).

One more elastic dynamic bulk modulus is expressed by:

$$= (K_d E_d)/(3-6\nu_d)$$

Note that $G_d = 0$ for a fluid, as fluids, cannot support shear, and the maximum value of Poisson's ratio $\nu_d = 0.5$; $\nu_d \approx 0.05$ for very hard rocks, $\nu_d \approx 0.45$ for loose, unconsolidated sediments, average ≈ 0.25.

The ranges of seismic velocities in the soils and rocks of Israel were presented in section 4.2.2 (from Gorstein and Ezersky (2015)). It was shown that ranges of the both V_p and V_s seismic velocities in soils and rocks generally agree with those measured by other researchers in shallow subsurface (Jakosky 1957, Press 1966,

Reynolds 2011). However, the V_P values in rocks reported in the literature are higher (for instance, $V_P = 5800$ m s^{-1} in granite, $V_P = 6100$ m s^{-1} in limestone) than those measured in Israel. The higher V_P values could be explained by the high stress and better rock quality in deep measurement conditions. The shear-wave velocities measured throughout Israel vary by a wide range: from 100 m s^{-1} in sands and LCs to more than 600 m s^{-1} in gravels, and from 400 m s^{-1} in conglomerates to more than 2400 m s^{-1} in basalts and dolomites. The variability of V_s and thickness of soft sediments overlying hard rock on the one side, and very limited availability of the densely distributed geotechnical information such as V_s at depth call for less expensive and less time-consuming methods to provide the required parameters for site effect assessment.

Here we are intended to present mainly seismic velocities (V_P and V_S) of the soluble rocks in comparison with the non-soluble ones. The velocity ranges measured in the field (*in situ*) are presented in figure 4.196(a). The velocity ranges measured in the laboratory conditions are shown in figure 4.196(b).

The compilation of rock properties ranges indicates the variability of velocities from author to author following the time of surveys, place, type of rocks, the experience of the researcher. However, we carried out an analysis of all factors including where measurements were carried out in the field or the laboratory, collected different data, and compiled tables with all new understanding. We have organized the table by giving priority to rocks subjected to karstification. Unfortunately, there is not so much data for gypsum and anhydrite.

These data we selected from the separate studies like that of Rucker *et al* (2005), Ahmed (1989) (denoted by 1 and 2 asterisks in figure 4.196). Rucker *et al* (2005) selected a dissolving gypsum layer using the low values of V_P and V_s between hard

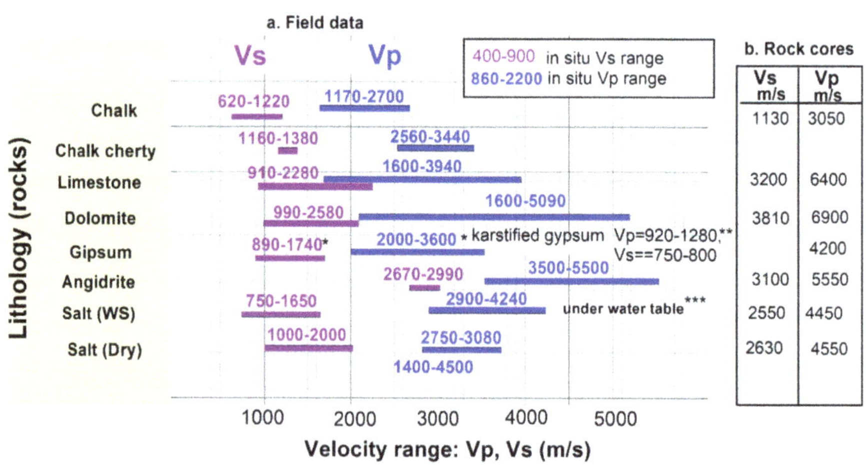

Figure 4.196. A broad range of soluble rock's V_P and V_s velocity worldwide. (a) field data, (b) rock samples. Data have been compiled from different researches. Asterisks denote *Rucker *et al* (2005), **Ahmed (1989), ***Ezersky (2006), ****Ezersky and Goretsky (2014). The salt velocity of more than 4550 m s^{-1} is explained by impurity (Jenyon 1986), rock core velocities—from Mavko *et al* (2009).

gypsum layers (three asterisks in figure 4.196). Salt properties are given in Ezersky (2006) and Ezersky and Goretsky (2014) (three and four asterisks, respectively).

We separated ranges of the seismic velocities measured in the field and rock samples and one can see that there are essential differences. Reynolds (2011) explains this circumstance by the fact that *in situ* rock is usually heavily fractured. The refraction velocities measure both the rock and the discontinuities together, whereas a laboratory measurement on a solid sample cannot, by the virtue of the scale of the sample. This conclusion is explained well by the scale effect in rocks (see section 4.11.1 above).

4.11.3 Properties of the DS soils

Arkin and Gilat (2000) selected three main types of sinkholes: (1) gravel holes occurring in alluvial fans, (2) mud holes developing in the mudflats between fans, and (3) a combination of both types that are forming at the front of young alluvial fans where they overlap mudflats. Figure 4.197 shows the main sinkhole types formed in alluvial fans of Ein Gedi (a) and Ein Boqeq (b), and the DS mud of the southern Mineral Beach site (c).

Following these definitions, two main types of DS soils can be selected—mud (named also DS clay or LC) and sandy-gravel alluvium (Yechieli *et al* 2004, Frydman *et al* 2008, 2014).

The study of the sinkhole hazard problem requires knowledge of the soil's geotechnical and geophysical properties. These properties are a very important issue of the geotechnical modeling and calculations of the sinkhole site stability (Maimon *et al* 2005, Shalev *et al* 2006, Shalev and Lyakhovsky 2012, Al-Halbouni *et al* 2019).

The study is based on inter-relations available between the geophysical (elastic, electrical, etc) properties and the geotechnical ones (Ezersky and Livne 2013). The propagation of elastic waves through geomaterials provides information about the critical properties of engineering materials (saturation, state of stress, stiffness, diagenesis, and even soil type) and the spatial distribution of these properties (e.g. layers, anomalies, inclusions, water table, etc) (Stokoe *et al* 1994, Santamarina *et al* 2001). The development of the geophysical application is based on evaluation of near-surface geotechnical materials in two directions: (1) soil dynamics specifically, for designing dynamically loaded foundations where small-strain shear stiffness G_{max}

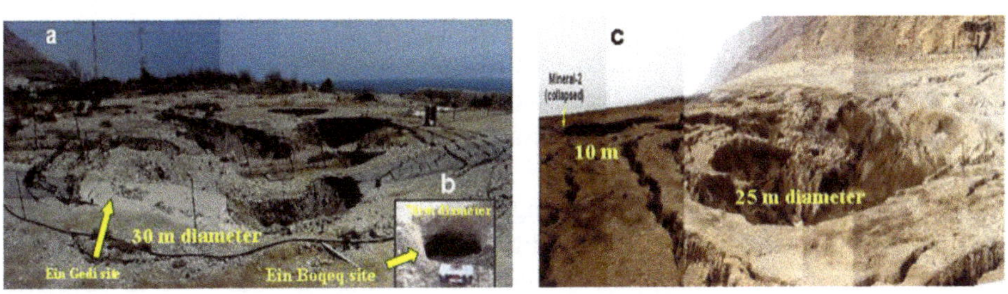

Figure 4.197. Two types of sinkholes (a, b) in sandy-gravel sediments of Ein Gedi and Ein Boqeq, respectively; and (c) in the DS mud (LCs) of the southern Mineral Beach site.

and G_0 (*hereafter, dynamic shear modul*us) is the key soil properties responsible for its strength; and (2) geotechnical earthquake engineering for site response analysis. In the latter case, measurements in the small-strain and nonlinear-strain ranges are required. This requirement necessitated the combined use of field dynamic (small-strain) and laboratory dynamic and static (small-to-large-strain) measurements. Resistivity also provides useful information about the fluid properties (Schon 2011, Legchenko *et al* 2009, Frumkin *et al* 2011, Ezersky and Livne 2013).

Different authors suggest existence of the relationship between the dynamic and geotechnical properties for cohesive (Blake and Gilbert 1997) and cohesionless (Cha and Cho 2007) soils. If such a correlation exists, the V_s values measured in the field would be used for evaluation *in situ* strength. The SPT and other borehole soundings (CPT, DSPT) methods are widely used for studying the geotechnical soil properties (Lacroix and Horn 1973, DIN 1980, Rogers 2006, Lutenneger 2008). In particularity, Terzaghi *et al* (1996) suggest that there is a correlation between the SPT N_{60} and unconfined undrained shear strength. The parameter *N*-blows in SPT measurements is obtained following ASTM D4633 (2016) and correction to N_{60} value is carried out following ASTM D1586/1586M-18 (2018). Recently a lot of researchers link the SPT *N*-blow parameter with shear wave velocities V_s of different soils (Jinan 1987, DeJong 2007, Hasancebi and Ulusay 2007, Schon 2011).

To investigate geophysical and geotechnical properties of the lithology of sinkhole sites and relationships between the geophysical and geotechnical rock and soil properties *in situ*, boreholes drilling was carried out in the DS sites composed of typical rocks (see section 4.8). In this section, we intend to summarise the properties of the DS soils and salt and their inter-relations.

The investigations carried out in boreholes have allowed studying properties of the main lithological types of the sinkhole sites along the DS western shore. Soil and salt samples collected from the boreholes enabled us to study the properties of these rocks in different scales of the laboratory and *in situ* (Livne 2012, Ezersky *et al* 2013c, Ezersky and Livne 2013). We have discovered some new properties of these soils that open new possibilities for the application of geophysical methods. Then we briefly consider these findings.

Lime carbonates. It is carbonate soil named by Frydman *et al* (2008) as '*lime carbonate*' that is composed predominantly of calcite and not clay minerals, whereas other Israeli clays are comprised largely of montmorillonite (smectite) clay minerals (Frydman 2000). The DS data lie consistently above those of other Israeli clays and correspond to average effective strength parameters: $C' = 0$ kPa and $\phi' = 34°$. As follows from the modified plasticity card (Casagrande 1932a, Terzaghi *et al* 1996) presented in figure 4.198 undisturbed specimens extracted from the borehole MN-5E are classified after the Unified Soil Classification System (USCS) as CL (Goretsky 2012, Livne 2012).

It should be medium to high dry strength, medium toughness at plastic limits, and no to a very slow reaction to the shaking test (Terzaghy *et al* 1996). LC consists of 94%–99% fines (<0.074 mm), has water content ranging from 40% to 48% with liquid limit, LL = 31–44 and plasticity index, PI = 8–18. It is a very fine material that was considered in past as the DS clay or mud (Arkin and Gilat 2000). The clastic

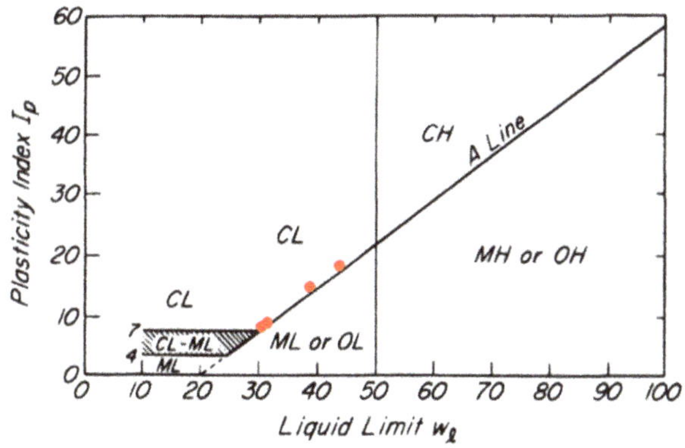

Figure 4.198. Properties of LCs put on the modified plasticity card for use with the USCS (after Terzaghy *et al* (1996)). Red points relate to lime carbonate (LC) samples. (CL—inorganic clay of medium plasticity; CH—inorganic clays of high plasticity; ML—inorganic silt of medium compressibility or organic silt (OL); MH—inorganic silt of high compressibility or organic clays (OH)). The card is based on a plasticity card (after Casagrande (1932a)).

detrital material was brought into the basin by seasonal floods. Areas of such soil were named mudflats. These areas when the brine is saturated are characterized by marshy consistency like marshland. Its characteristic feature is that the Earth's surface subsides underfoot. The upper soil layer when dried seems to be dense, but under this layer, the soil remains marshy. There were cases when at such sites the wheels of the 4×4 vehicle momentarily fell through such soil and the vehicle could only be taken out by a powerful tractor.

As mentioned above, despite the outer similarity to clay, this soil is not composed of clay minerals. The brine-saturated LCs (the DS mud) are characterized *in situ* by the low V_s values of 120–250 m s^{-1}. They do not exhibit cohesion (Ezersky and Livne 2013). Frydman (2000) explains it by the mineralogical composition of the DS material composed predominantly of calcite. This feature allows consideration of saturated LC as cohesionless soil like fine sands. LCs demonstrate one more peculiarity, i.e., very low cation exchange capacity (CEC) (Khlaifata *et al* 2010) that allows the use of the TEM resistivity method for determining porosity of both areas of sand and DS mud (Ezersky and Frumkin 2017). Note also that the significant difference of V_s allows the mapping of different lithology along with DS coastal areas.

Sandy-gravel sediments were tested in the Ben Gurion University (BGU) geotechnical laboratory by Livne (2012) following Frydman *et al*'s (2008) methodology. These sediments are classified as SW, SM, SP, e.g. coarse clastic sands and gravels (after USCS), where the water content is 15%–25%. The sands consist of carbonate particles with a total unit weight (dry) γ_t varying from 16.6 to 23 kN m^{-3} and fines. The pore fluid is the DS brine with a density of ρ_f = 1240 kg m^{-13}.

Carbonate soils of the western DS area are typical also (with some reserves) for its eastern (Jordanian) shores (see table 4.15) where both coarse sandy grave sediments (Taqieddin *et al* 2000) and mudflats composed of LCs (Khlaifata *et al* 2010) are found.

Table 4.15. Soil and DS brine properties. (Reprinted from Ezersky *et al* (2017).)

Lithology	Sandy-gravel sediments (SG)		LCs		Comments
Soils	Laboratory*	Field	Laboratory**	Field	
Classification	SW, SM, SP—coarse clastic sands and gravels		CL—LCs		after USCS
Fine content, %			94–99		(<0.074 mm),
Water content, W %	15–25		40–48		
Unit weight, γ_t N m^{-3}	16.6–23		11.27–13.99 / 16.68–18.34		dry / saturated
Liquid limits (LL)			31–44		
Plasticity index (PI)			8–18		
Void ratio dry, e			0.57–1.02		No pressure
Void ratio dry under pressure, e_σ			0.82–1.53		Confining 50–500 kPa
Porosity dry, Ø			0.45–0.60		No pressure
Porosity dry under pressure, $Ø_\sigma$			0.41–0.50		Confining 50–500 kPa
Density dry, D kg m^{-3}			1085–1427		min—max
Density wet, D kg m^{-3}			1628–1806		min—max
Density dry, D_σ kg m^{-3}			1301–1534		Confining 50–500 kPa
Density wet, D_σ kg m^{-3}			1806–1920		Confining 50–500 kPa
Friction angle, $\varphi°$			34		Saturated
Cohesion C, kPa			0		Saturated
Friction angle, φ^0			1		Naturally wet
Cohesion C, kPa			19–20		Naturally wet
V_s, m s^{-1}		300–650	90–350	120–250	Bender test at 0–500 kPa
V_p, m s^{-1}		700–1500 / 1540–2700	1770–1890*SC	1125–1400 / 1690–2390	dry / saturated
SPT, blows/30 cm		16–50		5–25	
Bulk resistivity ρ_x, Ωm		0.3–0.5	0.54–1.40,*SC	0.2–0.3	*Saturated, consolidated
Pore fluid density, DS brine, kg m^{-3}	1230/1240				Northern/southern basin
DS water	Laboratory*	Field			
Total dissolved solids (TDS), g L^{-1}	340/380				Northern/southern basin
V_p, m s^{-1}	1800	1820			

*Tests were carried out at the Ben Gurion University of Negev (BGU) geotechnical laboratory by E Livne (2012).

**Tests were carried out in the Building and Infrastructure Testing Lab. Ltd (Haifa) by Dr Ian Goretsky. Both testings of soils were carried out following Friedman *et al*'s (2008) methodology for the saline soils.

Table 4.16. Seismic velocities of the DS salt measured in the laboratory and *in situ* conditions. (Reprinted from Ezersky *et al* (2017).)

Salt	Laboratory*	Field	Comments
V_p, m s^{-1}	3700–4460	2900–4250	
V_s, m s^{-1}	1900–2620	750–1800	
Bulk resistivity ρ_x, Ωm	2–100	0.5–2.0	
Porosity, ∅	0.032–0.21	0.15–0.40*	*Evaluated
Hydraulic conductivity, *K*, m s^{-1}	2.3×10^{-9}–4.5×10^{-6}	1×10^{-6}–2.0×10^{-4}*	*Evaluated
Dry density, d_{dry}, kg m^{-3}	1635–2106		
Wet density, d_{wet}, kg m^{-3}	1681–2116		

* Ezersky and Goretsky (2014).

4.11.4 Properties of the DS salt

The DS salt was studied both *in situ* and under laboratory conditions. Under *in situ* conditions, numerous seismic refraction investigations were carried out that enabled us to estimate the seismic V_p velocities of salt and to establish the velocity criterion of salt (Ezersky 2006). Then, the MASW method allowed the evaluation of shear wave velocities (V_s) of salt (Ezersky *et al* 2013a). Ezersky and Legchenko (2014) have suggested the methodology of hydraulic conductivity *K* calculation based on the V_s velocity. *In situ* salt resistivity was measured by the TEM method (Ezersky *et al* 2011). Field (*in situ*) data acquisitions were carried out by a team of the Geophysical Institute of Israel (Lod).

Samples of the DS salt were studied by ultrasonic and resistance methods (Ezersky and Goretsky 2014). The preparation of salt samples and measurements of V_p and V_s velocities were carried out in the laboratory of the Geophysical Institute of Israel using ultrasonic equipment SonicViewer SX Model 5251C, specific resistivity was carried out using a Nilsson 400 resistance meter (Ezersky and Goretsky 2014).

Measurements of porosity ∅, hydraulic conductivity *K*, and density *d* (dry and wet) were carried out in the Building and Infrastructure Testing Lab. Ltd (Haifa) by Dr I Goretsky (Ezersky and Goretsky 2014).

Table 4.16 shows that elastic velocities of the DS salt vary in a wide range. That corresponds to figure 4.195 from 2900 to 4460 m s^{-1}.

References

Abelson M 2021 Hydrological and geological controls on the evolution of the Dead Sea Sinkholes *The Many Facets of Israel's Hydrogeology* ed U Kafri and Y Yechieli (Cham: Springer)

Abelson M, Kurzon I, Crouvi O, Wachs D and Yechieli Y 2002 Development of the Dead Sea sinkholes—detection from aerial photographs up to 2001: geological Survey of Israel *Report* GSI/30/2002 (in Hebrew with English abstract)

Abelson M, Baer G, Stivelman V, Wachs D, Raz E, Crouvi O, Kurzon I and Yechieli Y 2003 Collapse-sinkholes and radar interferometry reveal neotectonics concealed within the Dead Sea basin *Geophys. Res. Lett.* **30** 1545

Abelson M, Yechieli Y, Crouvi O, Baer G, Wacks D, Bein A and Shtivelman V 2006 Evolution of the Dead Sea sinkholes *New Frontiers in Dead Sea Paleoenvironmental Research* ed Y Enzel, A Agnon and M Stein 401 (Boulder, CO: Geological Society America) pp 241–53

Abelson M, Gabay R, Shalev E and Yechieli Y 2009 Sinkholes hazard around the evaporation ponds Dead Sea southern basin (https://govder/reports/abels.il/BlobFolon-et-al-report-2009-27/he/report-2009_GSI-27–2009.pdf)

Abelson M, Yechieli Y, Baer B, Lapid G, Behar N, Calvo R and Rosensaft M 2017 Natural versus human control on subsurface salt dissolution and development of thousands of sinkholes along the Dead Sea coast *J. Geophys. Res.: Earth Surf.* **122** 16

Abelson M, Yechieli Y, Bein Y, Shtivelman V, Crouvi A, Kalbo R, Steinberg G, Baer G and Waks D 2004 Classification and mapping of areas of likelihood of sinkhole formation along Dead Sea shores *GSI Report* GSI/13/2004 http://www.gsi.gov.il/eng/ (in Hebrew)

Abelson M, Aksinenko T, Kurzon I, Pinsky V, Baer G, Nof R and Yechieli Y 2018 Nanoseismicity forcasts sinkhole collapse in the Dead Sea coast years in advance *Geology* **46** 83–6

Abou Karaki N, Fiaschi S, Paenen K, Al-Awabdeh M and Closson D 2019 Exposure of tourism development to salt karst hazards along the Jordanian Dead Sea shore *Hydrol. Earth Syst. Sci.* **23** 2111–27

Abueladas A and Al-Zoubi A 2004 The application of a combined geophysical survey (GPR and seismic refraction) for mapping sinkholes in Ghor Al-Haditha area *Fall Meeting Supplement, Eos Transactions American Geophysical Union* Abstract GP11A-0825

AEMR 2018 TEM FAST 48 (Manual) v.8a (http://aemr.net/tem_res.htm)

AEMR 2018 TEM-FAST, TEM-RESEARCHER, TEM-3D-WIZARD, Manuals, http://aemr.net

Ahmed H 1989 Applications of mode-converted shear waves to rock-property estimation from vertical seismic profiling data. *Geophysics* **54** 478–85

Aksinenko T, Pinsky V, Polozov A, Giller V, Kadosh D and Levi D 2012 Operation and monitoring microseismic system and data processing for location of micro-events for the project 'Dead Sea sinkholes' *GII Report* No. 581/663/12 (Available in the Library of GSI)

Albouy V, Andrieux P, Descloitres V, Robain H, Vassal J and Sorrensen K 2001 Time domain electromagnetism. Comparison between 3 commercial systems *IRD Report* Centre de Recherche Geophysique of Garchy, France

Al-Halbouni D, Holohan E P, Saberi L, Alrshdan H, Sawarieh A, Closson D, Walter T R and Dahm, T 2017 Sinkholes, subsidence and subrosion on the eastern shore of the Dead Sea as revealed by a close-range photogrammetric survey *Geomorphology* **285** 305–24

Al-Halbouni D, Holohan E P, Tahery A, Schopfer M P J, Emam S and Dahm T 2018 Geomechanical modelling of sinkhole development using distinct elements: model verification for a single void space and application to the Dead Sea area *Solid Earth* **9** 1341–73

Al-Halbouni D, Holohan E P, Taheri A, Watson R, Polom U, Schöpfer P J, Emam S and Dahm T 2019 Distinct element geomechanical modelling of the formation of sinkhole clusters within large-scale karstic depressions *Solid Earth* **10** 1219–41

Al-Zoubi A and Ten Brink U S 2001 Salt diapirs in the Dead Sea basin and their relationship to quaternary extensional tectonics *Mar. Pet. Geol.* **18** 779–97

Al-Zoubi A, Abueladas A-R, Al-Ruzouq R, Camerlynck C, Akkawi E, Ezersky M, Abou-Hammateh Z, Ali W and Al-Rawashdeh S 2007 Use of 2D multi electrodes resistivity imagining for sinkholes hazard assessment along the eastern part of the Dead Sea, Jordan *Am. J. Environ. Sci.* **3** 229–33

Al-Zoubi A, Eppelbaum L, Abueladas A, Ezersky M and Akkawi E 2011 Preliminary magnetic data analysis in the eastern Dead Sea shore for identification of subsurface inhomogeneities and sinkholes *Proc. of the Near Surface 2011 Conf. (Leicester, UK)* p30

Al-Zoubi A, Eppelbaum L, Abueladas A, Ezersky M and Akkawi E 2013 Removing regional trends in microgravity in complex environments: testing on 3D model and field investigations in the eastern Dead Sea coast (Jordan) *Int. J. Geophys. Spec. Iss.* **2013** 341797

Arandjelovic D 1969 A possible way of tracing groundwater flows in karst *Geophys. Prospect.* **XII** 404–21

Arandjelovic D 1976 Geophysics in the karst *Institute for Geological, Geophysical, and Geotechnical Research—Geophysical Institute Beograd* **17** 1976

Arav R, Filin S and Avni Y 2020 Sinkhole swarms from initiation to stabilization based on in situ high-resolution 3-D observations *Geomorphology* **351** 106916

Archie G E 1942 Electrical resistivity as an aid in determining some reservoir characteristics *Trans. Am. Inst. Min. Eng.* **146** 54–62

Arkin Y and Gilat A 2000 Dead Sea sinkholes- an ever-developing hazard. *Environ. Geol* **39** 711–22

Arkin Y, Gilat A and Kofman L 2000 Developing Dead Sea sinkholes revealed by ground penetrating radar *GSI Curr. Res* **12** 25–30

ASTM 1996 ASTM Standard D 3999 1991(1996) Standard Test Methods for the Determination of the Modulus and Damping Properties of Soils using the Cyclic Triaxial Apparatus, Annual Book of ASTM Standards, 1996, vol 04.08.

ASTM 2006 ASTM D 4428/D 4428M-2014 Standard Test Method for Crosshole Seismic Testing

ASTM 2006 ASTM D 5777-700 (reapproved 2006) Standard Guide for Using the Seismic Refraction Method for Subsurface Investigation

ASTM 2007 ASTM D 7400-7 Standard Test Methods for Downhole Seismic Testing

ASTM 2008 Standard D 2845 Standard Test Method for Laboratory determination of pulse velocities and ultrasonic elastic constants of rock

ASTM 2012 Standard G 187 Test Method for Measurement of Soil Resistivity Using the Two-Electrode Soil Box Method.

ASTM 2016 ASTM D-4633. Standard Test Method for Energy Measurement for Dynamic Penetrometers.

ASTM 2018 D 1586/1586M-18 Standard Test Method for Standard Penetration Test (SPT) and Split-Barrel Sampling of Soils.

ASTM 2019 ASTM D 8295-19 Standard Test Method for Determination of Shear Wave Velocity and Initial Shear Modulus in Soil Specimens using Bender Elements

Avni Y *et al* 2016 Self-accelerated development of salt karst during flash floods along the Dead Sea Coast, Israel *J. Geophys. Res. Earth Surf.* **121** 17–38

Auken E and Christiansen A V 2004 Layered and laterally constrained 2D inversion of resistivity data *Geophysics* **69** 752–61

Baer G, Schattner U, Wachs D, Sandwell D, Wdowinski S and Frydman S 2002 The lowest place on Earth is subsiding—an InSAR (interferometric synthetic aperture radar) perspective *Geol. Soc. Am. Bull.* **114** 12–3

Baer G, Magen Y, Nof R N, Raz E, Lyakhovsky V and Shalev E 2018 InSAR measurements and viscoelastic modeling of sinkhole precursory subsidence: implications for sinkhole formation, early warning and sediment properties *J. Geophys. Res.* **123** 678–93

Banab K K and Motazedian D 2010 On the efficiency of the multi-channel analysis of surface wave method for shallow and semi-deep loose soil layers *Int. J. Geophys.* **2010** 403016

Barker R 1992 A simple algorithm for electrical imaging of the subsurface *First Break* **10** 53–62
Barsukov P O and Fainberg E B 2001 Superparamagnetic effect over gold and nickel deposits *Eur. J. Environ. Eng. Geophys.* **6** 61–72
Barsukov P, Fainberg E and Khabensky E 2015 Shallow investigations by TEM-FAST technique: methodology and examples *Electromagnetic Sounding of the Earth's Interior. Theory, Modeling, Practice* 2nd edn ed V Spichak (Amsterdam: Elsevier) pp 47–78
Barsukov P O and Fainberg E B 2016 On the geoelectric structure of the Sphinx Area in Egypt *Near Surf. Geophys.* **14** 39–45
Barton N 2006 *Rock Quality, Seismic Velocity, Attenuation and Anisotropy* (London: Taylor and Francis) p 756
Bartov Y, Agnon A, Enzel Y and Stein M 2006 Late quaternary faulting and subsidence in the central Dead Sea basin *Israel J. Earth Sci.* **55** 17–31
Bashirov A E, Eppelbaum L V and Mishne L R 1992 Improving Eötvös corrections by wide-band noise Kalman filtering *Geophys. J. Int.* **108** 193–7
Batayneh A T, Abueladas A A and Moumani K A 2002 Use of ground penetrating radar for assessment of potential sinkhole conditions: an example from Ghor al Haditha area, Jordan *Environ. Geol.* **41** 977–83
Beck A E 1981 *Physical Principles of Exploration Methods* (London: Macmillan)
Belfer I, Bruner I, Keydar S, Kravtsov A and Landa E 1998 Detection of shallow objects using refracted and diffracted seismic waves *J. Appl. Geophys.* **38** 155–68
Ben-Avraham Z 1997 Geophysical framework of the Dead Sea: structure and Tectonics *The Dead Sea: the Lake and Its Setting* ed T M Niemi, Z Ben-Avraham and J Gat (Oxford: Oxford University Press) pp 22–35
Benson A K 1995 Applications of ground-penetrating radar in assessing some geological hazards: examples of groundwater contamination, faults, cavities *J. Appl. Geop. R.S. Hys* **33** 177–93
Benson R C and Yuhr L 1993 Spatial sampling consideration and their applications to characterizing fractured rock and karst systems *Applied Karst Geology* ed B F Beck (Rotterdam: Balkema) pp 99–113
Beres M, Luetcher M and Paymond O 2001 Integration of penetrating radar and micro-gravimetric methods to map shallow caves *J. Appl. Geophys.* **46** 249–62
Bernab Y, Mok U and Evans B 2003 Permeability porosity relationship in rocks subjected to various evolution processes *Pure Appl. Geophys.* **160** 937–60
Bernaix J 1967 *Etude Geotechnique de la Roche de Malpasset* Paris (in French)
Beyth M 2007 The Red Sea and the Mediterranean–Dead Sea canal project *Desalination* **214** 364–70
Bichara M, Erling J C and Lakshmanan J 1981 Technique de mesure et d'interpretation minimisant les erreurs de mesure en microgravimetrie *Geophys. Prospect.* **29** 782–89
Bidyashwari H, Kushwahal R S, Chandra M and Okenndro M 2017 Physical properties of soil and its implication to slope stability of Nungbi Khunou, NH-150, Manipur *Int. J. Geosci.* **8** 1332–43
Blakely R J 1995 *Potential Theory in Gravity and Magnetic Applications* (Cambridge: Cambridge University Press)
Bieniawski Z T 1974 Geomechanical classification of rock masses and its application in tunneling *3rd Congress of the ISRM, Proc., Volume II, Part A (Denver, CO)* 27–32

Billi A, De Filippis L, Poncia P P, Sella P and Faccenna C 2016 Hidden sinkholes and karst cavities in the travertine plateau of a highly-populated geothermal seismic territory (Tivoli, central Italy) *Geomorphology* **255** 63–80

Bingham M H and Fry J M 2010 Regression: linear models in statistics *Undergraduate Math Series* (London: Springer)

Biot M A 1952 Propagation of elastic waves in a cylindrical bore containing a fluid *J. Appl. Phys.* **23** 997–1005

Blake W D and Gilbert R B 1997 Investigation of possible relationship between undrained shear strength and shear wave velocity for normally consolidated clays *Offshore Technology Conf. (Houston, TX, 5–8 May 1997)*

Bodet L, Van Wijk K, Bitri A, Abraham O, Côte P, Grandjean G and Leparoux D 2005 Surface-wave inversion limitations from laser-Doppler physical modeling *J. Environ. Eng. Geophys.* **10** 151–62

Bodet L, Abraham O and Clorennec D 2009 Near-offsets effects on Rayleigh-wave dispersion measurements: physical modelling *J. Appl. Geophys.* **68** 95–103

Bodet L, Galibert P Y, Dhemaied A, Camerlynck C and Al-Zoubi A 2010 Surface-wave profiling for sinkhole hazard assessment along the eastern dead sea shoreline, Ghor Al–Haditha, Jordan *72th EAGE Conf. & Exibition incorporating EUROPEC 2010 (Barcelona, 14–17 June)* M027

Bohlen T, Kugler S, Klein G and Theilen F 2004 1.5-D inversion of lateral variation of Scholte wave dispersion *Geophysics* **69** 330–44

Boiero D and Socco L V 2010 Retrieving lateral variations from surface wave dispersion curves analysis *Geophys. Prospect.* **58** 1365–2478

Bolshakov V D 1983 *Theory of Observation Mistakes* (Moscow: Nedra) (in Russian)

Booth A D, Clark R and Murray T 2010 Semblance response to a ground penetrating radar wavelet and resulting errors in velocity analysis *Near Surf. Geophys.* **8** 235–46

Borda M 2011 *Fundamentals in Information Theory and Coding* (Berlin: Springer)

Brace W F 1981 The effect of size on mechanical properties of rock *Geophys. Res. Lett.* **8** 651–52

Bristow C/S and Jol H M (ed) 2003 *Ground Penetrating Radar in Sediments* (London: Geological Society) pp 1–7 Special Publication, 211

Bruner I, Charach J and Medvedev B 2003 High resolution seismic reflection survey along dike 5/ 8, Dead Sea works *Joint GII-Ecolog-DSW Report* No. 320/336/0 3

Buker F, Gurtner M, Horstmeyer H and Green A G 1996 Three-dimensional mapping of glaciofluvial, and deltaic sediments in central Switzerland using ground penetrating radar *6th Int. Conf. on Ground Penetrating Radar (GPR'96) (Sendai, Japan)* 45–50

Buthens J, Filippi M, Asadi N, Zare M, Slechta S and Churackova Z 2009 Surficial deposits on salt diapirs (Zagros Mountains and Persian Gulf Platform, Iran): characterization, evolution, erosion and the influence on landscape morphology *Geomorphology* **107** 195–209

Camerlynck C, Al-Zoubi A, Alem Z, Bruner I, Ezersky M and Legchenko A 2006 New geophysical approach to prediction of sinkhole hazards related to the drop of the dead sea level and human activity. project sinkhole hazard *SfP* 981128

Cannaday F X 1964 Modulus of elasticity of a rock determined by four different methods *Report of Investigations* 6533 US Bureau of Mines

Cardarelli E, Di Filippo G and Tuccinardi E 2006 Electrical resistivity tomography to detect buried cavities in Rome: a case study *Near Surf. Geophys.* **4** 387–92

Casagrande A 1932a Researchon the Attenberglimits of soils *Public Roads* **13** 121–36

Carter M 1983 *Geotechnical Engineering Textbook* (London: Pentech Press) p 226

Casas A, Lazaro R, Vilas M and Busquet E 1996 Detecting karstic cavities with ground penetrating radar at different geological environments in Spain *Proc. 6th Int. Conf. GPR'96 (Sendai, Japan)* pp 455–60

Casto D W, Luke B, Calderon-Matias C and Kaufmann R 2009 Interpreting surface-wave data for a site with shallow bedrock *J. Environ. Eng. Geophys.* **14** 115–27

Cercato M 2009 Addressing non-uniqueness in linearized multichannel surface wave inversion *Geophys. Prospect.* **57** 27–47

Cha M and Cho G-C 2007 Shear strength estimation of sandy soils using shear wave velocity *Geotech. Test. J.* **30** 484–95

Chalikakis K, Plagnes V, Guerin R, Valoes R and Bosch F P 2011 Contribution of geophysical methods to karst-system exploration: an overview *Hydrogeol. J.* **19** 1169–80

Charrach J 2019 Investigations into the Holocene geology of the Dead Sea basin *Carbonates Evaporites* **34** 1415–42

Charrach J, Goretsky I and Frydman S 2007 Geotechnical studies of soils from an evaporateenvironment in Israel *Proceedings: 5th International Congress on Environmental Geotechnics, Cardiff, Wales 1* (London: Thomas Telford) pp 407–14

Chevalier A, Legchenko A, Girard J-F and Descloitres M 2014 3D Monte Carlo inversion of Magnetic Resonance measurements *Geophys. J. Int.* **198** 216–28

Closson D 2005 Structural control of sinkholes and subsidence hazards along the Jordanian Dead Sea coast *Environ. Geol.* **47** 290–301

Closson D, Abou Karaki N, Klinger Y and Hussein M J 2005 Subsidence and sinkhole hazard assessment in the southern DS area, Jordan *Pure Appl. Geophys.* **162** 221–48

Closson D and Abou Karaki N 2009 Salt karst and tectonics: sinkholes development along tension cracks between parallel strike-slip faults, Dead Sea, Jordan *Earth Surface Process Landf.* **34** 1408–21

Conyers L B 2013 *Ground-penetrating Radar for Archaeology* 3rd edn (Lanham, MD: Altamira Press, Rowman and Littlefield Publishers)

Conyers L B 2016 Ground-penetrating Radar for Geoarchaeology 1st edn (New York: Wiley), p 160

Crampin S, McGonigle R and Bamford D 1980 Estimating crack parameters from observations of P-wave velocity anisotropy *Geophysics* **45** 345–60

Cruz J C R, Hubral P, Tygel M, Schleicher J and Hocht G 2000 The common reflecting element (CRE) method revisited *Geophysics* **65** 979–93

Dal Moro G (ed) 2015 *Surface Wave Analysis for Nearsurface Applications* (Amsterdam: Elsevier) p 244

Daniels D J (ed) 2007 *Ground Penetrating Radar* 2nd edn (London: The Institution of Engineering and Technology)

Debeglia N, Bitri A and Thierry P 2006 Karst investigations using microgravity and MASW: application to Orléans, France *Near Surf. Geophys.* **4** 215–25

Deer D U 1968 Geological consideration *Rock Mechanics in Engineering Practice* ed K G Stagg and O C Zienkevich (London: Wiley) pp 1–20

DeGroot-Hedlin C and Constable S 1990 Occam's inversion to generate smooth, two-dimensional models form magneto telluric data *Geophysics* **55** 1613–24

DeJong J T 2007 Site characterization—guidelines for estimating vs based on in-situ tests *Stage I. Internal Report* (http://dap3.dot.ca.gov/shaketable/references/Vs30_correlations_US. Davis.pdf)

De Waele J, Picotti V, Zini L, Cucchi F and Forti P 2009 Karst phenomena in the Cordillera de la Sal (Atacama, Chile) *GeoActa Spec. Publ.* **2** 113–27

Dhemaied A 2007 Tomographie seismique d'une zone de subsidence (Sinkhole zone, Ghor Al-Haditha, Jordanie) *Unpublished MS Report* Paris 6 University, Paris (in French)

DIN 1980 DIN 4094 part 2. Dynamic and Static pentrometers, Application and evaluation *Unpublished reports of the Tamar Regional Council* Deutsches Institut für Normung DS sinkhole database 1998–2016

Dvorkin J 2008 Yet another Vs equation *Geophysics* **73** E35–9

Duba A, Piwinskii A J, Santor M and Weed H C 1978 The electrical conductivity of sandstone, limestone and granite *Geophys. J. R. Astron. Soc.* **53** 583–97

Dunn K J, Bergman D J and Latorraca G A 2002 *Nuclear Magnetic Resonance Petrophysical and Logging Applications* (London: Elsevier Science)

EJGE 1999 Vs estimator v.1.2 (http://ejge.com/ejgesoft/prog-vs.htm)

El-Isa Z, Rimawi O, Jarrar G, Abu-Karaki N, Taqieddin S, Atalah M, Abderahman N and Al-Saed A 1995 Assessment of the hazard of subsidence and sinkholes in Ghor Al-Haditha area *Unpublished Report to Jordan Valley Authority* University of Jordan, Amman 141

Eppelbaum L V 2007a Revealing of subterranean karst using modern analysis of potential and quasi-potential fields *Proc. 2007 SAGEEP Conference* vol 20 *(Denver, CO)* pp 797–810

Eppelbaum L V 2007b Localization of ring structures in Earth's environments *J. Arch. Soc. Slovak. Acad. Sci., Spec. Issue: Arch. Prosp.* **XLI** 145–8

Eppelbaum L V 2008 Remote operated vehicle geophysical survey using magnetic and VLF methods: proposed schemes for data processing and interpretation *Collections of Selected Papers of the 2008 SAGEEP Conf.* vol 21 *(Philadelphia, PA)* 938–63

Eppelbaum L V 2010 An advanced methodology for remote operation vehicle magnetic survey to delineate buried targets *Trans. of the 20th General Meeting of the Intern. Mineralogical Association (Budapest, Hungary)* 103 CH30G: Archaeometry (general session): Composition, technology and provenance of archaeological artifacts

Eppelbaum L V 2010 Archaeological geophysics in Israel: past, present and future *Adv. Geosci.* **24** 45–68

Eppelbaum L V 2011a Study of magnetic anomalies over archaeological targets in urban conditions *Phys. Chem. Earth.* **36** 1318–30

Eppelbaum L V 2011b Review of environmental and geological microgravity applications and feasibility of their implementation at archaeological sites in Israel *Int. J. Geophys.* **2011** 927080

Eppelbaum L V 2013 Interpretation of magnetic anomalies due to archaeological and environmental targets classified as 'quasi thick bed bodies' in complex physical–geological environments *Proc. 2013 SAGEEP Conf.* vol 26 *(Denver, CO)* 415–24

Eppelbaum L V 2015a Interpretation of microgravity and magnetic data for delineation of subsurface salt bodies in unfavourable S/N ratio conditions *Proc. 2015 SAGEEP Conf.* vol 28 *(Austin, TX)* 686–95

Eppelbaum L V 2015b Quantitative interpretation of magnetic anomalies from bodies approximated by thick bed models in complex environments *Environ. Earth Sci.* **74** 5971–88

Eppelbaum L V 2015c Quantitative interpretation of magnetic anomalies from thick bed, horizontal plate and intermediate models under complex physical–geological environments in archaeological prospection *Archaeol. Prospect.* **23** 255–68

Eppelbaum L V 2016 Remote operated vehicles geophysical surveys in air, land (underground) and submarine archaeology: general peculiarities of processing and interpretation *Trans. of the 12th EUG Meet., Geophysical Research Abstracts* vol 18 *(Vienna)* 1–7 EGU2016-10055

Eppelbaum L V 2019 *Geophysical Potential Fields: Geological and Environmental Applications* (Amsterdam: Elsevier)

Eppelbaum L, Ezersky M, Al-Zoubi A, Goldshmidt V and Legchenko A 2008 Study of the factors affecting the karst volume assessment in the Dead Sea sinkhole problem using microgravity field analysis and 3D modeling *Adv. Geosci.* **18** 97–115

Eppelbaum L V and Khesin B E 2004 Advanced 3-D modelling of gravity field unmasks reserves of a pyrite-polymetallic deposit: a case study from the Greater Caucasus *First Break* **22** 53–6

Eppelbaum L V and Khesin B E 2012 *Geophysical Studies in the Caucasus* (Heidelberg: Springer)

Eppelbaum L V, Khesin B E and Itkis S E 2001 Prompt magnetic investigations of archaeological remains in areas of infrastructure development: Israeli experience *Archaeol. Prospect.* **8** 163–85

Eppelbaum L V and Mishne L R 1988 Some theoretical aspects of optimal filtration of geophysical signals with dependent noises (in Russian) *Deposited by VINITI, USSR Academy of Sciences* **No. 6034-B88** 10

Eppelbaum L V and Mishne A R 2011 Unmanned airborne magnetic and VLF investigations: effective geophysical methodology of the near future *Positioning* **2** 112–33

Eppelbaum L V and Pilchin A N 2006 Methodology of Curie discontinuity map development for regions with low thermal characteristics: an example from Israel *Earth Planet. Sci. Lett.* **243** 536–51

Eppelbaum L V and Khesin B E 2004 Advanced 3-D modelling of gravity field unmasks reserves of a pyrite-polymetallic deposit: a case study from the Greater Caucasus *First Break* **22** 53–6

Ezersky M 2002a The CVES method as applied to geohydrological, environmental, and engineering surveys in Israel *Isr. J. Earth Sci.* **511** 29–34

Ezersky M 2002b Some features of the subsurface geoelectric structure in the sinkhole occurrence sites along the Dead Sea shore line *EEGS-ES 8th Meeting, Aveiro, Portugal* 343–5

Ezersky M 2003a Study of the shallow subsurface sinkhole hazard areas along the Dead Sea shore using geophysical methods, Part 1, Results *GII Report* No. 211/247/02-1

Ezersky M 2003b Study of the shallow subsurface sinkhole hazard areas along the Dead Sea shore using geophysical methods, Part 2 Interpretation *GII Report* No. 211/247/02-2

Ezersky 2006 The geophysical properties of the Dead Sea salt in application to sinkhole problem *J. Appl. Geophys.* **58** 45–58

Ezersky M 2008 Geoelectric structure of the Ein Gedi sinkhole occurrence site at the Dead Sea shore in Israel *J. Appl. Geophys.* **64** 56–69

Ezersky M 2011a Geophysical surveys at two drilling sites of the Dead Sea area: Ein-Boqeq and Shalem-2 (Mineral SPA) *Report* #322/562/11

Ezersky M 2011b Improvement of the seismic refraction methods for mapping of the buried salt layers along the Dead Sea shoreline *GII Report* No. 211/486/09 24

Ezersky M G and Eppelbaum L 2017 Geophysical monitoring of underground constructions and its theoretical basis *Int. J. Georesour. Environ.* **3** 56–72

Ezersky M and Frumkin A 2013 Faults–dissolution front relations and the DS sinkholes problem *Geomorphology* **201** 35–44

Ezersky M and Frumkin A 2017 Evaluation and mapping of Dead Sea coastal aquifers salinity using Transient Electromagnetic (TEM) resistivity measurements *C. R. Geosci.* **349** 1–11

Ezersky M G and Frumkin A 2020 Identification of sinkhole origin using surface geophysical methods, Dead Sea, Israel *Geomorphology* **364** 107225

Ezersky M G and Frumkin A 2021 Subaerial morphology affected by groundwater aggressiveness: sinkhole susceptibility above karstified salt, Dead Sea *Geomorphology* **375** 107525

Ezersky M G and Goretsky I 2014 Velocity-resistivity versus porosity-permeability inter-relations in Dead Sea salt samples *Eng. Geol.* **183** 96–115

Ezersky M G and Goretsky I 2015 The scale effect of Dead Sea salt velocities based on seismic-acoustic measurements *Mechanical Behavior of Salt VIII* ed L Roberts, K Mellegard and F Hansen (Rapid City, SD/London: CRC/Balkema) pp 137–44

Ezersky M G and Legchenko A 2008 Sinkhole hazards. Part I. Mapping of the salt in sinkhole hazardous sites along western Dead Sea shore using Seismic Refraction method. NATO Programme Science for Peace, Project *SfP 981128. Report* No. 232/361/08. GII-IRD 63

Ezersky M G and Legchenko A 2014 Quantitative assessment of in-situ salt karstification using shear wave velocity, Dead Sea *Geomorphology* **221** 150–63

Ezersky M G and Legchenko A 2015 Mapping of salt consolidation and permeability using MASW method in the Dead Sea sinkhole problem *Engineering Geology for Society and Territory* ed G Lollino, A Manconi, F Guzzetti, M Culshaw, P Bobrowsky and F Luino vol 5 (Berlin: Springer) pp 465–69

Ezersky M and Livne E 2013 Geotechnical and geophysical properties of soils in the Dead Sea sinkhole problem *19th EAGE Near Surface Geoscience (Bohum, Mo P13, 9–12 September 2013)* 4pp, on CD

Ezersky M and Sobolevsky L 2002 Estimates of the rock mass quality using geophysical methods *Second stage. G.I.I. Report* No. 263/228/02

Ezersky M G, Ilyin M M, Yakovlev B P and Kolichko A V 1991 Specifics of rock mass unloading during construction of the Rogun hydropower station with the use of geophysical data *7th Congress on ISRM, September, 1991, Aahen* (Rotterdam/Boston, MA: A.A. Balkema) pp 1099–104

Ezersky M G, Rudyak M S and Zhdanov V V 1993 Forecast and monitoring of stability of the tunnel Zhinvali Project on basis of geophysical data *Int. on Assessment and Prevention of Failure Phenomena in Rock Engineering (Istambul, April, 1993)* 485–92

Ezersky M, Bruner I, Keydar S, Trachtman P and Rybakov M 2006 Integrated study of the sinkhole development site on the western shores of the Dead Sea using geophysical methods *Near Surf. Geophys.* **4** 335–43

Ezersky M G, Legchenko A, Camerlynck C and Al-Zoubi A 2007 The Salt formation edge as a major indicator of the sinkhole hazard in the Dead Sea Western Coast *13th EAGE Near Surface Meeting & Exhibition 2007 (Istambul, 1–5 Sep 2007)* P67 (4pp) on CD

Ezersky M G, Legchenko A and Camerlynck C 2008 Sinbkhole hazards Part III. TEM mapping of shallow aquifers and study of their link with sinkhole development in the Dead Sea coastal area of Israel NATO Programme Science for Peace, Project *SfP 981128. Report* No. 232/362/08 GII Paris 6 University-IRD

Ezersky M, Legchenko A, Camerlynck C and Al-Zoubi A 2009 Identification of sinkhole development mechanism based on a combined geophysical study in Nahal Hever South area (Dead Sea coast of Israel *Environ. Geol.* **58** 1123–41

Ezersky M, Legchenko A, Camerlynck C, Al-Zoubi A, Eppelbaum L, Keidar S, Baucher M and Chalikakis K 2010 The Dead Sea sinkhole hazard—new findings based on a multidisciplinary geophysical study *Z. Geomorphol.* **54** 69–90

Ezersky M G, Legchenko A, Al-Zoubi A, Levi E, Akkawi E and Chalikakis K 2011 TEM study of the geoelectrical structure and groundwater salinity at the Nahal Hever sinkhole site, Dead Sea shore, Israel *J. Appl. Geophys.* **75** 99–112

Ezersky M, Keydar G, Gorstein S, Kalmanovich M and Perelman, N. M 2012 Combining H/V spectral ratio from ambient noise with the multichannel analysis of surface waves (MASW)

method to map subsurface: test in the Dead Sea area *Unpublished GII Report. Resume of the first year of the study. Publication* No. ES-14-2012. No. 572/680/12

Ezersky M, Bodet L, Akkawi E, Al-Zoubi A, Camerlynck C, Dhemaied A and Galibert P-Y 2013a Seismic surface-wave prospecting methods for sinkhole hazard assessment along the Dead Sea shoreline *J. Environ. Eng. Geophys.* **18** 233–53

Ezersky M G, Eppelbaum L V, Al-Zoubi A, Keydar S, Abueladas A-R, Akkawi E and Medvedev B 2013b Geophysical prediction and following development sinkholes in two Dead Sea areas, Israel and Jordan *Environ. Earth Sci.* **70** 1463–78

Ezersky M G, Al-Zoubi A, Eppelbaum L V and Keydar S 2013c Sinkhole hazard assessment of the dead sea area in Israel and Jordan: multidisciplinary study *Final Technical Report* MERC_M27-050 p 142

Ezersky M, Gorstein G, Kalmanovich M and Perelman, N. M 2013d Combining H/V spectral ratio (HVSR) technique with method using surface waves (MASW) to map the subsurface: implementation in the Dead Sea area *Unpublished GII Report 025/759/13*

Ezersky M, Gorstein G and Kalmanovich M 2014 Combination of active and passive MASW with HVSR methods for improving the accuracy and reliability of vs model (In site response assessment) *Publication No. ES-54-14, Unpublished GII Report* No. 080/802/14 (https://earthquake.co.il/docs/Reports/2014-080-802-14.pdf)

Ezersky M G, Gorstein M and Kalmanovich M 2015 Combination of active and passive MASW with HVSR methods for improving the accuracy and reliability of vs model (In site response assessment) *GII Report* 080/824/15 (https://earthquake.co.il /docs/ 2015 -080-824-15.pdf)

Ezersky M G, Legchenko A, Eppelbaum L and Al-Zoubi A 2017 Overview of the geophysical studies in the Dead Sea coastal area related to evaporite karst and recent sinkhole development *Int. J. Speleol.* **46** 277–302

Ezersky M G, Eppelbaum L, Legchenko A, Al-Zoubi A and Abueladas A 2021 Salt layer characteristics in the Ghor Al-Haditha area, Jordan: comprehensive combined reprocessing of geophysical data *Environ. Earth Sci.* 80 80–124

Filin S, Baruch A, Avni Y and Marco S 2011 Sinkhole characterization in the Dead Sea area using airborne laser scanning *Nat. Hazards* **58** 1135–54

Filin S, Baruch A, Morik S, Avni Y and Marko S 2012 Use of airborne laser scanning to characterize land degradation processes—the Dead Sea as a case *Surv. Rev.* **44** 84–90

Fitterman D V, Meeks J A C and Ritsema I L 1988 Equivalence behavior of three electrical sounding methods as applied to hydrogeological problems *50th Annual Meeting and Technical Exhibition of the European Association of Exploration Geophysics (The Hague, June 6–10, 1988)* (PROTEM TDEM sounding, Geonics Limited) 16

Flathe H 1976 The role of a geological concept in geophysical research work for solving hydrogeological problems *Geoexploration* **14** 195–206

Ford D C and Williams P W 2007 *Karst Geomorphology and Hydrology* (New York: Chapman and Hall)

Frumkin A 1994 Hydrology and denudation rates of halite karst *J. Hydrol.* **162** 171–89

Frumkin A 1995 Rapid entrenchment of stream profiles in the salt caves of Mount Sedom, Israel *Earth Surf. Process. Landforms* **20** 139–52

Frumkin A and Raz E 2001 Collapse and subsidence associated with salt karstification along the Dead Sea *Carbonates Evaporites* **16** 117–30

Frumkin A, Kofman L and Ezersky M 2008 Improvement of the reliability of subsurface void detection, including sinkhole development, at the Dead Sea shore area by means of Ground

Penetration Radar (GPR) *Intermediate Report* 283\311\07 Hebrew University—TECHNION Research and Development Foundation LTD—Geophysical Institute of Israel

Frumkin A, Kofman L and Ezersky M 2009 Improvement of the reliability of subsurface void detection, including sinkhole development, at the Dead Sea shore area by means of Ground Penetration Radar (GPR) *Report* No. MNI_ES-36-2008 Hebrew University—TECHNION Research and Development Foundation LTD—Geophysical Institute of Israel

Frumkin A, Ezersky M, Al-Zoubi A, Akkawi E and Abueladas A-R 2011 The Dead Sea hazard: geophysical assessment of salt dissolution and collapse *Geomorphology* **134** 102–17

Frydman S 2000 Shear strength of Israeli soils *Isr. J. Earth Sci* **49** 55–64

Frydman S, Charrash J and Goretsky I 2008 Geotechnical properties of evaporate soils on the Dead Sea area *Eng. Geol.* **101** 236–44

Frydman S, Charrash J and Goretsky I 2014 A geotechnical study of evaporitic, lacustrine sedimentsin the saline environment of the Dead Sea area *Eng. Geol.* **181** 309–22

Gabriels P, Snieder R and Nolet G 1987 In situ measurements of shear-wave velocity in sediments with higher-mode Rayleigh waves *Geophys. Prospect.* **35** 187–96

Gadirov V and Eppelbaum L V 2015 Density-thermal dependence of sedimentary associations calls to reinterpreting detailed gravity surveys *Ann. Geophys.* **58** 1–6

Galve J, Gutiérrez F, Lucha P, Guerrero J, Remondo J, Bonachea J and Cendrero A 2009a Probabilistic sinkhole modelling for hazard assessment *Earth Surf. Processes Landf.* **34** 437–52

Galve J P, Gutiérrez F, Remondo J, Bonachea J, Lucha P and Cendrero A 2009b Evaluating and comparing methods of sinkhole susceptibility mapping in the Ebro Valley evaporite karst (NE Spain *Geomorphology* **111** 160–72

Gelchinsky B 1988 Common reflecting element (CRE) method *Explor. Geophys.* **19** 71–5

Gibson P J, Lyle P and George D M 2004 Application of resistivity and magnetometry geophysical techniques for near-surface investigations in karstic terranes in Ireland *J. Cave Karst Stud.* **66** 35–8

Glover M J 1992 Void detection using standing wave analysis *Ground Penetrating Radar* ed J Pilon vol 90-4 (Ottawa: Geological Survey of Canada) pp 63–73

Goldman M, Gilad D, Ronen A and Melloul A 1991 Mapping of sea water intrusion into the coastal aquifer of Israel by the time domain electromagnetic method *Geoexploration* **28** 153–74

Goldman M, du Plooy A and Eckard M 1994 On reducing ambiguity in the interpretation of transient electromagnetic sounding data *Geophys. Prospect.* **42** 3–25

Goldman M and Ezersky M 1999 Geoelectrical measurements at the Nahal Hever site *Unpublished GII* Rep 841/92/98 (in Hebrew)

Goldman M, Shtivelman V, Wachs D and Yechieli Y 2000 Shallow integrated geophysical study in the sinkhole development area at the Dead Sea coast, Israel *6th EEGS Meeting (Bochum, Germany)* Expanded Abstracts, CH04

Goldman M and Kafri U 2021 Geoelectric, geoelectromagnetic, and combined geophysical methods in groundwater exploration in Israel *The Many Facets of Israel's Hydrogeology Springer Hydrogeology* ed U Kafri and Y Yechieli (Cham: Springer) pp 299–394

Goldshleger N, Basson U, Fastig S and Azaria I 2018 Using combined close-range active and passive- remote sensing methods to detect sinkholes *J. Remote Sens. GIS* **7** 1000222

Goretsky I 2012 Report on Results of Laboratory Soil Testing, Dead Sea area, July, 2012 *Project* 606132/2011

Gorstein M and Ezersky M 2015 Combination of HVSR and MASW methods to obtain shear wave velocity model of subsurface in Israel *Int. J. Green Econ.* **1** 20–41

GPR-SLICE v7.0 2012 User's manual (www.gpr-survey.com)

Grandjean G and Bitri A 2006 2M-SASW: Multifold multichannel seismic inversion of local dispersion of Rayleigh waves in laterally heterogeneous subsurface: application to the Super-Sauze Earth Flow *Near Surf. Geophys* **4** 367–75

Grasmuck M and Green A G 1996 3-D georadarmapping: looking into subsurface *Environ. Eng. Geosci.* **11** 1–6

Greenfield R J 1979 Review of geophysical approaches to the detection of karst *Bull. Assoc. Eng. Geol.* **XVI** 393–408

Greenwood W 2015 In situ shear wave velocity measurements in rocks (https://geoengineer.org/education/web-class-projects/cee-544-soil-site-improve-winter-2015/assignments/in-situ-shear-wave-velocity-measurements-in-rocks)

Grinsted A, Moore J C and Jevrejeva S 2004 Application of the cross wavelet transform and wavelet coherence to geophysical time series *Nonlinear Process. Geophys.* **11** 561–6

GSI 2012 Sinkholes hazard map along the Dead Sea coast (1:100000) *Report* Geological Survey of Israel Updated 27.02.2020 (in Hebrew) (https://gov.il/en/departments/general/feasibility-to-create-sinkholes)

Gutiérrez F, Cooper H A and Johnson K S 2008 Identification, prediction, and mitigation of sinkhole hazards in evaporate karst areas *Environ. Geol.* **53** 1007–22

Gutiérrez F, Galve J P, Lucha P, Castañeda C, Bonachea J and Guerro J 2011 Integrating geomorphological mapping, trenching, InSAR and GPR for the 1 identification and characterization of sinkholes: a review and application in the 2 mantled evaporite karst of the Ebro Valley (NE Spain) *Geomoirphology* **134** 144–56

Gutiérrez F, Parise M, De Waele J and Jourde H 2014 A review on natural and human induced geohazards and impacts in karst *Earth Sci. Rev.* **138** 61–88

Gutiérrez F, Zarroca M, Linares R, Carles Roqué C, Carbonel D, Guerrero J, McCalpind J P, Comas X and Cooper A H 2018 Identifying the boundaries of sinkholes and subsidence areas via trenching *Eng. Geol.* **233** 255–68

Hardin B O and Drnevich V P 1972 Shear modulus and damping in soils: design equations and curves *Proc. Am. Soc. Civil Eng., J. Soil Mech. Found. Div.* **98** 667–92

Hardy H R and Belesky R M 1987 Use of acoustic emission and seismic tomography techniques to study sinkhole development *Jpn. J. Appl. Phys.* **26** 3–8

Hasancebi N and Ulusay R 2007 Empirical correlations between shear wave velocity and penetration resistance for ground shaking assessments *Bull. Eng. Geol. Environ.* **66** 203–13

Haskell N 1953 The dispersion of surface waves on multilayered media *Bull. Seismol. Soc. Am.* **43** 17–34

Hoover R A 2003 Geophysical choices for karst investigations ed B F Beck *Sinkholes and the Engineering and Environmental Impacts of Karst* (Reston, VA: American Society of Civil Engineers) pp 529–38

Huang N E, Shen Z, Long S R, Wu M C, Shih H H, Zheng Q, Yen N-C, Tung C C and Liu H H 1998 The empirical mode decomposition and the Hilbert spectrum for non linear and non-stationary time series analysis *Proc. R. Soc. A: Math. Phys. Eng. Sci.* **454** 903–95

Interpex Ltd 2019 ZIX1D v.3.6 inversion software (http://interpex.com/SoftwareIndex.htm)

IRIS Instruments 2010 http://iris-instruments.com)

Ivanov J, Leitner B, Shefchik W, Shwenk J T and Peterie S L 2013 Evaluating hazards at salt cavern sites using multichannel analysis of surface waves *Leading Edge* **32** 298–305

Ivanov J, Miller B, Feigenbaum R D, Morton D, Peterie S L C and Dunbar, J.B. S L 2017 Revisiting levees in southern Texas using Love-wave multichannel analysis of surface waves with the high-resolution linear Radon transform *Interpretation* **5** T287–98

Iyisan R 1996 Correlation between shear wave velocity and in-situ penetration test results *Digest* **96** 371–4

Jaeger C 2009 *Rock Mechanics and Engineering* 2nd edn (Cambridge: Cambridge University Press)

Jakosky J 1957 *Exploration Geophysics* (Los Angeles, CA: Trija Publishing Company) p 1195 (clays)

Janelid I and Kvapil R 1965 Sublevel caving *Int. J. Rock Mech. Min. Sci.* **3** 129–53

Jenyon M K 1986 *Salt Tectonics* (London: Elsevier)

Jinan Z 1987 Correlation between seismic wave velocity and number of blow of SPT and depth. *Selected Papers from the Chin. J. Geotech. Eng, Am. Soc. Civil Eng.* 92–100

Joswig M 2005 *Second Stuttgart Summer School on Nanoseismic Monitoring, Course Tutorial and SparseNet Software* (http://geophys.unistuttgart.de/sss05.html)

Kafri U and Goldman M 2005 The use of the time domain electromagnetic method to delineate saline groundwater in granular and carbonate aquifers and evaluate their porosity *J. Appl. Jeophys.* **57** 167–78

Kafri U, Goldman M and Lang B 1997 Detection of subsurface brines, freshwater bodies and the interface configuration in-between by the time domain electromagnetic method in the Dead Sea rift, Israel *Environ. Geol.* **31** 42–9

Kafri U and Goldman M 2006 Are the lower subaquifers of the Mediterranean coastal aquifer blocked to seawater intrusion? Results of a TDEM (time domain electromagnetic) study *Israel Earth Sci.* **55** 55–68

Kamenetsky F 1997 *Electromagnetic Geophysical Research by the Method of Transient Processes* (Moscow: GEOS) p 162

Kamenetsky F M, Stettler E H and Trigubovich G M 2010 *Transient Geo-electromagnetics* (Munich: GEOS) p 304

Kanasevich E R and Phadke S M 1988 Imaging discontinuities on seismic sections *Geophysics* **53** 334–45

Karous M and Hjelt S E 1983 Linear-filtering of VLF dip-angle measurements *Geophys. Prospect.* **31** 782–894

Kaufman A 1978 Resolving capabilities of the inductive methods of electroprospecting *Geophysics* **43** 1392–98

Keller G V and Frischknecht F C 1966 *Electrical Methods in Geophysical Prospecting* (Oxford: Pergamon) p 519

Keydar S 2004 Homeomorphic imaging using path integrals *66th EAGE Conf. & Exhibition (Paris)* Expanded Abstract P078

Keydar S and Shtivelman V 2005 Imaging zero-offset section using multipath summation *First Break* **23** 21–4

Keydar S and Mikenberg M 2008 A newtime correction formula in three-dimensional media as a function of wavefront attributes *J. Seism. Explor.* **17** 349–69

Keydar S and Mikenberg M 2010 Prestack time migration using the Kirchhoff sum along a new approximation of the reflection travel time curve *Proc. 72nd European Association of Geoscientists and Engineers Conf. and Exhibition (EUROPEC'10) (Barcelona, June 2010)* 4916–20

Keydar S, Gelchinsky B and Berkovitch A 1996 Common shot-point stacking and imaging method *J. Seismic Explor.* **5** 261–74

Keydar S, Pelman D and Ezersky M 2010 Application of seismic diffraction imaging for detecting near-surface inhomogeneities in the Dead Sea area *J. Appl. Geophys.* **71** 47–52

Keydar S, Bodet L, Camerlynck C, Dhemaied A, Galibert P Y, Ezersky M and Akkawi E 2011 A new approach for shallow subsurface imaging and its application to the Dead Sea sinkhole problem *Proc. 73rd EAGE Conf. and Exhibition (Vienna, April 2011)* 1–4

Keydar S, Medvedev B, Ezersky M and Sobolevsky L 2012a Imaging shallow subsurface of Dead Sea area by Common Shot Point stacking and diffraction method using weighted multipath summation *J. Civil Eng. Sci.* **1** 75–9

Keydar S, Medvedev B, Al-Zoubi A and Ezersky M 2012b Another look of imaging of shallow subsurface: real examples from the Dead Sea sinkhole development areas, in EGU General Assembly *Geophys. Res. Abst.* **14** 1432

Keydar S, Medvedev B, Al-Zoubi A, Ezersky M and Akkawi E 2013 3D Imaging of Dead Sea area using weighted multipath summation: a case study *Int. J. Geophys.* **2013** 692452

Khlaifata A, Al-Khashmanb O, Qutoba H and Oil W 2010 Physical and chemical characterization of Dead Sea mud *Mater. Charact.* **61** 564–68

Khmelevskoy V K 1979 *Short Course of Exploration Geophysics* (Moscow: Moscow University Press) (in Russian)

Kidanu S T, Torgashov E V and Varnavina A V *et al* 2016 ERT-based investigation of a sinkhole in Greene County, Missouri *AIMS Geosci* **2** 99–115

Kidanu S, Varnavina A, Anderson N and Torgashov E 2020 Pseudo-3D electrical resistivity tomography imaging of subsurface structure of a sinkhole—a case study in Greene County, Missouri *AIMS Geosci.* **6** 54–70

Kiro Y, Yechieli Y, Shalev E, Lyakhovsky V and Starinsky A 2008 Time response of the water table and saltwater transition zone to a base level drop *Water Resour. Res.* **44** 1–15

Klem-Musatov K 1994 Theory of seismic diffraction ed F Hron and L Lines (Tulsa, OK: Society of Exploration Geophysicists) p 409

Knopoff L 1952 On Rayleigh wave velocities *Bull. Seismol. Soc. Am.* **42** 307–8

Kofman L 1994 Use of ground penetrating radar in geotechnical applications *Proc. 12th Conf. of the Israel Association for the Advancement of Mineral Engineering (Eilat)* 158–61

Kofman L 1996 Use of georadar for detection of underground structural irregularities *IGS Ann. Meet* 51

Kofman L, Ronen A and Frydman S 2006 Detection of model voids by identifying reverberation phenomena in GPR records *J. Appl. Geophys.* **59** 284–99

Krawczyk C M, Polom U, Trans S and Dahm T 2012 Sinkholes in the city of Hamburg—New urban shear-wave reflection seismic system enables high-resolution imaging of subrosion structures *J. Appl. Geophys.* **78** 133–43

Krawczyk *et al* 2015 New process model for the Dead Sea sinkholes at Ghor Al Haditha, Jordan, derived from shear-wave reflection seismics *EGU General Assembly 2015 (Vienna)* ID 5761

Kumar P and Foufoula-Georgiou E 1997 Wavelet analysis for geophysical applications *Rev. Geophys.* **35** 385–412

Kumar S S, Krishna A M and Dey A 2013 Parameters influencing dynamic soil properties: a review treatise *National Conf. Recent Advances in Civil Engineering (November 15–16, 2013)* 1–10

Lacroix Y and Horn H 1973 Direct determination and indirect evaluation of relative density and its use on earthwork construction projects: in evaluation of relative density and its role in geotechnical projects involving cohesionless soils *ASTM Spec. Tech. Publ.* **523** 251–80

Landa E and Keydar S 1998 Seismic monitoring of diffraction images for detection of local heterogeneities in the shallow subsurface *Geophysics* **63** 1093–105

Landa E, Shtivelman V and Gelchinsky B 1987 A method for detection of diffracted waves on common-offset sections *Geophys. Prospect.* **35** 359–73

Landa E, Fomel S and Moser T J 2006 Path-integral seismic imaging *Geophys. Prospect.* **54** 491–506

Lankston R 1990 High-resolution refraction seismic data acquisition and interpretation *Geotechnical and Environmental Geophysics Investigations in Geophysics No. 5* ed Reviewer, Tutorial and S Ward vol 1 (Tulsa, OK: Society of Exploration Geophysicists) pp 45–73

Legchenko A 2013 *Magnetic Resonance Imaging for Groundwater* (London: Wiley-ISTE) p 157

Legchenko A V and Shushakov O A 1998 Inversion of surface NMR data *Geophysics* **63** 75–81

Legchenko A, Ezersky M, Camerlynck C and Al-Zoubi A 2007 MRS study of the evolution of aquifer in sinkhole hazardous area (Dead Sea coast) *13th EAGE Near Surface Meeting & Exhibition 2007 (Istanbul, 3–5 Sep 2007)* B29

Legchenko A, Ezersky M, Boucher M, Camerlynk C, Al-Zoubi A and Chalikakis K 2008a Pre-existing caverns in salt formations could be the major cause of sinkhole hazards along the coast of the Dead Sea *Geophys. Res. Lett.* **35** L19404

Legchenko A, Ezersky M, Camerlink C, Al-Zoubi A, Chalikakis K and Girard J-F 2008b Locating water-filled karst caverns and estimating their volume using magnetic resonance soundings *Geophysics* **73** 51–61

Legchenko A, Ezersky M, Girard J-F, Baltassat J-M, Camerlynck C and Al-Zoubi A 2008c Interpretation of the MRS measurements in rocks with high electrical conductivity *J. Appl. Geophys.* **66** 118–27

Legchenko A, Ezersky M, Camerlynck C, Al-Zoubi A and Chalikakis K 2009 Joint use of TEM and MRS method in the complex geological setting *C. R. Geosci.* **341** 908–17

Legchenko A, Descloitres M, Vincent C, Guyard H, Garambois S, Chalikakis K and Ezersky M 2011 Three-dimensional magnetic resonance imaging for groundwater *New J. Phys.* **13** 025022

Lehmann-Horn J A, Hertrich M, Greenhalgh S A and Green A G 2011 Three-dimensional magnetic field and NMR sensitivity computations incorporating conductivity anomalies and variable-surface topography *IEEE Trans. Geosci. Remote Sens.* **49** 3878–91

Levi E, Goldman M, Hadad A and Gvirtzman H 2008 Spatial delineation of groundwater salinity using deep time domain electromagnetic geophysical measurements: a feasibility study *Water Resour. Res* **44** W12404

Levi E, Ezersky M, Al-Zoubi A and Eppelbaum L 2011 Study of the salt layer environments using TEM method in the Dead Sea sinkhole problem *EAGE-ES 17th Meeting (Leicester, 12–14 September 2011)*

Livne E 2012 Evaluation of the liquefaction potential along the dead-sea western shores—regional screening and geotechnical analysis of two boreholes *MSc Thesis* Ben Gurion University of the Negev

Loke M H and Barker R D 1996 Rapid least-squares inversion of apparent resistivity pseudosections by a quasi-Newton method *Geophys. Prospect.* **44** 131–52
Loke M H 2004 Electrical imaging surveys for environmental and engineering studies *A Practical Guide to 2D and 3D Surveys* Unpublished
Louie J N 2001 Faster, better: shear-wave velocity to 100 meters depth from refraction microtremor arrays *Bull. Seism. Soc. Am.* **91** 347–64
Lowe D and Waltham T 1995 A dictionary of karst and caves: a brief guide to the terminology and concepts of cave and karst science *Cave Studies Series Number 6* (London: British Cave Research Association) p 41
Lucha P, Cardona F, Gutiérrez F and Guerrero J 2008 Natural and human-induced dissolution and subsidence processes in the salt outcrop of the Cardona Diapir (NE Spain) *Environ. Geol.* **53** 1023–35
Lutenneger A 2008 The standard penetration test—more than just a one number test *Geotechnical and Geophysical Site Characterization* ed Huang and Mayne (London: Taylor and Francis Group) pp 481–5
Lyachovitsky F M, Chmelevskoy V K and Yaschenko Z G 1989 *Engineering Geophysics* (Moscow: Nedra) (in Russian)
Maimon O, Lyakhovsky V, Agnon A and Abelson M 2005 Stability of cavities and formation of sinkholes along the Dead Sea coast *Geological Survey of Israel Report* GSI/19/05 (in Hebrew) Abstract in English
Maimon O, Abelson M, Rybakov M and Lyakhovsky V 2009 An examination whether microgravity reflects underground cavities along the Dead Sea shoreline: 3-D numerical model of rock mass stability *Report* GSI/09/2009 Geological Survey of Israel
Mala 2010 ProEx—professional explorer control unit *Operating Manual* (Mala) 60
Massonet D and Feigle K 1998 Radar interferometry and its application to changes in the Earth's surface *Rev. Geophys.* **36** 441–500
Mavko G, Mukerji T and Dvorkin J 2009 *The Rock Physics Handbook* 2nd edn (Cambridge: Cambridge University Press)
McNeill J D 1980a Electromagnetic terrain conductivity measurement at low induction numbers *Technical Note TN-6* Geonics Ltd, Mississauga, Ontario
McNeill J D 1980b Applications of transient electromagnetic techniques *Technical Notes TN-7* Geonics Ltd, Mississauga, Ontario
McNeill J D 1980c Electrical conductivity of soils and rocks. *Technical Notes TN-5* Geonics Ltd, Mississauga, Ontario
Mellet J S 1995 Ground penetrating radar applications in engineering, environmental management, and geology *J. Appl. Geophys.* **33** 157–66
Mendecki A J (ed) 1997 *Seismic Monitoring in Mines* (London: Chapman and Hall) p 262
Mendonca A F, Pires A C B and Barros J G C 1993 Pseudo-sinkholes in laterritic terrains, Brasilia, Brazil *Applied Karst Geology* ed Beck (Rotterdam: Balkema) pp 43–9
Milanović P T 2000b *Geological Engineering in Karst: Dams, Reservoirs, Grouting, Groundwater Protection, Water Tapping, Tunneling* (Belgrade: Zebra Publishing Ltd) p 347
Miller R D, Xia J, Park C B and Ivanov J 1999 Multichannel analysis of surfaces waves to map bedrock *Lead. Edge* **18** 1337–448
Nakamura Y 1989 A method for dynamic characteristics estimation of subsurface using microtremor on the ground surface *QR RTRI* **30** 25–33

Nelson R G and Haigh J H 1990 Geophysical investigations of sinkholes in lateric terrains *Geotechnical and Environmental Geophysics Investigations in Geophysics No 5* ed S Ward vol III (Tulsa, OK: Society of Exploration Geophysicists) pp 133–53

Nof R, Baer G, Ziv A, Eyal Y, Raz E, Atzori S and Salvi S 2013 Sinkhole precursors along the Dead Sea, Israel, revealed by SAR interferometry *Geology* **41** 1019–22

Nof R N, Abelson M, Raz E, Magen Y, Atzori S, Salvi S and Baer G 2019 SAR interferometry for sinkhole early warning and susceptibility assessment along the Dead Sea, Israel *Remote Sens.* **11** 89

Ohta Y and Goto N 1978 Empirical shear wave velocity equations in terms of characteristic soil indexes *Earthquake Eng. Struct. Dyn.* **6** 167–87

Okada H 2003 The microtremor survey method *Geophysical Monograph Series* (Tulsa, OK: Society of Exploration Geophysicists (SEG))

O'Neill A 2003 Full-waveform reflectivity for modelling, inversion and appraisal of seismic surface wave dispersion in shallow site investigations *PhD Thesis* The University of Western Australia, School of Earth and Geographical Sciences

O'Neill A, Kampbell T and Matsuoka T 2008 Lateral resolution and lithological interpretation of surface wave profiling *Lead. Edge* **27** 1550–63

Orhan O, Yakar M and Ekercin S 2020 An application on sinkhole susceptibility mapping by integrating remote sensing and geographic information systems *Arab. J. Geosci.* **13** 886

Palmer D 1986 Refraction seismics *The Lateral Resolution of Structure and Seismic Velocity* (London: Geophysical Press Ltd)

Palmstrom A 1995 RMi—a rock characterization system for rock engineering purposes *PhD Thesis* Oslo University

Parise M and Lollino P 2011 A preliminary analysis of failure mechanisms in karst and man-made underground caves in southern Italy *Geomorphology* **134** 132–43

Park C B, Miller R D and Xia J 1999 Multichannel analysis of surface waves (MASW) *Geophysics* **64** 800–8

Park C B, Miller R D, Xia J and Ivanov J 2004 Imaging dispersion curves of passive surface waves: SEG expanded abstracts *Soc. Explor. Geophys.* (NSG 1.6)

Park C B and Miller R D 2006 Roadside Passive MASW *Proc. of SAGEEP (Seattle, WA, April 2–6, 2006)*

Park C B, Miller R D, Xia J and Ivanov J 2007 Multichannel analysis of surface waves (MASW) —active and passive methods *Lead. Edge* 60–4

Park C B, Ivanov J and Brohammer M 2008 *SurfSeis2, User's Manual v.2.0* ed M Adkins-Heljeson (Lawrence, KS: Kansas Geological Survey)

Pasquet S and Bodet L 2015 SWIP: an integrated workflow for surface-wave dispersion inversion and profiling *Geophysics* **82** WB47–61

Perelman N and Zaslavsky Y 2001 Analysis of seismic signals in frequency domain, (SEISPECT) *GII Report* 569/345/01. 2001

Polom U, Alrshdan H, Al-Halbouni D, Holohan E P, Dahm T, Sawarieh A, Atallah M Y and Krawczyk C M 2018 Shear wave reflection seismic yields subsurface dissolution and subrosion patterns: application to the Ghor Al-Haditha sinkhole site, Dead Sea, Jordan *Solid Earth* **9** 1079–98

Press F 1966 Seismic velocities *Handbook of Physical Constants* ed S P Clark (New York: The Geological Society of America) p 195

Rabinovich M 1995 Errors of 1-D interpretation of 3-D TDEM data in the application of mapping saltwater/freshwater contact *J. Appl. Geophys.* **34** 23–34

Raiche A P, Jupp D L B, Rutter H and Vozoff K 1985 The joint use of coincident loop transient electromagnetic and Shlumberger sounding to resolve layered structure *Geophysics* **50** 1618–27

Rajab J A 2021 Mapping the near-surface geoelectrical structure of the Mottled Zone using the very low frequency–electromagnetic method *J. Appl. Geophys.* **184** 1–12

Rao M V M S and Lakshmi K J P 2003 Shear-wave propagation in rocks and other lossy media: an experimental study *Curr. Sci.* **85** 1221–5

Rawlings J O, Pantula S G and Dickey D A 1998 *Applied Regression Analysis: A Research Tool* 2nd edn (New York: Springer)

Raz E 2000 Formation of sinkholes in the Dead Sea area—a surface survey: geological survey of Israel *Report* GSI/31/2000 (in Hebrew)

Reppert M, Vichabian Y, Sogade J and Morgan F D 1999 Cave detection using GPR *SAGEEP, March* 287–92

Reynolds J M 2011 *An introduction to Applied and Environmental Geophysics* 2nd edn (New York: Wiley)

Rogers J D 2006 Subsurface exploration using the standard penetration test and the cone penetrometer test *Geolog. Soc. Am. Environ. Eng. Geosci.* XII 161–79

Ronen A, Ezersky M, Beck A, Gatenio B and Simhiov R B 2019 Use of GPR method for prediction of sinkholes formation along the Dead Sea Shores, Israel *Geomorphology* **328** 28–43

Renalier F, Jongmans D, Savvaidis A, Wathelet M, Endrun B and Cornou C 2010 Influence of parameterization on inversion of surface wave dispersion curves and definition of an inversion strategy for sites with a strong V_s contrast *Geophysics* **75** B197

Rix G J and Leipski A E 1991 Accuracy and resolution of surface wave inversion *Recent Advances in Instrumentation, Data Acquisition and Testing in Soil DynamicsGeotechnical Special Publication No. 28* ed S K Bhatia and G W Blaney (Reston, VA: American Society of Civil Engineers) pp 17–23

Roche 0, Druitt H T and Merle O M 2000 Experimental study of caldera formation *J. Geophys. Res.* **105** 395–416

Rodriguez V, Gutierrez F, Green A G, Carbonel D, Horstmeyer H and Schmelbach C 2014 Characterizing sagging and collapse sinkholes in a mantled karst by means of ground penetrating radar (GPR) *Environ. Eng. Geosci.* XX 109–32 (the Geological Society of America)

Ronen A and Beck A 1997 Ground penetrating radar and FDEM surveys between kilometers 243-45 of Route #90, Ein Gedi, Dead Sea *GII Report No. 214/58/97* (in Hebrew)

Ronen A, Tahan S, Ziner I and Beck A 2015 GPR survey for detection and mapping underground voids at the Ein Gedi site *Geotec Report* No: B2015/10/02 (in Hebrew)

Reuter F and Stoyan D 1993 Sinkholes in carbonate, sulphate, and chloride karst regions: principles and problems of engineering geological investigations and predictions, with comments for the construction and mining industries Applied Karst Geology ed B F Beck (Rotterdam: Balkema) pp 3–25

Roth M J S, Mackey J R, Mackey C and Nyquist J E 1999 A case study of the reliability of multi-electrode earth resistivity testing for geotechnical investigations in karst terrains *Hydrogeology and Engineering Geology of Sinkholes and Karst* ed B F Beck, J Pettit and J G Herring (Rotterdam: Balkema) pp 247–52

Rucker M L, Crum G, Meyers R and Lommler J C 2005 Geophysical identification of evaporite dissolution structures beneath a highway alignment *Sinkholes and the Engineering and Environmental Impacts of Karst* (Reston, VA: ASCE)

Rybakov M 2003 Microgravity survey along the Ein Gedi bypass *GII Report* No. 860/240/0

Rybakov M 2005 Microgravity and micromagnetic solution for Dead Sea sinkholes problem *Near Surface 2005* B020 *(Palermo, 4–7 September 2005)*

Rybakov M, Goldschmidt V, Fleischer L and Rotstein Y 2001 Cave detection and 4-D monitoring: a microgravity case history near the Dead Sea *Lead. Edge* **20** 896–900

Rybakov M, Rotstein Y, Shirman B and Al-Zoubi A 2005 Cave detection near the Dead Sea—a micromagnetic feasibility study *Lead. Edge* **24** 585–90

Rybakov M 2007 Sinkhole Hazards. Part V, Detection of the underground water filled cavities at the sinkhole hazardous sites using microgravity method *Report of the GII* No. 232/284/0

Sandberg S K 1993 Examples of resolution improvement in geoelectrical soundings applied to groundwater investigations *Geophys. Prospect.* **41** 207–27

Sandmeier K J 2020 REFLEXW Version 9.5, Sandmeier Geo, Karlsruhe, Germany (https://sandmeier-geo.de/index.html)

Santamarina J C, Klein K A and Fam M A 2001 *Soils and Waves* (Chichester: Wiley) p 488

Savich A I, Koptev V I, Nikitin B N and Yaschenko Z G 1969 *Seismic-Acoustic Methods for Study of the Rock Masses* ed A M Epinatieva (Moscow: Nedra) p 239 (in Russian)

Salameh E and El-Naser H 2000 Changes in the Dead Sea level and their impacts on the surrounding groundwater bodies *Acta Hydrochem. Hydrobiol.* **28** 2–33

Savich A I, Iliin M M, Koptev V I, Ezersky M G and Zamahaev A M 1979 *Opyt and perspectivy dolgovremennykh seismo-akusticheskih nabludeniy na uchastke Inguri GES.Sbornik seismicheskoe prosvechivanie ochagovykh zon v prognoze zemletryaseniy I geodinamike* (Moscow: Nauka Publisher) pp 49–70

Savich A I, Iliin M M, Ezersky M G and Kalinin N I 1983 Long-term geophysical observations on the Inguri Damrock foundation *Bull. of the IAEG, No. 26-27, Paris* 315–9

Savich A I 1984 Generalized relations between static and dynamic indicex of rock deformobility *Hydrotechn. Construct.* **18** 394–400

Schleicher J and Costa J C 2009 Migration velocity analysis by double path-integral migration *Geophysics* **74** WCA225–31

Schneider J A, Hoyos L Jr., Mayne P W, Macari E J and Rix G J 1999 Field and laboratory measurements of dynamic shear modulus of Piedmont residual soils *Behavioral Characteristics of Residual Soils, GSP 92* (Reston, VA: ASCE) pp 12–25

Schon J H 2011 *Physical Properties of Rocks. A Workbook* (Amsterdam: Elsevier)

Seevers D O 1966 A nuclear magnetic method for determining the permeability of sand-stones *Annual Logging Symp. Transactions: Society of Professional Well Log Analysts* paper L

Serra O 1984 Fundamentals of well-log interpretation *Acquisition of Logging Data* (Amsterdam: Elsevier) p 423

SESAME User Guide 2006 Guidelines for the implementation of the H/V Spectral Ratio technique on ambient vibration measurements, processing and interpretation (http://sesame-sim.sourceforge.net/docs/SESAME/SESAME_Users_Guide.html)

Shalev E, Lyakhovsky V and Yechieli Y 2006 Salt dissolution and sinkhole formation along the Dead Sea shore *J. Geophys. Res.* **111** B03102

Shalev E, Yechieli Y, Bein A and Halamish N 2005 Prohibiting groundwater flux into the Dead Sea evaporation ponds—hydrogeological alternatives *Israel Geol. Surv. Report* TR-GSI/11/ 2005 (in Hebrew)

Shalev E and Lyakhovsky V 2012 Viscoelastic damage modeling of sinkhole formation *J. Struct. Geol.* **42** 163–70

Shapira A and Avirav V 1995 PS-SDA operation manual *Unpublished IPRG Report* No. Z1/567/79

Shatterjee S and Sadi A S 1996 *Regression Analysis by Example* (New York: Wiley)

Sharma R S and Bakkapatham A T 2008 An investigation of unsaturated soil stiffness *The 12th Int. Conf. of IACMAG (Goa, India, 1–6 October, 2008)* pp 2042–8

Shtivelman V 1995 Kinematical inversion of first arrivals of refracted wave—a combined approach *Geophysics* **61** 509–19

Shtivelman V, Goldman M, Ronen A and Ezersky M 1999 Shallow geophysical surveys in the sinkhole development area at the Nahal Hever site *GII Report* No. 823/87/98 (in English)

Shtivelman V and Keydar S 2005 Imaging subsurface inhomogeneities using multipath summation *First Break* **23** 39–42

Shtivelman V, Keydar S and Mikenberg M 2009 Imaging near surface heterogeneties using weighted multipath summation *Near Surf. Geophys.* **7** 171–7

Shviro M, Haviv I and Baer G 2017 High-resolution InSAR constraints on flood-related subsidence and evaporate dissolution along the Dead Sea shores: interplay between hydrology and rheology *Geomorphology* **293** 53–68

Silaeva O I and Shamina O G 1958 The distribution elastic pulses in celindrical specimens *USSR Acad. Sci. Izv. Geophys. Ser.* 32–43 (English edn 1958 **1** 17–24)

Sjögren B 2001 A brief study of the generalized reciprocal method and some limitations of the method *Geophys. Prospect.* **48** 815–34

Skempton A W 1986 Standard penetration test procedures and the effects in sands of overburden pressure, relative density, particle size, aging and overconsolidation *Geotechnique* **36** 425–47

Smith D L and Randazzo A F 1993 Abandoned mine detection and stability assessment: similarities to subsidenc e prediction for dissolution cavities *Applied Karst Geology* ed B F Beck (Rotterdam: Balkema) pp 135–42

Socco L V and Strobia C 2004 Surface-wave method for near-surface characterization: a tutorial *Near Surf. Geophys.* **2** 165–85

Socco L V, Boiero D, Foti S and Wisén R 2009 Laterally constrained inversion of ground roll from seismic reflection records *Geophysics* **74** G35–45

Socco L V, Foti S and Boiero D 2010 Surface-wave analysis for building near-surface velocity models—established approaches and new perspectives *Geophysics* **75A** 83–102

Spangenberg E, Spangenberg U and Heindorf C 1998 An experimental study of transport properties of porous rock salt *Phys. Chem. Earth.* **23** 367–71

Stein M, Torfstein A, Gavrieli I and Yechieli Y 2010 Abrupt aridities and salt deposition in the post-glacial Dead Sea and its north Atlantic connection *Quat. Sci. Rev.* **29** 567–75

Stevens K, Lodha G, Holloway A and Soonawala N 1995 The application of groundpenetrating radar formapping fracture zones in plutonic rocks within the Whiteshell Research Area, Pinawa, Manitoba, Canada *J. Appl. Geophys.* **33** 125–41

Stilke G 1959 On elastic surface waves at a cylindrical hole in an infinite solid *Geophys. Prospect.* **7** 273–86

Stiller M, Yechieli Y and Gavrieli I 2016 Rates of halite dissolution in natural brines: Dead Sea solutions as a case study *Chem. Geol.* **447** 161–72

Stokoe K H II, Wright G W, James A B and Jose M R 1994 Characterization of geotechnical sites by SASW method *Geophysical* Characterization of Sites ed R D Woods (International Science Publisher)

Stokoe K H II and Santamarina J S 2000 Seismic-wave-based testing in geotechnical engineering. Invited paper *Int. Conf. on Geotechnical and Geological Engineering, GeoEng 2000 (Melbourn, Australia)* 1490–536

Stoyer C 2010 Equivalence analysis of DC and EM data for layered models using the resolution matrix *SAGEEP 173-82 842*

Strakhov V N 1976 On the theory of filtration and transformation of potential fields under conditions of different prior information on noises in input data *Izvestiya Acad. Sci. USSR, Ser. Phys. Earth* 60–8 (in Russian); in English see *Phys. Solid Earth*

Strobbia C, Laake A, Vermeer P and Glushchenko A 2011 Surface waves: use them then lose them. Surface wave analysis, inversion and attenuation in land reflection seismic surveying *Near Surf. Geophys.* **9** 1–12

Stagg K G and Zienkevich O C (ed) 1968 *Rock Mechanics in Engineering Practice* (London: Wiley)

SurfSeis v.6.1 2017 www.KGS.ku.edu/software/surfseis

Sykora D W 1987 Creation of a data base of seismic shear wave velocities for correlation analysis *Geotech lab. Miscellaneous paper GL-87-26, U.S.Army Engineer Waterways experiment station, Vicksburg, MS*

Szczepański T 2008 Soil small strain parameters derived from wave velocity measurements *Geologia* **50** 85–9

AEMR 2017 TEM-researcher v.8 (http://aemr.net/tem_res.htm)

AEMR n.d. TEM-3D Wizard program (http://aemr.net/ tem_res.htm)

Taqieddin S A, Abderahman N S and Atallah M 2000 Sinkhole hazard along the eastern Dead Sea shoreline area, Jordan: a geological and geotechnical consideration *Environ. Geol.* 39 1237–53

Telford W M, Geldart L P and Sheriff R E 1993 *Applied Geophysics* (Cambridge: Cambridge University Press)

Ten Brink U S, Ben-Avraham Z, Bell R E, Hassouneh M, Coleman D F, Andreasen G, Tibor G and Coakley B 1993 Structure of the Dead Sea full-apart basin from gravity analyses *J. Geophys. Res.* **98** 21877–94

Terentiev V A 1993 Compact field impulsive ultrasonic meter S-70 *USSR Acad. Sci. Izv.Phys. Earth Ser.* **11** 54–7

Terzaghi K, Peck R B and Mesri G 1996 *Soil Mechanics in Engineering Practice* 3rd edn (New York: Wiley)

Thomas B and Roth M J S 1999 Evaluation of site characterization methods for sinkholes in Pennsylvania and New Jersey *Eng. Geol.* **52** 147–52

Thomson W T 1950 Transmission of elastic waves through a stratified solid medium *J. Appl. Phys.* **21** 89–94

Van Sambeek L L 1996 Dissolution-induced mine subsidence at the Retsof Salt Mine *Proc. of the SMRI Fall Meeting (Solution Min. Res. Inst., Cleveland, OH)* 289–309

van Shoor M 2002 Detection of sinkholes using 2D electrical resistivity imaging *J. Appl. Geophys.* **50** 393–9

Vapnik Y, Sokol E, Murashko M and Sharygin V 2006 The enigma of Hatrurim *Mineral Obser.* **10** 68–77

Veselov K E and Sagitov M Y 1968 *Gravity Prospecting* (Moscow: Nedra) (in Russian)

Vucetic M and Dobry R 1991 Effect of soils plasticity on cyclic response, *J. Goetech. Eng., ASCE* **117** 898–907

Wathelet M, Jongmans D and Ohrnberger M 2004 Surface-wave inversion using a direct search algorithm and its application to ambient vibration measurements *Near Surf. Geophys.* **2** 211–21

Wathelet M 2008 An improved neighbourhood algorithm: parameter conditions and dynamic scaling *Geophys. Res. Lett.* **35** L09301

Weber N, Antler G, Lazar B, Stein M, Yechieli Y and Gavrieli I 2021 Hydrological and thermodynamic controls on late Holocene gypsum formation by mixing saline groundwater and Dead Sea brine *Geochim. Cosmochim. Acta* **316** 363–83

Werner E 1984 Review of electrical resistivity methods *Proc. of the 1st Multidisciplinary Conf. on Sinkholes (Orlando, FL, 15–17 October)* pp 231–4

Wichura M J 1988 Algorithm AS 241: the percentage points of the normal distribution *Appl. Stat.* **37** 477–84

Whiteley R and Eccleston P 2006 Comparison of shallow seismic refraction interpretation methods for regolith mapping *2006 18th ASEG Geophysical Conf. July, 2006*

Wust-Bloch G H and Joswig M 2006 Pre-collapse identification of sinkholes in unconsolidated media at Dead Sea area by 'nanoseismic monitoring' (graphical jackknife location of weak sources by few, low-SNR records) *Geophys. J. Int.* **167** 1220–32

Xia J, Miller R D and Park C B 1999 Estimation of near-surface shear-wave velocity by inversion of Rayleigh waves *Geophysics* **64** 691–700

Xia J, Miller R D and Park C B 2003 Inversion of high-frequency surface waves with fundamental and higher modes *J. Appl. Geophys.* **52** 45–57

Yechieli Y 1993 The effects of water level changes in closed lakes (Dead Sea) on the surrounding groundwater and country rocks *PhD Thesis* Weizmann Institute of Science, Rehovot, Israel

Yechieli Y, Ronen D, Berkovitz B, Dershovitz W S and Hadad A 1995a Aquifer characteristics derived from the interaction between levels of a terminal lake (Dead Sea) and adjacent aquifer *Water Resour. Res.* **31** 893–902

Yechieli Y, Ronen D and Berkowitz B 1995b Are sedimentary salt layers always impermeable? *Geophys. Res. Lett.* **22** 2761–4

Yechieli Y 2000 Fresh-Saline ground water interface in the western Dead Sea area *Ground Water* **38** 615–23

Yechieli Y, Kafri U, Goldman M and Voss C I 2001 Factors controlling the configuration of the fresh-saline water interface in the Dead Sea coastal aquifers: synthesis of TDEM surveys and numerical groundwater modeling *Hydrol. J.* **9** 367–77

Yechieli Y, Wachs D, Shtivelman V, Abelson M, Onn C, Raz E and Baer G 2002 Formation of sinkholes along the shore of the Dead Sea—summary of the first stage of investigation *Geol. Surv. Israel. Curr. Res.* **13** 1–6

Yechieli Y, Abelson M, Bain A, Shtivelman V, Crouvi O, Wacks D, Bar G, Kalbo R and Lyachovski V 2004 Sinkhole phenomena along the Dead Sea shoreline *GSI Report* No. GSI/ 21/04 (in Hebrew)

Yechieli Y, Abelson M, Bein A and Crouvi O 2006 Sinkhole 'swarms' along the Dead Sea coast: reflection of disturbance of lake and adjacent groundwater systems *GSA Bull.* **118** 1075–87

Yechieli Y 2007 Geochemical monitoring of groundwater in boreholes at the DS coastal area–project of sinkholes *Unpublished GSI Report* TR-GSI/18/2007 (in Hebrew)

Yechieli Y, Shalev E, Wollman S, Kiro Y and Kafri U 2010 Response of the Mediterranean and Dead Sea coastal aquifers to sea level variations *Water Resour. Res.* **46** W12550

Yechieli Y, Abelson M and Baer G 2016 Sinkhole formation and subsidence along the Dead Sea coast, Israel *Hydrogeol. J.* **24** 601–12

Yilmaz O 1987 *Seismic Data Processing* ed S M Doherty (Littleton, CO: Society of Economic Geologists) p 239

Zaslavsky Y, Shapira A, Gorstein M, Kalmanovich M, Giller V, Perelman N, Livshits I, Giller D and Dan I 2005 Site response from ambient vibrations in the towns Lod and Ramle (Israel) and earthquake hazard assessment *Bull. Earthquake Eng.* **3** 355–81

Zaslavsky Y *et al* 2008 Microzoning of site response parameters in the towns of Dimona and Bet Shean (Israel) *Boll. Geofis., Teor. Appl.* **49** 109–30

Zaslavsky Y, Shapira A, Ataev G, Gorstein M, Aksinenko T, Kalmanovich M, Perelman N and Hofstetter R 2009 Using ambient noise measurements in the process of assessing earthquake hazards in urban areas: examples from Israel ed T Miura and Y Ikeda *Earthquake Engineering: New Research* (Hauppauge, NY: Nova Publisher) pp 155–218

Zhdanov M S and Keller G V 1994 *The Geoelecrical Methods in Geophysical Exploration* (Amsterdam: Elsevier)

Zhou W, Beck B F and Stephenson J B 1999 Investigation of ground water flow in karst areas using component separation natural potential measurements *Environ. Geol.* **37** 19–25

Chapter 5

Joint use of different geophysical methods

5.1 Reconcilability of the different geophysical methods (experimental data from the Dead Sea)

5.1.1 A brief preamble

Parts of this chapter have been reprinted from Ezersky *et al* (2006), Legchenko *et al* (2009), Ezersky and Frumkin (2020).

The complexity of geophysical methods sensing for the karst phenomenon localization and ambiguity of geophysical observations analysis call for an integration of different geophysical fields and in general their integration with geological methods (surface mapping, drilling, geochemical analyses, etc). On the other hand, researchers should be sure that different geophysical methods based on sensibility to different parameters of medium (resistivity, density, elastic velocities, dielectric constant, etc) show the same results in the mapping of those parameters, revealing similar anomalies, e.g. that those different methods are reconcilable one to other.

First, we have to check the reconcilability of different geophysical fields in the same areas. Then, we will consider the integration of different methods to reveal some parameters and properties that cannot be obtained using separate methods.

We have carried out two international projects to understand the potential of geophysical methods in the environment and geological setting of the karstic areas. A lot of different geophysical methods were tested by different researchers, teams, institutions. Those were discussed in section 4.1 (table 4.1). As result, we have tested different methodologies and obtained the different types of results in 1D, 2D, and 3D modes in six sites from the north to the south of the western Dead Sea (DS) side. In this section, we can compare the results of different methods obtained in the same sites.

doi:10.1088/978-0-7503-3635-2ch5

Sinkhole formation is a very complicated process. It has geological (structure and lithology), hydrogeological and tectonic aspects and it develops in space and time. This process causes changes in the structure and properties of the subsurface.

The methods used to study this process must therefore be adequate, e.g. capable of covering all these aspects. The present assumption is that void formation at depth of tens of meters causes changes in the medium structure and properties (conductivity, density, velocity, etc). These changes occur in the shallow subsurface above the cavity extending to the surface, as well as encompassing gradual long-term subsurface soil decompaction, subsidence, and/or collapse (Bruner *et al* 2003, Ezersky 2003a, 2003b). In our studies, this assumption is corroborated by a methodology that uses a combination of high-resolution seismic reflection and diffraction, seismic refraction, electrical resistivity tomography (ERT), and ground penetrating radar (GPR), and the microgravity and micromagnetic surveys (Rybakov *et al* 2001, 2005, Ezersky *et al* 2006). Then we will consider the results of different geophysical methods (maps, graphs, etc) in the sinkhole development sites presented in figure 5.1.

5.1.2 Examples of reconcilability different geophysical results obtained through the same DS areas

The studies carried out by different geophysical methods through the same areas of the DS have shown good reconcilability of results.

5.1.2.1 Nahal Hever south (NHS) (Site III in figure 5.1)

As an example, the residual gravity and TEM 1 Ωm interface maps through the NHS are shown in figure 5.2.

The residual gravity anomaly map (figure 5.2(a)) shows the mass deficit of the subsurface caused by the karstic cavities in the salt layer (Rybakov *et al* 2001), whereas variations of the 1 Ωm, and especially its low elevation depression in the center of the study area (figure 5.2(b)), has been caused by the accumulation of less saline groundwater in the sinkhole formation zone having potential to dissolve salt (Ezersky *et al* 2009).

5.1.2.2 Ein Gedi (Site II in figure 5.1)

A similar character of correlation takes place in the Ein Gedi site (figure 5.3).

Figure 5.3 shows even without special analysis, that the maps are similar in their main details (especially in the segment away from the road). The lilac zones on both maps show soil decompaction: a high resistivity anomaly on the resistivity map (figure 5.3(a)) and a negative gravity anomaly on the microgravity map (figure 5.3(b)). Since the ERT measurements were carried out before the road reconstruction and microgravity surveys were performed during and after reconstruction (dynamic compaction), there is a notable difference in the results just along the road segment: the microgravity shows a positive anomaly (soil compaction) while the ERT shows

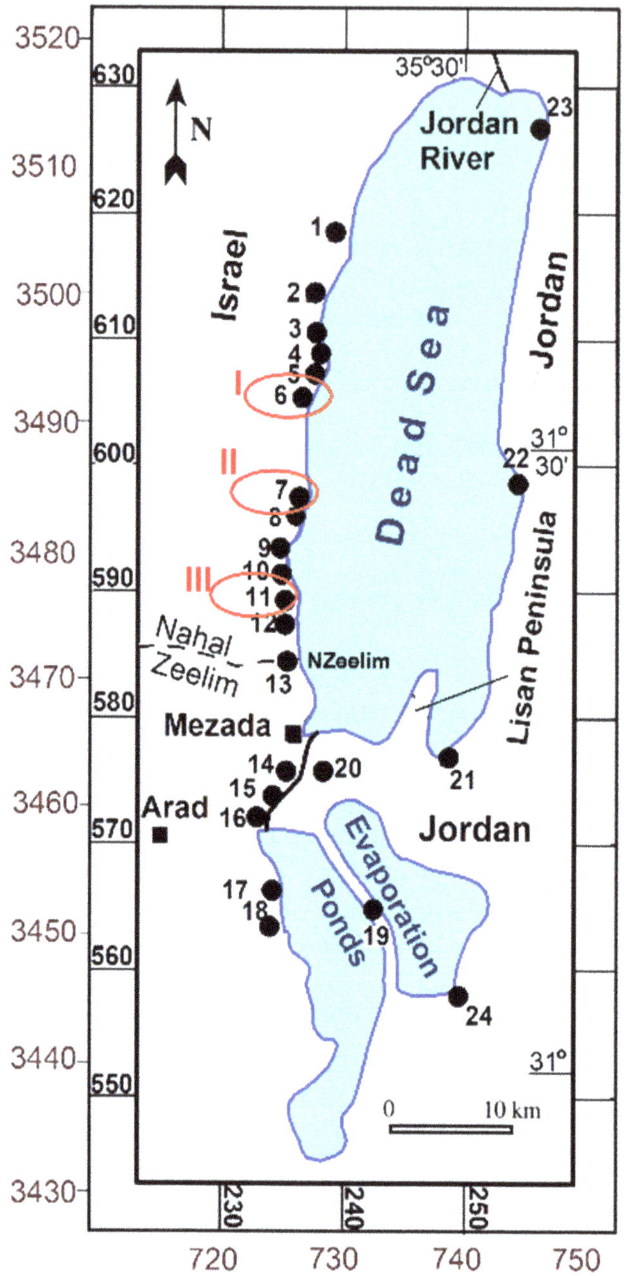

Figure 5.1. Location map of the study sites. Black points—sinkhole sites; red ellipses—study sites, 6—Mineral Beach (Site I); 7—Ein Gedy (Site II); 11—Nahal Hever south (Site III). The complete list of sinkhole sites is shown in figure 4.1 (the coordinates in the frame are the new Israel Mercator grid in km, the outer coordinates are the universal Transverse Mercator grid, in km).

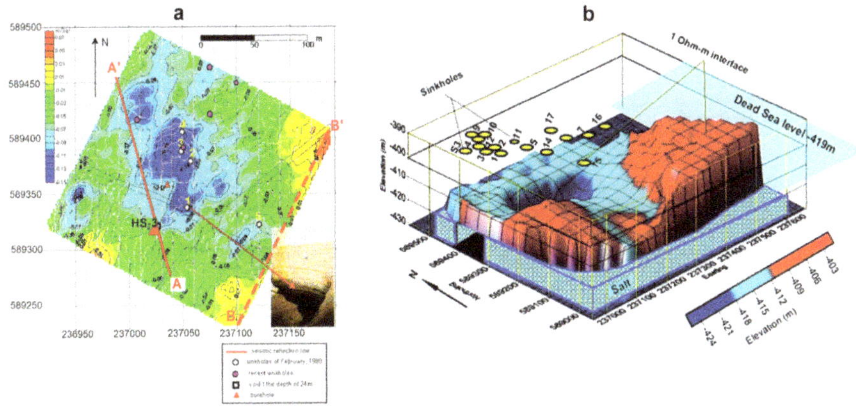

Figure 5.2. NHS. (a) Residual microgravity map and (b) TEM 1 Ωm interface map in 3D presentation. (Reprinted from (a) Rybakov *et al* (2001), with permission of SEG; (b) Ezersky *et al* (2009), with permission of Springer.) (The coordinates are the new Israel Mercator grid in km.)

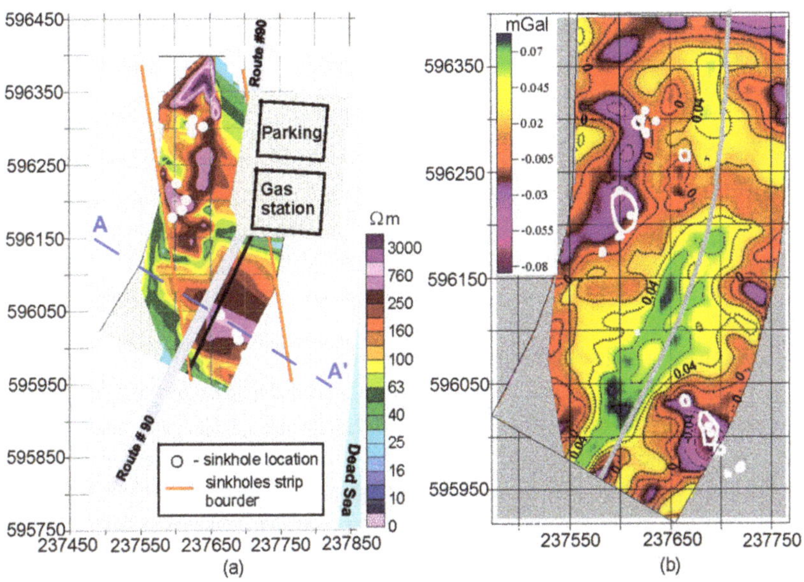

Figure 5.3. Ein Gedi site. (a) Resistivity map at the 17.5 m depth and (b) residual gravity anomaly map (Rybakov 2003). (The coordinates are the new Israel Mercator grid in km.)

slight decompaction under the road. Combined results along line A–A′ (see figure 5.3(a) for location) are presented lower in figure 5.4. There are resistivity section and residual gravity anomaly graphs along the section crossing the sinkhole strip in an approximately west–east direction.

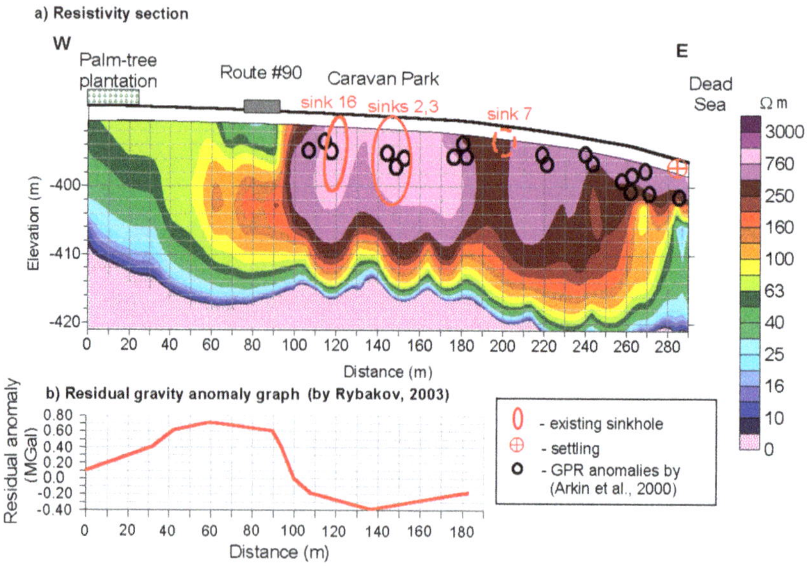

Figure 5.4. Ein Gedi site. (a) Resistivity 2D cross-section along with line A–A′ shown in figure 5.3a; and (b) residual gravity anomaly graph (b) along the same line.

The negative residual gravity anomaly correlates with a high resistivity anomaly zone involving underground cavities and sinkholes.

Ezersky (2008) has compared GPR detected void distribution through the Ein Gedi area (figure 5.5(b)) (Arkin and Gilat 2000, Ezersky 2008).

Numerous GPR anomalies related to the void presence were mapped by L Kofman at the Ein Gedi area. Results of this mapping are represented in figure 5.5. The radar image of the sinkhole is shown in figure 5.5(a) (Arkin and Gilat 2000).

The distribution of the detected GPR anomalies through the Ein Gedi area (red circles in figure 5.5(b)) is shown in figure 5.5(b) in comparison with the high resistivity anomaly shown in figure 5.3(a). The depth of the anomalies was in the range of 2–8 m. It can be seen in figure 5.5(b), that a good correlation exists between GPR anomalies (circles) and high resistivity area (gray-black) (Ezersky 2008). The image of this area (near the storeroom) 6 years later (19 March 2006) is shown in figure 5.5(c). Great subsidence (over 10 m deep) with multiple sinkholes has developed at the previous anomalies area. There is evidence that embryos of sinkholes detected by L Kofman in 1997/98 have collapsed.

5.1.2.3 Mineral Beach (Site I in figure 5.1)

A similar comparison of the ERT survey with GPR results in the Mineral Beach site is shown in figure 5.6.

One can see a high resistivity anomaly at the middle of the ERT line (figure 5.6(a)) corresponds to the GPR anomaly in figure 5.6(b).

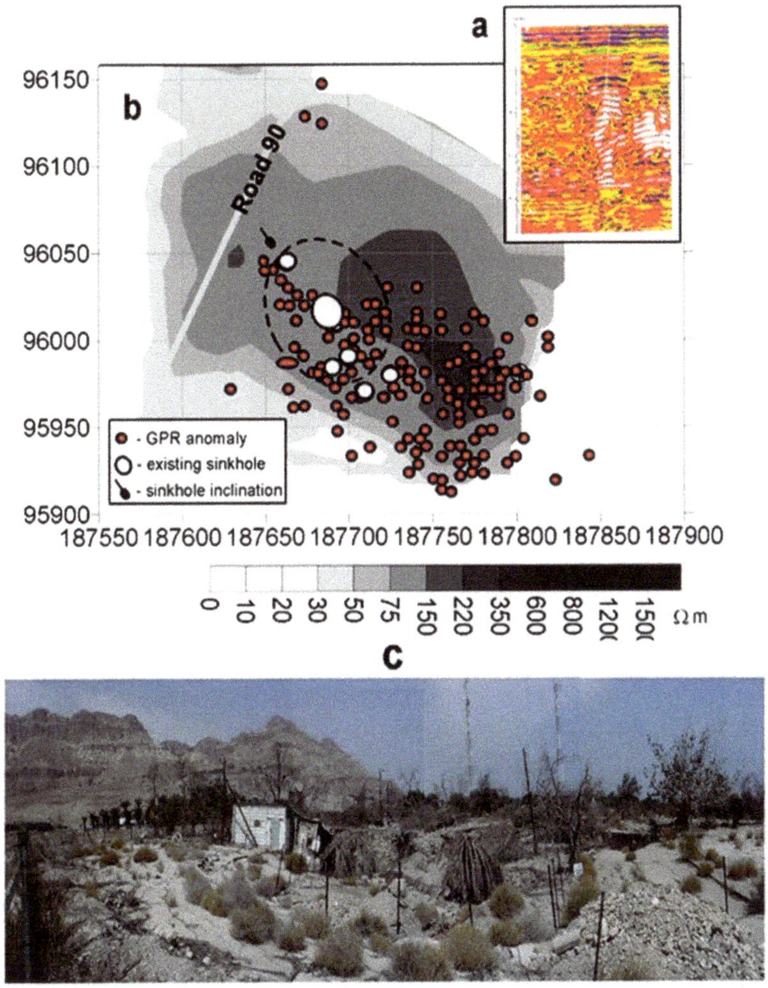

Figure 5.5. (a) GPR anomaly image (after Arkin and Gilat (2000), Kofman (1994)); (b) their distribution through Ein Gedi Caravan Park area; and (c) view of this site in March 2006. (Reprinted from Ezersky (2008), copyright (2008) with the permission of Elsevier.)

5.1.2.4 Nahal Hever south (Site III in figure 5.1)

Finally, Ezersky *et al* (2006) presented the depth sections measured by different methods (figure 5.7) along the same line A–A′ in the NHS site (see figure 5.2(a) for line location).

The combined results of the four methods are shown in figure 5.7. There are (a) the ERT resistivity—depth section, (b) seismic diffraction, (c) seismic reflection depth sections, and (d) the residual gravity anomaly graph. The presumed geological section based on geophysical data is shown in figure 5.7(e).

Two impressive features are presented in the resistivity section (figure 5.7(a)). First is a heightened resistivity anomaly of 40–60 Ωm (compared with a background

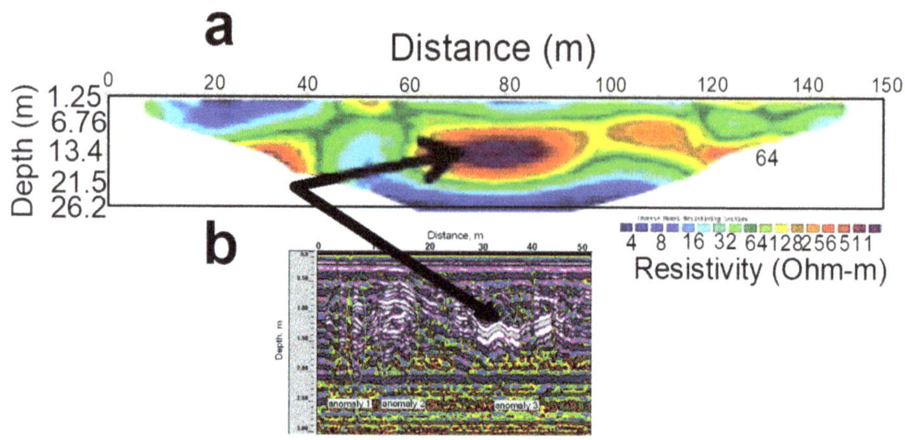

Figure 5.6. Mineral Beach site. (a) ERT and (b) GPR anomalies near the bus park. (a) Levi *et al* (2012), (b) Frumkin *et al* (2009).

of 20 Ωm) at the HS-3 borehole location, The second feature is a considerably higher resistivity zone 120–150 Ωm (in the north-western part of the section in the upper part of the subsurface). The second anomaly is higher than the background by a factor of 6–12. The zone extends down to a depth of 20 m, which is approximately the level of the water table. In figure 5.7(b), the diffraction stack image overlying the depth section is shown. There are visible anomalies at distances of 45 m (CDP 90) and 135 m (CDP 270) indicating the fault zones. These anomalies coincide with the discontinuities on the reflection depth section in figure 5.7(c) where structural features such as faults may be observed at approximately the same locations as increased resistivity anomalies in figure 5.7(a). The reflector has a funnel-shaped relief at depths of 20–25 m. The funnels correspond to the resistivity anomalies at 20–50, 60–80, and 110–150 m. In figure 5.7(d) two graphs are shown extended along with two lines from the residual gravity anomaly map (figure 5.2(a)) of Rybakov *et al* (2001). The first graph is plotted along reflection line MF-2 (A–A′), while the second is along the background line B–B′ located at the very eastern boundary of the mapping area away from the sinkhole sites. Comparing these graphs one can see a considerable difference in the levels of these two graphs (figure 5.7(d)). The background curve (blue) has a mean level of around 0 mGal or higher.

The entire A–A′ (MF-2) curve (red), however, shows a considerable negative residual anomaly along the entire line, starting from the HS-3 borehole location. Note that there is a good agreement with the results acquired at a different time from 1999 (microgravity) to 2002 (Bruner *et al* 2003). The geological section along line MF-2 has been shown in figure 5.7(e). It was obtained from the velocity map. The section is based on the seismic velocity of the refractor along line MF-2. According to our criterion, refractor velocities of more than 2900 m s^{-1} characterize the salt layer (Ezersky 2006). The salt layer is overlain by alluvium. Presuming that the lateral structure of the reflection section (the tortuous red line in figure 5.7(c)) reflects the real topography of the reflector, it is combined with the

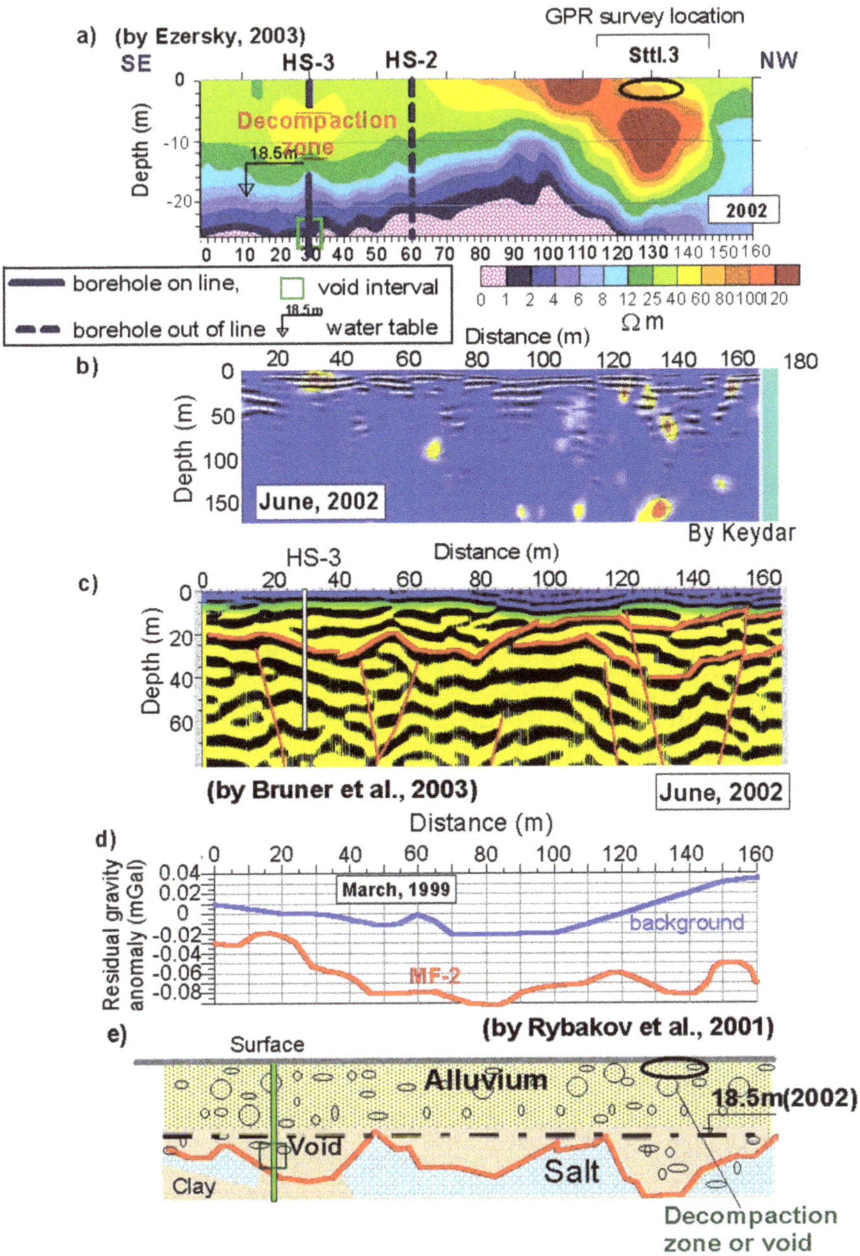

Figure 5.7. Multidisciplinary study in the NHS site. (a) ERT, (b) seismic diffraction and (c) reflection depth sections, (d) residual gravity anomaly graphs (blue line—background variations; red line—variations along with MF-2 (A–A′)), and (e) geological model derived from the geophysical results of the NHS site. See figure 5.2(a) for line location. The scales in the *Y*-direction in (a), (c), and (e) are different. (Reprinted from Ezersky *et al* (2006), with the permission of EAGE.)

geological section (the red line in figure 5.7(e)). Note some features in figure 5.7: (a) the salt layer has a funnel-shaped surface, which can be associated with dissolution processes (figure 5.7(b)). The funnel forms within the salt unit are presuming cavities, formed as a result of complete salt dissolution. (b) Borehole HS-3 has verified the presence of a cavity filled with brine to the southeast of the salt unit at the depth interval 23–28 m. Above the salt unit, one can observe the bell-shaped resistivity anomaly mentioned previously figure 5.7(a). (c) The north-western funnel in the geological section coincides with the large high-resistivity anomaly in figure 5.7(a). (d) There is some similarity between the reflector shape (figure 5.7(c) and microgravity graph figure 5.7(d)), especially in the central and right-hand parts of the section.

Resuming: figure 5.7 presents a combination of electrical resistivity tomography (ERT), high-resolution seismic diffraction, reflection, refraction, and microgravity methods allowing studying the structure and properties of the shallow subsurface at the sinkhole development site in the NHS area (on the shore of the DS) in Israel. These methods have different resolutions and penetration depths. The methods are complementary to each other and reveal useful information about the subsurface. Many methods show similar anomalies of different parameters (resistivity, residual gravity, velocity, etc) confirming their rightness. At the same time, when combined they provide additional information about the subsurface.

5.2 Practical integration of different geophysical methods in the Dead Sea area

5.2.1 Introduction

In this section we will consider the history cases when the use of several geophysical methods allows resolving of whatever problem, revealing of any specific parameter that cannot be obtained by one of them. Such an approach is often used in the borehole logging methodology (Serra 1984). Such an approach was considered in section 4.9 where estimating possible site effect is resolved by the joint implementation of horizontal-to-vertical spectral ratios (HVSR, named also Nakamura's) technique, which is based on ambient noise measurements and seismic methods such as S-wave refraction or multichannel analysis of surface waves (MASW) method. This problem is an integral part of the evaluation of the seismic hazard and reduction of earthquake damages (SESAME 2006, Ezersky *et al* 2014, Gorstein and Ezersky 2015). Another example we have considered in section 2.2.1.5 where integration of the GPR and surface nuclear magnetic resonance sounding (SNMR) methods was applied in the Tête Rousse Glacier (French Alps) to determine the status of the glacier safety. They included the following problems: (1) to study present conditions of the glacier; (2) to drain this subglacial water pocket, and (3) to monitor its evolution. The GPR method was intended to determine the geometry of the cavity within a glacier, whereas the SNMR method must evaluate the volume of water within the glacier and monitor its variation during a time, (Legchenko *et al* 2011, Vincent *et al* 2015, Garambois *et al* 2016).

The seismic diffraction method was integrated with seismic reflection to detect the upper edge of faults (figure 4.40) (Keydar *et al* 2012).

5.2.2 Some examples of the integration of geophysical methods

5.2.2.1 TEM and SNMR combination (Legchenko *et al* 2008c, 2009)
Two aspects of integrated use of the SNMR and transient electromagnetics (TEM) methods were considered: (a) measurements in conditions of high salinity subsurface and (b) distinguishing lithology of the subsurface.

(a) Measurements in conditions of high salinity subsurface

Trushkin *et al* (1995) have shown that for interpreting SNMR data the electrical conductivity of rocks has to be taken into account. The DS coastal areas are characterized by very saline groundwater. In these conditions, the MRS signal can be subjected to attenuation and a phase shift that may influence the efficiency of the SNMR method. That is why it was proposed to measure the conductivity/resistivity distribution using responding geophysical methods. Goldman *et al* (1994) have shown that when working in areas with high electrical conductivity of the subsurface, transient electromagnetic (TEM) measurements can be efficiently combined with SNMR. However, it is necessary to estimate the influence of TEM equivalency on SNMR results. Legchenko *et al* (2008c) have demonstrated using numerical modeling that SNMR has a relatively low sensitivity to possible variations in TEM results caused by the equivalence problem. Hence, TEM can be considered as a reliable tool for providing the resistivity distribution that is necessary for correct inversion of MRS data. Therefore, in the DS area, SNMR can and must be carried out in combination with the TEM method

(b) The combined use of TEM and SNMR methods to distinguish lithology of the subsurface

The combined use of TEM and SNMR methods allows investigating variations in groundwater salinity and the lithology of water-saturated rocks. Stephens *et al* (1998) have shown that total porosity decreases monotonically from fine to coarse fractions. For example, it could be clay or silt and gravel or cobbles, respectively. Bulk electrical conductivity (resistivity) depends on total porosity and fines content among other factors. From gravel to clay the porosity changes more than twofold, whereas according to Archie's law, the resistivity can change fivefold even if the resistivity of the groundwater does not change. Also increasing clay content causes a resistivity decrease even higher than that predicted by Archie's law.

TEM measures the electrical conductivity of the saline aquifer, but it is sensitive to both the salinity of the groundwater and the lithology of the rocks (porosity, or size of grain). SNMR, however, is more sensitive to the groundwater amount and the lithological variations in the subsurface than to the electrical conductivity.

Legchenko *et al* (2009) have shown that under Nahal Hever conditions, SNMR contributes to resolving the fundamental uncertainty in the TEM interpretation, which is caused by the equivalence between groundwater resistivity and lithology. The maps of the distribution of TEM and SNMR parameters through the NHS area are shown in figure 5.8. There are shown also TEM and SNMR stations.

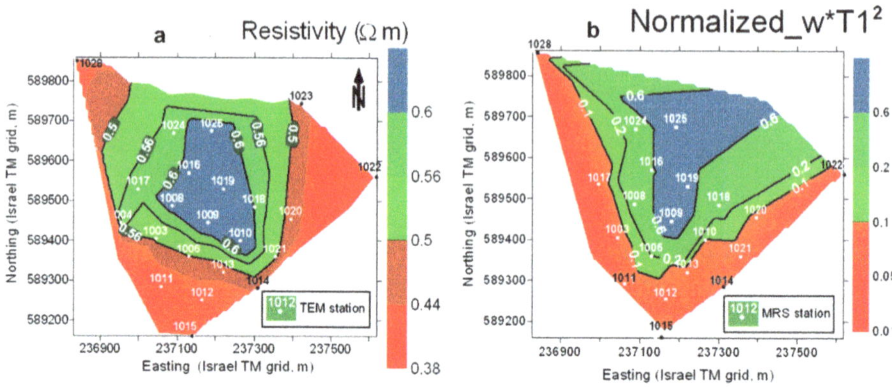

Figure 5.8. NHS site. Maps of the distribution of (a) bulk resistivity and (b) normalized SNMR hydraulic conductivity. White points are numbered TEM and SNMR stations. (Reproduced from Legchenko *et al* (2009) copyright 2009 published by Elsevier Masson SAS. All rights reserved. Joint use of TEM and MRS method in the complex geological setting.

(1) At the north, the area is characterized by high values of SNMR relaxation time and SNMR water content (and consequently high SNMR hydraulic conductivity) and a resistivity of the aquifer of $0.6 < \rho_x < 1.0$ Ωm. It suggests that the area is composed of permeable rock (sand and probably karstic salt formation) saturated with very saline water (DS brine).

(2) In the south of the area, low relaxation time T1 < 100 ms and low water content were revealed by the SNMR method. Such parameters correspond to low permeable material (clay, silt, compact salt). Resistivity $\rho_x < 0.4$ Ωm is also typical for clayey material saturated with DS brine. Maybe that a solid salt formation is enveloped by clay.

(3) Some TEM measurements reveal relatively high values of ρ_x in the range of 0.4–0.6 Ωm (and even 1.0 Ωm), at the same time, SNMR estimation of the hydraulic conductivity suggests a low permeability formation (for example soundings 1010, 1018). We interpret this formation as solid salt surrounded by relatively low clay content materials.

Figure 5.9 shows inter-relations between bulk TEM resistivities and SNMR parameters measured at the same point. The part of the site is also shown in the figure.

Variation of the TEM and SNMR parameters from the south to the north of the NHS area are presented in the table 5.1.

From the figures and table, one can see that in the NHS the TEM shows an increase of resistivity from the south to the north of the area. These variations are caused primarily by changes in the rock material—marl at the south and sandy sediments at the north of the area.

The resistivity of the subsurface increases from the south to the north testifying to the salinity decreasing. However, the variations in lithology observed with SNMR allow us to explain the observed increase in resistivity assuming unchanging salinity

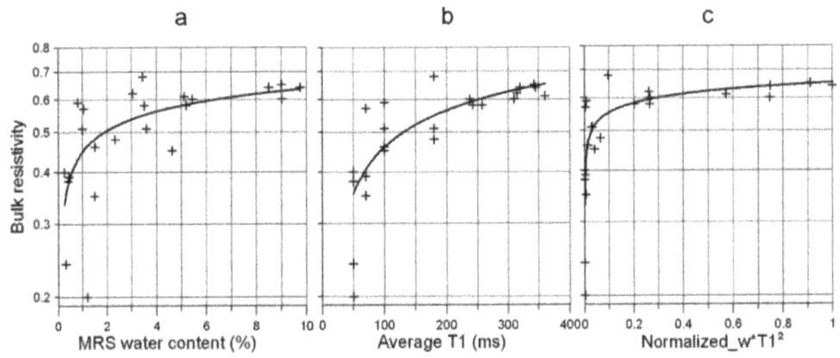

Figure 5.9. Experimental correlations between TEM and SNMR results in the NHS site. (a) bulk resistivity versus SNMR water content; (b) bulk resistivity versus SNMR relaxation time T1; (c) bulk resistivity versus normalized MRS estimator of the hydraulic conductivity denoted near experimental data: N—north of the site, S—south, C—center, W—west, E—east. Signs + are SNMR stations. (Reproduced from Legchenko *et al* (2009), copyright 2009 published by Elsevier Masson SAS. All rights reserved. Joint use of TEM and MRS method in the complex geological setting.)

Table 5.1. Change of the TEM and SNMR parameters from the south to the north of the NHS area.

Part of the site	ρ_x (Ωm)	SNMR water content%	SNMR T1 (ms)	Normalized_w*T1^2
South of the site (S)	0.2–0.6	<2.5	<100	<0.05
Center of the site (C)	>0.6	2–5	100–200	0.05–0.5
North of the site (N)	0.6–1.0	>5	200–400	0.5–1

of the groundwater. Thus, our results suggest that groundwater salinity in Nahal Hever does not change significantly around the area.

Thus, combining TEM and SNMR methods, we can show that the DS aquifer in the zone of sinkhole formations is characterized by a bulk resistivity of $\rho_x > 0.4$ Ωm, whereas the zones of silt and marl detected with SNMR in the south of the area are characterized by a bulk resistivity of $\rho_x < 0.4$ Ωm.

It seems to be that so low resistivity cannot explain the salt dissolution in the southern part of the NHS area. However, Ezersky and Frumkin (2021, table 1a) have shown that groundwater with resistivity 0.3–0.4 Ωm is slightly-to-medium aggressive concerning salt and able to dissolve up to 110 g salt L^{-1}.]

5.2.2.2 Integration of MASW and SNMR methods

DS sinkholes along with both Israeli and Jordanian shores are linked to karst cavities that form through slow salt dissolution (Frumkin *et al* 2011). Quantitative characteristics of salt karstification are its porosity and hydraulic conductivity (Shalev *et al* 2006). Measured *in situ*, these parameters would be an important indicator of sinkhole hazard. To reveal the parameter of hydraulic conductivity of the *in situ* salt layer, Ezersky and Legchenko (2014) suggested using the integrated methodology of MASW and SNMR methods. Both shear wave velocity (V_s), and

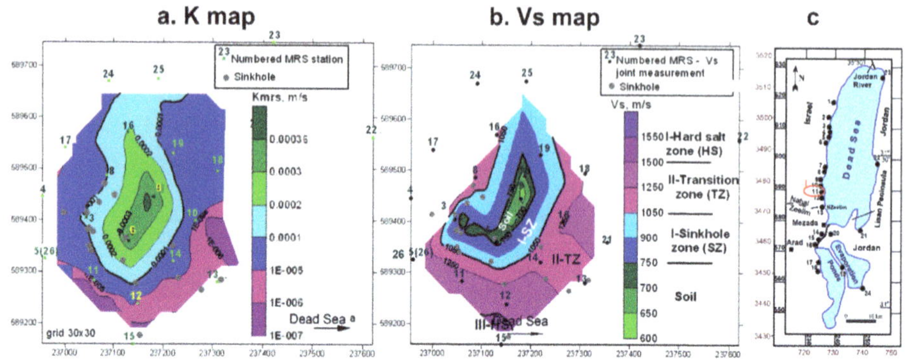

Figure 5.10. Maps of hydraulic conductivity K_{SNMR} (a), and shear wave velocity Vs (b) within the salt layer at the NHS area (see (c) for NHS site location). (Reprinted from Ezersky and Legchenko (2014), with permission from Elsevier.)

hydraulic conductivity (K), depend on porosity (Shalev *et al* 2006, Dvorkin 2008). Shear wave *in situ* conditions Vs can be measured using the MASW method (Ezersky *et al* 2013a), whereas K is measured using the SNMR technique (Vouillamoz *et al* 2003). At that, hydraulic conductivity measured by the SNMR method K_{SNMR} is calibrated to similar hydrogeological parameters K using a pumping test with empirical constant C_p that depends on the rock material (Lachassagne *et al* 2005). In this case $C_p = 7.5 \times 10^{-8}$ for the sand gravel aquifer.

The study was carried out in the NHS site characterized by the low level of electromagnetic noise (figure 5.10(c)). The distribution of hydraulic conductivity K and V_s through the NHS area is shown in figures 5.10(a) and (b), respectively.

The maps show a similar distribution of both parameters K and Vs through the NHS area. It means both parameters are defined by the same factors and we can use the maps in figures 5.10(a) and (b) for generating K versus Vs relationships. These relationships are shown in figure 5.11 where red circles correspond to *in situ* values of Vs (ranged 1600–600 m s^{-1}) and K corresponds to the permeability range 10^{-6} to 4×10^{-4} m s^{-1}. Field relations were augmented by similar inter-relations obtained in the laboratory on samples of DS salt blue triangles in figure 5.11 (Ezersky and Goretsky 2014). Thus, relationships between Vs (totally ranged from 600 m s^{-1} to 2500 m s^{-1}) and K (from 10^{-8} to 4×10^{-4}) cover the entire range of K versus Vs values measured in both field and laboratory conditions.

Both field and laboratory K and Vs values fit the exponential function

$$\ln(K) = -0.0044 \times Vs - 5.416 \quad (5.1)$$

with a determination coefficient (R^2) of 0.88 and a p-value that is <0.000 01 testifying that the result is significant.

A classification based on V_s and K can be generated now for salt conditions using that suggested by Holting (1989) (see the left column in figure 5.11). The

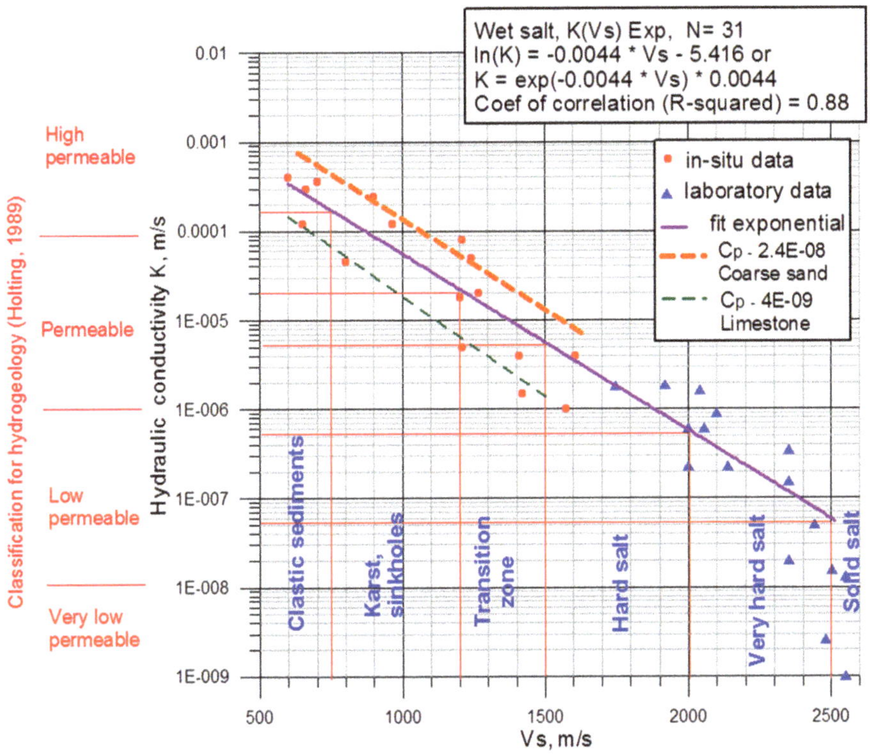

Figure 5.11. Relationship *Vs*–*K* based on calibration of elastic shear wave velocities to hydraulic conductivity at different measuring scales. Triangles are paired with laboratory ultrasonic and permeability measurements on brine-saturated salt samples. Circles are *in situ K* MRS and *Vs* MASWmeasurements in the water-saturated salt layer using a 100 × 100 m^2 loop and a 120 m-long line. Alternative graphs for different C_p are given by Nielsen *et al* (2011). Holting's (1989) classification of hydraulic conductivity used in hydrogeology is shown at the left. (Reprinted from Ezersky and Legchenko (2014), with permission from Elsevier.)

corresponding grades of sinkhole hazard (vertical red lines in figure 5.11) are shown in table 5.2.

This classification was verified in the Mineral Beach sinkhole development area. The mapping of sinkhole sites shows that they form within highly conductive zones with $K \geqslant 5.5 \times m\ 10^{-5}$. It is suggested that this methodology, with some modification, can be used for evaluating the conductive properties of karstified rock and associated sinkhole hazards (Ezersky and Legchenko 2014).

5.2.2.3 Integration of seismic refraction and TEM

The integration of seismic refraction (SRFR) and TEM methods in the DS area is intended to map the salt border and to determine groundwater salinity and aggressiveness near the salt to evaluate sinkhole susceptibility potential along with the salt border (named also dissolution front) (Ezersky 2011, Ezersky and Frumkin 2017, 2021)

Table 5.2. Classifications of salt conditions based on shear wave velocity (*Vs*). (Reprinted from Ezersky and Legchenko (2014), with permission from Elsevier.)

Classification by salt conditions	Karstified salt	Transition zone	Hard salt	Very hard salt	Solid salt
Classification by sinkhole hazard	Sinkhole hazard zone	Future sinkholes	No sinkholes	No sinkholes	No sinkholes
Classification by permeability[1]	High permeable-to-permeable	Permeable	Permeable to-low-permeable	Low permeable	Low permeable to very low permeable
Classification by hydraulic conductivity (m s^{-1})	1.5×10^{-4}–2×10^{-5}	2×10^{-5}–5×10^{-6}	5×10^{-6}–5×10^{-7}	5×10^{-7}–5×10^{-8}	$<5 \times 10^{-8}$
Porosity[2]	>40	40–29	29–15	15–8.0	<8.0%
Velocity *Vs* (m s^{-1})	760–1200	1200–1500	1500–2000	2000–2500	>2500
Velocity *Vp* (m s^{-1})	2500–2900	2900–3300	3300–3800	3800–4300	>4300

[1] HPerm—High permeability; Perm—Permeability; LPerm—Low permeability; VLPerm—Very low permeability.

[2] Porosity has been calculated from permeability using their inter-relationships (Ezersky and Goretsky 2014).

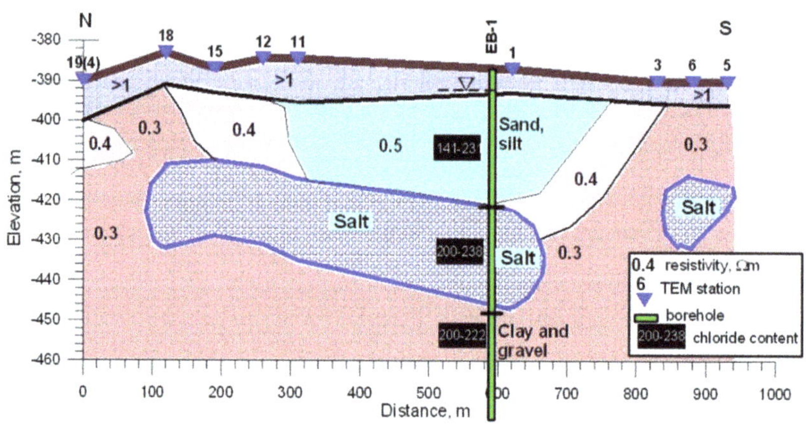

Figure 5.12. TEM resistivity section along the salt layer edge (Ezersky 2011).

The methodology of salt border mapping is described in Ezersky (2006). The classification of groundwater is based on the calibration of 28 boreholes drilled by the Geological Survey of Israel (GSI) along the DS western shore from the north to south (Ezersky and Frumkin 2017). Finally, Ezersky and Frumkin (2021) have suggested the classification of groundwater aggressiveness for two main sediments of the DS western shore sandy-gravel and lime carbonates (marl) tables 4.7(a) and (b), respectively. The TEM resistivity map along the salt border in the Ein Boqeq site is shown in figure 5.12.

Figure 5.13. Embryos of sinkholes were developed in the Ein Boqeq amphitheater site in 2008 (Ezersky *et al* (2013c), with permission of Tamar Regional Council.)

As followed from the Ein Boqeq-1 lithological section the subsurface is composed here of sandy-gravel sediments whose aggressiveness is defined in table 4.7(a). One can see that salt is enveloped mainly by aquifer with bulk resistivity ρ_x= 0.3–0.4 Ωm. Only the uppermost part of the salt in contact with sediments is characterized by ρ_x= 0.5 Ωm. Following table 4.7(a) such resistivity relates to a class of slightly-to-medium aggressive. Also, the groundwater level in this place is located at a high level on the surface. Thus, sinkhole susceptibility here is of low probability. The confirmation of such a conclusion can be some small cavity embryos discovered by the drilling in the subsurface during the site reconstruction works in 2008 (figure 5.13).

Similar integration of SRFR method with TEM method was described in section 4.3.2. The SRFR method helps us in the salt border mapping, whereas the TEM method was used to map bulk resistivity in the salt area and out of the salt area. In the limits of the 'salt area' resistivity (after Archie's law) is a measure of the salt porosity (Frumkin *et al* 2011), whereas, in the 'no salt area' resistivity is interpreted in terms of groundwater salinity based on resistivity-salinity calibrations generated for the DS region (Ezersky and Frumkin 2017).

5.2.2.4 Integration of different geophysical methods to reveal sinkhole susceptibility (Ezersky and Frumkin 2020)

We have considered above different examples of integration of different geophysical methods for resolving separate problems. However, there is the most important case of some method integration to identify the origin of the sinkhole formation (Ezersky and Frumkin 2020). Arkin and Gilat (2000) have considered a piping model, based on washing out fines by underground flows in the Newe Zohar site (figure 5.14). Their model was based on signs of piping and subsurface flows revealed from surface observations. Similar signs were observed in other sites too.

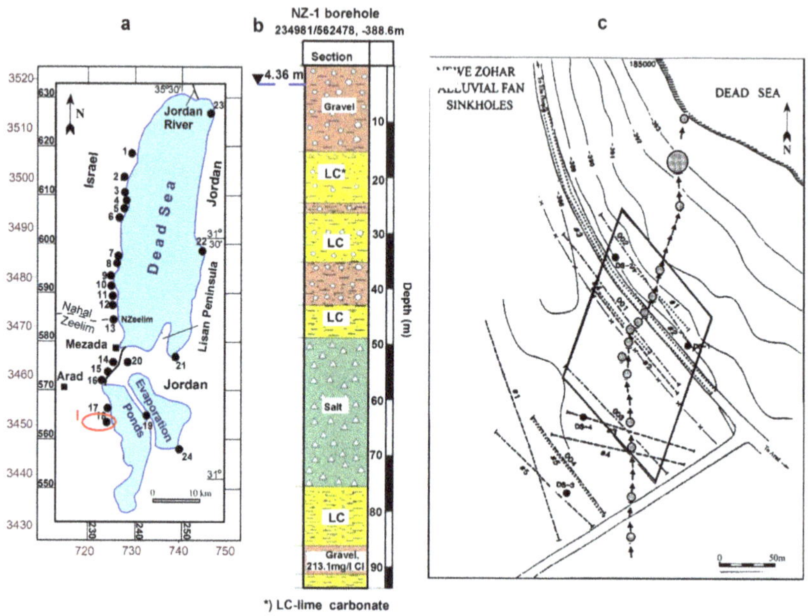

Figure 5.14. Newe Zohar case history. (a) Site location map. (b) NZ-1 borehole lithology section (compilation from Ezersky *et al* (2010)). (c) Piping model of sinkhole formation from Arkin and Gilat (2000). Explanation in text. (Modified from Ezersky and Frumkin (2020), with permission from Elsevier.)

Ezersky and Frumkin (2020) show that robust identification of sinkhole origin can be achieved when proper geophysical methodologies and their application are used.

They have considered the Newe Zohar site located in the southern part of the DS (figure 5.14(a)) to analyze different signs of sinkhole formation models using various geophysical methods. A line of sinkholes was extended along with a subsurface flow line in the frontal area of the Newe Zohar alluvial fan (figure 5.14(c)). The diameter of sinkholes is ranged from 3 to 5 m, and down to 5 m in depths. Sometimes the sinkhole floor was wet. Three boreholes of 10–15 m were drilled at the site that discovered loose alluvium. The authors identified the mechanism of the sinkhole as a piping one. Later, borehole NZ-1 revealed a 30 m thick salt layer with the top at 50 m depth (figure 5.14(b)) (Yechieli *et al* 2004).

Geophysical methods were carried out through the area to study different aspects of sinkhole formation (figure 5.15).

The SRFR method, allows discovering of the salt layer based on longitudinal wave velocity Vp (figure 5.15(a)); the MASW method allows us to determine the salt layer properties (rigidity) based on shear wave velocity Vs (figure 5.15(c)); finally, the TEM method (figure 5.15(d)) allows us to evaluate the degree of aggressiveness of groundwater concerning salt, based on bulk resistivity values. The results show that:

The sinkhole area is underlain by a salt layer edge that is extended along the sinkhole line (dashed black line in figure 5.15(a)); (a) MASW results have

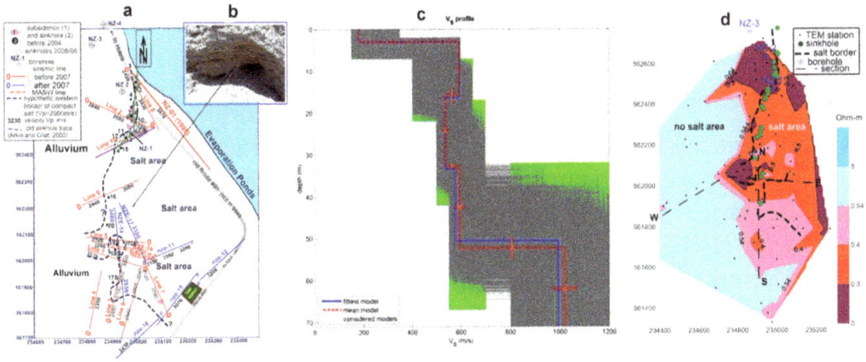

Figure 5.15. Newe Zohar case history. (a) Salt layer map based on compressional wave velocity *V*p (dashed black line), the solid green line is sinkhole trace from figure 5.14(c); (compilation from Ezersky *et al* (2010)); dashed red line is MASW measurement; (b) sinkhole No 19s with visible water channels at the bottom; (c) shear wave velocity *V*sv versus depth profile of the dispersion inversion: salt layer with *V*sv = 1170 m s^{-1} is located at a depth of 50 m. (d) The TEM resistivity map at the salt top in the Newe Zohar site. Four resistivity zones correspond to the zonation in table 4.7(b) within the light carbonate lithology. (Reprinted from Ezersky and Frumkin (2020), with permission of Elsevier.)

shown that velocity V_{sv} = 1170 m s^{-1} (with respect to table 5.2) characterizes a karstified and permeable salt layer with a porosity of 40%, i.e. sinkhole hazardous zone.

TEM resistivity map (figure 5.15(d)) shows that sinkholes distributed along with resistivity ρ_x = 0.3–0.4 Ωm zone. Groundwater, having such bulk resistivity in the lime carbonate lithology following table 4.7(b) relates to slight-to-medium aggressive. Such groundwater contains 105–224 g L^{-1} chloride and can dissolve up to 119 g L^{-1} halite (Ezersky and Frumkin 2017). It means, that the groundwater has the potential to slowly dissolve salt in the contact zone, at least (Ezersky and Frumkin 2017).

Ezersky and Frumkin (2020) have suggested the following mechanism of sinkhole formation. Firstly, salt is dissolved at the depth causing subsequent piping in the uppermost subsurface. Slight salt dissolution causes overlying sediments to subside, increasing their permeability. This allows shallow infiltrating water from the Hever Zohar alluvial fan to flow through subsurface sediments with increased porosity. This is supported by the fact that the process takes place during ascending DS ponds level. In the future, increase in the level of the pond may decrease the piping process.

Thus, integrating the use of different geophysical methods allows the identification of sinkhole formation origin. The underlying cavities along the DS are shown *to form primarily by salt dissolution, with some cases of additional piping*. Applying the right geophysical parameters for groundwater and salt sediments classification proves to be crucial for understanding subsurface geomorphology.

5.3 Integrated gravity–magnetic wavelet methodology for localization of subsurface karst

5.3.1 Essence of the problem

It is known that the increasing number of indicators of the desired target requires additional amounts of standard (background) information (Duda and Hart 1973). At the same time, a necessary number of such standards are available only in well-studied areas. However, in these areas, geological–geophysical prediction is less urgent. Therefore, a geophysical set should contain a minimum possible number of methods (Eppelbaum 2014a, 2020).

Usually, a number of geophysical methods applied for geological mapping or for searching different types of economic minerals do not exceed two or three (Parasnis 1997, Telford *et al* 1993, Khesin *et al* 1996, Kearey *et al* 2002, Eppelbaum and Khesin 2012, Dentith and Mudge 2014, Eppelbaum 2020).

Therefore, we select for buried karst recognition two often employed geophysical methods (applied in both shores of the DS): gravity and magnetic. Advanced peculiarities of new contemporary gravimetric and magnetometric equipment enable registering earlier inaccessible low-amplitude anomalies. Simultaneous improvement of observation methodology and development of new methodologies for potential geophysical field processing and interpretation have triggered manifold application of the mentioned fields in karst areas (e.g., Waltham *et al* 2004, Butler 2005, Debeglia *et al* 2006, Eppelbaum 2007, Eppelbaum *et al* 2008, Mochales *et al* 2008, Jacob *et al* 2010, Chalikakis *et al* 2011, Kaufmann *et al* 2011, Ezersky *et al* 2013c, 2021).

Wavelet methodology is an essential tool employed in geophysical signal processing for denoising and enhancement of signals, and distinguishing signals with closed characteristics (Kumar and Foufoula-Georgiou 1997, Alperovich *et al* 2001, 2003, Grinsted *et al* 2004, Wapenaar *et al* 2005, Yates *et al* 2007, Fedi *et al* 2010, Eppelbaum *et al* 2011, Averbuch *et al* 2014, 2016, 2019).

A non-conventional *three-stage methodology* of combined gravity–magnetic (GM) localization of karst terranes has been presented in Eppelbaum *et al* (2014). The *first stage* includes 3D modeling of potential geophysical effects from an extended set of typical subsurface karst terranes. We analyze here high-precise gravity and magnetic effects from different kinds of buried karst terranes. The *second stage* consists of the development of non-conventional signal processing methods to analyze profile GM observations. Finally, at the *third stage*, a novel methodology of geophysical data combined interpretation based on the modern developments in wavelet data mining is presented.

Thus, it is suggested that the advanced wavelet combined methodology will allow localizing the hidden karst terranes occurring in complex geological environments under small values of the signal-to-noise ratio.

5.3.2 Short description of the applied methodology

Review of numerous publications (e.g., Yechieli *et al* 2004, Chalikakis *et al* 2011, Eppelbaum 2007, Legchenko *et al* 2008a, 2008c, Ezersky *et al* 2010, 2013b, 2021,

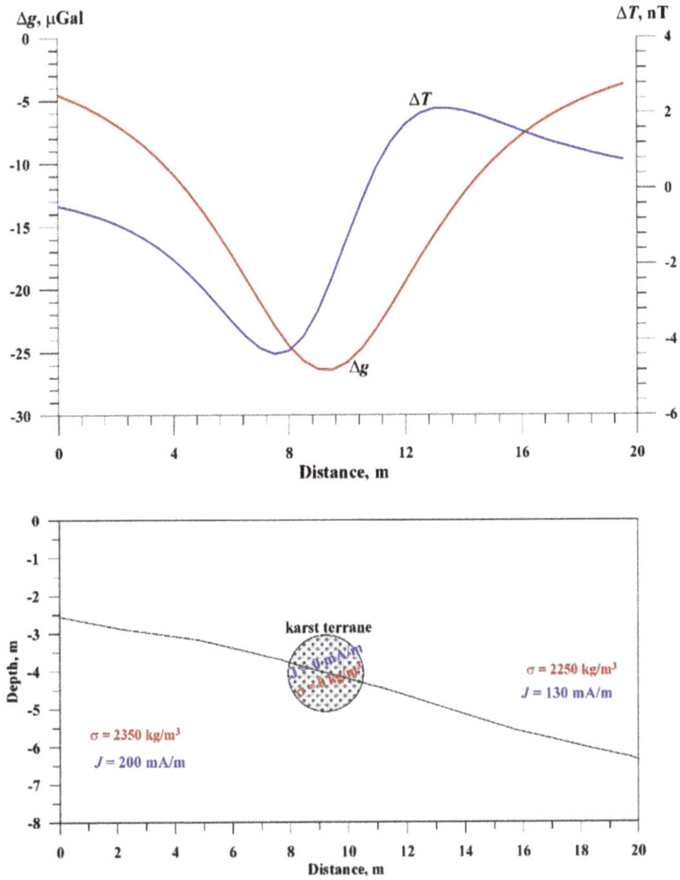

Figure 5.16. 3D GM effects from the PGM with the karst cavity only (**Model 1a**). Legend is shown in figure 5.17 (after Eppelbaum *et al* (2014), copyright Azerbaijan National Academy of Science 1995–2022).

Alperovich *et al* 2013a, Yechieli *et al* 2016, Kafri and Yechieli 2020) and the authors' personal experience, allowed development of 30 various physical–geological models (PGMs) of karst terranes typical for the DS region. The same sections were computed without the anomalous target—the karst cavity. Thus, altogether sixty PGMs were utilized. Figure 5.16 shows gravity and magnetic curves computed (for this, 3D modeling GSFC-M software (Eppelbaum 2007, Eppelbaum *et al* 2008) was applied) from the PGM 1a (from the underground cavity only). Figures 5.17 and 5.18 display PGM 1b (underground cavity + geological section), and PGM 1c (geological section only, without cavity), respectively. The PGM presented in figure 5.16 was utilized as GM 'useful signal learning'. Figures 5.19 and 5.20 show gravity and magnetic effects computed from PGMs 19a and 28a (both with the cavity), respectively. By such a manner were developed sixty (and with only karst cavities effects—ninety) PGMs.

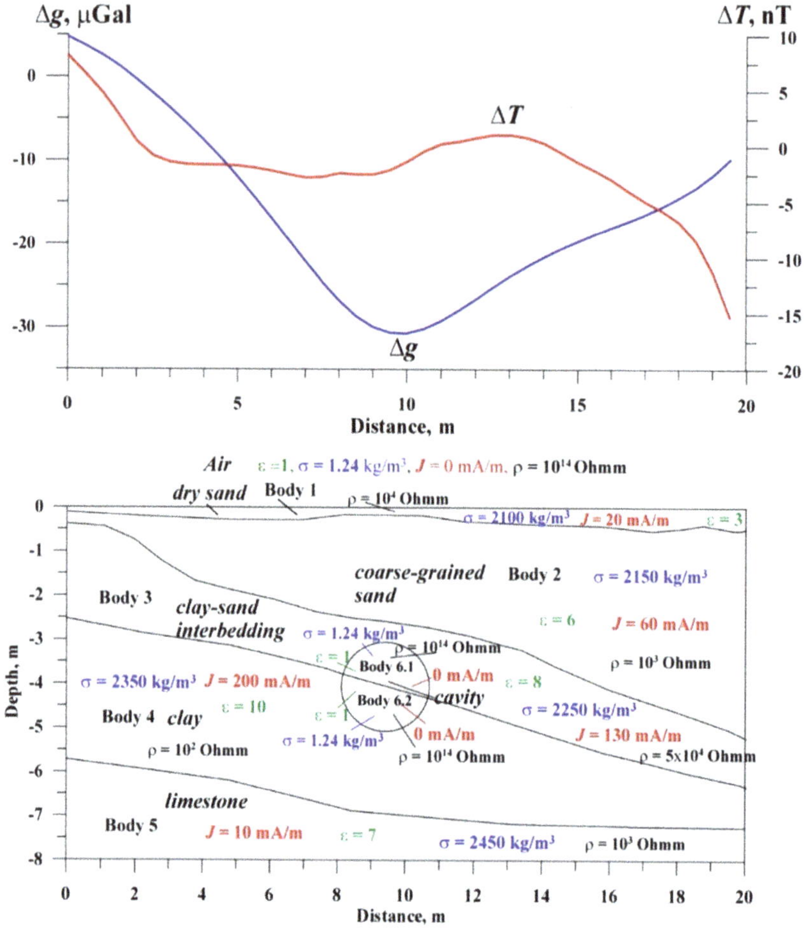

Figure 5.17. 3D GM effects from the PGM with the karst cavity (**Model 1b**). Here σ is the density, J is the magnetization, ρ is the electric resistivity, and ε is the dielectric permittivity (last two parameters are intended for the further GPR and ERT modeling). Gravity curve is presented in microGal (1 μGal = 10^{-8} m s^{-2}) and magnetic curve—in nanoTesla (nT) (after Eppelbaum *et al* (2014), copyright Azerbaijan National Academy of Science 1995–2022).

The anomalous geophysical effects from the buried karst cavities were specially modeled as significantly small for reliable testing of the developed wavelet algorithms. Several last PGM (27–30) were complicated by GM disturbing effects from fault and some metallic targets (e.g., bodies 3 and 4, respectively, in figure 5.18). Such complicated PGMs, undoubtedly, reflects some real classes of karst terranes occurring in the unfavorable for GM searching conditions. Anomalous effects from the two aforementioned targets significantly disturb the combined GM anomalies and impede the wavelet methodology application for underground karst terrane delineation.

The developed wavelet approach to recognition of hidden karst terranes consists of advanced processing of each geophysical method (G and M) and nonconventional

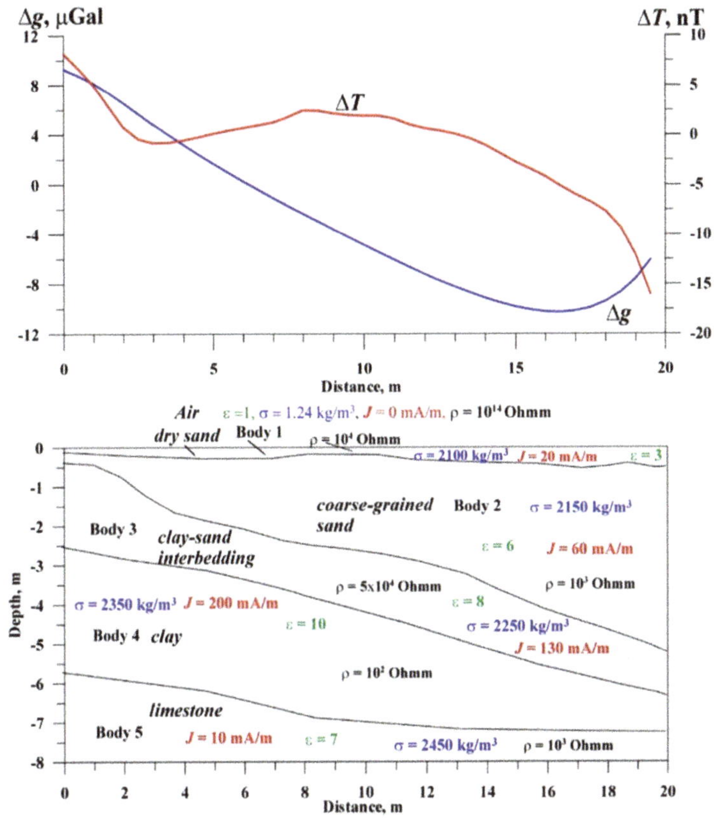

Figure 5.18. 3D GM effects from the PGM without the karst cavity (**Model 1c**). Legend is shown in figure 5.17 (after Eppelbaum *et al* (2014), copyright Azerbaijan National Academy of Science 1995–2022).

integration of the geophysical methods between themselves. This approach is based on the modern developments in wavelet theory and data mining (Averbuch *et al* 2014, 2016, 2019). The methodology is applied for derivation of enhanced (e.g., coherence portraits) and combined images of computed GM fields. It is based on the matching pursuit with the wavelet packet dictionaries enabling the extraction of desired signals even from strongly noised data (Averbuch *et al* 2014, Eppelbaum *et al* 2014).

The typical physical–mathematical problem is the extraction of significant features from available data disturbed by random noise and non-relevant background (e.g., Averbuch *et al* 2014). If the signal significant structure consists of several sine waves then we may represent it via trigonometric basis (Fourier analysis). In this case, we compare the signal with a set of sinusoids and extract consistent ones. An indicator of the presence of a sine wave in a studied signal $f(t)$ is the Fourier coefficient:

$$\int f(t) \sin wt \quad dt. \tag{5.2}$$

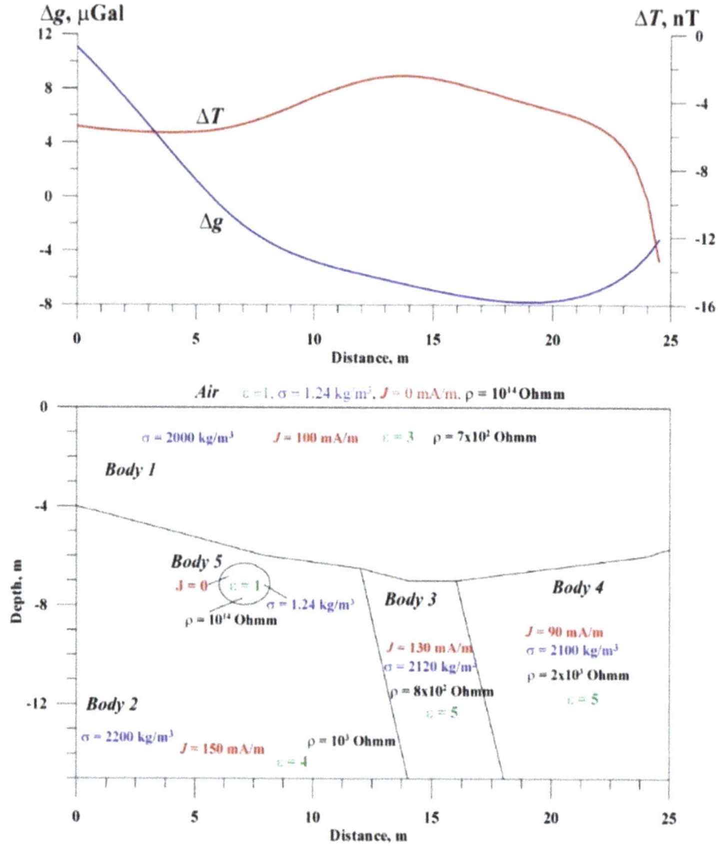

Figure 5.19. 3D GM effects from the PGM with the karst cavity (**Model 19a**). A description of physical properties is given in figure 5.17 (after Eppelbaum *et al* (2014), copyright Azerbaijan National Academy of Science 1995–2022).

Wavelet analysis provides a rich library of waveforms and fast and numerous procedures for the representation of signals and selection of relevant waveforms (e.g., Averbuch *et al* 2014, 2016). The basic assumption enabling an application of wavelet analysis (WA) is that the essential structure of the analyzed signal consists of a limited number of various waveforms. The most promising way to reveal this structure is a representation of the studied signal by a set of basic elements containing waveforms coherent to the signal. At the same time, small coefficients are expecting for the noise and structures incoherent to all basic waveforms.

Wavelets ranging from functions of arbitrary smoothness to fractal ones. Wavelet procedure involves two main aspects: (1) a *decomposition*, i.e. breaking up a signal to obtain the wavelet coefficients, and (2) a *reconstruction*, which consists of reassembling the signal from the coefficients.

There are many modifications to the WA. Note, first of all, so-called *continuous WA* in which signal $f(t)$ is tested for the presence of waveforms $\Psi\left(\frac{t-b}{a}\right)$.

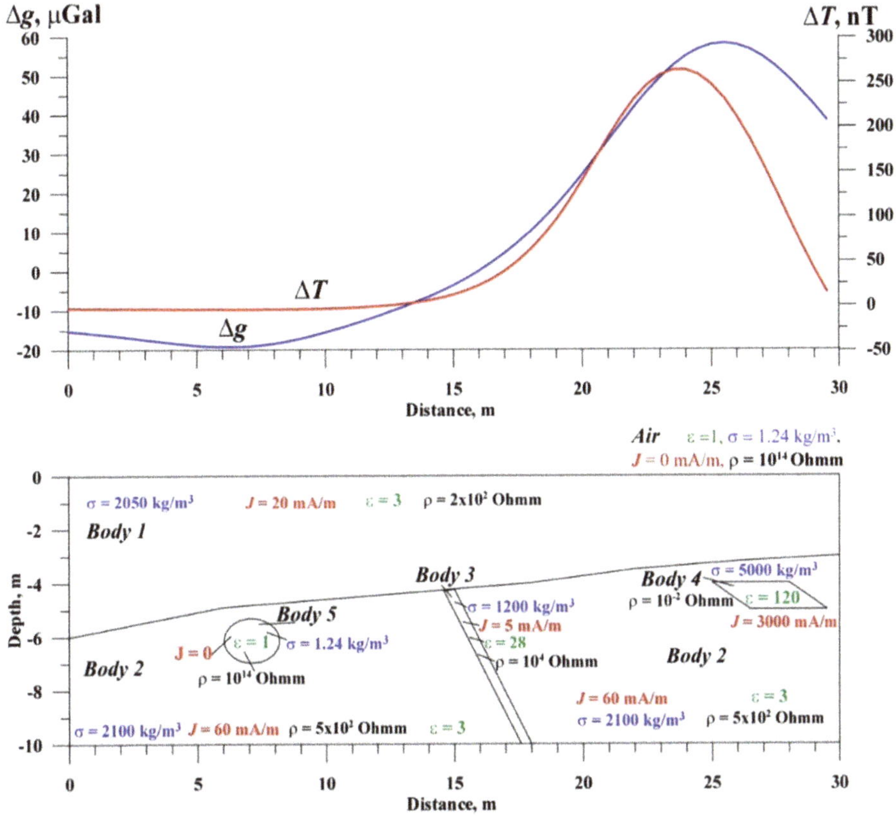

Figure 5.20. 3D GM effects from the PGM with the karst cavity (**Model 28a**). A description of physical properties is given in figure 5.17. (After Eppelbaum *et al* (2014), copyright Azerbaijan National Academy of Science 1995–2022.)

Here, a is the scaling parameter (dilation), b determines the location of the wavelet $\Psi\left(\frac{t-b}{a}\right)$ in a signal $f(t)$. The integral

$$(W_\Psi)(b, a) = \int f(t)\Psi\left(\frac{t-b}{a}\right) \tag{5.3}$$

is the continuous wavelet transform.

When parameters a and b intake some discrete values, we have the discrete wavelet transform (generally speaking most geophysical observations relate to the discrete wavelet transform). A common scheme of the wavelet decomposition tree is presented in figure 5.21.

The obtained signal is compared with the available testing signal on each level. The estimated wavelet coefficients enable the reconstruction of the first approximation of A_i and D_i. On the next level, wavelet transform is applied to

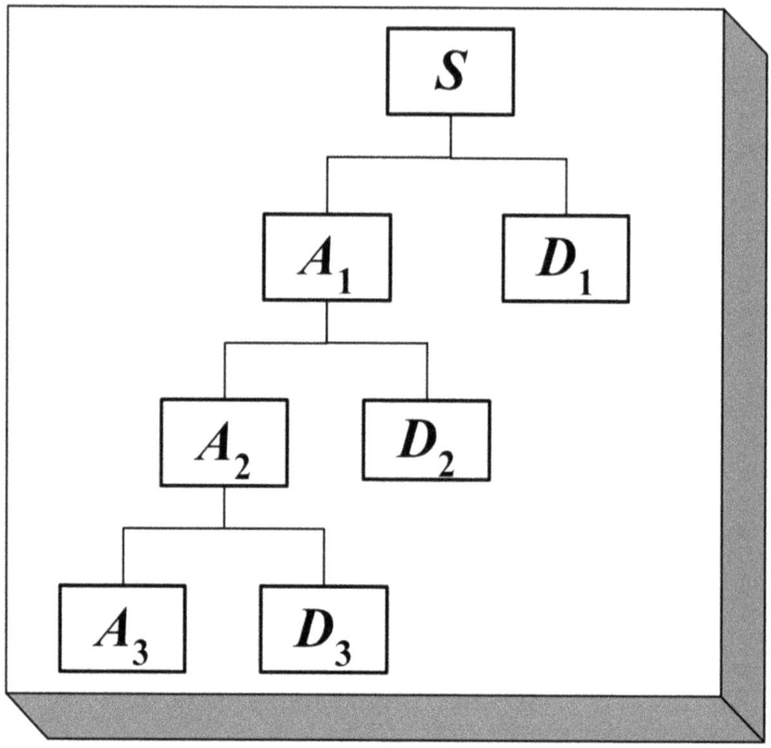

Figure 5.21. Principal block diagram of WA. A_i is the approximation, D_i is the details (after Eppelbaum *et al* (2014), copyright Azerbaijan National Academy of Science 1995–2022).

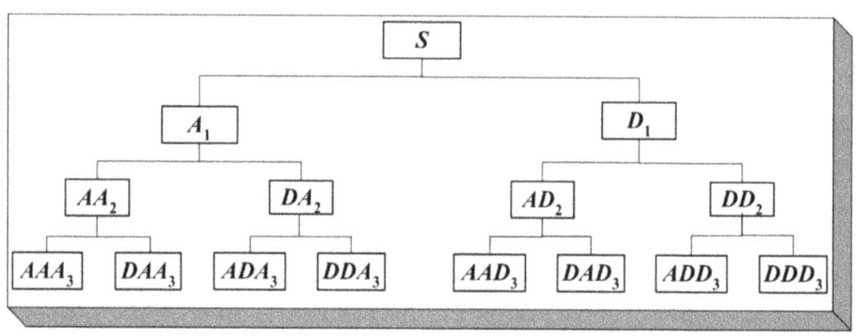

Figure 5.22. Wavelet packet tree (after Eppelbaum *et al* (2014), copyright Azerbaijan National Academy of Science 1995–2022).

the approximation. Then, we can present A_1 as $A_2 + D_2$, etc. Thus, if S—signal, A—approximation, D—details, then

$$S = A_1 + D_1 = A_2 + D_2 + D_1 = A_3 + D_3 + D_2 + D_1. \tag{5.4}$$

Wavelet packet transform is employed for both low-pass results (approximations) and high-pass results (details) (figure 5.22).

The goal of the modern geophysical data examination is to detect the geophysical signatures of buried targets at significantly noisy areas via the analysis of observed physical parameters with a minimal number of false alarms and miss-detections (Eppelbaum *et al* 2011, Eppelbaum 2020). The recently developed technique of diffusion clustering combined with the aforementioned wavelet methods was applied to integrate the GM data and detect karst cavities at a depth.

For analyzing the GM data, we applied a technique based on the algorithm to characterize combined GM images by a limited number of parameters (Eppelbaum *et al* 2012, Eppelbaum 2017). The developed algorithm consists of the following main phases: (a) accumulation of the database, (b) characterization of the GM images, and (c) reduction of dimensionality. Then, each GM image is characterized by the histogram of the coherency directions (Alperovich *et al* 2013b, Eppelbaum *et al* 2014). As a result of the previous steps, we obtain two sets (domains) **K** and **N** of the signatures vectors for images containing karst cavity (**K**) and non-karst subsurface (**N**), respectively. The obtained 3D domain of the data representatives can be utilized as a reference set for the classification of newly arriving GM data. Thus, we can receive the advanced procedure 'Learning with a teacher'.

5.3.3 Gravity and magnetic data processing

As a result of 3D GM modeling, 30 pairs of data vectors $\mathbf{g}_k$ and $\mathbf{m}_k$ were computed, $k = 1,\ldots,30$, respectively, from the subsurface section without a cavity (**N** set) and the same number of pairs retrieved from the subsurface section with a karst cavity (**K** set). Besides this, 30 PGMs were computed consisting of GM effects from the karst cavity only (these effects were utilized as reference ones).

The interim goal is to extract characteristic features of sets $\mathbf{g}_k$ and $\mathbf{m}_k$ such that the maps of these sets can be reliably separated. When this goal is achieved, these maps will be used as reference sets for the construction of classifiers, which automatically select newly arrived data to contain buried cavities and not contain them.

For achieving this goal, several steps are necessary. Either of the GM data vectors $\mathbf{g}_k$ and $\mathbf{m}_k$, $k = 1,\ldots,30$ comprise 40–60 computation points (step of computation was 0.5 m), which, being plotted look like smooth curves. Then, we merge these two vectors into one combined data vector. For further analysis, we will employ data vectors, whose lengths are a power of 2. Each vector is symmetrically extended from its left and right ends to the length 64. The window is applied to the extended vector to avoid discontinuities when merging (figure 5.23) (Eppelbaum *et al* 2014).

Finally, the extended and mirrored GM data vectors are merged into one combined vector of length 128. The window is applied to the extended vector to avoid leaps when merging. Figure 5.23 displays the main steps of the wavelet preprocessing: extension of the initial GM observational curves to 2^n–point curves, with the following smoothing by a sliding window. Figure 5.24 displays merged GM data curves for models (PGMs) without a karst cavity (**N**), and figure 5.25—merged GM data curves for the PGMs with a modeled karst cavity (**K**).

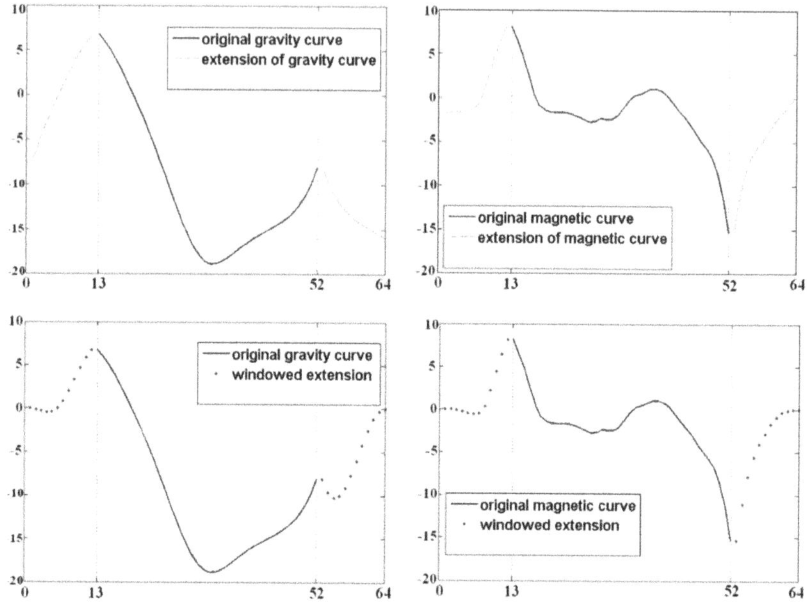

Figure 5.23. Preprocessing a GM pair of data vectors. The upper panel respects the case of 'no karst' (**N**), and the lower one shows the case of 'karst' presence (**K**) (after Eppelbaum *et al* (2014), copyright Azerbaijan National Academy of Science 1995–2022).

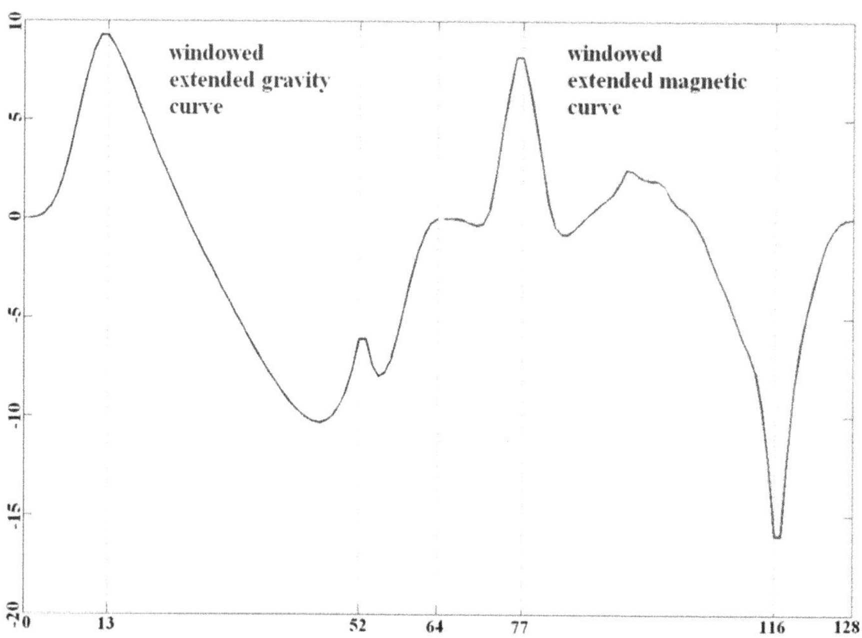

Figure 5.24. Merged combined GM curves from the PGMs without buried karst (after Eppelbaum *et al* (2014), copyright Azerbaijan National Academy of Science 1995–2022).

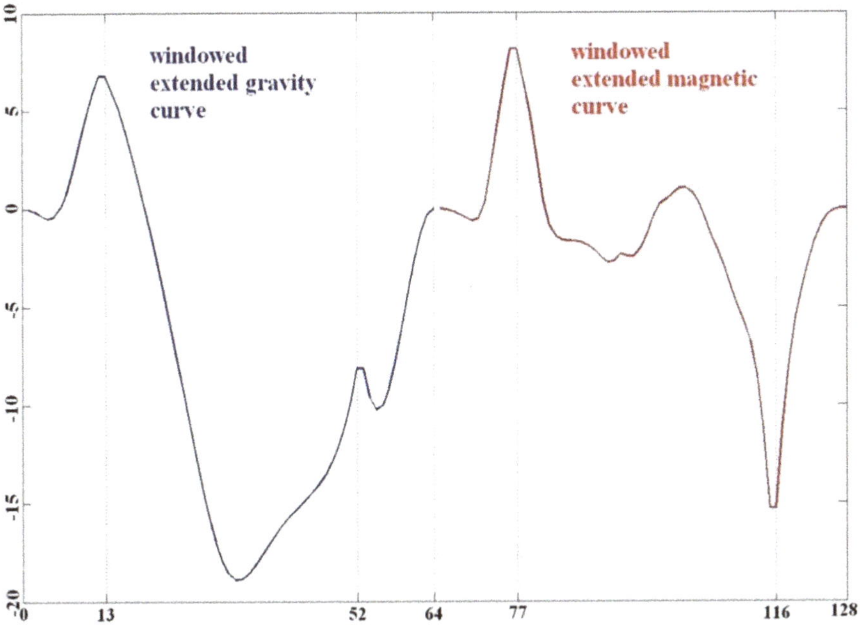

Figure 5.25. Merged combined GM curves from the PGMs with the available buried karst (**K**) (after Eppelbaum *et al* (2014), copyright Azerbaijan National Academy of Science 1995–2022).

The GM curves differ by some subtle geometrical details. To reveal these differences and use them for discrimination of **N** versus **K** sets we employ wavelet packet transforms and dimensionality reduction by diffusion maps (Averbuch *et al* 2014, 2016, Eppelbaum *et al* 2014).

5.3.4 Extraction of characteristic features using wavelet packet transforms

We select to characterize the geometric peculiarities in the merged data via the representation of the curve in the Fourier domain. The assumption is based on that each of the **K** and **N** data sets is characterized by a few dominating frequency bands. The most optimal tool to reveal these characteristic frequency bands gives the wavelet packet (WP) analysis (Averbuch *et al* 2001, 2010a). Once implemented, the WP transform of useful signal yields greatly redundant partitions of the frequency domain.

The transform consists of a successive application to a signal (curve) a pair (low-pass/high-pass) of quadrature mirror filters. Let us, **L** and **H** be low-pass and high-pass quadrature mirror filters, respectively. Their magnitude responses divide the investigated frequency domain into two halves. The one-scale WP transform of a signal **f** of length $N = 2^j$ uses the filters **L** and **H**, which is followed by downsampling by a factor of 2. As a result, we get two coefficient arrays **Lf** and **Hf** of length **N/2** each. The transform is an invertible one. The array **Lf** contains all necessary information needed to restore the low-frequency component of the signal **f**, while the high-frequency component can be restored from **Hf**. Since **L** and **H** are quadrature

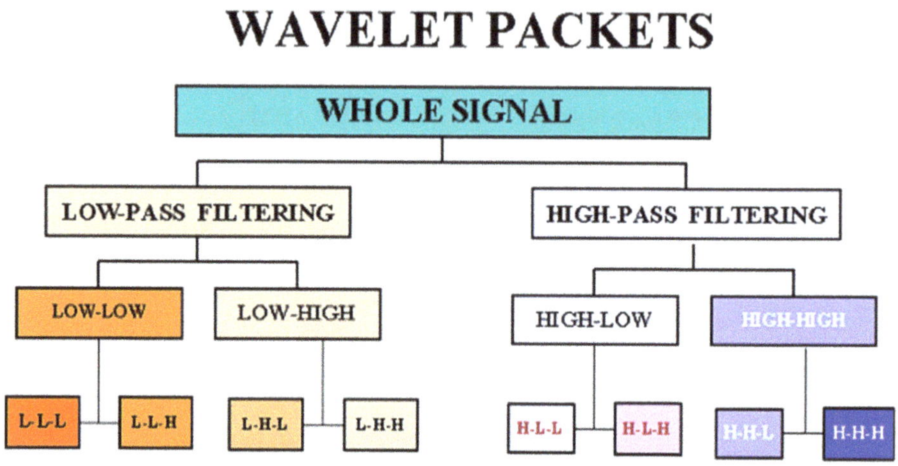

Figure 5.26. Typical diagram of WP transform (after Eppelbaum *et al* (2014), copyright Azerbaijan National Academy of Science 1995–2022).

mirror filters, the energy (sum of squares) in the coefficient arrays **Lf** and **Hf** are equal to the energies in the low—and high-frequency components of the signal **f**, respectively.

The second-scale transform consists of the application of the filters **L** and **H** to either of the arrays **Lf** and **Hf** and subsequent downsampling. As a result, we get four coefficient arrays **LLf**, **LHf**, **HLf**, and **HHf** of length $N/4$ each, relating to quarters 3 of the frequency domain. Similarly, to pass to the next scale, the filters **L** and **H** are applied to either of the four second-scale arrays. After downsampling, eight subarrays are produced, and so on. Figure 5.26 presents the diagram of the WP transform. Each block of the WP coefficients is related to a certain frequency band. Blocks from different scales correspond to frequency bands of different widths. Here it is necessary to note that the set of energies within the coefficients blocks produces a 'spectrum' of frequency bands of the desired signal rather than the spectrum of frequencies.

Extraction of the characteristic features of the **N** and **K** sets are by the following way. The data curves from **N** ('no cavity') and **K** ('karst cavity') are gathered into two matrices $\mathbf{M}_N$ and $\mathbf{M}_K$, respectively. Both matrices are of size 30 ×128. Then we applied the WP transform down to a fourth decomposition scale on each row of the matrices $\mathbf{M}_N$ and $\mathbf{M}_\mathrm{K}$. This step includes the calculation of energies (sums of squares) of each of 31 blocks of the transform coefficients. These energies are gathered into matrices $\mathbf{E}_N$ and $\mathbf{E}_K$ of size 30 × 31, respectively.

Then, we average the energy matrices $\mathbf{E}_N$ and $\mathbf{E}_K$ along with the columns. We regard two vectors $\overline{E}_N = \text{mean}(E_N)$ and $\overline{E}_K = \text{mean}(E_K)$ of length 31 as signatures of the classes **N** and **K**, respectively. The difference $D_{KN} = \overline{E}_K - \overline{E}_N$ of the signature vectors is calculated. The difference testifies to how different blocks of the transform coefficients discriminate between the two data sets. The next step is to select C (8–12) blocks, which best discriminate between two data sets. Let $\mathbf{I} = i_1,\ldots,i_k$ be a list of

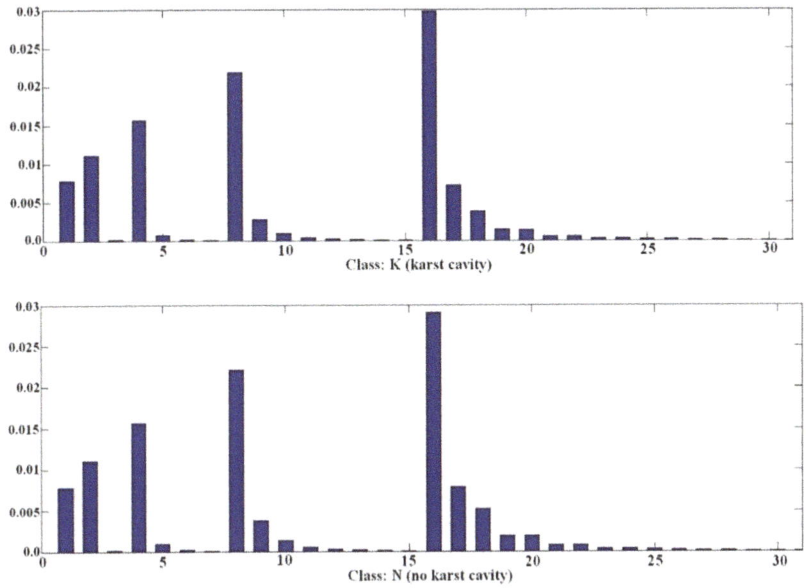

Figure 5.27. Signature vectors $\overline{E}_K$ and $\overline{E}_N$ of the classes **K** and **N** (after Eppelbaum *et al* (2014), copyright Azerbaijan National Academy of Science 1995–2022).

indices of the best discriminating blocks. And finally, we reduce the energy matrices $\mathbf{E}_N$ and $\mathbf{E}_K$ to the feature matrices $\mathbf{F}_N$ and $\mathbf{F}_K$ of size $30 \times C$ by discarding all the columns, whose indices are not included in the list **I**. Thereby, each merged data curve is represented by C features.

Figures 5.27 and 5.28 display the signature vectors $\overline{E}_K$ and $\overline{E}_N$, and figure their difference $\mathbf{D}_{KN}$, respectively. This difference, taking into consideration low values of anomalous GM effects is sufficiently small but quite detectable.

5.3.5 Reduction of dimensionality and obtained results

Reducing the data sets by embedding feature vectors into the 3D Euclidean space is intended to employ the so-called diffusion map. This map allows us to reveal the internal structure of the data sets **K** and **N** and to reliably separate them. For this aim, a matrix of the diffusion distances for the combined feature matrix $\mathbf{F} = \mathbf{F}_N \cup \mathbf{F}_C$ of size $60 \times C$ is constructed (Coifman and Lafon 2006, Averbuch *et al* 2010b). Then, each row of the matrices $\mathbf{F}_N$ and $\mathbf{F}_C$ is projected onto three first eigenvectors of the matrix **D(F)**. As a result, each data curve is represented by a 3D point in the Euclidean space formed by eigenvectors of **D(F)**. The Euclidean distances between these 3D points are necessary to show the similarity of the data curves. Figure 5.29 displays scattered projections of the data curves onto the diffusion eigenvectors. As a result of the aforementioned operations, we embedded the original data into 3D space where data set related to the **K** (karst cavity) are well separated from the **N** (non-cavity) data set. This 3D set of the data representatives will be used as a reference set for the estimation (classification) of newly arriving data. It is easy to see

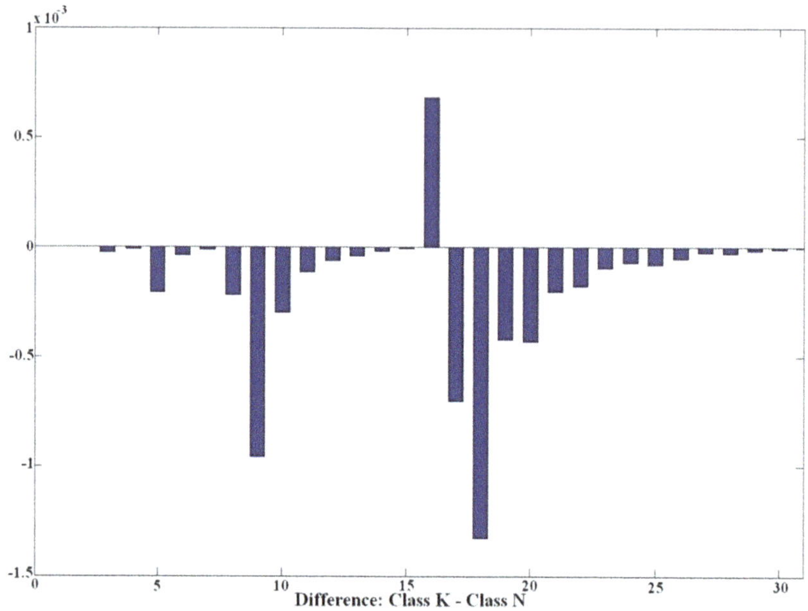

Figure 5.28. The difference $\mathbf{D}_{KN}$ for the classes **K** and **N** (after Eppelbaum *et al* (2014), copyright Azerbaijan National Academy of Science 1995–2022).

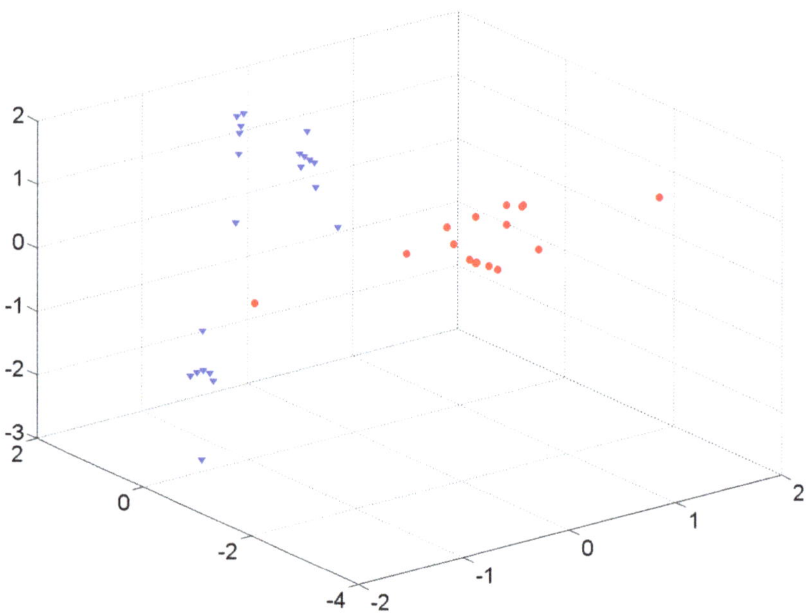

Figure 5.29. Scattered projections of the data curves onto the diffusion eigenvectors. Blue triangles and red circles represent the **N** data curves (non-karst areas) and **K** (karst areas), respectively (after Eppelbaum *et al* (2014)), copyright Azerbaijan National Academy of Science 1995–2022).

from figure 5.29 that the score of blue triangles (absence of karst) is unambiguously separated from the score of red circles (presence of karst). Geophysically it means the studied area can be reliably divided into karst-containing and 'free'-from-karst terranes districts.

Thus, we demonstrated the real possibilities for the classification of geophysical anomalies (on the example of series of GM PGMs) to detecting hidden karst terranes in complex physical–geological environments. Such computations could be performed for other geophysical method integrations (for instance, ERT + GPR, ERT + magnetics, VLF + ERT, etc). Some advanced wavelet methodologies were effectively applied for the combined subsurface GPR—ERT detailed examination in the Montagnole testing site, Franch Alps (Alperovich *et al* 2013a, 2013b).

5.4 Compatibility of the different geological–geophysical methods for buried karst localization

5.4.1 A brief preamble

Usually, the number of geophysical methods applied in geological mapping for solving different problems do not exceed two or three (Parasnis 1997, Telford *et al* 1993, Khesin *et al* 1996, Kearey *et al* 2002, Eppelbaum and Khesin 2012, Dentith and Mudge 2014, Eppelbaum 2019).

Let us consider the application of two methods (e.g., gravity and magnetic surveys). They are classified as follows: 1 is the negative field, 2 is the positive field, 3 is the roughly zero field, 4 is the alternating field, and 5 is the high-gradient field (table 5.3).

As follows from this table, four combinations of two parameters can represent four classes of desired targets. The number of possible combinations of two parameters divided into five categories is 25 (it follows from the elementary combinatorics). The geological features (each class) are characterized by one of these 25 available combinations. Here is necessary to underline, that it is easy to assume a mistake: e.g., instead of 'marl' with the characteristic of '13' may be selected 'gabbroid' with the characteristic '22'.

Table 5.3. An example of combined interpretation for two gravity and magnetic methods.

Typical geophysical field combination										
Magnetic					Gravity					Geological class of targets
1	2	3	4	5	1	2	3	4	5	
*							*			Marl
	*					*				Gabbroid
*					*					Salt
			*						*	Fracture zone

The number of applied combinations can be increased by increasing secondary features (downward and upward continuation, second and third horizontal and vertical derivatives, different transformations, etc).

It is well known that modern geophysical–environmental monitoring depends on the effectiveness of combined geophysical multilevel observations (underground, surface investigations, and remote sensing techniques) (e.g., Proto *et al* 2010). In turn, geophysical monitoring depends on the efficiency of applied algorithms to integrated analysis and revealing relationships between the employed fields and their associations with geological and environmental features.

One of the most important stages is the formulation of a problem 'what kind of new information can be derived from geophysical monitoring'. The generally available geological problem should comply with: (1) measuring capabilities of the selected geophysical method(s), and its applicability for measuring the field containing the information required; (2) physical properties of the medium under study, and their ability to generate detectable geophysical signals (anomalies); (3) methods for data processing and interpretation, namely, their possibility to extract information from the measured field containing the effects of the desired geological (or environmental) targets. A geological (environmental or technical) problem is considered to be well defined and solvable when all three conditions are satisfied (Khesin and Eppelbaum 1997).

Since the solution of geological problems calls for the integration of different geophysical methods, the observation values, petrophysical–morphological description, and interpretation data should be expressed in the same terms. Only after this operation can some concrete particular problems be formulated and classified.

In some cases, only the comparison of fields measured by various geophysical methods leads to an 'informational jump' (Khesin *et al* 1996), which results in revealing the desired targets. Sometimes, only numerical calculation of the amount of the corresponding information contained in observations may substantiate the presence of an object of the desired class (e.g., Svetov 1992, Khesin *et al* 1996). These estimations are especially convenient for the combined analysis, as they provide means for operating the data obtained by even principally different methods and expressed in the common (informational) unit. The target is singled out by maximum or minimum information values (depending on the concrete PGM) obtained from the integration of the geophysical methods.

Geophysical data processing is mainly intended to reduce and eliminate noises. However, it should be noted that the solutions to these problems are closely intermingled and mainly based on the interpreter's model conceptions.

5.4.2 Formalization of geological–geophysical process

All types of geological–geophysical investigations oriented to localization of the buried karst phenomena are applied in a definite succession in time and space. For this study, different geoscience means are employed (geophysical investigations, geological mapping, geochemical analyses, drilling, excavations, etc). The final aim of the application of geosciences is the best recognition (the most accurate and

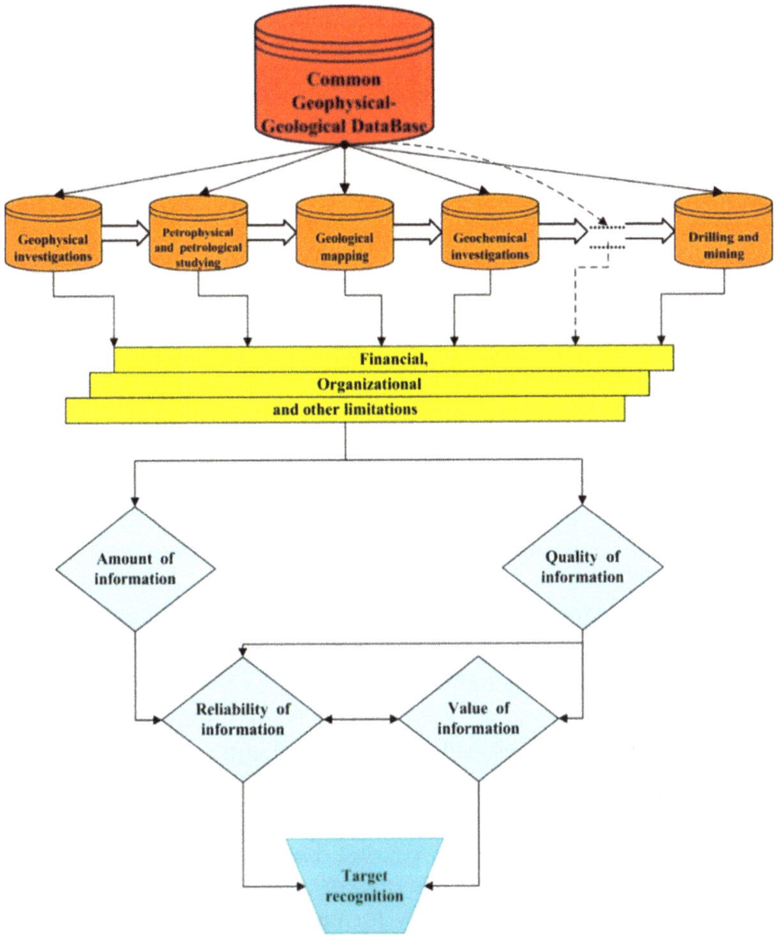

Figure 5.30. Tools to estimate combined geophysical–geological information (revised after Eppelbaum *et al* (2003a), copyright 2003 edited and published by Japan Society of Geoinformatics Production services Kasuga Co. Ltd).

low-cost identification of targets) in the studied area accompanied by several restrictions presented in figure 5.30.

The problem of geological means (tools) rational integration can be solved based on the following criteria:

(1) necessary financial expenditures for the realization of the geoscience integration (cost criterion *C*);
(2) necessary time for the realization of the geoscience integration (time criterion *T*);
(3) informativeness of the geoscience integration (informational criterion Π).

Criterions C and T can be easily determined by a direct calculation whereas estimation of the criterion Π is a complex research problem (Eppelbaum *et al* 2003a).

A simplified algorithm can be written as:

$$\Omega = \Gamma \bigcup C \bigcup T, \tag{5.5}$$

where $\cup$ is the symbol of unification.

All available geological/environmental and geophysical information can be presented in the classic three-level variant: (a) syntactical—number of information; (b) semantic—the substance of information, and (c) pragmatic—the value of information (Eppelbaum *et al* 2003a).

A basic logical-heuristic model of the geoscience information can be presented in the following form (Eppelbaum *et al* 2003a):

$$\Gamma = Q \bigcup R \bigcup V, \tag{5.6}$$

where Q is the quantitative estimation of information, R is the estimation of informational reliability according to the semantic criterion, V is the estimation of an informational value corresponding to the pragmatic criterion, and $\cup$ is the symbol of unification.

This algorithm contains the fundamental terms of the theory of information and is integrated with the structural (hierarchical) approach. This approach enables constructing each geological–geophysical indicator as a feature reflecting a set of typical situations. After this stage, the depth of recognition searching is estimated and calculated using the developed informational approach (Eppelbaum 2020). Realization of the proposed strategy provides quantitative calculation and effective monitoring of geophysical–environmental studies.

In infield conditions, many probabilistic factors affect the numerical data obtained by a set of geophysical tools (means). One of the essential problems is the impossibility to obtain a satisfactory formalized description of various factors influencing to results of local determinations. Similar situations are known in the theory of 'decision-making' where often a full mathematical formalization of the research problem is complicated. Our experience enables us to propose that the application of expert conclusions in many situations will be more optimal.

We consider the reliability of geophysical prospecting at the level of local determination and examine the reliability of information obtained by a separate mean or set of means (Eppelbaum *et al* 2003a, Eppelbaum 2020). The main aim of such a study is a problem of determination of the set of geophysical means by assumed reliability of these means (in contrast to Khesin and Eppelbaum (1997), where mainly statistical criteria were analysed). Solving this problem will enable us to find the most optimal integrations of geophysical–geological means (tools) under different physical–geological conditions. Determination of reliability of separate means may be obtained by results of repeated observations or may be derived from the expert estimations.

A flow-chart of integrated geophysical–geological comparative study is shown in figure 5.31.

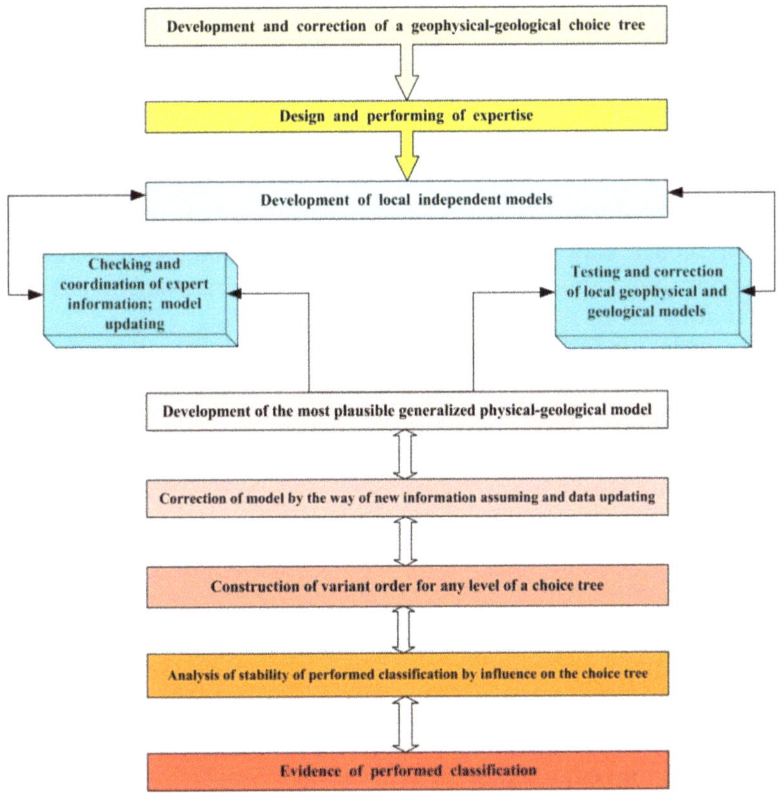

Figure 5.31. Flow-chart of geological investigation process using comparative informational analysis (revised after Eppelbaum *et al* (2003a), copyright 2003 edited and published by Japan Society of Geoinformatics Production services Kasuga Co. Ltd).

5.4.3 Multimodel approach to geophysical data analysis (on the example of magnetic data examination)

The multimodel approach to geophysical data analysis may be illustrated in the example of detailed magnetic prospecting. Quantitative analysis of magnetic anomalies (as well as other geophysical methods) is traditionally oriented to a sole interpretation model of hidden target identification. If there are several hypotheses relating to parameters of the anomalous body usually only one model is distinguished describing the target in the domain Ω_x of k-dimensional space of the physical–environmental factors. However, as a rule, many geological and environmental features are damaged by human, industrial and geodynamic activity as well as different geological/environmental processes.

Additional noise affecting geophysical analysis is caused by the influence of rugged terrain relief, oblique polarization of geological objects, and heterogeneous host medium. As a consequence, the response function Γ_t—geophysical field—may ambiguously represent the anomalous target. Therefore, domain Ω_x may be divided into several subdomains Ω_1, Ω_2, ..., Ω_m, in each of which a single model will dominate (Eppelbaum 2010). Then we could develop m physical–environmental models of the same object, and each will correct for separate subdomains Ω_1, Ω_2, …, Ω_m.

For instance, the remnants of an ancient wall in definite geological conditions may be recognized as a thick bed (magnetic field), *sphere* (gravity field), and *horizontal plate* (resistivity). The multimodel approach may also be realized based on various levels of geophysical field registration. As a result, different interpretation (explanation) models may be used in the process of quantitative analysis. Combining several response functions Γ_l, we obtain a more accurate and reliable physical–environmental model of the buried target.

A simple model presented in (Eppelbaum 2010) exemplifies the utilization of two different interpretations of the same hidden target by carrying out a magnetic survey at two different levels (0.1 and 3.0 m over the Earth's surface, respectively). Indeed, from the magnetic survey at the 0.1 m levels, it is a typical thick bed (TKB) model, and at the 3.0 m level, the anomalous body may be interpreted as a horizontal circular cylinder (HCC). Results of the TKB model interpretation were used to localize a center of the upper edge of the anomalous body, and the HCC model was applied for the determination of a center of the HCC (Eppelbaum 2010). Integrating these two models (we have two response functions Γ_1 and Γ_2 from the subdomains Ω_1 and Ω_2, respectively), we can develop a generalized model of the anomalous body.

As a common note, combining different geophysical methods with their multi-level observations (including different depth of electrodes or geophones grounding) (Eppelbaum 2008, 2016) we could receive a great number of combinations, which may be used for extracting additional useful information from observed geophysical data.

5.4.4 On the correlation between the useful signal and noise

Usually in geophysical observations, it is supposed that the noise is normally distributed and not correlated with the geophysical signal. Analysis of various geophysical fields indicates that these assumptions often do not clamp (Eppelbaum and Mishne 1988, 2011). For example, a principal scheme of correlating a useful signal with delaying wide-band noise was considered in Bashirov *et al* (1992).

Let us consider a simple case. Sometimes a correlation between the useful signal and noise dispersion is observed. In this case, a cross-correlation function B_{sn} between the useful signal and noise may be presented as

$$B_{\text{sn}}(m) = E[(S_m - E(S_m))(n_m - E(n_m))], \tag{5.7}$$

where E is the statistical expectation, S_m and n_m are the expected useful signal and noise, respectively.

Equation (5.7) may be transformed as

$$B_{\text{sn}}(m) = E[S_m(n_m - E(n_m))]. \tag{5.8}$$

Considering the properties of dispersion as a random quantity (e.g., Triola 2007), equation (5.7) may be rewritten in the following form

$$B_{sn}(m) = E\left[S_m\left(n_m - E(n_m) \pm \sqrt{D_n^2}\right)\right], \tag{5.9}$$

where D_n is the dispersion.

It is taken into account that

$$E\left[S_m\left(n_m - E(n_m) - \sqrt{D_n^2}\right)\right] = 0.$$

Equation (5.7) may be represented in the following form

$$B_{sn}(m) = E\left(S_m\sqrt{D_n^2}\right). \tag{5.10}$$

Thus, equation (5.10) describes the cross-correlation function between the useful signal and noise dispersion. Such a relationship permits the utilization of these peculiarities in the corresponding models of geophysical observations.

5.4.5 Geophysical data transformation as an effective tool for delineation of desired targets

Sometimes, for removing the geological disturbances expressed as quasi-white noise, it is optimal to convert observed geophysical fields to some informational measures. Eppelbaum *et al* (2003b) proposed to employ the following expression:

$$J_i = U_i\left(\sum_i^n U_i\right)^{-1} \lg\left[U_i\left(\sum_i^n U_i\right)^{-1}\right]K, \tag{5.11}$$

where U_i is the geophysical observation at ith point ($U_i > 0$) at the area under study, n is the total number of observations, and K is some coefficient (K application (integer) is necessary for improving the developed J_i map visualization).

This algorithm has been successfully applied in large, medium, and small scales for the processing of gravity, magnetic, self-potential, and VLF data. An example of gravity field processing in the NHS (western DS shore) by using equation (5.14) is shown in figure 5.32. Here the negative (green) areas indicate the presence of buried sinkholes.

5.4.6 Estimation of integration efficiency by revealing weak anomalies

In some cases, a simple model of the anomalous target/host media can be applied. If geophysical method integration is focused to study some independent features of equal value, the reliability of anomaly detection γ (Eppelbaum *et al* 2001) can be described by an error function (probability integral) (Eppelbaum 2014a) as:

$$\gamma = \mathrm{erf}\left(\frac{\sqrt{\sum_i \upsilon_i}}{2}\right), \tag{5.12}$$

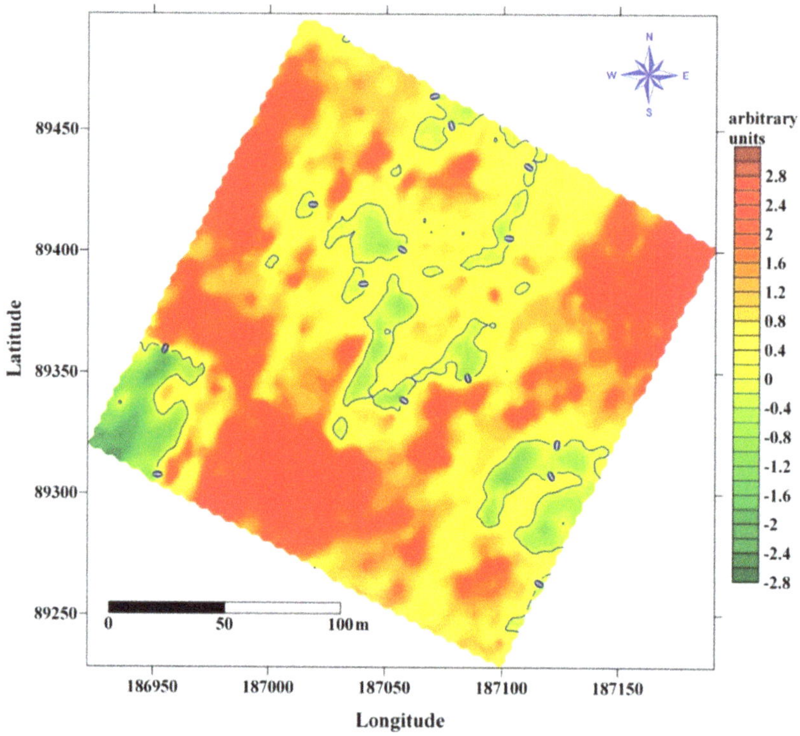

Figure 5.32. Map of the information parameter J_i computed for the NHS site (Israel DS coast) (revised after Eppelbaum *et al* (2008), with permission of EGU, Creative Commons Attribution 3.0 License.

where erf(x) is the error function, and υ_i is the ratio of the anomaly square to the noise dispersion for each ith geophysical field.

Let us assume that the geophysical anomaly is described by three points, and the mean square of the anomaly for each geophysical field is equal to noise dispersion. For a single method, the reliability of the detection of an anomaly of a known form and amplitude is expressed by the known Kotelnikov's criterion ('irrelevant observer') (e.g., Borda 2011) erf($\sqrt{\upsilon_i}/2$). Then, the reliability value for single methods is 0.61 and those of a set of two and three methods are 0.77 and 0.87, respectively (according to equation (5.12)) (Eppelbaum 2014a). This means that the risk of an erroneous solution (q value) at the integration of two or three methods decreases by the factors of 1.7 and 3, respectively.

Eppelbaum (2020) has utilized a solution of the well-known cartographic 'four colors' problem for selecting a set of two independent geophysical methods as more optimal numbers (geophysical methods are *a priori* assumed as equal ones). However, from the theoretical and practical point of view, the employment of different geophysical methods has no equal value.

A comparison of the risk versus expenditures enables one to find an optimum set of methods. For instance, in our case—*localization of the hidden karst funnels*—obviously ERT method prevails on the self-potential method application. In this situation, we need to apply a more complex expression for a calculation of optimal geophysical method integration (with an *a priori* assumed percent of geophysical method 'usefulness') (Eppelbaum 2014b).

5.4.7 Computing integrated geophysical and related parameters

To reveal geophysical anomalies observed in noisy conditions, different correlation techniques are employed, and various kinds of filters. In self-setting filtering, the anomaly parameters are not specified *a priori* but determined one way or another during the data processing. The areas, where geophysical anomalies obtained by different methods form identical combinations, are joined by common contours. After that, certain geological content is assigned to each selected contour. This method usually provides an enormous amount of dispassionate information. Nevertheless, being based on such delineation, it has definite limitations. Some anomalies revealed by this technique can be of no interest as far as predicting and prospecting problems are troubled.

Among the automatic classification methods, two large groups may be distinguished: (1) hierarchical, and (2) cluster techniques. They indicate inter-group and intra-group indicators of similarity, respectively. As a similarity measure, one can take a distance in a multi-dimensional space, a cosine of the angle between the vectors of indicators. This method is known as a general procedure for constructing a differentiating surface by the data on standard images (learning 'with a teacher'). Automatic classification of objects according to their typical indicators leads to forming some opinion as to their value.

The results of the application of the automatic classification methodologies depend to a great extent on the changes in the indicator set, particularly if their number is great. Moreover, the analysis of the content of selected groups is entrenched, as in the case of filtration-correlation techniques.

5.4.8 Revealing of objects (classes of objects) with expected properties

Let us envisage a case, where all physical properties of the targets under study are exactly known. Then, observations and analysis have been conducted without visible errors and it is possible to perform an unambiguous classification. Therefore, a deterministic solution to the problem of target recognition is a special case of a more general probabilistic solution. Employing the aforementioned techniques, except for the statistical ones, we can formulate an unmistakable judgment about the assignment of the desired target to one or another class of geological structures. Theoretical error probability in determining the target's nature is taken to be zero. In this connection, some researchers call such techniques logical (deterministic). However, under definite assumptions to anomalies and various kinds of noise, one can give a resolving rule with an estimate of the combined interpretation reliability.

When it is necessary to recognize, for example, the significant environmental features against the background of noise: location of faults, buried trenches, karst cavities, or rockslides, these above-mentioned techniques prove to be highly promising since they use most mutual indicators for different classes of the desired targets.

In environmental geophysics, it was suggested to employ the ‘reduction’ of geophysical method results, alongside the computation of an integrated index (‘index of contrast’), (‘generalized function’) (Khesin *et al* 1996). The latter is obtained as a sum of anomalies and normalized noise for each indicator under study. In many cases, the application of this logical approach enables obtaining reliable results. The most accurate definition of this approach is the summation or cross-correlation of data obtained by different geophysical methods (Khesin and Eppelbaum 1997).

5.5 Theoretical aspects of geophysical method integration based on information theory

Information theory (Shannon 1948) is the most optimal to express the results of geological–geophysical studies, i.e. the process of acquisition and analyzing the geoscience information. Therefore, despite the probabilistic calculus of the amount of information, the informational approach employs some deterministic features. The logical character of these interpretation methods is based on the information–statistical (logical–statistical) approach (Khesin *et al* 1996). Furthermore, this approach embraces such probabilistic methods as correlation, filtering, and statistical decision-making, which are widely accepted for different kinds of data processing and interpretation.

Since the solution of geological problems (we aim to localize buried karstic cavities) calls for the integration of different geophysical methods, the aforementioned aspects (geophysical observations, petrophysical–morphological properties, and interpretation) should be expressed in the same terms.

Sometimes, only estimation of the amount of the corresponding information contained in geophysical observations may authenticate the presence of the target of the desired class. These estimations are specifically expedient for the combined interpretation, as they provide means for operating data obtained by different methods and expressed in the common (informational) units. The target is singled out by maximum (sometimes minimum, depending on the constructed physical–geological model) information achieved from the geophysical method integration.

The processing of geophysical data is generally envisioned to reduce and eliminate different kinds of noise. The main problem of the qualitative interpretation is to recognize the target of the desired class, while the quantitative interpretation is intended to determine the geometrical parameters of the anomalous target. The solutions to these problems are closely interwoven and strongly associated with the interpreter’s model conceptions.

The most geophysical methods applied for searching and localization of hidden karst funnels are based on the contrast petrophysical properties between the target under study and host media. So, the study of the density, electric, thermal, acoustic, and other properties occupies a significant place in this investigation. Comprehensive analysis of these properties will enable increasing the reliability of combined geophysical monitoring.

5.5.1 Quantitative estimation of integration efficiency

The selection of certain geophysical methods to be integrated and the concrete number of the methods can be quantified. For this purpose reliable information, statistical and other criteria can be used. The informational criterion utilizes the quantity of information obtained by an individual geophysical method or by a set of various methods. The statistical criterion outlines the risk of decision-making following from results of geophysical surveys. We consider the informational criterion to be preferable because geophysical prospecting is a permanent process of acquisition and processing of different kinds of information. Shannon's (1948) and Brillouin's (1962) works have been triggering a wide application of information methodologies in various branches of science. A detailed review of the informational approach application in geophysics was given in (Eppelbaum 2020).

Khesin and Eppelbaum (1997) have indicated that the informational and statistical approaches represent two aspects of the same common approach.

Let us assume that there is only one hidden target (sinkhole) that occurs in the area under study. Let us divide this area into N equal cells. Assume further that each cell is characterized by the same probability to contain the sinkhole. Thus, the probability of the target's appearance is equal to/in each cell. Then, the entropy of the experiment β (discovery of the sinkhole) is log N. The entropy is determined using the following formula:

$$H(\beta) = -\sum_{i=1}^{N} P(B_i) \cdot \log_2 P(B_i), \tag{5.13}$$

where $P(B_i)$ is the probability of B_i outcome (B is a multitude of outcomes for the experiment β).

In equation (5.13) $H(\beta)$ value is a maximum of the possible uncertainty. Experiment α (geophysical observations) provides additional information about the hidden target. A value is a multitude of outcomes for the experiment α.

The difference between uncertainties in β results before and after experiment α allowing estimation of the information of α relating to β:

$$I(\alpha, \beta) = H(\beta) - H(\beta \mid \alpha), \tag{5.14}$$

where $H(\beta(\alpha))$ is the conditional entropy for the experiment β (under the condition that the experiment α has been realized).

Conditional entropy is an average value of random variable taking value $H(\beta_i \mid \alpha_i)$ with the probability of $P(A_i)$:

$$H(\beta_i \mid \alpha_i) = \sum_{i=1}^{N} P(A_i) H(B_i \mid A_i). \tag{5.15}$$

5.5.2 Estimation of information by indicator (field) gradations

Gradations of various indicators can be also applied for obtaining useful information about the existence of near-surface anomalous targets: sinkholes, faults, and contact zones. Any physical field changing, their derivatives or transformations, etc, maybe considered as indicators. Let us denote $P(A_i \mid B)$ *a posteriori* the probability of finding the ith gradation of the indicator A over the targets B, $P(A_i)$ is the prior probability of finding the same gradation in the area under survey. Then the partial information on the presence of target $I_{A_i \to B}$ contained in the registered A_i gradation can be expressed as:

$$I_{Ai \to B} = \log_2 \left[\frac{P(A_i \mid B)}{P(A_i)} \right], \tag{5.16}$$

since the uncertainty in A_i registration before the survey is log $P(A_i)$ and after the survey—$\log_2 P(A_i \mid B)$.

Similarly,

$$I_{Ai \to \overline{B}} = \log_2 \left[\frac{P(A_i \mid \overline{B})}{P(A_i)} \right]. \tag{5.17}$$

Statistically, the probabilities (to be more precise, relative frequencies) are expressed by ratios of areas: S is the total area under study, S_i is the area occupied by the ith interval of isolines A_i, S_p is the total area occupied of the studied targets' roofs (upper edges of targets), S_{pi} is the common area occupied simultaneously by the interval A_i and S_p. In our case, the target roof is the upper edge of the hidden sinkhole.

However, it should take into account that the effect of inclined magnetization (polarization) may distort the vertical projection of the anomalous object's roof to the Earth's surface (e.g., Eppelbaum 2010).

Based on the aforementioned expressions we can write

$$P(A_i) = \frac{S_i}{S}, \quad P(A_i \mid B) = \frac{S_{pi}}{S_p}, \quad P(A_i \mid \overline{B}) = \frac{S_i - S_{pi}}{S - S_p}. \tag{5.18}$$

When the areas are calculated (probability estimates), one may obtain the increment of information contained in A_i gradation:

$$I'_{A \to B} = \sum_i [P(A_i) \cdot I_{A_i \to B}]. \tag{5.19}$$

5.5.3 Estimation of individual method efficiency

When geophysical methods are selected for integration, it is practical to determine the amount of information provided by each method.

Suppose we have a well-investigated area, and 1/50 of it contains a hidden sinkhole. It's known that in the latter part (1/50) applied geophysical field (for instance, gravity) is always negative, while in 49/50 of the area it may be either positive or negative with equal probability. Then we can conclude *a priori* that 2% and 98% of the area contains sinkhole and does not contain it. The result of the experiment β will be designated as follows: B—part of this area containing sinkhole, $\overline{B}$—part of the area that does not contain sinkhole. Thus, $P(B) = 0.02$, $P(\overline{B}) = 0.98$. According to expression (5.13), $H(\beta) \cong 0.14$.

The result of the geophysical experiment α will be designated as follows: A is negative gravity field and $\overline{A}$ is positive gravity field. The relative partial entropy (after recording the negative gravity field, associated with the hidden karst terrane) can be calculated the following way:

$$P(A) = 0.51; P(\overline{A}) = 0.49; P(B \mid A) = 2/51,\ P(\overline{B} \mid A) = 1 - P(B \mid A) = \frac{49}{51}.$$

Thus, $H(\beta(\alpha)) \cong 0.24$.

According to the recorded positive gravity field, the district certainly does not contain the sinkhole: $H(B \mid \overline{A}) = 0$. Consequently, partial entropy for the experiment β under the conditions of α realization, following expression (5.15) is: $H(\beta(\alpha)) \cong 0.12$. Thus, the uncertainty of the determination of the hidden karst decreased after the gravity field measurements from 0.14 to 0.12 bit (i.e., $\approx$ on 14.5%).

Pay attention to the cost of each geophysical measurement, we can estimate the effectiveness of the corresponding methods. For this purpose, it is enough to compare the values (costs) of information units supplied by each method. On this basis, a valid solution for the reasonable integration of methods can be made (Eppelbaum 2020).

5.5.4 Estimation of geophysical integration efficiency based on probabilities of errors of first and second kinds

Quantitative estimation of the classification efficiency can be estimated not only for separate methods but also for a set of geophysical methods by calculating the reliability of identification geological targets. The probability of error of the second kind (M_2) is expressed as the relative frequency of an erroneous diagnosis for objects from sampling B (geological targets). The probability error of the first kind (M_1) is expressed as the relative frequency of an erroneous diagnosis for objects from the sampling $\overline{B}$ (the rest of the targets). Values of these errors are employed to determine a total unconditional error of the separation q between B and $\overline{B}$ classes (i.e., the risk of an erroneous solution):

$$q = M_2P(B) + M_1P(\overline{B}), \tag{5.20}$$

where $P(B)$ and $P(\overline{B})$ are prior probabilities of the appearance of the targets of the first and second classes, respectively.

If $P(B) = P(\overline{B}) = 0.5$, then q value corresponds to the intersection area of distribution densities $P(X(B))$ and $P(X \mid \overline{B})$, where X is the separation index. It can represent field amplitude, the value of the complex indicator, and some other parameters. The separation reliability (γ) is:

$$\gamma = 1 - q. \tag{5.21}$$

A total empirical error is compared with a theoretical one. The close values of the errors attest to a correct assumption and high reliability of identification. Using logical-informational methods (Khesin *et al* 1996), the classification reliability is estimated merely by empirical errors. The errors due to referring geophysical data to either class (containing some class of targets or not) can be determined as follows. The absence of anomalies of the complex indicator in a well-investigated area (containing the desired target) is an error of the second kind, or 'omission of target'. The presence of these anomalies in the 'empty' part of the investigated area is an error of the first kind, or 'false alarm' (Khesin and Eppelbaum 1997).

Comparing the results obtained by integration of rapid geophysical methods with the outcomes of more complex geophysical integration gives the possibility to estimate the respective errors and the reliability of classification.

Let us consider one practical example. For 12 prospective areas, their dangerous nature (karst terranes) was confirmed by more complex geophysical integration for 11 of them. New areas without the karst cavities have not been revealed by a new geophysical integration of complex geophysical methods. Thus, $M_1 = 0$, $M_2 = 1/12$. Assuming that $P(B) = P(\overline{B}) = 0.5$, and taking into account expressions (5.20) and (5.21), then $q = 0.04$ and $\gamma = 0.96$. Thus, this kind of estimation also approves the reliability of rapid geophysical method integration.

5.5.5 Computing integrated geophysical and related parameters

Difficulties in the choice of reference targets led to the development of methodologies based on the analysis of observed geophysical fields, anomaly correlation, and differentiation between the obtained groups of analogous combinations.

Let us imagine a situation, where all the physical characteristics of targets under study are known exactly. In this case, observations and analysis of their results have been accomplished without a single error, which gives a possibility to perform an unambiguous segregating into *a priori* selected classes. Therefore, a generalized deterministic approach to the recognition of hidden objects is a separate case of the more common probabilistic one. Employing the aforementioned methodologies, except for the statistical ones, we obtain an unmistakable judgment about the assignment of the target to one or another class. Theoretical error probability in defining the kind of hidden target is taken to be zero. Therefore, some researchers call such methodologies logical (deterministic). At the same time, under certain assumptions both to anomalies and noise, we receive a solving rule with an estimate of the combined interpretation reliability.

When it is necessary to recognize, for example, the significant environmental features against the noise background: karst cavities or rockslides, location of faults, buried trenches, these techniques of geophysical data analysis prove to be highly auspicious since they employ most mutual indicators for different classes of the desired targets.

In environmental geophysics, it was suggested to apply the 'reduction' of petrophysical or geophysical exploration results, alongside the computation of an integrated index ('index of contrast'), and 'generalized function'. The latter is obtained as a sum of anomalies and quasi-normalized distributed noise for each indicator under examination. In many cases, reliable results have been obtained using this logical approach. The most precise quantitative content of this approach is due to the summation or cross-correlation of information obtained by different methods (Eppelbaum and Khesin 2012).

5.5.6 Informational integrated indicator

At each observation point, the amount of geophysical information J_i due to the application of the ith geophysical method is

$$I_i = -\log_2 P_j [\text{bit}] \tag{5.22}$$

or

$$I_i \cong \log_2 \left(\frac{U_i}{\Delta U_i} \right), \tag{5.23}$$

where P_j is the relative frequency of the jth interval of the ith indicator in the histogram of its distribution, U_i and ΔU_i are the amplitude and the error of this indicator's determination, respectively.

In a common case, ΔU_i should be considered as a mean square error (MSE). The MSE might be calculated by different techniques (Bolshakov 1983, Parasnis 1997, Telford *et al* 1993). It should be borne in mind that some researchers recommend employing a double MSE value (e.g., Bolshakov 1983, Khesin *et al* 1996).

It should be underlined that even such simple calculations of information measure can be an effective tool by operating with different physical fields.

After summing up the information elements, different kinds of random noise and components, caused by various geological and environmental features, are suppressed. To avoid recognition of fictitious objects by calculation of $\frac{1}{n}\sum_{i=1}^{n} J_i$ which is possible with a great amount of information contained in only one or two methods, an additional integrated criterion is of essential importance. It allows one to justify different physical values and depends on the number of significant indicators:

$$I_{\text{integr}} = \sum_{k=1}^{\frac{n(n-1)}{2}} \frac{(J_p)_k}{(J_p)_{\max}} [\text{arbitrary units}], \tag{5.24}$$

where J_p is determined using pairwise combinations of results of n geophysical methods (I_1 and I_2 are pairwise geophysical observations from the total number n of applied geophysical methods):

$$I_p = (I_1 + I_2)\frac{I_1}{I_2}, \quad (I_1 \leqslant I_2), \tag{5.25}$$

To avoid missing the deeply located target detection, it is expedient in some situations to use relative frequencies of average values (or average field estimates) on a sliding averaging interval instead of P_j and U_i values.

Along with a simplified version based on summing up informants obtained from equation (5.25), the method has been realized as an 'integration' program enabling the calculation of the sums of informants by equation (5.24). Unlike other methods, J_{integr} allows one to reveal the objects characterized by an unlimited number of indicators of different amplitudes and, at the same, to avoid missing a target, which for some reasons did not manifest itself by any geophysical indicator. Such a combination of indicators permits making significant interpretative conclusions.

In geophysical practice, J_i is usually replaced by the relative amount of information, called also a coefficient of informativity (Eppelbaum and Khesin 2012)

$$K_i = \frac{J_i}{\overline{J_i}}. \tag{5.26}$$

The value $\overline{J_i}$ determines the information obtained when the result of U_j falls into the x_j interval of the histogram at an equal probability of falling into any of the R histogram's intervals. It is equal to average (complete) information contained in the measurements performed by a single method (Ventsel 1983):

$$\overline{J_i} = \log_2 R. \tag{5.27}$$

The application of parameter K_i enables one to consider differences in the ranges of different geophysical fields. However, the application of expressions (5.22)–(5.27) maybe not be effective for the cases of insufficient numbers of geophysical observations.

References

Alperovich L, Averbuch A, Eppelbaum L and Zheludev V 2013a Delineation of karst terranes in complex environments: application of modern developments in the wavelet theory and data mining *Transactions of the 9th EUG Meeting* vol 15 (Vienna: Geophysical Research Abstracts), 1–2 EGU2013-11063

Alperovich L, Eppelbaum L, Zheludev V, Dumoulin J, Soldovieri F, Proto M, Bavusi M and Loperte A 2013b A new combined wavelet methodology applied to GPR and ERT data in the Montagnole experiment (French Alps) *J. Geophys. Eng.* **10** 025017

Alperovich L, Zheludev V and Hayakawa M 2001 Application of a wavelet technique to detection of earthquake signatures in the geomagnetic field *J. Nat. Hazards Earth Syst. Sci.* **1** 75–81

Alperovich L, Zheludev V and Hayakawa M 2003 Use of wavelet analysis for detection of seismogenic ULF emissions *Radio Sci.* **38** 1093

Arkin Y and Gilat A 2000 Dead Sea sinkholes—an ever-developing hazard *Environ. Geol* **39** 711–22

Averbuch A Z, Hulata E, Zheludev V A and Kozlov I 2001 A wavelet packet algorithm for classification and detection of moving vehicles *Multidimension. Syst. Signal Process.* **12** 9–31

Averbuch A Z, Neittaanmäki P and Zheludev V A 2014 *Spline and Spline Wavelet Methods with Applications to Signal and Image Processing. Vol I: Periodic Splines* (Berlin: Springer)

Averbuch A Z, Neittaanmäki P and Zheludev V A 2016 *Spline and Spline Wavelet Methods with Applications to Signal and Image Processing. Vol II: Non-periodic Splines* (Berlin: Springer)

Averbuch A Z, Neittaanmäki P and Zheludev V A 2019 *Spline and Spline Wavelet Methods with Applications to Signal and Image Processing. Vol III: Selected Topics* (Berlin: Springer)

Averbuch A, Zheludev V, Neittaanm P and Koren J 2010a Block based deconvolution algorithm using spline wavelet packets *J. Math. Imaging Vis.* **38** 197–225

Averbuch A, Hochman K, Rabin N, Schclar A and Zheludev V 2010b A diffusion frame-work for detection of moving vehicles *Digital Signal Process.* **20** 111–22

Bashirov A E, Eppelbaum L V and Mishne L R 1992 Improving Eötvös corrections by wide-band noise Kalman filtering *Geophys. J. Int.* **108** 193–97

Bolshakov V D 1983 *Theory of Observation Mistakes* (Moscow: Nedra) (in Russian)

Borda M 2011 *Fundamentals in Information Theory and Coding* (Berlin: Springer)

Brillouin L N 1962 *Science and Information Theory* 2nd edn (New York: Academic)

Bruner I, Ezersky M, Keydar S and Trachtman P 2003 An integrated study of sinkhole development sites using geophysical methods *GII Report* No. 728/314/0 3

Butler D K (ed) 2005 *Near-Surface Geophysics(Investigations in Geophysics, No. 13)* (Houston, TX: Society of Exploration Geophysicists)

Chalikakis C, Plagnes V, Guerin R, Valois R and Bosch F R 2011 Contribution of geophysical methods to karst-system exploration: an overview *Hydrol. J.* **19** 1169–80

Coifman R R and Lafon S 2006 Diffusion maps, applied and computational harmonic analysis *Appl. Comput. Harmonic Anal.* **21** 5–30 (Special issue on diffusion maps and wavelets)

Debeglia N, Bitri A and Thierry P 2006 Karst investigations using microgravity and MASW: application to Orléans, France *Near Surf. Geophys.* **4** 215–25

Dentith M and Mudge S 2014 *Geophysics for the Mineral Exploration Geoscientist* (Cambridge: Cambridge University Press)

Dvorkin J P 2008 Yet another Vs equation *Geophysics* **73** E35–9

Duda R O and Hart P E 1973 *Pattern Classification and Scene Analysis* (New York: Wiley)

Eppelbaum L V 2007 Revealing of subterranean karst using modern analysis of potential and quasi-potential fields *Coll. of Papers of the 2007 SAGEEP Conf. (Denver, CO)* 797–810

Eppelbaum L V 2008 Remote-operated vehicle geophysical survey using magnetic and VLF methods: proposed schemes for data processing and interpretation *Collections of Selected Papers of the 2008 SAGEEP Conf.* vol 21 *(Philadelphia, PA)* 938–63

Eppelbaum L V 2010 Archaeological geophysics in Israel: past, present and future *Adv. Geosci.* **24** 45–68

Eppelbaum L V 2014a Four color theorem and applied geophysics *Appl. Math.* **5** 358–66

Eppelbaum L V 2014b Geophysical observations at archaeological sites: estimating informational content *Archaeol. Prospect.* **21** 25–38

Eppelbaum L V 2016 Remote operated vehicles geophysical surveys in air, land (underground) and submarine archaeology: general peculiarities of processing and interpretation *Transactions of the 12th EUG Meeting* vol 18 (Vienna: Geophysical Research Abstracts), 1–7 EGU2016-10055

Eppelbaum L V 2017 Delineating buried karst terranes on the basis of modern wavelet theory and data mining *Proc. of the 2017 SAGEEP Conf.* vol 30 *(Denver, COA)* 343–46

Eppelbaum L V 2019 *Geophysical Potential Fields: Geological and Environmental Applications* (Amsterdam: Elsevier)

Eppelbaum L 2020 Theories of probability, information and graphs in applied geophysics ed K Kyamakya *Prime Archives in Applied Mathematics* (Hyderabad: Vide Leaf), 1–35

Eppelbaum L V and Khesin B E 2012 *Geophysical Studies in the Caucasus* (Berlin: Springer)

Eppelbaum L V and Mishne L R 1988 Some theoretical aspects of optimal filtration of geophysical signals with dependent noises (in Russian) *Deposited by VINITI, USSR Academy of Sciences* **No. 6034-B88** 10

Eppelbaum L V, Khesin B E and Itkis S E 2001 Prompt magnetic investigations of archaeological remains in areas of infrastructure development: Israeli experience *Archaeol. Prospect.* **8** 163–85

Eppelbaum L, Eppelbaum V and Ben-Avraham Z 2003a Formalization and estimation of integrated geological investigations: informational approach *Geoinformatics* **14** 233–40

Eppelbaum L, Ben-Avraham Z and Itkis S 2003b Ancient Roman remains in Israel provide a challenge for physical–archaeological modeling techniques *First Break* **21** 51–61
Eppelbaum L V, Ezersky M G, Al-Zoubi A S, Goldshmidt V I and Legchenko A 2008 Study of the factors affecting the karst volume assessment in the Dead Sea sinkhole problem using microgravity field analysis and 3D modeling *Adv. GeoSci.* **19** 97–115
Eppelbaum L V, Alperovich L, Zheludev V and Pechersky A 2011 Application of informational and wavelet approaches for integrated processing of geophysical data in complex environments *Proc. of the 2011 SAGEEP Conf.* vol 24 *(Charleston, SC)* 24–60
Eppelbaum L V, Alperovich L S, Zheludev V, Ezersky M, Al-Zoubi A and Levi E 2012 Integration of potential and quasipotential geophysical fields and GPR data for delineation of buried karst terranes in complex environments *Transactions of the 8th EUG Meeting* vol 14 (Vienna: Geophysical Research Abstracts), 1–3 EGU2012-3418
Eppelbaum L V, Zheludev V and Averbuch A 2014 Diffusion maps as a powerful tool for integrated geophysical field analysis to detecting hidden karst terranes *Izvestiya Acad. Sci. Azerb. Rep., Ser.: Earth Sci.* **1–2** 36–46
Ezersky M 2003a,b Study of the shallow subsurface sinkhole hazard areas along Dead Sea shore using geophysical methods: part I. Results and Part II. Interpretation *G.I.I. Reports* No. 211/247/02-1,2
Ezersky 2006 The geophysical properties of the Dead Sea salt in application to sinkhole problem *J. Appl. Geophys.* **58** 45–58
Ezersky M 2008 Geoelectric structure of the Ein Gedi sinkhole occurrence site at the Dead Sea shore in Israel *J. Appl. Geophys.* **64** 56–69
Ezersky M 2011 Improvement of the seismic refraction methods for mapping of the buried salt layers along the Dead Sea shoreline Lod, Israel *GII Report* No. 211/486/09 24
Ezersky M and Frumkin A 2013 Faults–dissolution front relations and the DS sinkholes problem *Geomorphology* **201** 35–44
Ezersky M G and Goretsky I 2014 Velocity-resistivity versus porosity-permeability inter-relations in Dead Sea salt samples *Eng. Geol.* **183** 96–115
Ezersky M G and Legchenko A 2014 Quantitative assessment of in-situ salt karstification using shear wave velocity, Dead Sea *Geomorphology* **221** 150–63
Ezersky M and Frumkin A 2017 Evaluation and mapping of Dead Sea coastal aquifers salinity using Transient Electromagnetic (TEM) resistivity measurements *Comptes Rendus Geosci.* **349** 1–11
Ezersky M G and Frumkin A 2020 Identification of sinkhole origin using surface geophysical methods, Dead Sea, Israel *Geomorphology* **364** 107225
Ezersky M G and Frumkin A 2021 Subaerial morphology affected by groundwater aggressiveness: sinkhole susceptibility above karstified salt, Dead Sea *Geomorphology* **375** 107525
Ezersky M, Bruner I, Keydar S, Trachtman P and Rybakov M 2006 Integrated study of the sinkhole development site on the western shores of the Dead Sea using geophysical methods: near surface *Geophysics* **4** 335–43
Ezersky M, Legchenko A, Camerlynck C and Al-Zoubi A 2009 Identification of sinkhole development mechanism based on a combined geophysical study in Nahal Hever South area (Dead Sea coast of Israel) *Environ. Geol.* **58** 1123–41
Ezersky M, Legchenko A, Camerlynck C, Al-Zoubi A, Eppelbaum L, Keidar S, Baucher M and Chalikakis K 2010 The Dead Sea sinkhole hazard—new findings based on a multidisciplinary geophysical study . *Z. Geomorphol.* **54** 69–90

Ezersky M, Bodet L, Akkawi E, Al-Zoubi A, Camerlynck C, Dhemaied A and Galibert P-Y 2013a Seismic surface-wave prospecting methods for sinkhole hazard assessment along the Dead Sea shoreline *J. Environ. Eng. Geophys.* **18** 233–53

Ezersky M G, Eppelbaum L V, Al-Zoubi A, Keydar S, Abueladas A-R, Akkawi E and Medvedev B 2013b Geophysical prediction and following development sinkholes in two Dead Sea areas, Israel and Jordan *Environ. Earth Sci.* **70** 1463–78

Ezersky M G, Al-Zoubi A, Eppelbaum L V and Keydar S 2013c Sinkhole hazard assessment of the Dead Sea area in Israel and Jordan: multidisciplinary study *Final Technical Report* MERC_M27-050 142

Ezersky M, Gorstein G and Kalmanovich M 2014 Combination of active and passive MASW with HVSR methods for improving the accuracy and reliability of Vs model (In site response assessment) *Publication No.ES-54-14, Unpublished GII Report* No.080/802/14 (https://earthquake.co.il/docs/Reports/2014_080_802_14.pdf)

Fedi M, Cella F, Quarta T and Villani A V 2010 2D continuous wavelet transform of potential fields due to extended source distributions *Appl. Comput. Harmon. Anal.* **28** 320–37

Frumkin A, Kofman L and Ezersky M 2009 Improvement of the reliability of subsurface void detection, including sinkhole development, at the Dead Sea shore area by means of Ground Penetration Radar (GPR) *Report* No. MNI_ES-36-2008 Hebrew University—TECHNION Research and Development Foundation LTD—Geophysical Institute of Israel

Frumkin A, Ezersky M G, Al-Zoubi A, Akkawi E and Abueladas A-R 2011 The Dead Sea hazard:geophysical assessment of salt dissolution and collapse *Geomorphology* **134** 102–17

Garambois S, Legchenko A, Vincent C and Thibert A 2016 Ground-penetrating radar and surface nuclear magnetic resonance monitoring of an englacial water-filled cavity in the polythermal glacier of Tête Rousse *Geophysics* **81** WA131–46

Goldman M, Rabinovich B, Rabinovich M, Gilad D, Gev I and Schirov M 1994 Applications of the integrated NMR-TDEM method in groundwater exploration in Israel *J. Appl. Geophys.* **31** 27–52

Gorstein M and Ezersky M 2015 Combination of HVSR and MASW methods to obtain shear wave velocity model of subsurface in Israel *IJGE* **1** 20–41

Grinsted A, Moore J C and Jevrejeva S 2004 Application of the cross wavelet transform and wavelet coherence to geophysical time series *Nonlinear Processes Geophys.* **11** 561–6

Holting B 1989 *Hydrogeologie* 3rd edn (Stuttgart: Ferdinand Enke)

Jacob T, Bayer R, Chery J and Le Moigne N 2010 Time-lapse microgravity surveys reveal water storage heterogeneity of a karst aquifer *J. Geophys. Res.* **115** 1–18

Kafri Y and Yechieli Y (ed) 2020 *The Many Facets of Israel's Hydrogeology* (Berlin: Springer)

Kaufmann G, Romanov D and Nielbock R 2011 Cave detection using multiple geophysical methods: Unicorn cave, Harz Mountains, Germany *Geophysics* **76** B71–7

Kearey P, Brooks M and Hill I 2002 *Introduction to Geophysical Exploration* (Oxford: Blackwell Science)

Keydar S, Medvedev B, Ezersky M and Sobolevsky L 2012 Imaging shallow subsurface of Dead Sea area by Common Shot Point stacking and diffraction method using weighted multipath summation *J. Civil Eng. Sci.* **1** 75–9

Khesin B E and Eppelbaum L V 1997 The number of geophysical methods required for target classification: quantitative estimation *Geoinformatics* **8** 31–9

Khesin B E, Alexeyev V V and Eppelbaum L V 1996 *Interpretation of Geophysical Fields in Complicated EnvironmentsModern Approaches in Geophysics* (Dordrecht: Kluwer Academic Publishers)

Kofman L 1994 Use of ground penetrating radar in geotechnical applications *Proc. 12th Conf. of the Israel Association for the Advancement of Mineral Engineering (Eilat)* 158–61

Kumar P and Foufoula-Georgiou E 1997 Wavelet analysis for geophysical applications *Rev. Geophys.* **35** 385–412

Lachassagne P, Baltassat J-M, Legchenko A and de Gramont H M 2005 The links between MRS parameters and the hydrogeological parameters *Near Surf. Geophys.* **3** 259–65

Legchenko A, Ezersky M, Boucher M, Camerlynck C, Al-Zoubi A and Chalikakis K 2008a Pre-existing caverns in salt formations could be the major cause of sinkhole hazards along the coast of the Dead Sea *Geophys. Res. Lett.* **35** 1–5

Legchenko A, Ezersky M, Girard J-F, Baltassat J-M, Camerlynck C and Al-Zoubi A 2008c — Interpretation of the MRS measurements in rocks with high electrical conductivity *J. Appl. Geophys.* **66** 118–27

Legchenko A, Ezersky M, Camerlynck C, Al-Zoubi A and Chalikakis K 2009 Joint use of TEM and MRS method in the complex geological setting *C. R. Geosci.* **341** 908–17

Legchenko A, Descloitres M, Vincent C, Guyard H, Garambois G, Chalikakis K and Ezersky M 2011 Three-dimensional magnetic resonance imaging for groundwater *New J. Phys.* **13** 025022

Levi E, Abueladas A-R, Al-Zoubi A, Akkawi E and Ezersky M 2012 Application of geoelectrical methods in the DS sinkhole problem, Israel and Jordan *Geophys. Res. Abst.* **14** EGU2012–4695

Mochales T, Casas A M, Pueyo E L, Pueyo O, Roman M T, Pocovi A, Soriano M A and Anson D 2008 Detection of underground cavities by combining gravity, magnetic and ground-penetrating radar surveys: a case study from the Zaragoza area, NE Spain *Environ. Geol.* **53** 1067–77

Nielsen M R, Hagensen T F, Chalikakis K and Legchenko A 2011 Comparison of transmissivities from MRS and pumping tests in Denmark *Near Surf Geophys* **9** 211–23

Parasnis D S 1997 *Principles of Applied Geophysics* 5th edn (London: Chapman and Hall) revised and suppl.

Proto M *et al* 2010 Transport infrastructure surveillance and monitoring by electromagnetic sensing: the ISTIMES project *Sensors* **10** 10620–39

Rybakov M 2003 Microgravity survey along the Ein Gedi bypass *GII Report* No. 860/240/0 2

Rybakov M, Goldschmidt V, Fleischer L and Rostein Y 2001 Cave detection and 4-D monitoring: a microgravity case history near the Dead Sea *Lead. Edge* **20** 896–900

Rybakov M, Rotstein Y, Shirman B and Al-Zoubi A 2005 Cave detection near the Dead Sea—a micromagnetic feasibility study *Lead. Edge* **24** 585–90

Serra O 1984 Fundamentals of well-log interpretation *Acquisition of Logging Data* (Amsterdam: Elsevier), p 423

SESAM User Guide 2006 Guidelines for the implementation of the H/V Spectral Ratio technique on ambient vibration measurements, processing and interpretation (http://sesamesim.sourceforge.net/docs/SESAME/SESAME_Users_Guide.html)

Shalev E, Lyakhovsky V and Yechieli Y 2006 Salt dissolution and sinkhole formation along the Dead Sea shore *J. Geophys. Res.* **111** B03102

Shannon C E 1948 A mathematical theory of communication *Bell Sist. Techn.* **27** 379–432

Stephens D B, Hsu K S, Prieksat M A, Ankeny M D, Blanford N, Roth T L, Kelsey J A and Whitworth J R 1998 A comparison of estimated and calculated effective porosity *Hydrol. J.* **6** 156–65

Svetov B S 1992 *Information Theory Basis of Geophysics* (Moscow: Bookseries: Advanced Geophysics. Russian Experience. Issue No.1, Electromagnetic Research Centre)

Telford W M, Geldart L P and Sheriff R E 1993 *Applied Geophysics* (Cambridge: Cambridge University Press)

Triola M F 2007 *Essentials of Statistics* 4th edn (Reading, MA: Addison-Wesley)

Trushkin D V, Shushakov O A and Legchenko A V 1995 Surface NMR applied to an electroconductive medium *Geophys. Prospect.* **43** 623–33

Ventsel E S 1983 *The Probability Theory* 5th edn (Moscow: Nauka), p 534

Vincent C, Thibert E, Gagliardini O, Legchenko A, Gilbert A, Garambois S, Condom T, Baltassat J-M and Girard J-F 2015 Mechanisms of subglacial cavity filling in Tête Rousse glacier *J. Glaciol.* **61** 609–23

Vouillamoz J M, Legchenko A, Albouy Y, Bakalowicz M, Baltassat J M and Al-Fares W 2003 Localization of saturated karst aquifer with magnetic resonance sounding and resistivity imagery *J. Ground Water* **41** 578–87

Waltham T, Bell F G and Culshow M G 2004 *Sinkholes and Subsidence: Karst and Cavernous Rocks in Engineering and Construction* (Berlin: Springer)

Wapenaar K, Ghose R, Toxopeus G and Fokkema J 2005 The wavelet transform as a tool for geophysical data integration *Integr. Comput.-Aided Eng.* **12** 5–23

Yates T T, Si B C, Farrell R E and Pennock D J 2007 Time, location, and scale dependence of soil nitrous oxide emissions, soil water, and temperature using wavelets, cross-wavelets, and wavelet coherency analysis *J. Geophys. Res.* **112** 1–15

Yechieli Y, Abelson M, Bain A, Shtivelman V, Onn C, Wacks D, Bar G, Kalbo R and Lyachovski V 2004 Sinkhole phenomena along the Dead Sea shore line *Unpublished Report* GSI-21/04

Yechieli Y, Abelson M and Baer G 2016 Sinkhole formation and subsidence along the Dead Sea coast, Israel *Hydrogeol. J.* **24** 601–12

IOP Publishing

Applied Geophysics for Karst and Sinkhole Investigation
The Dead Sea and other regions
Michael Ezersky, Lev V Eppelbaum and Anatoly Legchenko

Chapter 6

Discussion, conclusions, and recommendations

6.1 Place of geophysics in the karst problem

The place of geophysical methods in the karst problem has been evaluated by both geologists and geophysicists (e.g., Arandjelović 1969, 1976, Milanović 1979, 2000b, 2005, Milanović *et al* 2019, Savich and Kujundjić 1990, Telford *et al* 1990, Hoover 2003, Chalikakis *et al* 2011, Benson and Yuhr 2016) as well as by normative documentation like ASTM D 6429-99 (2006), ISRM (2004).

Geologists especially note the complicated karst environments (e.g., Gutiérrez *et al* 2008, Chalikakis *et al* 2011, De Waele *et al* 2011, Parise *et al* 2015). Karst areas are characterized by a specific type of morphology formed in the carbonate and evaporitic rocks, mainly under different types of dissolution processes (Benson and Yuhr 2016). The karst areas are the subjects of many geomorphological, geological, hydrogeological, geotechnical, environmental, archaeological, etc explorations. The structure of the karst systems and the related aquifers can be extremely heterogeneous (Bakalowicz 1995, Ford and Williams 2007, White 2007). Conventional geological, hydrogeological, and geomorphological studies provide partial information on the degree of karstification development, but cannot determine the internal structure of the karst system. The knowledge of geometry and the structure of the different parts of a karst system (epikarst, infiltration zones, karst conduits, cavities, etc) involves many uncertainties because of the complex nature of the karst areas.

The basic purpose of geophysical methods is an investigation of those parts of karstified carbonate or evaporite rocks that cannot be observed directly. Exploration boreholes are valuable and not only present unique geological data, but also provide the application of geophysical and other techniques for evaluation of the deeper portions of the karst terranes. However, in many cases, the obtained information characterizes only the borehole vicinity. Results obtained from geological boreholes separated only by several meters, may be quite different, specifically when related to the karst hydrogeological and engineering characteristics. Geophysical methods enable the investigation of a particular volume of the rock mass outside the borehole

doi:10.1088/978-0-7503-3635-2ch6

space and delineate zones containing discontinuities which are always important in the karst investigations (Milanović 2005).

Geophysics aims to map characteristic anomalies of geophysical parameters between the boreholes (when they exist) or to detect such anomalies when the boreholes are absent and then check out them (ISRM 2004).

The geophysical methods include a wide diversity of surface, marine, airborne, and borehole measurements. Several geophysical surveys can be carried out from an aircraft such as electromagnetic (EM), magnetic, thermal imagery, and radiometric measurements. Remote sensing geophysical measurements are cost-effective when coverage of large areas is required, especially when the area is not easily accessible for ground surveys. We mean upon the surface geophysical methods, which represent an extremely powerful set of instruments helping to characterize the complex subsurface geological and hydrological conditions, typically in the range of 1 to 30 m, or so. Surface geophysics have a wide range of applications including engineering, environmental, water, and mineral resources, and even archeological applications which are referred to as applied engineering or environmental geophysics (Benson and Yuhr 2016).

Geophysics in the Karst. Arandjelović (1976) has been handled in such a way as to provide insight into the possible uses to which geophysical methods can be put to solve diverse geological, hydrogeological, engineering–geological, and geotechnical problems in karst. Milanović (2005) stated that no geophysical method has been developed yet for reliably determining the spatial position of underground cavities and water flows, especially in the cases when the ratio of the 'karst occurrence/karst size' is small. The latter conclusion is right for the surface methods because the resolution of geophysical methods decreases with the depth. However, taking into account that a cavity is a geological object affected by different tectonic and erosion processes during its development, the detection depth of the desired object can be increased to the ratio 'karst occurrence/karst size' ~10 (see overview in section 2.1.1) (Benson and Yuhr 1993). Also, the quick development of geophysical technologies such as electrical resistivity tomography (ERT), seismic tomography, surface nuclear magnetic resonance sounding (SNMR), etc, took place at the beginning of the 2000s. A lot of corresponding surveys were carried out during recent decades when various researchers have enriched defined success in this field. There have appeared different geophysical methodologies trying to resolve the problem of underground cavity detection. A second important feature of geophysical surveys is the combination of different methods allowing increase of the reliability and effectiveness of the integrated analysis: gravity, magnetic and ground-penetrating radar (GPR) (Mochales *et al* 2008, Abdullah *et al* 2019), ERT, and multichannel analysis of surface waves (MASW) (Bansan and Anderson 2018, Pasquet and Bodet 2017), GPR, and MRS (Garambois *et al* 2016), gravity, ERT and induced polarization (IP) (Martinez-Moreno *et al* 2018), seismic refraction tomography and ERT (Brixova *et al* 2018). Finally, the application of 3D techniques caused by the development of powerful computers and modern software also enabled the improvement of geophysical methodologies (Legchenko *et al* 2011, Keydar *et al* 2013, Eppelbaum *et al* 2014, Rodriguez *et al* 2014, Intrieri *et al* 2018, Xie *et al* 2018).

Undoubtedly, we need to note the combined techniques linking surface geophysics with borehole tomography (Zhang *et al* 2019).

We have considered geophysical methods in sections 6.2 and 6.4. Note here the main features of new geophysical technologies.

SNMR (named also, magnetic resonance sounding (MRS)). The maximum depth of investigation with SNMR is about 100 m. A karst aquifer may be represented as a water-saturated limestone matrix with a water-filled cavity. In terms of the SNMR method, the matrix is characterized by low water content (approximately 1%) and relatively short relaxation times (100–150 ms), whereas the cavity shows much higher water content as well as a longer relaxation time (>400 ms). Field results demonstrated that in favorable conditions (a low EM noise environment and a relatively shallow, large target) the SNMR method, used with a coincident transmitter/receiver loop can be an effective tool for locating a water-filled karst conduit (Boucher *et al* 2006, Legchenko *et al* 2008a). Apart from this, SNMR allows the monitoring of the water volume variation within the cave for a long time (some years) (Legchenko *et al* 2011, 2014).

GPR is a conventional method that has been used in 2D and 3D modes to detect karstic cavities down to 15–20 m (Mochales *et al* 2008, Pueyo-Anchuela *et al* 2009a, 2009b, Rodriguez *et al* 2014, Ronen *et al* 2019, Hussain *et al* 2020). The penetration depth of the GPR signal depends on antennae frequency and dielectric constant of the underground that in turn depends on humidity and soil salinity. There is a known case of penetration depth of 50 m from the surface in the polythermal Tête Rousse glacier (Garambois *et al* 2016, see section 2.2.1.5 of this book). Such success was reached with the 100 MHz antennae because the dielectric constant ε_r was of very low value (1–3).

ERT found application for cavity detection from the beginning of the 2000s (Park *et al* 2013, Billi *et al* 2016, Bansan and Anderson 2018). The results of the numerical ERT modeling confirm that of the four electrode arrays tested, the resistivity distribution obtained using pole–dipole array gives the best images (Park *et al* 2013). The field survey confirms this conclusion. Bansan and Anderson (2018) evaluate the penetration depth of the ERT method as 30 m at least. Some researchers denote that the use of the ERT method alone does not permit the unique detection of caves (Ezersky *et al* 2000). That is why usually geophysicists use additional methods like GPR, microgravity, MASW, etc.

Microgravity. The success of the microgravity method is caused by the high sensitivity of modern gravimeters allowing calculating residual anomalies and different types of noise that can exceed residual gravity effects by some tenfold (Al-Zoubi *et al* 2013). The microgravity method can be successful when there is a contrast in the density of the cavity filling and its environment. This fact as well as the size of the cavity influence the depth of the cavity detection (Rybakov *et al* 2001, Eppelbaum *et al* 2008). The success of the method is increased by combination with other techniques. Martinez-Moreno *et al* (2018) have used microgravity, ERT, and IP methods; Debeglia *et al* (2006) combined microgravity with MASW method, etc.

Borehole tomography. Zhang *et al* (2019) have used combined technology of seismic and electrical cross-borehole tomographies to predict the covered karst along with the metro trace in the urban territory in Chine down to 50 m.

The method of underground detection using standing waves has been suggested and tested by Kolesnikov and Fedin (2018).

The examples given show the progress in the application of geophysical methodologies in the cavity detection problem. Similar progress has been reached also in other fields of the karst investigation problem.

The geophysical methods can be used during two main stages of the characterization of site conditions. At the initial stage of the surveys, when the main goal is to detect the site anomalous peculiarities, the density of sampling usually increases. At this stage, the employment of the geophysical method has often a qualitative character. Later, during the subsequent stages of investigations (designing and construction), the geophysical data may be applied more quantitatively to determine the specific engineering or hydrological parameters, e.g. P-wave velocity for rippability, calculating elastic constants from the P-wave and S-wave velocities, locating the fresh-seawater interface, and estimating the size and dimensions of the anomalous conditions such as fractures faults and cave systems (Greenhouse *et al* 1995, ISRM 2004).

6.2 Principles of the karst investigations

Our personal experience and review of the numerous publications enable us to formulate three important principles of karst investigations using geophysical methods.

1. The scale of measurements. It is necessary to identify the expected volume of subsurface and to be able to compare the size of the desired features with the resolution of the applied geophysical methods. It means that the scale of the applied geophysical method must be selected following the geometrical parameters of the karstic target under study. It should be taken into account as one of the primary requirements by the planning of geophysical surveys (Benson and Yuhr 1993, Khesin *et al* 1996).
2. Any geophysical measurements, must begin and end with the geological factors and geological laws and patterns. Such measurements are only an aid in helping to visualize the geological conditions that cannot be directly seen. It is always necessary to interpret geophysical data in conjunction with the known geological peculiarities from the outcrops, boreholes, or any other features such as 'benchmarks'. Geophysical measurements can be very effective by allowing interpolation between such sources, or careful extrapolation away from them. However, in every case, the knowledge derived from geophysical measurements must be eventually reconfirmed against the known geological conditions (from McNeill 1980).
3. Finally, the planning of geophysical investigations in the engineering–geological zone and their subsequent interpretation requires the development of corresponding engineering–geological (physical–geological) models (Khesin *et al* 1996). Ogilvy *et al* (1980) proposed that *a priori* geological–geophysical information of the main peculiarities of engineering–geological and hydrogeological conditions in the region under study can be used as a preliminary

model at the first stage of the investigation. The size, shape, and location depth of anomalous targets are of high practical importance. We have considered some theoretical models of the sinkhole formation in section 2.1.2. From the above analysis, follow two significant conclusions:

(1) the void formation at a depth causes as a rule vertical and horizontal fractures and faults at shallow subsurface;
(2) the void formation occurring at some depth, generates the collapse processes propagating during a prolonged time to the earth's surface.

Both these conclusions allow being optimistic with the respect to the possibility to detect the buried void (karst cavity) and the subsidence of the shallow subsurface before the expected collapse will happen.

Moreover, the used geophysical methods have to satisfy the available geological model. We can suggest as an example the seismic refraction general reciprocal method (GRM) technique applied to the mapping of the buried salt edge in the Dead Sea (DS) shores (Ezersky 2006). A heterogeneous refractor consisting of the mixture of the salt and the loose sediments (geological model) exactly corresponds to the GRM technique which is intended to study the inhomogeneous refractor (Palmer 1986). Such a combination gives very sure results repeating from site to site.

Reynolds (2011) suggested applying some principles to increase the reliability of geophysical method employment in the karst problem resolving.

Applied geophysics contains a wide range of effective and powerful tools which, will produce a large volume of useful information subject to their correct use in the right situations. However, all the tools, if misused or abused, will not work effectively. It should be briefly explained how the applied geophysical methods can be employed appropriately and to highlight the advantages and disadvantages of the various techniques (Savich and Kujundjić 1990, Reynolds 2011, Benson and Yuhr 2016).

(1) If geophysical methods are applied to resolve any private problem, for instance, to detect a subsurface cavity, this information should be verified by drilling
(2) Geophysics, as a rule, is part of the engineering–geological project, and thus geophysicists should always present their data in a format understandable to engineers–geologists or the Chief Engineer of the project. It means, if the geophysicists produce results that are solely in terms of the specific geophysical methods (for instance, seismic velocity variations), they need to transform the velocity data into an engineering index (modulus of elasticity, strength, rippibility, etc) with which engineers would be familiar.
(3) Often, geophysical solutions are not unique, and they may be geologically nonsensical. If the geophysical model is not geologically plausible (applicable), then the constructed geophysical model has to be rejected and a new one must be developed which does provide a reasonable geological solution.
(4) In other cases, the geological model proves to be inconsistent with the geophysical interpretation, then it may require the updated geological

information to be re-evaluated. One of the authors of this book for many years was working in the field of engineering–geological designing of high dams. The experience shows that geologists relate to geophysical information with great attention and respect and are always ready to discuss possible alternative decisions.

(5) It is important to understand that the various geophysical methods are based on different principles and are sensitive to the different physical properties of the medium. Therefore, the appropriate techniques must be used for a given type of investigation. For example, gravity methods are sensitive to the density contrasts between the cavities and surrounding rocks. GPR method is sensitive to the dielectric constants and will map well objects with these contrasting parameters, etc. The same can be said concerning the electrical methods sensitive to the electrical resistivities of layered media, etc.

6.3 Conclusions

The great volume of measurements was carried out in the DS shore areas in Israel and Jordan. A lot of geophysical methods were tested, and the enhanced methodologies of the studies were suggested (table 6.1).

1. Application of surface geophysical methods allowed us to: delineate the location of the salt deposits, map groundwater salinity, perform *in situ* and laboratory measurements of the shear and longitudinal wave velocities (*Vp and Vs*) corresponding to different geological formations typical for the DS area, detecting water-saturated caverns and estimating their volumes (see table 6.1).
2. The properties of the DS soils and salts were studied *in situ* and laboratory conditions (tables 4.13 and 4.14, in sections 4.11.3 and 4.11.4, respectively). This additional knowledge may contribute to more accurate numerical modeling of the sinkhole development in the DS coast.
3. Combined analysis of geophysical data suggests that the sinkholes preferentially appear along the salt edge, exposed to brackish groundwater flow. This conclusion was confirmed (with small variations) in all investigated sites both in Israel and Jordan (Ezersky *et al* 2013b).
4. The existence of different mechanisms of sinkhole development points to the complexity of the phenomenon and our geophysical results support some of the models. For instance, the 'salt edge' model of the DS sinkhole formation is based on seismic refraction findings (Ezersky 2006, Frumkin *et al* 2011). Moreover, the seismic reflection method allowed reconciling the 'salt edge' model with the 'structural' one (Ezersky and Frumkin 2013). Also, geophysical methods enabled us to confirm the salt edge model instead of 'piping' that (Ezersky and Frumkin 2020).
5. An optimal integration of geophysical methods significantly decreases the risk of erroneous solution that is shown on the probabilistic-information examples (Ezersky and Frumkin 2020).

Table 6.1. Results of geophysical studies in the DS sinkhole problem (Ezersky *et al* (2017), with additions).

Method	Main results	References
SRFR	1. Sinkholes are formed along the salt edge (dissolution front). 2. Criterion of salt was substantiated. 3. The *in situ* *P*-wave velocity (*Vp*) varies between 2900 m s^{-1} (rarely, 2850 m s^{-1}) to 4250 m s^{-1}. The zone of compressional velocity Vp_{min} = 2900 m s^{-1} is the lower limit of the salt layer. Salt edge was delineated along western DS shore and in Ghor Al-Haditha in Jordan. 4. The distribution of sinkhole sites along with the salt layer was shown. 5. Model of pre-existing sinkholes was suggested.	Ezersky (2006), Ezersky *et al* (2007, 2010), Legchenko *et al* (2008a)
MASW	1. Search of underground voids in soil. 2. Measurements of shear wave velocities (*Vs*) *in situ* salt layers; *in situ* salt *Vs* are varied between 750–800 m s^{-1} and 1800 m s^{-1}. 3. Salt classification based on *Vs* was suggested. 4. Salt layer parameters in the Ghor Al-Haditha were calculated based on MASW field data.	Bodet *et al* (2010), Ezersky *et al* (2013a, 2021), Ezersky and Goretsky (2014), Ezersky and Legchenko (2014)
SRFL_P	1. 2D, 3D improvement of fault detection using multipass summation technique. 2. Suggested combined model of sinkhole formation based on faults-dissolution front relations.	Keydar *et al* (2012, 2013), Ezersky and Frumkin (2013)
SDFR	1. Investigated diffraction anomalies in the DS sinkhole sites. 2. Combination of SDFR with SRFL allowed sure detection of concealed faults.	Keydar *et al* (2010, 2011)
Transient electromagnetic (TEM, TDEM)	1. Localization of brackish-salt water interface. 2. Evaluation of DS water salinity based on bulk resistivity, mapping of sinkhole development sites; estimation of salt layer thickness. 3. Suggested methodology of evaluation of groundwater aggressiveness concerning salt.	Ezersky *et al* (2011), Frumkin *et al* (2011), Ezersky and Legchenko (2015), Ezersky and Frumkin (2017, 2021)

(*Continued*)

Table 6.1. (*Continued*)

Method	Main results	References
Electric resistivity tomography (ERT)	1. Study of resistivity structure of subsurface. 2. Detection of sinkhole sites located on high resistivity alluvium.	Al-Zoubi *et al* (2007), Ezersky (2008), Frumkin *et al* (2011)
Surface nuclear magnetic resonance (SNMR) + 3D-SNMR (MRT) (Former name MRS)	1. Determination of water content (*Wi*), relaxation time (T_1), and hydraulic conductivity (*K*) of the subsurface, evaluation of water-saturated caves volume. In practice, a karst aquifer generally is composed of karst channels and caverns. In the DS conditions, a long apparent relaxation time T_{1ap}>450 ms is an unambiguous signature of karst. The signal generated by water in a porous medium is characterized by a shorter relaxation time T_1 < 400 ms. 2. It has been shown that SNMR can be used jointly with the TEM method because the equivalence of resistivity structure does not affect SNMR results. At the same time, TEM allows correction of the SNMR inversion	Legchenko *et al* (2004, 2007, 2008a, 2008b, 2008c, 2011), Ezersky *et al* (2010)
SNMR + TEM	Investigation of subsurface lithology: sandy gravel and silty sediments were identified.	Legchenko *et al* (2009)
SNMR+MASW	Evaluation of the soil hydraulic conductivity (*K*) based on the correlation between *Vs* and *K*. Hydraulic conductivity (*K*) in the same zones varies between about 10^{-4} m s^{-1} to slightly over 10^{-8} m s^{-1}. Sinkholes form within highly conductive zones with $K \geqslant 2 \times 10^{-5}$.	Ezersky and Legchenko (2014)
HVSR+MASW	1. Suggested methodology of joint generating of deep Vs structure model for site effect and spectral acceleration evaluation; and 2. Detection of a deep salt structure.	Ezersky *et al* (2012, 2013b, 2014, 2015), Gorstein and Ezersky (2015)
Ground-penetrating radar (GPR)	Investigation of the shallow structure of sinkhole sites, detection of subsurface cavities.	Arkin *et al* (2000), Kofman *et al* (2006), Frumkin *et al* (2011), Ronen *et al* (2019)
Microgravity	1. Detection of caverns and evaluation of their volume. 2. Suggested model of cavity detection based on depth, volume, and density contrast of medium and filling material. 3. Removing of regional trends methodology.	El-Isa *et al* (1995), Eppelbaum *et al* (2008), Ezersky *et al* (2013b), Eppelbaum (2015), Rybakov *et al* (2001), Al-Zoubi *et al* (2013)

Micromagnetic	Detection of underground cavities based on magnetic anomaly interpretation.	Rybakov *et al* (2005), Eppelbaum (2015)
Interferometric synthetic aperture radar (InSAR)	1. Subsidence of surface measurement. 2. Detection of sinkholes throughout hazardous areas.	Baer *et al* (2002), Closson *et al* (2003) Nof *et al* (2013, 2019), Baer *et al* (2018)
Nano seismicity and microseismic activity	Detection of sinkhole hazardous sites using the nanoseismic activity: two sites were investigated: Nahal Hever South; Mineral Beach.	Wust-Bloch and Joswig (2006), Abelson *et al* (2018)
Multidisciplinary geophysics	Integration of different geophysical methods allowed identify the origin of the Newe Zohar site	Ezersky and Frumkin (2021)
Laboratory testing of DS soil and salt	Samples of soil and salt collected in Ein Boqeq and Mineral Beach sites were tested, allowing carrying out of model geotechnical and geophysical calculations.	Livne (2012), Ezersky and Livne (2013), Ezersky and Goretsky (2014, 2015)

6. 3D combined modeling of geophysical fields (first of all, gravity and magnetic, SNMR, GPR, etc) essentially increases the reliability of the developed physical–geological models. Employment of the modern developments in the wavelet analysis combined with geophysical field modeling allowed creation of an advanced semi-automatic searching geophysical system with a ‘teacher’.
7. Applying elements of the theory of information enables the formalization of many geological and geophysical factors and presents these data as unified information units.

6.4 Recommendations

Our experience has indicated the following.

At present, the most efficient methods to detect cavities are the GPR, micro-gravity, ERT methods that should be used in combination with other methods. The seismic diffraction method is now under development and can be successful in the nearest future. Note that the effect of stoping—gradual ascending of the cavity to surface in loose sediments—allows detection of the cavity above saline groundwater and a prolonged time before the surface collapse.

We recommend using the 3D-SNMR method for locating water-filled caverns and estimation of their volume.

To map the salt dissolution front, the seismic refraction method using the GRM technique is effective and exact. Its combination with the MASW method makes the evaluation of the salt layer karstification and rigidity possible.

The TEM method can be used to evaluate the salinity and groundwater aggressiveness depending on the salt solution composition.

The seismic reflection method, especially based on the S-wave source, can be very useful to detect faults and topography of the salt layer in the shallow subsurface, whereas the P-wave reflection can be helpful at greater depths. The reflection method is effective in combination with the seismic diffraction method.

The remote sensing methods (InSAR, laser, and LiDAR) can be very effective to map the surface subsidence and appearance of the sinkhole signs through large areas.

References

Abdullah *et al* 2019 Characterization of subsurface cavities using gravity and ground penetrating radar *J. Environ. Eng. Geophys.* **24** 265–76

Abelson M, Aksinenko T, Kurzon I, Pinsky V, Baer G, Nof R and Yechieli Y 2018 Nanoseismicity forcasts sinkhole collapse in the Dead Sea coast years in advance *Geology* **46** 83–6

Al-Zoubi A, Abueladas A-R, Al-Ruzouq R, Camerlynck C, Akkawi E, Ezersky M, Abou-Hammateh Z, Ali W and Al-Rawashdeh S 2007 Use of 2D multi electrodes resistivity imagining for sinkholes hazard assessment along the eastern part of the Dead Sea, Jordan *Am. J. Environ. Sci.* **3** 229–33

Al-Zoubi A, Eppelbaum L, Abueladas A, Ezersky M and Akkawi E 2013 Removing regional trends in microgravity in complex environments: testing on 3D model and field investigations in the eastern Dead Sea coast (Jordan) *Int. J. Geophys. Spec. Iss.* **2013** 341797

Arandjelović D 1969 A possible way of tracing groundwater flows in karst *Geophys. Prospect.* **XII** 404–21

Arandjelović D 1976 Geophysics in the karst. Institute for geological, geophysical, and geotechnical research *Geophys. Inst. Beograd* **17** 1976

Arkin Y, Gilat A and Kofman L 2000 Developing Dead Sea sinkholes revealed by ground penetrating radar *GSI Curr. Res.* **12** 25–30

ASTM D 2006 6429-99 reapproved 2006 *Standard Guide for Selecting Surface Geophysical Methods* (West Conshohocken, PA: ASTM) p 11

Baer G, Schattner U, Wachs D, Sandwell D, Wdowinski S and Frydman S 2002 The lowest place on earth is subsiding—an InSAR (interferometric synthetic aperture radar) perspective *Geol. Soc. Am. Bull.* **114** 12–23

Baer G, Magen Y, Nof R N, Raz E, Lyakhovsky V and Shalev E 2018 *InSAR Measurements and Viscoelastic Modeling of Sinkhole Precursory Subsidence: Implications for Sinkhole Formation, Early Warning and Sediment Properties.* (Washington, DC: American Geophysical Union)

Bakalowicz M 1995 La zone d'infiltration des aquifères karstiques: méthodes d'étude—structure et fonctionnement [Infiltration zones in karst aquifers: methods of study—structure and functioning] *Hydrogéologie* **4** 3–21

Bansan K and Anderson N 2018 Multichannel analysis of surface waves: estimating depth to bedrock and acoustic properties in Karst Terrain *ARMA17* **82** 9

Billi A, De Filippis L, Poncia P P, Sella P and Faccenna C 2016 Hidden sinkholes and karst cavities in the travertine plateau of a highly-populated geothermal seismic territory (Tivoli, central Italy) *Geomorphology* **255** 63–80

Boucher M, Legchenko A, Girard J F, Baltassat J M, Dorflinguer N and Chalikakis K 2006 Using 2D inversion of MRS soundings to locate a water-filled karst conduit *J. Hydrol.* **330** 413–21

Benson R C and Yuhr L 1993 Spatial sampling consideration and their applications to characterizing fractured rock and karst systems *Applied Karst Geology* ed B F Beck (Rotterdam: Balkema) pp 99–113

Benson R C and Yuhr L B 2016 *Site Characterization in Karst and Pseudokarst Terranes. Practical Strategies and Technology for Practicing Engineers, Hydrologists and Geologists* (Berlin: Springer) p 421

Bodet L, Galibert P Y, Dhemaied A, Camerlynck C and Al-Zoubi A 2010 Surface-wave profiling for sinkhole hazard assessment along the eastern dead Sea Shoreline, Ghor Al-Haditha, Jordan *72th EAGE Conf. & Exibition Incorporating EUROPEC 2010 (Barcelona, 14–17 June, 2010) M027 4*

Brixova B, Mosna A and Putiska R 2018 Applications of shallow seismic refraction measurements in the Western Carpathians (Slovakia): case studies *Contrib. Geophys. Geod.* **48** 1–21

Chalikakis K, Plagnes V, Guerin R, Valois R and Bosch F P 2011 Contribution of geophysical methods to karst-system exploration: an overview *Hidrol. J.* **19** 1169–80

Closson D, Abou Karaki N, Hansen H, Derauw D, Barbier C and Ozer A 2003 Space-borne radar interferometric mapping of precursory deformations of a dyke collapse—Dead Sea area—Jordan *Int. J. Remote Sens.* **24** 843–9

Debeglia N, Bitri A and Thierry P 2006 Karst investigations using microgravity and MASW: application to Orléans, France *Near Surf. Geophys.* **4** 215–25

De Waele J, Gutiérrez F, Parise M and Plan L 2011 Geomorphology and natural hazards in karst areas: a review *Geomorphology* **134** 1–8

Eppelbaum L V, Zheludev V and Averbuch A 2014 Diffusion maps as a powerful tool for integrated geophysical field analysis to detecting hidden karst terranes *Izv. Acad. Sci. Azerb. Rep., Ser.: Earth Sci.* **1–2** 36–46

Eppelbaum L V 2015 Interpretation of microgravity and magnetic data for delineation of subsurface salt bodies in unfavourable S/N ratio conditions *Proc. 2015 SAGEEP Conf. (Austin, TX)* vol 28 686–95

Eppelbaum L V, Ezersky M G, Al-Zoubi A S, Goldshmidt V I and Legchenko A 2008 Study of the factors affecting the karst volume assessment in the Dead Sea sinkhole problem using microgravity field analysis and 3D modeling *Adv. GeoSci.* **19** 97–115

El-Isa Z, Rimawi O, Jarrar G, Abu-Karaki N, Taqieddin S, Atalah M, Abderahman N and Al-Saed A 1995 Assessment of the hazard of subsidence and sinkholes in Ghor Al-Haditha area *Unpublished Report to Jordan Valley Authority* University of Jordan, Amman

Ezersky 2006 The geophysical properties of the Dead Sea salt in application to sinkhole problem *J. Appl. Geophys.* **58** 45–58

Ezersky M 2008 Geoelectric structure of the Ein Gedi sinkhole occurrence site at the Dead Sea shore in Israel *J. Appl. Geophys.* **64** 56–69

Ezersky M and Frumkin A 2013 Faults—dissolution front relations and the DS sinkholes problem *Geomorphology* **201** 35–44

Ezersky M and Frumkin A 2017 Evaluation and mapping of Dead Sea coastal aquifers salinity using Transient Electromagnetic (TEM) resistivity measurements *C. R. Geosci.* **349** 1–11

Ezersky M G and Frumkin A 2020 Identification of sinkhole origin using surface geophysical methods, Dead Sea, Israel *Geomorphology* **364** 107225

Ezersky M G and Frumkin A 2021 Subaerial morphology affected by groundwater aggressiveness: sinkhole susceptibility above karstified salt, Dead Sea *Geomorphology* **375** 107525

Ezersky M G and Goretsky I 2014 Velocity-resistivity versus porosity-permeability inter-relations in Dead Sea salt samples *Eng. Geol.* **183** 96–115

Ezersky M G and Goretsky I 2015 The scale effect of Dead Sea salt velocities based on seismic-acoustic measurements *Mechanical Behavior of Salt* ed L Roberts VIII, K Mellegard and F Hansen (Rapid City, SD: A.A. Balkema) pp 137–44

Ezersky M G, Legchenko A, Camerlynck C and Al-Zoubi A 2007 The salt formation edge as a major indicator of the sinkhole hazard in the Dead Sea western coast *13th EAGE Near Surface Meeting & Exhibition 2007 (Istambul, Turkey, 1–5 september 2007)* P67 (4pp) on CD

Ezersky M G and Legchenko A 2014 Quantitative assessment of in-situ salt karstification using shear wave velocity, Dead Sea *Geomorphology* **221** 150–63

Ezersky M G and Legchenko A 2015 Mapping of salt consolidation and permeability using MASW method in the Dead Sea sinkhole problem ed G Lollino, A Manconi, F Guzzetti, M Culshaw, P Bobrowsky and F Luino *Engineering Geology for Society and Territory* vol 5 (Berlin: Springer) pp 465–9

Ezersky M and Livne E 2013 Geotechnical and geophysical properties of soils in the Dead Sea sinkhole problem *19th EAGE Near Surface Geoscience (Bohum, Mo, 9–12 September 2013)* P13, 4pp, on CD

Ezersky M G, Goldman M, Bar-Matthews M and Ayalon A 2000 Geoelectric investigation of the Soreq Cave area in Israel *6th Annual EAGE-lEEGS Meeting (Bohum, September 2000)* 4 cp-51-00126

Ezersky M G, Legchenko A, Camerlynck C and Al-Zoubi A 2007 Multidisciplinary geophysical study of the sinkhole hazard sites in the Dead Sea coastal areas *EAGE 69th Conf. and Exhibition (London, 11–14 June 2007) E-011 4 on CD*

Ezersky M, Legchenko A, Camerlynck C, Al-Zoubi A, Eppelbaum L, Keidar S, Baucher M and Chalikakis K 2010 The Dead Sea sinkhole hazard—new findings based on a multidisciplinary geophysical study *Z. Geomorphol.* **54** 69–90

Ezersky M G, Legchenko A, Al-Zoubi A, Levi E, Akkawi E and Chalikakis K 2011 TEM study of the geoelectrical structure and groundwater salinity at the Nahal Hever sinkhole site, Dead Sea shore, Israel *J. Appl. Geophys.* **75** 99–112

Ezersky M, Bodet L, Akkawi E, Al-Zoubi A, Camerlynck C, Dhemaied A and Galibert P-Y 2013a Seismic surface-wave prospecting methods for sinkhole hazard assessment along the Dead Sea shoreline *J. Environ.Eng. Geophys.* **18** 233–53

Ezersky M G, Eppelbaum L V, Al-Zoubi A, Keydar S, Abueladas A-R, Akkawi E and Medvedev B 2013b Geophysical prediction and following development sinkholes in two Dead Sea areas, Israel and Jordan *Environ. Earth Sci.* **70** 1463–78

Ezersky M G, Legchenko A, Eppelbaum L and Al-Zoubi A 2017 Overview of the geophysical studies in the Dead Sea coastal area related to evaporite karst and recent sinkhole development *Int. J. Speleol.* **46** 277–302

Ezersky M G, Keydar S, Gorstein M, Kalmanovich M and Perelman N 2012 Combining H/V spectral ratio from ambient noise with the multichannel analysis of surface waves (MASW) method to map subsurface: test in the Dead Sea area. Resumé of the first year of the study *Unpublished GII* Report No. 572/680/12 Publication No. ES-14-2012

Ezersky M G, Gorstein M, Kalmanovich M and Perelman N 2013b Combining H/V spectral ratio (HVSR) technique with method using surface waves (MASW) to map the subsurface: implementation in the Dead Sea area *Unpublished GII Report* 025/759/13

Ezersky M G, Gorstein M and Kalmanovich M 2014 Combination of active and passive MASW with HVSR methods for improving the accuracy and reliability of Vs model (in site response assessment) *Unpublished GII Report* No. 080/802/14 Publication No. ES-54-14, https://earthquake.co.il/docs/Reports/2014_080_802_14.pdf (accessed 30 March 2021)

Ezersky M G, Gorstein M and Kalmanovich M 2015 Combination of active and passive MASW with HVSR methods for improving the accuracy and reliability of Vs model (in site response assessment) *GII Report* 080/824/15 https://earthquake.co.il/docs/2015_080-824-15.pdf (accessed 30 March 2021)

Ezersky M G, Eppelbaum L, Legchenko A, Al-Zoubi A and Abueladas A 2021 Salt layer characteristics in the Ghor Al-Haditha area, Jordan: comprehensive combined reprocessing of geophysical data *Environ. Earth Sci.* **80** 80–124

Ford D C and Williams P W 2007 *Karst Geomorphology and Hydrology* (New York: Chapman and Hall) p 601

Frumkin A, Ezersky M, Al-Zoubi A, Akkawi E and Abueladas A-R 2011 The Dead Sea hazard: geophysical assessment of salt dissolution and collapse *Geomorphology* **134** 102–17

Garambois S, Legchenko A, Vincent C and Thibert A 2016 Ground-penetrating radar and surface nuclear magnetic resonance monitoring of an englacial water-filled cavity in the polythermal glacier of Tête Rousse *Geophysics* **81** WA131–46

Gorstein M and Ezersky M 2015 Combination of HVSR and MASW methods to obtain shear wave velocity model of subsurface in Israel *Int. J. Green Econ.* **1** 20–41

Greenhouse J *et al* 1995 An introduction to application of surface geophysics in environmental investigations *A Short Course Presented at the Symp. on the Application of Geophysics to Environmental and Engineering Problems, Environmental and Engineering Geophysical Society (Orlando, FL)*

Gutiérrez F, Cooper H A and Johnson K S 2008 Identification, prediction, and mitigation of sinkhole hazards in evaporate karst areas *Environ. Geol.* **53** 1007–22

Hoover R A 2003 Geophysical choices for karst investigations *Sinkholes and the Engineering and Environmental Impacts of Karst* ed B F Beck (Reston, VA: American Society of Civil Engineers) pp 529–38

Hussain Y, Uagoda R, Borges W, Nunes J, Hamza O, Condori C, Aslam K, Dou J and Cardenas-Soto M 2020 The potential use of geophysical methods to identify cavities, sinkholes and pathways for water infiltration: a case study from Mambai, Brazil *ESS Open Arch.* (preprint)

Intrieri E, Giovanni G and Gracchi T *et al* 2018 Application of an ultra-wide band sensor-free wireless network for ground monitoring *Eng. Geol.* **228** 1–14

ISRM 2004 Suggested methods for land geophysics in rock engineering *Final Report of ISRM Commission*

Keydar S, Pelman D and Ezersky M 2010 Application of seismic diffraction imaging for detecting near-surface inhomogeneities in the Dead Sea area *J. Appl. Geophys.* **71** 47–52

Keydar S, Bodet L, Camerlynck C, Dhemaied A, Galibert P Y, Ezersky M and Akkawi E 2011 A new approach for shallow subsurface imaging and its application to the Dead Sea sinkhole problem *Proc. 73rd EAGE Conf. and Exhibition (Vienna, April 2011)* pp 1–4

Keydar S, Medvedev B, Ezersky M and Sobolevsky L 2012 Imaging shallow subsurface of Dead Sea area by common shot point stacking and diffraction method using weighted multipath summation *J. Civil Eng. Sci.* **1** 75–9

Keydar S, Medvedev B, Al-Zoubi A, Ezersky M and Akkawi E 2013 3D Imaging of Dead Sea area using weighted multipath summation: a case study *Int. J. Geophys.* **2013** 692452

Khesin B E, Alexeyev V V and Eppelbaum L V 1996 *Interpretation of Geophysical Fields in Complicated Environments. Series: Modern Approaches in Geophysics* (Boston, MA: Kluwer Academic Publishers)

Kofman L, Ronen A and Frydman S 2006 Detection of model voids by identifying reverberation phenomena in GPR records *J. Appl. Geophys.* **59** 284–99

Kolesnikov Y U and Fedin K V 2018 Detecting underground cavities using microtremor data: physical modeling and field experiment *Geophys. Prospect.* **66** 342–53

Legchenko A, Baltassat J-M, Bobachev A, Martin C, Robin H and Vouillamoz J-M 2004 Magnetic resonance sounding applied to aquifer characterization *J. Ground Water* **42** 363–73

Legchenko A, Ezersky M, Camerlynck C and Al-Zoubi A 2007 MRS study of the evolution of aquifer in sinkhole hazardous area (Dead Sea coast) *13th EAGE Near Surface Meeting & Exhibition 2007 (Istanbul, Turkey, 3–5 September 2007)* B29

Legchenko A, Ezersky M, Boucher M, Camerlynk C, Al-Zoubi A and Chalikakis K 2008a Pre-existing caverns in salt formations could be the major cause of sinkhole hazards along the coast of the Dead Sea *Geophys. Res. Lett.* **35** L19404

Legchenko A, Ezersky M, Camerlink C, Al-Zoubi A, Chalikakis K and Girard J-F 2008b Locating water-filled karst caverns and estimating their volume using magnetic resonance soundings *Geophysics* **73** 51–61

Legchenko A, Ezersky M, Girard J-F, Baltassat J-M, Camerlynck C and Al-Zoubi A 2008c Interpretation of the MRS measurements in rocks with high electrical conductivity *J. Appl. Geophys.* **66** 118–27

Legchenko A, Ezersky M, Camerlynck C, Al-Zoubi A and Chalikakis K 2009 Joint use of TEM and MRS method in the complex geological setting *Comptes Rendus Geosci.* **341** 908–17

Legchenko A, Descloitres M, Vincent C, Guyard H, Garambois G, Chalikakis K and Ezersky M 2011 Three-dimensional magnetic resonance imaging for groundwater *New J. Phys.* **13** 025022

Legchenko A *et al* 2014 Monitoring water accumulation in a glacier using magnetic resonance imaging *Cryosphere* **8** 155–66

Livne E 2012 Evaluation of the liquefaction potential along the dead-sea western shores—regional screening and geotechnical analysis of two boreholes *MSc Thesis* Ben-Gurion University of the Negev, Beersheba

Martínez-Moreno F J, Monteiro Santos F A, Galindo-Zaldívar F A, González-Castillo L, Pedrera A, Bernando I, Gonçalves S and Ribeiro J A 2018 Characterization of a cave by means of microgravity and electrical resistivity 3D-inversions: Zé de Braga cave (Mira de Aire, Portugal) *First Break* **36** 29–34

McNeill J D 1980 Applications of transient electromagnetic techniques *Technical Notes TN-7* (Mississauga, ON: Geonics Ltd), 17

Milanović P 1979 *Hydrogeologija karsta I metode istrazivanja* (Trebinje: 'HE 'Trebishnica'', Institute za karstenje I zastitu voda na krsu)

Milanović P T 2000b *Geological Engineering in Karst: Dams, Reservoirs, Grouting, Groundwater Protection, Water Tapping, Tunneling* (Belgrade: Zebra Publishing Ltd) p 347

Milanović P 2005 *Water Resources Engineering in Karst* (Boca Raton, FL: CRC Press) p 328

Milanović P, Maximovich N and Meshcheryakova O 2019 *Dams and Reservoirs in Evaporites.* (Berlin: Springer)

Mochales T, Casas A M, Pueyo E L, Pueyo O, Román M T, Pocoví A, Soriano M A and Ansón D 2008 Detection of underground cavities by combining gravity, magnetic and ground penetrating radar surveys: a case study from the Zaragoza area, NE Spain *Environ. Geol.* **53** 1067–77

Nof R, Baer G, Ziv A, Eyal Y, Raz E, Atzori S and Salvi S 2013 Sinkhole precursors along the Dead Sea, Israel, revealed by SAR interferometry *Geology* **41** 1019–22

Nof R N, Abelson M, Raz E, Magen Y, Atzori S, Salvi S and Baer G 2019 SAR interferometry for sinkhole early warning and susceptibility assessment along the Dead Sea, Israel *Remote Sens.* **11** 89

Ogilvy A A, Bogoslovsky V A and Kuzmina E N 1980 Geoelectric models in engineering geophysics *Geophys. Prospect.* **28** 945–55

Palmer D 1986 *Refraction seismics. The Lateral Resolution of Structure and Seismic Velocity* (London: Geophysical Press Ltd)

Parise M, Closson D, Gutierrez F and Stefanović Z 2015 Anticipating and managing engineering problems in the complex karst environment *Environ. Earth Sci.* **74** 7823–35

Pasquet S and Bodet L 2017 SWIP: an integrated workflow for surface-wave dispersion inversion and profiling *Geophysics* **82** WB47–61

Park M K, Park S, Yi M-J, Kim C, Son J-S, Kim J-H and Abraham A A 2013 Application of electrical resistivity tomography (ERT) technique to detect underground cavities in a karst area of South Korea *Environ. Earth Sci.* **71** 2797–806

Pueyo-Anchuela O, Juan A P, Soriano M A and Casas-Sainz A M 2009a Characterization of karst hazards from the perspective of the doline triangle using GPR: examples from Central Ebro Basin (Spain) *Eng. Geol.* **108** 225–36

Pueyo-Anchuela O, Casas Sainz A M, Soriano M A and Pocoví Juan A 2009b Mapping subsurface karst features with GPR: results and limitations *Environ. Geol.* **58** 391–9

Reynolds J M 2011 *An Introduction to Applied and Environmental Geophysics* 2nd edn (New York: Wiley)

Rodriguez V, Gutiérrez F, Green A G, Carbonel D, Horstmeyer H and Schmelbach C 2014 Characterizing sagging and collapse sinkholes in a mantled karst by means of ground penetrating radar (GPR) *Environ. Eng. Geosci.* **20** 109–32

Ronen A, Ezersky M, Beck A, Gatenio B and Simhiov R B 2019 Use of GPR method for prediction of sinkholes formation along the Dead Sea Shores, Israel *Geomorphology* **328** 28–43

Rybakov M, Goldschmidt V, Fleischer L and Rotstein Y 2001 Cave detection and 4-D monitoring: a microgravity case history near the Dead Sea *Lead. Edge* **20** 896–900

Rybakov M, Rotstein Y, Shirman B and Al-Zoubi A 2005 Cave detection near the Dead Sea—a micromagnetic feasibility study *Lead. Edge* **24** 585–90

Savich A I and Kujundjić B D (ed) 1990 *Complex Engineering—Geophysical Explorations at the Hydrotechnical Construction* (Moscow: Nedra) (in Russian)

Telford W M, Geldart L P, Sheriff R E and Keys D A 1990 *Applied Geophysics* 2nd edn (Cambridge: Cambridge University Press)

White W B 2007 A brief history of karst hydrogeology: contributions of the NSS *J Cave Karst Stud.* **69** 13–26

Wust-Bloch G H and Joswig M 2006 Pre-collapse identification of sinkholes in unconsolidated media at Dead Sea area by 'nanoseismic monitoring' (graphical jackknife location of weak sources by few, low-SNR records) *Geophys. J. Int.* **167** 1220–32

Xie P, Wen H, Fu J, Li R and Hu J 2018 A methodology for three-dimensional modeling of subsurface geologic structure in mantled karst area *J. Comput. Methods Sci. Eng.* **18** 667–82

Zhang J, Liu S, ChenZg Q, Wang B and Ren C 2019 Application of cross-borehole integrated geophysical methods for the detailed investigation of karst in urban metro construction *J. Environ. Eng. Geophys.* **24** 525

www.ingramcontent.com/pod-product-compliance
Ingram Content Group UK Ltd.
Pitfield, Milton Keynes, MK11 3LW, UK
UKHW051911100926
8556IPUK00005B/846

9 780750 336338